Problems and Solutions in Many-Particle Systems

Online at: https://doi.org/10.1088/978-0-7503-6447-8

Problems and Solutions in Many-Particle Systems

**Pradeep Kumar Sharma MSc, Mtech, MPhil, PhD, AMIE,
CEng(I) MIET, MInstP, SMIEEE**
Consultant physicist and researcher, (India)

IOP Publishing, Bristol, UK

ISBN   978-0-7503-6447-8 (ebook)
ISBN   978-0-7503-6443-0 (print)
ISBN   978-0-7503-6444-7 (myPrint)
ISBN   978-0-7503-6446-1 (mobi)

DOI   10.1088/978-0-7503-6447-8

Version: 20250201

IOP ebooks

British Library Cataloguing-in-Publication Data: A catalogue record for this book is available from the British Library.

Published by IOP Publishing, wholly owned by The Institute of Physics, London

IOP Publishing, No.2 The Distillery, Glassfields, Avon Street, Bristol, BS2 0GR, UK

US Office: IOP Publishing, Inc., 190 North Independence Mall West, Suite 601, Philadelphia, PA 19106, USA

This book is dedicated to my wife Mrs Usha Sharma, my son Mr Hayagreeva Sharma, my brother-in-laws Mr Rajendra Kumar Dash and Mr Rabindra Kumar Dash for their constant support and motivation.

Contents

Foreword   x

Preface   xi

Acknowledgements   xiii

Author biography   xv

1   System of particles   **1-1**

1.1   Introduction   1-1
1.2   Definition of center of mass   1-2
1.3   Characteristics of center of mass   1-6
1.4   Calculation of position of center of mass (finding center of mass)   1-14
1.5   Work–energy theorem for a system of particles   1-44
1.6   Impulse and momentum equation for a system of particles   1-52
1.7   Advantages of center of mass frame   1-64
1.8   Application of impulse-momentum equation and work–energy   1-66
      theorem for a two-particle system
1.9   Application of impulse-momentum equation and work–energy   1-73
      theorem for many-particle system

2   Collisions   **2-1**

2.1   Introduction   2-1
2.2   Definition   2-1
2.3   Impulsive force and impulse   2-3
2.4   Types of collision   2-5
2.5   Conservation of linear momentum in a collision   2-9
2.6   Energy consideration in collisions   2-11
2.7   Elastic collision   2-17
2.8   Elastic-oblique collision   2-20
2.9   Complete inelastic collision ($e = 0$)   2-22
2.10 General solution for head-on collision between two bodies   2-28
2.11 General solution for oblique inelastic impact of two smooth bodies   2-36

3   Gravitation   **3-1**

3.1   Introduction   3-1
3.2   Kepler's laws   3-1

3.3	Newton's law of universal gravitation	3-3
3.4	Inertial and gravitational mass	3-9
3.5	Gravitational field and field intensity, superposition of gravitational field	3-11
3.6	Calculation of gravitational field intensity	3-13
3.7	Work done by gravity	3-32
3.8	Gravitational potential energy between two particles	3-34
3.9	Gravitational potential	3-35
3.10	Calculation of gravitational potential	3-37
3.11	Gravitational potential energy of a group of particles	3-48
3.12	Earth's gravitational field	3-54
3.13	Variation of $|\vec{g}_{\text{eff}}|$ and apparent weight	3-56
3.14	The motion of planets	3-60
3.15	Motion of planets and satellites in circular orbit	3-63
3.16	Escape velocity	3-67
3.17	Orbital velocity and nature of orbits of a satellite	3-72
3.18	Weightlessness	3-73
3.19	Earth as an inertial reference frame	3-75
4	**Fluid statics**	**4-1**
4.1	Introduction	4-1
4.2	Fluids and solids	4-1
4.3	Definition of fluid statics	4-5
4.4	Density	4-6
4.5	Specific gravity (relative density)	4-6
4.6	Hydrostatic force	4-9
4.7	Hydrostatic pressure	4-12
4.8	Pressure in a non-accelerating liquid	4-16
4.9	Pressure due to many non-accelerating liquid layers	4-25
4.10	Hydrostatic pressure in a vertically accelerating liquid	4-34
4.11	Hydrostatic pressure in a horizontally accelerating liquid	4-37
4.12	Hydrostatic force on a flat surface	4-51
4.13	Hydrostatic force on a curved surface	4-62
4.14	Archimedes' principle	4-77
4.15	Application of archimedes' principle	4-80
4.16	Buoyant force in accelerating liquid	4-99
4.17	Hydrostatic torque	4-106
4.18	Pascal's law and its application	4-110

5   Fluid dynamics   5-1

5.1   Definition of hydrodynamics   5-1
5.2   General characteristics of fluid flow   5-3
5.3   Streamline motion   5-6
5.4   Equation of continuity   5-7
5.5   Flux of $\vec{v}$-field (ϕ_v)   5-9
5.6   Flux density   5-13
5.7   Equation of state of fluid motion   5-15
5.8   Bernoulli's theorem   5-17
5.9   Applications of Bernoulli's theorem and equation of continuity   5-22

6   Properties of matter   6-1

6.1   Introduction   6-1
6.2   Elastic property   6-1
6.3   Hooke's law   6-10
6.4   Calculation of deformation   6-19
6.5   Elastic energy   6-25
6.6   Calculation of elastic energy   6-26
6.7   Viscosity   6-28
6.8   Newton's law of viscosity   6-29
6.9   Stoke's law   6-31
6.10 Surface tension   6-34
6.11 Molecular theory of surface tension   6-35
6.12 Measuring surface tension   6-36
6.13 Surface energy   6-40
6.14 Relation between surface tension and surface energy   6-42
6.15 Pressure difference across a liquid surface   6-44
6.16 Angle of contact   6-48
6.17 Capillary action   6-49

Foreword

It is my pleasure to write a foreword for this book authored by Dr Pradeep Kumar Sharma. The author of this book knew me while he was a lecturer of Physics in Brilliant Tutorials (BT), Chennai and I was a professor of Physics in Indian Institute of Technology (IIT), Madras. I also know the author for his five books—*GRB Understanding Physics*. The author gained a vast experience and expertise in training the best students for various competitive examinations such as IIT-JEE, Olympiads, etc. In 2020, the author personally visited my house in Chennai and requested me to edit this book. Although I could not edit, I spent some time in reviewing the matter and gave some suggestions to improve the overall quality and content of this book.

This book is well-written and well-balanced with the theories and problems. All concepts are covered in-depth. The explanations have been presented in detail in a simple and lucid style. Most appropriate examples are given by the author based on his vast experience of teaching physics in reputed institutes across the country such as BT, FIIT-JEE, etc. The book has an impressive layout and excellent quality of its production. In each chapter, the author has included systematic theories with the best examples and scholarly set solved problems. This will assist both teachers and students in building and strengthening the concepts of physics.

This book will be immensely useful to all college students preparing for entrance examinations such as IIT-JEE, Physics Olympiads, etc. Furthermore, this book could be useful for Physics GRE and PhD Qualifying examinations of top universities across the world. Teachers can also use this book for enhancing their subject knowledge so as to impart a better physics education. Each topic is dealt scrupulously so as to enhance the student's understanding of the subject to a great extent. I thank the author for this splendid work as the result of his hard work and determination. I wish him all the best for his forthcoming books.

Dr Jagabandhu Majhi
Professor of Physics (Retired)
Indian Institute of Technology, Madras

Preface

Overview of the series

From my studenthood to date I have been in search of a series of books on *College Physics* that carries every concept in depth along with the scholarly problems to cater for the needs of potential students preparing for top entrance examinations. I found that students and scholars were spending hundreds of hours gathering information from different sources, websites and books on each concept and chapter. This practice still continues today. This is due to the lack of a desired college physics book series. We have a lot of books in the market; each has its own merits and limitations. Practically, it is not possible to get everything in a single book. However, a potential student expects everything in a book such as varieties of adequate problems with detailed theories and examples and solved problems.

Students are confused by the scattering and variety of sources for the problems, as well as their provenance and validity of solutions. So, not just helpless students but also teachers need concise and precise books where maximum concepts are covered systematically. The proposed series of books is a sincere effort to minimize all the above limitations of existing books and resources and maximise the potential of the books in terms of content, quality, rigor, and depth in theories, questions, and problems.

There are no such books available in the market balanced with problems, theories, quality, and quantity of problems. In this series, we have attempted to balance the books with theories, examples, and problems using a systematic approach for concept building. By virtue of my experience and expertise, as well as suggestions and recommendations of my colleagues and hundreds of my gifted students, the proposed series of books is an outcome of my strong desire to transform physics-phobia into a physics-loving attitude for the students.

Readership

At present, sophomores prepare for entrance examinations to get into premier universities like Oxford, Cambridge, etc, in the United Kingdom; MIT, Princeton, etc, in the Unites States of America. Millions of students all over the world appear for Physics Olympiads (both national and international) and other international physics examinations such as Physics GRE, etc. In India, millions of students appear for national-level examinations for prestigious Indian Institutes of Technology (IITs) through a toughest Joint Entrance Examination called IIT-JEE with the lowest success rate in the world.

Moreover, this book is also useful for students preparing for the PhD qualifying examinations of top universities and Physics GRE examinations. In India, students preparing for UPSC examinations and semester examinations and physics majors in BSc and BEng (or any branch of engineering study) can use this book as a handbook of physics problems. This book will be ultimately useful for Indian students preparing for JEE (Mains and Advanced).

To educate potential students, teachers should be strong in concepts and problem-solving ability. I strongly hope that this text could be a valuable reference book for JEE-educators in India due to its content and quality, such as familiar problems, theories, creative problems, and problems with detailed solutions along with impressive coloured diagrams throughout. I have handpicked many problems from various sources. Each problem is observed being asked in different ways in various examinations all over the world. I spent adequate time in analysing each problem's most original version and then put it in my series. Furthermore, I prefer to put all possible questions linked with the original problems in one place so that the student does not have to waste their time thinking too much and wandering.

About this book

Mechanics of many-particle system (center of mass, collision, gravitation, fluid mechanics, properties of matter). This book contains six chapters:
1. System of particles
2. Collision
3. Gravitation
4. Fluid statics
5. Fluid dynamics
6. Properties of matter.

The first chapter introduces the concept of center of mass in order to apply Newton's laws and work–energy theorem for a system of particles. The end of the first chapter deals with the impulse-momentum equations for a system of particles. The second chapter deals with the basics of collision. Generally, we use Newton's impact or empirical formula (NIF) and conservation of momentum for a system particles that forms the basis of explaining the other chapters such as gravitation, fluid mechanics (static and dynamics), and properties of matter (elasticity, surface tension, capillary action, and viscosity).

How to use this book

Students should first complete the given theory with the examples and then try to solve the problems in their own way. Sometimes students find better methods for solving the problems and should not rely solely on the methods provided. However, if a student falters after repeated attempts, they can refer to the given solutions. Although I tried to create error-free text, some unavoidable mistakes might appear in some subtle form each time. So there will always be scope for improvement of the content. I request my readers to critically analyze my work and give their valuable comments and suggestions for the overall improvement of this book to make it error-free in due course.

Acknowledgements

It goes without saying that a teacher is worth millions of books. An ideal teacher is built by the association of potential students. I acquired my experience and expertise by the association of gifted students of premier institutes. So, first of all, I would like to express my gratitude to the directors of all the institutes where I worked. Most illustrious are Brilliant Tutorials Private Limited, Madras (Chennai), FIIT-JEE Ltd, New-Delhi, and Narayana Group of Educational Institutions, Hyderabad.

My high school science and mathematics teacher Mr Ramesh Chandra Behera imparted an ever-lasting style of solving problems and presented the theory in a handy and student-friendly way. All the attributes of my gifted teachers (many of them have left their mortal bodies) and students (who are now global leaders in their own fields) are reflected in these books. So, I express my sincere thanks to all my revered teachers (gurus) and past students.

I am thankful to Professor T Surya Kumar, Professor Bhanumati and Professor G N Subramanian of BT who was well-versed with both physics and mathematics helped me to solve the standard problems in mechanics.

In FIIT-JEE Limited, New-Delhi, I thank my seniors Er Srikant Kumar and Er P K Mishra for imparting a standard subject knowledge enhancement programme that had a great impact on this book.

I thank the late Mr Prakash Chand Bathla for offering me to write five books under his nationally leading publication house (G R Bathla & Sons). Based upon my experience as a national level author and educator, I could attain a chance to write for an international publishing house (IOPP).

While I headed the department of Physics in Narayana Group of educational institutions, I would like to thank all my previous top students who edited my books. Furthermore, I express my deepest gratitude to the principals and deans of Nayayana Group, Hyderabad especially Mr Krishna Reddy and Mr Ramalinga Reddy, with whom I worked major portion of my professional career, for giving me operational freedom, status, stability and respect. There, I could complete some theories and examples of the present series in rudimentary form. My sincere thanks and admiration to some leading physics educators such as Er Aditya Sachan, Er L N Prusty, Er Sekhar Somnath, Mr Monoj Pandey and Mr S K Singh. I am also thankful to Professor Kundal Rao and Professor Raghunath of Narayana IIT and Professor Srinivasa Chary of Sri-Mega for their suggestions and inspirations for my publishing works.

I would like to express my profound gratitude to my wife Usha in supporting and bearing me in the pandemic in 2020 when I started conceptualizing this book-series.

I remain obliged to the commissioning editor of IOP Publishing Mr John Navas for his insightful comments, suggestions, and expertise, which have enhanced the rigor and depth of this work. Furthermore, I thank Mr David McDade and Phoebe Hooper, who streamlined the publication work of this book.

I express my gratitude to my Ex-publisher Mr Monoj Bathla who suggested me to accept the offer of IOPP realising the suitability of the publisher with my work.

I sincerely thank Mr Bismay Parida (Readers institute, Balasore, Odisha State) for his continuous effort in typing the manuscript in time.

I thank Professor Peter Dobson (Oxford University) who taught me how to do the things with perfection and I am applying this idea in the present publication. I am deeply indebted to Dr Benjamin Hourahine, Professor Yu Chen of University of Strathclyde for imparting a standard knowledge of nanoscience so that I could include this fastest growing field in my problem book series.

One of my notable friends is Mr Rajinder Sehra, director of S&RJ Ltd and Foot Print Media Production near Glasgow. His constant motivation for writing this series is also praiseworthy.

Furthermore, I would like to thank a potential Physics educator Mr Mithilesh for finding time to review some of the problems in my book. At last, my sincere thanks to Er K K Khandelwal (a graduate from IIT, an ex-baeurocrat and a gifted senior Physics educator) for reviewing some of the controversial problems in my book.

This book would not have been possible without the collective contributions and support of all those mentioned above. Their guidance and encouragement have been instrumental in the completion of this significant milestone in my authorship. Taking this as a blessing of the Almighty, I pray for the attainment of knowledge that would be an ultimate solution to all problems of human being and other living entities.

Pradeep Kumar Sharma
17 September 2024

Author biography

Pradeep Kumar Sharma

He is a well-known physics educator in India possessing more than three decades of experience in physics education and research in training the aspirants of the joint entrance examination conducted by prestigious Indian Institutes of Technology, popularly known as IIT-JEE. Many of his students also won gold and silver medals in national and international physics Olympiads. His vast experience as a potential teacher, team leader and head of the department in some premier institutes like Brilliant tutorials, (Chennai), FIIT-JEE Ltd (New Delhi), Narayana Group (Andhra and Telangana) etc., made him extend his service as a consultant physicist to mentor both students and teachers of reputed groups in India. He has authored bestselling study materials and five books known as GRB Understanding Physics for the entrance examinations. He has been associating as a research scholar of physics education, nanoscience, metaphysics and management in some Indian and foreign universities such as Oxford University, Strathclyde University, Sofia University, Indian Institutes of Technology, Patna etc. Furthermore, he is continuing his research while affiliated with various national and international organizations such as IEEE (USA), IET (UK), IE(I), IOP(UK) etc. He has published dozens of papers in national and international journals like IEEE-Scopus journals and journals published by Institute of Physics (UK). He is currently busy in completing the problems and solutions of a series of six books which will be ready to publish very shortly. Also, he is planning to design a unique interactive study material in the mode of Active Teaching and Active Learning (ATAL) that will make the physics easier for the students to learn.

IOP Publishing

Problems and Solutions in Many-Particle Systems

Pradeep Kumar Sharma

Chapter 1

System of particles

1.1 Introduction

In the mechanics (kinematics and dynamics) of a particle, we find the position or displacement from velocity. The expression of velocity can be derived using two different methods:

1. From acceleration of a particle by using Newton's laws of motion
2. By using the work–energy theorem

For both methods, we start with drawing a free-body diagram (FBD) of the particle and measure the net force acting on it. Then by dividing the net force $\vec{F}$ by the mass of the particle, we find the acceleration given as

$$\vec{a} = \frac{\vec{F}}{m}$$

Integrating the acceleration, we find velocity. In the other method, in general, we can use the work–energy theorem. If the forces acting on the particle are conservative in nature, we can use the principle of conservation of energy.

Now let us try to understand the mechanics of an extended object—a *system (or group) of particles* in order to define and analyse the linear motion of the object. Let an object (or a system of particles) experience many forces from outside the system—external forces at its different points. If we add all external forces to get a net force F, say, then we can write Newton's 2nd law as

$$\vec{F} = m\vec{a},$$

where F = Net force acting on the object, which is easy to understand. Now a natural question arises, what is $\vec{a}$? In fact, it is an acceleration, but acceleration of which point of the object? Thus, we have to develop a notion of a special point called *center of mass.* It plays a significant role in understanding the mechanics of all states of matter such as solid, liquid, and gas.

1.2 Definition of center of mass

For the sake of simplicity, let us consider an object consisting of a system of *two* particles and try to apply the laws of motion. Let us assume that external forces F_1 and F_2 act on the particles of masses m_1 and m_2, respectively. If there are forces interacting between the particles we call them internal forces. Let the internal forces acting on m_1 and m_2 be F_{12}, F_{21}, respectively. Then the net force acting on m_1 is the sum of the external and internal forces acting on it, which can be given as

$$(\vec{F}_{net})_1 = \vec{F_1} + \vec{F}_{12} \tag{1.1}$$

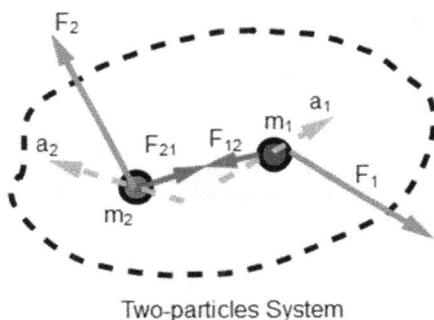

Two-particles System

Applying Newton's law on the article m_1, we have

$$(\vec{F}_{net})_1 = m_1 \vec{a_1} \tag{1.2}$$

where a_1 = acceleration of m_1.

Substituting $(\vec{F}_{net})_1$ from equation (i) in equation (ii), we have

$$\vec{F_1} + \vec{F}_{12} = m_1 \vec{a_1} \tag{1.3}$$

Similarly, by application of Newton's second law on the 2nd particle m_2, we have

$$\vec{F_2} + \vec{F}_{21} = m_2 \vec{a_2} \tag{1.4}$$

Summing up the last two equations, the net force F_{net} acting on the system is given as

$$\vec{F}_{net} = (\vec{F_1} + \vec{F_2}) + (\vec{F}_{12} + \vec{F}_{21}) = m_1 \vec{a_1} + m_2 \vec{a_2} \tag{1.5}$$

According to Newton's third law, action and reaction forces are equal in magnitude and opposite in direction. So, net internal force turns out to be zero as they appear as action-reaction pairs. Then, we have

$$(\overrightarrow{F_{12}} + \overrightarrow{F_{21}}) = 0 \qquad (1.6)$$

Using the last two equations, the net force for a system of particles can be given as For a system of two particles, we have

$$\overrightarrow{F}_{\text{net}} = \overrightarrow{F_1} + \overrightarrow{F_2} = m_1\overrightarrow{a_1} + m_2\overrightarrow{a_2}$$

For a system of n particles, in a nut shell, we can write

$$\overrightarrow{F}_{\text{net}} = \sum \overrightarrow{F}_i = \sum m_i \overrightarrow{a_i},$$

where $\overrightarrow{F_i}$ = external force acting on ith particle of mass m_i and $\overrightarrow{a_i}$ = acceleration of m_i. The above equation can be directly used for any system comprising point masses.

Example 1 A system of three particles of masses m_1, m_2, and m_3 interconnected by two light springs experience forces $\overrightarrow{F_1} \overrightarrow{F_2}$, and $\overrightarrow{F_3}$ respectively. If the accelerations of m_2 and m_3 are a_2 and a_3, respectively, find the acceleration a_1 of m_1. Assume $F_1 = 3N$, $F_2 = 1N$, $F_3 = 2N$, $a_2 = 2$ m s^{-2}, and $m_1 = m_2 = m_3 = 1$ kg.

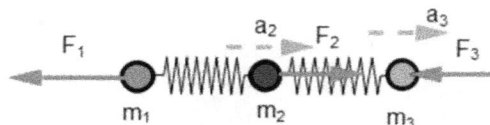

Solution
Since the spring forces are internal forces of the system of three particles, the net force acting on the system is $\overrightarrow{F}_{\text{net}} = \overrightarrow{F_1} + \overrightarrow{F_2} + \overrightarrow{F_3}$, which is equal to the vector sum of the product of mass and acceleration of individual particles given as

$$\sum m_i \overrightarrow{a_i} = m_1\overrightarrow{a_1} + m_2 a_2 + m_3\overrightarrow{a_3}$$

Substituting $\overrightarrow{F_1} = -3\hat{i}\,\text{N}$, $\overrightarrow{F_2} = 1\hat{i}\,\text{N}$, and $\overrightarrow{F_3} = -2\hat{i}\,\text{N}$, we have

$$\overrightarrow{F}_{\text{net}} = (-3 + 1 - 2)\hat{i} = -4\,\text{N}$$

$$\Rightarrow m_1\overrightarrow{a_1} + m_2\overrightarrow{a_2} + m_3\overrightarrow{a_3} = -4$$

Putting $\overrightarrow{a_2} = +2\hat{i}$ ms^{-2}, $\overrightarrow{a_3} = +\hat{i}$ ms^{-2} and $m_1 = m_2 = m_3 = 1$ kg in the last equation, we have

$$(1)(\overrightarrow{a_1}) + (1)(2\hat{i}) + (1)(+\hat{i}) = 0$$

$$\Rightarrow \overrightarrow{a_1} = -3\hat{i} \text{ m s}^{-2}$$

Hence, m_1 accelerates toward the left with a magnitude of 3 m s^{-2}.

Example 2 A system of two oscillating particles of masses m_1 and m_2 interconnected by an ideal spring is released in gravity. If the acceleration of m_1 is $\vec{a}$, find the acceleration of m_2.

Solution

In this system of two particles (m_1 and m_2), the spring forces are internal forces so we can write

$$(\vec{F_{12}} + \vec{F_{21}}) = 0$$

Then, the net force acting on the system is

$$\vec{F_{net}} = \vec{F_1} + \vec{F_2},$$

which is equal to the vector sum of the product of mass and acceleration of individual particles given as

$$\sum m_i \vec{a_i} = m_1 \vec{a_1} + m_2 a_2$$

$$\vec{F_{net}} = \vec{F_1} + \vec{F_2} = m_1 \vec{a_1} + m_2 a_2$$

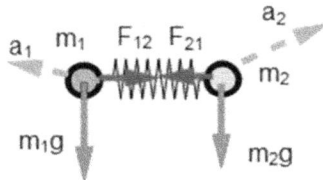

As the system is falling in gravity, the external forces are gravity forces (weights). Then, substituting $\vec{F_1} = m_1 \vec{g}$, $\vec{F_2} = m_2 \vec{g}$ and $\vec{a_1} = \vec{a}$ in the last equation, we have

$$\vec{a_2} = \frac{m_1 \vec{g} + m_2 \vec{g} - m_1 \vec{a}}{m_2} \quad \text{Ans.}$$

The previous discussion gives us the idea that for a system of particles the sum of the product of mass and acceleration of particles of the system is equal to the net force, that is, net external force acting on the system. Symbolically $\vec{F_{net}} = \sum m_i \vec{a_i}$.

After being familiar with the expression $\vec{F_{net}} = \sum m_i \vec{a_i}$, let us now divide both sides by total mass of the system, that is, $\sum m_i$. This gives

$$\frac{\vec{F_{net}}}{\sum m_i} = \frac{\sum m_i \vec{a_i}}{\sum m_i}$$

The RHS term is the ratio of 'total force' and 'total mass', which gives us an 'acceleration' $\vec{a}$, say. Then, we can write

$$\frac{\vec{F_{net}}}{\sum m_i} = \vec{a}, \text{ where } \vec{a} = \frac{\sum m_i \vec{a_i}}{\sum m_i}$$

This can also be given as

$$\overrightarrow{F_{net}} = \left(\sum m_i\right)\overrightarrow{a}$$

Here, we have three terms $\overrightarrow{F_{net}}$, $\sum m_i$, and $\overrightarrow{a}$. Let us look at each term. The first term $\overrightarrow{F_{net}}$ signifies the net (total) force acting on the system, second term $\sum m_i$ gives us the total mass of the system, but what about the third term $\overrightarrow{a}$? Well, it is an acceleration, but acceleration at which point? Furthermore, you can also ask: where does the force $\overrightarrow{F_{net}}$ act? Let us try to answer these questions.

As the force $\overrightarrow{F_{net}}$ is the sum of the external forces acting at different point masses of the system, it makes no sense to say that $\overrightarrow{F_{net}}$ acts on any particle of the system. Thus, it is essential to imagine a 'special point' at which the net force acts so as to move it with an acceleration $\overrightarrow{a}$, which is equal to net force divided by total mass of the system. This hypothetical point moves with an acceleration $\overrightarrow{a}$ as if it carries the entire mass $(\sum m_i)$ of the system experiencing the net force $\overrightarrow{F_{net}}$ acting on the system. Hence, it is natural to imagine that the entire mass of the system is centered at that 'special point', which can be logically called 'center of mass'. As this is just a 'hypothetical point' by definition, it is not essential that center of mass should carry some mass.

Example 3 Find the acceleration of center of mass of the system of two particles of masses m_1 and m_2 connected by a light inextensible string that passes over a smooth fixed pulley.
   Solution
   Method 1:
   Here the system is $(m_1 + m_2)$.

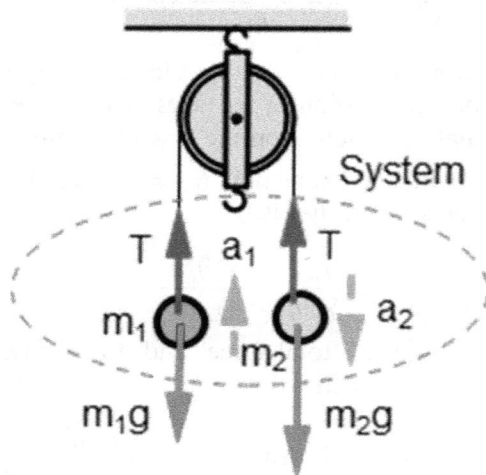

The net force acting on the system is $\vec{F} = -(m_1 g + m_2 g - 2T)\hat{j}$, where $T = \frac{2m_1 m_2 g}{m_1 + m_2}$ as derived earlier:

$$\Rightarrow \vec{F} = -\frac{(m_1 - m_2)^2}{m_1 + m_2} g\hat{j} = \frac{(m_1 - m_2)^2}{m_1 + m_2}\vec{g} \tag{1.7}$$

Then, the acceleration of center of mass of the system is given as

$$\vec{a}_{CM} = \frac{\vec{F}}{m_1 + m_2} \tag{1.8}$$

Substituting $\vec{F}$ from equation (1.7) in equation (1.8), we have

$$\vec{a}_{CM} = \left|\frac{m_1 - m_2}{m_1 + m_2}\right|^2 \vec{g}$$

Method 2:

Applying Newton's second law on each particle we find their accelerations as

$$-\vec{a_1} = \vec{a_2} = \frac{m_2 - m_1}{m_1 + m_2}\vec{g}$$

Then substituting $\vec{a_1}$ and $\vec{a_2}$ in the expression

$$\vec{a}_{CM} = \frac{m_1 \vec{a_1} + m_2 \vec{a_2}}{m_1 + m_2}, \text{ we have}$$

$$\vec{a}_{CM} = \left|\frac{m_2 - m_1}{m_1 + m_2}\right|^2 \vec{g} \quad \text{Ans.}$$

N.B.: If we take the pulley into account, assuming mass of the pulley as M, you can prove that the acceleration of pulley + particle system will be equal to

$$\frac{(m_1 - m_2)^2 g}{(m_1 + m_2)(M + m_1 + m_2)}$$

The momentum of center of mass may roughly mean the product of mass of the system and velocity of center of mass. Similarly, kinetic energy (KE) of center of mass roughly means $\frac{1}{2}$ (mass of the system) $\times$ (CM)^2. In this way, using the powerful concept of center of mass, we will derive the expressions for KE, momentum, and other parameters of the system of particles.

1.3 Characteristics of center of mass

(i) **Acceleration of center of mass**: After realising the requirement of the concept of center of mass, let us discuss all the factors that define the center of mass. As you know, in kinematics, a point is characterised by its position, velocity and acceleration. Since we find the acceleration from force equation, then velocity from acceleration and position from velocity, let us define the center of mass by using the idea of force as follows.

Center of mass is a point that moves with an acceleration equal to the net force acting on the system divided by the mass of the system. Symbolically

$$\vec{a}_{CM} = \frac{\vec{F}}{\sum m_i} = \frac{\sum m_i \vec{a}_1}{\sum m_i}$$

Basically, we have two equations; one is

$$\vec{a}_{CM} = \frac{\vec{F}}{\sum m_i}$$

And the other is

$$\vec{a}_{CM} = \frac{\sum m_i \vec{a}_1}{\sum m_i}$$

Note that both equations give the same result, but the approach may be different. This will be clear in the following example.

Example 4 A flexible chain of mass m is rotating about a vertical axis passing through the point P by an inextensible light string. If the string makes a constant angle θ with the vertical, find the (a) radius of the circular path traced by the center of mass of the chain (b) tension in the string.

Solution

The forces acting on the chain are tension T and weight $mg\downarrow$. Let us write the force equation of the center of mass, which moves in a horizontal circle of radius r. Let the center of mass have acceleration a_c, which is radially inward as it moves in a horizontal circle.

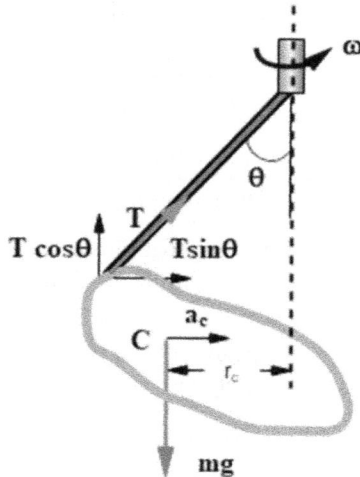

$$F_r = ma_r,$$

where $F_r = T \sin \theta$ and $a_r = r\omega^2$. This gives

$$T \sin \theta = mr\omega^2 \qquad (1.9)$$

The net vertical force acting on the chain is

$$F_y = ma_y.$$

Since the center of mass moves in a horizontal circle,

$$a_y = 0.$$

$$\Rightarrow F_y = T \cos \theta - mg = 0$$

$$\Rightarrow T \cos \theta = mg \qquad (1.10)$$

Eliminating T from equations (1.9) and (1.10), we have

$$r = \frac{g \tan \theta}{\omega^2} \text{ Ans.}$$

(b) Using the equation (1.10) the tension in the string is

$$\Rightarrow T = mg \sec \theta \text{ Ans.}$$

N.B.: All forces T and mg are imagined to act at the center of mass. If you take the radius of the ring or radius of the circular path traced by any other point of the chain (such as Q), you will get the wrong answer. That means, in $F = ma$, a is the acceleration of center of mass and F is the net force acting on the body (center of mass).

(ii) Velocity of center of mass: The change in velocity of center of mass from $t = t_1$ and $t = t_2$ can be given as

$$\Delta \vec{v}_C = \int_{t_1}^{t_2} \vec{a}_C dt$$

By substituting the acceleration of the center of mass

$$\vec{a}_C = \frac{\sum m_i \vec{a}_i}{\sum m_i},$$

In the last equation, we have

$$\Delta \vec{v}_C = \frac{\sum m_i \int_{t_1}^{t_2} \vec{a}_i dt}{\sum m_i},$$

where $\int_{t_1}^{t_2} \vec{a}_i dt = \Delta \vec{v}_i$

$$\Rightarrow \Delta \vec{v}_C = \frac{\sum m_i (\Delta \vec{v}_i)}{\sum m_i}$$

Since the change in velocity of the center of mass is

$$\Delta \vec{v_C} = \vec{v_C} - \vec{u_C} \text{ and } \sum m_i(\Delta \vec{v_i}) = \sum m_i \vec{v_i} - \sum m_i \vec{u_i}$$

Comparing both sides we have,

$$\vec{v_C} = \frac{\sum m_i \vec{v_i}}{\sum m_i}$$

Example 5 Find the velocity of center of mass of the system of two moving particles of mass m and $2m$, as shown in figures (i), (ii), and (iii).

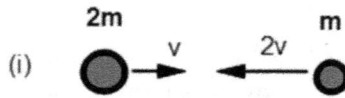

Solution
(i) Let us use the equations

$$\vec{v_C} = \frac{\sum m_i \vec{v_i}}{\sum m_i},$$

$$\Rightarrow \vec{v_C} = \frac{(m)(2v\,\hat{i}) + (2m)(-v\,\hat{i})}{m + 2m} = 0$$

(ii) $\vec{v_C} = \frac{m(2v\,\hat{i}) + 2m(v\,\hat{i})}{m + 2m} = \frac{4v}{3}\,\hat{i}$

(iii) $\vec{v_C} = \frac{m(-2v\,\hat{i}) + 2m(v\,\hat{j})}{m + 2m} = \frac{2v}{3}(\hat{i} - \hat{j})$

Example 6 Find the velocity of center of mass of the system of three particles of mass m, $2m$, and $3m$ placed at the vertices of an equilateral triangle.

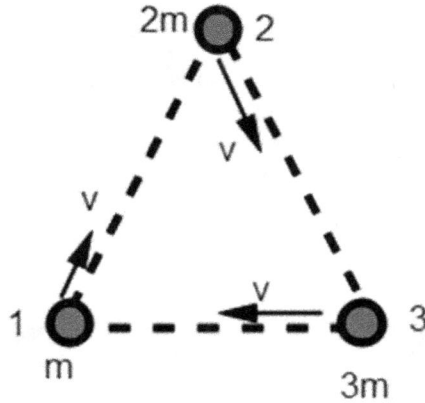

Solution

(i) Let us use the equation

$$\vec{v}_C = \frac{\sum m_i \vec{v}_i}{\sum m_i} = \frac{m_1 \vec{v}_1 + m_2 \vec{v}_2 + m_3 \vec{v}_3}{m_1 + m_2 + m_3}$$

$$\Rightarrow \vec{v}_C = \frac{(m)(\frac{1}{2}v\,\hat{i} + \frac{\sqrt{3}}{2}v\,\hat{j}) + (2m)(\frac{1}{2}v\,\hat{i} - \frac{\sqrt{3}}{2}v\,\hat{j}) + (3m)(-v\,\hat{i})}{m + 2m + 3m}$$

$$= -\frac{1}{4}(\hat{i} + \frac{1}{\sqrt{3}}\hat{j})\,v \text{ Ans.}$$

(iii) Displacement of center of mass: Integrating velocity of center of mass with time, the displacement of center of mass is

$$\vec{s}_C = \int \vec{v}_C dt,$$

where $\vec{v}_C = \frac{\sum m_i \vec{v}_i}{\sum m_i}$

$$\Rightarrow \vec{s}_C = \frac{\sum m_i \int_{t_1}^{t_2} \vec{v}_i dt}{\sum m_i},$$

where

$$\int_{t_1}^{t_2} \vec{v}_i dt = \vec{s}_i.$$

Then, we have

$$\vec{s}_C = \frac{\sum m_i \vec{s}_i}{\sum m_i}$$

Example 7

Four particles of mass m, $2m$, $3m$, and $4m$ are placed at the vertices of a rectangle $ABCD$ as shown in the figure. Let the particles move (a) from A to B, B to Cm C to D, and D to A, respectively (b) to the center of the rectangle. In this process find the displacement of center of mass of the system of four particles. Put $AB = l$ and $BC = 2l$ and assume A as the origin of the coordinate system.

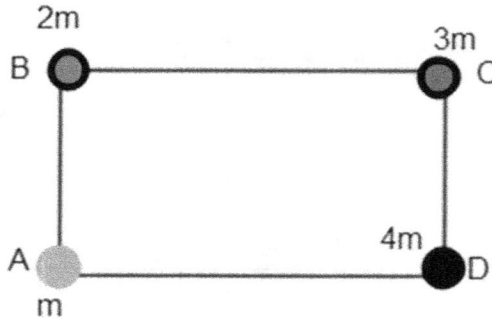

Solution

Referring to the formula $\vec{s_C} = \frac{\sum m_i \vec{s_i}}{\sum m_i}$ for the system of four particles, we have

$$\vec{s_C} = \frac{m_1 \vec{s_1} + m_2 \vec{s_2} + m_3 \vec{s_3} + m_4 \vec{s_4}}{m_1 + m_2 + m_3 + m_4},$$

where $m_1 = m$, $m_2 = 2m$, $m_3 = 3m$, $m_4 = 4m$,

$$\vec{s_1} = l\hat{j}, \; \vec{s_2} = 2l\hat{i}, \; \vec{s_3} = -l\hat{j} \text{ and } \vec{s_4} = -2l\hat{i},$$

$$\Rightarrow \vec{s_C} = \frac{(m)(l\hat{j}) + 2m(2l\hat{i}) + 3m(-l\hat{j}) + 4m(-2l\hat{i})}{m + 2m + 3m + 4m}$$

$$= -\frac{1}{5}\hat{i} - \frac{21}{5}\hat{j} \text{ Ans.}$$

(b)

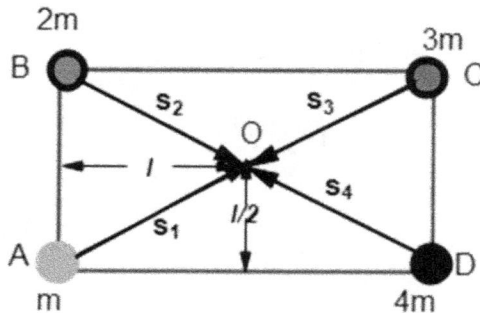

Putting $m_1 = m$, $m_2 = 2m$, $m_3 = 3m$, $m_4 = 4m$ in the equation

$$\vec{s_C} = \frac{m_1 \vec{s_1} + m_2 \vec{s_2} + m_3 \vec{s_3} + m_4 \vec{s_4}}{m_1 + m_2 + m_3 + m_4}, \text{ we have}$$

$$\vec{s_C} = \frac{m \vec{s_1} + (2m) \vec{s_2} + (3m) \vec{s_3} + (4m) \vec{s_4}}{m + 2m + 3m + 4m}$$

$$\vec{s_C} = \frac{\vec{s_1} + 2\vec{s_2} + 3\vec{s_3} + 4\vec{s_4}}{10}$$

Then, we can put $\vec{s_1} = l(\hat{i} + 0.5\hat{j})$, $\vec{s_2} = l(\hat{i} - 0.5\hat{j})$, $\vec{s_3} = l(-\hat{i} - 0.5\hat{j})$ and $\vec{s_4} = l(-\hat{i} + 0.5\hat{j})$ and simplify the factors to obtain

$$\vec{s_C} = -0.4l\hat{i} \text{ Ans.}$$

Example 8 Two particles 1 and 2 are attached with an inextensible light string that passes over a smooth pulley. Another particle 3 (an insect, say) sits on the particle 2 such that the system $(1 + 2 + 3)$ is in equilibrium. (a) If the total mass of the system is M and the mass of the insect is m, find the displacement of the center of mass of the system when the insect crawls with an upward distance 'y' relative to the string. (b) Why should the center of mass move up? (c) What is the tension in the displacements of 1, 2, $(2 + 3)$?

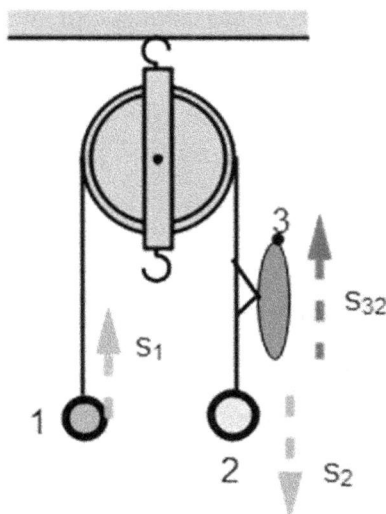

Solution

The system has three particles 1, 2, and 3. The displacement of the system can be given as

$$\vec{s}_{CM} = \frac{m_1 \vec{s_1} + m_2 \vec{s_2} + m_3 \vec{s_3}}{m_1 + m_2 + m_3}$$

where $\vec{s_1} = -\vec{s_2}$ (because the string is inextensible) and $\vec{s_3} = \vec{s_{32}} + \vec{s_2}$. Then, we have

$$\vec{s}_{CM} = \frac{m_1 \vec{s_1} + m_2 \vec{s_2} + m_3(\vec{s_{32}} + \vec{s_2})}{m_1 + m_2 + m_3}$$

$$= \vec{s}_{CM} = \frac{m_1 \vec{s_1} + (m_2 + m_3)\vec{s_2} + m_3 \vec{s_{32}}}{m_1 + m_2 + m_3}$$

Since, $m_1 = m_2 + m_3$ and $\vec{s_1} + \vec{s_2} = 0$, we have

$$m_1 \vec{s_1} + (m_2 + m_3)\vec{s_2} = 0$$

$$\Rightarrow \vec{s}_{CM} = \frac{m_3 \vec{s_{32}}}{m_1 + m_2 + m_3}$$

Putting $m_3 = m$, $m_1 + m_2 + m_3 = M$ and $\vec{s_{32}} = y\hat{j}$ in the last equation, we have

$$\vec{s}_{CM} = \frac{m}{M} y\hat{j}$$

(b) The upward impulse changes the momentum of the system in an upward direction.

(c) The net force acting on the system is
$2T - Mg = Ma_c$,

$$\text{where } \vec{a_c} = \frac{d^2 \vec{s}_{CM}}{dt^2} = \frac{m}{M}\frac{d^2 y}{dt^2}\hat{j}$$

This gives us

$$T = \frac{1}{2}\left(Mg + m\frac{d^2 y}{dt^2}\right) = \frac{1}{2}(Mg + ma_r)$$

(iv) Position of the center of mass: After calculating the velocity and acceleration, we want to find the location of center of mass given by the position vector $\vec{r_C}$. Thus, the center of mass is characterised by its position, velocity, and acceleration. You can see that $\vec{r_C}$ comes from $\vec{v_C}$, $\vec{v_C}$ comes from $\vec{a_C}$ and $\vec{a_C}$ comes from $\vec{F}_{net}$. By using the expression

$$\vec{S_C}(=\Delta \vec{r_C}) = \frac{\sum m_i \Delta \vec{r_i}}{\sum m_i},$$

1-13

And comparing both sides, we have

$$\vec{r_C} = \frac{\sum m_i \Delta \vec{r_i}}{\sum m_i}$$

Example 9 Locate the center of mass of a system of three particles 1, 2, and 3 each of mass m situated at the vertices of a right-angled triangle. Put $\theta = 60°$.

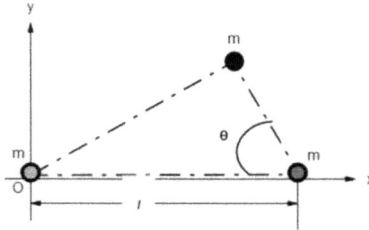

Solution

The position of the center of mass is $\vec{r_C} = \frac{m_1 \vec{r_1} + m_2 \vec{r_2} + m_3 \vec{r_3}}{m_1 + m_2 + m_3}$.

Substituting $m_1 = m_2 = m_3 = m, \vec{r_1} = 0, \vec{r_2} = \frac{3l}{4}\hat{i} + \frac{\sqrt{3}}{4}l\hat{j}$ and $\vec{r_3} = l\hat{i}$, we have

$$\vec{r_C} = \frac{m(\frac{3l}{4}\hat{i} + \frac{\sqrt{3}}{4}l\hat{j}) + m(l\hat{i})}{m + m + m}$$

$$\Rightarrow \vec{r_C} = \frac{7l}{12}\hat{i} + \frac{\sqrt{3}l}{12}\hat{j} \text{ Ans.}$$

N.B.: If the particles have different masses, the centroid of the triangle and center of mass of the system do not coincide. It may be possible when masses are adjusted properly so that $m_1 \vec{r_1} + m_2 \vec{r_2} + m_3 \vec{r_3} = 0$, where $\vec{r_1}, \vec{r_2}$, and $\vec{r_3}$ are taken from the geometrical center of the triangle.

Let us summarize the above discussion.

The center of mass of a system of particles is characterised by its position, displacement, velocity, and acceleration, given as

$$\vec{r_C} = \frac{\sum m_i \vec{r_i}}{\sum m_i}, \quad \Delta \vec{r_C} = \frac{\sum m_i \Delta \vec{r_i}}{\sum m_i}, \quad \vec{v_C} = \frac{\sum m_i \vec{v_i}}{\sum m_i} \text{ and } \vec{a_C} = \frac{\sum m_i \vec{a_i}}{\sum m_i} \left(= \frac{\vec{F_{net}}}{\sum m_i} \right) \text{ respectively.}$$

1.4 Calculation of position of center of mass (finding center of mass)

Now let us find the location of center of mass of various systems of particles. In some cases the arrangement of the particles are discrete and in other cases the particles are continuously arranged. Generally, a solid or liquid is practically composed of continuously arranged particles like molecules. But when we observe under microscope, the molecules do not touch each other; these are microscopically just discrete particle distributions. However, it appears as a continuous mass distribution in macroscopic scale (gross matter). Furthermore, if you go deeper into the atomic scale, you will find that the molecules we consider as point masses are composed of tinier atoms. An atom can also be thought as a point mass when we compare a molecule, but it has many subatomic particles like neutron, proton, and nucleus. This discussion emphasizes the fact that microscopically matter is quantised. This means that matter has discrete (discontinuous) mass distribution at microscopic level, whereas macroscopically matter appears to be continuous.

Let us discuss the methods of finding the position of center of mass of a system comprising (a) discrete mass distribution and (b) continuous mass distribution.

(A) Discrete mass distribution

Example 10 Find the center of mass of a system of eight particles placed at the corners of a cube of edge l.

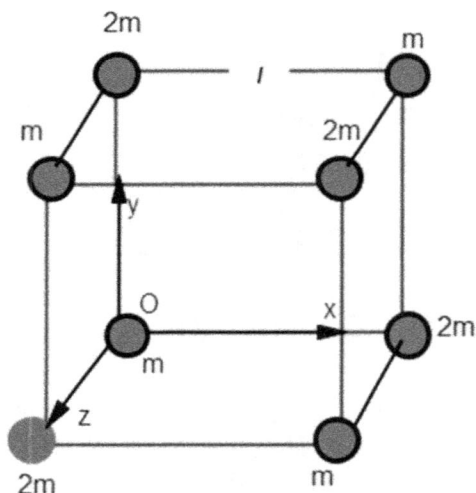

Solution
There are eight particles; $n = 1 \rightarrow 8$. Following the general formula

$$\vec{r_C} = \frac{\sum m_i \vec{r_i}}{\sum m_i}, \text{ we can write}$$

$$\vec{r_C} = \frac{\displaystyle\sum_{i=1}^{i=8} m_i \vec{r_i}}{\displaystyle\sum_{i=1}^{i=8} m_i} \tag{1.11}$$

Substituting the masses and position vectors of the particles, we have

$$\sum_{i=1}^{i=8} m_i = 13m \tag{1.12}$$

The sum of product of mass and corresponding position vectors is

$$\sum_{i=1}^{i=8} m_i \vec{r_i} = m(0\hat{i} + 0\hat{j} + 0\hat{k}) + 2m(l\hat{i} + 0\hat{j} + 0\hat{k}) + 2m(0\hat{i} + l\hat{j} + 0\hat{k}) + 2m(0\hat{i} + 0\hat{j} + l\hat{k})$$

$$+ m(l\hat{i} + l\hat{j} + 0\hat{k}) + m(0\hat{i} + l\hat{j} + l\hat{k}) + m(l\hat{i} + 0\hat{j} + l\hat{k}) + 3m(l\hat{i} + l\hat{j} + l\hat{k})$$

$$\sum_{i=1}^{i=8} m_i \vec{r_i} = 7ml(\hat{i} + \hat{j} + \hat{k}) \tag{1.13}$$

Using last three equations, we have

$$\vec{r_C} = \frac{7ml(\hat{i} + \hat{j} + \hat{k})}{13m} = \frac{7l}{13}(\hat{i} + \hat{j} + \hat{k}) \text{ Ans.}$$

Example 11 Find the center of mass of a system of four particles placed at the vertices of a regular tetrahedron of side l.

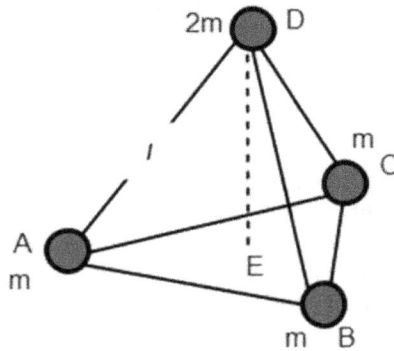

Solution
Since equal masses are placed at the vertices of the equilateral triangular base ABC, using the principle of symmetry, the center of mass of the base must lie at the point E. So, we can replace the triangle ABC by a particle of mass $3m$ placed at point E. Now we have two particles of mass $2m$ and $3m$ placed at D and E, respectively. Then, the center of mass O of the system is located at a height OE given as

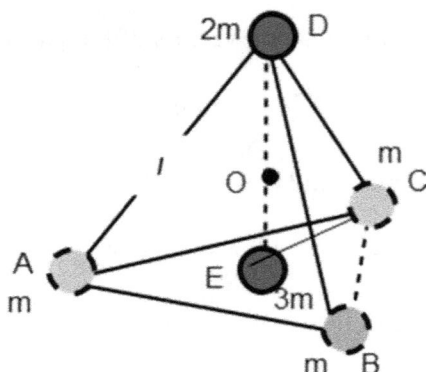

$$OE = \frac{2m}{2m + 3m}DE = \frac{2}{5}DE,$$

In the right-angled triangle DEC,

$$DE = \sqrt{CD^2 - CE^2} = \sqrt{l^2 - \left(\frac{l}{\sqrt{3}}\right)^2} = \sqrt{\frac{2}{3}}\,l$$

Then, we have

$$OE = =\frac{2}{5}\sqrt{\frac{2}{3}}\,l$$

This means that the center of mass of the given system of particle is located at a height of $\frac{2}{5}\sqrt{\frac{2}{3}}l$ from the base, on the vertical line DE passing through the apex D of the tetrahedron. Ans.

(B) Continuous mass distribution

For continuous mass distribution it is difficult to distinguish ith practice from jth particle. In that case, we have to consider an elementary segment of mass 'dm'. This behaves as a point mass. Hence, we can write 'dm' instead of 'm_i'. Assuming the position vector of dm as $\vec{r}$, we can substitute $\vec{r_i}$ as $\vec{r}$. Finally, substituting '$\int$' for '$\sum$', we can find the formula for position of center of mass for continuous mass distribution, which can be given as

$$\vec{r_C} = \frac{\int \vec{r}\,dm}{\int dm}$$

(a) **Linear mass distribution:** If the object is linear, $dm = \lambda dr$. Then, we have

$$r_C = \frac{\int \lambda r\,dm}{\int \lambda dm}$$

Linear mass distribution can be straight or curved. Let us discuss the center of mass of straight and curved linear mass in the following examples.

(i) Straight line

Example 12 Find the position of center of mass of a thread of uniform cross-section whose linear mass density varies with x as $\lambda = \lambda_0(1 + \frac{x}{l})$.

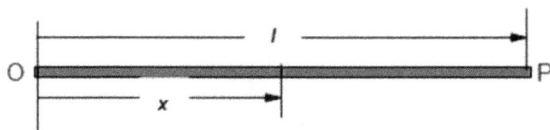

Solution

The formula for the position of the center of mass is

$$r_C = \frac{\int \lambda r\, dr}{\int \lambda\, dr},$$

where $\lambda = \lambda_0(1 + \frac{x}{l})$,

$$\Rightarrow x_C = \frac{\int_0^l \lambda_0(1 + \frac{x}{l})x\, dx}{\int_0^l \lambda_0(1 + \frac{x}{l})dx} = \frac{5}{9}l \text{ Ans.}$$

N.B.: In the above example we found that the center of mass and the geometrical center (mid-point) of the thread are different. In which case both coexist if the thread is uniform.

In the formula $r_C = \frac{\int \lambda r\, dr}{\int \lambda\, dr}$, we cannot take λ out of the integral and cancel in both numerator and denominator before evaluating the integral when $\lambda = f(r)$ in the case of non-uniform mass distribution. However, we can do the same when $\lambda = c$ for uniform mass distribution. In that case, the center of mass and mid-point of a straight uniform object coincide. Another important thing we need to remember is that, in the above formula, we must treat 'r' as 'vector' even though we mention the formula in terms of scalars. In other words, r is an algebraic scalar. If the position lies in $-$ve x-, y-, and z-directions (axes), we write it negative and *vice versa*. Lastly, we note that $m = \lambda l$ is valid for uniform objects only; however, $m = \int \lambda\, dl$ is a general expression.

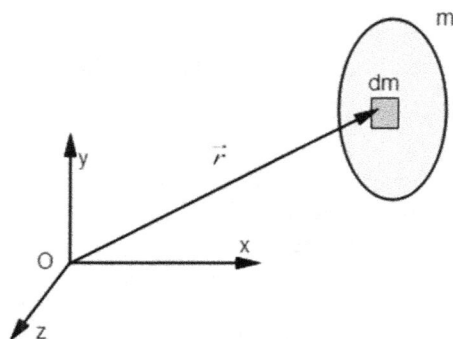

Sometimes we come across the systems of straight rods. In that case, how do we locate the center of mass of the system? Let us see through the following example.

Example 13 Four uniform rods AB, BC, CD, and DA of linear mass densities λ, 2λ, 3λ and 4λ are joined end to end to form a rectangle. The point A of the rectangle coincides with the origin O of the coordinate system. Find the position of center of mass of the system of four rods.

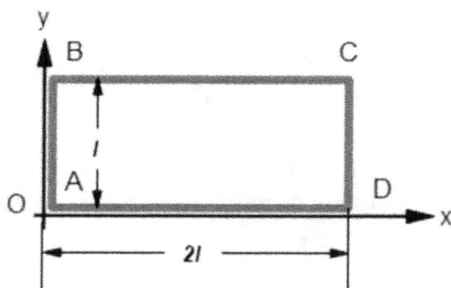

Solution
Since each rod is uniform, the center of mass of each rod lies at its mid-point. Hence, we replace the rods by point masses of magnitude equal to the mass of the respective rods and place the point masses at the mid-points of the rods. Then using the formula for discrete particle distribution, we can locate the center of mass of the system.

Following the above logic, we have

$$\vec{r_C} = \frac{m_1 \vec{r_1} + m_2 \vec{r_2} + m_3 \vec{r_3} + m_4 \vec{r_4}}{m_1 + m_2 + m_3 + m_4},$$

where $m_1 = \lambda l$, $m_2 = (2\lambda)(2l) = 4\lambda l$,

$$m_3 = (3\lambda)(l) = 3\lambda l, \quad m_4 = (4\lambda)(2l) = 8\lambda l, \quad \vec{r_1} = \frac{1}{2}\hat{j},$$

$$\vec{r_2} = l(\hat{i} + \hat{j}), \quad \vec{r_3} = l(2\hat{i} + \frac{\hat{j}}{2}) \text{ and } \vec{r_4} = l\hat{i}$$

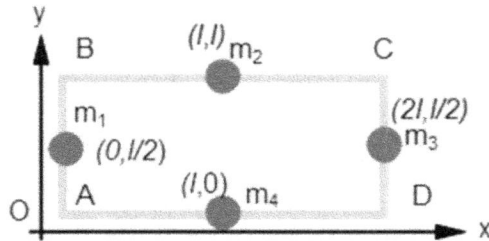

After substituting the values of mass,

$$\vec{r_C} = \frac{(\lambda l)\vec{r_1} + (4\lambda l)\vec{r_2} + (3\lambda l)\vec{r_3} + (8\lambda l)\vec{r_4}}{\lambda l + 4\lambda l + 3\lambda l + 8\lambda l}$$

$$= \frac{\vec{r_1} + 4\vec{r_2} + 3\vec{r_3} + 8\vec{r_4}}{16}$$

Then, substituting the values of position vectors,

$$\vec{r_c} = \frac{\frac{l}{2}\hat{j} + 4l(\hat{i} + \hat{j}) + 3l(2\hat{i} + \frac{1}{2}\hat{j}) + 8(l\hat{i})}{16}$$

After simplifying the factors, we have

$$\vec{r_c} = \frac{3(3\hat{i} + \hat{j})l}{8} \text{ Ans.}$$

Example 14 A uniform rod of mass m and length l is bent to form a right-angled triangle as shown in the figure. Locate the center of mass of the rod. Put $\theta = 30°$.

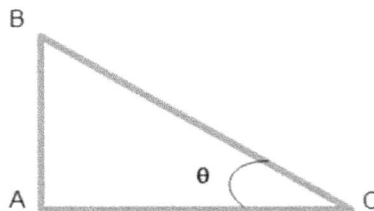

Solution

Since the rods are uniform, the center of mass of each rod lies at its mid-point. Hence, by replacing the rods by point masses of magnitude equal to the mass of the respective rods and placing the point masses at the mid-points of the rods, we have the following diagram:

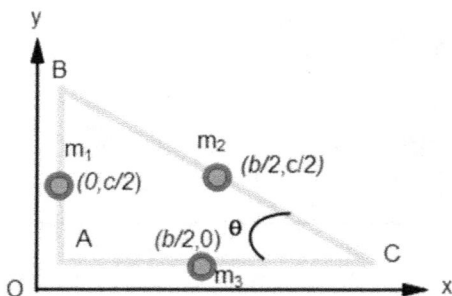

Then using the formula for discrete particle distribution, we can locate the center of mass of the system. Let $BC = a$, $AC = b$, and $AB = c$.

Following the above logic, we have

$$\overrightarrow{r_C} = \frac{m_1 \overrightarrow{r_1} + m_2 \overrightarrow{r_2} + m_3 \overrightarrow{r_3}}{m_1 + m_2 + m_3},$$

where $m_1 = \lambda c = \lambda a \cos \theta$, $m_2 = \lambda a$,

$$m_3 = \lambda b = \lambda a \cos \theta, \quad \overrightarrow{r_1} = \frac{c}{2}\hat{j} = \frac{a \sin \theta}{2}\hat{j}$$

$$\overrightarrow{r_2} = \left(\frac{b\hat{i}}{2} + \frac{c\hat{j}}{2}\right) = \frac{a}{2}(\cos \theta \hat{i} + \sin \theta \hat{j}) \text{ and } \overrightarrow{r_3} = \frac{b}{2}\hat{i} = \frac{a \cos \theta}{2}\hat{i}$$

$$\Rightarrow \overrightarrow{r_C} = \frac{(\lambda a \sin \theta)(\frac{a \sin \theta}{2}\hat{j}) + (\lambda a)(\frac{a \sin \theta}{2}\hat{i} + \frac{a \sin \theta}{2}\hat{j}) + (\lambda a \cos \theta)(\frac{a \cos \theta}{2}\hat{i})}{\lambda a \sin \theta + \lambda a + \lambda a \cos \theta}$$

$$= \frac{\cos \theta(1 + \cos \theta)\hat{i} + \sin \theta(1 + \sin \theta)\hat{j}}{2(1 + \sin \theta + \cos \theta)}a$$

$$\overrightarrow{r_C} = \frac{l}{4(3 + \sqrt{3})}\{(2\sqrt{3} + 3)\hat{i} + 3\hat{j}\}$$

relative to A. Ans.

Summarizing the above facts:

When a system of rods is given, find the center of mass of each rod and make an equivalent system of point masses comprising the center of mass of the rods. Then using the formula $\vec{r_C} = \frac{\sum m_i \vec{r_i}}{\sum m_i}$, find the center of mass of the system of rods. Keep in mind that for a uniform straight rod, we can directly substitute it by a point mass equal to the mass of the rod placed at the mid-point of the rod. However, in a non-uniform case, we have to follow the basic procedure to find the center of mass.

In the previous discussion we learned that whenever a rod is bent, its shape and position of the particles change. As a consequence, the center of mass of the system changes its position.

Let us now bend the rods into curves and try to find their center of mass. Since a circle is the simplest curve, let us take an example of a circular arc.

(ii) Circular arc

Example 15 A uniform rod of length l is bent into a circular arc of radius R. Locate the center of mass of the rod.
  Solution
  Let us draw the x-axis such that it passes through the center 'O' of curvature of the circular arc and divides the circular arc into two equal halves.
$\vec{r_C} = \frac{\int \vec{r}\, dm}{\int dm}$, we have

$$\vec{r_C} = \frac{\int \vec{r}\, d\phi}{\int d\phi}, \quad \text{where } \vec{r} = x\hat{i} + y\hat{j}$$

Substituting $x = R\cos\phi$ and $y = R\sin\phi$, we have

$$\vec{r_C} = \frac{R\int \cos\phi\, d\phi}{\int d\phi}\hat{i} + \frac{R\int \sin\phi\, d\phi}{\int d\phi}\hat{j}$$

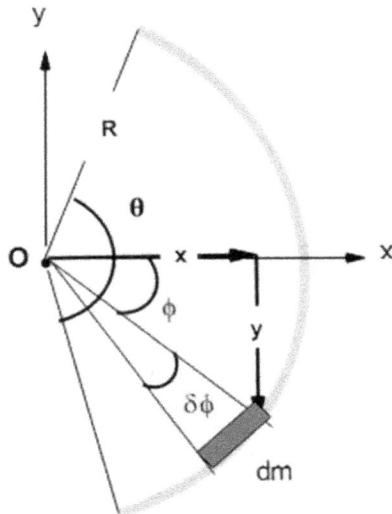

If the arc subtends and angle $\theta(=\frac{1}{R})$, which is symmetrical about the x-axis, we put a lower limit of $\phi = -\frac{\theta}{2}$ and an upper limit of $\phi = \frac{\theta}{2}$ to obtain

$$\overrightarrow{r_C} = \frac{R \int_{-\theta/2}^{\theta/2} \cos \phi d\phi}{\int_{-\theta/2}^{\theta/2} d\phi} \hat{i} + \frac{R \int_{-\theta/2}^{\theta/2} \sin \phi d\phi}{\int_{-\theta/2}^{\theta/2} d\phi} \hat{j}$$

After evaluating the integrations, you can see that the second term of the RHS will vanish giving us $\overrightarrow{r_C} = \frac{2R \sin \frac{\theta}{2}}{\theta} \hat{i}$.

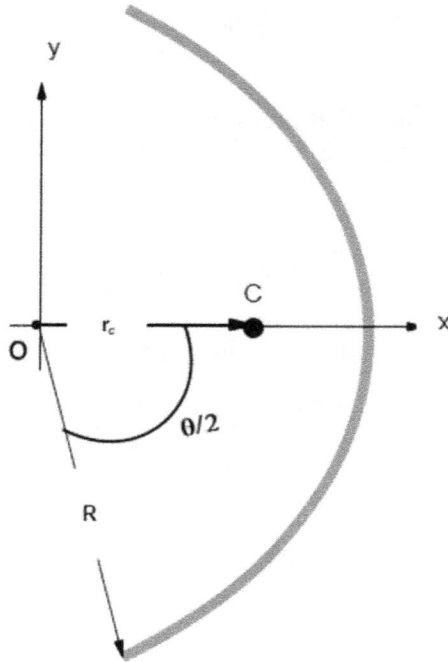

Substituting $\theta = \left(\frac{l}{R}\right)$, we have

$$\overrightarrow{r_C} = \frac{2R^2}{l} \sin\left(\frac{l}{2R}\right) \hat{i} \text{ Ans.}$$

N.B.: The center of mass of an arc of radius R subtending an angle θ at the center of its curvature lies at a distance of $r = \frac{2R \sin \frac{\theta}{2}}{\theta}$ from its center of curvature on the radial line passing through the mid-point of the arc. As the uniform arc is symmetrical about the x-axis, logically we can conclude that the y-coordinate of $\overrightarrow{r_C}$ must be zero.

Example 16 A uniform rod of length l is bent into a circular arc MQN. If the ends of the rod subtends an angle MNQ $= \theta = 30°$ at the point Q as shown in the figure, find the position r_c of the center of mass of the rod.

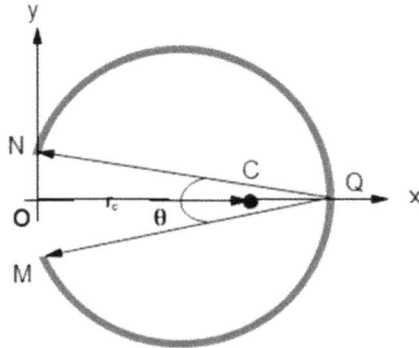

Solution
The angle made by the end M and N of the arc at its center of curvature P is given as

$$\alpha = 2\beta = 2\{2(\theta/2\} = 2\theta$$

So, the angle subtended by the arc MQN at P is

$$\phi = \pi - 2\theta. \tag{1.14}$$

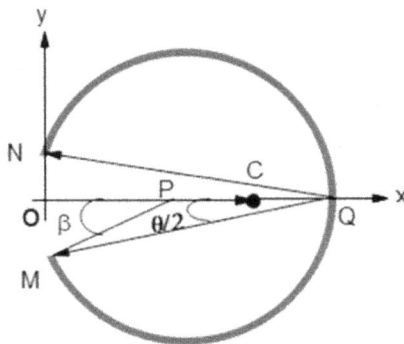

The position of the center of mass of the circular arc from its center of curvature P is

$$r'_C = PC = \frac{2R \sin \frac{\phi}{2}}{\phi} \tag{1.15}$$

Using last two equations, we have

$$r'_C = \frac{2R \sin \frac{\pi - 2\theta}{2}}{\pi - 2\theta} = \frac{R \cos \theta}{\pi/2 - \theta}, \tag{1.16}$$

where the length of the arc MQN is

$$l = R(\pi - 2\theta) \tag{1.17}$$

Putting the value of r from equation (1.17) in equation (1.16), finally we have

$$r'_C = \frac{l/(\pi - 2\theta)\cos\theta}{\pi/2 - \theta} = \frac{2l\cos\theta}{(\pi - 2\theta)^2} \tag{1.18}$$

Then, $OC = r_c = OP + PC = OP + r'_c$, where $OP = R\cos\theta$.
So, we have

$$r_c = R\cos\theta + r'_c \tag{1.19}$$

Using the last two equations,

$$r_C = \frac{2l\cos\theta}{(\pi - 2\theta)^2} + R\cos\theta \tag{1.20}$$

Using equations (1.17) and (1.20), finally we have

$$r_C = \frac{2l\cos\theta}{(\pi - 2\theta)^2} + \frac{l}{(\pi - 2\theta)}\cos\theta$$

$$\Rightarrow r_C = \frac{l\cos\theta}{(\pi - 2\theta)}\left\{\frac{2}{(\pi - 2\theta)} + 1\right\}$$

$$\Rightarrow r_C = \frac{9l(3 + \pi)}{4\pi^2} \text{ Ans.}$$

(b) Surface (areal) mass distribution: Physically or practically, continuous surface mass distribution means thin sheets (laminae). When we consider a surface mass distribution, two types of surfaces come to mind; one is a plane surface and the other is a curved surface. We can discuss in detail for each surface later on. At present let us try to derive a general expression for locating the center of mass. For this consider an elementary area dA containing an elementary mass dm at the point P having position vector $\vec{r}$. Assuming the surface mass distribution at the position P, we can write $dm = \sigma dA$.

Thin sheet

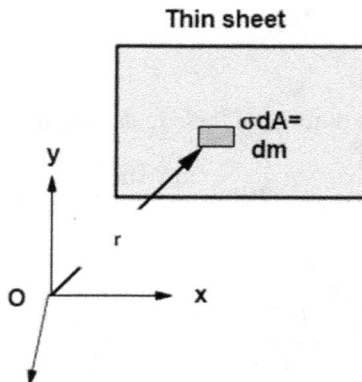

Substituting this in the basic formula,

$\vec{r_C} = \frac{\int \vec{r} \, dm}{\int dm}$, we have

$$\vec{r_C} = \frac{\int_S \vec{r} \, dA}{\int_S \sigma \, dA},$$

where '$\int_S$' is the surface (double) integral. We will talk about the surface integral in a user-friendly way. So do not panic when you look at this strange looking integral. You will understand in a moment. For uniform surface mass distribution, σ is a constant. In that case, we can write $m = \sigma A$. Then, we can take σ out of the integral and cancel it from both the numerator and denominator. This gives,

$$\vec{r_C} = \frac{\int_S \vec{r} \, dA}{A}$$

This formula is undoubtedly simpler in the absence of σ, which can also be a function of distance for non-uniform distribution of mass.

(i) Flat sheets: Let us try to apply the above formula for flat surfaces. First of all we will use the Cartesian coordinate system. For this, let us imagine that the sheet lies in the x-y plane. Then the elementary area dA at a point P having position vector $\vec{r}$ can be given as

$$dA = dx. \ dy$$

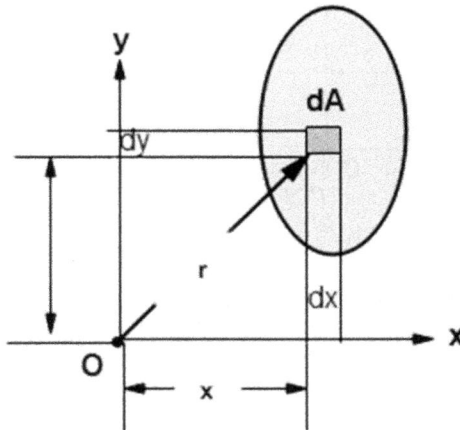

Substituting $dA = dx. \ dy$ and $\vec{r} = x\hat{i} + y\hat{j}$ in the formula $\vec{r_C} = \frac{1}{A} \int_S \vec{r} \, dA$, we have

$$\vec{r_c} = \left(\frac{l}{A}\int x\ dx\ dy\right)\hat{i} + \left(\frac{l}{A}\int y\ dx\ dy\right)\hat{j},$$

where $\vec{r_c} = x_c\hat{i} + y_c\hat{j}$

Then, we have

$$x_C = \frac{1}{A}\iint x\ dx\ dy,\ \ y_C = \frac{1}{A}\iint y\ dx\ dy$$

Let me explain the physical significance of the surface integral. The surface integral '$\int_S$' can be written as a double integral '$\iint$'.

It may seem mathematically a bit different from line integral '$\int_l$' what we call the simple integtral.

When we analyse the double integral $\iint x\ dx\ dy$, we can see that $\int dx\ dy (=dA)$ is the sum of elementary areas. Hence,

$$\iint x\ dx\ dy = \int x \int dx\ dy$$

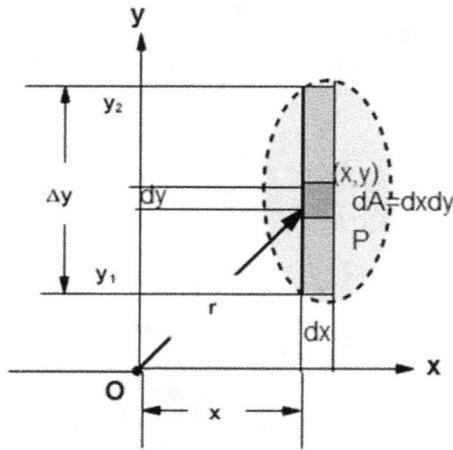

Which means that the elementary areas are summed up keeping 'x' constant. Substituting $\int dx\ dy = \int dA = dA'$, we have

$$\iint x\ dx\ dy = \int x\ dA'$$

where dA'= area of the vertical strip of thickness dx at constant x. Similarly, we can write

$$\iint y\ dx\ dy = \int y \int dA = \int y\ dA',$$

where dA'= area of the horizontal strip of thickness dy at a constant y.

Now substituting $\iint x \, dx \, dy = \int x \, dA'$ and $\iint y \, dx \, dy = \int y \, dA'$ in the above formula, we have $x_C = \frac{1}{A} \int x \, dA'$

$$y_C = \frac{1}{A} y \, dA'$$

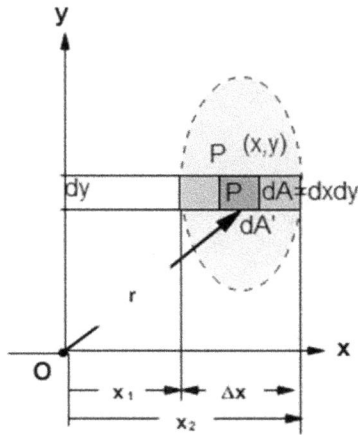

Note that $dA'=$ area of the elementary strips parallel to the y- and x-axes keeping x and y constants as shown in the above figure.

After simplifying the expressions to simple integrals, now you can apply them more conveniently in the following examples. However, you are always encouraged to use the basic (original) formula in more complex problems.

Example 17 Locate the center of mass of a triangular lamina.

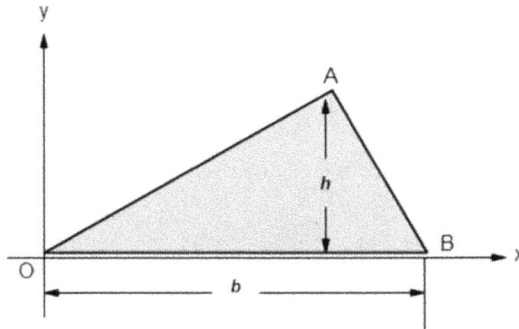

Solution

To find the y-coordinate of center of mass we will have to take an elementary strip of length l at a constant y parallel to the base of the triangle which lies on x-axis. The area of the strip is given as

$$dA' = l \, dy, \tag{1.21}$$

Using the properties of similar triangles OAB and CAD
CD/OB $= (h{-}y)/h$

$$\Rightarrow \frac{l}{b} = \frac{h - y}{h} \tag{1.22}$$

Using last two equations, we have

$$dA' = \frac{b(h - y)}{h} dy \tag{1.23}$$

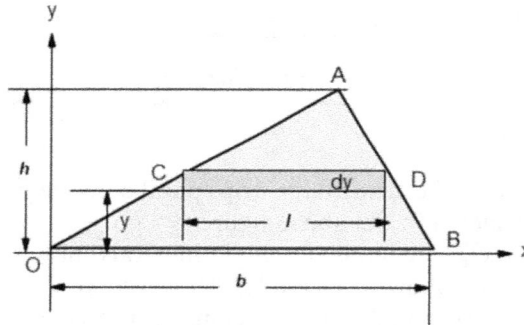

The y-coordinate of the center of mass is

$$y_C = \frac{1}{A} \int y \, dA' \tag{1.24}$$

Using last two equations, we have

$$y_C = \frac{1}{A} \int_0^h \frac{by}{h}(h - y)dy \tag{1.25}$$

Substituting $A = \frac{1}{2}bh$ in the above equation and evaluating the integral, we have $y_C = \frac{h}{3}$. Ans.

We can find x_C by using the formula

$$x_C = \frac{1}{A} \int x \, dA' \tag{1.26}$$

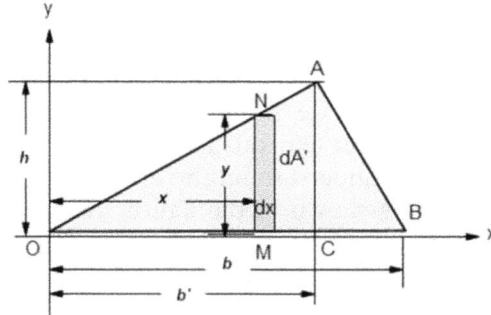

To find the x-coordinate of center of mass we will have to take an elementary vertical strip of length y at a distance x from the origin. The area of the strip is given as

$$dA' = y\,dx \tag{1.27}$$

Using the properties of similar triangles OAC and ONM, we have
NM/AC = OM/OC

$$\frac{y}{h} = \frac{x}{b'} \tag{1.28}$$

Using the last two equations, we have

$$dA' = \frac{xh}{b'}dx \tag{1.29}$$

Using the equations (1.26) and (1.29), we have

$$x_C = \frac{1}{A}\left\{\int_0^{b'} \frac{x^2 h}{b'}dx + \int_{b'}^{b} \frac{x(b-x)h}{b-b'}dx\right\}$$

Evaluating the integral and putting $A = bh/2$, and simplifying the factors, we have

$$x_c = \left(\frac{b+b'}{3}\right)$$

N.B.: The coordinates of center of mass of a uniform triangular lamina are independent of any coordinate system because it is situated at a distance of one third of the altitudes of the triangle from the respective bases of the triangle. The coordinates of C are given as
$x_C = (b + b')/3$ and $y_C = h/3$.

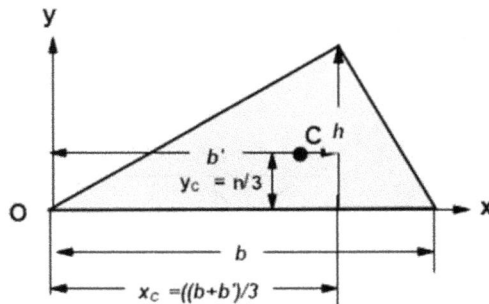

Calculation of $\vec{r_C}$ by cylindrical coordinate system: Now let us find the center of mass of a circular sector. For this we have many methods. Let us first do it by using a basic method. For this we need to be familiar with the 'cylindrical coordinate' system. If you do not fear this name 'cylindrical coordinate system' and simply go with the basic understanding with the formation and calculation of the elementary area dA as shown in the figure, you will tackle the problem easily.

Taking an elementary area dA at P, we have

$$dA = (r \, d\phi)(dr) = r \, dr \, d\phi$$

Then putting dA in the formula,

$$x_C = \frac{\int_s x \, dA}{A}$$

we have

$$x_C = \frac{\iint x \, r \, dr \, d\phi}{A},$$

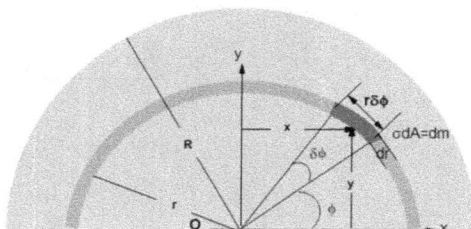

Putting $x = r \cos \phi$ in the last expression, we have

$$x_C = \frac{\iint (r \cos \phi)(r \, dr)d\phi}{A}$$

Similarly, putting dA in the formula

$y_C = \frac{\int_s y \, dA}{A}$, we have

$$y_C = \frac{\iint (r \sin \phi)(r dr)d\phi}{A},$$

where A = area of the given sector.

In a cylindrical coordinate system, the coordinates of the center of mass of a uniform lamina can be given as

$$x_C = \frac{1}{A} \int r^2 dr \int \cos \phi \, d\phi$$

$$y_C = \frac{1}{A} \int r^2 dr \int \sin \phi \, d\phi$$

Example 18 Locate the center of mass of a uniform semicircular plate of radius R.
Solution

For the semicircular plate $r_1 = 0$, $r_2 = R$, $\phi_1 = 0$, $\phi_2 = \pi$, and $A = \frac{\pi R^2}{2}$. Putting these values in the obtained expressions

$$y_C = \frac{1}{A} \int r^2 dr \int \sin \phi \, d\phi, \quad x_C = \frac{1}{A} \int r^2 dr \int \cos \phi \, d\phi$$

we have

$$x_C = \frac{2}{\pi R^2} \left(\int_0^R r^2 \, dr \int_0^\pi \cos \phi \, d\phi \right) = 0$$

and

$$y_C = \frac{2}{\pi R^2} \left(\int_0^R r^2 \, dr \int_0^\pi \sin \phi \, d\phi \right)$$

$$= \frac{2}{\pi R^2} \frac{R^3}{3} (\cos 0 - \cos \pi) = \frac{4R}{3\pi}$$

Alternative method (1): Applying the Cartesian coordinate system we derive the formula

$$y_C = \frac{1}{A} \int y \, dA',$$

where $y = \sqrt{R^2 - x^2}$ and $dA' = 2x \, dy$

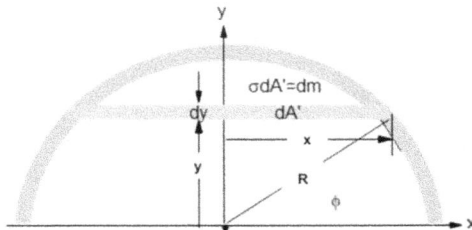

Substituting

$$dy = -\frac{x\,dx}{\sqrt{R^2 - x^2}}$$

We have

$$dA' = \frac{2x^2 dx}{\sqrt{R^2 - x^2}}$$

Then, we have

$$y_C = \frac{1}{A} \int_0^R \left(\sqrt{R^2 - x^2}\right)\left(\frac{2x^2 dx}{\sqrt{R^2 - x^2}}\right)$$

$$= \frac{2R^3}{3A}, \text{ where } A = \frac{\pi R^2}{2}$$

Finally, this gives

$$y_C = \frac{4R}{3\pi}$$

Alternative method (2): Let us take a thin sector OAB of radius R, which subtends an elementary angle $d\phi$ at the center of curvature O.

This sector behaves as a triangle whose center of mass 'C' is located at a distance $OC = \frac{2R}{3}$ from its apex O, as proved earlier. Then the mass dm of the elementary sector OAB can be imagined at a distance

$$y\left(=\frac{2R\sin\phi}{3}\right).$$

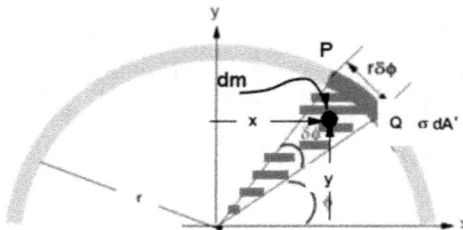

Then, following the basic formula,

$$y_C = \frac{\int y\,dm}{\int dm} = \frac{\int y\,dA'}{A}$$

1-33

and substituting
$y = \frac{2R}{3} \sin \phi$, we have

$$y_C = \frac{1}{A} \int \left(\frac{2R}{3} \sin \phi\right) dA',$$

where

$$dA' = \text{ area of } OAB = \frac{1}{2}(R\, d\phi)R = \frac{R^2 d\phi}{2}$$

and $A = $ area of the semicircle $= \frac{\pi R^2}{2}$.

This gives

$$y_C = \frac{2R}{3\pi} \int_0^\pi \sin \phi \, d\phi = \frac{4R}{3\pi}$$

and

$$x_C = \frac{2R}{3\pi} \int_0^\pi \cos \phi \, d\phi = 0$$

Alternative method (3): Let us imagine that the semicircular sector is the combination of many elementary sectors.

Since each elementary sector OAB behaves as a triangle having area $dA = \frac{1}{2}R^2 d\phi$, its center of mass is at a distance $\frac{2R}{3}$ from O. Hence, the semicircular disc can be imagined as a semicircle which gives the locus of center of mass of each elementary sector having angular position ranging from $\phi = 0$ to $\phi = \pi$.

Since the total mass of the elementary sectors is equal to mass of the semicircular plate, the center of mass of the semicircular arc produced by the center of mass of the differential sectors gives us the center of mass of the semicircular plate y_c.

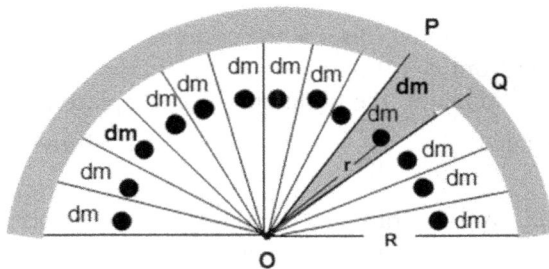

As we know
$y_C = \frac{2r}{\pi}$ for the semicircular ring,
where $r = \frac{2}{3}R$,

Then, we have $y_C = \frac{4R}{3\pi}$.

Alternative method (4): Let us take an elementary semicircular ring of radius r and thickness dr. The mass of the ring is

$$dm = \sigma dA_{ring}, \text{ where } dA_{ring} = 2\pi r dr$$

This gives $dm = 2\pi \sigma r dr$.

Since the center of mass of the ring is situated at a distance $y' = \frac{2r}{\pi}$ from O, $\frac{\int y' \, dm}{\int dm}$ will be equal to the y_C of the semicircular disc.

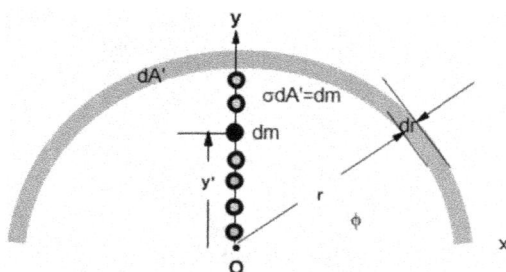

The elementary semicircular strip of radius r, area
dA' and mass dm is equivalent to a point mass dm
placed at a radial distance y'=2π/r.

Substituting $y' = \frac{2r}{\pi}$, $dm = 2\pi \sigma r dr$, we have

$$y_C = \frac{\int_0^R \left(\frac{2r}{\pi}\right)(2\pi \sigma r dr)}{\left(\frac{1}{2}\pi R^2\right)} = \frac{4R}{3\pi} \text{ Ans.}$$

Since the semicircular lamina (disc) is the combination of concentric elementary rings of radii ranging from $r = 0$ to $r = R$, the entire lamina can be thought of as point masses dm (= mass of the differential ring) placed one after the other respective ring, we have $y_C = \frac{\int y' \, dm}{\int dm} = \frac{\int y' \, dm}{M}$, which will give the center of mass of the lamina.

N.B.: If the circular sector of radius R (lamina) is uniform and subtends an angle θ at the center of curvature and it is placed symmetrical with the x-axis, its center of mass will be situated on the x-axis at a distance

$$r_C(=x_C) = \frac{4R \sin\left(\frac{\theta}{2}\right)}{3\theta}$$

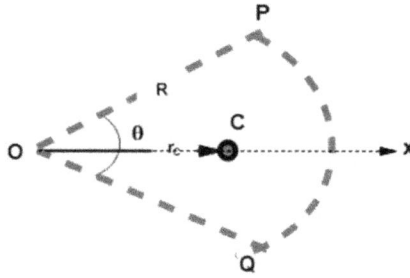

Example 19 Following the idea of this coaxial ring and using basic formula $y_C = \frac{\int y \, dm}{M}$, prove that the center of mass of a uniform thin hollow cone of height H is located at a height $\frac{H}{3}$ from the base on its axis.

Solution

As we know, a hollow cone is the combination of a continuous array of coaxial thin rings of radii ranging from R to zero.

If dm = mass of a thin ring, for a thickness dl of the strip, $dm = \sigma dA'$, where dA' = area of the stuff of the thin ring = $2\pi\sigma r dl$. As the ring is uniform, its center of mass is located at its geometrical center C. So the ring is substituted by a point mass dm.

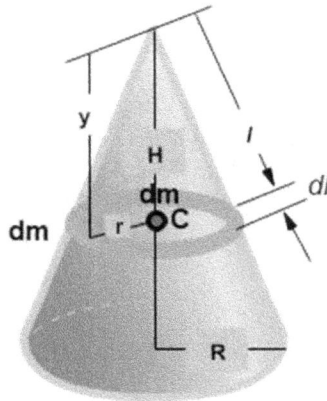

The area of the ring is

$$dA' = 2\pi r dl \tag{1.30}$$

Then, the center of mass of the cone is

$$y_C = \frac{1}{A} \int y \, dA' \tag{1.31}$$

The curved surface area of the cone is

$$A = \pi r l \tag{1.32}$$

Using last three equations, we have

$$y_C = \frac{1}{A} \int y \, dA' = \frac{1}{\pi R L} \int y \, (2\pi r \, dl)$$

$$\Rightarrow y_c = \frac{2}{RL} \int y \, r \, dl \tag{1.33}$$

Geometrically, we have

$$l/L = r/R = y/H \tag{1.34}$$

Using the last two equations, we have

$$\Rightarrow y_c = \frac{2}{RL} \int_0^L \left(\frac{lH}{L}\right)\left(\frac{lR}{L}\right) dl$$

$$\Rightarrow y_c = \frac{2}{RL}\left(\frac{HR}{L^2}\right) \int_0^L l^2 \, dl = \frac{2}{3}H$$

Thus, the center of mass of a hollow cone is situated at a height of $H/3$ from the base, where $H =$ height of the cone.

N.B.: When the thin curved surface is symmetrical to any axis, the y-axis (say), first of all take a differential (elementary) ring at a height y. Then find its area dA. After that, its mass can be given as $dm = \sigma dA$, where $\sigma =$ surface mass density. Since the ring is symmetrical about the y-axis, its center of mass is located at the same height y on the y-axis. Finally, the center of mass of the given surface can be given as $y_C = \frac{\int y \, dm}{M}$. For uniform surface $M = \sigma A$, where $A =$ area of the differential ring, we have

$$y_C = \frac{1}{A} \int y \, dA'$$

Express y and dA' in terms of given parameters by using geometry, to find the position of center of mass of the system of particles.

Curved surface
If the surface is curved, we can find the elementary area dA as shown as the shaded patch by using spherical coordinate system.

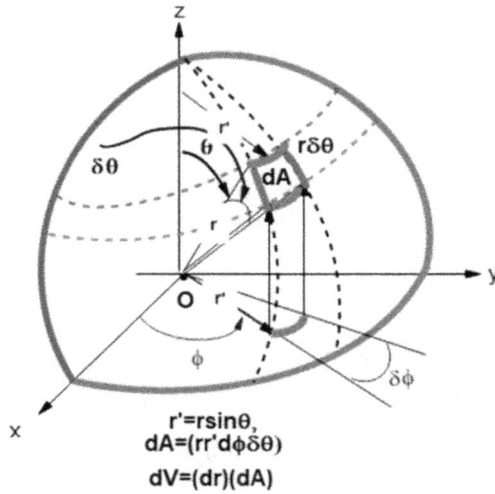

r'=rsinθ,
dA=(rr'dφδθ)
dV=(dr)(dA)

The area of the elementary patch is

$$dA = (r'd\phi)(r\ d\theta), \qquad (1.35)$$

where $r' = r \sin \theta$.
So, we have

$$dA = r^2 \sin \theta d\theta d\phi$$

Now we can substitute dA in the formula
$\overrightarrow{r_C} = \frac{1}{A} \iint \overrightarrow{r}\ dA$ to obtain

$$\overrightarrow{r_C} = \frac{r^2}{A} \iint \overrightarrow{r} \sin \theta d\theta d\phi$$

However, in simpler problems such as hemispherical surface, we easily find $\overrightarrow{r_C}$ by ranging θ from 0 to $\frac{\pi}{2}$ and ϕ from 0 to 2π, as described in the following example.

Example 20 Find the center of mass of a thin hemispherical bowl of radius R.
Solution
Following the basic formula

$$\overrightarrow{r_C} = \frac{r^2}{A} \iint \overrightarrow{r} \sin \theta d\theta d\phi$$

we understand that one component of $\overrightarrow{r}$ is directed along the y-axis, that is, what we roughly call the 'axis of the hemisphere'. Since the hemisphere is symmetrical about the y-axis, logically we can conclude that its center of mass is located at the y-axis. Then we can write,

$$r_C = y_C = \frac{R^2}{A} \int y \sin \theta d\theta d\phi,$$

where $y = R \cos \theta$ and $A = 2\pi R^2$.

This gives

$$y_C = \frac{R}{2\pi} \int_0^{\pi/2} \sin \theta$$

$$= \frac{R}{2} \int_0^{\pi^2} \sin 2\theta d\theta = \frac{R}{2}$$

Alternative method: After considering an elementary ring of thickness $Rd\theta$ and radius $r(=R \sin \theta)$ at a height y, we can find its mass as $dm = \sigma dA'$, where $dA' =$ area of the ring.

Substituting $dA' = (2\pi r')Rd\theta = 2\pi(R \sin \theta)Rd\theta = 2\pi R^2 \sin \theta d\theta$, we have

$$dm = 2\pi R^2 \sigma \sin \theta d\theta$$

Since the ring is symmetrical about the y-axis, its center of mass will be located at the y-axis at $y = R \cos \theta$.

As the hollow hemisphere is a combination of coaxial elementary rings of radii ranging from $r = 0$ to $r = R$ and their center of mass lies on the y-axis, we can imagine a continuous arrangement (array) of point masses 'dm' on the y-axis spreading from $y = 0$ to $y = R$.

Hence, the center of mass of the hollow hemisphere is given as

$$y_C = \frac{\int y \, dm}{M}$$

Substituting $y = R \sin \theta$, $dm = 2\pi R^2 \sigma \sin \theta d\theta$ and $M = 2\pi R^2 \sigma$, we have

$$y_C = R \int_0^{\pi/2} \sin \theta \cos \theta d\theta$$

which will give the previous result:

$$y_C = \frac{R}{2}$$

Volume–mass distribution: In volume–mass distribution, mass is distributed throughout the given volume (space) of matter. The density of volume–mass distribution at any point is given as

$$\rho = \frac{dm}{dV}$$

This gives us the mass of the elementary segment if the density of the material is given as a constant or as a function of distance r. After finding the elementary mass $dm = \rho dV$, substitute this in the basic equation

$$\vec{r_C} - = \frac{\int \vec{r} \, dm}{\int dm}$$

Non-uniform distribution
This gives

$$\vec{r_C} = \frac{\int_v \rho \vec{r} \, dv}{\int \rho \, dV},$$

where $\int_v$ = volume (or triple) itegral and $\rho = f(r)$.

Uniform distribution
If the object is 'uniform', we can take 'ρ' out of the integral and cancel it in both numerator and denominator. Then, we have

$$\vec{r_C} = \frac{\int \vec{r} \, dV}{V}$$

The elementary volume can be found with the help of different coordinate systems. In the Cartesian coordinate system, $dV = dx \, dy \, dz$. In the cylindrical coordinate system, $dV = R \, dr \, d\phi \, dz$, and in the spherical coordinate system, $dV = R^2 dr \sin \theta \, d\theta d\phi$.

If you are not familiar with the given ideas, we will consider some simpler cases such as uniform cone, hemisphere, paraboloid, etc, where we can locate the center of mass, avoiding the above rigorous mathematical expressions. However, the above expressions of 'dV' are useful if the density 'ρ' varies with the coordinates or the given object does not possess any symmetry about any axis.

Let us discuss a common procedure to locate the center of mass of simple objects.

If we have an object symmetrical about the y-axis (say), we will take a thin plate (elementary strip) at a height y parallel to the xz-plane. Then find its volume. As it is uniform and symmetrical about the y-axis, the center of mass of the differential (elementary) strip lies at its geometrical center, that is, at the same height on the y-axis. As a consequence, the given object can be imagined as a continuous line mass distribution in the y-axis. Now we can find the center of mass of the line distribution by using the formula

$$y = \frac{\int y \, dm'}{\int dm'} = \frac{\int y \, dm}{M},$$

where $dm' = \rho dV'$ and $M = \rho V$.

Note that dm' and dV' are the mass and volume of the elementary strip (but not that of the elementary segment of the elementary strip).

This gives

$$y_C = \frac{\int y \, dV'}{V}$$

Let us use this formula.

Example 21 Locate the center of mass of a uniform solid hemisphere.

Solution

Take an elementary strip of radius r and thickness dy parallel to the xz-plane at a distance y. The volume of the strip is given as

$$dV' = \pi r^2 dy$$

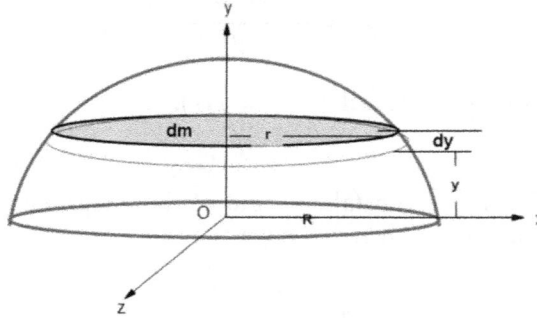

Then substituting dV' in the equation

$$y_C = \frac{\int y \, dV'}{V},$$

we have

$$y_C = \frac{1}{V} \int y(\pi r^2 dy),$$

where $r^2 = R^2 - y^2$ and $V = \frac{2}{3}\pi R^3$.

This gives

$$y_C = \frac{3}{8}R$$

Example 22 Locate the center of mass of a solid cone of height H following the above method.

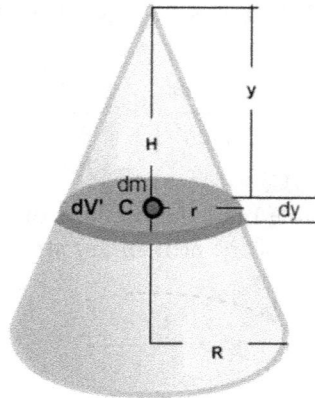

Solution

Take an elementary strip of radius r and thickness dy parallel to the xz-plane at a height y. The volume of the strip is given as

$$dV' = \pi r^2 dy$$

Then substituting dV' in the equation

$$y_C = \frac{\int y \, dV'}{V},$$

we have

$$y_C = \frac{1}{V} \int y(\pi r^2 dy),$$

where $r = \frac{y}{H}R$ and $V = \frac{1}{3}\pi R^2 H$.

After evaluating the integral and simplifying the factors, we have

$y_C = \frac{3}{4}H$ from the apex on its axis of symmetry. **Ans.**

Let us now tabulate the important results of the previous discussion.

N.B.:

Area of the elementary rings

Each elementary ring of radius r behaves as a rectangular strip of length equal to the perimeter of the ring, that is, $2\pi r$. The width of the ring is dr for a disc $Rd\theta$ for the sphere and dl for the cone.

Then

$$dA = (2\pi r)(dr)$$

For a disc

$= (2\pi r)(R \, d\theta)$ for sphere, where $r = R \sin\theta$

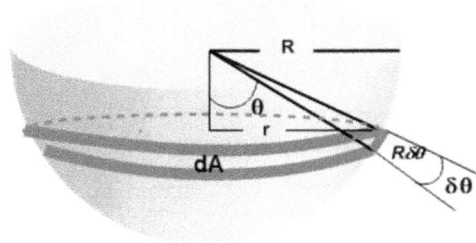

$= (2\pi r). \, dl$ for a cone, where $dl = dr \, \text{cosec} \, \theta$ put $\tan\theta = \frac{R}{H}$.

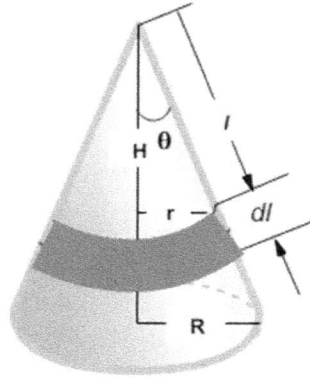

Volume of the thin (elementary) strips

Each strip behaves as a cylinder of (thickness) height dy and base area πr^2. Then the volume of the cylinder is given as

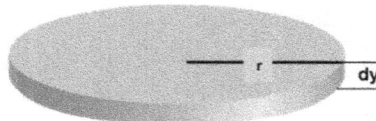

Thin Disc

$$dV = \pi r^2 dy,$$

where

$$r^2 = R^2 - y^2$$

for the sphere as shown in the last figure.

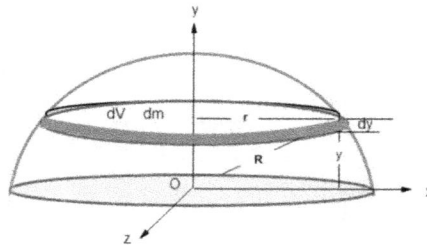

and $r = \left(\frac{H-y}{H}\right)R$ for the cone.

After realising the importance of center of mass, for a system of particles, let us use the concept of work and energy on the system of particles.

1.5 Work–energy theorem for a system of particles

Work done by external forces: Let us recast the system of n particles on which external forces F_1, F_2...F_n act on particles of masses m_1, m_2...m_n, respectively, as stated at the beginning of the chapter. The work done by the force F_1 during a small displacement ds_1 of the particle m_1 is given as

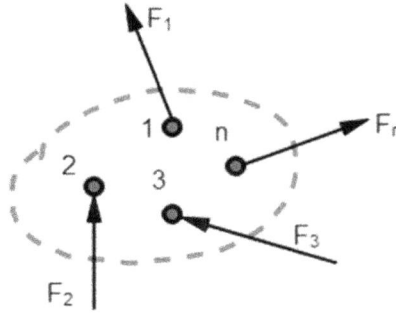

$dW_1 = \vec{F_1} \cdot \vec{ds_1}$, where $\vec{ds_1} = \vec{ds_{1C}} + \vec{ds_C}$.

$\vec{ds_{1C}}$ = displacement of m_1 relative to center of mass and $\vec{ds_C}$ = displacement of center of mass relative to ground.

This gives $dW_1 = \vec{F_1} \cdot \vec{ds_{1C}} + \vec{F_1} \cdot \vec{ds_C}$

Similarly, work done by other external forces can be given as

$$dW_2 = \vec{F_2} \cdot \vec{ds_{2C}} + \vec{F_2} \cdot \vec{ds_C}, \quad dW_3 = \vec{F_3} \cdot \vec{ds_{3C}} + \vec{F_3} \cdot \vec{ds_C}, \quad dW_n = \vec{F_n} \cdot \vec{ds_{nC}} + \vec{F_n} \cdot \vec{ds_C}$$

Summing up the works due to all external forces, we have

$$W_{ext} = \int \vec{F_1} \cdot \vec{ds_{iC}} + \int \left(\sum \vec{F_1}\right) \cdot \vec{ds_C}$$

The first term signifies the sum of work done by the external forces relative to the center of mass frame. The second term gives the work done by the net force, that is, $\vec{F_{ext}}(=\sum\vec{F_1})$ on the center of mass even though any force may not act on the center of mass.

Work done by internal force: Now let us consider the work done by the internal forces. Following the above-derived equation substituting $\vec{F_1}$ by $\vec{f_1}$ (internal forces), the total work done by the internal forces can be given as

$$W_{int} = \int \vec{f_i} \cdot \vec{ds_{iC}} + \int \left(\sum \vec{f_i}\right) \cdot \vec{ds_C}$$

Since the net internal force is zero, that is, $\sum\vec{f} = 0$, we have

$$W_{int} = \int \vec{f_i} \cdot \vec{ds_{iC}}$$

For non-rigid objects consisting of a group of particles, in general, the displacement of different particles relative to center of mass will be different. This gives non-zero work due to the internal forces. However, the total work done by internal forces will be zero for the rigid bodies as there is no relative displacement between center of mass and the particles along the line of their separation.

Example 23 Derive an expression for work done by internal forces on the system of two interacting particles.

Solution

The work done by $\vec{F_1}$ is

$$dW_1 = \vec{F_1} \cdot d\vec{r_1} = \vec{F} \cdot d\vec{r_1}$$

The work done by $\vec{F_2}$ is

$$dW_2 = -\vec{F} \cdot d\vec{r_2}$$

The total work done is

$$dW = \vec{F} \cdot d\vec{r_1} - \vec{F} \cdot d\vec{r_2}$$

$$\vec{F} = (d\vec{r_1} - d\vec{r_2}) = \vec{F} \cdot d\vec{r_{12}} = -\vec{F} \cdot d\vec{r_{21}}$$

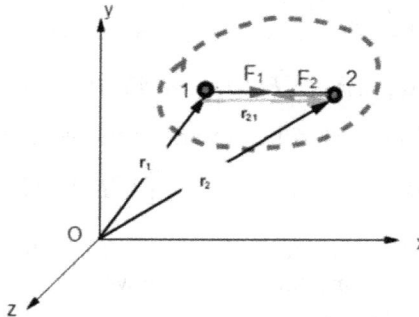

Since $d\vec{r_{21}} = d\vec{r_{21_\parallel}} + d\vec{r_{21_\perp}}$, we have

$$dW = -\vec{F} \cdot d\vec{r_{21_\parallel}} - \vec{F} \cdot d\vec{r_{21_\perp}}$$

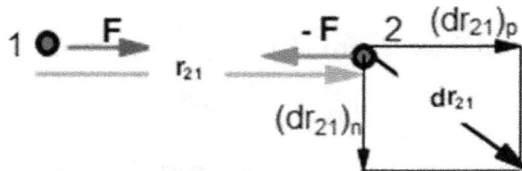

Since $\vec{F} \perp d\vec{r}_{21_\perp}$, we have

$$dW = -\vec{F}.d\vec{r}_{21_\parallel}$$

Since $d\vec{r}_{21_\parallel}$ and $\vec{F}$ are parallel, we have

$$dW = -Fdr_{21}$$

Substituting $dr_{21} = dr$ we have

$$dW = -Fdr$$

Then, the total work done by the internal forces is

$$W = \int dW = -\int Fdr,$$

where $F =$ magnitude of the internal force and $r =$ distance of separation between the particles.

N.B.: Let's note the following points:

1. The total work done by the internal forces does not depend on the reference frame, which can be given as $W_{int} = \int Fdx$, where $x =$ relative distance of separation.
2. If x increases, work done is negative; when x decreases work done will be positive. If x remains constant no work will be done.
3. Total work done by internal forces on center of mass is zero because $\sum F_{int} = 0$.

Potential energy: The total energy of a system is basically defined as the sum of potential and kinetic energy.

We have many sources of potential energy. If the interaction is gravitational, the potential energy is gravitational; if the interaction between the particles is electrostatic, potential energy is termed as electrostatic potential energy, and so on.

For a system of particles we have different formulae for potential energy. In the chapter 'Gravitation' we will derive the gravitational potential energy of interaction of a system of particles such as

$U_{gr} = -\frac{1}{2}\sum V_i m_i$ for a discrete particle system, where $V_1 =$ gravitational potential at ith point due to all masses except m_1 and $m_1 =$ mass of ith paticle,

For continuous mass distribution, if gravitational potential V at any point of the system is given, the gravitational potential energy of the system is given as $U = \frac{1}{2}\int Vdm$.

We have similar sets of formulae for potential (energy of interaction for discreate and continuous charge distribution) in electrostatics.

As we know, potential energy expression is valid only when the forces are conservative; no energy must be wasted in the form of radiant energy (heat, light, and sound). For this, we can find the work done by all conservative forces using the expression as given in the previous section. Then using the expression

$W_{\text{con}} = -\Delta U = (U_{\text{final}} - U_{\text{initial}})$ and setting the value of U_{initial} we can find the expression for U_{final}.

Potential energy of interaction is equal to negative of work done by internal conservative forces, which does not depend on the reference frame.

Sometimes we come across a system of particles such as water bottle, sand bag, any rigid or non-rigid objects falling under the earth's gravitational field. In this case how to find the potential energy of the system? Let us see in the following example.

Example 24 Derive an expression for the gravitational potential energy of a system of particles due to the earth's gravity.

Solution

For any system of discrete distribution of particles, let us take the ith particle of mass m_1, which is situated at a height y_i (say), and gravitational potential energy of m_i is given as $U_i = m_i g y_i$

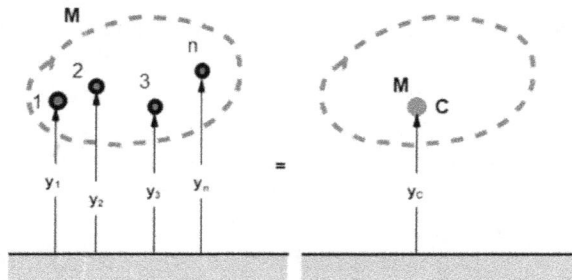

Summing up the potential energy of all particles, we have the total potential energy.

$$U = \sum U_i = \sum m_i g y_i = g \sum m_i y_i$$

Since $\sum m_i y_i = (\sum m_i) y_C$, by substituting $\sum m_i (=M)$ we have

$$U = Mg y_C$$

N.B.: Note that $U = Mg y_C$ is strictly valid only when the size of the object (system of particles) is much smaller than the radius of the earth, so that we can assume a constant value of g (uniform gravitational field) for all particles.

After understanding the meaning of potential energy of a system of interacting particles, let us derive the expression of kinetic energy of a system of particles.

Kinetic energy of a system of particles: Total kinetic energy of a system of n particles (say) is equal to the sum of kinetic energy of all particles relative to ground.

Total kinetic energy $K = \sum \frac{1}{2} m_i v_i^2$, where $\vec{v_i} = \vec{v_{iC}} + \vec{v_C}$,

v_i = velocity of m_1, $\vec{v_{iC}}$ = velocity of m_1 relative to center of mass, and $\vec{v_C}$ = velocity of center of mass.

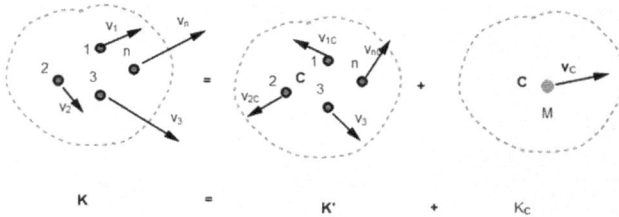

This gives

$$K = \frac{1}{2}\sum m_i \, |\vec{v_{iC}} + \vec{v_C}|^2 = \frac{1}{2}\sum m_i (v_{iC}^2 + v_C^2 + 2\vec{v_{iC}} \cdot \vec{v_C}) = \frac{1}{2}\sum m_i v_{iC}^2 + \frac{1}{2}\left(\sum m_i\right) v_C^2 + \sum m_i \vec{v_{iC}} \cdot \vec{v_C}$$

Since $\sum m_i \vec{v_{iC}} = 0$ (sum of momenta of all particles relative to the center of mass of the system is zero) substituting

$$\sum m_i \vec{v_{iC}} \, \vec{v_C} = \vec{v_C}\left(\sum m_i \vec{v_{iC}}\right) = 0$$

we have

$$K = \sum \frac{1}{2} m_i c_{iC}^2 + \frac{1}{2}\left(\sum m_i\right) v_C^2,$$

where $\sum m_i = M$ = total mass of the system and $\sum \frac{1}{2} m_i v_{iC}^2$ = sum of kinetic energies of the particles relative to center of mass of the system, which is termed as internal kinetic energy or kinetic energy relative to center of mass frame ($=K'$, say).

Then, we have

$$K = K' + \frac{M v_C^2}{2}$$

The above expression tells us that the KE of a system of particles is equal to the KE of the system relative to center of mass plus the KE of the center of mass ($\frac{M}{2} v_C^2$). Hence, the KE of a system is minimum relative to the frame attached with center of mass, which is equal to K'. The kinetic energy relative to the center of mass K' is called internal KE. This does not depend on any reference frame. Thus, K' is an internal property of the system.

Example 25 Find the KE of a system of two particles of masses m_1 and m_2 having velocities $\vec{v_1}$ and $\vec{v_2}$ relative to the center of mass of the system.

Solution

The kinetic energy of the system relative to center of mass is

$$K' = K - \frac{Mv_C^2}{2},$$

where $\vec{v}_C = \frac{m_1\vec{v_1} + m_2\vec{v_2}}{m_1 + m_2}$, $M = (m_1 + m_2)$ and $K = \frac{1}{2}m_1v_1^2 + \frac{1}{2}m_2v_2^2$.

Then, we have $K' = \frac{1}{2}m_1v_1^2 + \frac{1}{2}m_2v_2^2 - \frac{1}{2}(m_1 + m_2)|\frac{m_1\vec{v_1} + m_2\vec{v_2}}{m_1 + m_2}|$.

Expanding the square and simplifying the factors, we have

$$K' = \frac{m_1m_2\,|\vec{v_1} - \vec{v_2}|^2}{2(m_1 + m_2)}$$

Alternative method: $K' = \frac{1}{2}m_1v_{1C}^2 + \frac{1}{2}m_2v_{2C}^2$, where $\vec{v}_{1C} = \vec{v_1} - \vec{v}_C$ and $\vec{v}_{2C} = \vec{v_2} - \vec{v}_C$

Substituting

$$\vec{v}_C = \frac{m_1\vec{v_1} + m_2\vec{v_2}}{m_1 + m_2}$$

we have $K' = \frac{1}{2}m_1\,|\vec{v_1} - \frac{m_1\vec{v_1} + m_2\vec{v_2}}{m_1 + m_2}|^2 + \frac{1}{2}m_2\,|\vec{v_2} - \frac{m_1\vec{v_1} + m_2\vec{v_2}}{m_1 + m_2}|^2$

$$= \frac{1}{2}\frac{m_1m_2}{m_1 + m_2}\,|\vec{v_1} - \vec{v_2}|^2$$

Example 26 Two particles of mass $3m$ and $2m$ are moving towards each other with speeds v_0 and $2v_0$, respectively. If the particles are interconnected by a massless spring of stiffness k and the spring is compressed by a length x, find the (a) total mechanical energy of the system and (b) the total internal energy of the system.

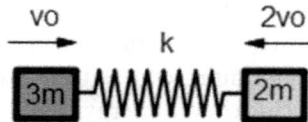

Solution

(a) Total mechanical energy is given as

$$E = U + K$$

where $U = \frac{k}{2}x^2$ and $K = \frac{1}{2}(3m)v_0^2 + \frac{1}{2}(2m)(2v_0)^2$

Then, we have

$$E = \frac{11}{2}mv_0^2 + \frac{k}{2}x^2$$

(b) The total internal energy is $E_{int} = E - \frac{Mv_C^2}{2}$, where $M = (3m + 2m)$ and

$$v_C = \frac{(3m)(v_0) + (2m)(-2v_0)}{3m + 2m} = -\frac{v_0}{5}$$

This gives $E_{int} = \frac{27}{5}mv_0^2 + \frac{k}{2}x^2$.

Work–Energy Theorem: Now we have all expressions in our hand such as work done by internal and external forces and KE of system of particles. Applying the work–energy theorem on the system of particles, we have

$$W_{int} + W_{ext} = \Delta K$$

where W_{int}= sum of work done by all internal forces, W_{ext}= sum of work done by all external forces and ΔK= sum of change in KE of all particles of the system.

The above expression tells us that for a system of particles, the sum of the work done by all forces (internal and external) acting on each element or particles of the system is equal to the sum of the change in KE of all particles of the system. This is what we call the work–energy theorem for a system of particles.

Example 27 Two blocks of masses m_1 and m_2 interconnected by a light spring of stiffness k are kept on a horizontal surface. The coefficient of friction between the blocks and horizontal surface is μ. If a horizontal force $F(>\mu m_2 g)$ acts on m_2, the block m_2 slides. In consequence the spring elongates by a length x. Find the work done by (i) internal forces, (ii) external force F, and (iii) all forces on the system of the blocks and spring.

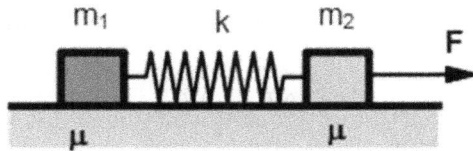

Solution

(i) For the system $(m_1 + spring + m_2)$, spring force is the internal force. The work done by the spring is given as

$$W_{sp} = -\int F \, dr$$

where $F = kx$ and $dr = dx$.

This gives $W_{int} = W_{sp} = -\frac{k}{2}x^2$.

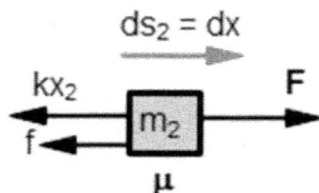

(ii) We have two external forces, that is, F and the frictions acting on m_1 and m_2. If you assume m_1 to be stationary, friction on m_1 cannot do work. Then the total work done by external force is

$$W_{\text{ext}} = \int \vec{F} \cdot d\vec{s_2} + \int \vec{f} \cdot d\vec{s_2}$$

where $d\vec{s_2} = dx\hat{i}$ and $\vec{f} = -\mu m_2 g \hat{i}$.

Then, $W_{\text{ext}} = F \int_0^x dx - \mu m_2 g \int_0^x dx$
This gives $W_{\text{ext}} = Fx - \mu m_2 g x$

(iii) The total work done is $W = W_{\text{int}} + W_{\text{ext}}$, where $W_{\text{int}} = \frac{1}{2}kx^2$ and $W_{\text{ext}} = Fx - \mu m_2 g x$.

Then, we have $W = -\frac{kx^2}{2} + Fx - \mu m_2 g x$.

Conservation of mechanical energy: If the external and internal forces acting on the system of particles are conservative, we can write

$$W_{\text{ext}} + W_{\text{int}} = -\Delta U,$$

where $W_{\text{ext}}=$ total work done by all external forces, $W_{\text{int}}=$ total work done by all internal forces and $\Delta U=$ change in the potential energy of the system.

Substituting $-\Delta U = W_{\text{ext}} + W_{\text{int}}$ in the work–energy theorem $W_{\text{int}} + W_{\text{ext}} = \Delta K$ we have

$$-\Delta U = \Delta K,$$

This gives $\Delta U + \Delta K = 0$.

The above expression tells us that when the forces (external and internal) involved in the group of particles are conservative, the total mechanical energy of the system remains conserved.

Example 28 Discuss the effect of stopping a gas jar filled with an ideal gas of mass M and moving with a speed v.

Solution

When the gas jar stops, the center of mass of the gas will stop $(v_{\text{center of mass}})_2 = 0$.

Since the collision of gas particles with the wall of the gas jar is elastic, we can conserve the KE of the system (gas but not gas jar). As there is no attraction between the ideal gas particles, the potential energy of interaction is zero. Neglecting the effect of gravity of the earth, the total internal energy is equal to the internal KE of the gas. Hence, we can conserve the KR of the system of gas particles to obtain

$$\frac{M(v_C^2)_1}{2} + (K')_1 = \frac{M(v_C^2)_2}{2} + (K')_2,$$

where $(v_C^2)_1 = v$ and $(v_C^2)_2 = 0$.

Then, the internal kinetic energy K'_2 just after stopping the gas jar increases by $\frac{Mv^2}{2}$. As we will explain in the chapter 'Kinetic Theory of Gas', due to increase in internal kinetic energy, the temperature of the gas will rise.

The KE of a system of two particles relative to the center of mass of the system can be given as

$$K' = \frac{1}{2}\mu v_{rel}^2,$$

where $\mu = \frac{m_1 m_2}{m_1 + m_2}$ (called reduced mass because $\mu < m_1$ and $\mu < m_2$) and $v_{rel} = |\vec{v_1} - \vec{v_2}|$.

Total mechanical energy of a system of particles: Once we have the KE and PE of a system of particles, just add them to obtain the total energy, which can be expressed as

$$E = U + K,$$

where U = potential energy and K = kinetic energy of the system of particles.

Total internal energy: When we subtract the KE of center of mass from the total mechanical energy of the system, we will get total internal energy, because the KE of the center of mass is controlled by external forces. Hence, the total internal energy is given as

$$E_{\text{int}} = E - \frac{Mv_C^2}{2}$$

1.6 Impulse and momentum equation for a system of particles

Momentum of a particle: Newton's second law for a particle tells us that a particle changes its velocity at a rate $\vec{a}(=\frac{d\vec{v}}{dt})$ in the direction of a force $\vec{F}$ applied on it. According to Newton's second law, applied for a particle of mass m, we have

$$\vec{F} = m\vec{a} -= \frac{md\vec{v}}{dt}$$

Since 'm' is a constant, take it into the derivative.

This gives $\vec{F} = \frac{d}{dt}(m\vec{v})$.

This expression tells us that the force changes the vector quantity '$m\vec{v}$', which is defined as 'momentum of the particle' and denoted by the symbol $\vec{P}$.

$$\vec{P} = m\vec{v}$$

Newton describes the momentum as quantity of motion.

Single force acting on a particle: In other words, linear momentum of a particle is defined as the product of mass and velocity of the particle. Since mass is a scalar and velocity is a vector quantity, momentum $\vec{P}$ is a vector quantity. Therefore, we can express the 'force as a rate of change in linear momentum'.

$$\vec{F} = \frac{d\vec{P}}{dt}$$

The above expression tells us that force is equal to the rate of change of linear momentum, which is a standard expression of Newton's second law of motion.

Many forces acting on a particle: When a number of forces $\vec{F_1}$, $\vec{F_2}$, ...$\vec{F_n}$, say, act on a particle, each force will tend to change the linear momentum of the particle. As a consequence, the net change in momentum is caused by the combined effect (action) of all forces. Then, the net (or resultant) force acting on the particle is given as.

$$\vec{F}\left(=\vec{F}_{net} = \sum \vec{F_1}\right) = \frac{d\vec{P}}{dt}$$

Hence, Newton's second law can be expressed as

$$\vec{F} = \frac{d\vec{P}}{dt}$$

where $\vec{F}$ = net force and $\vec{P} = m\vec{v}$.

The net force acting on a particle is equal to the rate of change of linear momentum of the particle, where linear momentum of the particle is defined as the product of its mass and velocity. Newton calls linear momentum 'quantity of motion of matter'. In this way, force causes the change in quantity of motion. The total change in momentum is attributed to the net (total) force but not individual (component) force.

Let us see how the net force can change the momentum in the following example.

Example 29 A particle experiences two forces; one of them is a constant force $\vec{F_1} = (2\hat{i} + 3\hat{j})N$. If the particle changes its momentum with time according to the relation $\vec{P} = t^3\hat{i} + t^2\hat{j}$, find the other force $\vec{F_2}$.

Solution

The net forces are given as

$$\vec{F} = \frac{d\vec{P}}{dt}$$

where $\vec{P} = t^3\hat{i} + t^2\hat{j}$.

Then, we have $\vec{F} = 3t^2\hat{i} + 2t\hat{j}$.

Since $\vec{F} = \vec{F_1} + \vec{F_2}$, $\vec{F_2}$ can be given as $\vec{F_2} = \vec{F} - \vec{F_1}$, substituting $\vec{F} = 3t^2\hat{i} + 2t\hat{j}$ and $\vec{F} = (2\hat{i} + 3\hat{j})$ we have $\vec{F_2} = (3t^2 - 2)\hat{i} + (2t - 3)\hat{j}$.

Momentum of a system of particles: For a system of n particles, the momentum $\vec{P}$ of the system is equal to the vector sum of the momenta of all particles of the system.

$$\vec{P}_{\text{system}}(=\vec{P}) = \sum \vec{P_1},$$

where $\vec{P_i} = m_i \vec{v_i}$ (momentum of ith particle).

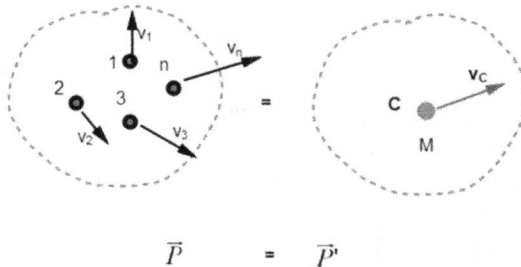

$$\vec{P} \quad = \quad \vec{P'}$$

This gives $\vec{P} = \sum_{i=1}^{i=n} m_i \vec{v_i}$.

Since $m_i \vec{v_i} = (\sum m_i)\vec{v_C}$,

we have $\vec{P} = (\sum m_i)\vec{v_C}$.

Substituting $\sum m_i = M$,

we have $\vec{P} = M\vec{v_C}$,

where $M =$ mass of the system and $\vec{v_C} =$ velocity of center of mass of the system.

The above discussion tells us that the total momentum of a system of particles is equal to the vector sum (or resultant) of momenta of all particles of the system, which can also be equal to the product of total mass and the velocity of the center of mass of the system.

Example 30 What is the momentum of a system comprising a gas jar of mass M_0 moving with velocity $\vec{v}$ and the gas of mass m? If the gas jar is stopped, what will be the momentum of the system?
Solution

$$\vec{P}_{\text{system}} = M\vec{v}_C$$

where $M = M_0 + m$ and $v_C = v$.

Then, we have $\vec{P}_{\text{system}} = (M_0 + m)\vec{v}$. Ans.

When the center of mass of a system of particles (gas + gas jar in the last example) does not move, we can say that the momentum of the system is zero, even though each gas particle has some velocity. You should remember that the total momentum is equal to the vector sum of individual momentum, but not the sum of the magnitudes of momenta of all particles.

Symbolically $|\vec{P}| = |\sum \vec{P_i}|$ but not $|\vec{P}| = \sum |\vec{P_i}|$

However, the KE of the system need not be zero even though the center of mass of the system does not move. For instance, in the last example, we have

$$K = K' + \frac{mv_C^2}{2},$$

where $v_C = 0$ after stopping the gas jar. Then, we have $K = K' =$ sum of the KE of all particles relative to center of mass, which can be equal to the sum of the KE of all particles relative to the ground because $v_C = 0$ (gas jar is stopped).

Conservation of linear momentum: For particle of m moving with a velocity v, Newton's second law can be written as

$$\vec{F} = \frac{d\vec{P}}{dt},$$

where $\vec{P} = m\vec{v}$ and $\vec{F} = \vec{F}_{net}$.

For a group of particles, Newton's 2nd law can be given as

$$\vec{F} = \frac{d\vec{P}}{dt}$$

where $\vec{P} = \sum m_i \vec{v_i} = M\vec{v}_C$ and $\vec{F} = \vec{F}_{net} = \sum \vec{F_i}$; $\vec{F_i} =$ external force acting at ith point.

If there is 'no net force' acting on a particle or a system of particles, its momentum remains constant.

When $\overrightarrow{F} = 0$, we have $\frac{d\overrightarrow{P}}{dt} = 0$.

Hence, $\overrightarrow{P} = $ constant.

That is what we call the law of conservation of linear momentum.

In other words, whenever no net force acts on a system of one or more particles, the linear momentum of the system remains constant (conserved). Since the internal force is zero, it (the total internal force) cannot change the linear momentum of the system, even through the individual momentum of the particles of the system may change.

Example 31 Discuss the possibility of conservation of linear momentum of a block moving on a rough inclined plane if $\mu = \tan \theta$.

Solution

Let us consider the motion of the block in the x-direction. The net force acting on the block is

$$F_x = mg \sin \theta - f,$$

where $f = \mu mg \cos \theta$.

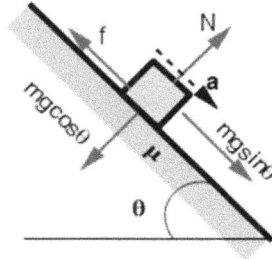

This gives $F_x = mg(\sin \theta - \mu \cos \theta)$.

Since $\mu = \tan \theta$, $F_x = 0$. Hence, we can conserve the momentum of the block down the plane.

However, you can not conserve the block's momentum when it is pushed upwards along the plane in the negative x-direction because of the net force acting on the block down the slant, which is given as $F_{net} = 2mg \sin \theta \searrow$.

The above example tells us that if the net force acting on a particle in a certain direction is zero, the momentum of the particle will remain constant in that direction.

Sometimes we observe a particle in an accelerating frame. In that case, in addition to the real forces we must take the pseudo-forces into account. If the net effect of $\overrightarrow{F}_{real}$ and $\overrightarrow{F}_{pseudo}$ is zero, in that case we can conserve the momentum of the particle (relative to the accelerating or non-inertial frame).

In other words, in accelerating the (non-inertial) frame, if the resultant of all forces ($\overrightarrow{F}_{pseudo}$ and $\overrightarrow{F}_{real}$) acting on a particle is zero (in any direction), the momentum of the particle remains constant relative to that frame (in that direction).

Let us use the above idea in the following example.

Example 32 Can you conserve the linear momentum of a block placed on a smooth accelerating wedge? Explain.

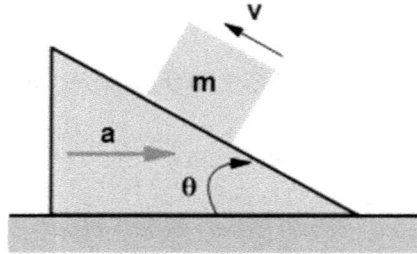

Solution

Imposing pseudo-force on the block as observed from the accelerating wedge, we have the net force

$$F = m\overrightarrow{g} + \overrightarrow{N} + \overrightarrow{F}_{ps} \text{ where } \overrightarrow{F}_{ps} = m\overrightarrow{a} -$$

Resolving the forces in x- and y-axes, we have $F_x = mg \sin \theta - ma \cos \theta$
$F_y = -mg \cos \theta - ma \sin \theta + N$

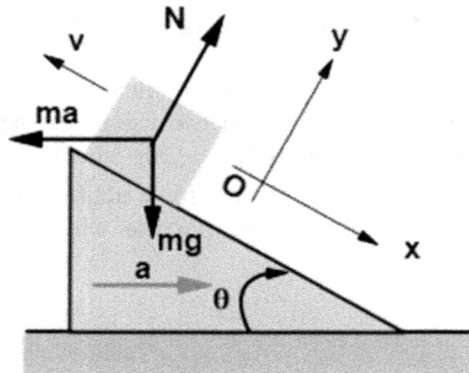

Since the block does not lose contact with the wedge, we can say that the block does not move relative to the wedge in the y-direction (normal to the wedge). In other words we can say that the acceleration of the block is zero relative to the wedge in the y-direction. In this way, we have $F_y = 0$ on the block so as to remain with zero momentum along the y-axis relative to the wedge. That means, we can conserve the 'zero momentum' of the block relative to the wedge in the y-direction (normal to the slant of the wedge) for any acceleration (including zero acceleration) of the wedge. Now let us analyse the effect of force in the x-direction. Since $F_x (= mg \sin \theta - ma \cos \theta)$, F_x will be zero only when $a = g \tan \theta$. So, if the wedge accelerates towards the right having acceleration $a = g \tan \theta$, the net force acting on the block relative to the wedge will be zero (in both x- and y-direction). Hence, the momentum of the block remains constant relative to the wedge when $\vec{a} = g \sin \theta \hat{i}$. If we push the block of mass m with a velocity $\vec{v}$ relative to the wedge along the x-direction, it will go on moving with the same momentum '$m\vec{v}$' relative to the wedge.

Let's now look at some examples of conservation of linear momentum of a system containing two or more particles.

Example 33 Can you conserve the linear momentum of the following system?
  (a) A system of two particles interconnected by a light spring is released in a smooth (ii) horizontal plane (ii) inclined plane.

  (b) The motion of a smooth block on a smooth prismatic wedge placed on a horizontal surface.

Solution
  (a)
   (i) The net external force acting on the system $A + B$ is zero because the internal spring forces are equal and opposite. Hence, the momentum $\vec{P}$ of the system remains constant, which can be given as

$$\vec{P} = (m)(-v_0 \hat{i}) + (2m)(2v_0 \hat{i}) = 3mv_0 \hat{i}$$

   (ii) Down the inclined plane, the net force acting on the system is non-zero because the component of the weight 3 mg of the system parallel to the inclined plane is non-zero. So, we cannot conserve the momentum of the system.

(b) The normal reactions N and $-N$ between the wedge and block are equal opposites. Hence, there will be no effect of the internal forces of the system $(M + m)$. Then, we have the vertical external forces such as gravity forces $-Mg\hat{j}$ and $-mg\hat{j}$ and normal reaction $N_1\hat{j}$ offered by the ground on the wedge. As there is no horizontal external force acting on the system $(M + m)$, we can conserve the 'momentum of the wedge + block system' horizontally. Since the block accelerates vertically downwards with $a_y = a \sin \theta$, where $a=$ acceleration of m relative to M, the net force acting on the system $(M + m)$ is not zero. Then the net vertical force acting on the system $(M + m)$ is $F_y = [N_1 - (M + m)g] \neq 0$.

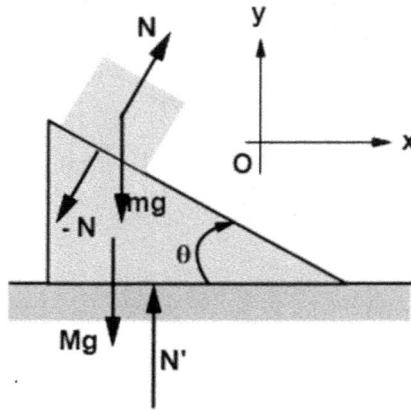

Hence, we cannot conserve the vertical momentum of the wedge-block system.

Example 34 Can you conserve the linear momentum of the following system?
   (a) Particles interconnected by a spring projected arbitrarily.

(b) The particle m is projected on the curved surface of the smooth wedge.

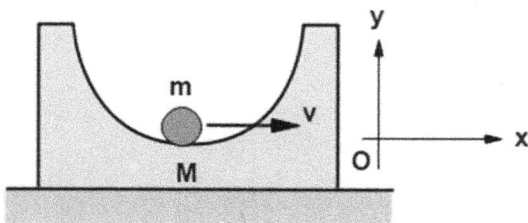

Solution

(a) Yes, because $F_{net} = 0$ on the spring–particle system as the spring force is an internal force in this system. Ans.

(b) P_x = constant because $F_x = 0$ but $P_y \neq$ constant because $F_y \neq 0$. Ans.

Linear impulse: As discussed earlier, the effect of a force F can be given as $\vec{F} = \frac{d\vec{P}}{dt}$.

Then, the change in momentum of a particle under the action of a single force is given as

$$d\vec{P} = \vec{F}\,dt$$

The right-hand term '$\vec{F}\,dt$' is called the 'linear impulse' of force $\vec{F}$ during time dt. Impulse is a vector quantity because a vector $\vec{F}$ is multiplied with a scalar dt.

If we want to find the impulse of the force F over a finite time interval $\Delta t = t_2 - t_1$, we will have to integrate (sum up) all elementary impulses.

Impulse-momentum equation: The net impulse during a time interval $\Delta t = t_2 - t_1$ can be given as

Imp $= \int_t^{t_2} \vec{F}\,dl$, where $\vec{F}\,dt = d\vec{P}$

This gives Imp $= \int_{P_1}^{P_2} d\vec{P} = \vec{P_2} - \vec{P_1} = \Delta\vec{P}$ (say).

Then, we have Imp $= \int \vec{F}\,dt = \Delta\vec{P}$.

The impulse of a force $\vec{F}$ during a time interval Δt is numerically equal to the change in momentum of the particle during the time interval Δt.

When many forces act on a particle, the sum of impulses of all forces is equal to the impulse of the net force. Symbolically, $\sum \vec{F_i}\,dt = \int (\sum \vec{F_i})\,dt$.

In that case, the net impulse but not the impulse of individual forces over any interval is equal to the change in momentum of the particle during that time interval.

When many forces act on a particle, impulse of the net force during any time interval is equal to the change in momentum of the particle during the given interval.

Remember that the 'impulse-momentum' equation is just a time integral form of Newton's 2nd law. Similar to this, the 'work–energy theorem' is a distance integral of force, derived from Newton's 2nd law. Hence, we should not treat 'impulse-momentum' and 'work–energy' relation as totally different concepts. The basis of each concept lies in Newton's 2nd law. We recast the integral and differential forms of Newton's 2nd law and work–energy theorem as follows:

$$\text{Imp} = \int \overrightarrow{F} \, dt = \Delta \overrightarrow{P}$$

$$\overrightarrow{F} = \frac{d\overrightarrow{P}}{dt}$$

$$\text{Work} = W = \int F. \; ds = \Delta K$$

$$F = \frac{dK}{dx}$$

In this way, we have learned that the net force F is a time derivative of momentum and space (distance) derivative of energy.

In other words, impulse-momentum relation relates force and momentum and work–energy theorem relates work and energy.

Graphical significance of impulse-momentum equation
The impulse of a force F during a time interval $\Delta t = t_2 - t_1$ is given as

$$\text{Imp} = \int_{t_1}^{t_2} \overrightarrow{F} \, dt,$$

where '$\overrightarrow{F} \, dt$' graphically represents the area of the thin vertical strip under the F–t curve during an elementary time dt.

Substituting $\overrightarrow{F} \, dt$ (= Area of the elementary strip of height F and breadth dt) = dA, we have

$$\text{Imp} = \int_{t_1}^{t_2} Fdt = \int dA = A$$

where A = area generated by Ft curve with time axis between the times $t = t_2$ and $t = t_2$.

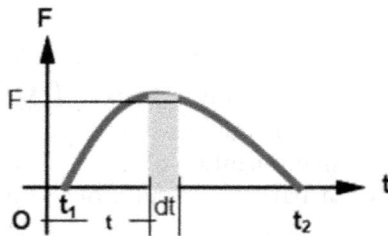

The above discussion tells us that

$$\text{Imp} = \int \overrightarrow{F} \, dt$$

which represents the area under F–t graph with time axis.

If the area lies above the t-axis, it is positive. It means the change in momentum $\Delta\vec{P}$ (caused by $\vec{F}$ only) is positive. The 'positive $\Delta\vec{P}$' points in positive directions of the axes, but you should not misunderstand that the magnitude of final linear momentum is greater than the magnitude of initial linear momentum.

Similarly, if the area lies below the time axis, the impulse is negative. Then $\Delta\vec{P}$ (due to $\vec{F}$ only) is negative, which signifies that $\Delta\vec{P}$ points in negative x-, y-, and z-directions in coordinate axes.

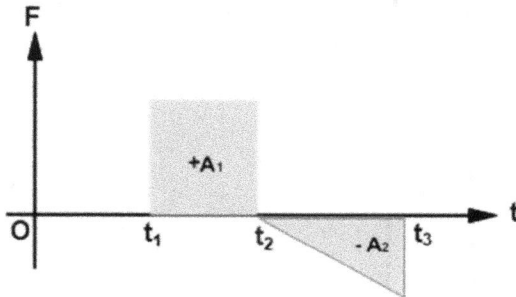

Sometimes a force may reverse its direction from positive to negative. A positive force gives a positive impulse, changing the momentum in the positive direction. Likewise, a negative force causes a negative impulse, which changes the momentum in the negative direction. In this way, during a time interval $\Delta t(=t_2 - t_1)$ say, we can have both positive impulse during some time intervals and negative impulse for the other time intervals. In other words, the area under the F–t graph will be sometimes positive and sometimes negative.

In this case, let us sum up all areas algebraically. The total impulse is Imp = Total area, A (say). Then, $A = (A_1) + (-A_2) =$ (magnitude of positive area) $-$ (magnitude of negative area) $=$ (magnitude of area lying above t-axis) $-$ (magnitude of area lying below t-axis).

If the total area A is positive, we have a net positive impulse and hence a positive $\Delta\vec{P}$. If the net area A is zero, $\Delta\vec{P} = 0$; there is no net change in linear momentum in the given time interval. However, $\Delta\vec{P} = 0$ does not tells us that the momentum of the particle remains constant for all instants between the time interval. When A is negative, we have a net negative impulse. This signifies a change in momentum in negative directions.

Recapitulating:

1. The area under the above F–t graph gives the impulse of the force $\vec{F}$, which is equal to the change in momentum 'due to that force'. Hence, the area under any force-time graph can be equated to the total change in momentum of a particle due to that force.

2. The net impulse = net area of net force–time curve.

3. If impulse is positive the change in momentum due to that impulse [but not necessarily the total (or actual) change in momentum] is directed in a positive direction and *vice versa*.

4. If net impulse is zero, $\Delta\vec{P} = 0$, but this does not conclude that $\vec{P}$ remains conserved.

Example 35 A particle of mass 1 kg moves with a velocity $v_0 = 2$ m s^{-1} in the positive x-direction at $t = 0$. The particle experiences a force F acting along the x-axis, which varies with time as shown in the figure. Find the momentum of the particle at the end of the 5th second.

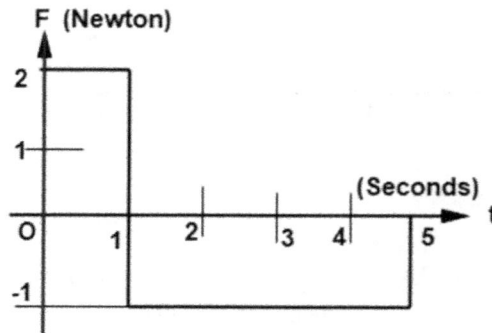

Solution

The net impulse $= \int F dt =$ Area under F–t graph $= A_1 - A_2$, where $A_1 = 2 \times 1 = 2$ and $A_2 = 1 \times 4 = 4$.

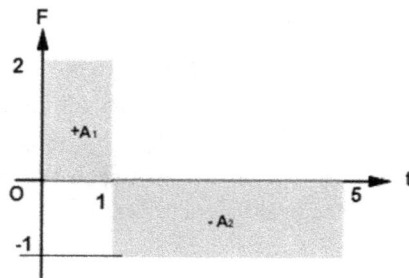

Then, we have $\int F dt = 2 - 4 = -2$.

Since $Fdt = \Delta\vec{P}$, we have $\Delta\vec{P} = -2\hat{i}$.

Substituting $\Delta\vec{P} = \vec{P} - \vec{P_0}$, we have $\vec{P} = \vec{P_0} - 2\hat{i}$, where $P_0 = mv_0$. Substituting $m = 1$ kg and $v_0 = 2$ ms^{-1} we have $\vec{P} = 0$.

Role of Newton's 3rd law, conservation of linear momentum of system of particles: We have discussed how Newton's 2nd law takes part in explaining the principle of conservation of linear momentum of a system of particles. Now we will discuss the role of Newton's 3rd law in explaining the principle of conservation of linear momentum of a system of particles.

Newton's 3rd law tells us that internal forces appear as action-reaction pairs. Let's consider a pair of two interacting particles A and B experiencing forces F and -F, respectively. Then the sum of impulses of these two forces will be equal to zero. This means that, according to Newton's 3rd law, the equal and opposite action-reaction pairs of forces as a whole contribute no net impulse. Then adding the impulses of all possible action-reaction pairs of internal forces, we have no net impulse. As a result, no change in momentum takes place due to internal forces as a whole. Hence, when the net effect of external forces acting on the system of particles is zero, we can conserve the linear momentum of the system.

You can summarize the above matter in the following way. Since the net internal force is equal to zero according to Newton's 3rd law, when the net external force acting on a system is zero, the momentum of the system is conserved in accordance with Newton's 2nd law.

Example 36 Considering the entire universe as a closed system, discuss the plausibility of conservation of linear momentum of the universe.

Solution

Since particles interact with each other, action-reaction pairs cannot produce any net impulse. Newton's law deals only with the interaction between material particles. Sometimes photons interact with matter. In that case, we need to modify Newton's laws using the modern ideas of relativity and quantum theories and try to conserve the momentum of the universe as a vector sum of momenta of all material and photon particles. Needless to say, no net force is acting on the universe when we assume it as an isolated system. Hence, the linear momentum of the universe is a 'constant quantity'. Note that, '*conservation of linear momentum*' *is an independent principle*.

1.7 Advantages of center of mass frame

The following properties of center of mass makes the problem solving easier:

1. Net force acting on the system relative to center of mass is zero, whereas the net force acting on center of mass (relative to ground) is equal to the resultant of all external forces acting on the system.

2. Momentum of the system relative to center of mass is zero. This means, neither internal forces nor external forces can change the momentum of the system relative to center of mass.

3. Kinetic energy of a system is minimum relative to center of mass.

4. The work done by pseudo (inertial) force in center of mass frame is zero for a translating system.

5. The work done by gravity is zero relative to center of mass (for the bodies having the size negligible compared to the radius of the earth).

6. The torques about the center of mass point and about any other moving or stationary points coinciding with the center of mass are equal.

7. The net torque produced by the internal forces is zero about the center of mass.

8. Gravity cannot produce a torque about the center of mass of a body whose size is negligible compared to the radius of the earth.

9. The torque about the center of mass causes rotation of a system of particles (rigid bodies).

10. The angular momentum about the center of mass defines the rotation of a system of particles.

11. The linear momentum of a system is equal to the linear momentum of the center of mass of the system.

12. The internal angular momentum of a system is equal to the angular momentum of the system relative to its center of mass.

13. The kinetic energy of a system is (generally) greater than the KE of the system about the center of mass of the system. However, both are equal in center of mass frame.

14. For a two-particle system $K_{int} = K_{system} = \frac{1}{2}\mu v_{rel}^2$ relative to the center of mass, where μ= reduced mass of the system and v_{rel}= relative velocities between the particles.

15. For a two-particle system, $\vec{P}_{system} = \frac{1}{2}\mu\vec{v}_{rel}$ relative to the center of mass.

16. For a two (or many) particle system $\sum m_i \vec{r}_{iC} = 0$, $\sum m_i \vec{v}_{iC} = 0$ and $\sum m_i \vec{a}_{iC} = 0$-. That means, the sum of (i) moment of masses, (ii) linear momentam, and (iii) product of mass and acceleration of the particles relative to center of mass are zero individually.

17. For a two-particle system, momenta of the particles are equal and opposite relative to center of mass.

18. For a two-particle system, the ratio of KE of the particles relative to the center of mass is equal to the inverse ratio of their masses.

19. For a two-particle system, the ratio of velocities, displacements, and accelerations of the particles is equal to the inverse ratio of their masses.

20. If $F_{ext} = 0$, the center of mass moves with constant velocity. Hence, the center of mass (but not necessarily other points of the system that may be accelerating) can be treated as an inertial frame (point).

21. The net effect of internal forces is zero on the center of mass, whereas it is not zero (in general) for other points of the system. As a consequence, the

center of mass moves with a constant velocity whereas other points may move with same accelerations when $F_{ext} = 0$.

22. Angular momentum (and rate of change in angular momentum) of a system of particles relative to center of mass point and pf any stationary or moving point coinciding with the center of mass is equal.

23. For any system, the internal KE is equal to KE of the system relative to the center of mass, which becomes an internal property of the system.

24. For rigid bodies the KE of rotation is equal to the KE of the particles due to their motion perpendicular to the line joining the center of mass with the corresponding particles of the system.

25. The KE of vibration of a two-particle system is the KE of the particles due to their motion along the line joining the center of mass of the system with the particles.

26. Internal forces as a whole cannot change the KE of the center of mass of the system whereas external force can change the KE of the center of mass of the system.

27. When the net work done by the internal forces is non-zero, the KE of the system about the center of mass changes due to the internal forces.

28. Internal forces cannot change the momentum of the center of mass of the system.

Some points regarding the center of mass will be covered in the forthcoming chapters. Anyway, keeping all the above points in mind, we will proceed to the next section where we will clear some of the above points (concepts) through suitable examples.

1.8 Application of impulse–momentum equation and work–energy theorem for a two-particle system

Work done by internal forces: In the previous sections we learned that, for a two-particle system, work done by internal force is given as

$$W_f = -\int f \, dr,$$

where f = internal force and r = distance of separation between the particles m_1 and m_2.

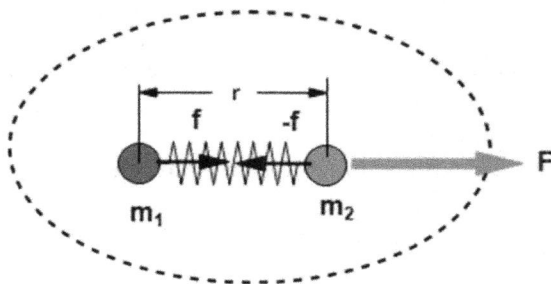

(a) If f = spring force, we have $f = kx$ (by Hooke's law) and $dr = dx$.
This gives $W_f = -k \int_{x_i}^{x_f} x\,dx = -\frac{k}{2}(x_f^2 - x_i^2)$.

(b) If f = gravitational force, we have

$$f = \frac{Gm_1 m_2}{r^2} \text{ and } dr = dx$$

This gives $W_{gr} = Gm_1 m_2\left(\frac{1}{r_f} - \frac{1}{r_i}\right)$.

(c) If f = electrostatic force, we have $f = \frac{q_1 q_2}{r\pi\varepsilon_0 r^2}$ and $dr = dr$.
Then, we have

$$W_{el} = -\frac{q_1 q_2}{4\pi\varepsilon_0}\left(\frac{1}{r_f} - \frac{1}{r_i}\right).$$

(d) If f = static friction, we have

$$f = f_s(\leqslant \mu_s N) \text{ and } dr = 0$$

[because there is no relative displacement (sliding) between the contacting suefaces].
This gives $W_s = 0$.

(e) If f = kinetic friction, we have $f = f_k = \mu_k N$ and $dr = dx$ [= relative elementary distance moved between the contacting surfaces (objects)].
This gives $W_k = -f_k x$, where $f_k = \mu_k N$ (assuming N= constant).

(f) If f = tension or normal reactions (constraint forces), put $f = T$ (or N) and $dr = 0$ (because there is no relative displacement between the objects conncted by string, rods, etc).

This gives $W_{constraint} = 0$.
In all the above cases, we observe that $W_{internal}(=W_f)$ does not depend on the choice of reference frame, as it deals with the magnitude of force of interaction 'f' which, in general, is a function of relative distance of separation of the particles.

Work done by external forces: As discussed earlier, work done by external forces can be given as

$$W_{ext} = \sum \int \vec{F_i} \cdot d\vec{r_{iC}} + \int \vec{F}.d\vec{r_C},$$

where $\vec{F}$ = net external force and $\vec{F_i}$ = external force acting at ith mass (or point), $d\,\vec{r_{iC}}$ = displacement of ith point relative to center of mass, and $d\,\vec{r_C}$ = elementary displacement of center of mass.

ΔK: Applying the KE expression for the two-particle system relative to ground, we found kinetic energy of the system as

$$K = \frac{1}{2}\mu v_{rel}^2 + \frac{1}{2}Mv_C^2$$

where $\mu = \frac{m_1 m_2}{m_1 + m_2}$, $v_{rel} = |\vec{v_1} - \vec{v_2}|$ and $M = (m_1 + m_2)$.

Then, the change in kinetic energy of the system is

$$\Delta K = \frac{1}{2}\mu\left[(v_{rel})_f^2 - (v_{rel})_i^2\right] + \frac{1}{2}M(v_{C_f}^2 - v_{C_f}^2)$$

Work–energy theorem relative to ground: Apply the work–energy theorem $W = \Delta K$, where $W = W_{ext} + W_{int}$. Then substitute the derived values of ΔK, W_{int} using the data of the question.

Work–energy theorem relative to center of mass:
Substituting $W_{ext} = \sum \int \vec{F_i} \cdot d\,\vec{r_{iC}} + \int \vec{F} \cdot d\,\vec{r_C}$, $\Delta K = \Delta K' + \Delta K_C$ we have

$$W_{int} + \sum \int \vec{F_i} \cdot d r_{iC} + \int \vec{F} \cdot d\,\vec{r_C} = \Delta K' + \Delta K_C.$$

Comparing both sides, we have the work–energy theorem relative to center of mass, which can be given as
where $\Delta K' = \frac{1}{2}\mu(v_{rel_f}^2 - v_{rel_i}^2)$
This gives

$$W_{int} + \sum \int \vec{F_i} \cdot d\,\vec{r_{iC}} = \frac{1}{2}\mu(v_{rel_f}^2 - v_{rel_i}^2)$$

If $F_{ext} = 0$, we can write

$$W_{int} = \frac{1}{2}\mu(v_{rel_f}^2 - v_{rel_i}^2)$$

When we use the work–energy theorem relative to an accelerating center of mass, we need to consider the work done by the pseudo-forces $-m_1\vec{a_C}$ and $-m_2\vec{a_C}$, where $a_C = \frac{F}{M}$. Then we find the W_{int} and W_{ext} relative to CM'. Finally, find the change in kinetic energy $(\Delta K')$ relative to center of mass: $W_{ext} = \sum \int \vec{F_i} \cdot d\,\vec{r_{iC}}$ and $W_{int} = -\sum f \cdot dr$. Note that pseudo-forces as a whole cannot perform work relative to the center of mass. Hence, you may disregard the effect of pseudo-force when solving the two-particle system relative to their center of mass.

Example 37 Two constant forces F_1 and F_2 act on two smooth particles (blocks) of masses m_1 and m_2 interconnected by a light spring of stiffness k. Find the maximum elongation of the spring using the work–energy theorem.

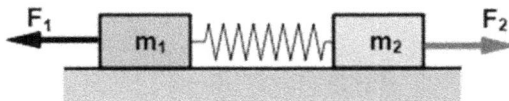

Solution

Let us assume that after some time, x_1 and x_2 are the position vectors of the particles m_1 and m_2, respectively. During a small (elementary) time interval dt, let the particles move through m_2, respectively. During a small (elementary) time interval dt, let the particles move through elementary distances dx_1 and dx_2, respectively.

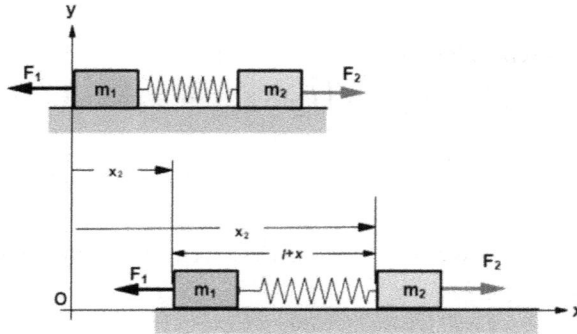

The work done by external force F_1 is

$$W_{F_1} = -F_1 x_1$$

Work done by external force F_2 is

$$W_{F_2} = -F_2 x_2$$

Work done by the internal force (spring forces) is $W_{sp} = -\frac{k}{2}x^2$, where $x = $ elongation of the spring.

Then, the total work done is

$$W = W_{F_1} + W_{F_2} + W_{sp}$$

$$= -F_1 x_1 + F_2 x_2 - \frac{k}{2}x^2$$

The change in kinetic energy of the system until the spring has maximum elongation is

$$\Delta K = \frac{1}{2}(m_1 + m_2)v^2$$

where $v = $ speed of the particles because both m_1 and m_2 will move with the same velocity at the time of maximum separation between them which corresponds to maximum elongation of the spring.

Now applying the work–energy theorem $W = \Delta K$, we have

$$-F_1 x_1 + F_2 x_2 - \frac{k}{2}x^2 = \frac{1}{2}(m_1 + m_2)v^2$$

Substituting $x_1 = (x_{1C} + x_C)$ and $x_2 = x_{2C} + x_C$, we have

$$(-F_1 + F_2)x_C + F_2 x_{2C} - F_1 x_{1C} - \frac{1}{2}kx^2 = \frac{1}{2}(m_1 + m_2)v_C^2$$

Comparing the terms on both sides, we have the following two equations:

$$F_2 x_{2C} - F_1 x_{1C} - \frac{k}{2}x^2 = 0 \tag{1.36}$$

$$(-F_1 + F_2)x_C = \frac{1}{2}(m_1 + m_2)v^2 \tag{1.37}$$

Substituting $x_{1C} = \frac{-m_2 x}{m_1 + m_2}$ and $x_{2C} = \frac{-m_1 x}{m_1 + m_2}$ in equation (1.36) we have

$$F_2\left(\frac{xm_1}{m_1 + m_2}\right) - F_2\left(\frac{-m_2 x}{m_1 + m_2}\right) = \frac{1}{2}kx^2$$

This gives

$$x = \frac{2(m_1 F_2 + m_2 F_1)}{k(m_1 + m_2)} \text{ Ans.}$$

Alternative method (center of mass frame):

$$W_{\text{int}} + \sum \int \vec{F_i} \cdot d\,\vec{r_{iC}} = \frac{1}{2}\mu(v_{rel_f}^2 - v_{rel_i}^2)$$

Since $v_{rel_f} = 0$ and $v_{rel_i} = 0$, we have $W_{\text{int}} + \sum \vec{F_i} \cdot d\,\vec{r_{iC}} = 0$
Putting $W_{\text{int}} = \frac{kx^2}{2}$, we have $\sum \vec{F_i} \cdot d\,\vec{r_{iC}} = \frac{kx^2}{2}$, where

$$\sum \int \vec{F_i} \cdot d\,\vec{r_{iC}} = \int \vec{F_1} \cdot d\,\vec{r_{iC}} + \int \vec{F_2} \cdot d\,\vec{r_{2C}}$$

Then, we have

$$\int_0^{\Delta \vec{r_{iC}}} \vec{F_1} \cdot d\,\vec{r_{1C}} + \int_0^{\Delta \vec{r_{2C}}} \vec{F_2} \cdot d\,\vec{r_{2C}} = \frac{k}{2}x^2$$

Since F_1 and F_2 are constant forces, we have

$$F_1 \Delta r_{1C} + F_2 \Delta r_{2C} = \frac{k}{2}x^2 \tag{1.38}$$

Since $-m_1 \Delta r_{1C} + m_2 \Delta r_{2C} = 0$ and $\Delta r_{1C} + \Delta r_{2C} = \frac{m_1 x}{m_1 + m_2}$

Substituting Δr_{1C} and Δr_{2C} in equation (1.38), we have

$$F_1\left(\frac{m_2 x}{m_1 + m_2}\right) + F_2\left(\frac{m_1 x}{m_1 + m_2}\right) = \frac{kx^2}{2}$$

This gives

$$x = \frac{2(m_1 F_2 + m_2 F_1)}{(m_1 + m_2)k} \text{ Ans.}$$

Conservation of linear momentum: If $F_{\text{ext}} = 0$ on the two-particle system, external impulse is zero. Then, we can conserve the linear momentum of the system; $\vec{P_i} = \vec{P_f}$. This means that $\vec{v_C} =$ constant and $\Delta K_C = 0$. Hence, $\Delta K = \Delta K'$.

Substituting $\Delta K_C = 0$ in the work–energy theorem $W_{\text{int}} + W_{\text{ext}} = \Delta K_C + \Delta K'$, we have

$$W_{\text{int}} + W_{\text{ext}} = \Delta K', \text{ where}$$

$$\Delta K' = \frac{1}{2}\mu\left[(v_{\text{rel}})_f^2 - (v_{\text{rel}})_i^2 \right]$$

Let us execute the given ideas in the following examples.

Example 38 A block of mass m_1 is projected with a velocity v_0 so that it climbs onto the smooth wedge of mass m_2. If the block does not leave the wedge, find the maximum height attained by the block.

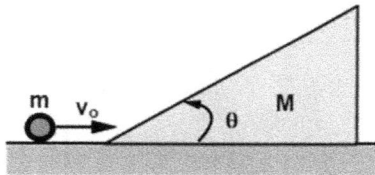

Solution
The block attains a maximum height when the final relative velocity v_{rel_f} between block and wedge becomes zero.
That means, $(v_{\text{rel}})_f = 0$.
The initial relative velocity between block and wedge is given as

$$(v_{\text{rel}})_i = v_0$$

Substituting $(v_{\text{rel}})_f = 0$ and $(v_{\text{rel}})_i = v_0$ in the formula $\Delta K = \frac{1}{2}\mu[(v_{\text{rel}})_f^2 - (v_{\text{rel}})_i^2]$ we have $\Delta K = -\frac{1}{2}\mu v_0^2$.

While ascending through a height h, the external force, that is, the earth's gravity, does work on the block m_1, which can be given as

$$W_{\text{ext}}(= W_{\text{gr}}) = -m_1 g h$$

The internal (constraint) forces (normal reactions) between m_1 and m_2 as a whole do not perform work; $W_{int} = 0$. Finally, putting $\Delta K = -\frac{1}{2}\mu v_0^2$, $W_{ext} = -m_1gh$, and $W_{int} = 0$ in the work–energy theorem $W_{ext} + W_{int} = \Delta K'$, we have

$$-m_1gh + 0 = -\frac{1}{2}\mu v_0^2,$$

where $\mu = \frac{m_1 m_2}{m_1 + m_2}$.

This gives

$$h = \frac{v_0^2}{2(1 + \frac{m_1}{m_2})g} \quad \text{Ans.}$$

Ground frame method: If you want to proceed from the basic, then write the following equations separately.

P = C: Since no external force acts horizontally, $P_x = C$

$$(\vec{P_x})_f = (\vec{P_x})_i,$$

where $\vec{P_x}i = m_1v_0\hat{i}$ and $\vec{P_{x_f}} = (m_1 + m_2)v\,\hat{i}$

(because, finally at the maximum height, both m_1 and m_2 have the same velocity as the relative sliding between them stops).

Then, we have

$$(m_1 + m_2)v = m_1v_0 \tag{1.39}$$

The work–energy theorem:

$$W_{ext} + W_{int} = \Delta K$$

where $W_{ext} = -m_1gh$, $W_{int} = W_{constraint} = 0$ and $\Delta K = \left[\left(\frac{m_1v^2}{2} + \frac{m_2v^2}{2}\right) - \frac{m_1v_0^2}{2}\right]$.

This gives

$$-m_1gh = \frac{1}{2}(m_1 + m_2)v^2 - \frac{m_1v_0^2}{2} \tag{1.40}$$

Substituting v from equation (1.39) in equation (1.40), we have

$$h = \frac{v_0^2}{2\left(1 + \frac{m_1}{m_2}\right)g} \quad \text{Ans.}$$

N.B.: By substituting $W_{gr} = -\Delta U_{gr}$ you can also apply the principle of conservation of energy, which originates from the work–energy theorem as a special case.

1.9 Application of impulse-momentum equation and work–energy theorem for many-particle system

(a) Constant mass system: Suppose there are two holes, 1 and 2, in a container filled with a liquid. In hole 1 and 2 an equal amount of liquid flows such that the mass of liquid in the container remains constant. Let the velocities of incoming water at hole 1 and outgoing water at hole 2 be v_1 and v_2, respectively. If 'dm' is the amount of liquid that enters into the container with velocity v_1 and the same amount of liquid escapes from the container with velocity v_2, the change in momentum of the system is equal to the change in momentum of the liquid, which can be given as

$$d\vec{P} = dm\,\vec{v_2} - dm\,\vec{v_1} = dm(\vec{v_2} - \vec{v_1})$$

The impulse $\int F_{ext}\,dt$ of the external forces acting on the system (container + liquid) must be equal to change in momentum of the system. Writing the impulse-momentum equation, we have

$$\vec{F}_{ext}dt = d\vec{P}$$

This gives $\vec{F}_{ext} = \frac{d\vec{P}}{dt}$, where $d\vec{P} = dm(\vec{v_2} - \vec{v_1})$.

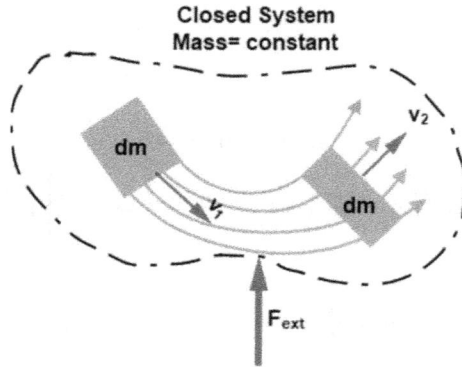

Closed System
Mass= constant

Then, we have

$$\vec{F}_{ext} = \frac{dm}{dt}(\vec{v_2} - \vec{v_1})$$

Note that here $\frac{dm}{dt}$ is not the rate of gain or rate of loss of mass. It is just a rate of flow, $\frac{dm}{dt} = A_1\rho v_1 = A_2\rho v_2$, according to the equation of continuity (mass conservation).

Example 39 An L-shaped tube of uniform cross-section 'a' carries a flowing liquid. If the liquid moves with a speed v and changes its direction by $90°$, find the external force F required to keep the tube in equilibrium. Neglect friction.

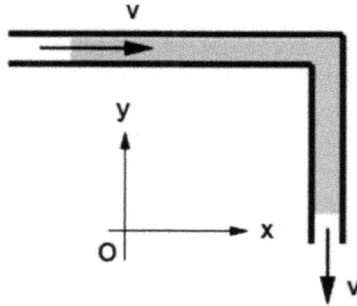

Solution
Following the derived expression

$$\vec{F}_{ext} = \frac{dm}{dt}(\vec{v_2} - \vec{v_1})$$

and substituting $\vec{v_1} = v\,\hat{i}$, $\vec{v_2} = -v\,\hat{j}$ and $\frac{dm}{dt} = a\rho v$, we have

$$\vec{F}_{ext} = a\rho v[(-v\,\hat{j}) - (v\,\hat{i})] = -a\rho v^2(\hat{i} + \hat{j})$$

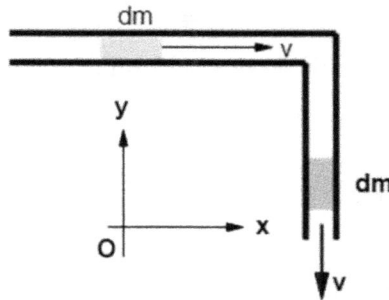

The magnitude of the force is equal to $\sqrt{2}\,a\rho v^2$ and it is applied at an angle of $135°$ with positive x- direction.

Example 40 A smooth piston of area A is pushed by a constant horizontal force with a velocity v_1. This results in escaping the incompressible non-viscous liquid of density ρ, with a velocity v_2. Find the magnitude of F.

Solution

When we push an elementary segment of liquid of mass dm with a velocity v_1 through a distance dx, say, an equal amount of liquid will be escaped from the other end of the tube, during the same time dt, with a velocity v_2.

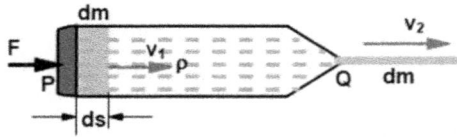

Then, the change in kinetic energy ΔK of the liquid = change in kinetic energy of the elementary segment 'dm' because the other portions of the liquid do not change their KE.

Since the mass 'dm' changes its KE from $\frac{dm}{2}v_1^2$ to $\frac{dmv_2^2}{2}$, we have

$$dK = \frac{dm}{2}(v_2^2 - v_1^2)$$

The external force F performs work, which can be given as

$$dW_{\text{ext}} = Fdx$$

The internal forces as a whole do not perform any work.
Hence,

$$W_{\text{int}} = 0$$

Hence, the net work done is

$$dW = Fdx$$

Using the work–energy theorem, we have $dW = dK$, where $dW = Fdx$ and $dK = \frac{dm}{2}(v_2^2 - v_1^2)$.
Then, we have

$$F = \frac{1}{2}\left(\frac{dm}{dx}\right)(v_2^2 - v_1^2)$$

Substituting $dm = \rho A dx$, we have

$$F = \frac{\rho A}{2}(v_2^2 - v_1^2) \text{ Ans.}$$

(b) Variable mass system
(i) *Constant mass approach*: Let's take an example of a rocket that moves with a velocity $\overrightarrow{v_1}$ after ejecting a gas of mass m_2. Let's assume that the ejected gas moves with a velocity $\overrightarrow{v_2}$. If the mass of the rocket + the remaining gas inside the rocket is equal to m_1, $(m_1 + m_2)$, that is, the total mass of the system (rocket + gas) remains constant. The momentum of the above system is

$$\overrightarrow{P} = m_1 \overrightarrow{v_1} + m_2 \overrightarrow{v_2}$$

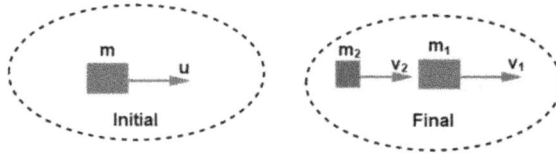

This rate at which the momentum of the system changes is given as

$$\frac{d\overrightarrow{P}}{dt} = \frac{d(m_1 \overrightarrow{v_1} + m_2 \overrightarrow{v_2})}{dt}$$

This gives

$$\frac{d\overrightarrow{P}}{dt} = \frac{dm_1 \overrightarrow{v_1}}{dt} + \frac{m_1 d\overrightarrow{v_1}}{dt} + \frac{dm_2 \overrightarrow{v_2}}{dt} + \frac{m_2 d\overrightarrow{v_2}}{dt} \tag{1.41}$$

Since $(m_1 + m_2) =$ constant, we have

$$\frac{dm_1}{dt} = -\frac{dm_2}{dt} \tag{1.42}$$

Substituting $\frac{dm_1}{dt} = -\frac{dm_2}{dt}$ from equation (1.42) in equation (1.41), we have

$$\frac{d\overrightarrow{P}}{dt} = \frac{dm_1}{dt}(\overrightarrow{v_1} - \overrightarrow{v_2}) + m_1 \frac{d\overrightarrow{v_1}}{dt} + m_2 \frac{d\overrightarrow{v_2}}{dt} \tag{1.43}$$

After leaving the 'system m_1' (rocket + residual gas), the 'system m_2', that is, ejected gas, moves with constant velocity $\overrightarrow{v_2}$.

Hence, $\frac{d\overrightarrow{v_2}}{dt} = 0$

Substituting $\frac{d\overrightarrow{v_2}}{dt} = 0$ in equation (1.43), we have

$$\frac{d\overrightarrow{P}}{dt} = (\overrightarrow{v_1} - \overrightarrow{v_2})\frac{dm_1}{dt} + m_1 \frac{d\overrightarrow{v_1}}{dt} \tag{1.44}$$

Since the change in the momentum of the system $(m_1 + m_2)$ is caused by the external force $\overrightarrow{F_{ext}}$, using Newton's 2nd law law of motion for the system of particles, we have

$$\vec{F}_{ext} = \frac{d\vec{P}}{dt} \tag{1.45}$$

Substituting $\frac{d\vec{P}}{dt}$ from equation (1.44) in equation (1.45), we have

$$\vec{F}_{ext} = (\vec{v_1} - \vec{v_2})\frac{dm_1}{dt} + m_1\frac{d\vec{v_1}}{dt}$$

where $(\vec{v_2} - \vec{v_1})(=\vec{v}_{rel}$, say) is the velocity of the escaping gas relative to the variable mass system 'm_1' (rocket + residual gas).

In general, if m is the mass of the variable mass system moving with a velocity $\vec{v}$ at any instant and $\vec{F}_{ext}$ is the net external force acting on it (excluding the reaction force of the incoming or outgoing mass), putting $\vec{v_1} - \vec{v_2} = -\vec{v}_{rel}$, $m_1 = m$, and $\vec{v_1} = \vec{v}-$, Newton's 2nd law can be written as

$$\vec{F}_{ext} = -\vec{v}_{rel}\frac{dm}{dt} + m\frac{d\vec{v}}{dt},$$

where '$\vec{v}_{rel}\frac{dm}{dt}$' is called the reaction or impact force $\vec{F}_{imp}$ or $\vec{F}_{reaction}$. Then, we can write

$$\vec{F}_{ext} + \vec{F}_{imp} = m\frac{d\vec{v}}{dt},$$

where $\vec{F}_{imp} = \vec{v}_{rel}\frac{dm}{dt}$

N.B.: Let's try to write Newton's 2nd law for the variable mass system (VMS) in the following way.

Suppose the VMS has a mass m and the velocity of its center of mass is $\vec{v_1}$ at any instant. Then, the momentum of the VMS can be given as

$$\vec{P} = m\vec{v_1}$$

Applying Newton's 2nd law

$$\vec{F}_{ext} = \frac{d\vec{P}}{dt},$$

we have $\vec{F}_{ext} = \frac{d}{dt}(m\vec{v_1})$.

This gives

$$\vec{F}_{ext} = m\frac{d\vec{v}}{dt} + \frac{\vec{v_1}dm}{dt}$$

which is totally different from the obtained expression, $\vec{F}_{ext} = -\vec{v}_{rel}\frac{dm}{dt} + m\frac{d\vec{v_1}}{dt}$!

So where is the mistake? Probably ignoring the rate of change in the momentum of little escaping stuff, that is, $\vec{v_2}\frac{dm_2}{dt}$, which makes all the difference! However, you may consider the above (generally wrong) formula correct only when $v_2 = 0$ (the special case when the mass leaves VMS with zero velocity).

(ii) *Impact force method*: Let's take an elementary segment of mass dm fixed with VMS. By applying suitable force (may be a variable force) somehow we keep the velocity of the VMS constant, that is, $\vec{v_1}$ (say). Suppose the elementary mass leaves the VMS during time dt with a relative velocity $\vec{v}_{rel}$ and starts moving with velocity $\vec{v_2}$ (say).

Hence, $\vec{v_2} = \vec{v_1} + \vec{v}_{rel}$.

Then the change in momentum of the elementary segment dm during time of impact dt is

$$d\vec{P} = dm\,\vec{v_2} - dm\,\vec{v_2} = dm(\vec{v_2} - \vec{v_1})$$

Substituting $\vec{v_2} - \vec{v_1} = \vec{v}_{rel}$ we have

$$d\vec{P} = dm\,\vec{v}_{rel}$$

If the elementary segment experiences a force $\vec{F}_{imp}$ to change its momentum during time dt, by using Newton's 2nd law or impulse-momentum equation, we can write

$$\vec{F}_{imp}dt = dm\,\vec{v}_{rel}$$

This gives

$$\vec{F}_{imp} = \vec{v}_{rel}\frac{dm}{dt}$$

Impact force (on an elementary mass or on VMS) depends on the (i) rate at which the VMS gains or loses mass and (ii) the velocity (magnitude and direction) of the incoming or outgoing mass relative to the VMS. If the VMS loses mass, $\frac{dm}{dt}$ is $-$ve. If it gains, $\frac{dm}{dt}$ is $+$ve.

Example 41 A trolley car of mass M_0 initially moving in the rain with a velocity $\vec{v_0}$ towards the right experiences a constant horizontal force $\vec{F}$ $-$. If the rain falls vertically such that the rainwater collected by the trolley car per unit time is μ, find the velocity of the trolley car after a time t measured from the instant of beginning the interaction (collection) of rainwater with the trolley car. Neglect friction at the ground.

Solution

Method 1 (constant mass method)

Let's take total rainwater plus trolley car as a system. The net horizontal force acting on the system is $\vec{F}_{ext} = F\hat{i}$.

1-78

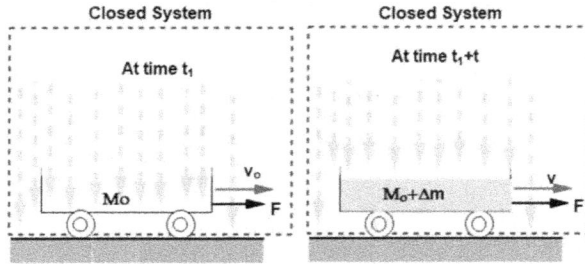

During a time t, the impulse of $\overrightarrow{F}_{\text{ext}}$ is

$$\text{Imp} = \int \overrightarrow{F}_{\text{ext}} dt = Ft\hat{i} \qquad (1.46)$$

Let the rainwater of mass Δm collected in the trolley car during time t move horizontally with a velocity v.

This causes a change in momentum of the system from $\overrightarrow{P_1}$ to $\overrightarrow{P_2}$ in horizontal during time t, which can be given as

$\Delta\overrightarrow{P} = \overrightarrow{P_2} - \overrightarrow{P_1}$, where $\overrightarrow{P_2} = (M_0 + \Delta m)v\,\hat{i}$ and $\overrightarrow{P_2} = M_0 v_0\hat{i}$

Then, we have

$$\Delta\overrightarrow{P} = (M_0 + \Delta m)v\,\hat{i} - M_0 v_0\hat{i}$$

Since the rainwater is collected at a constant rate of 'μ', $\Delta = \mu t$ substituting $\Delta m = \mu t$, we have

$$\Delta\overrightarrow{P} = (M_0 + \mu t)v\,\hat{i} - M_0 v_0\hat{i} \qquad (1.47)$$

By equating the impulse from equation (1.46) and $\Delta\overrightarrow{P}$ from equation (1.47) using the impulse-momentum equation, we have

$$Ft\hat{i} = (M_0 + \mu t)v\,\hat{i} - M_0 v_0\hat{i}$$

This gives $v = \frac{Ft + M_0 v_0}{M_0 + \mu t}$ Ans.

Method 2 (impact force method)

We can solve the above example by impact force method as follows. As the raindrops stick to the container, it gets a horizontal momentum just after the impact. So, a forward impact (reaction) force acts on each colliding raindrop. As a consequence, an equal and opposite (backward) impact force $\overrightarrow{F}_{\text{imp}}$ acts on the car by the colliding raindrops. Let's calculate it by using the derived formula

$$\overrightarrow{F}_{\text{imp}} = \overrightarrow{v}_{\text{rel}}\frac{dm}{dt}$$

Here, $\frac{dm}{dt} = \mu$ because the VMS gains mass and $(\vec{v}_{rel}) =$ horizontal velocity of raindrops relative to the trolley car. If the trolley car has a velocity $\vec{v}$ at any instant, $\vec{v}_{rel} = (\vec{v}_r)_x - (\vec{v}_t)_x$, where $(\vec{v}_r)_x = 0$ as the rain is falling vertically and $(\vec{v}_t)_x = v\,\hat{\imath}$.

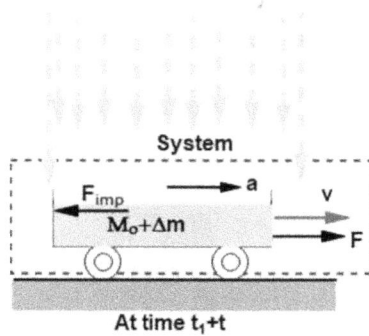

System

F_{imp} a   v

$M_0 + \Delta m$

F

At time $t_1 + t$

Then, we have $\vec{v}_{rel} = -v\,\hat{\imath}$.

Substituting $\frac{dm}{dt} = \mu$ and $\vec{v}_{rel} = -v\,\hat{\imath}$ in the formula $\vec{F}_{imp} = \vec{v}_{rel}\frac{dm}{dt}$, we have

$$\vec{F}_{imp} = -\mu v\,\hat{\imath}$$

Negative signifies that $\vec{F}_{imp}$ acts backward on the trolley car. After finding F_{imp}, apply Newton's 2nd law on the VMS to obtain

$F_{net}(= F_{ext} - F_{imp}) = ma$, where $a = \frac{dv}{dt}$, $F_{ext} = F$ and $F_{imp} = -\mu v$. This gives

$$F - \mu v = m\frac{dv}{dt}$$

Substituting $m = M_0 + \mu t$ and rearranging the terms, we have

$$\frac{dt}{M_0 + \mu t} = \frac{dv}{F - \mu v}$$

If the velocity of the trolley car changes from v_0 to v during time t, integrating both sides we have

$$\int_0^t \frac{dt}{M_0 + \mu t} = \int_{v_0}^v \frac{dv}{F - \mu v}$$

Evaluating the integration, we have

$$v = \frac{Ft + M_0 v_0}{M_0 + \mu t} \text{ Ans.}$$

N.B.: Using the impact force method, find $\vec{F}_{imp}$ by considering the signs of $\frac{dm}{dt}$ and $\vec{v}_{rel}$; then add the impact force $\vec{F}_{imp}$ with the given applied forces on VMS. Finally, the force $\vec{F}_{net}(=\vec{F}_{app} + \vec{F}_{imp})$ is equated with $m\vec{a}$, where $\vec{a} = \frac{d\vec{v}}{dt}$ and $m = $ mass of VMS $= M_0 + \mu t$. Then do the mathematical work by solving the differential equation. Using the impact force method you will get $\vec{a}$, then $\vec{v}$, but using the constant mass approach, you will get $\vec{v}$ and then $\vec{a} = \frac{d\vec{v}}{dt}$.

Example 42 A chain of uniform linear mass density μ is released from rest from vertical position as shown in the figure. Find the reaction force offered by the ground after the chain falls through a distance y.

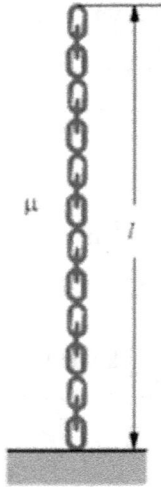

Solution

Method 1 (constant mass approach)

Let's consider the total chain as a system of constant mass M. If the mass of the chain fallen is m' and the mass of the falling part of the chain is m, we have

$$m + m' = M.$$

Hence, we have

$$\frac{dm}{dy} = -\frac{dm'}{dy}$$

If the velocity of the falling part is $\vec{v}(=-v\,\hat{j})$ its momentum is

$$\vec{P}_1 = -mv\,\hat{j}$$

Since the fallen part does not move, its momentum P_2 is zero. Then, the total momentum of the system (total chain) is

$$\overrightarrow{P} = \overrightarrow{P_1} + \overrightarrow{P_2} = -mv\,\hat{j}$$

The rate of change in momentum of the system is

$$\frac{d\overrightarrow{P}}{dt} = \frac{d(m\overrightarrow{v})}{dt} = \frac{md\,\overrightarrow{v}}{dt} + \overrightarrow{v}\frac{dm}{dt} \qquad (1.48)$$

Since the falling part has a free acceleration

$$\frac{d\,\overrightarrow{v}}{dt} = -g\hat{j} \qquad (1.49)$$

By using equations (1.48) and (1.49), we have

$$\frac{d\overrightarrow{P}}{dt} = -\left(v\frac{dm}{dt} + mg\right)\hat{j}$$

Substituting $\frac{dm}{dt} = \frac{dm}{dy} \cdot \frac{dy}{dt} = \mu v$, we have $\frac{dP}{dt} = \mu v^2 + mg$.
Since the chain falls freely through a distance y, $v^2 = 2gy$.
Then,

$$\frac{d\overrightarrow{P}}{dt} = -\{\mu(2gy) + mg\}\hat{j}$$

where m = mass of the falling part.

Equating $\frac{d\overrightarrow{P}}{dt}$ with the net force $\overrightarrow{F}$, we have $\overrightarrow{F} = -(mg + 2\mu gy)\hat{j}$, where
$\overrightarrow{F} = \{-(mg + m'g) + N\}\hat{j}$ (referring to the free-body diagram)
This gives

$$-mg - m'g + N = -(mg + 2\mu gy)$$

Then, we have $N = m'g - 2\mu gy$, where $m' = \mu y$.
This gives

$$N = 3\mu gy \text{ Ans.}$$

Method 2 (impact force method)
Since each falling link collides with the fallen links with a relative velocity
$\overrightarrow{v}_{\text{rel}} = -v\,\hat{j}$, where v = velocity of the falling chain, we have the impact formula
for the VMS 'm' given as $\overrightarrow{F}_{\text{imp}} = \overrightarrow{v}_{\text{rel}}\frac{dm'}{dt}$, where $\overrightarrow{v}_{\text{rel}} = -v\,\hat{j}$ and $\frac{dm'}{dt} = \mu v$ (positive,
as mass m' is increasing).
This gives

$$\overrightarrow{F}_{\text{imp}} = -\mu v^2\hat{j},$$

where $v = \sqrt{2gy}$ because the chain falls freely.
Hence, we have

$$\overrightarrow{F}_{\text{imp}} = -2\mu gy\hat{j}$$

1-82

Now considering the external forces $\overrightarrow{N}$ and $m'\overrightarrow{g}$ and adding $\overrightarrow{F}_{imp}$, the net force on m' is

$$\overrightarrow{F}_{imp} = (-m'g - 2\mu gy + N)\hat{j}$$

Since the acceleration of the fallen part is zero, substitute $a = 0$ in the formula $F_{net} = m'a$ to obtain

$$N - m'g - 2\mu gy = 0, \text{ where } m' = \mu y. \text{ This gives}$$

$$N = 3\mu gy \text{ Ans.}$$

N.B.: Since each elementary segment dm of the falling mass changes its velocity from $-v\,\hat{j}$ to zero, the change in momenta of dm can be given as $\overrightarrow{dP} = dmv\,\hat{j}$ during time dt.

Hence, the reaction (impact force acting on the mass dm) is

$$\overrightarrow{F}'_{imp} = \frac{\overrightarrow{dP}}{dt} = \frac{dm}{dt}v\,\hat{j}, \text{ where } \frac{dm}{dt} = \mu v.$$

This gives $\overrightarrow{F}'_{imp} = \mu v^2\hat{j}$.

Since the chain is continuously falling, the lowest colliding link experiences an impact (reaction) force F'_{imp} and the heap (fallen portion) receives an equal and opposite force $\overrightarrow{F}_{imp} = \overrightarrow{F}'_{imp} = -\mu v^2\hat{j}$ in the vertically downward direction. Since the chain is flexible, the lowest link of the falling part will be automatically detached from the link, which collides and falls dead just after the collision. Hence, the falling part of the chain cannot receive any reaction (impulsive or impact) force $\overrightarrow{F}_{imp}$. However, the interacting link (the link that is neither fallen nor falling) will receive an impact force F_{imp} from the fallen part and gives an equal and opposite impact force to the fallen part. As a consequence the falling part moves with freefall acceleration g.

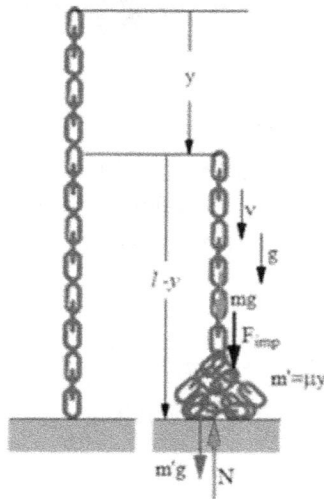

Problem 1 Calculation of position of center of mass
Point (discrete mass distribution)
1. Four small balls are placed at the vertices of a square of diagonal of length $L = 2\sqrt{2}$ m. Taking 2 kg mass as the origin, if the coordinates of center of mass of the system of four particles are 1.5, y, find the value of m and y.

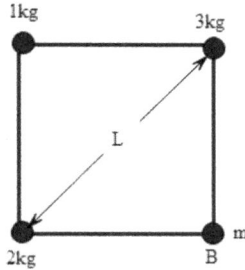

Solution
Referring to the figure given below, the x-coordinate of the center of mass is

$$x_C = \frac{2(0) + m(2) + 3(2) + 1(0)}{2 + 1 + 3 + m}$$

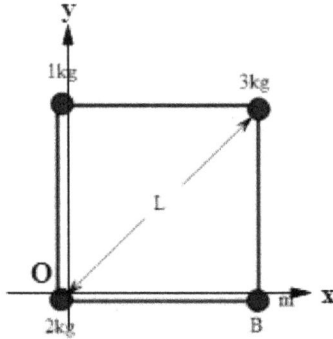

$$x_c = \frac{2m + 6}{m + 6}$$

$$1.5 = \frac{2m + 6}{m + 6} \Rightarrow \frac{3}{2}m + 9 = 2m + 6$$

$$\Rightarrow \frac{m}{2} = 3 \Rightarrow m = 6 \text{ Ans.}$$

Then $y_c = \dfrac{2(0) + 1(2) + 3(2) + m(0)}{m + 6}$

$$= \frac{8}{6 + m} = \frac{8}{6 + 6} = \frac{8}{12} = \frac{2}{3} \text{ Ans.}$$

Problem 2 Four particles A, B, C, and D of masses 3, 1, 2, and 4 kg, respectively, are placed at the vertices of a square ABCD of side $L = 2\,m$. (a) Find the center of mass of the system of particles. (b) If the particles move to their nearest vertices in anticlockwise sense with a same speed v, find the (a) displacement and (b) velocity of the center of mass of the system. Take the point A as the origin of the coordinate system.

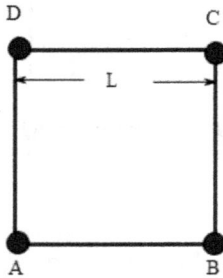

Solution

(a) Referring to the last figure, the displacement of the center of mass is

$$\vec{s_C} = \frac{3 \times 2\hat{i} + 1 \times 2\hat{j} + 2(-2\hat{i}) + 4(-2\hat{j})}{10}$$

$$= \frac{2\hat{i} - 6\hat{i}}{10} = (\hat{i} - 0.6\hat{j})m \text{ Ans.}$$

(b) The velocity of center of mass is $\vec{v_C} = \frac{(3)v\,\hat{i} + (1)v\,\hat{j} + 2(-v\,\hat{i}) + 4(-v\,\hat{j})}{10}$

$$= \frac{v}{10}\{\hat{i} - 3\hat{j}\} \text{ Ans.}$$

Problem 3 A thin rod of length l and linear mass density varies linearly from μ at A to $\eta\mu$ at B. Find the center of mass of the rod.

Solution

Referring to the figure below, the value of linear mass density at a distance x is

$$\mu' = \mu\left\{1 + \frac{n-1}{L}x\right\}$$

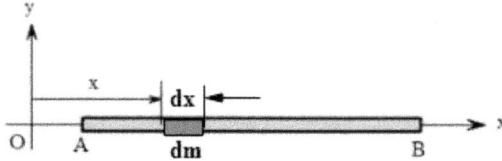

The position of the center of mass is

$$x_c = \frac{\int x\,dm}{\int dm} = \frac{\int x\mu' A\,dx}{\int \mu' A\,dx} = \frac{\int x\mu'\,dx}{\int \mu'\,dx}$$

$$= \frac{\int x\,\mu\left\{1 + \frac{n-1}{L}x\right\}dx}{\mu\int\left\{1 + \frac{n-1}{L}x\right\}dx}$$

$$= \frac{\left.\frac{x^2}{2} + \frac{n-1}{L}\frac{x^3}{3}\right|_0^L}{\left.x + \frac{n-1}{L}\frac{x^2}{2}\right|_0^L}$$

$$= \frac{\frac{L^2}{2} + \frac{n-1}{L}\frac{L^3}{3}}{L + \frac{n-1}{L}\frac{L^2}{2}}$$

$$= L\left\{\frac{\frac{1}{2} + \frac{n-1}{3}}{1 + \frac{n-1}{2}}\right\} = \frac{3 + 2n - 2}{\frac{6(2+n-1)}{2}}L$$

$$= \frac{(2n+1)}{3(n+1)}L = \frac{2n+1}{3(n+1)}L \quad \text{Ans.}$$

Problem 4 Find the center of mass of the T-shaped rod if the limbs OA and BC have lengths a and b, respectively. Take O as the origin.

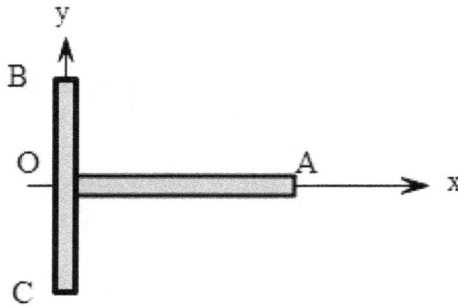

Solution

Let's reduce the uniform rods into the point masses placed at the centers of the rods because they are uniform. Now, the position of the center of mass of a two-particle system is

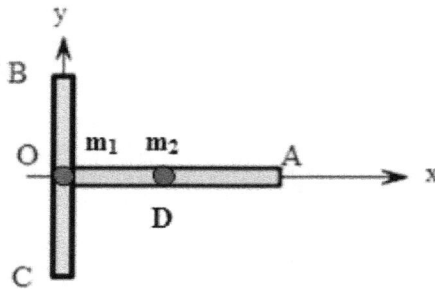

$$x_c = \frac{m_1(0) + m_2\left(\frac{l_2}{2}\right)}{m_1 + m_2}$$

$$= \frac{l_2}{2\left(1 + \frac{m_1}{m_2}\right)}$$

Since the rods are uniform, mass is directly proportional to length; thus, the mass ratio is equal to the ratio of length.

$$x_c = \frac{b}{2\left(1 + \frac{a}{b}\right)} = \frac{b^2}{2(a + b)}$$

Problem 5 Two plates A and B are joined to form a T-shaped plate. Assume the length and breadth of the plates are $l_1 = 12$ m, $b_1 = 2$ m, and $l_2 = 10$ m, $b_2 = 1.8$ m, respectively. Find the center of mass of the T-shaped rod.

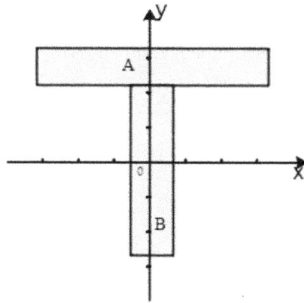

Solution

Due to symmetry of the system about the y-axis, $x_c = 0$. The position of center of mass of the system is

$$y_c = \frac{m\left(\frac{l_2 + b_1}{2}\right) + M(0)}{M + m}$$

$$= \frac{l_2 + b_1}{2\left(\frac{M}{m} + 1\right)},$$

where $\frac{M}{m} = \frac{l_2 b_2}{l_1 b_1}$

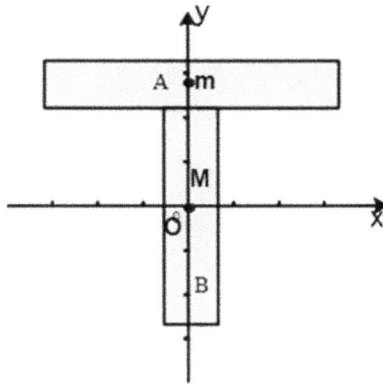

$$\Rightarrow y_c = \frac{l_2 + b_1}{2\left(\frac{l_2 b_2}{l_1 b_1} + 1\right)}$$

$$= \frac{10 + 2}{2\left(\frac{10 \times 1.8}{12 \times 2} + 1\right)} = \frac{12}{2(7/4)} = \frac{24}{7} \quad \text{Ans.}$$

Problem 6 The uniform plate ABC is connected to two uniform rods AD and CD of masses m_1 and m_2 and linear mass density $x = 2$ kg m^{-1}. A particle of mass m is welded at the point D. Find the center of mass of the combination relative to the corner B. Put $M = 6$ kg, $AD = a = 2$ and $AB = b = 1$ and $m = 5$.

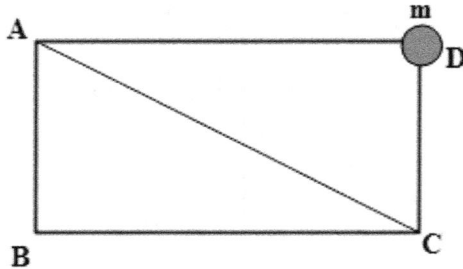

Solution

The center of mass of the triangular plate is at $\frac{a}{3}, \frac{b}{3}$. The center of mass of the rods will be at its geometrical centers. Then we have

$$X_c = \frac{M(\frac{a}{3}) + m_1 \frac{a}{2} + m(a) + m_2(a)}{M + m_1 + m_2 + m}$$

$$= \frac{(\frac{M}{3} + \frac{m_1}{2} + m_2 + m)}{M + m_1 + m_2 + m} a$$

$$= \frac{\frac{M}{3} + \frac{xa}{2} + xb + m}{M + xa + xb + m} a$$

$$= \frac{\frac{M}{3} + \frac{x}{2}(a + 2b) + m}{M + x(a + b) + m} a$$

$$= \frac{\frac{6}{3} + \frac{2}{2}(2 + 2 \times 1) + 5}{6 + 2(2 + 1) + 5} a$$

$$= \frac{2 + 4 + 5}{6 + 6 + 5} a = \frac{11}{17} a$$

$$11 \times \frac{2}{17} = \frac{22}{17} \text{ m}$$

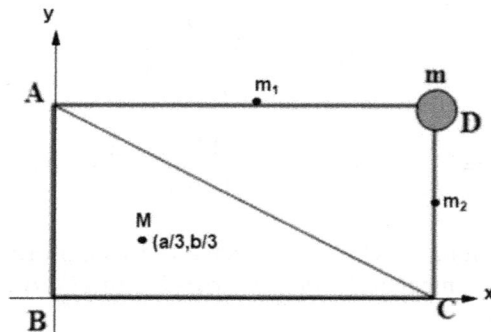

Similarly, $y_c = \dfrac{M(\frac{b}{3}) + m_1 b + mb + m_2(\frac{b}{2})}{M + m_1 + m + m_2}$

$$= \dfrac{m\frac{b}{3} + \left(m_1 + \frac{m_2}{2}\right)b + mb}{M + m_1 + m_2 + m}$$

$$= \dfrac{\frac{M}{3} + \left(m_1 + \frac{m_2}{2}\right) + m}{M + m_1 + m_2 + m}b$$

$$= \dfrac{\frac{M}{3} + x\left(a + \frac{b}{2}\right) + m}{M + x(a + b) + m}$$

$$= \dfrac{\frac{6}{3} + 2\left(2 + \frac{2}{2}\right) + 5}{6 + 2(2 + 1) + 5}b$$

$$= \dfrac{2 + 6 + 5}{6 + 6 + 5}(1) = \dfrac{13}{17} \text{ Ans.}$$

Problem 7 Infinite numbers of blocks are piled as shown in the figure. Each succeeding block has half the length and breadth of its preceding block. Taking the constant density of all the bricks, calculate the x-component of the center of mass of the system of blocks.

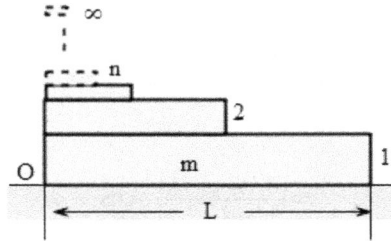

Solution

If the length and breadth of a block become less by u and v, respectively, than that of its immediate lower block, the corresponding mass of the block will be uv times less than that of the lower block. Then, $m_1 = m$, $m_2 = \dfrac{m}{uv}$, $m_3 = \dfrac{m}{(uv)^2}$, $m_4 = \dfrac{m}{(uv)^3}$, and so on. Then the x-coordinate of the center of mass is

$$x_c = \dfrac{m_1 x_1 + m_2 x_2 + m_3 x_3 + \cdots}{m_1 + m_2 + m_3 + \cdots}$$

$$= \dfrac{(m)\left(\frac{l}{2}\right) + \left(\frac{m}{uv}\right)\left(\frac{l/u}{2}\right) + \dfrac{m}{(uv)^2}\dfrac{(l/u^2)}{2} + \cdots}{m + \dfrac{m}{uv} + \dfrac{m}{(uv)^2} + \cdots}$$

$$= \frac{m\frac{l}{2}\left\{1 + \frac{1}{u^2v} + \frac{1}{(u^2v)^2} + \cdots\cdots\right\}}{m\left\{1 + \frac{1}{uv} + \frac{1}{(uv)^2} + \cdots\cdots\right\}}$$

$$= \frac{1 + \frac{1}{u^2v} + \frac{1}{(u^2v)^2} + \cdots}{1 + \frac{1}{uv} + \frac{1}{(uv)^2}} \frac{l}{2}$$

$$= \frac{\frac{1}{1 - \frac{1}{u^2v}}}{\frac{1}{1 - \frac{1}{uv}}} \frac{l}{2} = \frac{l}{2}\left\{\frac{u^2v}{u^2v - 1} \times \frac{uv - 1}{uv}\right\}$$

$$= \frac{ul(uv - 1)}{2(u^2v - 1)} \text{ Ans.}$$

The y-coordinate of the center of mass is

$$y_c = \frac{m_1 y_1 + m_2 y_2 + m_3 y_3 + \cdots\cdots}{m_1 + m_2 + m_3 + \cdots\cdots}$$

$$= \frac{m\frac{l}{2} + \left(\frac{m}{uv}\right)\left(\frac{l/2}{v}\right) + \left(\frac{m}{(uv)^2}\right)\left(\frac{l/2}{v^2}\right)}{m + \frac{m}{uv} + \frac{m}{(uv)^2} + \cdots\cdots}$$

$$= \frac{l}{2}\left[\frac{1 + \frac{1}{uv^2} + \frac{1}{(uv^2)^2} + \cdots}{1 + \frac{1}{uv} + \frac{1}{(uv)^2} + \cdots}\right]$$

$$= \frac{vl(uv - 1)}{2(uv^2 - 1)} \text{ Ans.}$$

Problem 8 There are n numbers of identical planks of mass m and length l that are piled as shown in the figure. If the system of all planks will remain in equilibrium and a constant overhang of magnitude b is maintained, find the maximum value of N.

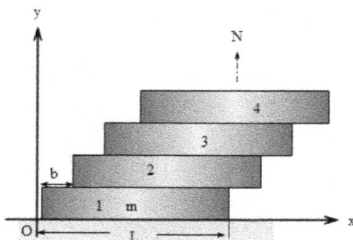

Solution

If b = horizontal distance between two consecutive bricks, then the x-coordinate of the center of mass of the system of blocks from the end of the first block is

$$x_c = \frac{m\left(\frac{l}{2} + b\right) + m\left(\frac{l}{2} + 2b\right)}{m(n-1)}$$

$$= \frac{(n-1)m\frac{l}{2} + mb\{1 + 2 + \dots + (n-1)\}}{m(n-1)}$$

$$= \frac{l}{2} + \frac{mb(n-1)n}{2(n-1)m}$$

$$x_c = \frac{l}{2} + \frac{bn}{2} \qquad (1.50)$$

For no toppling of the $(n-1)$ blocks, its center of mass must pass through the edge P of the first blocks. Then, the center of mass of $(n-1)$ blocks is given as

$$x_c \leqslant l \qquad (1.51)$$

Solving these two equations (1.50) and (1.51), we have

$$b \leqslant \frac{l}{n} \Rightarrow n \geqslant \frac{l}{b} \text{ Ans.}$$

Problem 9 A circular portion of radius r is cut from a uniform circular plate of radius R as shown in the figure. Find the position of the center of mass of the remaining portion.

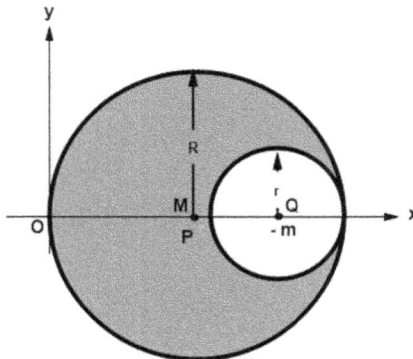

Solution

As the portion of disc of radius r is removed, the mass of the removed portion can be imagined as a negative mass m, say. Let the mass of the complete disc of radius R before removal be M. Due to the uniformity of the disc, we can substitute these discs by two point masses M and $-m$ placed at the centers P and Q, respectively, of the respective discs.

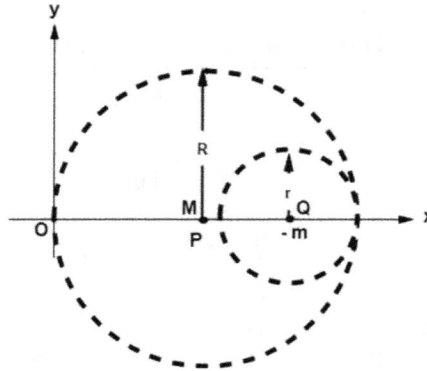

The center of mass of the two-particle system is

$$x_c = \frac{MR + m(2R - r)}{M + (-m)}$$

$$x_c = \frac{MR - m(2R - r)}{M + (-m)} = \frac{MR - mR - mR + mr)}{M - m}$$

$$= \frac{MR - mR - mR + mr)}{M - m}$$

$$= R - \frac{m(R - r)}{M - m} = R - \frac{(R - r)}{M/m - 1}$$

Putting $M/m = (R/r)^2$, we have

$$x_c = R - \frac{(R - r)}{(R/r)^2 - 1}$$

$$= R - \frac{r^2}{R + r} = \frac{R^2 + Rr - r^2}{R + r} \quad \text{Ans.}$$

Problem 10 A circular portion is removed from the square lamina as shown in the figure. Find the center of mass of the remaining portion of the lamina relative to the point (a) O if $AB = \eta R$. Put $\eta = 4$.

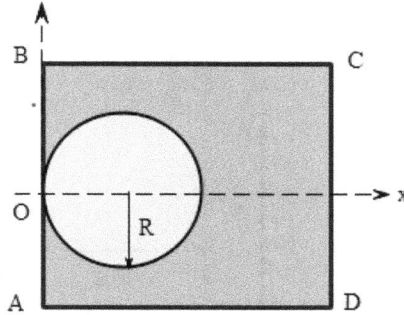

Solution

Following the logic of the previous problem,

$$x_c = \frac{m_1 x_1 - m_2 x_2}{m_1 - m_2}$$

$$= \frac{(\eta R)^2 \sigma \cdot \frac{\eta R}{2} - \pi R^2 \sigma \cdot R}{\eta^2 R^2 \sigma - \pi R^2 \sigma} = \frac{\eta^3 - 2\pi}{2(\eta^2 - \pi)} R$$

$$= \frac{\left(\frac{\eta^3}{2} - \pi\right) R}{(\eta^2 - \pi)} = \frac{(\eta^3 - 2\pi)R}{2(\eta^2 - \pi)}, \text{ where } \eta = 4$$

$$\text{Then, } x_c = \frac{32 - \pi}{16 - \pi} R \text{ from O}$$

Problem 11 Two plates A and B are joined to form a L-shaped plate. Assume the length and breadth of the plates are $l_1 = 10$ m, $b_1 = 1.5$ m, and $l_2 = 12$ m, $b_2 = 2$ m, respectively. Find the center of mass of the L-shaped rod. Assume that the ratio of the masses of A and B $= 2$.

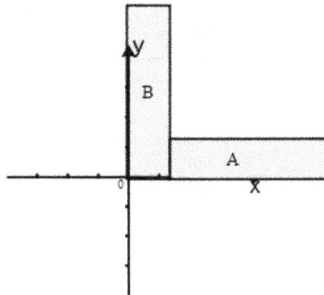

Solution

Each plate can be substituted by the point mass placed at the center of mass A and B of the plates. Then, the center of mass of the two-particle system can be given by our usual method as follows:

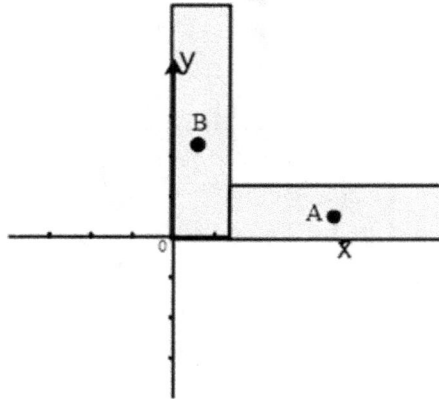

The x-coordinate of the center of mass is

$$x_c = \frac{m_A\left(\frac{l_1}{2} + b_2\right) + m_B\frac{b_2}{2}}{m_A + m_B}$$

$$= \frac{\frac{m_A}{m_B}\left(\frac{l_1}{2} + b_2\right) + \frac{b_2}{2}}{\frac{m_A}{m_B} + 1}$$

$$= \frac{2\left(\frac{10}{2} + 2\right) + \frac{2}{2}}{2 + 1} = 5$$

The x-coordinate of the center of mass is

$$y_c = \frac{m_B\frac{l_2}{2} + m_A\frac{b_1}{2}}{m_A + m_B}$$

$$= \frac{\left\{\frac{m_A}{m_B}b_1 + \frac{l_2}{1}\right\}}{2\left(\frac{m_A}{m_B} + 1\right)}$$

1-95

$$= \frac{2 \times 1.5 + 12}{2(2 + 1)} = \frac{15}{3 \times 2} = 2.5$$

Problem 12 A hemispherical bowl of mass M and radius R is completely filled with water of mass m. Find the center of mass of the system $(M + m)$. Put $M/m = 2$.

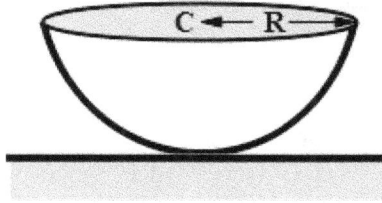

Solution

There are two bodies; one is the hemispherical bowl of mass m and radius R whose center of mass is at a height $y_1 = \frac{R}{2}$. The other is the hemispherical portion of water of mass M and radius R whose center of mass is at height $y_2 = \frac{5R}{8}$, from ground. Then the center of mass of the system can be given as

$$y_c = \frac{m_1 y_1 + m_2 y_2}{m_1 + m_2}$$

$$= \frac{m\frac{R}{2} + M\frac{5R}{8}}{m + M}$$

$$= (4m + 5M)\frac{R}{8(M + m)}$$

Putting $\frac{M}{m} = 2$, we have

$$yc = \frac{4 + 5(M/m)R}{8(1 + M/m)}$$

$$= (4 + 5 \times 2)\frac{R}{8(1 + 2)} = \frac{7R}{12} \text{ Ans.}$$

Problem 13 Find the center of mass of the shaded portion sphere of radius a and thickness $h = a/2$ relative to the center of the sphere.

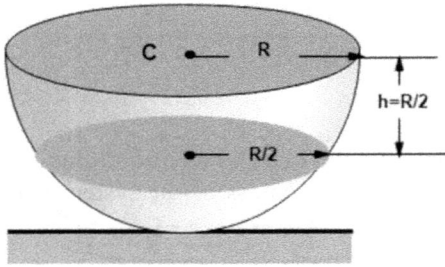

Solution

The formula for the y-position of center of mass is

$$y_c = \frac{\int y \, dm}{\int dm} = \frac{\int y \rho \pi r^2 dy}{\int \rho \pi r^2 dy}$$

$$= \frac{\int y r^2 dy}{\int r^2 dy} = \frac{\int R \sin \theta R^2 \cos^2 \theta R \cos d\theta}{\int R^2 \cos^2 \theta R \cos d\theta}$$

$$= R \frac{\int_0^\theta \sin \theta \cos^3 \theta d\theta}{\int \cos^2 \theta \cos \theta d\theta}$$

$$= R \frac{\int t^2 dt}{(1 - t^2) dt} = \frac{|(R/4) \cos^4 \theta|_0^{30^0}}{\sin \theta - \frac{\sin^3 \theta}{3} \Big|_0^{30^0}}$$

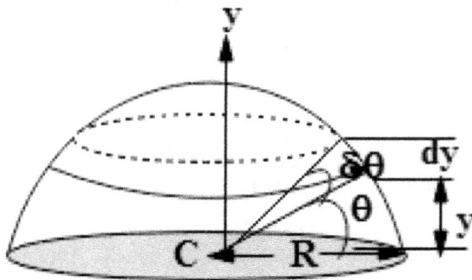

$$= \frac{R}{4} \frac{|\cos^4 30^0 - \cos^4 0|}{\sin 30^0 - \frac{\sin^3 30}{3}}$$

$$=\frac{R}{4}\frac{\left|\left(\frac{\sqrt{3}}{2}\right)^4-1\right|}{\frac{1}{2}-\frac{1}{24}}=\frac{R}{4}\frac{\left|\frac{9}{16}-1\right|}{\frac{11}{24}}$$

$$=\frac{R}{4}\frac{7}{16}\times\frac{24}{11}=\frac{21}{88}R \text{ from C.}$$

N.B.: Proof of the Integral:

$$\int \cos^3\theta d\theta = \int \cos^2\theta\cos\theta d\theta$$

$$=\int (1-\sin^2\theta)\cos\theta d\theta,\ \text{put } \sin\theta = t,\ \cos\theta d\theta = dt$$

$$=\int (1-t^2)dt$$

$$=t-\frac{t^3}{3} = \sin\theta - \frac{\sin^3\theta}{3}\Big|_0^{30^0}$$

$$=\frac{1}{2}-\frac{1}{24} = \frac{11}{24}$$

Problem 14 A square plate of mass M length l is kept by the side of another plate of mass m and height $l/2$ as shown in the figure.

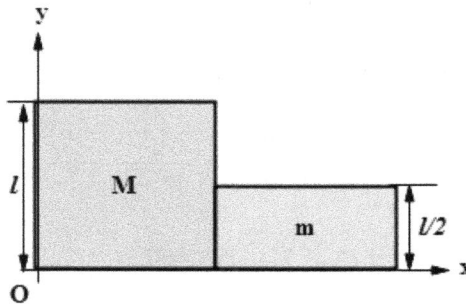

Solution

(a) If $m_1 = m = $ mass of the square when we take half of the square its mass will be $m_2 = \frac{m}{2}$. Then the center of mass of the square and half square will be $(\frac{l}{2}, \frac{l}{2})$ and

$(\frac{3l}{2}, \frac{l}{4})$, respectively. After converting the given system to the system of two particles, let's find the center of mass using the general procedure.

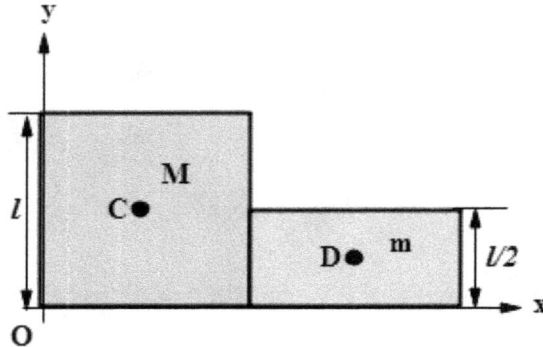

The x-component of the center of mass of the given section is

$$x = \frac{m_1 \frac{l}{2} + m_2 \frac{3l}{2}}{m_1 + m_2}$$

$$= \frac{l}{2} \frac{\frac{m_1}{m_2} + 3}{\frac{m_1}{m_2} + 1}$$

Putting $\frac{m_1}{m_2} = 2$, we have

$$x_c = \frac{l}{2}\left(\frac{2+3}{2+1}\right) = \frac{5l}{6} \text{ Ans.}$$

The y-component of center of mass of the section is

$$y_c = \frac{m_1 \frac{l}{2} + m_2 \frac{l}{4}}{m_1 + m_2} = \frac{l}{4} \frac{\left(\frac{2m_1}{m_2} + 1\right)}{\frac{m_1}{m_2} + 1}$$

$$= \frac{l}{4} \frac{(2 \times 2 + 1)}{2+1} = \frac{5l}{12} \text{ Ans.}$$

(b) Then, the initial potential energy of the segment is

$$U = (m_1 + m_2)gy_c = \left(m + \frac{m}{2}\right)g5l$$

$$= \frac{3m}{2}g\frac{5}{12} = \frac{5mgl}{8}$$

Problem 15 In figure (a), (b), and (c), two smooth blocks of mass M and m are placed side by side. In figure (d), the block m is placed on the block of mass M. A horizontal force F acts in each case. Find the acceleration of center of mass.

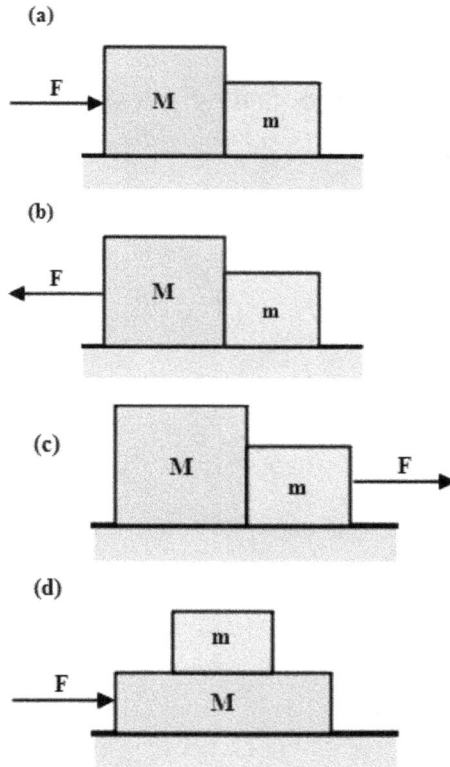

(a)

(b)

(c)

(d)

Solution
(a) When pushed, the blocks do not lose contact with each other. They move as a combined block. Then the center of mass of the system moves with an acceleration equal to the acceleration of the blocks.

$$a_c = \frac{F}{M + m} = a_1 = a_2$$

(b) However, when the block M is pulled by the force F, it loses contact with the block m. So, these accelerations will be different. Since the net force acting on the block M is $-F\hat{i}$, its acceleration is

$$\vec{a_M} = -\frac{F}{M}\hat{i}$$

Since no net force acts on the block m, its acceleration is zero.

$$\vec{a_2} = 0$$

But the center of mass of $M + m$ will move with an acceleration

$$\vec{a_c} = \frac{-F\hat{i}}{M + m} \text{ Ans.}$$

(c) If the force F pulls the block m, it will lose contact with the block M. Then,

$$a_M = 0$$

$$a_m = \frac{F}{m}\hat{i}$$

$$\vec{a_c} = \frac{F}{M + m}\hat{i}$$

(d) As the net force F acts on M, the acceleration of M is

$$\vec{a_M} = \frac{F}{M}\hat{i}$$

Since no net force acts on m, its acceleration is

$$\vec{a_m} = 0$$

The net force acting on $(M + m)$ is $F\hat{i}$. So, its center of mass of moves with an acceleration

$$\vec{a_c} = \frac{F}{M + m}\hat{i}$$

Problem 16 A smooth block of mass m is in contact with the accelerating trolley car of mass M when the trolley car is moved with an acceleration a. Find the acceleration of the center of mass of the system $(M + m)$.

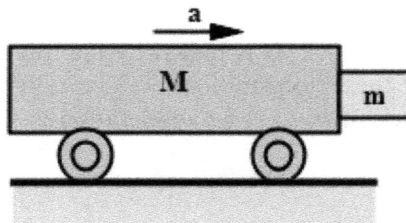

Solution

As the trolley car accelerates, the block m does not lose contact as a normal reaction acts between the cart and block exists. So, the x-component of acceleration of M and m will be equal to $a\hat{i}$. The net vertical force acting on the block m is mg. So, its vertical acceleration is $\overrightarrow{a_y} = -g\,\hat{j}$, whereas the vertical acceleration of the car is zero because its gravity is balanced by the normal reason offered by the ground. Then, the center of mass of the system $(M + m)$ moves with an acceleration whose x-component is

$$\overrightarrow{a_x} = \frac{m_1 \overrightarrow{a_{x_1}} + m_2 \overrightarrow{a_{x_2}}}{m_1 + m_2}$$

$$= \frac{Ma\hat{i} + ma\hat{i}}{M + m} = a\hat{i}$$

The y-component of the acceleration of the center of mass of the system is

$$\overrightarrow{a_y} = \frac{m_1 a_{y_1} + m_2 a_{y_2}}{m_1 + m_2}$$

$$= \frac{M(0) + m(-g\,\hat{j})}{M + m}$$

$$\Rightarrow \overrightarrow{a_y} = \frac{-mg}{M + m}\hat{j}$$

Then, the net acceleration of center of mass is

$$\overrightarrow{a} = \overrightarrow{a_x} + \overrightarrow{a_y} = a\hat{i} - \frac{mg}{M + m}\hat{j} \text{ Ans.}$$

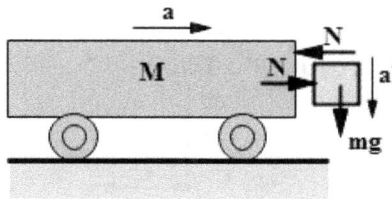

Problem 17 A flexible loop of total mass m is connected to a light rod that rotates with an angular speed ω making an angle θ with the axis of rotation, as shown in the figure. If the distance between the center of mass of the loop and axis is $r_c = r$, find the (a) tension in the rod and (b) angle θ made by the rod with vertical.

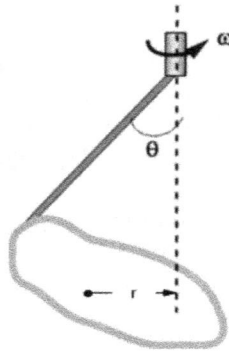

Solution

(a) The center of mass of the chain moves in a horizontal circle. So, its acceleration is horizontal and centripetal. Let r = radius of the circle made by center of mass C. Then, the acceleration of the center of mass of the chain is

$$a_c = \frac{F_{net}}{m}$$

$$\Rightarrow a_c = \frac{T \sin \theta}{m} \tag{1.52}$$

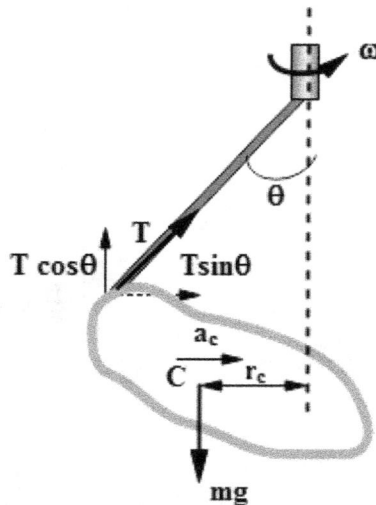

The net vertical force acting on the chain is

$$F_y = T \cos \theta - mg \tag{1.53}$$

Since the center of mass of the chain does not move vertically, its vertical acceleration is zero.

Putting $a_y = 0$ in the last equation, we have

$$T \cos \theta - mg = 0 \tag{1.54}$$

The centripetal acceleration of the center of mass is

$$a_c = r\omega^2 \tag{1.55}$$

Using the last three equations, we have

$$T \sin \theta = mr\omega^2 \tag{1.56}$$

$$T \cos \theta = mg \tag{1.57}$$

Using equations (1.56) and (1.57)

$$T = m\sqrt{(r\omega^2) + g^2}$$

$$= m\sqrt{r^2\omega^4 + g^2}$$

(b) Referring to the free-body diagram, we have $\tan \theta = \frac{r\omega^2}{g}$

$$\Rightarrow \theta = \tan^{-1}\left(\frac{r\omega^2}{g}\right) \text{ Ans.}$$

Problem 18 A man of mass m walks on a wedge with a constant magnitude of its velocity v relative to the wedge of mass M. If he starts from A and reaches the point B of the wedge, assuming smooth ground, find the (a) velocity of the wedge, (b) time taken by the man to reach the top point B of the wedge, and (c) displacement of the wedge.

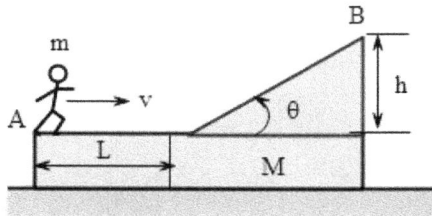

Solution

From AC, let the velocity of the wedge be v_M. The velocity of the man relative to the wedge is

$$v_m = v - v_M$$

Conserving horizontal momentum (because $F_x = 0$ on the system $M + m$),

$$-Mv_M + m(v - v_M) = 0$$

$$\Rightarrow v_M = \frac{mv}{M + m} \tag{1.58}$$

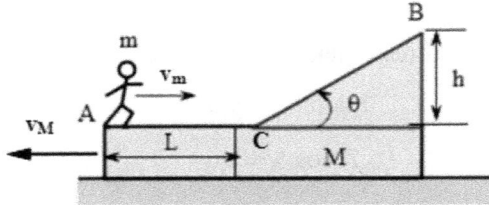

From C to B, let the velocity of the wedge be v'_m. The horizontal velocity of the man is

$$v_m = v \cos \theta - v_M$$

Then conserving the horizontal momentum of $M + m$,

$$-Mv'_M + m(v \cos \theta - v'_M) = 0$$

$$\Rightarrow v'_M = \frac{mv \cos \theta}{M + m} \tag{1.59}$$

(b) The time taken by the man to travel from A to B is

$$t = \frac{AC + CB}{v}$$

$$= \frac{L + \left(\frac{h}{\sin \theta}\right)}{v}$$

$$\Rightarrow t = \frac{L \sin \theta + h}{v \sin \theta}. \text{ Ans.}$$

(c) The total distance covered by the wedge in backward direction is

$$x = v_m t_{AC} + v'_M t_{CB}$$

$$= \frac{mv}{M + m}\left(\frac{L}{v}\right) + \left(\frac{mu \cos \theta}{M + m}\right)\left(\frac{h}{v \sin \theta}\right)$$

$$= \frac{ML}{M + m}\left[1 + \frac{h}{L} \cot \theta\right]. \text{ Ans.}$$

Problem 19 Two men of mass m_1 and m_2 are standing at the ends A and B of the boat of mass M. If they walk up to the mid-point of the boat, find the (a) displacement of the boat, (b) final position of the boat, (c) final position of m_1, and (d) final position of m_2. Assume that l_0 = initial distance of the end A of the boat from the shore.

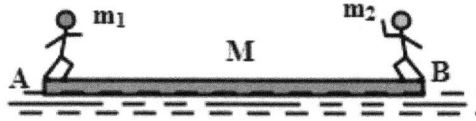

Solution

(a) Let the displacement of the boat be $\vec{s}$. As the displacement of man 1 relative to the boat is
$$\vec{s_{1b}} = \frac{l}{2}\hat{i} \text{ and that of 2 relative to the boat is } \vec{s_{2b}} = -l\hat{i}, \text{ these displace-}$$
ments relative are
$$\vec{s_1} = \vec{s_{1b}} + \vec{s_b} = \vec{s_{1b}} + \vec{s} \text{ and}$$

$$\vec{s_i} = \vec{s_{2b}} + \vec{s_b} = \vec{s_{2b}} + \vec{s}$$

Putting all these value in the equation
$$m_1\vec{s_1} + m_2\vec{s_2} + M\vec{s} = (M + m_1 + m_2)\vec{s_c} = 0$$

We have,
$$m_1(\vec{s_{1b}} + \vec{s}) + m_2(\vec{s_{2b}} + \vec{s}) + M\vec{s} = 0$$

$$\Rightarrow \vec{s} = -\frac{m_1\vec{s_{1b}} + m_2\vec{s_{2b}}}{M + m_1 + m_2}$$

$$\Rightarrow \vec{s_b} = \vec{s} = -\left\{\frac{m(\frac{l}{2}\hat{i}) + 2m(-l\hat{i})}{M + m + 2m}\right\}$$

$$= \frac{3ml}{2(M + 3m)}\hat{i} \text{ Ans.}$$

(b) The final position of the boat is
$$\vec{r_b} = \vec{s_b} + \vec{r_b}$$

$$= \left\{l_0 + \frac{l}{2} + \frac{3ml}{2(M + 3m)}\right\}\hat{i}$$

$$= \left[x_0 + \frac{l}{2} \left\{ 1 + \frac{3m}{M + 3m} \right\} \right] \hat{i}$$

(c) The final position of m_1 is

$$\vec{r_1} = \vec{r_1} + (\vec{s_{1b}} + \vec{s_b})$$

$$= x_0 \hat{i} + \frac{l}{2} \hat{i} + \frac{3ml}{2(M + 3m)} \hat{i}$$

$$= \left\{ x_0 + \frac{l}{2}(1 + \frac{3m}{M + 3m}) \right\} \hat{i}$$

(d) The final position of m_2 is

$$\vec{r_2} = \vec{r_{2,0}} + \vec{s_{2b}} + \vec{s_b}$$

$$= (l + x_0)\hat{i} + (-l\hat{i}) + \frac{3ml}{2(M + 3m)} \hat{i}$$

$$\left\{ x_0 + \frac{2M + 9m}{2(M + 3m)} l \right\} \hat{i}$$

Problem 20 A cube of ice floats on a cubical water beaker. If the ice melts, what happens to the (b) level of water and the (b) position of the center of mass of the system (ice + water)?

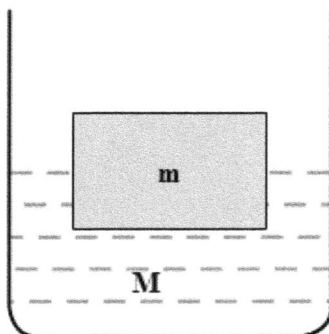

Solution

(a) Let $y =$ depth of immersion of the ice cube. Since the ice is floating, the buoyant force is

$$F_b = mg$$

$$\Rightarrow Vpg = mg$$

$$\Rightarrow Ay\, \rho g = (Al\sigma)g$$

$$\Rightarrow y = l\frac{\sigma}{\rho} \tag{1.60}$$

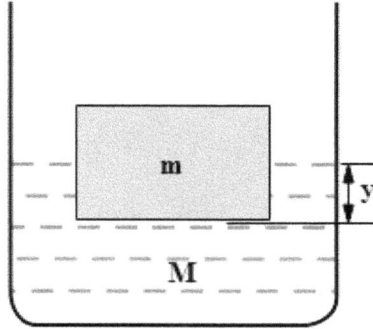

If the ice cube melts, let it form a rectangular structure of water having the base equal to that of the ice. If y' is the height of the rectangular structure of water, obeying the conservation of mass, we have

$$m_\omega = m_{ice}$$

$$\Rightarrow Ay'\rho = Al\sigma \tag{1.61}$$

Using equations (1.60) and (1.61), we have

$$y' = y \tag{1.62}$$

This means that after melting of ice the level of water does not increase.

(b) Since the surrounding water does not change its center of mass and the mass of ice remains same as the mass of water formed, due to the decrease in volume, the resulting height of the center of mass decreases. Then, the decrease in height of the center of mass of the ice cube of mass m when it melts is

$$\Delta h_c = h_2 - h_1 = \frac{y'}{2} - \frac{l}{2} \tag{1.63}$$

Using equations (1.62) and (1.63), we have

$$\Delta h_c = \frac{l\sigma}{2\rho} - \frac{l}{2} = \frac{l}{2}\left(\frac{\sigma}{\rho} - 1\right)$$

So, the decrease in height of the center of mass of the system $(M + m)$ is

$$\Delta h'_c = \frac{m\Delta h_c + M(0)}{M + m}$$

$$= -\frac{ml}{2(M + m)}\left(1 - \frac{\sigma}{\rho}\right) \text{ Ans.}$$

Problem 21 Two persons A and B each of mass m walk from the top of the wedge of mass M at time $t = 0$ and reach the bottom of the wedge simultaneously at time $t = T$. If the ground is smooth and the wedge is movable, find the (a) displacement of the wedge during the time T (b) the displacement of the center of mass of the system $(A + B + \text{wedge})$.

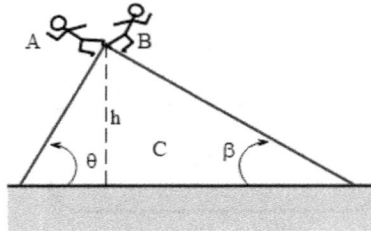

Solution

Let s_{1w}, s_{2w} be the displacements of the persons relative to the wedge, respectively, and $\vec{s_w} = $ displacement of the wedge. Then, the displacement of the center of mass of the wedge + balls system is

$$\vec{s_c} = \frac{m_1 \vec{s_1} + m_2 \vec{s_2} + m_3 \vec{s_3}}{m_1 + m_2 + m_3} \tag{1.64}$$

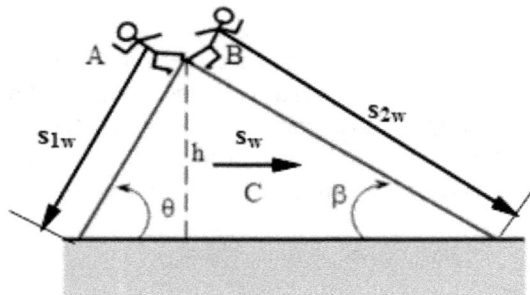

Using Galilean transformation of displacements, we have

The displacement of $A = \vec{s_1} = \vec{s_{1w}} + \vec{s_w}$ \hfill (1.65)

$$\text{The displacement of B} = \overrightarrow{s_2} = \overrightarrow{s_{2\omega}} + \overrightarrow{s_\omega} \tag{1.66}$$

$$\text{The displacement of the wedge} = \overrightarrow{s_3} = \overrightarrow{s_\omega} \tag{1.67}$$

Using the last four equations,

$$\overrightarrow{s_c} = \frac{m_1(\overrightarrow{s_{1\omega}} + \overrightarrow{s_\omega}) + m_2(\overrightarrow{s_{2\omega}} + \overrightarrow{s_\omega}) + m_3\overrightarrow{s_\omega}}{M + m_1 + m_2}$$

$$\overrightarrow{s_c} = \frac{m_1\overrightarrow{s_{1\omega}} + m_2\overrightarrow{s_{2\omega}} + (m_1 + m_2 + m_3)\overrightarrow{s_\omega}}{M + m_1 + m_2}$$

Since no net horizontal force acts on the system $(M + 2m)$, $a_{c_x} = 0$. So, $v_{c_x} =$ constant, which will be equal to zero because all the bodies were at rest initially:
Putting $\overrightarrow{s_c} = \overrightarrow{s_{c_x}} + \overrightarrow{s_{c_y}}$ and $\overrightarrow{s_{c_x}} = 0$, we have

$$m_1\overrightarrow{s_{1\omega_x}} + m_2\overrightarrow{s_{2\omega_x}} + \overrightarrow{s_\omega}(m_1 + m_2 + M) = 0$$

Putting $\overrightarrow{s_{1\omega_x}} = -h\cos\theta\,\hat{i}$ and $\overrightarrow{s_{2\omega_x}} = h\cos\beta\,\hat{i}$, we have $\overrightarrow{s_\omega} = \frac{h(\cos\theta - \cos\beta)\hat{i}}{M + 2m}$ Ans.
(b) Since $\overrightarrow{s_{c_x}} = 0$, $\overrightarrow{s_c} = \overrightarrow{s_{c_y}}$

$$= \frac{m_1\overrightarrow{s_{1_y}} + m_2\overrightarrow{s_{2_y}} + m_3\overrightarrow{s_{3_y}}}{m_1 + m_2 + m_3}$$

$$= \frac{m(-h\hat{j}) + m(-h\hat{j}) + 0}{M + 2m} = -\frac{2mh}{M + 2m}\hat{j} \text{ Ans.}$$

Problem 22 In an ideal pulley-particle system, the system is released from rest from the initial positions of the bodies as shown in the figure. Find the (a) position, (b) velocity, and (c) acceleration of the center of mass of the system as a function of (i) distance x travelled by each body and (ii) time.

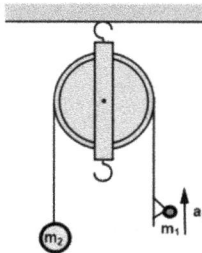

Method 1

Let a = acceleration of m_1 relative to the string. If the string touching the mass m_1 accelerates down with a, the other mass m_2 moves with an upward acceleration a. Then the accelerations of m_1 and m_2 are

$$a_1 = (a_r - a) \text{ and } a_2 = a$$

Hence, the acceleration of center of mass of the system $m_1 + m_2$ is

$$\vec{a}_c = \frac{m_1 \vec{a}_1 + m_2 \vec{a}_2}{m_1 + m_2}$$

$$= \frac{m_1(a_r - a)\hat{j} + m_2(a\hat{j})}{m_1 + m_2}$$

$$\Rightarrow \vec{a}_c = \frac{m_1}{m_1 + m_2} a_r \hat{j} + \frac{m_2 - m_1}{m_1 + m_2} a\hat{j} \tag{1.68}$$

The force equation on m_1:

$$T - m_1 g = m_1 a_1 = m_1(a_r - a) \tag{1.69}$$

The force equation on m_2:

$$T - m_2 g = m_2 a_2 = m_2 a \tag{1.70}$$

Solving equations (1.69) and (170)

$$m_1 g + m_1(a_r - a) = m_2(g + a) \Rightarrow (m_1 + m_2)a = (m_1 - m_2)g + m_1 a_r$$

$$\Rightarrow a = \frac{m_1 - m_2}{m_1 + m_2} g + \frac{m_1 a_r}{m_1 + m_2} \tag{1.71}$$

Putting the value of a form equation (1.71) in equation (1.68), we have

$$a_c = \frac{m_1 a_r}{m_1 + m_2} + \frac{m_2 - m_1}{m_1 + m_2} \left\{ \frac{m_1 - m_2}{m_1 + m_2} g + \frac{m_1 a_r}{m_1 + m_2} \right\}$$

$$= \frac{m_1 a_r}{m_1 + m_2} \left(1 + \frac{m_2 - m_1}{m_1 + m_2} \right) - \left\{ \frac{m_2 - m_1}{m_1 + m_2} \right\}^2 g$$

$$= \frac{2 m_1 m_2 a_r}{m_1 + m_2} - \left(\frac{m_2 - m_1}{m_1 + m_2} \right)^2 g$$

If the insect m_1 stops accelerating relative to the string, putting $a_r = 0$, we have

$$a_c = -\left(\frac{m_2 - m_1}{m_1 + m_2} \right) g \text{ as expected.}$$

Method 2

The net force acting on the system $M + m$ is

$$2T - (m_1 + m_2)g = (m_1 + m_2)a_c \qquad (1.72)$$

Using equations (1.70) and (1.71), we have

$$T = m_2(g + a)$$

$$= m_2 \left\{ g + \frac{m_1 a_r}{m_1 + m_2} + \frac{m_1 - m_2}{m_1 + m_2} g \right\}$$

$$= m_2 \left\{ \frac{m + a_r}{m_1 + m_2} \right\}$$

$$\Rightarrow T = \frac{m_1 m_2}{m_1 + m_2}(2g + a_r) \qquad (1.73)$$

Using equations (1.72) and (1.73),

$$a_c = \frac{2T}{m_1 + m_2} - g$$

$$= \frac{2}{m_1 + m_2} \left\{ \frac{m_1 m_2}{m_1 + m_2}(2g + a_r) \right\} - g$$

$$= \frac{2m_1 m_2(2g + a_r)}{(m_1 + m_2)^2} - g$$

$$= \frac{2m_1 m_2 a_r}{(m_1 + m_2)^2} - g \left\{ 1 - \frac{4m_1 m_2}{(m_1 + m_2)^2} \right\}$$

$$= \frac{2m_1 m_2 a_r}{(m_1 + m_2)^2} - (\frac{m_1 - m_2}{m_1 + m_2})^2 g$$

Problem 23 In the pulley-particle system, in one side we have a bunch of bananas hanging and on the other side a monkey whose mass M is equal to the mass of the bananas moves with an upward acceleration a_r relative to the string in order to grab the banana. (a) What is the acceleration of center of mass of the monkey-bananas system? (b) Can the monkey succeed in grabbing the banana?

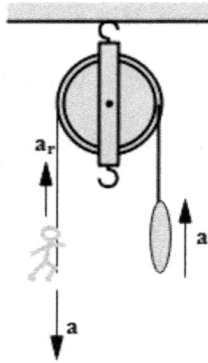

Solution

If $m_1 = m_2$ using equation (1.71), we have

$$a = \frac{m_1 a_r}{m_1 + m_2} = \frac{a_r}{2}$$

Then, the acceleration of center of mass is

$$a_c = \frac{2m_1 m_2}{(m_1 + m_2)^2} a_r = \frac{2 \times m \times m}{(m + m)^2} a_r = \frac{a_r}{2}$$

The acceleration of m_1 and m_2 are equal. So the relative acceleration between man and monkey is

$$a_{rel} = |a_1 - a_2|$$
$$= |a_r - a - a|$$
$$= |a_r - 2a|$$
$$= \left| a_r - 2\frac{a_r}{2} \right| = 0$$

In this case, the relative distance between the bodies remains constant because the relative displacement is

$$S_{rel} = \frac{1}{2} a_{rel} t^2 = 0$$

Then the monkey cannot catch the bananas.

Problem 24 In the following figure, a ladder of length L with a boy hangs from one end of the string and the other end of the string is connected to a block of mass M. Initially the system is at rest. Let the boy climb the ladder from its bottom to top with a constant average velocity v_r. Find the (a) acceleration of the center of mass of the system, (b) time after the boy reaches the top of the ladder, (c) maximum displacement of the center of mass, (d) maximum velocity of the center of mass, and (e) tension on the string.

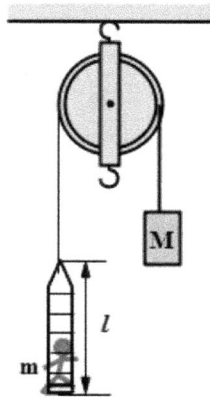

Solution

(a) The acceleration of the center of mass of the system is

$$2M\vec{a_c} = m(a_r - a)\hat{j} + (M - m)(-a)\hat{j} + Ma\hat{j}$$

$$\Rightarrow \vec{a_c} = -\frac{ma_r}{2M}\hat{j}$$

(b) Then the time after the boy reaches the top of the ladder is

$$t = \sqrt{\frac{2l}{a_r}} = \sqrt{\frac{2l}{\frac{2M}{m}a}} = \sqrt{\frac{ml}{Ma}}$$

(c) The displacement of the center of mass is

$$\vec{S_c} = \frac{a_c t^2}{2}\hat{j} = \frac{m}{2M}\left(\frac{a_r t^2}{2}\right)\hat{j}$$

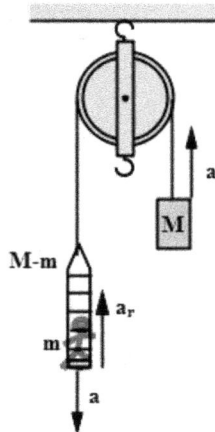

Putting $\frac{a_r t^2}{2} = l$, we have

$$\vec{S_c} = \frac{ml}{2M}\hat{j} \text{ Ans.}$$

(d) The velocity of the center of mass is

$$\vec{v_c} = a_c t = \left(\frac{2M}{m}a\right)\sqrt{\frac{ml}{Ma}} = 2\sqrt{\frac{Mal}{m}}$$

(e) The tension on the string is

$$2T - (M + m)g = 2Ma_c$$

$$\Rightarrow T = 2M(g + a_c)$$

Putting $a_c = \frac{ma_r}{2M}$, we have

$$T = 2M\left(g + \frac{ma_r}{2M}\right) \text{ Ans.}$$

Problem 25 The ball of mass m is released from a height h so that it moves down the smooth wedge. Let the block reach the bottom of the wedge, but it does not touch the ground. Find (a) the KE of the system if the velocity of the ball relative to the wedge is v_r at the bottom, (b) value of v_r, (c) velocity of the wedge, and (d) fraction of energy carried by the wedge.

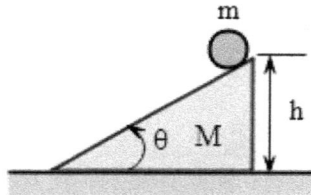

Solution

(a) If v = real velocity of the wedge, the horizontal velocity of the ball is

$$v_{b_x} = (v_r \cos\theta - v)$$

The vertical velocity of the ball is

$$v_{b_y} = v_r \sin\theta$$

Then, the KE of the ball-wedge system is

$$K = \frac{1}{2}Mv^2 + \frac{1}{2}m\{(v_r \cos\theta - v)^2 + (v_r \sin\theta)^2\} \Rightarrow K = \frac{1}{2}(M + m)v^2 + \frac{1}{2}m(v_r^2 - 2v_r v \cos\theta) \quad (1.74)$$

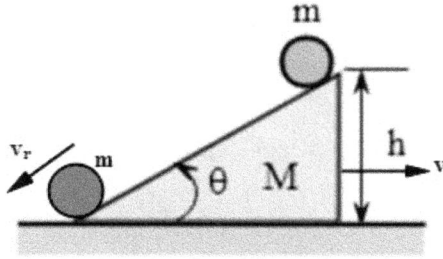

Considering horizontal momentum,
$$-Mv + m(v_r \cos\theta - v) = 0$$

$$\Rightarrow v = \frac{mv_r \cos\theta}{M+m} \tag{1.75}$$

Putting u from equation (1.75) in equation (1.74),

$$K = \tfrac{1}{2}(M+m)\left(\frac{mv_r\cos\theta}{M+m}\right)^2 + \tfrac{1}{2}mv_r^2$$
$$- \tfrac{1}{2}m2v_r\left(\frac{mv_r\cos\theta}{M+m}\right)\cos\theta$$

$$\Rightarrow K = \frac{m^2 v_r^2 \cos^2\theta}{2(M+m)} + \tfrac{1}{2}mv_r^2\left(1 - \frac{2m\cos^2\theta}{M+m}\right)$$

$$= \frac{mv_r^2}{2}\left[\frac{m}{M+m}\cos^2\theta - \frac{2m\cos^2\theta}{M+m} + 1\right]$$

$$= \frac{mv_r^2}{2}\left[1 - \frac{m\cos^2\theta}{M+m}\right]$$

$$\Rightarrow K = \frac{m(M + m\sin^2\theta)v_r^2}{2(M+m)} \quad \text{Ans.}$$

(b) Conserving energy
$$\Delta K + \Delta U = 0$$

$$(K - 0) + (-mgh) = 0$$

$$\Rightarrow K = mgh$$

Putting the above value of K,
$$\frac{m(M + m\sin^2\theta)v_r^2}{2(M+m)} = mgh$$

$$\Rightarrow v_r = \sqrt{\frac{2(M+m).\,gh}{(M+m\sin^2\theta)}} \quad \text{Ans.}$$

(c) Then, the velocity of the wedge can be given as by equation (1.75) as

$$v = \frac{mv_r\cos\theta}{M+m}$$

$$= \frac{m\cos\theta}{M+m}\sqrt{\frac{2(M+m)gh}{M+m\sin^2\theta}}$$

$$= \sqrt{\frac{2m^2gh\cos^2\theta}{(M+m)(M+m\sin^2\theta)}} \quad \text{Ans.}$$

(d) The fraction of energy carried by the wedge is

$$\eta = \frac{K_M}{mgh} = \frac{\frac{1}{2}Mv^2}{mgh} = \frac{Mv^2}{2mgh}$$

$$= \frac{(2m^2gh\cos^2\theta)M}{2(M+m)(M+m\sin^2\theta)(mgh)}$$

$$= \frac{Mm\cos^2\theta}{(M+m)(M+m\sin^2\theta)} \quad \text{Ans.}$$

Problem 26 The ball of mass m is released from a height h so that it moves down the smooth wedge. Let the block reach the ground. Find the (a) velocity of the ball relative to ground at the bottom of the wedge, (b) fraction of energy lost by the ball after launching onto the ground without bouncing, and (c) velocities of ball and wedge after the ball reaches the ground.

Solution

(a) If $v_{mM} = v_{\text{rel}}$ at the bottom of the wedge, the KE of the wedge particle system is

$$K = \frac{m(M+\sin^2\theta)v_r^2}{2(M+m)} \tag{1.76}$$

Conserving energy of $(M+m)$, we have

$$\Delta K + \Delta u = 0,$$

where $\Delta K = K - 0 = K$ and $\Delta U = -mgh$

$$\Rightarrow K = mgh \tag{1.77}$$

1-117

Using equations (1.76) and (1.77),

$$v_r = \sqrt{\frac{2(M+m)gh}{M + m \sin^2 \theta}} \tag{1.78}$$

Then the velocity of the ball is

$$\vec{v_m} = (-v + v_r \cos \theta)\hat{i} = v_r \sin \theta \hat{j} \tag{1.79}$$

Conserving horizontal linear momentum

$$-Mv + m(v - v_r \cos \theta) = 0$$

$$\Rightarrow v = \frac{mv_r \cos \theta}{M + m} \tag{1.80}$$

Using equations (1.79) and (1.82)

$$\vec{v_m} = \left(v_r \cos \theta - \frac{mv_r \cos \theta}{M + m}\right)\hat{i} - v_r \sin \theta \hat{j}$$

$$= \frac{mv_r \cos \theta}{M + m}\hat{i} - v_r \sin \theta \hat{j}$$

$$= v_r \left(\frac{M \cos \theta}{M + m}\hat{i} - \sin \theta \hat{j}\right) \tag{1.81}$$

Using equations (1.78) and (1.81)

$$\vec{v_m} = \sqrt{\frac{2(M+m)gh}{M + m \sin^2 \theta}}\left(\frac{M}{M + m} \cos \theta \hat{i} - \sin \theta \hat{j}\right)$$

(b) As the collision is inelastic, $c = 0$ along the vertical. Since the ground is smooth, we can still conserve horizontal momentum of $M + m$. So, the vertical velocity $-v_r \sin \theta \hat{j}$ will be last. Then, the portion of energy lost will be

$$\frac{\Delta K}{K} = \frac{\frac{1}{2}mv_r^2 \sin^2 \theta}{\frac{m}{2}\left(\frac{M + m \sin^2 \theta}{M + m}\right)v_r^2} \tag{1.82}$$

$$= \frac{(M + m)\sin^2 \theta}{M + m \sin^2 \theta} \tag{1.83}$$

After putting $\frac{\Delta K}{K} = \frac{\frac{1}{2}mv_0^2 \sin^2 \theta}{mgh} = \frac{v_r^2}{2gh} \sin^2 \theta$ in equation (1.83), we have

$$\frac{\Delta K}{K} = \left\{\frac{2(M+m)gh}{M + m \sin^2 \theta}\right\}\left(\frac{1}{2gh}\right)(\sin^2 \theta) = \frac{(M + m)\sin^2 \theta}{M + m \sin^2 \theta}$$

(c) Conserving linear momentum
$Mv_2 = mv_1$(b)
Balancing energy

$$\frac{M}{2}v_2^2 + \frac{1}{2}mv_1^2 + \Delta K = mgh \qquad (1.84)$$

Using equations (1.53) and (1.84)

$$\Delta K + \frac{M}{2}\left(\frac{mv_1}{2}\right)^2 + \frac{1}{2}mv_1^2 = mgh$$

$$\Rightarrow \Delta K + \frac{mv_1^2}{2}\left(\frac{m}{M} + 1\right) = mgh$$

$$\Rightarrow \frac{mv_1^2}{2}\left(\frac{M+m}{M}\right) = mgh - \Delta K \qquad (1.85)$$

Using equations (1.82) and (1.83)

$$\frac{mv_1^2}{2}\left(\frac{M+m}{M}\right)$$

$$= mgh - mgh\left(\frac{M+m}{M+m\sin^2\theta}\right)$$

$$= mgh\left\{\frac{M + m\sin^2\theta - m\sin^2\theta - M\sin^2\theta}{M+m\sin^2\theta}\right\}$$

$$\Rightarrow \frac{(M+m)v_1^2}{2M} = gh\left\{\frac{M\cos^2\theta}{M+m\sin^2\theta}\right\} \Rightarrow v_1^2 = \frac{2M^2\cos^2\theta gh}{(M+m)(M+m\sin^2\theta)}$$

$$\Rightarrow v_1 = M\cos\theta\sqrt{\frac{2gh}{(M+m)(M+m\sin^2\theta)}}$$

Then, $v_2 = \frac{mv_1}{M}$

$$= m\cos\theta\sqrt{\frac{2gh}{(M+m)(M+m\sin^2\theta)}} \quad \text{Ans.}$$

Problem 27 A ball of mass m is given a velocity v_o such that it climbs onto a wedge of mass M without bouncing. How high will it go?

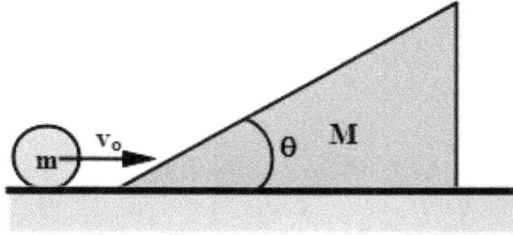

Solution

Let $v =$ velocity of the wedge just after the ball begins to slide over it. Since collision is absent along the slope, we can assume the linear momentum of the ball to obtain its velocity $v_0 \cos \theta$ along the slope. For inelastic collision (no bouncing), putting $e = 0$, we have $v_1 = v \sin \theta$. Then the horizontal momentum of $M + m$ is

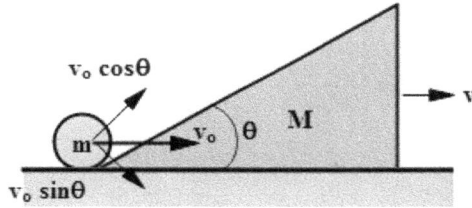

$$mv_0 = P_x = m(v_0 \cos \theta \cos \theta + v_1 \sin \theta) + Mv$$

Putting $v_1 = v \sin \theta$, we have

$$m(v_0 \cos^2 \theta + v \sin^2 \theta) + Mv = mv_0$$

$$\Rightarrow v(M + m \sin^2 \theta) = mv_0 \sin^2 \theta$$

$$\Rightarrow v = \frac{mv_0 \sin^2 \theta}{M + m \sin^2 \theta} \quad \text{Ans.}$$

The velocity of the ball relative to the wedge is given as

$$v_r = v_0 \cos \theta - v \cos \theta = (v_0 - v)\cos \theta$$

Putting the value of v

$$v_r = \left(v_0 - \frac{mv_0 \sin^2 \theta}{M + m \sin^2 \theta} \right)\cos \theta$$

$$\Rightarrow v_r = \frac{Mv_0 \cos \theta}{M + m \sin^2 \theta} \quad \text{Ans.}$$

At the highest position both M and m in have same velocity v'.

Then, $(M + m)v' = mv_0$

$$\Rightarrow v' = \frac{mv_0}{M + m}$$

Conserving energy

$$(KE_m)_i + (KE_M)_i$$

$$= (KE_m)_f + (KE_M)_f + (PE_M)_f$$

$$\Rightarrow \frac{1}{2}m(v_0^2 \cos^2\theta + v^2 \sin^2\theta) + \frac{M}{2}v^2 = \frac{(M + m)}{2}v'^2 + mgh$$

Putting $v' = \frac{mv_0}{M+m}$ and $v = \frac{mv_0 \sin^2\theta}{M + m \sin^2\theta}$ in the last equation, we have

$$\frac{1}{2}mv_0^2 \cos^2\theta + \frac{1}{2}(M + m\sin^2\theta)\left(\frac{mv_0 \sin^2\theta}{M + m\sin^2\theta}\right)^2 = \frac{M + m}{2}\left(\frac{mv_0}{M + m}\right)^2 + mgh$$

$$\Rightarrow \frac{1}{2}mv_0^2\left\{\cos^2\theta + \frac{m\sin^4\theta}{M + m\sin^2\theta}\right\} = \frac{mv_0^2}{2}\frac{m}{M + m} + mgh$$

$$\Rightarrow \frac{1}{2}mv_0^2\left\{\cos^2\theta + \frac{m\sin^4\theta}{M + m\sin^2\theta} - \frac{m}{M + m}\right\} = mgh$$

$$\Rightarrow h = \frac{v_0^2}{2g}\left\{\frac{M\cos^2\theta + m\sin^2\theta(\cos^2\theta + \sin^2\theta)}{M + m\sin^2\theta} - \frac{m}{M + m}\right\}$$

$$\Rightarrow h = \frac{v_0^2}{2g}\left\{\frac{M\cos^2\theta + m\sin^2\theta}{M + m\sin^2\theta} - \frac{m}{M + m}\right\}$$

$$= \frac{v_0^2}{2g}\frac{\left(\begin{array}{c}M^2\cos^2\theta + Mm\sin^2\theta + Mm\cos^2\theta \\ + m^2\sin^2\theta - Mm - m^2\sin^2\theta\end{array}\right)}{(M + m)(M + m\sin^2\theta)}$$

$$= \frac{v_0^2}{2g}\frac{M^2\cos^2\theta}{(M + m)(M + m\sin^2\theta)}$$

Problem 28 The ball of mass m is released from a height h so that it moves down the smooth wedge. Let the block reach the ground smoothly without losing energy in collision. Find the velocity of the (a) ball and wedge relative to ground, (b) center of mass of the wedge-ball system after the ball reaches the ground, and (c) fraction of energy transferred to the wedge.

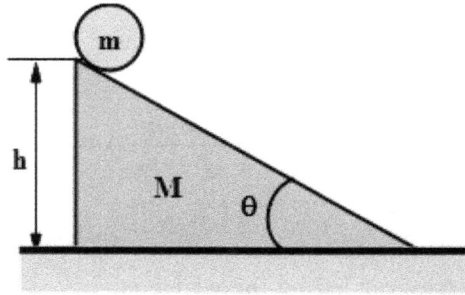

Solution

(a) Let v_1 = recoil velocity and v_2 be the velocity of m. Then, conserving the linear momentum

$$-Mv_1 + mv_2 = 0$$

$$\Rightarrow Mv_1 = mv_2 \tag{1.86}$$

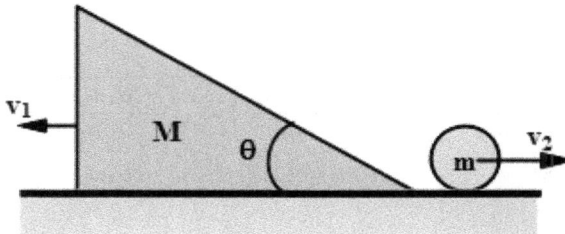

Conserving energy

$$-mgh + \frac{M}{2}v_1^2 + \frac{1}{2}mv_2^2 = 0$$

$$\Rightarrow \frac{Mv_1^2}{2} + \frac{mv_2^2}{2} = mgh \tag{1.87}$$

Solving equations (1.86) and (1.87)

$$\frac{Mv_1^2}{2} + \frac{m}{2}\left(\frac{M}{m}v_1\right)^2 = mgh$$

$$\Rightarrow \frac{Mv_1^2}{2}\left(1 + \frac{M}{m}\right) = mgh$$

$$\Rightarrow v_1 = \sqrt{\frac{2mgh}{M+m}} \quad \text{Ans.}$$

$$v_2 = \frac{M}{m} v_1 = \sqrt{\frac{2Mgh}{M+m}} \quad \text{Ans.}$$

(b) The velocity of the center of mass is $v_c = \frac{-Mv_1 + mv_2}{M+m} = 0$ Ans.

(c) The portion energy carried by the wedge is

$$\eta = \frac{\frac{1}{2}Mv_1^2}{mgh} = \frac{Mv_1^2}{2mgh} = \frac{M}{2mgh}\left(\sqrt{\frac{2mgh}{M+m}}\right)^2$$

$$= \frac{M}{M+m} \quad \text{Ans.}$$

(d) The relative velocity between the ball and wedge is

$$v_{rel} = v_1 + v_2 = v_1 + \frac{Mv_1}{m}$$

$$= v_1\left(1 + \frac{M}{m}\right)\sqrt{\frac{2mgh}{M+m}}\left(\frac{M+m}{m}\right) = \sqrt{\frac{2(M+m)gh}{m}} \quad \text{Ans.}$$

Problem 29 The bob of mass m is suspended from a block of mass M by an ideal string of length L as shown in the figure. The block can slide along a smooth horizontal bar. If the bob is given a horizontal velocity v_o, find the (a) vertical velocity of the bob when the string becomes horizontal (b) minimum value of v_o so that the string becomes horizontal (c) tension in the string (d) horizontal acceleration of M (e) total acceleration of m.

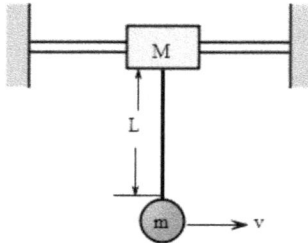

Solution

(a) Let v_x= horizontal velocity of M and m and v_y= vertical velocity of m.

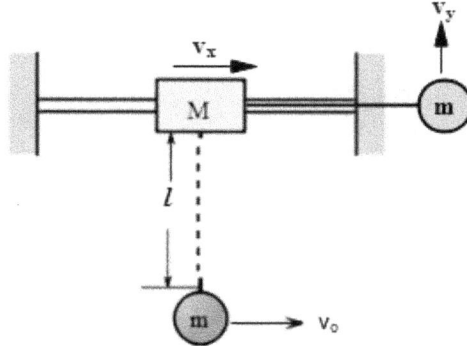

Then the KE of the system is

$$K = \frac{1}{2}(M + m)v_x^2 + \frac{1}{2}mv_y^2 \tag{1.88}$$

Conserving the linear momentum of the system $(M + m)$

$$(M + m)v_x = mv_0$$

$$\Rightarrow v_x = \frac{mv_0}{M + m} \tag{1.89}$$

Using equations (1.88) and (1.89)

$$K = \frac{1}{2}(M + m)\left(\frac{mv_0}{M + m}\right)^2 + \frac{1}{2}mv_y^2$$

$$= \frac{m^2v_0^2}{2(M + m)} + \frac{1}{2}mv_y^2$$

The conservation of energy gives

$$K + U = \frac{1}{2}mv_0^2$$

$$\Rightarrow \frac{m^2v_0^2}{2(M + m)} + \frac{1}{2}mv_y^2 + mgl = \frac{1}{2}mv_0^2$$

$$\Rightarrow \frac{1}{2}mv_y^2 + mgl = \frac{mv_y^2}{2}\left(1 - \frac{m}{M + m}\right)$$

$$\Rightarrow \frac{1}{2}mv_y^2 = \frac{Mmv_0^2}{2(M+m)} - mgl$$

$$\Rightarrow v_y = \sqrt{\frac{Mv_0^2}{(M+m)} - 2gl}$$

(b) For $v_y \geqslant 0$, $v_0 \geqslant \sqrt{\frac{2(M+m)gl}{M}}$

(c) The tension in the string is given as

$$T = Ma_x \tag{1.90}$$

$$T = m\left(\frac{v_y^2}{l} - a_x\right) \tag{1.91}$$

Eliminating a_x between equation (1.90) and (1.91)

$$\frac{T}{M} + \frac{T}{m} = \frac{v_y^2}{l}$$

$$\Rightarrow T = \frac{Mmv_y^2}{(M+m)l} \tag{1.92}$$

Putting v_y in equation (1.92)

$$T = \frac{Mm}{(M+m)l}\left[\frac{Mv_0^2}{M+m} - 2gl\right] \text{ Ans.}$$

(d) Then the horizontal acceleration of M is

$$\vec{a}_M = \frac{\vec{T}}{M}$$

$$= \frac{m}{M+m}\left[\frac{Mv_0^2}{(M+m)l} - 2g\right]\hat{i} \text{ Ans.}$$

(e) The vertical acceleration of $m = (\vec{a}_m)_y = -g\hat{j}$
The horizontal acceleration of m is

$$(\vec{a}_m)_x = -\frac{T}{m}\hat{i}$$

$$= -\frac{M}{M+m}\left[\frac{Mv_0^2}{(M+m)l} - 2g\right]$$

Then, the total acceleration of m is

$$\vec{a}_m = \vec{a}_{m_x} + \vec{a}_{m_y}$$

$$= -\left[\frac{M}{M+m} \left\{ \frac{Mv_0^2}{(M+m)l} - 2g \right\} \hat{i} + g\hat{j} \right]$$

Problem 30 A sphere of radius r is released from rest from the top of a smooth wedge so that the sphere slides along the hemispherical surface of the wedge. When the sphere reaches its lowest position, find the (a) velocity of the sphere relative to the wedge, (b) velocity of the wedge, (c) displacement of the wedge, and (d) the normal reaction acting on the system $(M + m)$ by the ground.

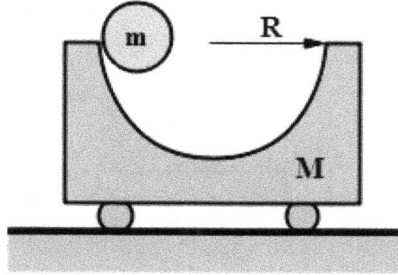

Solution

(a) Let v_r be the velocity of m relative to M at the bottom. Then the change in KE of $(M+m)$ is

$$\Delta K = \frac{1}{2} \frac{Mm}{M+m} v_r^2 \tag{1.93}$$

As the ball slides down, the change in potential energy of $M + m$ is

$$\Delta U = -mg(R - r) \tag{1.94}$$

Consuming energy

$$\Delta K + \Delta U = 0 \tag{1.95}$$

Using the last three equations

$$\frac{Mmv_r^2}{2(M+m)} - mg(R - r) = 0$$

$$\Rightarrow v_r = \sqrt{\frac{2(M+m)g(R-r)}{M}}$$

(b) Conserving linear momentum

$$-Mv + m(v_r - v) = 0$$

$$\Rightarrow v = \frac{mv_r}{M + m} \tag{1.96}$$

Putting v_r in equation (1.96)

$$v = \frac{m}{M + m}\sqrt{\frac{2(M + m)g(R - r)}{M}}$$

$$= \sqrt{\frac{m^2 g(R - r)}{(M + m)m}} \quad \text{Ans.}$$

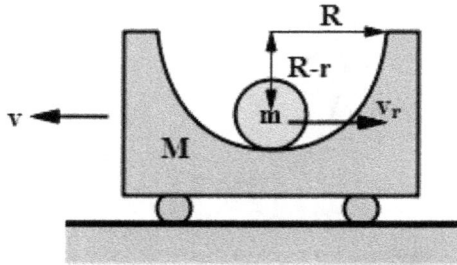

(c) If the displacement of the wedge is equal to x to the left of the displacement of the ball relative to ground is,

$$x_m = (R - r) - x$$

Then, the displacement of the center of mass is

$$\vec{x_c} = \frac{M\vec{x_M} + m\vec{x_m}}{M + m}$$

Since $\vec{x_c} = 0$, we have

$$M\vec{x_M} + m\vec{x_m} = 0$$

$$\Rightarrow M(-x) + m[(R - r) - x] = 0$$

$$\Rightarrow x = \frac{m(R - r)}{M + m} \quad \text{Ans.}$$

(d) The net vertical force acting on $M+m$ is

$$\{N - (M + m)g\}\hat{j} = M\vec{a_{My}} + m\vec{a_{my}}$$

Since $a_{M_y} = 0$ and $a_{m_y} = a_{mM_y} = \frac{v_r^2}{R-r}$, we can write

$$N = (M + m)g + \frac{mv_r^2}{R - r}$$

Putting the value of v_r, we have

$$N = (M + m)g + \frac{m}{R - r}\left\{\frac{2(M + m)g}{M}(R - r)\right\}$$

$$= (M + m)g + 2(M + m)\frac{mg}{M}$$

$$= (M + m)g\left(1 + \frac{2m}{M}\right)$$

$$= \left(1 + \frac{2m}{M}\right)(M + m)g \text{ Ans.}$$

Problem 31 A small bob of mass m is released from rest from the given position of the smooth wedge of mass M so that the bob slides down the spherical surface of radius R inside the wedge. The wedge is kept by the side of the vertical wall as shown in the figure. (a) How high will the bob go? (b) What is the maximum velocity of center of mass of the system $(M + m)$ when the bob reaches its lowest position?

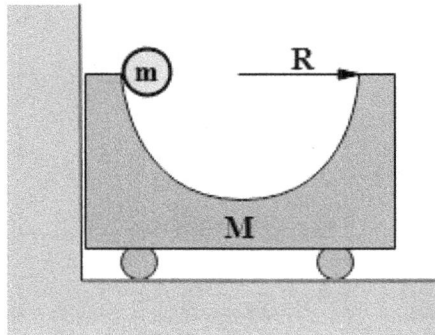

Solution

(a) Due to the presence of the wall, the wedge does not recoil util the ball reaches the bottom.

Conserving energy, we have

$$\Delta U + \Delta K = 0$$

$$-mgR + \frac{1}{2}mv^2 = 0$$

$$\Rightarrow v = \sqrt{2gr} \qquad (1.97)$$

As the ball attains the height position, let both ball and wedge move with same velocity v' conserving horizontal momentum,

$$mv = (M + m)v' \qquad (1.98)$$

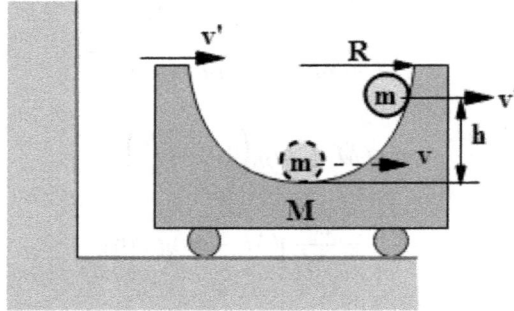

Conserving energy:

$$\frac{1}{2}mv^2 = \frac{1}{2}(M + m)v'^2 + mgh \qquad (1.99)$$

Using equations (1.98) and (1.99)

$$\frac{Mmv^2}{2(M + m)} = mgh$$

$$\Rightarrow h = \frac{Mv^2}{2(M + m)g} \qquad (1.100)$$

Using equations (1.97) and (1.100), we have

$$h = \frac{MR}{M + m} \quad \text{Ans.}$$

(b) The horizontal velocity of center of mass is

$$v_c = \frac{mv}{M + m}, \quad \text{where } v = \sqrt{2gR}$$

$$\Rightarrow v_c = \frac{m\sqrt{2gR}}{M + m} \quad \text{Ans.}$$

Problem 32 The ball of mass m is released from the top of the hemispherical wedge with a small velocity v_o so that it slides along the smooth hemispherical wedge of mass M and radius R. (a) Find the speed of the ball and wedge relative to the wedge as the function of the angle revolved by the ball relative to the center of the hemisphere (b). At what angle will the ball leave the wedge? Put $m/M = 0{,}25/7$.

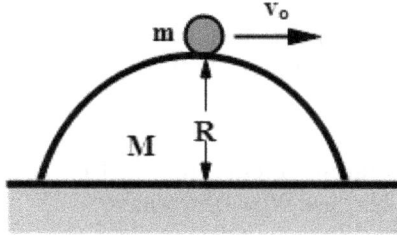

Solution

(a) If $u = v_{mM}$ and $v_M = v$, conserving horizontal momentum,

$$-Mv + m(u \cos \theta - v) = 0$$

$$\Rightarrow v = \frac{mu \cos \theta}{M + m} \tag{1.101}$$

Conservation of energy gives

$$\Delta K + \Delta U = 0$$

$$\Rightarrow \frac{1}{2}m(u^2 + v^2 - 2uv \cos \theta) + \frac{Mv^2}{2} - mgR(1 - \cos \theta) = 0$$

$$\Rightarrow \frac{1}{2}(M + m)v^2 + \frac{1}{2}mu^2 - muv \cos \theta = mgR(1 - \cos \theta) \tag{1.102}$$

Using equations (1.101) and (1.102)

$$\frac{1}{2}\frac{(M + m \sin^2 \theta)}{M + m}u^2 = gR(1 - \cos \theta)$$

$$\Rightarrow u = \sqrt{\frac{2gR(1 - \cos \theta)(M + m)}{M + m \sin^2 \theta}} \quad \text{Ans.}$$

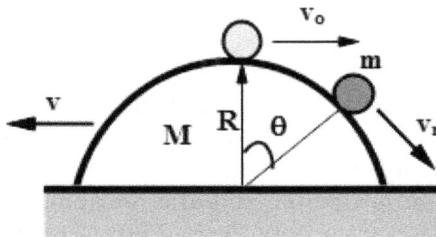

Putting the value of u in equation (1.101), we have

$$v = \sqrt{\frac{2m^2gR(1 - \cos\theta)}{(M + m)(M + m\sin^2\theta)}} \quad \text{Ans.}$$

(b) When the ball loses contact, $N = 0$, then the ball is in a projectile motion. The centripetal force acting on the ball is

$$-(N + mA\sin\theta) + mg\cos\theta = ma_r$$

where $a_r = \frac{v^2}{r}(r = $ radius of curvature at P).
Putting $A = 0$ because $N = 0$, we have

$$mg\cos\theta = \frac{mu^2}{R}$$

$$\Rightarrow u = \sqrt{gR\cos\theta}$$

Putting the value of u,

$$\frac{2gR(1 - \cos\theta)(M + m)}{M + m\sin^2\theta} = gR\cos\theta$$

$$\Rightarrow \frac{2(1 - \cos\theta)(M + m)}{M + m\sin^2\theta} = \cos\theta$$

Putting $\frac{m}{M} \to 0$, we have $\cos\theta = \frac{2}{3}$
$$\Rightarrow \theta = \cos^{-1}\frac{2}{3} \text{ (as expected)}$$
Putting $\frac{m}{M} = \frac{25}{7}$, we have

$$\frac{2(1 - \cos\theta)\left(1 + \frac{25}{7}\right)}{1 + \frac{7}{25}\sin^2\theta} = \cos\theta$$

Solving this equation, we have

$$\theta = \cos^{-1}\frac{4}{5} = 37°$$

Problem 33 A small ball of mass m is released from rest from the top of the smooth wedge of mass M. The wedge has a circular track of radius R along which the ball slides. Find the (a) velocity of m relative to M, (b) acceleration of the ball just before escaping from the wedge, (c) velocity of the ball just after escaping from the wedge, (d) change in acceleration just before and after escaping from the wedge, and (e) radius of curvature of the trajectory of the ball just after escaping from the wedge.

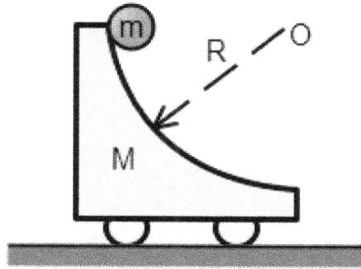

Soluion

(a) If $u = v_{mM}$= velocity of m relative to M, the total KE of the system ($M + m$) can be given as

$$K = \frac{Mmu^2}{2(M + m)} \qquad (1.103)$$

Conserving the energy, we have

$$\Delta K + \Delta U = 0$$

$$\Rightarrow (K - 0) + (-mgR) = 0$$

$$\Rightarrow K = mgR \qquad (1.104)$$

Using equations (1.103) and (1.104)

$$\frac{Mmu^2}{2(M + m)} = mgR$$

$$\Rightarrow u = \sqrt{\frac{2gR(M + m)}{M}} \qquad (1.105)$$

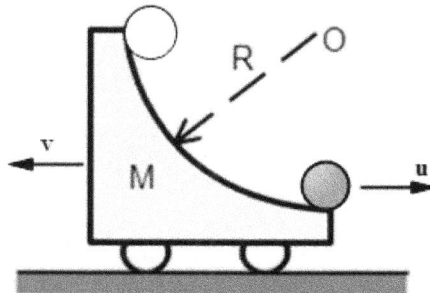

(b) The acceleration of m is

$$\vec{a_m} = \frac{u^2}{R}\hat{j} \qquad (1.106)$$

Using equations (1.105) and (1.106)

$$\vec{a_m} = \frac{2(M+m)g}{m}\hat{j} \text{ Ans.}$$

(c) Just after escaping from the wedge, the velocity of the ball is given as $v_m = (u - v)$, where $v = \frac{mu}{M+m}$

$$\Rightarrow v_m = \frac{Mu}{M+m} \qquad (1.107)$$

Putting the value of u from equation (1.105) in equation (1.107)

$$v_m = \frac{M}{M+m}\sqrt{\frac{2gR(M+m)}{M}}$$

$$= \sqrt{\frac{2MgR}{M+m}}$$

(d) Since the ball moves in a projectile path, its acceleration will be

$$\vec{a'}_m = -g\hat{j}$$

$\Rightarrow$ The change in acceleration is

$$\vec{a'}_m - \vec{a_m} = -g\hat{j} - \left\{ \frac{2(M+m)}{M}g\hat{j} \right\}$$

$$= -g\hat{j}\left(\frac{3M+2m}{M}\right) = -\left(3 + \frac{2m}{M}\right)g\hat{j} \text{ Ans.}$$

(e) The radius of curvature of the trajectory of the ball just after escaping from the wedge is

$$r = \frac{v_m^2}{g} = \frac{1}{g}\left(\sqrt{\frac{2MgR}{M+m}}\right)^2 = \frac{2MR}{M+m} \text{ Ans.}$$

Problem 34 A small ball of mass m is released from rest from the top of the smooth wedge of mass M. The wedge has a curved track along which the ball slides. Find the (a) maximum range of the ball after escaping from the wedge and the (b) velocity of the ball striking the ground.

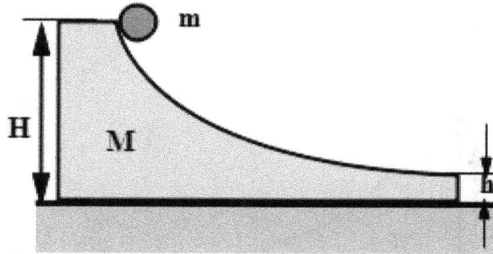

Solution

(a) Let v_1, v_2 be the velocities of wedge and ball, respectively.
Conserving momentum

$$-Mv_2 + mv_1 = 0 \qquad (1.108)$$

Conserving energy

$$\frac{M}{2}v_2^2 + \frac{1}{2}mv_1^2 = mgh \qquad (1.109)$$

Solving equations (1.108) and (1.109) $\frac{M}{2}\left(\frac{mv_1}{M}\right)^2 + \frac{1}{2}mv_1^2 = mgh$

$$\Rightarrow \frac{mv_1^2}{2}\left(\frac{m}{M} + 1\right) = mgh$$

$$\Rightarrow v_1 = \sqrt{\frac{2Mgh}{M + m}} \qquad (1.110)$$

The horizontal range is

$$R = u_1 t, \quad \text{where } t = \sqrt{\frac{2(H - h)}{g}}$$

$$\Rightarrow R = u_1 \sqrt{\frac{(H - h)}{g}} \qquad (1.111)$$

Using equations (1.110) and (1.111)

$$R = \left(\sqrt{\frac{2ghM}{M + m}}\right)\left(\sqrt{\frac{2(H - h)}{g}}\right) R = 2\sqrt{\frac{M}{M + m}h(H - h)}$$

For R to be maximum

$$\frac{d}{dh}(R^2) = 0$$

$$\Rightarrow -2h + H = 0 \Rightarrow h = \frac{H}{2}$$

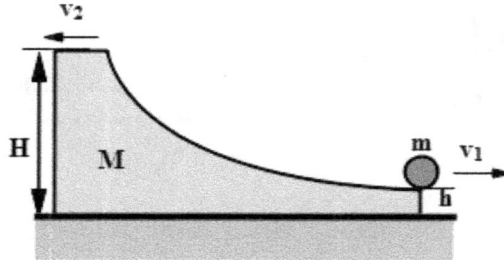

Therefore, $R_{\max} = 2\sqrt{\frac{M}{M+m}\left(\frac{H}{2}\right)\left(H - \frac{H}{2}\right)} = \sqrt{\frac{H}{M+m}}\, H$ Ans.

(b) The velocity of striking of the ball is

$$\frac{1}{2}m(v'^2 - v_m^2) = mg(H - h)$$

$$v' = \sqrt{v_m^2 + 2g(H - h)} \tag{1.112}$$

Putting $v_m = v_1 = \sqrt{\frac{2Mgh}{M+m}}$ from equation (1.110) in equation (1.112), we have

$$v' = \sqrt{\frac{2Mgh}{M+m} + 2g(H-h)} = \sqrt{2g\left\{\left(\frac{M}{M+m} - 1\right)h + H\right\}} = \sqrt{2g\left\{H - \frac{mh}{M+m}\right\}}$$

Problem 35 A block of mass M is free to slide along the horizontal rod. A bob of mass m is connected with the bob by a light inextensible string of length l. The bob is released from rest when the string is just taut and horizontal. When the string becomes vertical, (a) find the velocity of the bob relative to the block and velocity of the block, (b) acceleration of the bob, and (c) acceleration of the center of mass.

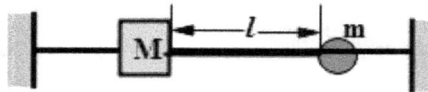

Solution

(a) The net horizontal force is zero.
Conserving the horizontal momentum

$$mv_1 + Mv_2 = 0 \qquad\qquad (1.113)$$

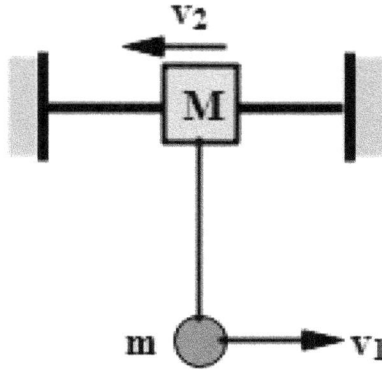

Conserving the energy

$$\frac{1}{2}mv_1^2 + \frac{1}{2}Mv_2^2 - mgh = 0$$

$$\Rightarrow \frac{1}{2}mv_1 + \frac{1}{2}M\left(\frac{mv_1}{M}\right)^2 = mgh$$

$$\Rightarrow \frac{m}{2}v_1^2\left(1 + \frac{m}{M}\right) = mgh$$

$$\Rightarrow u_1 = \sqrt{\frac{2Mgh}{M+m}}\,\hat{i}$$

$$v_2 = -\frac{mv_1}{M} = -\frac{m}{M}\sqrt{\frac{2Mgh}{M+m}}\,\hat{i}$$

(b) The acceleration of the bob is

$$\vec{a_1} = \vec{a_{12}} + \vec{a_2}, \text{ where } a_2 = 0$$

$$\Rightarrow \vec{a_1} = \vec{a_{12}}, \text{ where } a_{12} = \left(\frac{1}{l}\right)v^2 = \frac{(v_1 + v_2)^2}{l}$$

$$\Rightarrow \vec{a_1} = \frac{(v_1 + v_2)^2}{l}\hat{j} \qquad\qquad (1.114)$$

Putting $v_1 = \sqrt{\frac{2Mgh}{M+m}}$ and $v_2 = +\frac{m}{M\sqrt{M+m}}\sqrt{2Mgh}$ in the equation

$(1.114), \vec{a_1} = \{\sqrt{\frac{2Mgh}{M+M}}(1 + \frac{m}{M})\}^2\frac{1}{l}\hat{j}$

$$= \frac{2Mgh}{(M+m)l}\left(\frac{M+m}{M}\right)^2$$

$$= \frac{2(M+m)gh}{Ml}\hat{j} \quad \text{Ans.}$$

(c) The acceleration of the center of mass is

$$\vec{a_c} = \frac{m_1\vec{a_1} + m_2\vec{a_2}}{m_1 + m_2} = \frac{m\vec{a_1}}{m+M}$$

$$\Rightarrow \vec{a_c} = \frac{m}{M+m}\frac{2(M+m)gh}{Ml}\hat{j}$$

$$= \frac{2mgh}{Ml}\hat{j} \quad \text{Ans.}$$

Problem 36 In the last problem, let the bob be given a velocity v when the string was vertical and the block was at rest. Find the (a) acceleration of the bob, (b) acceleration of the center of mass of the system $(M + m)$, (c) tension in the string, (d) velocity of the center of mass, at the initial position, and (e) maximum height attained by the bob.

Solution

(a) Let a_{12} = acceleration of 1 relative to 2 and a_1 = acceleration of 1 relative to ground. Then, $\vec{a_1} = \vec{a_{12}} + \vec{a_2}$

Since the block 2 does not accelerate vertically as it moves along the horizontal direction, we put $a_2 = 0$ to obtain

$$\vec{a_1} = \vec{a_{12}} \tag{1.114a}$$

Since $\vec{a_{12}} = \frac{v^2}{l}\hat{j}$ we have $\vec{a_1} = \frac{v^2}{l}\hat{j}$

(b) The acceleration of center of mass is

$$\vec{a_c} = \frac{m_1\vec{a_1} + m_2\vec{a_2}}{m_1 + m_2}$$

$$= \frac{m\left(\frac{v^2}{l}\hat{j}\right) + m_2(0)}{m + M} = \frac{mv^2}{(M+m)l}\hat{j}$$

(c) The tension is

$$T - mg = ma_1$$

$$T = m(g + a_1)$$

$$\Rightarrow T = m\left(g + \frac{v^2}{l}\right) \text{ Ans.}$$

(d) The velocity of center of mass

$$\vec{v_c} = \frac{m_1 \vec{v_1} + m_2 \vec{v_2}}{m_1 + m_2}$$

$$= \frac{mv \hat{i} + m(0)}{M + m} = \frac{mv}{M + m} \hat{i} \text{ Ans.}$$

(e) Let v' = velocity of the combination M and m when m will be at the highest position consuming the horizontal momentum,

$$(M + m)v' = mu$$

$$\Rightarrow v' = \frac{mu}{M + m} \tag{1.115}$$

Conserving the energy,

$$\frac{1}{2}(M + m)v'^2 + mgh = \frac{1}{2}mv^2 \tag{1.116}$$

Solving equations (1.115) and (1.116),

$$\frac{1}{2}(M + m)\left(\frac{mv}{M + m}\right)^2 + mgh = \frac{1}{2}mv^2$$

$$\Rightarrow \frac{Mmv^2}{2(M + m)} = mgh$$

$$\Rightarrow h = \frac{Mv^2}{2(M + m)g} \text{ Ans.}$$

Problem 37 The ball of mass m is projected with a velocity v_o so that it climbs onto the wedge of mass M and height h. Find the (a) velocity with which it reaches the top of the wedge, (b) v_o so that the ball will reach up to the top of the wedge, (c) velocity with which the ball returns to the ground.

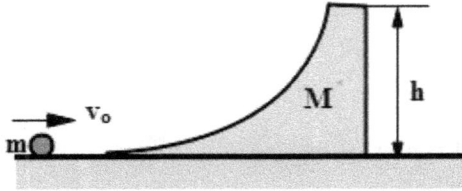

Solution

(a) At the height h let v_x be the horizontal velocity of both m and M and v_y be the vertical velocity of M.

Conserving horizontal momentum, between the position A and B

$$mv_0 = (M + m)v_x \tag{1.117}$$

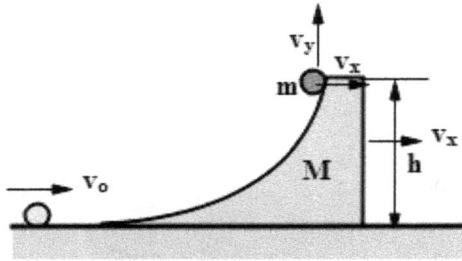

Conserving energy between the positions A and B,

$$\frac{1}{2}mv_0^2 = \frac{1}{2}Mv_x^2 + \frac{1}{2}m(v_x^2 + v_y^2) + mgh$$

$$\Rightarrow \frac{1}{2}mv_0^2 - \frac{1}{2}(M + m)v_x^2 - mgh = \frac{1}{2}mv_y^2 \tag{1.118}$$

Putting $v_x = \frac{mv_0}{M + m}$ from equation (1.117) in equation (1.118),

$$\frac{1}{2}mv_0^2 - \frac{1}{2}(M + m)\left(\frac{mv_0}{M + m}\right)^2 - mgh = \frac{1}{2}mv_y^2$$

$$\Rightarrow \frac{1}{2}mv_y^2 = \frac{Mmv_0^2}{2(M + m)} - mgh$$

$$\Rightarrow v_y = \sqrt{\frac{Mv_0^2}{M+m} - 2gh} \text{ Ans.}$$

(b) At the maximum height, $v_y = 0$

then $h_{\max} = \dfrac{Mv_0^2}{2(M+m)g}$ Ans.

(c) Since the horizontal velocity of both M and m are equal, the ball seems to move vertically up and down when observed from the wedge. However, with respect to ground-frame, the ball moves in a parabolic (projectile) path after escaping from the wedge. Its time of flight is

$$T = \frac{2v_y}{g} = \frac{2}{g}\sqrt{\frac{Mv_0^2}{M+m} - 2gh}$$

During this time, the wedge moves through a horizontal distance.

$$R = v_x T$$

$$= \frac{mv_0}{M+m}\frac{2}{g}\sqrt{\frac{Mv_0^2}{M+m} - 2gh}$$

$$= \frac{2mv_0}{(M+m)g}\sqrt{\frac{Mv_0^2}{M+m} - 2gh}$$

This means that the ball will climb the wedge, and move up and down in air under gravity after leaving the wedge. Then it launches onto the wedge again and comes down and leaves the wedge at its bottom moving to left with a velocity

$v_m = \frac{M-m}{M+m}v_0$. Thereafter, the wedge will move with a velocity $v_M = \frac{2m}{M+m}v_0$. This result can be obtained by conserving the horizontal momentum and KE of the wedge-ball system. So, it is just like the head-on elastic collision of a body with a stationary body.

Problem 38 Two smooth wedges are $M_1 = M$ and $M_2 = \eta M$, respectively, and each of the wedges has height H. If we release the ball of mass m from the given position, it slides down the first wedge and climbs onto the second wedge leaving the first wedge behind. Disregard the friction between all contacting surfaces. Find the (a) velocity of the ball at the lowest position and (b) maximum height attained by the ball.

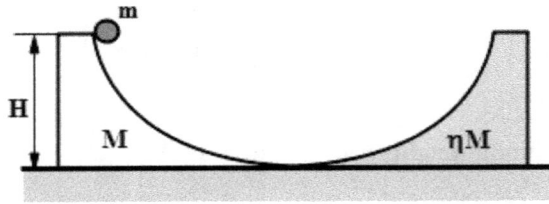

Solution

(a) When the ball reaches the bottom of the wedge, conserving linear (horizontal) momentum of the system $(M + m)$,

$$Mv_1 - mv_0 = 0 \tag{1.119}$$

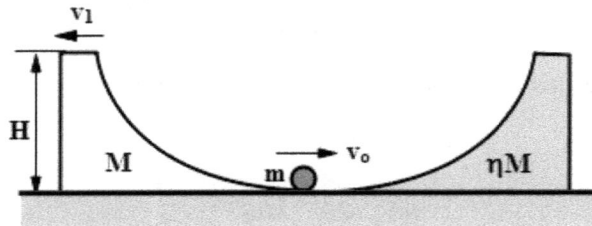

Energy conservation:

$$mgH = \frac{M}{2}v_1^2 + \frac{1}{2}mv_0^2 \tag{1.120}$$

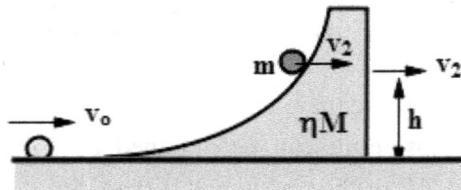

Solving equations (1.119) and (1.120),

$$mgH = \frac{M}{2}\left(\frac{mv_0}{M}\right)^2 + \frac{1}{2}mv_0^2$$

$$\Rightarrow v_0 = \sqrt{\frac{2gH}{1 + \frac{m}{M}}} \tag{1.121}$$

(b) This means that the ball leaves the wedge M and begins to climb onto the wedge of mass ηM with the velocity v_0.

Conserving horizontal momentum of the system $(m + \eta M)$, we have

$$mv_0 = (\eta M + m)v_2 \tag{1.122}$$

Conserving energy, we have

$$\frac{1}{2}mv_0^2 = \frac{1}{2}(\eta M + m)v_2^2 + mgh \tag{1.123}$$

Solving equations (1.122) and (1.123)

$$\frac{1}{2}mv_0^2 = \frac{1}{2}(\eta M + m)\left(\frac{mv_0}{\eta M + m}\right)^2 + mgh$$

$$\Rightarrow \frac{1}{2}\frac{\eta M m v_0^2}{(\eta M + m)} = mgh$$

$$\Rightarrow h = \frac{\eta M v_0^2}{2(\eta M + m)g} \tag{1.124}$$

Putting the value of v_0^2 from equation (1.121) in equation (1.124)

$$h = \frac{\eta M}{2(\eta M + m)g}\left(\frac{2mgH}{1 + \frac{m}{M}}\right)$$

$$= \left\{\frac{\eta M}{(\eta N + m)}\right\}\left\{\frac{M}{M + m}\right\}H$$

$$h = \frac{\eta M^2 H}{(M + m)(\eta M + m)}$$

Problem 39 A ball of mass m is released from rest from the top of a smooth wedge of mass M such that the ball compresses the spring of stiffness k after leaving the wedge. Find the (a) maximum velocity of the ball and (b) maximum compression of the spring.

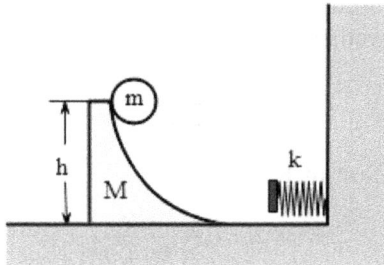

Solution

(a) Referring to the previous problem, the velocity of the ball is

$$v_0 = \sqrt{\frac{2gh}{1 + \frac{m}{M}}} \tag{1.125}$$

(b) If the ball compresses the spring through a distance x, say, we can conserve the energy to obtain

$$\frac{1}{2}mv_0^2 = \frac{1}{2}kx^2$$

$$\Rightarrow x = \sqrt{\frac{m}{k}}\, v_0 \tag{1.126}$$

Using equations (1.125) and (1.126)

$$x = \sqrt{\frac{m}{k}\left(\frac{2gh}{1 + \frac{m}{M}}\right)} \quad \text{Ans.}$$

Problem 40 A ball of mass m is released from rest from the top of a smooth wedge of mass M, which is fitted with a spring of stiffness k. Find the maximum (a) velocity of the (i) ball relative to wedge (ii) wedge and ball, relative to ground, (b) compression of the spring, (c) relative acceleration, and (d) time of contact of the ball with the spring.

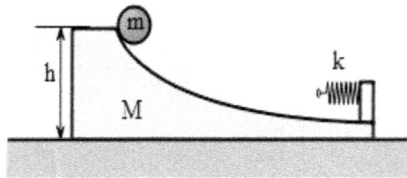

Solution

(a)

    (i) If the maximum velocity of the ball relative to the wedge is v_{rel}, the KE of the wedge particle system is

$$K = \frac{1}{2}\frac{Mm}{M + m}v_r^2 \tag{1.127}$$

The increase in KE is

$$\Delta K = K = \frac{Mm}{2(M + m)}v_r^2$$

The decrease in PE is

$$-\Delta U = mgh$$

$$\Delta K + \Delta U = 0$$

Since

$$-mgh + \frac{1}{2}\frac{Mm}{M+m}v_r^2 = 0$$

$$\Rightarrow v_r = \sqrt{\frac{2(M+m)gh}{M}} \quad \text{Ans.}$$

(ii) If v_1 and v_2 are the velocities of ball and wedge, respectively, conservation of horizontal momentum gives

$$mv_1 = Mv_2$$

$$\Rightarrow \frac{v_1}{v_2} = \frac{M}{m}$$

$$\Rightarrow \frac{v_1}{v_1+v_2} = \frac{M}{M+m}$$

$$\Rightarrow v_1 = \frac{M}{M+m}(v_1+v_2)$$

$$\Rightarrow v_1 = \frac{M}{M+m}v_r = \frac{M}{M+m}\sqrt{\frac{2(M+m)gh}{M}}$$

$$= \sqrt{\frac{2Mgh}{M+m}} \quad \text{Ans.}$$

Then, $v_2 = \frac{mv_1}{M} = \frac{m}{M}\sqrt{\frac{2Mgh}{M+m}}$

$$= \frac{2m^2gh}{M(M+m)} \quad \text{Ans.}$$

(b) At the time of maximum compression of the spring $v_r = 0$. Then both M and m will come to rest instantaneously. Now, conserving the energy, we have

$$\Delta U + \Delta K = 0$$

$$\Rightarrow -mgh + \frac{k}{2}x^2 + 0 = 0$$

$$\Rightarrow x = \sqrt{\frac{2mgh}{k}} \ \text{Ans.}$$

(c) The acceleration of m is

$$a_1 = \frac{kx}{M}$$

The acceleration of M is

$$a_2 = \frac{kx}{M}$$

Then the relative acceleration is

$$a_{\text{rel}} = a_1 + a_2$$

$$= kx\left(\frac{1}{m} + \frac{1}{M}\right) = \frac{(M+m)k}{Mm}x$$

Putting $x = \sqrt{\frac{2Mgh}{k}}$, we have

$$a_{\text{rel}} = \frac{M+m}{Mm}k\sqrt{\frac{2Mgh}{k}}$$

$$= \frac{M+m}{M}\sqrt{\frac{2ghk}{m}} \ \text{Ans.}$$

(d) The time of contact of the ball with the spring is

$$t_{\text{contact}} = \frac{T}{2}, \ \text{where } T = \frac{2\pi}{\omega_{\text{osc}}}$$

Putting $\omega = \sqrt{\frac{k}{\mu}}$, where $\mu = \frac{Mm}{M+m}$, we have $\omega = \sqrt{\frac{k(M+m)}{Mm}}$

$$\Rightarrow t_{\text{contact}} = \pi\sqrt{\frac{Mm}{(M+m)k}}$$

Problem 41 A block of mass m is tied to a small ideal spring of stiffness k. The spring is rigidly fitted with the smooth wedge of mass M at the point P as shown in the figure. Let the string be cut at time $t=0$. Find the (a) velocity of the block relative to the wedge, (b) differential equation of the relative motion between the bodies, (c) angular frequency of oscillation, (d) amplitude of oscillation, (e) maximum elongation of the spring, (f) maximum displacement of the wedge, and (g) amplitude of the wedge.

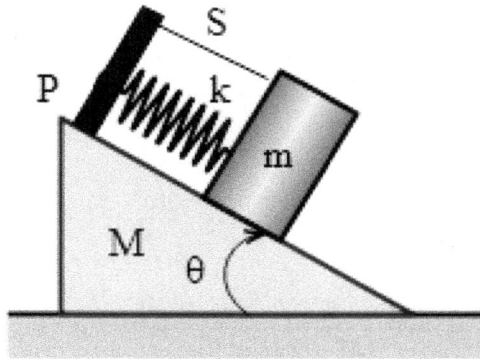

Solution

(a) If v_1 = velocity of the block relative to the wedge, following the previously discussed procedure, the increase in KE of the system $(M + m)$ can be given as

$$\Delta K = \frac{1}{2} \frac{(M + m \sin^2 \theta)}{M + m} v_r^2 \qquad (1.128)$$

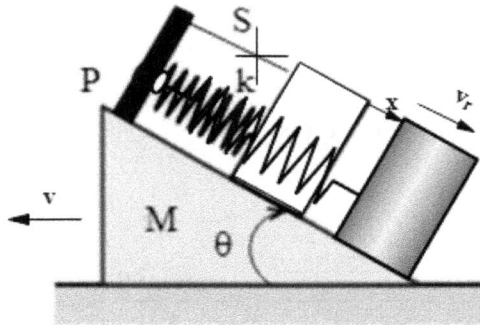

As the block falls through a vertical distance $x \sin \theta$ where x = elongation of the spring, the potential PE changes by $-mgx \sin \theta$ and the spring potential energy by $+\frac{1}{2}kx^2$. So the total change in PE is

$$\Delta U = -mgx + \frac{k}{2} x^2 \qquad (1.129)$$

According to conservation of energy

$$\Delta U + \Delta K = 0 \qquad (1.130)$$

Using equations (1.128), (1.129), and (1.130)

$$-mgx + \frac{1}{2}kx^2 + \frac{1}{2}\frac{(M + m \sin^2 \theta)}{M + m}v_r^2 = 0$$

$$\Rightarrow v_{\text{rel}} = \sqrt{\frac{(2mgx \sin \theta - kx^2)(M + m)}{M + m \sin^2 \theta}}$$

(b) Differentiating v_r with respect to x, $-mgx \sin \theta + \frac{1}{2}(2kx)) +$

$$\frac{1}{2}\frac{(M + m \sin^2 \theta)}{2(M + m)}2v_r\frac{dv_r}{dx} = 0$$

$$\Rightarrow v_r\frac{dy}{dx} = \frac{(mg \sin \theta - kx)(M + m)}{(M + m \sin^2 \theta)} = g' \text{ (say)} \qquad (1.131)$$

Since $v_r = \frac{dx}{dt}$, we have

$$\frac{d^2x}{dt^2} + \frac{k(M + m)x}{(M + m \sin^2 \theta)} = \frac{mg \sin \theta(M + m)}{M + m \sin^2 \theta}$$

(c) Comparing the above differential equation with $\frac{d^2x}{dt^2} + \omega_{0sc}^2 x = g'$, we have the frequency of oscillation, given as

$$\omega_{osc} = \sqrt{\frac{k(M + m)}{M + m \sin^2\theta}}$$

(d) The amplitude of oscillation is

$$A = \frac{g'}{\omega_{osc}^2}$$

$$= \frac{\frac{(mg \sin \theta)(M + m)}{M + m \sin^2 \theta}}{\frac{K(M + m)}{M + M \sin^2 \theta}} = \frac{mg \sin^2 \theta}{k} \text{ Ans.}$$

Alternate method: Putting $\frac{v_r dv_r}{dx} = 0$ in equation (1.131) we have

$$\frac{(Mg \sin \theta - kx)(M + m)}{M + m \sin^2 \theta} = 0$$

$$\Rightarrow x = A = \frac{mg \sin \theta}{k}$$

It is because at the mean position

$$v_r = , \text{ maximum and } a_r = \frac{v_r dv_r}{dx} = 0$$

(e) The maximum elongation of the spring can be found by putting $v_{rel} = 0$ to obtain $x_{max} = \frac{2mg \sin \theta}{k}$ Ans.

So, the amplitude of oscillation of m relative to M is

$$A = x_{max} - x_{mean} = \frac{2mg \sin \theta}{k} - \frac{mg \sin \theta}{k} = \frac{mg \sin \theta}{k}$$

(f) The maximum distance moved by the wedge is given as

$$-mx_1 + m(x_{max} \cos \theta - x_1) = 0$$

$$\Rightarrow (x_1)_{max} = \frac{mx_{max}}{M + m} \cos \theta$$

$$= \frac{m \cos \theta}{M + m} \left(\frac{2mg \sin \theta}{k} \right)$$

$$\Rightarrow (x_1)_{max} = \frac{2m^2g \sin \theta \cos \theta}{(M + m)k} \quad \text{Ans.}$$

(g) If the amplitude of oscillation of m relative to wedge M is A, then the amplitude of oscillation of the wedge is

$$A' = \frac{MA \cos \theta}{M + m} = \frac{M \cos \theta}{M + m} \left(\frac{mg \sin \theta}{k} \right) = \frac{M \cos \theta}{M + m} \left(\frac{mg \sin \theta}{k} \right)$$

$$= \frac{m^2g \sin \theta \cos \theta}{(M + m)k} \quad \text{Ans.}$$

Problem 42 A smooth bead of mass m is projected with a minimum speed v_o as shown in the figure so that it reaches the highest point P of the semi-circular cut of the wedge of mass M. Find the minimum value of v_o (d) recoil velocity of the wedge after the block leaves the wedge for the minimum value of v_o (d) value of v_o so that the wedge will break off at the time of the block.

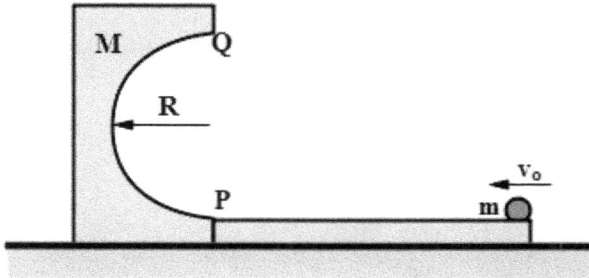

Solution

The net horizontal force acting on the system $(M + m)$ is zero. Hence, we can consume the horizontal momentum of the system.

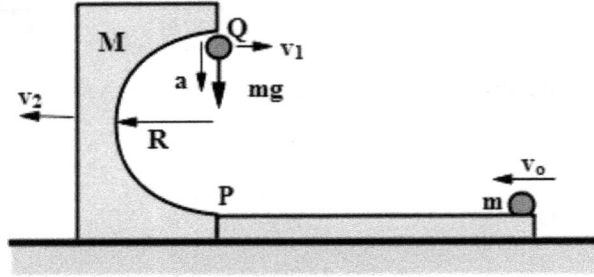

(a) If v_1 and v_2 are the velocities of the ball and wedge, respectively, conserving horizontal momentum, we have

$$mv_0 = Mv_2 - mv_1 \tag{1.132}$$

Conserving energy, we have

$$\frac{1}{2}mv_0^2 = \frac{M}{2}v_2^2 + \frac{1}{2}mv_1^2 + mg(2R) \tag{1.133}$$

The net vertical force acting on the ball is

$$N + mg = m(a_y) \tag{1.134}$$

The vertical acceleration of the ball is

$$a_y = \frac{(v_1 + v_2)^2}{R} \tag{1.135}$$

For minimum value of v_0, $N = 0$; then

$$(v_1 + v_2)^2 = gR$$

$$\Rightarrow v_1 + v_2 = \sqrt{gR} \tag{1.136}$$

Using equations (1.132) and (1.136)

$$mv_0 = M\left(\sqrt{gR} - v_1\right) - mv_1$$

$$\Rightarrow v_1 = \frac{M\sqrt{gR} - mv_0}{M + m} \tag{1.137}$$

Using equations (1.133) and (1.136)

$$\frac{1}{2}mv_0^2 = \frac{M}{2}(\sqrt{gR} - v_1)^2 + \frac{1}{2}mv_1^2 + 2mgR$$

$$= \frac{M}{2} \left\{ gR + v_1^2 - 2(\sqrt{gR})v_1 \right\} + \frac{1}{2} m v_1^2 + 2mgR$$

$$\frac{1}{2} m v_0^2 = (\frac{M}{2} + 2m)gR + \frac{(M+m)}{2} v_1^2 - m\sqrt{gR} \cdot v_1$$

$$m v_0^2 = (M + 4m)gR + (M + m)v_1^2 - 2M\sqrt{gR}\, v_1 \qquad (1.138)$$

Using equations (1.137) and (1.138)

$$m v_0^2 = (M + 4m)gR + (M + m)\left(\frac{M\sqrt{gR} - mv_0}{M + m} \right)^2$$

$$- 2M\sqrt{gR}\left(\frac{m\sqrt{gR} - mv_0}{M + m} \right)$$

$$= (M + 4m)gR + \frac{M^2 gR + m^2 v_0^2 - 2Mm\sqrt{gR}\, v_0}{M + m}$$

$$-\frac{2M^2 gR}{M + m} + \frac{2Mm\sqrt{gR}\, v_0}{M + m}$$

$$\Rightarrow m v_0^2 = (M + 4m)gR + \frac{m^2 v_0^2 - M^2 gR}{M + m} \Rightarrow m v_0^2 \left(1 - \frac{m}{M + m} \right) = gR\left(M + 4m - \frac{M^2}{M + m} \right)$$

$$\Rightarrow \frac{Mm v_0^2}{M + m} = \frac{M^2 + 4m^2 + 5Mm - M^2}{M + m} gR$$

$$\Rightarrow M v_0^2 = \frac{4m + 5M}{M} gR$$

$$\Rightarrow v_0 = \sqrt{\left(5 + \frac{4m}{M} \right) gR} \text{ Ans.}$$

N.B.: When the wedge is heavy or it is fixed with earth, putting $m/M \to 0$, we have $v_0 = \sqrt{5gR}$ as expected.

Alternative method: The relative velocity between the wedge and ball at the bottom is $u_r = v_0$. Then, the KE of the system $(M + m)$ relative to the center of mass frame is

$$K'_1 = \frac{Mm u_r^2}{2(M + m)} = \frac{Mm v_0^2}{2(M + m)} \qquad (1.139)$$

The KE of the system relative to the center of mass frame when the ball is at the top is

$$K'_f = \frac{Mm v_r^2}{2(M + m)} \qquad (1.140)$$

The velocity of the bead relative to the wedge at the top is given as

$$v_r^2 = gR \tag{1.141}$$

By using equations (1.140) and (1.141)

$$K'_f = \frac{MmgR}{2(M+m)} \tag{1.142}$$

Conserving energy relative to the center of mass frame is

$$K'_f + v'_f = K'_i + v'_f$$

$$\frac{MmgR}{2(M+m)} + mg(2R) = \frac{Mmv_0^2}{2(M+m)} + 0$$

$$\Rightarrow v_0^2 = \frac{4(M+m)}{M}gR + gR = \left(5 + \frac{4m}{M}\right)gR$$

$$\Rightarrow v_0 = \sqrt{\left(5 + \frac{4m}{M}\right)gR} \text{ Ans.}$$

Problem 43 One end of a spring is rigidly connected to a smooth cart of mass M and its other end is connected with a block of mass m as shown in the figure. Initially the spring is compressed by a distance x_o while the cart is very close to a vertical wall. If the system $(M+m)$ is released from rest, find the (a) maximum velocities of the cart and block and (b) distance covered by the center of mass of the system during the time in which the block hits the end of the cart. Put $AB = d$.

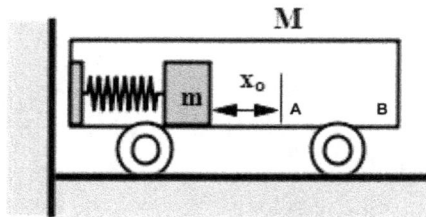

Solution

(a) Let v_1, v_2 be the velocities of the ball and cart, respectively.

$$\text{Since } F_h = 0, \quad P_h = C.$$

$$\Rightarrow -Mv_2 + mv_1 = 0 \tag{1.143}$$

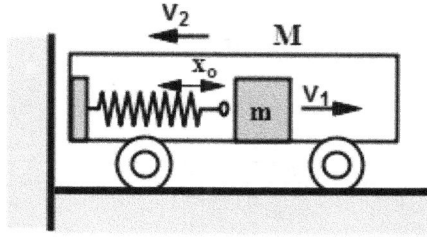

The cart rebounds with the same speed v_2. The relative velocity between m and M is

$$v_{rel} = v_1 - v_2$$

Then, the time after which the ball hits the other end of the cart is

$$t = \frac{d}{v_1 - v_2}$$

During this time, the center of mass of the system (cart-ball) moves through a distance

$$D = v_C t$$

$$\Rightarrow D = \left(\frac{mv_1 + Mv_2}{M + m} \right) \left(\frac{d}{v_1 - v_2} \right) \tag{1.144}$$

Conserving energy we have

$$\frac{1}{2}mv_1^2 + \frac{1}{2}Mv_2^2 = \frac{k}{2}x^2 \tag{1.145}$$

Using equations (1.143) and (1.145), we have

$$\frac{1}{2}mv_1^2 + \frac{1}{2}M\left(\frac{mv_1}{M}\right)^2 = \frac{kx^2}{2}$$

$$\Rightarrow \frac{mv_1^2}{2}\left(1 + \frac{m}{M}\right) = \frac{kx^2}{2}$$

$$\Rightarrow v_1 = \sqrt{\frac{Mk}{(M + m)m}}\, x$$

$$v_2 = \sqrt{\frac{mk}{(M + m)M}}\, x$$

(b) Putting the values of v_1 and v_2 in equation (1.144),

$$D = \frac{1}{M + m}\left\{ m\sqrt{\frac{Mk}{(M + m)m}}\, x_0 + Mx_0\sqrt{\frac{mk}{(M + m)M}} \right\}$$

$$\times \frac{d}{\sqrt{\frac{Mk}{(M+m)m}}\, x_0 - \sqrt{\frac{mk}{(M+m)M}}\, x_0}$$

$$\text{So, } D = \frac{2Mmd}{(M+m)(M-m)} = \frac{2Mmd}{M^2 - m^2}$$

Problem 44 A ball of mass m is placed at one end of a smooth tube of mass M. The ball is given a velocity v_o so that it will collide with the other end of the tube and then return back to the initial end. If the coefficient of restitution of collision is e, for a to-and-fro journey of the ball relative to the tube, find the (a) average speed and velocity relative to the tube, (b) velocities of the ball and tube after first collision.

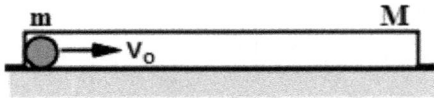

Solution

(a) The relative velocity decreases by a factor e in each collision. When it comes back to its initial position, the time taken is

$$t = \frac{d}{v_0} + \frac{d}{ev_0}$$

$$\Rightarrow t = \frac{d(e+1)}{v_0 e} \tag{1.146}$$

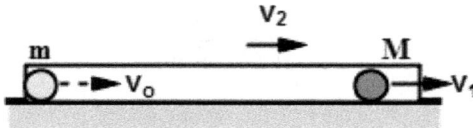

The average speed of the ball relative to the sledge is

$$v_{\text{av}} = \frac{d+d}{t} = \frac{2d}{t} \tag{1.147}$$

Using equations (1.146) and (1.147),

$$v_{av} = \frac{2d}{\frac{d(e+1)}{ev_0}}$$

$$\Rightarrow v_{av} = \frac{2ev_0}{e+1} \text{ Ans.}$$

Since the displacement of the ball relative to the tube is zero, its average velocity relative to the tube will be zero. Ans.

(b) The velocity of M and m just after first collision be v_1 and v_2. Then, conserving linear momentum,

$$mv_0 = mv_1 + Mv_2 \qquad (1.148)$$

$$-ev_0 = v_1 - v_2 \qquad (1.149)$$

Solving equations (1.148) and (1.149),

$$mv_0 - Mev_0 = (M+m)v_1$$

$$v_1 = \frac{m - eM}{M + m}v_0$$

$$|v_1| = \left| -\frac{(eM - m)}{M + m}v_0 \right|$$

$$\Rightarrow v_1 = \frac{eM - mv_0}{M + m}$$

From equation (1.149),

$$v_2 = v_1 + ev_0$$

$$= \frac{m - eM}{M + m}v_0 + ev_0$$

$$= v_0 \left\{ \frac{m - eM + eM + em}{M + m} \right\}$$

$$v_2 = \frac{mv_0(1 + e)}{M + m}$$

Problem 45 A block of mass m_1 is connected with another block of mass m_2 by a light spring of stiffness k.

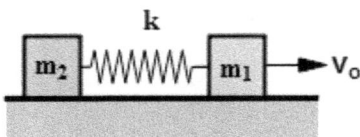

If the block m_1 is given a velocity v_o, disregarding friction, find the (a) displacement of center of mass as the function of time t, (b) the deformation x as the function of time t.

Solution

(a) The initial position of the center of mass of the system $(M + m)$ is

$$x_0 = \frac{m_1 l}{m_1 + m_2} \tag{1.150}$$

Let x_1, x_2 be the positions of the blocks m_1 and m_2. After a time t, the center of mass travels through a distance

$$s_c = v_c t, \tag{1.151}$$

where v_c = velocity of center of mass is given as

$$v_c = \frac{m_1 v_0}{m_1 + m_2} \tag{1.152}$$

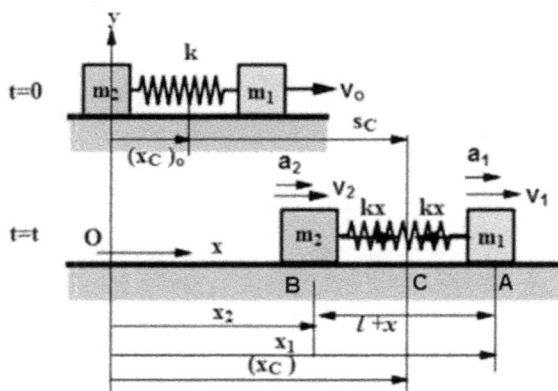

Using equations (1.151) and (1.152)

$$s_c = \frac{m_1 v_0 t}{m_1 + m_2} \tag{1.153}$$

1-155

(b) Let $x =$ elongation of the spring, which is given as

$$x_1 - x_2 = x \tag{1.154}$$

Differentiating with time, we have

$$\frac{d^2x_1}{dt^2} - \frac{d^2x_2}{dt^2} = \frac{d^2x}{dt^2}$$

$$a_1 - a_2 = \frac{d^2x}{dt^2} \tag{1.155}$$

Newton's second law on m_1:

$$a_1 = \frac{-kx}{m_1} \tag{1.156}$$

Newton's second law on m_2:

$$a_2 = \frac{kx}{m_2} \tag{1.157}$$

Using equations (1.155), (1.156), and (1.157),

$$-\frac{kx}{m_1} - \frac{kx}{m_2} = \frac{d^2x}{dt^2}$$

$$\Rightarrow \frac{d^2x}{dt^2} + k\left(\frac{1}{m_1} + \frac{1}{m_2}\right)x = 0$$

If we put $k(\frac{1}{m_1} + \frac{1}{m_2}) = \omega^2$, we have

$$\frac{d^2x}{dt^2} + \omega^2 x = 0 \tag{1.158}$$

The solution of this differential equation is given as

$$x = A \sin \omega t, \tag{1.159}$$

where $A =$ amplitude of oscillation which is equal to the maximum elongation of the spring

$$\Rightarrow \frac{dx}{dt} = A\omega \cos \omega t$$

At $t = 0$, $\frac{dx}{dt} = v_{rel} = v_0$.
Then, $v_0 = A\omega \cos 0$

$$\Rightarrow A = \frac{v_0}{\omega} \tag{1.160}$$

Using equations (1.159) and (1.160)

$$x = \frac{v_0}{\omega} \sin \omega t,$$

where $\omega = \sqrt{\dfrac{k(m_1 + m_2)}{m_1 m_2}}$ Ans.

Problem 46 A block of mass m_1 is connected with another block of mass m_2 by a light spring of stiffness k. If the block m_1 is pulled by a force F, disregarding friction, find the (a) position of the of center of mass as the function of time t, (b) deformation x as the function of time t, (c) minimum and maximum elongation of the spring, and (d) position of the blocks as the function of time.

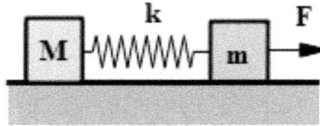

Solution

(a) Let x_1, x_2 be the positions of m and M, respectively. If x is the position of the center of mass of the system $M + m$ and $x_0 =$ initial position of center of mass, we can write

$$x = (x_c)_0 + s_c, \qquad (1.161)$$

where

$$(x_c)_0 = \frac{ml}{M + m} \qquad (1.162)$$

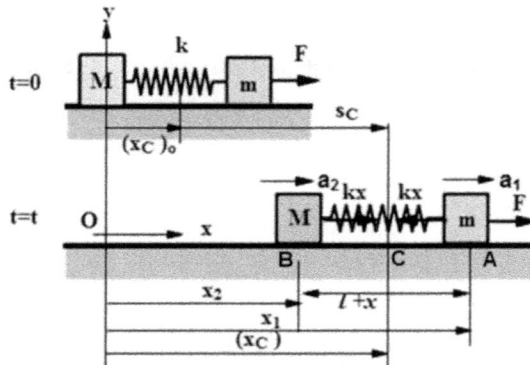

Furthermore, s_c= displacement of center of mass during time t

$$=\frac{1}{2}a_c t^2, \text{ where } a_c = \frac{F}{M+m}$$

$$\Rightarrow s_c = \frac{1}{2}\frac{F}{M+m}t^2 \tag{1.163}$$

Using equations (1.161) and (1.163)

$$x_c = \frac{ml}{M+m} + \frac{Ft^2}{2(M+m)} \tag{1.164}$$

(b) Applying NSL on m and M, we have

$$F - kx = ma_1 \tag{1.165}$$

$$kx = Ma_2 \tag{1.166}$$

Constraint equation:

$$x_1 - x_2 = x$$

$$\Rightarrow \frac{d^2x_1}{dt^2} - \frac{d^2x_2}{dt^2} = \frac{d^2x}{dt^2}$$

$$\Rightarrow a_1 - a_2 = \frac{d^2x}{dt^2} \tag{1.167}$$

Putting the values of a_1 and a_2 from equations (1.165) and (1.166) in equation (1.167).

$$\left(\frac{F}{m} - \frac{k}{m}x\right) - \frac{kx}{M} = \frac{d^2x}{dt^2}$$

$$\Rightarrow \frac{d^2x}{dt^2} + \frac{k(M+m)x}{Mm} = \frac{F}{m},$$

where g = constant.
Putting $\frac{k(M+m)}{Mm} = \omega^2$ and $\frac{F}{m} = g$, we obtain the differential equation

$$\frac{d^2x}{dt^2} + \omega^2 x = g \tag{1.168}$$

Its solution can be given as

$$x = \frac{g}{\omega^2}(1 - \cos \omega t), \tag{1.169}$$

where $\omega = \sqrt{\frac{k(M+m)}{Mm}}$

$$\Rightarrow x = \frac{\left(\frac{F}{m}\right) \times Mm}{k(M+m)}[1 - \cos \omega t]$$

$$\Rightarrow x = \frac{FM}{k(M+m)}(1 - \cos \omega t), \qquad (1.170)$$

where $\omega = \sqrt{\frac{k(M+m)}{Mm}}$

(c) The elongation will be maximum at the values of times when $\cos \omega t = -1$

$$t = \frac{\pi}{\omega}, \quad \frac{3\pi}{\omega} \quad \frac{(2n+1)\pi}{\omega}$$

$$\Rightarrow x_{\max} = \frac{2FM}{k(M+m)}$$

The elongation will be zero when $\cos \omega t = 1$

$$t = \frac{2n\pi}{\omega}$$

$$\Rightarrow x_{\min} = 0$$

(d) The position of A is

$$x_1 = x_i + s_c + AC$$

$$= \frac{ml}{M+m} + \frac{1}{2}a_c t^2 + \frac{M(l+x)}{M+m}$$

$$= l + \frac{1}{2}a_c t^2 + \frac{Mx}{M+m}$$

Problem 47 Two blocks being connected rigidly with a vertical spring are at rest. The spring is compressed by a distance x_o by pushing the upper block down by a string S as shown in the figure. Let the string be cut. (a) What is the minimum initial compression x_o of the spring so that the lower block will break off? (b) If $M = m$, what is the maximum height attained by the center of mass of the system if the system is released from an initial compression $x_i = 2 x_o$?

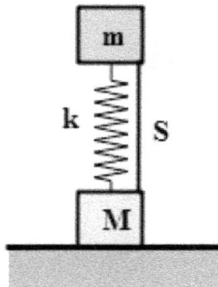

Solution

(a) If the spring is cut, the compressed spring will push the block m up and the block M down. As result, M will stay at rest and in begins to accelerate up. If the block m crosses the relaxed position of the spring ($x = 0$) and moves above this position, the spring will get elongated. Thereafter, the elongated spring pulls the block m down and the block M up. If the pulling force F_{sp} will be equal to Mg, the block will lose contact with the ground. Then, the elongation of the spring is

$$x = \frac{Mg}{k} \tag{1.171}$$

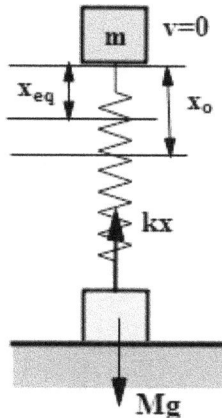

The change in spring potential energy between initial and final position x is

$$\Delta U_s = \frac{k}{2}(x^2 - x^2) \tag{1.172}$$

The change in gravitational potential energy as the block moves through a distance $(x + x_0)$ is

$$\Delta U = mg(x + x_0) \tag{1.173}$$

The change in KE of the block m (M is still at rest) is

$$\Delta K = \frac{1}{2}mv^2 \tag{1.174}$$

Conserving energy

$$\Delta U + \Delta K = 0$$
$$\Rightarrow \Delta U_g + \Delta U_s + \Delta K = 0$$

$$\Rightarrow mg(x + x_0) + \frac{k}{2}(x^2 - x_0^2) + \frac{m}{2}v^2 = 0$$

Putting $v = 0$ for minimum x_0, we have

$$mg(x + x_0) + \frac{k}{2}(x^2 - x_0^2) = 0 \Rightarrow mg + \frac{k}{2}(x - x_0) = 0 \qquad (1.175)$$

Putting $x = \frac{mg}{k}$ from equation (1.171) in equation (1.175)

$$mg + \frac{k}{2}\left(\frac{Mg}{k} - x_0\right) = 0$$

$$\Rightarrow \frac{kx_0}{2} = mg + \frac{Mg}{2}$$

$$\Rightarrow x_0 = \frac{(2m + M)g}{k} \quad \text{Ans.}$$

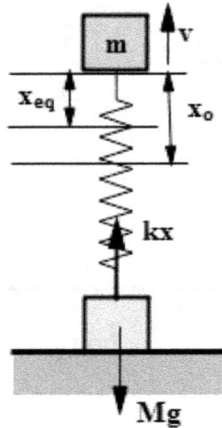

(b) Let us rewrite the equation

$$\frac{m}{2}v^2 + (x + x_0)\left\{mg + \frac{k}{2}(x - x_0)\right\} = 0$$

$$\Rightarrow v = \sqrt{\frac{2(x + x_0)}{m}\left\{\frac{k}{2}(x_0 - x) - mg\right\}}$$

$$\Rightarrow \sqrt{2(x + x_0)\left\{\frac{k}{2m}(x_0 - x) - g\right\}}$$

Putting $x = \frac{Mg}{k}$ and $x_0 = \frac{2(2m + M)}{k}$, we have

$$v = \sqrt{2\left\{\frac{Mg}{k} + \frac{2(2m + M)g}{k}\right\}\left\{\frac{k}{2m}\frac{(4m + M)g}{k} - g\right\}}$$

$$= \sqrt{2(4m + 3M)\frac{g}{k} \times \frac{(2m + M)g}{2m}}$$

$$= \left\{ \sqrt{\frac{(4m + 3M)(2m + M)}{mk}} \right\} g \text{ Ans.}$$

If $M = m$,

$$v = \left\{ \sqrt{\frac{21m}{k}} \right\} g \text{ Ans.}$$

The rise in the center of mass until the block M will break off is given as
$h_1 = \frac{M(0) + ms_m}{M + m} = \frac{s_m}{2}$ ($\because M = m$), where s_m = displacement of m

$$= (x + x_0)$$

$$= \frac{Mg}{k} + \frac{2(2m + M)g}{k}$$

$$= \frac{Mg}{k} + \frac{2(2m + m)g}{k} = \frac{7mg}{K}$$

Then, $h_1 = \frac{7mg}{2k}$

The velocity of center of mass at the time of breaking off is

$$v'_C = \frac{mv}{M + m} = \frac{Mv}{M + m} = \frac{v}{2}$$

Then the extra height moved by the center of mass is

$$h_2 = \frac{v'^2_C}{2g} = \frac{(\frac{v}{2})^2}{2g} = \frac{v^2}{8g}$$

Putting the obtained value of

$$v = \sqrt{\frac{21m}{k}} g, \text{ we have}$$

$$h_2 = \frac{v^2}{8g} = \frac{\frac{21m}{K}g^2}{8g} = \frac{21mg}{8k}$$

Hence, the total shift of the center of mass before it comes down is

$$h = h_1 + h_2 = \left(\frac{7}{2} + \frac{21}{8} \right)\frac{mg}{k} = \frac{49mg}{8k} \text{ Ans.}$$

Problem 48 A vertical spring-blocks system (two identical blocks each of mass m being connected by a light spring) is released from rest. After falling through a distance h, the lower block collides with the ground inelastically. Assume that $h = \eta x_s$, where $x_s =$ static deformation of the spring. Find the (a) ratio of maximum compression and static deformation of the spring and (b) total descent of the center of mass until the block A reaches its lowest position.

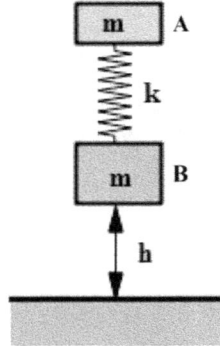

Solution

(a) Let $x =$ compression of the spring. The block A moves down through a vertical distance $h + x$. Then the change in gravitational potential energy is

$$\Delta U_g = -mg(h + x) \tag{1.176}$$

The change in spring potential energy is

$$\Delta U_s = \frac{1}{2}kx^2 \tag{1.177}$$

Since the block A will be at rest instantaneously at its lowest position, $v_f = 0$. As it was released from rest, $v_i = 0$. So, the change KE of the spring + block A is

$$\Delta K = 0$$

We can conserve the total energy of the system (spring + upper block), because the lower block collides inelastically and comes to rest just after the collision.

Conserving energy,

$$\Delta U + \Delta K = 0$$

$$\Rightarrow \Delta U_s + \Delta U_g = 0 \; (\because \Delta K = 0) \tag{1.178}$$

Using equations (1.176), (1.177), and (1.178)

$$\frac{k}{2}x^2 - mg(h + x) = 0$$

$$\Rightarrow kx^2 - 2mgx - 2mgh = 0$$

The solution of this quadratic equation is

$$x = \frac{2mg \pm \sqrt{(2mg)^2 + 4mghk}}{2k}$$

$$= \frac{mg}{k}\left[1 + \sqrt{1 + \frac{hk}{mg}}\right]$$

Putting $\frac{mg}{k} = x_s$ = static deformation, we have

$$x = x_s\left[1 + \sqrt{1 + \frac{hk}{mg}}\right]$$

$$x = x_s\left[1 + \sqrt{1 + \frac{h}{x_s}}\right]$$

If $h = \eta x_s$, we have

$$\frac{x}{x_s} = 1 + \sqrt{1 + \eta} \text{ Ans.}$$

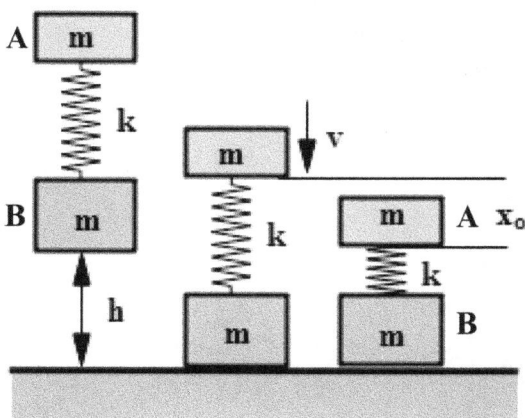

(b) The displacement of the center of mass until the block B hits the ground is $s_{c_i} = h = \eta x_s$ because both blocks A and B move through same distance h. The additional displacement of the center of mass until the block B comes to its lowest position is

$$S_{C_2} = \frac{m_A S_A}{m_A + m_B} = \frac{m S_A}{m + m} = \frac{S_A}{2}$$

where $s_A = x = x_s(1 + \sqrt{1 + \eta})$

$$\Rightarrow s_{C_2} = \frac{x_s}{2}\left(1 + \sqrt{1 + \eta}\right)$$

Then the total descent of the center of mass until the block A reaches its lowest position is

$$s_C = s_{C_1} + s_{C_2}$$

$$= \eta x_s + \frac{x_s}{2}\left(1 + \sqrt{1 + \eta}\right)$$

$$= \frac{2\eta + 1 + \sqrt{1 + \eta}}{2} \quad \text{Ans.}$$

Problem 49 Three identical particles B, C, and D are connected by two identical inextensible strings each of length l. The system of two strings and three particles is on a smooth horizontal plane. Let's give a velocity v_o to the another particle A of mass m. Find the (a) tension in the string just after the collision, (b) relative acceleration between C and D just after the collision, (c) fraction of energy lost in the collision, (d) velocity of C and D relative to the combination (A + B), (e) tension in the string, (f) accelerations of C and D, (g) accelerations of C and D relative to each other, and (h) acceleration of A, just before C and D collide.

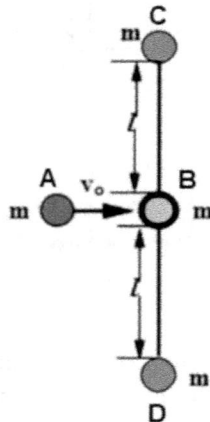

Solution

(a) Just before and after the collision between A and B, conserving linear momentum

$$mv_0 = (m + m)v$$

$$\Rightarrow v = \frac{v_0}{2} \tag{1.179}$$

As the combination (A + B) move with a velocity $v = \frac{v_0}{2}$ just after the collision when C and D are still stationary, the tension in the string is

$$T = \frac{mv_{\text{rel}}^2}{l} \tag{1.180}$$

Since $v_{\text{rel}} = v_{BC} = v_{BD} = \frac{v_0}{2}$ just after the collision, $T = \frac{mv_0^2}{4l}$ Ans.

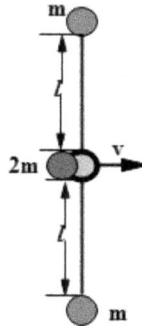

(b) Then the relative acceleration between C and D is $a_{CD} = 2a_C = 2a_D$ because both C and D accelerate towards the combined block (A + B).

Applying NSL, $a_C = a_D = \frac{T}{m} = \frac{\frac{mv_0^2}{4l}}{m} = \frac{v_0^2}{4l}$

$$\Rightarrow a_{CD} = 2a_C = 2 \times \frac{v_0^2}{4l} = \frac{v_0^2}{2l} \text{ Ans.}$$

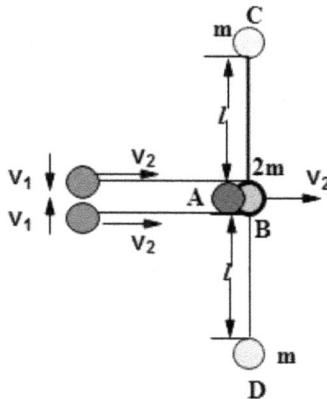

(c) Fraction of energy lost in the collision is

$$\eta = \frac{|\Delta K|}{K} = \frac{\frac{1}{2}mv_0^2 - \frac{1}{2}(m+m)v_1^2}{\frac{1}{2}mv_0^2},$$

where $v_1 = \frac{v_0}{2}$

$$\Rightarrow \eta = \frac{\frac{1}{2}mv_0^2 - \frac{1}{2}2m\left(\frac{v_0}{2}\right)^2}{\frac{1}{2}mv_0^2} = \frac{1}{2} \text{ Ans.}$$

(d) As the balls C and D revolve in the circles relative to the combination (A + B), let them have velocity v_1 and v_2 perpendicular and parallel to the string just before they meet.

Conserving the energy between the last two positions (just after the collision and just before C and D collide), we have

$$\Delta K = 0 \Rightarrow K_f = K_i$$

$$\Rightarrow \frac{1}{2}m(v_1^2 + v_2^2) \times 2 + \frac{1}{2}2mv_2^2 = \frac{1}{2}(2m)v^2$$

$$\Rightarrow 2(v_1^2 + v_2^2) + 2v_2^2 = 2v^2 \tag{1.181}$$

Using equations (1.179) and (1.181)

$$3v_2^2 + 2v_1^2 = 2\left(\frac{v_0}{2}\right)^2$$

$$\Rightarrow 3v_2^2 + 2v_1^2 = \frac{v_0^2}{2} \tag{1.182}$$

Conserving horizontal momentum along the strings between initial and final position of the system

$$mv_0 = 4mv_2$$

$$\Rightarrow v_2 = \frac{v_0}{4} \tag{1.183}$$

Using equations (1.182) and (1.183)

$$3\left(\frac{v_0}{4}\right)^2 + 2v_1^2 = \frac{v_0^2}{2}$$

$$\Rightarrow 2v_1^2 = \frac{v_0^2}{2} - \frac{3v_0^2}{16}$$

$$\Rightarrow 2v_1^2 = \frac{8v_0^2 - 3v_0^2}{16}$$

$$\Rightarrow v_1 = \frac{1}{4}\sqrt{\frac{5}{2}}\, v_0 \text{ Ans.}$$

(e) The tension in the string just before C and D collide is

$$T = \frac{mv'^2_1}{l} = \frac{mv^2_{AC}}{l} = m\frac{v^2_1}{l}$$

$$= \frac{m}{l}\left(\frac{v^2_0}{32}\right) = \frac{5mv^2_0}{32l} \text{ Ans.}$$

(f) The acceleration of C and D is

$$\vec{a_C} = \vec{a_D} = \frac{T}{m}\hat{i}$$

$$= \frac{5mv^2_0}{32ml}\hat{i} = \frac{5v^2_0}{32l}\hat{i}$$

(g) The acceleration of C of D relative A + B is

$$a_{\text{rel}} = \frac{v^2_{\text{rel}}}{l} = \frac{v^2_1}{l}, \text{ where } v_1 = \frac{1}{4}\sqrt{\frac{5}{2}}\, v_0$$

Then, we have $a_{\text{rel}} = \frac{5v^2_0}{32l}$ Ans.

(h) The acceleration of A just before C and D collide is $a' = 2T/2m = T/m$; put the obtained value of tension T (referring to (e)) at this instant.

Problem 50 A smooth chain of mass m and length l is kept stationary on the table. (a) At time $t = 0$, let's release the chain when it is overhanging length at the edge O is $x\,(= b)$. Find the (a) acceleration, (b) velocity of the chain as the function of x, and (c) time required by the chain to just touch the ground. Assume $h < l$.

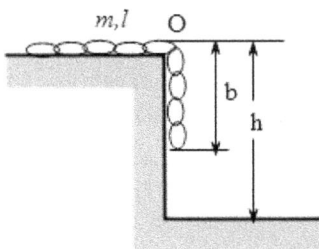

(a) For any overhang y the vertical pulling force acting on the chain is
F_{net} = weight of the hanging portion of the chain

$$=\left(\frac{m}{l}y\right)g = \frac{mg}{l}y$$

Then, the magnitude of tangential acceleration of the hanging chain is

$$a = \frac{F_{net}}{m} = \frac{mg}{l}y = \frac{g}{l}y$$

(b) Putting $a = \frac{vdv}{dy}$, we have

$$\frac{vdv}{dx} = \frac{g}{l}y$$

$$\Rightarrow vdv = \frac{g}{l}ydy$$

$$\Rightarrow \int_0^v vdv = \frac{g}{l}\int_b^y ydy$$

$$\Rightarrow \frac{v^2}{2} = \frac{g}{l}\left(\frac{y^2}{2} - \frac{b^2}{2}\right)$$

$$\Rightarrow v = \sqrt{g\frac{(y^2 - b^2)}{l}} \quad \text{Ans.}$$

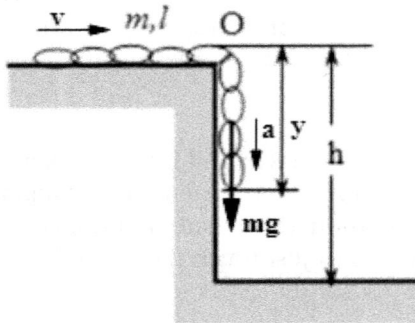

Alternative method: The change in gravitational potential energy relative to the table top when the lowest point of the chain touches the ground is

$$\Delta U = U_f - U_i$$

$$= (-\frac{my}{l}g\,\frac{y}{2}) - (m'g\frac{b}{2}), \text{ where } m' = \frac{m}{l}b$$

$$\Rightarrow \Delta U = -mg\frac{y^2}{2l} + \frac{m}{2l}gb^2$$

The change in KE is

$$\Delta K = \frac{1}{2}mv^2$$

Conserving energy,

$$\Delta U + \Delta K = 0$$

Using the last three equations, we have

$$\frac{mgy^2}{2l} - mg\frac{l}{2} + \frac{1}{2}mv^2 = 0$$

$$\Rightarrow v = \sqrt{\frac{g(y^2 - b^2)}{l}}$$

(c) Putting $v = \frac{dx}{dt} = \sqrt{\frac{g}{l}(y^2 - b^2)}$ and separating the variables, we have

$$\frac{dy}{\sqrt{y^2 - b^2}} = \sqrt{\frac{g}{l}} \, dt$$

Integrating both sides, we have $\displaystyle\int_0^l \frac{dy}{\sqrt{y^2 - b^2}} = \sqrt{\frac{g}{l}} \int_0^t dt$

$$\Rightarrow \ln\left(\frac{h + \sqrt{h^2 - b^2}}{h}\right) = \sqrt{\frac{g}{l}}\, t$$

$$\Rightarrow t = \sqrt{\frac{l}{g}} \ln\left(\frac{l + \sqrt{l^2 - b^2}}{h}\right) \quad \text{Ans.}$$

Problem 51 A uniform smooth chain AB of mass m is released from rest so that it slides along the circular track of radius R and then on the horizontal surface as shown in the figure. Find the (a) gravitational potential energy, (b) speed with which the chain leaves the circular track.

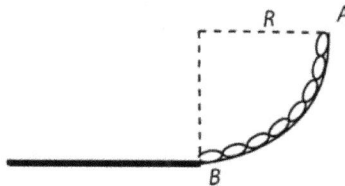

Solution

(a) The position of the center of mass C is

$$r_c = \frac{2R \sin \frac{90}{2}}{\frac{\pi}{2}}$$

$$= \frac{4R}{\pi} \times \frac{1}{\sqrt{2}} = \frac{2\sqrt{2}\,R}{\pi}$$

Relative to the center O of curvature of the arc, the coordinates of center of mass are

$$-\frac{r_c}{\sqrt{2}}, \quad -\frac{r_c}{\sqrt{2}} = -\frac{2R}{\pi}, \quad -\frac{2R}{\pi}$$

This means that the y-coordinate of center of mass is

$$y_c = -\frac{2R}{\pi}$$

Then, the gravitational potential energy is

$$U_i = mgy_c = mg\left(\frac{2R}{\pi}\right)$$

$$\Rightarrow U_i = -2mg\frac{R}{\pi}$$

Relative to ground

$$U = mgy_c = mg\left(R - \frac{2R}{\pi}\right) = mgR\left(1 - \frac{2}{\pi}\right) \text{ Ans.}$$

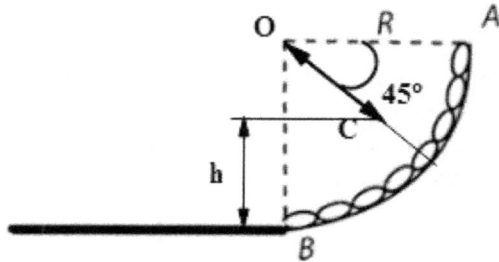

(b) When the chain slides to the ground the total chain will be below the point O by a distance R. Then, the gravitational potential energy relative to O is

$$U_f = -mgR$$

The conservation of energy gives

$$U_i = U_f + K$$

$$\Rightarrow -\frac{2mgR}{\pi} = -mgR + \frac{1}{2}mv^2$$

$$\Rightarrow mgR\left(1 - \frac{2}{\pi}\right) = \frac{1}{2}mv^2$$

$$\Rightarrow v = \sqrt{2gR(1 - \frac{2}{\pi})} \text{ Ans.}$$

Problem 52 A smooth rope ABC of mass m and length l is kept on the fixed disc of radius R. If the rope slides with a speed v, find the (a) net vertical force acting on the chain, (b) net acceleration of center of mass of the chain, (c) net momentum of the chain at the given position.

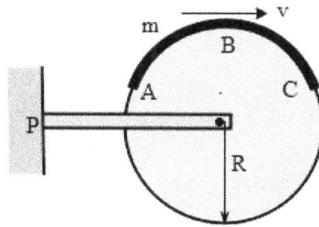

Solution

(a) The net radial force acting on the element is

$$dmg \cos \theta - dN = dm\frac{v^2}{R}$$

$$\Rightarrow dN = dm\left(g \cos \theta - \frac{v^2}{R}\right)$$

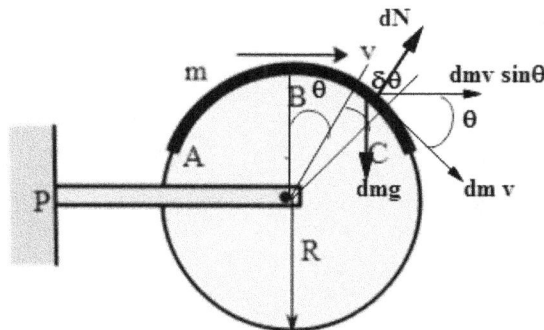

The net vertical force acting on the element is

$$dF_y = dN \cos\theta$$

$$= dm\left(g\cos^2\theta - \frac{v^2}{R}\cos\theta \right)$$

Then, the net vertical force acting on each quadrant of the chain is

$$F'_y = \int dF_y$$

$$= \int dm\left(g\cos^2\theta - \frac{v^2}{R}\cos\theta \right)$$

$$= \int \lambda R d\theta\left(g\cos^2\theta - \frac{v^2}{R}\cos\theta \right)$$

$$= \lambda R\left[\int_0^{\frac{\pi}{2}} g\cos^2\theta - \frac{v^2}{R}\int_0^{\frac{\pi}{2}} \cos\theta d\theta \right]$$

$$= \lambda R\left[\frac{g\pi}{2} - \frac{v^2}{R} \right] = \lambda R\left(\frac{\pi g}{2} - \frac{v^2}{R} \right)$$

Putting $\lambda = \frac{m}{\pi R}$, we have the vertical force acting on each quadrant of the chain given as

$$F'_y = \frac{m}{\pi R} \cdot R\left(\frac{\pi g}{2} - \frac{v^2}{R} \right)$$

$$= \frac{m}{\pi}\left(\frac{\pi g}{2} - \frac{v^2}{R} \right)$$

Since there are two equal half quadrants the net horizontal force $= 0$. So, the net horizontal force $= 0$. Then the net vertical force acting on the total chain is

$$F_y = 2F'_y = \frac{2m}{\pi}\left(\frac{\pi g}{2} - \frac{v^2}{R} \right)$$

$$\Rightarrow F_y = -mg\left[1 - \frac{2v^2}{\pi g R} \right]$$

Alternate method: Let $N =$ vertical reaction force N acting on the chain then, the net vertical force is

$$\overrightarrow{F}_{net} = mg - N = ma_C$$

$\Rightarrow N = m(g - a_C)$, where

$$a_C = \frac{v_c^2}{r_C} \text{ and } v_c = \frac{2v}{\pi}$$

Putting $r_C = \frac{2R}{\pi}$, we have

$$N = m\left[g - \frac{(\frac{2v}{\pi})^2}{(\frac{2R}{\pi})} \right]$$

$$\Rightarrow N = mg\left[1 - \frac{2v^2}{\pi g R} \right] \text{ Ans.}$$

(b) Hence, the net acceleration of center of mass of the chain is

$$\vec{a_C} = \frac{\vec{F}_{\text{net}}}{m}$$

$$= \frac{-\frac{2m}{\pi}\left(\frac{g}{2} - \frac{v^2}{R} \right)\hat{j}}{m}$$

$$= -\frac{2}{\pi}\left(\frac{g}{2} - \frac{v^2}{R} \right)\hat{j}$$

(c) Taking two elements symmetrically we can see that the net vertical momentum of the chain is zero. So, $v_y = 0$.

Then the velocity of center of mass is

$$v_x = \frac{\int dm v \cos \theta}{\int dm}$$

$$= \frac{\int_{-\frac{\pi}{2}}^{+\frac{\pi}{2}} \lambda R d\theta . \, v \cos \theta}{\int_0^\pi \lambda R d\theta}$$

$$= \frac{-v \int_{-\frac{\pi}{2}}^{\frac{\pi}{2}} \cos \theta \, d\theta}{\pi} = \frac{2v}{\pi}$$

$\Rightarrow$ The net momentum of the chain is

$$\vec{P} = m\frac{2v}{\pi}\hat{i}$$

The velocity of center of mass is

$$\vec{v_C} = \frac{\vec{P}}{m} = \frac{2v}{\pi}\hat{i}$$

Problem 53 A block of mass m is placed on a plank of mass M. The plank-block system is placed on a smooth horizontal surface. If the block is given a velocity v_o, the distance of relative sliding between M and m is x (let). Find the (a) total work done by friction in terms of x, (b) initial kinetic energy relative to center of mass of the system $(M + m)$, (c) maximum distance x covered by the block relative to the plank in terms of v_o. The coefficient of friction between the block and plank is μ.

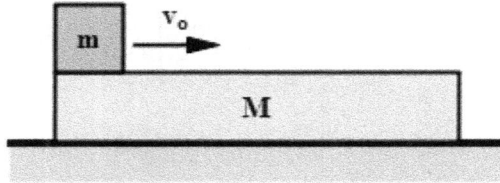

Solution

(a) Let x_1 and x_2 be the displacements of m and M relative to ground. Then, the work done by friction on m and M is given as

$$W_1 = -fx_1 \qquad (1.184)$$

$$W_2 = +fx_2 \qquad (1.185)$$

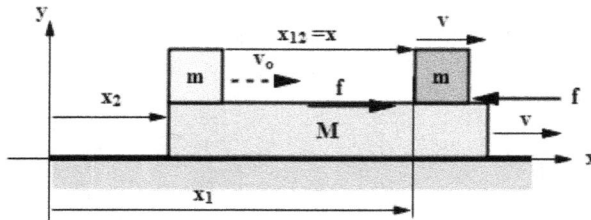

The net work by friction on the system $(M + m)$ is

$$W = W_1 + W_2$$

$$= -f(x_1 - x_2)$$

Putting $x_1 - x_2 = x$, we have

$$W_f = -fx \qquad (1.186)$$

where

$$f = \mu mg \qquad (1.187)$$

$$\Rightarrow W_f = -\mu mgx \qquad (1.188)$$

We can see that the total work done by friction does not depend on the reference frame.

(b) The velocity of center of mass of the system is

$$\vec{v_C} = \frac{mv_0}{M + m}\hat{i}$$

The initial KE of the system relative to the center of mass is

$$K' = \frac{1}{2}mv_{1C}^2 + \frac{1}{2}Mv_{2C}^2$$

$$= \frac{1}{2}m(v_1 - v_C)^2 + \frac{M}{2}(v_2 - v_C)^2$$

Putting $v_1 = v_0$ and $v_2 = 0$, we have

$$= \frac{1}{2}m(v_0 - v_C)^2 + \frac{1}{2}(Mv_C^2)$$

$$= \frac{1}{2}m\left(v_0 - \frac{mv_0}{M + m}\right)^2 + M\left(\frac{mv_0}{M + m}\right)^2$$

$$= \frac{m}{2}\left(\frac{Mv_0}{M + m}\right)^2 + M\left(\frac{mv_0}{M + m}\right)^2$$

$$= \frac{Mmv_0^2}{2(M + m)}(M + m)$$

$$= \frac{Mmv_0^2}{2(M + m)} \quad \text{Ans.}$$

(c) The net work done is

$W_{\text{net}} = W_f$(because work done by gravity and normal reaction will be zero)

$$\Rightarrow W_{\text{net}} = -\mu mgx$$

Since the change in kinetic energy of the center of mass $= \Delta K_C = 0$, the total change in KE is

$$\Delta K = \Delta K' = K'_f - K'_i$$

Since the block stops sliding on the plank,
putting $v'_{\text{rel}} = 0$, we have $K'_f = 0$

$$\Rightarrow \Delta K = -K'_i$$

$$\Rightarrow \Delta K = -\frac{Mmv_0^2}{2(M+m)}$$

The work–energy theorem states that

$$W_{net} = \Delta K$$

$$\Rightarrow -\mu mgx = -\frac{Mmv_0^2}{2(M+m)}$$

$$\Rightarrow x = \frac{Mv_0^2}{2\mu(M+m)g} \quad \text{Ans.}$$

Problem 54 A block of mass m is released from rest from the top of a smooth wedge of mass M_1 so that it slides along the smooth curved track and launches on to a plank of mass M_2 that is resting on a smooth ground as shown in the figure. The block slides on the plank and stops sliding at the other end of the plank. Find the (a) momentum delivered to both m and M_1, (b) velocities of m and M_1, (c) fraction of energy carried by the wedge M_1, (d) velocity of center of mass of $M_1 + m$, (e) velocity of center of mass of the system $(M_2 + m)$, (f) initial KE of the system $M_2 + m$ in center of mass frame, and (g) total distance travelled by the block relative to the plank M_2. Assume = coefficient of friction between the block and plank.

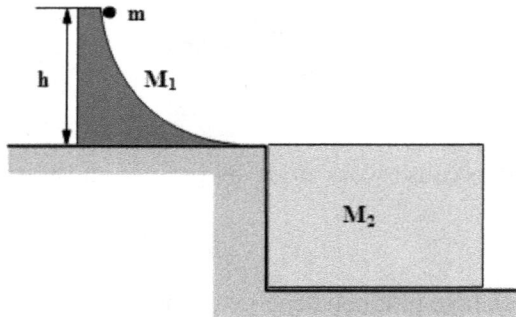

Solution

(a) Conserving linear momentum between the position 1 and 2 of the system $M_1 + m$,

$$p_1 = p_2$$

$$\Rightarrow M_1 v_1 = mv_2 \tag{1.189}$$

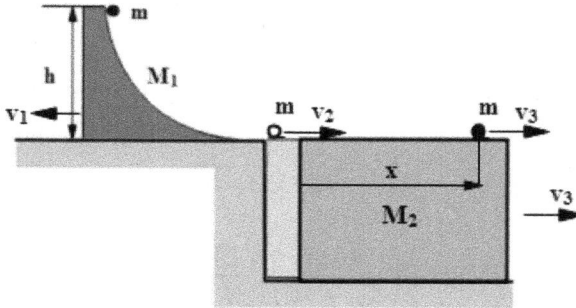

Conserving energy of $M_1 + m$

$$\frac{-p_1^2}{2M_1} + \frac{p_2^2}{2m} = mgh \tag{1.190}$$

Since $p_1 = p_2 = p_1$, say,

$$\frac{p^2}{2}\left(\frac{1}{M_1} + \frac{1}{m}\right) = mgh$$

$$\Rightarrow p = \sqrt{\frac{2M_1 m^2 gh}{M_1 + m}} \quad \text{Ans.}$$

(b) Then $v_1 = \frac{p}{M_1} = \sqrt{\frac{2m^2 gh}{M_1(M_1 + m)}}$ Ans.

Similarly, $v_2 = \frac{p}{m} = \sqrt{\frac{2M_1 gh}{(M_1 + m)}}$ Ans.

(c) Fraction of KE carried by m is

$$\eta = \frac{\frac{p^2}{2m}}{mgh} = \frac{2M_1 m^2 gh}{2(M + m)m \times mgh} = \frac{M_1}{M_1 + m}$$

(d) The horizontal velocity of center of mass of $M_1 + m$ is $v_{C_1} = 0$ because initially both M_1 and m are at rest.

(e) The velocity of center of mass of the system $(M_2 + m)$ is

$$v_C = \frac{Mv_2 + M_2(O)}{M + m}$$

$$= \frac{mv_2}{M_2 + m} = \frac{m}{M_2 + m}\sqrt{\frac{2M_1gh}{M_1 + m}}$$

(f) The initial KE of the system $M_2 + m$ in the center of mass frame is

$$K'_i = \frac{1}{2}\frac{M_2 mv_2^2}{(M_2 + m)}$$

$$= \frac{1}{2}\left(\frac{M_2 m}{M_2 + m}\right)\left\{\frac{2M_1gh}{M_1 + m}\right\}$$

$$= \frac{M_1 M_2 mgh}{(M_1 + m)(M_2 + m)} \quad \text{Ans.}$$

(g) The work done by friction till the block stops sliding relative to the plank M_2 is

$$W_f = -\mu mgx$$

which does not depend on the reference frame. The net work done by gravity and normal reaction is zero.

Using the work–energy theorem, we have

$$W_{\text{net}} = W_f = \Delta K = \Delta K' + \Delta K_C$$

$$\Rightarrow -\mu gx = K'_f - K'_i \, (\because \Delta K_C = 0)$$

Since $K'_f = 0$ as the final relative velocity $= 0$ and $K'_i = \frac{M_1 M_2 mgh}{(M_1 + m)(M_2 + m)}$, we have

$$-\mu mgx = -\frac{M_1 M_2 mgh}{(M_1 + m)(M_2 + m)}$$

$$\Rightarrow x = \frac{M_1 M_2 h}{(M_1 + m)(M_2 + m)} \quad \text{Ans.}$$

Problem 55 A smooth chain of mass m and length L is heaped on a table top. If we release the chain by slightly pulling its free end through the hole made on the table, the overhanging chain moves down by the effect of gravity. Find the (a) velocity, (b) acceleration of the chain, (c) time taken for the entire chain to slip down through the hole.

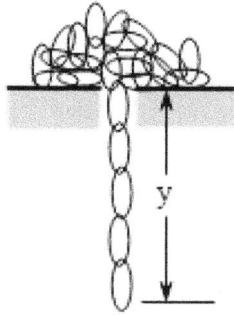

Solution

(a) If the open end of the chain is pulled by a small amount and slowly released, due to the effect of gravity the hanging part of the chain moves down with a velocity v and acceleration a, say.

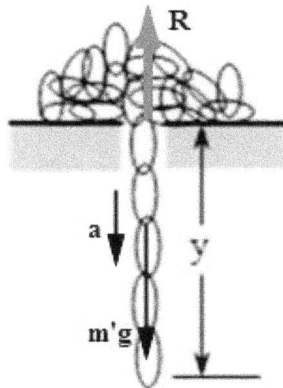

Let R = upward reaction offered by the leap as each link of the heap gains a downward velocity, given as

$$R = v_{rel}\frac{dm}{dt} = v\frac{dm}{dt} = v\,\lambda\frac{dx}{dt} = v\,\lambda.\,v$$

$$\Rightarrow R = \lambda v^2, \text{ where } \lambda = \frac{dm}{dx}$$

Then the net force acting on the chain is

$$m'g - R = m'a$$

$$m'g - \lambda v^2 = m'\frac{v\,dv}{dx},$$

$$\text{where } m' = \lambda x$$

$$\Rightarrow gx - v^2 = x\frac{v\, dv}{dx} \tag{1.191}$$

To solve this differential equation,
put $v^2 = kx$, then

$$\frac{2v\, dv}{dx} = k \tag{1.192}$$

Putting $\frac{v}{\frac{d}{dx}U} = \frac{K}{2}$ from equation (1.192) in equation (1.191), we have

$$gx - kx = x \cdot \frac{k}{2}$$

$$\Rightarrow 3\frac{kx}{2} = gx$$

$$\Rightarrow k = \frac{2g}{3}$$

Then, the velocity $v = \sqrt{kx} = \sqrt{\frac{2g}{3}x}$ Ans.

(b) The acceleration of the hanging chain is

$$a = \frac{v\, dv}{dx} = \frac{2g}{3} \text{ Ans.}$$

(c) Then the time taken for the entire chain to slip down through the hole is

$$t = \sqrt{\frac{2l}{a}} = \sqrt{\frac{2l}{\frac{2g}{3}}} = \sqrt{\frac{3l}{g}} \text{ Ans.}$$

(d) The velocity of the other end of the chain resting on the table to slip (escape) from the take is

$$v = \sqrt{\frac{2gx}{3}} \text{ where } x = l$$

$$= \sqrt{\frac{2gl}{3}} \text{ Ans.}$$

Problem 56 In the previous problem, let's replace the heaped chain by a smooth coil of a rope. If we slightly pull the free end of the rope through the hole made on the table, the overhanging rope moves down by the effect of gravity. Find the (a) velocity, (b) acceleration of the chain, (c) net force experienced by the heap due to the table as the function of the length y of the fallen part of the chain.

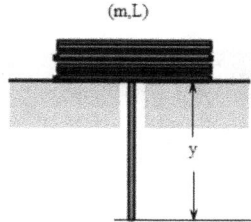

Solution

(a) As the overhang of the rope continuously pulls the immediate portion of the rope, the smooth transition of the elements of the coiled rope from rest to motion does not allow any impact. So, there will be no energy loss and we can conserve the mechanical energy of the rope. Conserving energy,

$$\Delta U + \Delta K = 0$$

$$-m'gy_C + \frac{1}{2}m'v^2 = 0$$

$$\Rightarrow -m'g\frac{y}{2} + \frac{m'v^2}{2} = 0$$

$$\Rightarrow u = \sqrt{gy}$$

(b) Then, the acceleration of the moving portion of the rope is

$$a = \frac{v\,dv}{dy} = \frac{1}{2}\frac{d}{dy}(v^2)$$

$$\frac{1}{2}\frac{d}{dy}(gy) = \frac{g}{2}$$

(c) The net force acting on the overhanging rope is

$$F_{net} = m'a$$

$$m'g - F_y = m'a$$

$$\Rightarrow F_y = m'(g - a) = \left(\frac{m}{l} y\right)\left(g - \frac{g}{2}\right)$$

$$\Rightarrow F_y = \frac{mgy}{2l} = \frac{\mu gy}{2}$$

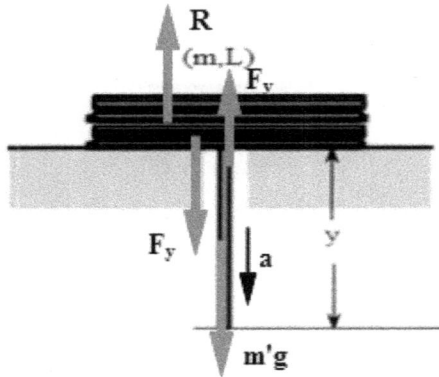

Problem 57 A uniform chain of mass m and length l hangs from a table top by a smooth tube such that the overhang touches the floor. If we release the chain from rest, it slides due to the effect of gravity. Find the (a) velocity of the chain when it leaves the table top, (b) velocity with which the free end of the chain touches the heap, (c) reaction offered by the ground on the heap when the last link is about to leave the point O.

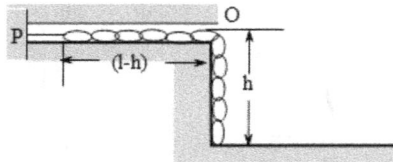

(a) The driving force of the chain is

$$F = mg = (\mu h)g = \mu gh$$

The acceleration of the chain is

$$a = -\frac{v\,dv}{dx} = \frac{\text{force}}{\text{mass}}$$

$$= \frac{\mu gh}{\mu(h + x)} = \frac{gh}{h + x}$$

$$-v\,dv = gh\frac{dx}{h+x}$$

$$-\int_0^v v\,dv = gh\int_{l-h}^0 \frac{dx}{(h+x)}$$

$$\Rightarrow -\frac{v^2}{2} = gh\,\ln(h+x)|_{l-h}^0$$

$$\Rightarrow -\frac{v^2}{2} = gh\{\ln h - \ln l\} = -gh\,\ln\frac{l}{h}$$

$$\Rightarrow v = \sqrt{2gh\,\ln\frac{l}{h}} = v_0,\ \text{say}$$

(b) Since the impact force given by the heap cannot act on the falling part as it acts on the lowest link of the falling part, which itself gets delinked just after the collision with the heap, then the falling chain falls freely. So, the velocity with which the force end of the chain touches the heap is given as

$$v^2 = v_0^2 + 2gh$$

$$= 2gh\,\ln\frac{l}{h} + 2gh$$

$$v = \sqrt{2gh\left(1 + \ln\frac{l}{h}\right)}\ \text{Ans.}$$

(c) The reaction offered by the ground on the heap is
$$R = \text{weight of the heap} + \text{Reaction force of the falling chain}$$

$$= \mu gh + \frac{dm}{dt}$$

$$= \mu gh + \mu v^2$$

Putting the obtained value of v, we have

$$R = \mu gh + \mu\left\{2gh\,\ln\frac{l}{h}\right\}$$

$$= \mu gh\left[1 + 2\ln\frac{l}{h}\right]\ \text{Ans.}$$

Problem 58 One end A of the uniform string is rigidly fixed with the ceiling. (a) If the free end C of the string is pulled up with a uniform velocity v, find the (a) force required to pull it at C as the function of y; (b) if the free end C of the string is pulled up with a constant force F, find the velocity of C as the function of y.

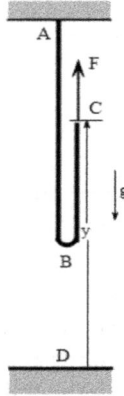

Solution

(a) Applying the work–energy theorem,

$$F = \frac{dE}{dy}$$

$$\Rightarrow F = \frac{d}{dy}(U + K) \tag{1.193}$$

The gravitational potential energy is

$$U = m_1 g y_{C_1} + m_2 g y_{C_2}$$

$$= \mu\left(l - \frac{y}{2}\right)g\left\{\frac{y}{2} + \left(\frac{l}{2} - \frac{y}{4}\right)\right\} + \left(\mu\frac{y}{2}\right)g\left(\frac{3y}{4}\right)$$

$$= \frac{\mu g}{2}\left[l^2 - \left(\frac{y}{2}\right)^2 + \frac{3y^2}{4}\right] = \frac{\mu g}{2}\left(l^2 + \frac{y^2}{2}\right)$$

The initial gravitational potential energy is

$$U_0 = mgy_0 = (\mu l)g\frac{l}{2} = \frac{\mu g l^2}{2}$$

Then,

$$\Delta U = U - U_0 = \frac{\mu g y^2}{4} \tag{1.194}$$

The change in the KE is

$$\Delta K = \frac{1}{2}m_2 v^2 = \frac{1}{2}\left(\mu\frac{y}{2}\right)v^2 \tag{1.195}$$

The work done by the force F is

$$W_F = F_y \tag{1.196}$$

$$W_F = \Delta U + \Delta K$$

$$= \frac{\mu g y^2}{4} + \frac{\mu y v^2}{4}$$

$$\Rightarrow W_F = \frac{\mu}{4}(gy^2 + v^2 y)$$

$$\Rightarrow F = \frac{dW_F}{dy} = \frac{\mu}{4}(2gy + v^2)\ \text{Ans.}$$

(b) If the force is kept constant

$$W_F = Fy$$

Then, $Fy = \Delta E = \Delta U + \Delta K$

$$= \frac{\mu}{4}(gy^2 + v^2 y)$$

$$\Rightarrow F = \frac{\mu}{4}(gy + v^2)$$

$$\Rightarrow v = \sqrt{\frac{4F}{\mu} - gy}$$

$$\Rightarrow v = \sqrt{\frac{4Fl}{m} - gy}$$

N.B.: For $v \geqslant 0$, $F \geqslant \dfrac{mg}{4}$

Problem 59 A uniform chain of mass m is heaped on a horizontal surface. The free end of the chain on the heap is pulled up with a constant velocity v. Find the (a) upward force acting on the chain (b) reaction force offered by the ground on the heap as the function of the length y of the chain above the heap.

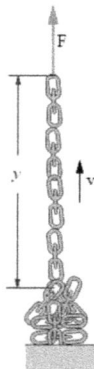

Solution
(a) In pulling the chain up, each link of the heap of the chain suddenly gets an upward momentum, it collides with the moving part of the chain causing an impact force F'.

The forces acting on the moving chain are gravity ($m_1 g$), upward applied force (F), and downward impact (reaction) force F'. Then the net force acting on the moving chain is

$$F - m_1 g - F' = m_1 a,$$

where $a = \frac{dv}{dt} = 0$ ($\because v = C$)

$$\Rightarrow F = m_1 g + F' \tag{1.197}$$

The reaction force F' is given as

$$F' = v \frac{dm}{dt} = v \frac{d}{dt}(\mu y) = \mu v \frac{dy}{dt}$$

$$\Rightarrow F' = \mu v^2 \tag{1.198}$$

Using equations (1.197) and (1.198)

$$F = m_1 g + \mu v^2$$

$$= (\mu y)g + \mu v^2$$

$$\Rightarrow F = \mu(v^2 + gy) \text{ Ans.}$$

(b) The net force acting on the heap is

$$F' + R - m_2 g = m_2 a,$$

Since the heap is at rest $a = 0$,

$$\Rightarrow R = m_2 g - F' \tag{1.199}$$

Using equations (1.198) and (1.199),

$$R = m_2 g - \mu v^2$$

$$= \{\mu(l - y)\}g - \mu v^2$$

$$= \mu\{gl - (v^2 + gy)\}$$

$$= mg - \mu(v^2 + gy) \text{ Ans.}$$

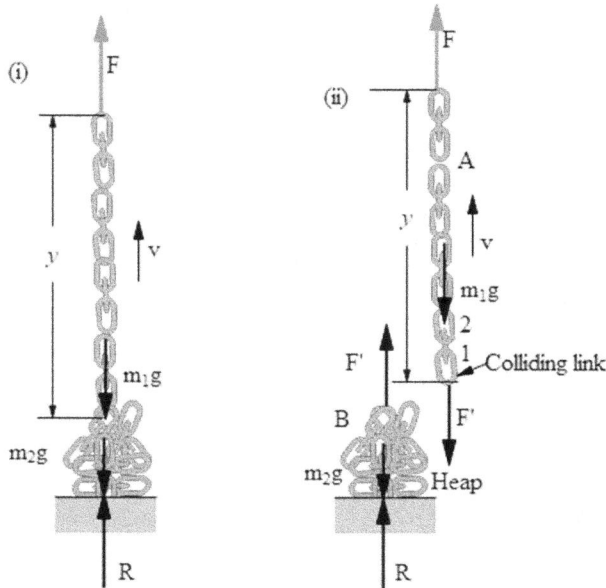

When the hanging part is moved up by the force F, the link 1 collides with the heap and imparts an upward force F' to the heap and receives equal and opposite downward force F' from the heap. In this case, as the links above the heap are moved up, they do not lose contact with each other.

Alternative solution
Applying Newton's second law on the total chain the net force is

$$F + R - mg = 0$$

$$\Rightarrow R = mg - F,$$

$$\text{where } F = \mu(v^2 + gy)$$

$$\Rightarrow R = mg - \mu(v^2 + gy) \text{ Ans.}$$

Problem 60 A uniform chain of mass m and length L hangs from a rigid support by a string S so that it just touches the horizontal floor. (i) The string is cut and then hold the top of the chain and lower it with a constant velocity v. The chain falls and forms a heap. Find the (a) force applied on the string and (b) force acting on the heap by the floor. (ii) If the string is cut and the chain is allowed to fall under gravity, find the reaction force exerted on the heap by the floor as the function of the length y of the falling part of the chain.

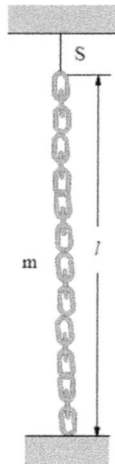

Solution

(i)

(a) While the chain moves down each lowest link is detached from this moving part with 'zero relative velocity'. So, the impact force acting on the moving part is zero. However, there will be an impact force F' acting downward on the heap as the detached link collides with the stationary heap with a non-zero relative velocity v. If dm = mass of the chain colliding with the heap during time dt with a velocity v, the downward impact force is

$$F' = \frac{dP}{dt} = v\,\frac{dm}{dt} = v\,(\mu v)$$

$$\Rightarrow F' = \mu v^2 \tag{1.200}$$

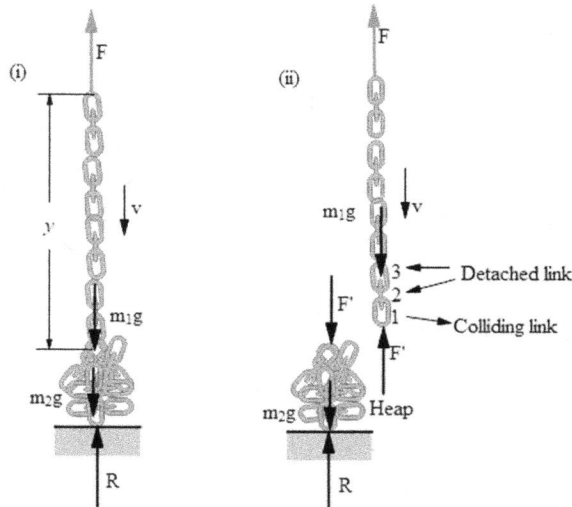

Link 1 collides with the heap and imparts a downward force F' on the heap and receives equal upward force F' from the heap. Just after the link 1 collides with the heap, the next immediate link 2 loses contact with link 1; so, the link 2 does not experience any impact force at this instant. This means that the falling link does not experience the impact force.

N.B.: The action-reaction pair each of magnitude F' acts on the heap and the colliding link F''; but the colliding link does not transfer any force to its neighbouring link of the falling part of the chain. Therefore, the forces acting on the falling part of the chain are gravity ($m_1 g$) and applied force F. So, the net force acting on m_1 is

$$F - m_1 g = 0$$

$$\Rightarrow F = m_1 g$$

$$\Rightarrow F = (\mu y)g = \mu g y \text{ Ans.}$$

(b) On the other hand, the forces acting on the heap are gravity ($m_2 g$), reaction (impact) force F', and the normal reaction force R offered by ground. So, the net force acting on m_2 (heap) is

$$F_{net} = R - F' - m_2 g = 0 (a = 0)$$

$$\Rightarrow R = F' + m_2 g$$

$$= \mu v^2 + \{\mu(l - y)\}g$$

$$= \mu[v^2 + g(l - y)]$$

$$= \frac{m}{l}[v^2 + g(l - y)]$$

$$= m\left[\frac{v^2}{l} + g\left(1 - \frac{y}{l}\right)\right]$$

(ii) If $F = 0$, the position of the chain above the heap will experience the gravity only. So, it falls freely with an acceleration g. Then, we have

$$v = \sqrt{2gl}$$

Putting this value of v in the last expression, we have

$$R = m\left[\frac{v^2}{l} + g\left(1 - \frac{y}{l}\right)\right]$$

$$= m\left[\frac{2gl}{l} + g\left(1 - \frac{y}{l}\right)\right]$$

$$= m\left[3g - g\frac{y}{l}\right]$$

$$= mg\left[3 - \frac{y}{l}\right] \text{ Ans.}$$

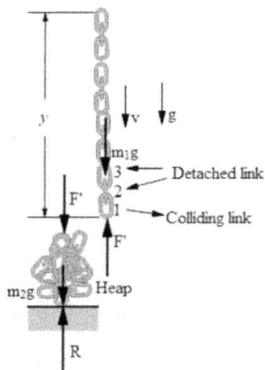

Impulse-momentum equation

Problem 61 A cart of mass M is pulled by a constant external force F on a smooth horizontal surface. Let the velocity of the cart be v_o just before it is being loaded by the sand falling from a stationary hopper. If the sand falls at a rate of μ kg s^{-1} and sticks to the cart, find the (a) velocity (b) (i) velocity (ii) acceleration of the cart as the function of time assuming $F = 0$ (c) (i) velocity (ii) acceleration when $v_o = 0$ (d) (i) force (ii) power (iii) rate of change of KE (iv) rate of loss of energy.

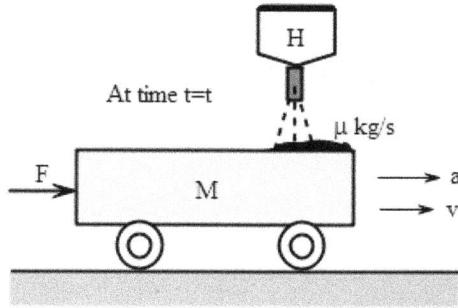

At time t=t

H

μ kg/s

F

M

a

v

Solution

(a) The change in horizontal momentum of the cart is

$$\Delta \vec{P} = \{(M + \mu t)v - Mv_0\}\hat{i} \qquad (1.201)$$

The impulse of the force is

$$\vec{I} = \vec{F}t = Ft\hat{i} \qquad (1.202)$$

Applying impulse-momentum equation,

$$\vec{I} = \Delta \vec{P} \qquad (1.203)$$

Using the equations (1.201), (1.202), and (1.203),

$$Ft\hat{i} = \{(M + \mu t)v - Mv_0\}\hat{i} \qquad (1.204)$$

$$\Rightarrow v = \frac{Ft + Mv_0}{M + \mu t} \quad \text{Ans.}$$

(b)

(i) If $F = 0$, $v = \frac{Mv_0}{M + \mu t}$ Ans.

(ii) $a = \frac{dV}{dt} = Mv_0 \frac{d}{dt}(M + \mu t)^{-1}$

$$\Rightarrow a = -\frac{\mu M v_0}{(M + \mu t)^2} \quad \text{Ans.}$$

(c)

    (i) If $v_0 = 0$, we have

$$v = \frac{Ft}{M + \mu t} \quad \text{Ans.}$$

    (ii) $a = \frac{dv}{dt} = F\frac{d}{dt}(\frac{t}{M+\mu t})$

$$= F\left\{\frac{(M + \mu t) - t(\mu)}{(M + mt)^2}\right\}$$

$$= \frac{MF}{(M + \mu t)^2} \quad \text{Ans.}$$

(d)

    (i) If $v_0 = v = u$, using equation (1.204)

$$Ft = (M + \mu t)u - Mu$$

$$\Rightarrow F = \mu v \quad \text{Ans.}$$

    (ii) The power delivered by the external agent is

$$P_{ext} = FV = \mu v.\, v = \mu v^2 \quad \text{Ans.}$$

    (iii) The rate of change in KE

$$\frac{dK}{dt} = \frac{d}{dt}\left(\frac{1}{2}mv^2\right)$$

$$= \frac{v^2}{2}(\frac{dm}{dt}) = \mu\frac{v^2}{2} \quad \text{Ans.}$$

    (iv) The rate of loss of energy

$$P_{LOSS} = P_{ext} - \frac{dK}{dt}$$

$$= \mu v^2 - \frac{\mu v^2}{2} = \frac{\mu v^2}{2} \quad \text{Ans.}$$

Problem 62 A trolley car is loaded with sand has mass M and is pulled by a constant force F. At $t = 0$, let the velocity of the car be v_o and sand starts leaking through a hole made at the bottom of the trolley car at a constant rate of μ kg s^{-1}. Find the (a) velocity and (b) acceleration of the trolley car as the function of time t.

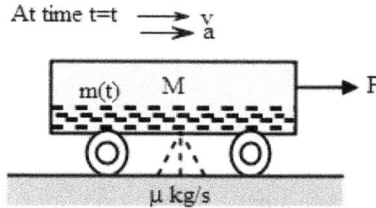

At time t=t $\longrightarrow$ v
$\Longrightarrow$ a

m(t)　　M　　　　　$\longrightarrow$ F

μ kg/s

Solution

(a) It is given that during time t, the water escapes at a constant rate of $\mu(=\frac{dm}{dt})$. So, the momentum of the trolley car is

$$P = (M - \mu t)v$$

Since initial momentum is zero ($v_0 = 0$), the change in momentum is

$$\Delta P = (M - \mu t)v$$

This is numerically equal to the impulse of the applied force F.

$$\Rightarrow Ft = \Delta P = (M - \mu t)v$$

$$\Rightarrow v = \frac{Ft}{M - \mu t} \quad \text{Ans.}$$

(b) The acceleration of the trolley car is

$$a = \frac{dv}{dt} = \frac{d}{dt}\left(\frac{Ft}{M - \mu t}\right)$$

$$= \frac{F\{(M - \mu t) - t(-\mu)\}}{(M - \mu t)^2}$$

$$\Rightarrow a = \frac{FM}{(M - \mu t)^2} \quad \text{Ans.}$$

Problem 63 The cylinder of length L and radius R is pushed by a constant force F at the time of entering into a vast region of stationary cosmic dust of uniform concentration of n particles per cubic meter. If the mass of each dust particle is m

and the dust particles stick to the cylinder, find the distance covered by the cylinder as a function of time.

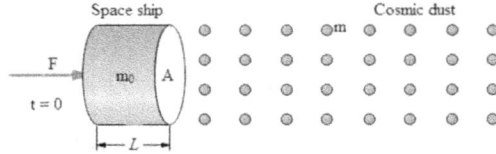

Solution

If n = number of dust particles per unit volume, in a volume $V = Ax$, there are $nV = nAx$ particles each of mass m. So, during time t, the extra mass deposited at the face of the cylinder is

$$m' = mnAx \tag{1.205}$$

The change in momentum of the system (cylinder + dust cloud) is

$$\Delta \vec{P} = (m_1 v - m_0 v_0)\hat{i}, \tag{1.206}$$

where m_0 = initial mass of the cylinder, v_0 = initial velocity of the cylinder, and $m_1 = m_0 + m'$. Applying the impulse-momentum equation,

$$\vec{F} t = \Delta \vec{P} \tag{1.207}$$

we have

$$Ft\hat{i} = (mv - m_0 v_0)\hat{i}$$

$$\Rightarrow v = \frac{Ft + m_0 v_0}{m} \tag{1.208}$$

Using equations (1.205) and (1.208),

$$v = \frac{Ft + m_0 v_0}{m_0 + m'}$$

$$v = \frac{Ft + m_0 v_0}{m_0 + Amnx} \tag{1.209}$$

Putting $v = \frac{dx}{dt}$, we have

$$\frac{dx}{dt} = \frac{Ft + m_0 v_0}{m_0 + Amnx}$$

$$\Rightarrow (m_0 + Amnx)dx = (Ft + m_0 v_0)dt$$

Integrating both sides, we have

$$\int_0^x (m_0 + Amnx)dx = \int_0^t (Ft + m_0 v_0)dt$$

$$\Rightarrow m_0 x + Amn\frac{x^2}{2} = \frac{Ft^2}{2} + m_0 v_0 t$$

$$\Rightarrow Amnx^2 + 2m_0 x - (Ft^2 + 2m_0 v_0 t) = 0$$

The solution of this quadratic equation is given as

$$x = \frac{-2m_0 \pm \sqrt{4m_0^2 + 4Amn(Ft^2 + 2m_0 v_0 t)}}{2Amn}$$

$$x = \frac{\sqrt{m_0^2 + Amn(Ft^2 + 2m_0 v_0 t)} - m_0}{Amn} \quad \text{Ans.}$$

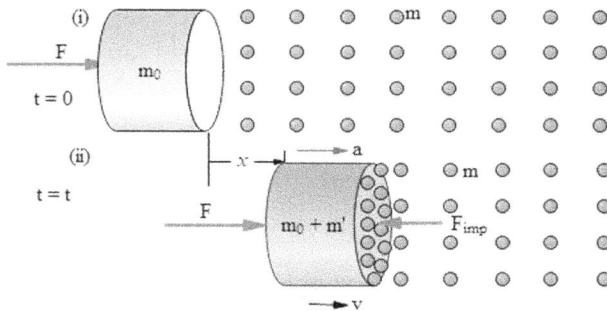

Problem 64 A rocket of initial mass m_o ejects the fuel at a relative velocity of u_{rel} near the earth's surface. The rocket starts from rest from ground level at time $t = 0$.
(a) Derive an expression for the velocity of the rocket after a time t from firing.
(b) With reference to (a), in gravity-free space find the velocity of the rocket after the fuel gets exhausted (put mass of the fuel $= \eta m_0$). (c) Find the total height attained by the rocket. Put $\frac{m_0}{m} = 2$ and $t = \frac{T}{2}$.

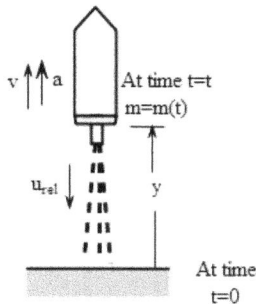

Solution

(a) The upward reaction force given by the exhausted fuel is given as

$$F = -u_r \frac{dm}{dt} \qquad (1.210)$$

The net force acting on the rocket is

$$F - mg = ma \qquad (1.211)$$

Using equations (1.210) and (1.211),

$$-u_r \frac{dm}{dt} - mg = ma = m\frac{dv}{dt}$$

$$\Rightarrow -u_r \frac{dm}{m} - gdt = dU$$

Integrating both slides, we have

$$-u_r \int_{m_0}^{m} \frac{dm}{m} - g \int_0^t dt = \int_0^v dv$$

$$\Rightarrow -u_r(\ln m - \ln m_0) - gt = v$$

$$\Rightarrow u = u_r \ln \frac{m_0}{m} - gt \text{ Ans.} \qquad (1.212)$$

(b) In the gravity-free space putting $g = 0$, we have

$$v = u_r \ln \frac{m_0}{m}$$

When the entire fuel is exhausted, putting
$m = m_0 - \eta m_0 = m_0(1 - \eta)$, we have

$$v = u_r \ln \frac{m_0}{m_0(1 - \eta)}$$

$$\Rightarrow v = -u_r \ln(1 - \eta) \text{ Ans.}$$

(c) Putting $u_{rel} = gt$, $\frac{m_0}{m} = 2$ and $t = \frac{T}{2}$ in the equation (1.212), we have

$$v_{max} = gT \ln 2 - \frac{gT}{2}$$

$$\Rightarrow v_{max} = g\frac{T}{2}(2 \ln 2 - 1) \qquad (1.213)$$

Just after all the fuel gets exhausted at height h_1, the rocket attains the maximum speed $v_{\max}$. Then, the maximum height attained by the rocket from that position is

$$h_2 = \frac{v_{\max}^2}{2g} = \frac{1}{2g}\left(\frac{gT}{2}\right)^2 (2\ln 2 - 1)^2$$

$$\Rightarrow h_2 = \frac{gT^2}{8}(2\ln 2 - 1)^2 \tag{1.214}$$

The height h_1 until all fuel gets exhausted is given as

$$h_1 = \int v\, dt$$

$$= v_{rel}\int_0^{\frac{T}{2}} \ln\left(\frac{m_0}{m_0 - \lambda t}\right)dt - \frac{gt^2}{2}\Big|_0^{\frac{T}{2}}$$

$$= -v_{rel}\int_0^{\frac{T}{2}} \ln\left(1 - \frac{\lambda t}{m_0}\right) - \frac{gT^2}{8} \tag{1.215}$$

Using the formula

$$\int \ln(a - bx) = -\frac{1}{b}[(a - bx)\ln(a - bx) + bx],$$

we can evaluate the integral of equation (1.215), and simplifying the factors, we have

$$h_1 = \frac{gT^2}{8}(3 - 4\ln 2) \tag{1.216}$$

Using (1.210) and (1.216), the total height attained by the rocket is

$$h = h_1 + h_2$$

$$= \frac{gT^{-2}}{8}(2\ln 2 - 1)^2 + \frac{gT^2}{8}(3 - 4\ln 2)$$

$$= \frac{gT^2}{8}(1 - \ln 2)^2 \text{ Ans.}$$

IOP Publishing

Problems and Solutions in Many-Particle Systems

Pradeep Kumar Sharma

Chapter 2

Collisions

2.1 Introduction

The collision is the most common phenomenon in which two particles may or may not touch physically. The colliding particles interact for a short time exchanging their momenta and energy. While collision between bat and ball is delightful, the collision of a vehicle is a horrible event. The pressure due to atmosphere or a fluid is the result of collision between the surrounding fluid molecules and the given surface. The chemical reactions are the result of the collisions between the atoms and ions of the reactants. The nuclear reactions occur due to the collision of bombarding particles like neutrons, α-particles, etc, with the nuclei. This Universe is said to be originated from a huge collision called the 'Big Bang' and sustained by continuous creation and annihilation of matter in the process of collision. Hence, the collision exists everywhere.

The collisions can be central, oblique, head-on, etc. The theory of collision is explained by the principle of impulse-momentum and conservation of momentum and energy.

2.2 Definition

The word collision comes from a Latin word which means 'to injure by striking. It is defined as a prompt interaction between two or more bodies that produces a considerable change in the momenta of the interacting bodies.

If you throw a ball vertically up with a speed $v = 10$ m s^{-1}, after a time

$$t = \frac{v}{g} = 1 \text{ s}$$

the ball will come to rest instantaneously at its highest position (assuming g $= 10$ m s^{-2} and ignoring the air resistance). This means that after a time of 1 s, the ball gives up all its momentum mv ($= 10$ kg m s^{-1}) to earth in the course of its interaction with earth. As the interaction time is not too small to change the momentum of the stone,

doi:10.1088/978-0-7503-6447-8ch2

it is not termed as a collision. The average downward force acting on the ball is $F = \frac{mv}{t} = 10$ N, which is numerically equal to the gravitational force (or weight) of the ball is called a non-impulsive force because does not arise from a collision.

If a ball of mass 1 kg collides with the ground with a velocity of 1 m s^{-1}, it may stop losing its total momentum of 10 kg m s^{-1} rapidly. The ball may also come to rest momentarily and then start bouncing up. Since the ball gives off its entire momentum very rapidly within a short time span (0.01 s, say) it will experience an average upward force of one hundred times the weight of the ball! Then, this upward force is impulsive force, which is different from the non-impulsive gravitational force of earth.

In general, impulsive forces are large and appear for a short time, whereas the non-impulsive forces can stay for ever.

Example 1 Steller collision: If gravitational force is not an impulsive force, then how can we explain the collision between two stars by their gravitational effect?

Answer: The collision between two stars, say, is called a stellar collision and lasts for a brief time of several micro- to milliseconds. For instance, a supernova event takes places in around 200 milliseconds but the after effect may take as long as several years to die down. However, this time span is infinitely small compared to the gigantic time span of millions of years of motion of the stars. Steller collision takes place in a high density of stars. In this Universe we have 400 billion stars. This collision takes place in an average time span of 10,000 years.

On an astronomical scale, the of time of motion of stars could be millions of years. Thus, a huge gravitational interaction between two approaching stars for several days or years is a non-impulsive force. However, the huge gravitational interaction between the astronomical bodies during a short time of interaction during the gravitational collapse of the cosmic bodies is defined 'collision'.

The above example clarifies the fact that the time of interaction should be very small compared to the time of motion (observation) of the interacting bodies. During this time of collision, the bodies must experience the forces much greater than the forces acting on them in their general (or normal) state of motion.

In the previous discussion we used an example of a collision between a ball and the ground. In this collision, physical contact takes place between the colliding bodies. This means the neutral bodies, in general, touch each other during their collision.

Example 2 Nuclear collision versus stellar or galactic collision: Is it always essential that the bodies touch each other during a collision? Justify your answer.

Answer. In Rutherford's famous Gold Foil Experiment the α-particles do not touch the nucleus because both α-particles and nucleus are positively charged. Since electric field is zero outside the gold atom, the α-particles experience zero force outside the gold atom. However, the α-particles experience tremendous electrostatic field (force) due to the interaction of nucleus inside the atom. As the α-particles move very fast, they remain inside the atom for a very short time. During the short time of journey inside the atom, the α-particles change their momenta very rapidly. It has been observed that some α-particles get deflected by large angles, and a few of them retrace their path by bouncing back after stopping at the closest distance from the nucleus. Hence, we state that α-particles collide with the gold atoms (nuclei) experiencing a large force during a short time, without touching the nuclei.

The above example clarifies the fact that in a collision the colliding bodies need not touch physically, and during the collision the bodies experience a very large force compared to the force acting on them in their general motion. The time of collision is negligible compared to the time of observation of the bodies.

2.3 Impulsive force and impulse

Let us assume that a colliding body changes its momentum by $\Delta\vec{P}$ during the time Δt of collision. Then the average change in momentum of the body is equal to the average force given as

$$\vec{F_{av}} = \frac{\Delta\vec{P}}{\Delta t},$$

This is the force acting on the body averaged over the time of impact.
Impulsive force: According to Newton's 2nd law, we have

$$\vec{F_{av}} = \frac{\Delta\vec{P}}{\Delta t}.$$

Since the collision time Δt is very small and the change in momentum $\Delta\vec{P}$ is finite, the body must experience a 'huge average force' during the collision. That is what we call 'average impact (or impulsive) force'.

Example 3 Collision force is an average force but *not* an instantaneous force: During a collision why is it convenient to use average force rather than the instantaneous force?
Solution
When two bodies collide, millions of molecules of the colliding bodies are pressed against each other. Hence, the impulsive force acting on the colliding bodies is a sum of millions of impulsive molecular interactions. The variation of molecular force between two isolated molecules follows the law given as

$$F = \frac{a}{r^6} - \frac{b}{r^{12}}$$

However, the net effect of millions of molecular forces is extremely complex during a collision. So, the exact variation of impulsive force with time is incredibly complex. This limits us use the differential expression of instantaneous force in Newton's 2nd law, given as

$$\vec{F} = \frac{d\vec{P}}{dt}$$

Now a question arises: when we do not know the exact variation of $\vec{F}$ as the function of time during collision, how can we find the final velocities (velocities of the bodies just after the collision)?

The straight answer to this basic question is to use an integral form of Newton's 2nd law called 'the impulse-momentum equation'.

Impulse-momentum equation: Let a body A collide with another body B in gravity. Ignoring the viscosity of air, apart from the non-impulsive force $\vec{F}_{ext}(=m\vec{g})$, the impulsive force N acts on A for a very short time $\Delta t(=t_2 - t_1)$ of collision. The impulsive force F varies abruptly without following any definite rule. So, it is impractical to use the instantaneous force to find any change in velocity or momentum of the body during the collision. Then, we have to introduce the concept of impulse of F (instead of instantaneous impulsive force F), which is given as

$$I = \int_{t_1}^{t_2} \vec{F}\, dt$$

which is equal to the change in momentum $\Delta\vec{P}$ of the body during the collision. Using the impulse-momentum equation, we have

$$\vec{j} = \int \vec{F}\, dt\left(=\vec{F}_{av}\Delta t\right) = \Delta\vec{P}.$$

Then, the average force is

$$\vec{F}_{av}\left(=\frac{\Delta\vec{P}}{\Delta t}\right)$$

This average force quantifies the 'strength of collision'. Since the time of impact is very small, the average impact (or impulsive) force is very large.

You can notice that in the F–t graph, a non-impulsive gravitational force continues to act during the collision. In general, the impulse of $\vec{F}_{gr}$ is

$$\text{Imp} = I = \int_{t_2}^{t_1} \vec{F}_{gr}\, dt\left(=F_{gr}\Delta t\right),$$

This is very small compared to $\int_{t_2}^{t_1} F\, dt$, so we can neglect the impulse of gravity.

Recapitulating, the net impulse on a colliding body is approximately equal to the impulse of the impulsive forces only during the collision, which is equal to the change in momentum of the colliding bodies during the collision. The impulsive force averaged over the time of impact is given as

$$\vec{F}_{av} = \frac{\Delta\vec{P}}{\Delta t}$$

Since $|\Delta\vec{P}|$ is finite and Δt is very small, generally $\vec{F}_{av}$ is a large force. The time period of collision depends on shape, size and velocities of the colliding bodies and their elastic properties. As the exact variation of impulsive forces with time is extremely uncommon, the idea of average impulsive force is generally more useful.

2.4 Types of collision

Line of impact: Let the bodies collide at a common point P. Then, we can draw a tangent at P and then draw a normal passing through this point. For the sake of simplicity let us assume that the colliding bodies 1 and 2 are smooth. So, no impulsive forces will act along the tangent. The impulsive force would be the normal reactions $\vec{N_1}$ and $\vec{N_2}$ only. In other words, we can say that collision or impact takes place along the normal. Hence, the normal line is known as *line of impact* or *line of action* of the smooth colliding bodies. As an action-reaction pair, $\vec{N_1}$ and $\vec{N_2}$ are equal opposite and collinear. Then, we can write

$$\vec{N_1} + \vec{N_2} = 0$$

This means that the net impulse of these forces

$$\vec{N_1} \text{ and } \vec{N_2} \text{ is zero.}$$

Tangential (frictional) impact force: If the bodies are rough, an action-reaction pair of tangential impulsive forces or impulsive frictions may act on the bodies along their common tangent.

Line of motion: The line in which the center of masses C_1 and C_2 of the bodies move are called the line of motion of the bodies 1 and 2, respectively.

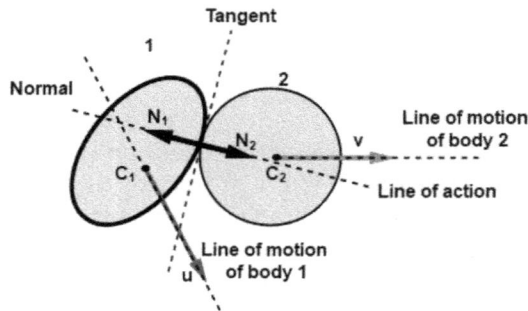

There are various types of collision depending on the position of the center of mass of the bodies relative to the line of impact given as follows.

Central impact: If the line of impact passes through the line joining the center of mass C_1 and C_2 of the bodies, it is called the 'central impact'.

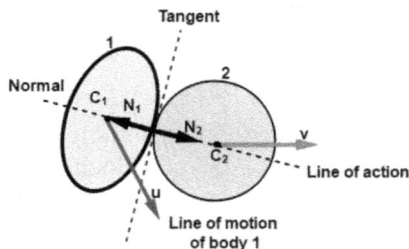

Oblique impact: If the line of the motion and line of impact does not coincide, it is known as an *oblique impact*.

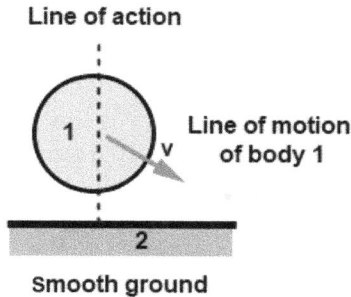

Line of action

Line of motion of body 1

Smooth ground

Eccentric impact: When the line joining the center of mass of the colliding bodies does not coincide with the line of impact, it is known as an eccentric impact. The impact of cricket ball and bat is the familiar example of an eccentric impact.

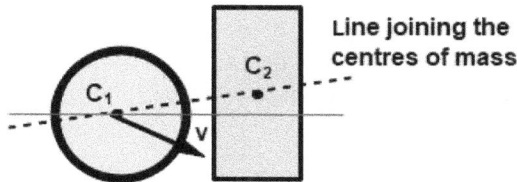

Line joining the centres of mass

Head-on impact (collision): If the line of motion and line of impact of the colliding bodies coincide, it is called a direct-central or head-on collision. The collision between two identical balls moving in the same line is a common example of a head-on collision.

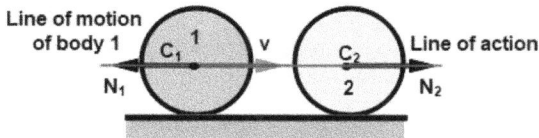

Line of motion of body 1

Line of action

Example 4 Describe the following impacts: (i) Head-on or direct-central, (ii) oblique-central, (iii) direct-eccentric, and (iv) oblique-eccentric.
Solution

(i) Line of impact, line of motion, and line joining the centers of mass are collinear. Hence, the collision is head-on (direct-central).

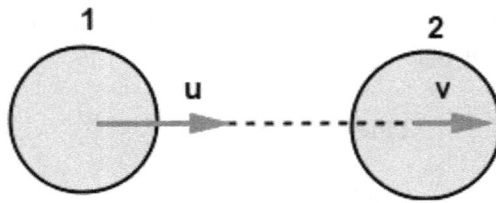

(ii) The line of motion is different from the line of impact and line joining the center of mass of the bodies. However, the center of mass of the bodies such as C_1 and C_2 lie on the line of impact. So this impact is said to be oblique-central.

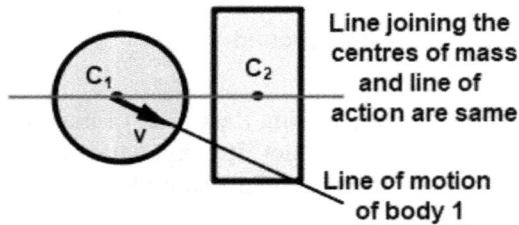

(iii) The line joining the center of mass C_1 and C_2 is different from the line of impact and line of motion. The lines of motion and line of impact are parallel. Hence, this collision is said to be direct-eccentric.

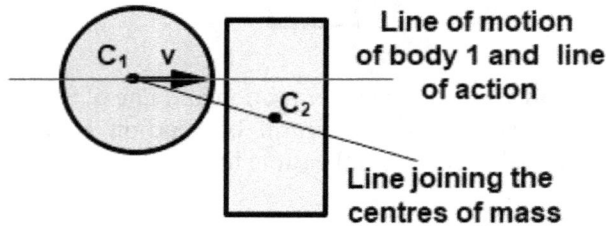

(iv) Since the line joining the center of mass of the bodies does not coincide with the line of impact, it is said to be an eccentric impact. Since the line of motion of the body does not coincide with (or is parallel to) the line of impact, you can call it oblique impact. A combination of these two is called oblique-eccentric impact.

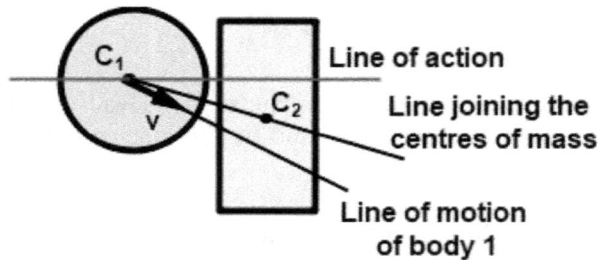

2-7

Recapitulating:

1. Impact forces between two colliding bodies are equal and opposite even through surfaces of the colliding bodies are tough. In addition to the normal impulsive forces, tangential impact forces also come into play.
2. The line of impact and line of motion must be parallel in direct impact.
3. The line of motion and line of impact are different in the oblique impact.
4. The line of impact and line joining the center of mass of the colliding bodies are different in the eccentric impact.

The above classification is done based on the geometry of the colliding bodies. Let us now classify the collisions based on the nature of the colliding bodies.

Scattering: When the composition and mass of the colliding particles (or bodies) remain unchanged during collision so that, before and after the collision, each colliding body remains identical, this type of collision is called scattering. For instance, collision between ideal gas molecules is an example of scattering.

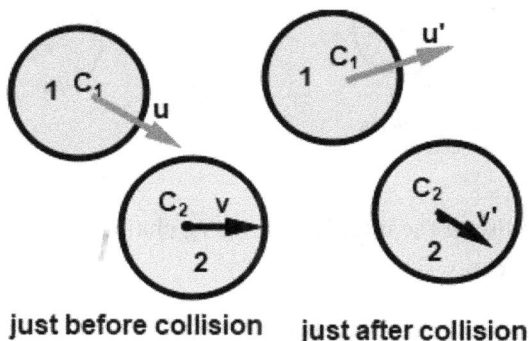

just before collision   just after collision

Reaction: Many times, a collision between two atoms/molecules A and B yields a couple of products (atoms/molecules) C and D, say. As the final particles are different the final particles (or bodies) of the colliding system are not identical with the initial particles, it is know as a 'reaction'. Thus, chemical and nuclear reactions are the consequences of collisions.

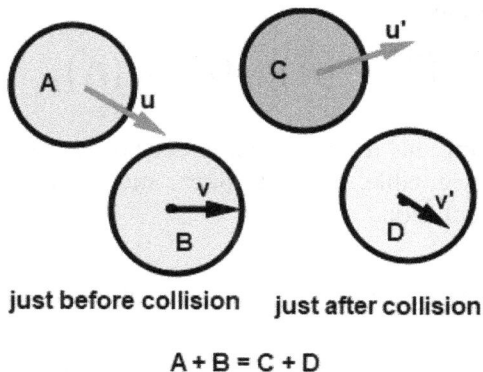

just before collision   just after collision

$$A + B = C + D$$

2.5 Conservation of linear momentum in a collision

Let us consider an isolated system of two colliding bodies A and B. Then, the net force acting on the system is zero. During the collision the impulse of impulsive force $\vec{F_A}$ changes the momentum of A from $(\vec{P_A})_i$ to $(\vec{P_A})_f$. Then, we have

$$\left(\vec{P_A}\right)_i + \int \vec{F_A} dt = \left(\vec{P_A}\right)_f \tag{2.1}$$

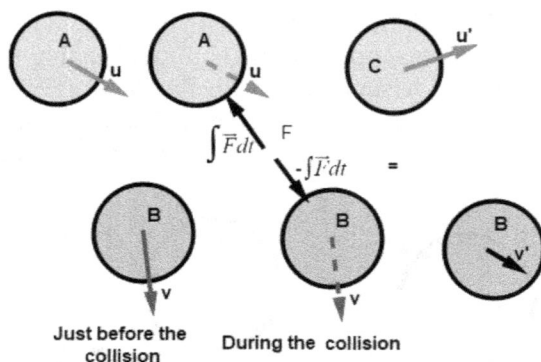

Just before the collision   **During the collision**

Likewise, the impulse of the impulsive force $\vec{F_B}$ changes the momentum of B from $(\vec{P_B})_i$ to $(\vec{P_B})_f$. So, we can write

$$\left(\vec{P_B}\right)_i + \int \vec{F_B} dt = \left(\vec{P_B}\right)_f \tag{2.2}$$

As $\vec{F_A}$ and $\vec{F_B}$ are action-reaction pairs,

$$\int \vec{F_A} dt + \int \vec{F_B} dt = 0 \tag{2.3}$$

Summing up equations (2.1) and (2.2), we have

$$\left(\vec{P_A}\right)_i + \left(\vec{P_B}\right)_i = \left(\vec{P_A}\right)_f + \left(\vec{P_B}\right)_f$$

The above expression tells us that when either no external impulse is present *or* external impulse are negligible, the total linear momentum of the system can be conserved during the collision.

Example 5 Discuss the validity of the principle of momentum of two iron balls colliding in mid-air.

Solution

For two iron balls in air as a system, gravitational forces (weights) of the balls are the external non-impulsive forces. But the impulse of gravity is negligible during the very short time of impact. Since the net impulse of impulsive forces acting on the balls as a whole is zero, it cannot change the momentum of the system during collision. Then, we can conserve the momentum of the system just before and after the impact. However, for a long time, the gravitational impulse will be considerably large and we cannot conserve the momentum of the system.

Example 6 A block of mass M is hanging from a light vertical spring of stiffness k. A bullet of mass m strikes the block with a velocity u. (a) Discuss the possibility of conservation of momentum. (b) What is the velocity of the bullet just after the collision if it remains inside the block after the collision?

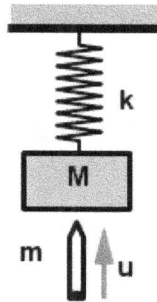

Solution

If we chose the system as bullet *plus* block, the external forces acting on the system are downward gravity $(M + m)g$ and upward spring force kx, where $x = $ elongation of the spring. As the block was hanging initially, the net force acting on the block just before the collision is zero. As the bullet is colliding, the net force acting on the system $(M + m)$ is equal to weight mg of the bullet. During the collision the impulsive reaction forces acting on the bullet and block are internal forces. So, these forces as a whole cannot change the linear momentum of the system. During the interaction between bullet and block the spring will be pushed up by a very small amount. So the impulse of this excess spring force over a very small time of interaction or collision is extremely small. Since the impulse of spring force and friction are very small, we can practically conserve the linear momentum of the system just before and after the collision.

N.B.:

1. Although the internal impulsive forces can change the momentum of the individual (colliding) bodies noticeably during the collision, the net internal impulsive force is zero and hence the momentum of the system remains practically conserved during the collision.

2. If an external non-impulsive force acts on a system, its impulse is very small during the impact. So, the momentum just before the impact is nearly equal to momentum just after impact.

2.6 Energy consideration in collisions

The need to apply the principle of energy conservation to find the velocities of the colliding bodies just after the collision.

Let us assume that two particles 1 and 2 carrying momenta $\vec{P_1}$ and $\vec{P_2}$ collide in a smooth horizontal plane. Just after the collision, let their momenta be $\vec{P}'_1$ and $\vec{P}'_2$, respectively. For an ideally or practically closed system, we can conserve the momentum of the system. Then, we have

$$\vec{P_1} + \vec{P_2} = \vec{P}'_1 + \vec{P}'_2$$

Now we have one equation and two unknown variables (i.e., $\vec{P}'_1$ and $\vec{P}'_2$). Then, we need another equation. Here we need the second equation which is the conservation of energy.

Example 7 Discuss the conservation of energy and momentum in the collision between a freely falling ball with the ground.
   Solution
   When we drop a ball from a certain height, it collides with the ground, bounces back and forth, and finally stops. The ball delivers a fraction of its momentum and energy to earth. If we take the combination of ball and earth as a closed system, the external force is zero because the gravitational force is a conservative and internal force for this sytem. Hence, we can conserve the total mass energy and momentum of the system. Furthermore, the ball loses some fraction of its mechanical energy during each collision in the form of heat, sound and light, etc. So, we cannot conserve the mechanical energy of the system. As the earth is extremely massive compared to the ball, the recoil velocity of earth is unnoticeable. So, the earth carries a negligible fraction of momentum of the ball. But the energy loss is considerable, which is divided between the ball and earth. If we take the ball as a system, neither its energy nor its momentum is conserved. Ans.
   If the colliding particles have tremendous speeds, their masses and momenta will be relativistic. For instance, in nuclear reactions, for the fast-moving particles, the relativistic effect comes into play, which says that mass increases with velocity.
   In 'all' kinds of collisions between the bodies, due to the effect of huge impulsive forces, the colliding bodies may change their internal structures, composition, and configuration of particles within them. So, it can change the 'internal energy' of the system, which depends on the configuration of the particles inside the colliding bodies and the kinetic energy of the particles (electrons, atoms, molecules, etc) of the colliding bodies relative to their centers of mass.
   When no external force acts on the system of colliding bodies, the total energy, that is, the kinetic energy of the center of mass of the bodies plus their internal

energy, remains conserved. In collisions, generally, the internal energy and the kinetic energy of the center of mass of the colliding bodies change.

Let us consider a system of two particles of masses m_1 and m_2 colliding in a region (shown as dotted circle) such that they interact without any physical contact. Just before the impact, let the masses of the colliding system be m_1, m_2, velocities $\vec{v_1}$, $\vec{v_2}$, and momenta $\vec{P_1}$, $\vec{P_2}$, respectively. Furthermore, let the masses of the systems just after the collision be m'_1 and m'_2, respectively, and their velocities and momenta be v'_1, v'_2 and $\vec{P}'_1$, $\vec{P}'_2$, respectively.

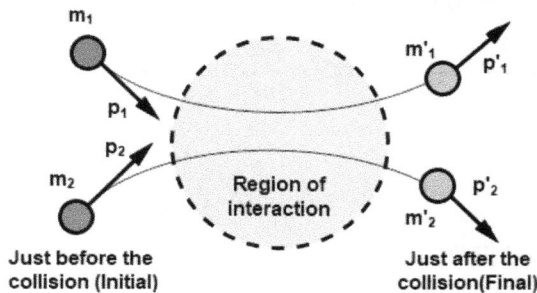

Just before the collision (Initial) · **Just after the collision(Final)**

The total energy of the systems just before the collision is

$$E_i = K_i + U_i \tag{2.4}$$

where K_i and U_i are the initial kinetic and potential energy of the colliding systems, respectively.

The initial KE of the system is

$$K_i = \frac{P_1^2}{2m_1} + \frac{P_2^2}{2m_2} \tag{2.5}$$

Using equations (2.4) and (2.5), we have

$$E_i = \frac{P_1^2}{2m_1} + \frac{P_2^2}{2m_2} + U_i \tag{2.6}$$

Similarly, the sum of total energy of the system just after the collision is

$$E_f = K_f + U_f \tag{2.7}$$

where K_f and U_f are the final kinetic and potential energy of the colliding system, respectively.

The final KE of the system (just after the collision) is

$$K_f = \frac{P_1'^2}{2m'_1} + \frac{P_2'^2}{2m'_2} \tag{2.8}$$

Using equations (2.7) and (2.8), we have

$$E_f = \frac{P_1'^2}{2m_1'} + \frac{P_2'^2}{2m_2'} + U_f \tag{2.9}$$

Using the principle of conservation of total energy just before and after the collision, we have

$$E_i = E_f \tag{2.10}$$

Using equations (2.6), (2.9), and (2.10), we have

$$\frac{P_1^2}{2m_1} + \frac{P_2^2}{2m_2} + U_i = \frac{P_1'^2}{2m_1'} + \frac{P_2'^2}{2m_2'} + U_f \tag{2.11}$$

Substituting

$$U_i - U_f = K_f - K_i = Q \tag{2.12}$$

in equation (2.11), we have

$$\frac{P_1'^2}{2m_1'} + \frac{P_2'^2}{2m_2'} = \frac{P_1^2}{2m_1} + \frac{P_2^2}{2m_2} + Q \tag{2.13}$$

where $Q =$ energy of reaction.

If $Q = 0$, the collision is said to be *perfectly elastic*; hence no change in KE takes place.

If $Q \neq 0$, the collision is said to be *inelastic* as KE changes after collision.

If $Q < 0$, KE decreases and hence the internal energy increases. Then, this inelastic collision is called *Endorgic*.

If $Q > 0$, KE increases, the internal energy decreases. That means, the KE of the particles of system increases at the expense of internal energy. Then, this inelastic collision is said to be *Exorgic*. Nuclear fission and some exothermic chemical reactions including explosions, etc, are the familiar examples of exorgic collision.

Example 8 Find the Q (energy of reaction) of the collision between an α-particle and a stationary N^{14} nucleus. The kinetic energy of bombarding of α-particle is $K_\alpha = 4\,\text{MeV}$ and proton is deflected with a kinetic energy $K_P = 2.09\,\text{MeV}$ at an angle $\theta = 60°$ to the direction of motion of α-particle.

Solution

Following the aforementioned general procedure, let us first conserve the momentum of the system. The initial and final momentum of the system can be given equated as

$$\vec{P_i} = \vec{P_f}, \quad \text{where } \vec{P_f} = \vec{P'}_1 + \vec{P'}_2 \text{ and } \vec{P_i} = \vec{P}.$$

Triangle law

Parallelogram law

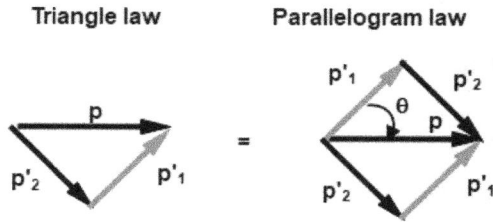

Substituting $\overrightarrow{P_i}$ and $\overrightarrow{P_f}$, we have

$$\overrightarrow{P} = \overrightarrow{P}'_1 + \overrightarrow{P}'_2 \tag{2.14}$$

$$\frac{P^2}{2m_1} = \frac{P'^2_1}{2m'_1} + \frac{P'^2_2}{2m'_2} - Q \tag{2.15}$$

From equation (2.14), we have $|\overrightarrow{P}'_2|^2 = |\overrightarrow{P} - \overrightarrow{P}'_1|^2$

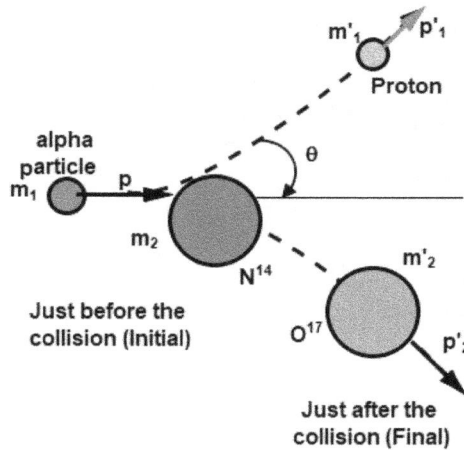

This gives

$$P'^2_2 = P^2 + P'^2_1 - 2PP'_1 \cos \theta \tag{2.16}$$

Substituting P'^2_2 from equation (2.16) in equation (2.15), we have

$$\frac{P^2}{2m_1} = \frac{P'^2_1}{2m'_1} + \frac{P^2 + P'^2_1 - 2PP'_1 - \cos \theta}{2m'_2} - Q$$

Substituting

$$\frac{P^2}{2m_1} = K_\alpha, \frac{P'^2_1}{2m'_1} = K_P$$

and simplifying the factors, we have

$$Q = K_\alpha \left(1 + \frac{m'_1}{m'_2}\right) - K_P\left(1 - \frac{m_1}{m'_2}\right) - \frac{2\sqrt{m_1 m'_1 K_\alpha K_P}}{m'_2} \cos\theta$$

where $m_1 = m_\alpha$, $m_2 = m_N$, $m'_1 = m_P$ and $m'_2 = m_0$, $\theta = 60°$.

Substituting the values of masses and kinetic energies,

$$K_\alpha = 4\,\text{MeV} \quad \text{and} \quad K_P = 2.09\,\text{MeV},$$

we have $Q = -1.2\,\text{MeV}$. Ans.

Negative Q signifies that the kinetic energy of α-particle is absorbed by the reaction so as to produce the products (proton and O^{17}) having greater internal energy.

Example 9 Two identical particles collide elastically ($Q = 0$). Find the angle of deviation between the deflected particles. Assume that one of the colliding particles is stationary.

Solution

Writing momentum conservation, we have $|\vec{P'}_1 + \vec{P'}_2| = |\vec{P_1}|$.
Squaring both sides, we have

$$P'^2_1 + P'^2_2 + 2.\,\vec{P'}_1.\,\vec{P'}_2 = P^2_1 \qquad (2.17)$$

Conserving energy, we have

$$\frac{P'^2_1}{2m'_1} + \frac{P'^2_2}{2m'_2} = \frac{P^2_1}{2m_1}$$

Since the collision is elastic and the particles are identical, $m'_1 = m'_2 = m_1$. Then, we have

$$P'^2_1 + P'^2_2 = P^2_1 \qquad (2.18)$$

Using equations (2.17) and (2.18), we have $\vec{P'}_1 + \vec{P'}_2 = 0$.
The above expression tells us that the particles will be deflected at an angle of $90°$ with each other.

Since $\vec{P'}_1.\,\vec{P'}_2 = 0$, either $P'_1 = 0$ or $P'_2 = 0$ or $\cos\theta = 0(\theta = \frac{\pi}{2}\text{rad})$. If $P'_1 = 0$, $P'_2 = P_1$; if $\cos\theta = 0$, $\vec{P'}_1 \perp \vec{P'}_2$. This means either the particles exchange their momenta during collision or the particles deflect making $90°$ with each other.

Example 10 Using the principle of conservation of momentum and energy, find the expressions for velocity and kinetic energy of a particle of mass m_1 colliding elastically with a momentum P_1 with the stationary mass m_2. Assume head-on collision.

Solution

By the conservation of momentum, we have

$$\vec{P'}_1 + \vec{P'}_2 = \vec{P_1}$$

Then squaring both sides, we have

$$P'^2_1 + P'^2_2 + 2P'_1 P'_2 \cos\theta = P^2_1$$

If we assume that the particles move after the collision in the direction of the colliding particle m_1, we have $\cos\theta = 1$.

Then, we have

$$P'_1 + P'_2 = P_1 \tag{2.19}$$

[You can also directly obtain the above equation (equation (2.19)) by assuming $\vec{P'}_1$, $\vec{P'}_2$ and $\vec{P_1}$ as unidirectional.]

Now conservation of energy gives

$$\frac{P^2_1}{2m_1} = \frac{P'^2_1}{2m_1} + \frac{P'^2_2}{2m_2} \tag{2.20}$$

Eliminating P'^2_2 from equation (2.20) by substituting $P'^2_2 = (P_1 - P'_1)^2$ from equation (2.19), we have

$$\frac{P^2_1}{2m_1} = \frac{P'^2_1}{2m_1} + \frac{P^2_1 + P'^2_1 - 2P_1 P'_1}{2m_2}$$

This gives

$$[P_1(m_1 - m_2) - P'_1(m_1 + m_2)]^2 = 0$$

Then, we have

$$P'_1 = \left(\frac{m_1 - m_2}{m_1 + m_2}\right) P_1$$

$$\Rightarrow K'_1 = \frac{P'^2_1}{2m_1} = \frac{P^2_1}{2m_1}\left(\frac{m_1 - m_2}{m_1 + m_2}\right)^2$$

$$\Rightarrow K'_1 = \left(\frac{m_1 - m_2}{m_1 + m_2}\right)^2 \frac{m_1^2 v_1^2}{2m_1}$$

$$= \left(\frac{m_1 - m_2}{m_1 + m_2}\right)^2 K_1,$$

where

$$K_1 = \frac{1}{2}m_1 v_1^2 = \frac{P^2_1}{2m_1} \text{ Ans.}$$

Similarly, we can find that

$$P'_2 = \frac{2m_2 P_1}{m_1 + m_2}, \; K'_2 = \frac{P_1^2}{2m_1}\left(\frac{m_1 - m_2}{m_1 + m_2}\right)^2 \; \text{Ans.}$$

2.7 Elastic collision

As discussed earlier, collision is said to be elastic when the total kinetic energy of a system of colliding particles (or bodies) remains conserved in the collision. For instance, if a ball collides with a wall elastically, it will bounce with the same speed, that is, the same kinetic energy. Another example is the collision of the gas particles (molecules) in a container. The elastic collision is subdivided into two parts: (i) direct-central (head-on) and (ii) indirect-central (oblique).

Head-on collision: As explained earlier, in a head-on collision, the line of motion, line of impact, and the center of mass of the colliding bodies lie on the same line.

Let us assume that two identical iron balls of mass m_1 and m_2 collide elastically with velocities u_1 and u_2, respectively. Let the particle move with velocities v_1 and v_2, respectively.

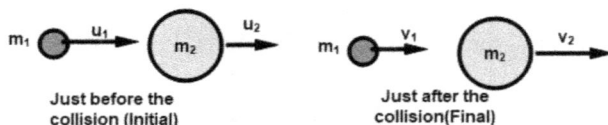

Just before the collision (Initial)   Just after the collision(Final)

Assuming all velocities are in the positive x-direction, let us write the equation of momentum and energy conservation.

Conserving momentum of the system $(m_1 + m_2)$, we have

$$m_1 u_1 + m_2 u_2 = m_1 v_1 + m_2 v_2$$

Conservation of kinetic energy of the system gives

$$\frac{1}{2}m_1 u_1^2 + \frac{1}{2}m_2 u_2^2 = \frac{1}{2}m_1 v_1^2 + \frac{1}{2}m_2 v_2^2 \tag{2.21}$$

From equation (2.21), we have

$$m_1(u_1 - v_1) = m_2(v_2 - u_2) \tag{2.22}$$

From equation (2.21), we have

$$m_1(u_1^2 - v_1^2) = m_2(v_2^2 - u_2^2) \tag{2.23}$$

Dividing equation (2.23) by equation (2.22), we have

$$u_1 + v_1 = v_2 = u_2$$

This gives

$$u_1 - u_2 = -(v_1 - v_2) \tag{2.24}$$

Solving equations (2.21) and (2.24), we have

$$v_1 = \frac{m_1 - m_2}{m_1 + m_2} u_1 + \frac{2m_2}{m_1 + m_2} u_2$$

$$v_2 = \frac{2m_1}{m_1 + m_2} u_1 + \frac{m_2 - m_1}{m_1 + m_2} u_2$$

Example 11 Explain the physical significance of the obtained result given as

$$u_1 - u_2 = -(v_1 - v_2).$$

Solution

The term $(u_1 - u_2)$ is the velocity of m_1 relative to m_2. If you stand on the body m_2, m_1 seems to approach m_2 with a relative velocity, called *velocity of approach*, given as

$$u_{\text{app}} = u_1 - u_2$$

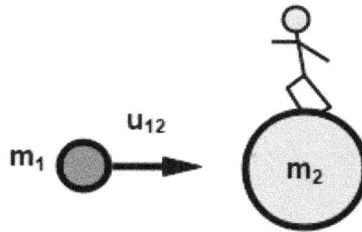

m_1 approaches
m_2 with u_{12}

The velocity of approach is the relative velocity between the colliding particles along the line of impact just before collision.

Similarly, the term $(v_1 - v_2)$ is the velocity with which m_1 moves away relative to m_2 just after the collision, known as *velocity of separation,* denoted as

$$u_{\text{sep}} = (v_1 - v_2)$$

As obtained earlier,

$$u_1 - u_2 = -(v_1 - v_2)$$

$$\Rightarrow v_{\text{app}} = -v_{\text{sep}}$$

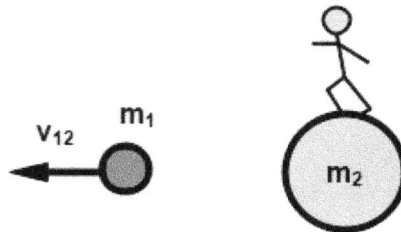

m_1 departs m_2
with v_{12}

It physically signifies that m_1 approaches (moves towards m_2) just before the collision with a speed $u_{app}(=u_1 - u_2)$ and will depart (move away from m_2) just after the collision with the same speed $v_{sep}(=v_1 - v_2)$, along the line of impact.

In a perfectly elastic collision

$$u_{12} = -v_{12}$$

Which states that velocity of approach $= -$velocity of separation.

The above expression is also valid in the center of mass frame of the colliding particles.

Example 12 Discuss different cases in a head-on elastic collision between two free particles of masses m_1 and m_2 colliding with speeds u_1 and u_2, respectively.

Solution

Case 1. $m_1 = m_2$

Substituting $m_1 = m_2$ in the formula

$$v_1 = \frac{m_1 - m_2}{m_1 + m_2}u_1 + \frac{2m_2}{m_1 + m_2}u_2,$$

we have

$$v_1 = u_2$$

Similarly substituting

$$m_1 = m_2 \text{ in the formula}$$

$$v_2 = \frac{2m_1}{m_1 + m_2}u_1 + \frac{m_2 - m_1}{m_1 + m_2}u_2$$

we have

$$v_2 = u_1$$

Since $v_1 = u_2$ and $v_2 = u_1$, the colliding particles just exchange their momenta.

Case 2. $m_2 \gg m_1$: When $m_2 \gg m_1$; $\frac{m_1}{m_2} \simeq 0$.

Then substituting $\frac{m_1}{m_2} \simeq 0$ in

$$v_1 = \frac{\frac{m_1}{m_2} - 1}{\frac{m_1}{m_2} + 1}u_1 + \frac{2}{1 + \frac{m_1}{m_2}}u_2$$

we have

$$v_1 \simeq -u_1 + 2u_2$$

Similarly, substituting $\frac{m_1}{m_2} \simeq 0$ in

$$v_2 = \frac{2\frac{m_1}{m_2}u_1}{\frac{m_1}{m_2} + 1} + \frac{1 - \frac{m_1}{m_2}}{1 + \frac{m_1}{m_2}}u_2,$$

we have

$$v_2 = u_2$$

The velocity of a massive body remains practically constant.

Case 3. $v_2 = 0$: When one of the bodies, m_2 say, is stationary, we have
$v_1 = \frac{m_1 - m_2}{m_1 + m_2} u_1$ and $v_2 = \frac{2m_1}{m_1 + m_2} u_1$.

(a) For $m_1 = m_2$, $v_1 = 0$ and $v_2 = u_2$, which means that the exchange of momentum between the bodies takes place. The body m_1 comes to rest and body m_2 carries the total momentum of m_1.

(b) When $m_1 < m_2$; v_1 is negative, m_1 will return back.

(c) When $m_1 > m_2$; v_1 is positive, m_1 will proceed forward.

(d) When $m_1 > > m_2$; $v_1 = u_1$ and $v_2 \cong 2u_1$, the body m_1 will move as it was moving and the stationary body m_2 receives twice the speed of the body 1. Thus, maximum momentum and velocity of m_1 can be delivered to m_2, when $\frac{m_1}{m_2} \to \infty$.

N.B.: From the foregoing discussion of head-on elastic collision, let us note the following points:

1. Head-on collision between identical particles leads to an exchange of velocity, energy, and momentum.

2. Relative velocity between the colliding points (along the line of impact) reverses after the collision without changing its magnitude.

3. When a massive body strikes a stationary light body, the velocity of the massive body remains approximately constant, and the lighter body moves with twice the velocity of the colliding body.

4. When a light body strikes a stationary massive body, the massive body remains practically at the same spot, whereas the light body retraces its path with nearly the same speed.

5.
    (i) When $u_2 = 0$, the maximum transfer of kinetic energy takes place, when $m_1 = m_2$.
    (ii) Maximum transfer of velocity takes place, when $m_1 \gg m_2$.
    (iii) Maximum momentum transfer takes place, when $m_2 \gg m_1$.

2.8 Elastic-oblique collision

For the sake of simplicity, let us consider the oblique collision between two balls of masses m_1 and m_2 on a smooth horizontal surface when m_2 is stationary. In the absence of friction, we can conserve the momentum of the system ($m_1 + m_2$) in any direction in the horizontal plane.

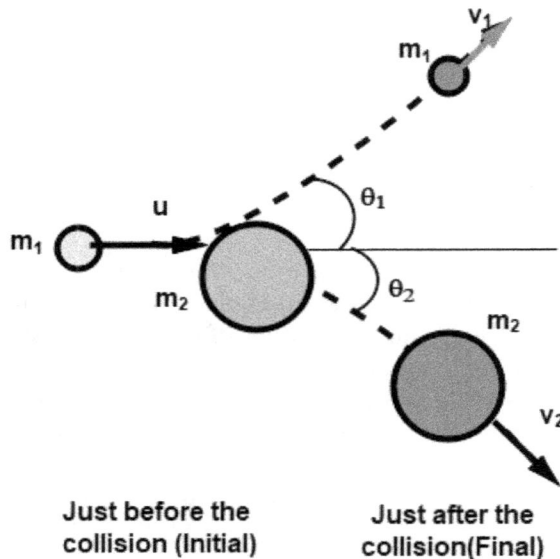

Just before the collision (Initial)   **Just after the collision(Final)**

Let the bodies move with velocities v_1 and v_2, respectively, just after the collision with angles θ_1 and θ_2 made relative to the x-axis. Conserving the momentum of the system in the x-direction, we have

$$m_1 u = m_1 v_1 \cos \theta_1 + m_2 v_2 \cos \theta_2 \qquad (2.25)$$

Conserving the momentum of the system in the y-direction,

$$-m_1 v_1 \sin \theta + m_2 v_2 \sin \theta_2 = 0 \qquad (2.26)$$

Conserving the kinetic energy of the system,

$$\frac{1}{2} m_1 u^2 = \frac{1}{2} m_1 v_1^2 + \frac{1}{2} m_2 v_2^2 \qquad (2.27)$$

Now we have three equations and four unknown quantities (i.e., v_1, v_2, θ_1 and θ_2). Hence, we need another equation that can be derived from the condition (data) given in the examples. To solve the equations let us look at the following example.

Example 13 Substituting $m_1 = 2m_2(=2m)$, $u = v_0$, and $\theta_1 = 30°$, find v_1, v_2, and θ_2 using the equations (2.25)–(2.27) derived above.
   Solution
Using equation (2.25), we have

$$2m v_0 = 2m v_1 \cos 30^0 + m v_2 \cos \theta_2$$

$$\Rightarrow v_2 \cos \theta_2 = 2v_0 - \sqrt{3} v_1 \qquad (2.28)$$

Using equation (2.26), we have

$$-(2m)v_1 \sin 30^0 + (m)v_2 \sin \theta_2 = 0$$

$$\Rightarrow v_2 \sin \theta_2 = v_1 \tag{2.29}$$

Using equation (2.27), we have

$$\frac{1}{2}(2m)v_0^2 = \frac{1}{2}(2m)v_1^2 + \frac{1}{2}(m)v_2^2$$

$$\Rightarrow 2v_1^2 + v_2^2 = 2v_0^2 \tag{2.30}$$

Squaring equations (2.28) and (2.29) and summing them, we have

$$(2v_0 - \sqrt{3}v_1)^2 + v_1^2 = v_2^2 \tag{2.31}$$

Eliminating v_2^2 from equations (2.30) and (2.31), we have

$$2v_1^2 + v_1^2 + (2v_0 - \sqrt{3}v_1)^2 = 2v_0^2$$

Then, we have

$$6v_1^2 - 4\sqrt{3}v_0v_1 + 2v_0^2 = 0$$

Solving the quadratic equation, we have

$$v_1 = \frac{4\sqrt{3}v_0 \pm \sqrt{(4\sqrt{3}v_0)^2 - (4)(6)(2v_0^2)}}{12} = \frac{v_0}{\sqrt{3}} \text{ Ans.}$$

Substituting $v_1 = \frac{v_0}{\sqrt{3}}$ in equation (2.30), we have $v_2 = \frac{2v_0}{\sqrt{3}}$.

Then substituting $v_2 = \frac{2v_0}{\sqrt{3}}$ and $v_1 = \frac{v_0}{\sqrt{3}}$ in equation (2.29), we have $\theta_2 = 30°$.

N.B.: In the foregoing example, if $m_1 = m_2$ and the collision is elastic, the maximum angle of deflection of m_1 relative to m_2 just after the collision is equal to 90°.

2.9 Complete inelastic collision ($e = 0$)

In a perfectly inelastic collision, both the colliding bodies move with the same velocity v, say, just after the collision, as a combined unit.

The initial momentum of the system is

$$\overrightarrow{P_i} = m_1\overrightarrow{v_1} + m_2\overrightarrow{v_2} \tag{2.32}$$

The final momentum of the system is

$$\overrightarrow{P_f} = m_1\overrightarrow{v} + m_2\overrightarrow{v} \tag{2.33}$$

Conserving the linear momentum of the system $(m_1 + m_2)$, we have

$$\vec{P_i} = \vec{P_f} \tag{2.34}$$

Using the last three equations,

$$(m_1 + m_2)\vec{v} = m_1\vec{v_1} + m_2\vec{v_2}$$

$$\Rightarrow \vec{v} = \frac{m_1\vec{v_1} + m_2\vec{v_2}}{m_1 + m_2} \tag{2.35}$$

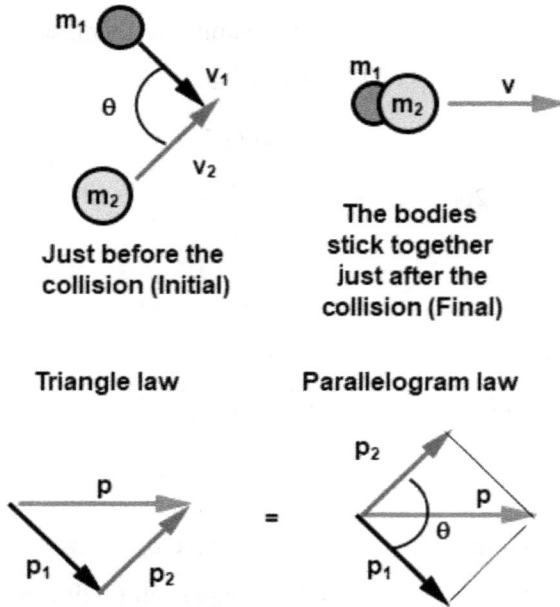

Just before the collision (Initial)

The bodies stick together just after the collision (Final)

Triangle law

Parallelogram law

The change in KE of the system is

$$\Delta K = \frac{1}{2}(m_1 + m_2)v^2 - \left(\frac{1}{2}m_1v_1^2 + \frac{1}{2}m_2v_2^2\right) \tag{2.36}$$

By substituting v from equation (2.35) in equation (2.36) and simplifying the terms, we have

$$\Delta K = \frac{m_1 m_2}{2(m_1 + m_2)}\,|\vec{v_1} - \vec{v_2}|^2$$

Example 14 A car of mass m with a velocity u strikes inelastically with a truck of mass M moving with a velocity v perpendicular to the direction of motion of the car.

If the truck and car stick together after the collision, find the (a) energy loss in the collision and (b) (i) time and (ii) distance of sliding of the combined mass, if the coefficient of friction between truck, car, and ground is μ.

Solution

(a) We have the formula

$$\Delta K = \frac{m_1 m_2}{2(m_1 + m_2)} |\vec{v_1} - \vec{v_2}|^2,$$

where $|\vec{v_1} - \vec{v_2}|^2 = v_1^2 + v_2^2 + 2v_1 v_2 \cos \theta$.

Substituting $m_1 = m$, $m_2 = M$, $v_1 = u$, $v_2 = v$ and $\theta = 90^0$, we have

(b)
$$\Delta K = \frac{Mm(u^2 + v^2)}{2(M + m)} \text{ Ans.}$$

(i) The time of sliding of the combined truck and car, after the collision, is given as

$$t = \frac{v'}{\mu g},$$

where $v' = $ velocity of the combination just after the collision, given as

$$v' = \frac{\sqrt{m^2 u^2 + M^2 v^2}}{(M + m)}$$

Using last two equations, we have

$$t = \frac{\sqrt{m^2 u^2 + M^2 v^2}}{\mu(M + m)g}$$

(ii) The distance of sliding of the combined truck and car, after the collision, is given as

$$s = \frac{(v')^2}{2\mu g}$$

Putting the obtained value of v', we have

$$s = \frac{m^2 u^2 + M^2 v^2}{2\mu(M + m)^2 g}$$

Sometimes the combined mass just after an inelastic collision is constrained to move in a 'definite line or plane'. In these cases, we need to conserve the momentum of the system $(m_1 + m_2)$ in the line of motion of the combined unit. We consider the following example to demonstrate this.

Example 15 A particle (a mud pallet, say) of mass m strikes a smooth stationary wedge of mass M with a velocity v_0, at an angle θ with horizontal. If the collision is perfectly inelastic, find the

    (a) Velocity of the wedge just after the collision

    (b) Change in KE of the system $(M + m)$ in collision

    (c) Impulse of the impact force on the wedge

    (d) Possibility of the conservation of linear momentum, if the ground is rough.

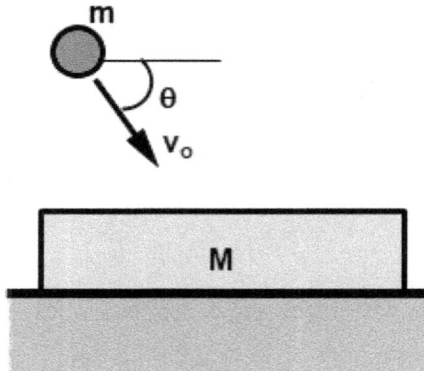

Solution

    (a) Let the system $(M + m)$ move as a single mass with a velocity v. Conserving the momentum of the system in horizontal, we have

$$mv_0 \cos \theta = (M + m)v$$

$$\Rightarrow v = \frac{mv_0 \cos \theta}{M + m} \qquad (2.37)$$

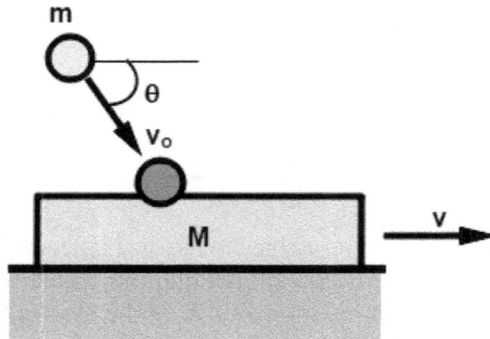

(b) The change in KE of the system is

$$\Delta K = \frac{1}{2}(M + m)v^2 - \frac{1}{2}mv_0^2 \qquad (2.38)$$

Using last two equations, we have

$$\Delta K = \frac{(M + m)}{2}\left(\frac{mv_0 \cos \theta}{M + m}\right)^2 - \frac{1}{2}mv_0^2$$

$$= \frac{-mv_0^2}{2}\left(1 - \frac{m}{M + m}\cos^2 \theta\right)$$

$$= -\frac{(M + m \sin^2 \theta)mv_0^2}{2(M + m)} \quad \text{Ans.}$$

(c) The impulse of the impact force on the wedge is

$$I = Mv = \frac{Mmv_0 \cos\theta}{M + m} \text{ pointing right.}$$

(d) If the ground is rough, the contact forces such as normal reaction and friction will be impulsive. Due to the impulsive friction acting during the collision, we cannot conserve the horizontal momentum of the system.

Example 16 Two particles of masses m_1 and m_2 are connected by a light and inextensible string that passes over a fixed pulley. Initially, the particle m_1 moves with a velocity v_0 when the string is not taut. Neglecting friction in all contacting surfaces, find the (a) velocities of the particles m_1 and m_2, (b) the change in momentum of the system $(m_1 + m_2)$ just after the string is taut, and (c) the possibility of conservation of linear momentum of the system $(m_1 + m_2 + \text{pulley})$.

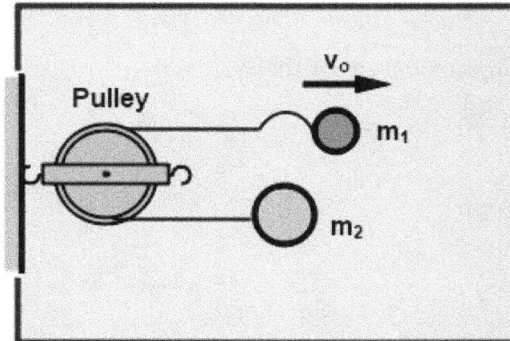

Solution

(a) During the impact, the string is taut as an impulsive tension develops. The leftward ($-$ve) impulse of tension T is equal to the change in momentum of m_1.

As the body m_1 changes its momentum from $m_1 v_o$ to mv, we can write

$$-\int T dt = m_1(v - v_0) \tag{2.39}$$

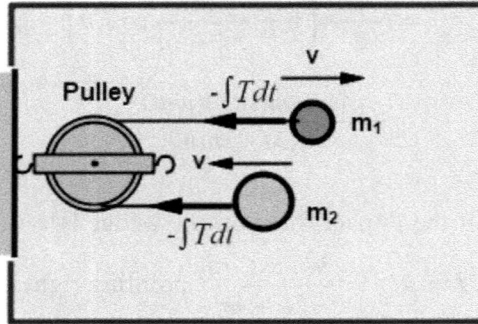

Again, due to the leftward impulse of tension T, m_2 changes its momentum from zero (as it was stationary) to $-mv$ as it moves towards the left after the string is taut. The impulse of the string is equal to the change in momentum of m_2, which is given as

$$-\int T dt = m_2(-v) \tag{2.40}$$

Using equations (2.39) and (2.40), we have

$$m_1(v_0 - v) = m_2 v$$

$$\Rightarrow v = \frac{m_1 v_0}{m_1 + m_2} \text{ Ans.}$$

(b) The change in momentum of the system $(m_1 + m_2)$ is given as

$$\overrightarrow{\Delta P} = (m_1 v - m_2 v)\hat{i} - m_1 v_o \hat{i}$$

$$= \frac{(m_1 - m_2)v_0}{m_1 + m_2}\hat{i} - m_1 v_o \hat{i}$$

$$= -\frac{2m_1 m_2 v_0}{m_1 + m_2}\hat{i} \text{ Ans.}$$

(c) We cannot conserve the momentum of the $(m_1 + m_2 + \text{pulley})$ because there is an external impact force acting on the pulley by the pivot that points towards the left.

N.B.: The student must remember that $\Delta\overrightarrow{P_1} + \Delta\overrightarrow{P_2} = \int \overrightarrow{T_1}\, dt + \int \overrightarrow{T_2}\, dt = \int (\overrightarrow{T_1} + \overrightarrow{T_2})\, dt \neq 0$ because $\overrightarrow{T_1} + \overrightarrow{T_2} \neq 0$; rather it is equal, $-2T\hat{i}$. Hence, in the above example we cannot apply the principle of conservation of linear momentum even though simply (or blindly) writing $\Delta P_1 = \Delta P_2$ you may get the answer. On the system $(m_1 + m_2)$, the net impulse (or net force) is not zero during the impact. Hence, the momentum of the system changes by an amount $\frac{2m_1 m_2 v_0}{m_1 + m_2}$ in the negative x-direction (towards the left).

2.10 General solution for head-on collision between two bodies

Time of deformation: Let two balls of masses m_1 and m_2 move in $+x$-direction with velocities u_1 and u_2, respectively. The collision occurs if $u_1 > u_2$. During the collision the bodies are pressed against each other due to the action of the impulsive forces $-F$ and F. Just before collision the relative velocity between the bodies is $u_1 - u_2$. This is known as 'velocity of approach'. The bodies push each other during the deformation phase of the collision by their impulsive forces. As a result, the relative velocity between the bodies will rapidly decrease to zero during a time interval called 'deformation time t_d'. At the end of the time interval t_d, both the bodies move with the same velocity v, say.

During the time of deformation, the body m_1 decelerates by changing its momentum from $m_1 u_1$ to $m_1 v$ by the action of the impulse $-\int F'dt$.

Then,

$$-\int_0^{t_d} F\,dt = m_1(v - u_1) \tag{2.41}$$

Meanwhile, body m_2 accelerates changing its momentum from $m_2 u_2$ to $m_2 v$ under the action of impulse $\int F\, dt$. Then we have

$$\int_0^{t_d} F\,dt = m_2(v - u_2) \tag{2.42}$$

Time of restitution: We have seen that at the end of the compression or deformation period the bodies will move with the same velocity v compressing each other to a maximum amount. Thereafter, the impulsive forces will still continue to push them until the bodies restitute (regain) their original shape and size. If the bodies take an additional time t_r to restitute and lose contact with each other, this period is called 'time of restitution'. It is practically evident that the impulse in a deformation phase is greater than the impulse of the forces in restitution phase. Let the instantaneous impulsive forces acting during restitution be $-F'$ and F'. During the time of restitution the body m_1 accelerates by $-F'$ from velocity v to v_1, say, by changing its momentum from mv to mv_1. Then the impulse of $-F'$ is

$$-\int_0^{t_r} F' \, dt = m_1 v_1 - m_1 v \tag{2.43}$$

Similarly, during the restitution phase, the body m_2 accelerates from velocity v to v_2, say, changing its momentum from mv to mv_2. Then, the impulse of F' is

$$\int_0^{t_r} F' \, dt = m_2 v_2 - m_2 v \tag{2.44}$$

Coefficient of restitution:
It is practically evident that the magnitude of change in momentum of the either body during the restitution phase is always less than that during the deformation phase. In other words, the magnitude of impulse of the deformation forces is always greater than the impulse of the restitution forces. Then, the ratio of magnitude of restitution impulse to the deformation impulse can be termed as the 'coefficient of restitution' denoted by the letter 'e', given as

$$\frac{\int_0^{t_r} F' \, dt}{\int_0^{t_d} F \, dt} = e \tag{2.45}$$

Using this definition of 'e' in equation (2.45) from equations (2.41) and (2.43) we have

$$\frac{\int_0^{t_r} F' \, dt}{\int_0^{t_d} F \, dt} = e = \frac{v_1 - v}{v - u_2} \tag{2.46}$$

Similarly, using equations (2.42) and (2.46), we have

$$e = \frac{v_2 - v}{v - u_2} \tag{2.47}$$

Using equations (2.46) and (2.47), we have $e = -\frac{v_1 - v_2}{u_1 - u_2}$

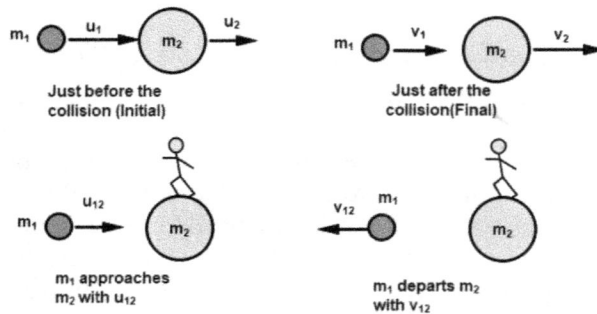

Just before the
collision (Initial)

Just after the
collision(Final)

m_1 approaches
m_2 with u_{12}

m_1 departs m_2
with v_{12}

Newton's law of restitution: The more familiar and convenient form of the above equation can be given as

$$v_1 - v_2 = -e(u_1 - u_2) \tag{2.48}$$

This formula is also called Newton's impact (or empirical) formula, which can be remembered as $-e$ (velocity of approach) = velocity of separation.

The coefficient of restitution depends on the nature of material, velocities, shape, and size of the colliding bodies.

Perfectly elastic collision: The value of e lies between 0 and 1; $0 \leqslant e \leqslant 1$. When $e = 1$, the impact is called perfectly elastic. In this collision, both deformation and restitution impulses are equal in magnitude. Putting $e = 1$, we have

$$u_1 - u_2 = -(v_1 - v_2)$$

This signifies that the relative velocity between the bodies just before the impact will be equal and opposite to that just after the impact. In other words, the velocity of approach is numerically equal to velocity of separation. You can realise this as follows:

You sit on either m_1 or m_2. Suppose you observe m_1 while sitting on m_2. You will find the velocity of m_1 relative to m_2 just before the collision as 'velocity of approach' $v_{\text{app}}[=(u_1 - u_2)]$. Just after the collision you will see that the body m_1 departs relative to m_2 (you) with the same speed $(u_1 - u_2)$ but in the opposite direction. Hence, $v_{\text{sep}} = -ev_{\text{app}}$, where $e = 1$ for elastic collision.

Perfectly inelastic collision: Putting $e = 0$, we have $v_{\text{sep}} = 0$. This means the restitution period is zero. Hence, the bodies stick together and move with the same velocity after the collision.

Conservation of linear momentum: Since the bodies in the isolated system $(m_1 + m_2)$ experience equal and opposite impulses during the collision, the net impulse be zero. Ignoring the impulses of external forces, we have

$$\vec{P_i} = \vec{P_f},$$

where

$$P_i = m_1 u_1 + m_2 u_2$$

and

$$P_f = m_1 v_1 + m_2 v_2$$

$$\Rightarrow m_1 u_1 + m_2 u_2 = m_1 v_1 + m_2 v_2 \tag{2.49}$$

Now we have two equations (2.48) and (2.49) for two unknown quantities such as v_1 and v_2. Solving these equations we have

$$v_1 = \frac{m_1 - em_2}{m_1 + m_2} u_1 + \frac{(1 + e)m_2}{m_1 + m_2} u_2$$

$$v_2 = \frac{(1 + e)m_1}{m_1 + m_2} u_1 + \frac{m_2 - em_1}{m_1 + m_2} u_2$$

Example 17 Derive an expression for coefficient of restitution using the center of mass reference frame.

Solution

The velocity of center of mass is

$$v_C = \frac{m_1 u_1 + m_2 u_2}{m_1 + m_2}$$

This is also equal to the common velocities of the bodies m_1 and m_2 at the end of the deformation phase.

The momentum of m_1 relative to center of mass frame is

$$\overrightarrow{P_{iC}} = m_1(u_1 - v_C) + m_2(u_2 - v_C)$$

At the end of the deformation phase, the initial momentum of $m_1 \overrightarrow{P_{iC}}$ impulse of F during t_d is

$$-\int_0^{t_d} F dt = \Delta P_{iC},$$

where

$$\Delta P_{iC} = -\frac{m_1 m_2}{m_1 + m_2}(u_1 - u_2)$$

This gives

$$\int_0^{t_d} F dt = \frac{m_1 m_2}{m_1 + m_2}(u_1 - u_2) \tag{2.50}$$

Similarly, in the restitution phase during the time t_r of restitution the momentum of m_1 increases from zero to $\frac{m_1 m_2 (v_1 - v_2)}{m_1 + m_2}$.

Hence, the impulse of $-F'$ is

$$-\int_0^{t_r} F' dt = \Delta P.$$

where

$$\Delta P = \frac{m_1 m_2 (v_1 - v_2)}{m_1 + m_2}$$

This gives

$$\int_0^{t_r} F' dt = \frac{m_1 m_2}{m_1 + m_2}(v_1 - v_2) \tag{2.51}$$

Now using equation (2.50) and (2.51), we have the same result:

$$e = \frac{\int_0^{t_r} F' dt}{\int_0^{t_r} F dt} = -\frac{(v_1 - v_2)}{u_1 - u_2} \quad \text{Ans.}$$

Example 18 Derive an expression for loss of kinetic energy in an inelastic head-on collision between two particles.

Solution

For an isolated system of two particles of masses m_1 and m_2, the KE is given by

$$K = K_C + \frac{1}{2}\mu u_{rel}^2$$

The velocity of center of mass of the system is

$$v_C = \frac{m_1 u_1 + m_2 u_2}{m_1 + m_2} = \frac{m_1 v_1 + m_2 v_2}{m_1 + m_2}$$

For an isolated system, the net external force is zero. So, v_C remains constant and the KE K_C of the center of mass must be constant. Then, the change in KE of the system is

$$\Delta K = \frac{1}{2}\mu\left[(v_{\text{rel}})_f^2 - (v_{\text{rel}})_i^2\right]$$

For head-on collision

$$(v_{\text{rel}})_f = -e(v_{\text{rel}})_i$$

$$\Rightarrow \Delta K = -\frac{1}{2}\mu(v_{\text{rel}})_i^2(1 - e^2),$$

where

$$(v_{\text{rel}})_i = |\vec{u_1} - \vec{u_2}|$$

and $\mu = \frac{m_1 m_2}{m_1 + m_2}$

Then, we have $\Delta K = -\frac{1}{2}\frac{m_1 m_2}{m_1 + m_2}(1 - e^2)(|\vec{u_1} - \vec{u_2}|)^2$.

Negative sign signifies that change in KE is negative.

This means that loss in KE is given as negative change in KE.

In a perfectly inelastic collision,

$$\Delta K = -\frac{m_1 m_2}{2m_1 + m_2}|\vec{u_1} - \vec{u_2}|^2$$

In a perfectly elastic collision,

$\Delta K = 0$ (because $e = 1$)

The relation

$$\Delta K = -\frac{1}{2}\mu(1 - e^2)u_{\text{rel}}^2$$

is valid for a one-dimensional head-on collision.

However, in a two-dimensional collision, if the colliding bodies are smooth, u_{rel} is equal to the relative velocity between the colliding bodies along the line of impact, just before collision.

For an isolated two-particle colliding system the net external force is zero. Then, the center of mass moves with a constant velocity and KE. In general, there is a loss of energy of the system. As the center of mass moves with a constant KE, the loss of KE must take place in the center of mass frame.

Example 19 Two iron balls of masses 1 and 2 kg are moving in the same direction. If they collide centrally with velocities 2 and 1 m s^{-1}, respectively. Find the (a) velocities of the bodies just after the collision, (b) velocity of center of mass of the system of the balls, and (c) change in KE of the system during collision. Assume $e = \frac{1}{2}$.

Solution

(a) Let us assume that the balls move with velocities v_1 and v_2 just after the collision.

The net external force acting on the system is $F_{ext} = 0$.

Conserving the linear momentum,

$$\left(\overrightarrow{P}\right)_f = \left(\overrightarrow{P}\right)_i,$$

where

$$\left(\overrightarrow{P}\right)_i = (1)(2) + 2(-1) = 0$$

and

$$(P)_f = m_1 v_1 + m_2 v_2 = (1)(v_1) + (2)(v_2) = v_1 + 2v_2$$

Then, we have

$$v_1 + 2v_2 = 0 \tag{2.52}$$

Writing Newton's empirical collision formula, we have

$$-e(u_1 - u_2) = v_1 - v_2$$

Substituting $u_1 = 2, u_2 = -1, v_1 = +v_1$ and $v_2 = +v_2$ and $e = \frac{1}{2}$, we have

$$-\frac{1}{2}[(2) - (-1)] = v_1 - v_2$$

$$\Rightarrow v_1 - v_2 = -\frac{3}{2} \tag{2.53}$$

Solving equations (2.52) and (2.53), we have $v_1 = -1$ ms^{-1} and $v_2 = \frac{1}{2}$ ms^{-1}. Ans.

(b) The velocity of the center of mass is

$$v_C = \frac{1(2) + 2(-1)}{1 + 2} = 0 \text{ Ans.}$$

(c) The change in KE of the system is $\Delta K = \frac{1}{2}(1)[(-1)^2 - (2)^2] + \frac{1}{2} \cdot 2[(\frac{1}{2})^2 - (-1)^2] = -\frac{1}{4}J$. Ans.

Example 20 Two spherical bodies of mass m_1 and m_2 fall freely through a distance h, before the body m_2 collides with the ground. If the coefficient of restitution of all collisions is e, find the velocity of m_1 just after it collides with m_2.

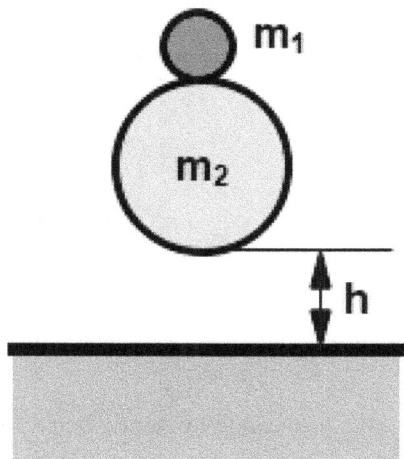

Solution

As the bodies m_1 and m_2 fall freely through a distance h, m_2 collides with ground with a velocity $v_0 = \sqrt{2gh}$. Hence, m_2 bounces up with a velocity ev_0 just after the collision with the ground. Then it collides with the freely falling body m_1. Just after the collision, let us assume that v_1 and v_2 are the velocities of m_1 and m_2, respectively. Ignoring the impulses due to the weights of the bodies during the period of collision, let us conserve the momentum of the system $(m_1 + m_2)$.

Conserving linear momentum.

$$P = -m_1v_0 + m_2ev_0 = m_1v_1 + m_2v_2 \qquad (2.54)$$

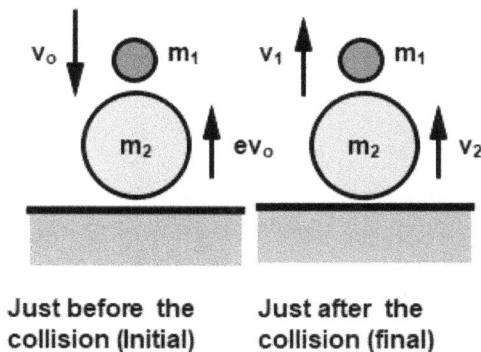

Just before the collision (Initial)   **Just after the collision (final)**

Newton's impact formula

$$-e[u_1 - u_2] = v_1 - v_2,$$

Putiing $u_1 = -v_0$ and $u_2 = ev_0$, we have

$$-e[(-v_0) - (ev_0)] = v_1 - v_2 \qquad (2.55)$$

Solving equations (2.54) and (2.55), we have

$$v_1 = \frac{-(m_1 - em_2)}{m_1 + m_2}v_0 + \frac{(1 + e)m_2}{m_1 + m_2}ev_0$$

Putting

$$m_1 = m_2, \ e = \frac{1}{2} \ \text{and} \ v_0 = \sqrt{2gh},$$

we have

$$v_1 = \sqrt{\frac{gh}{32}} \uparrow \ \text{Ans.}$$

N.B.: In the above example, you can show that when $e = 1$ and $m_1 = m_2$, we have $v_1 = \sqrt{2gh} \uparrow; v_2 = \sqrt{2gh} \downarrow$.

In the above example, at first m_2 collides with ground, then m_2 collides with freely falling m_1 while bouncing up. In this way, two collisions occur one by one. While freely falling, the reaction force between m_1 and m_2 is zero. However, during their collision, huge reaction (impulsive) forces act on them.

If m_2 is stationary just before the collision the other body m_1 collides with m_2 with speed u centrally. In this case, putting $u_2 = 0$ in the general expressions of v_1 and v_2, we have

$$v_1 = \frac{m_1 - em_2}{m_1 + m_2}u_1 \ \text{and} \ v_2 = \frac{(1 + e)m_1u}{m_1 + m_2},$$

where $u = u_1$.

Alternately, any problem of head-on collision can be solved from the basic method that relies on the principles of conservation of momentum and Newton's imperial collision formula.

Example 21 A particle of mass m_1 collides centrally with a stationary particle of mass m_2 with a velocity v_0. If the particle m_1 stops just after the collision, find the:
   (a) Coefficient of restitution
   (b) Change in kinetic energy of the system.

Solution

(a) Conserving momentum, we have

$$\vec{P_i} = \vec{P_f},$$

where $\vec{P_i} = m_1 v_0$ and $P_f = m_2 v_2 = m_2 v$

$$\Rightarrow m_2 v = m_1 v_0 \qquad (2.56)$$

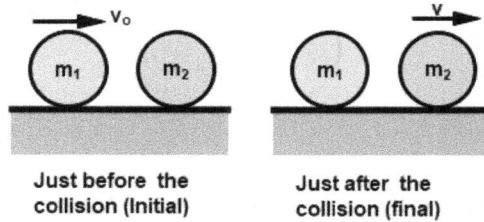

Just before the collision (Initial)   **Just after the collision (final)**

Applying NIF, we have

$$-e(u_1 - u_2) = v_1 - v_2,$$

$$\Rightarrow -e(v_0 - 0) = (0 - v)$$

$$\Rightarrow v = e v_0 \qquad (2.57)$$

Using equations (2.56) and (2.57), we have

$$e = \frac{m_1}{m_2} \text{ Ans.}$$

(b) The change in kinetic energy is

$$\Delta K = \frac{1}{2} m_2 v_2^2 - \frac{1}{2} m_1 v_0^2$$

Substituting $v_2 = e v_0$, we have

$$\Delta K = \frac{-\frac{1}{2} m_1 v_0^2 (m_2 - m_1)}{m_2} \text{ Ans.}$$

2.11 General solution for oblique inelastic impact of two smooth bodies

In oblique impact of smooth bodies, first of all we draw a tangent at the point of collision. Then draw a normal at the point of collision. Since the bodies are smooth, no impact force acts along the tangent. This means the impact (impulsive) action-reaction forces are $+N$ and $-N$, respectively. Since N and $-N$ are action-reaction forces, their net impulse is zero.

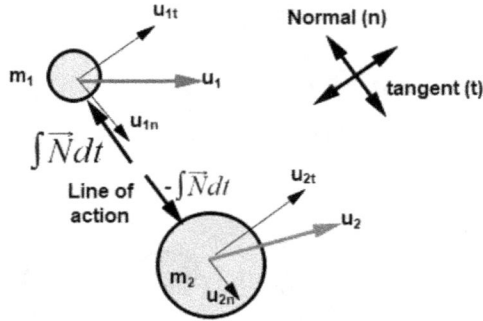

Case 1. $F_{ext} = 0$:

If no external force acts on the system $(m_1 + m_2)$, $\int F_{ext}\, dt = 0$. Even though some non-impulsive forces act on the system, the impulse of these forces during the short time of impact is negligible. Since the sum of impulses of all internal impulsive forces is zero, we can conserve the momentum of the system during the collision. This means that momentum just before collision = momentum during collision = momentum just after collision.

Resolving the momenta along the normal, we can write

$$m_1 \vec{u}_{1n} + m_2 \vec{u}_{2n} = m_1 \vec{v}_{1n} + m_2 \vec{v}_{2n} \tag{2.58}$$

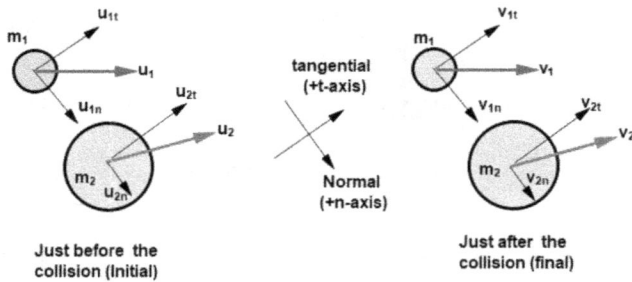

Just before the collision (Initial)

Just after the collision (final)

Since no tangential force acts on the bodies as they are smooth, individually we can conserve their tangential momenta.

$$m_1 \vec{u}_{1t} = m_1 \vec{v}_{2t} \tag{2.59}$$

$$m_2 u_{2t} = m_2 v_{2t} \tag{2.60}$$

Now, we have three equations and four unknown variables; i.e., v_{1t}, v_{2t}, v_{1n} and v_{2n}. So, we need another equation, which should be none other than Newton's empirical formula, which is always applied along the line of impact.

$$-e(\overrightarrow{u_{1n}} - \overrightarrow{u_{2n}}) = \overrightarrow{v_{1n}} - \overrightarrow{v_{2n}} \tag{2.61}$$

Now we have four equations and four unknown variables. You can solve these equations for all four unknown quantities specified earlier.

Example 22 A particle of mass $m_1 = m$ collides centrally with a smooth disc of mass $m_2 = M$. The disc moves vertically up and the particle moves horizontally to the right with equal speed v_0, just before the collision. Assuming the coefficient of restitution $e = \frac{1}{2}$ and $2m = M$, (a) write the four relevant equations following the process as described in the foregoing section and (b) find the speeds of the bodies just after collision.

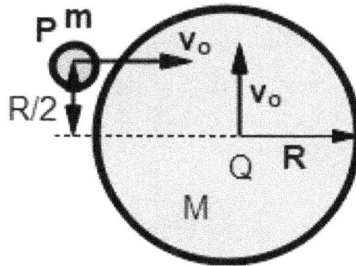

Solution
(a) Directly following the procedure described above, we conserve the momentum of the system $(m + M)$ along the normal.

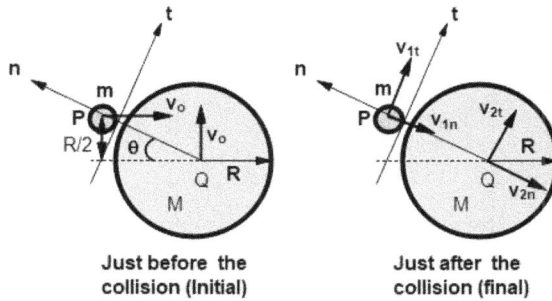

Just before the   Just after the
collision (Initial)   collision (final)

$$-mv_0 \cos\theta + Mv_0 \sin\theta = -mv_{1n} - Mv_{2n}$$

Substituting $2m = M$ and $\theta = 30°$, we have

$$2v_{2n} + v_{1n} = \frac{(\sqrt{3} - 2)v_0}{2} \tag{2.62}$$

Conserving tangential momentum of m, we have

$$mv_{1t} = mv_0 \sin\theta, \quad \text{where } \theta = 30°$$

$$\Rightarrow v_{1t} = \frac{v_0}{2} \tag{2.63}$$

Conserving tangential momentum of M, we have

$$Mv_0 \cos\theta = Mv_{2t}, \quad \text{where } \theta = 30°$$

Then we have

$$v_{2t} = \frac{\sqrt{3}\,v_0}{2} \tag{2.64}$$

Then use Newton's empirical formula

$$-e(u_{1n} - u_{2n}) = v_{1n} - v_{2n},$$

where $e = \frac{1}{2}$, $u_{1n} = -\frac{\sqrt{3}\,v}{2}$, $u_{2n} = +\frac{v_0}{2}$

$$\Rightarrow v_{1n} - v_{2n} = \left(\frac{(\sqrt{3}+1)}{4}\right)v_0 \tag{2.65}$$

(b) By solving equations (2.62) and (2.65), we have

$$v_{1n} = \left(\frac{(2\sqrt{3}-1)}{6}\right)v_0 \tag{2.66}$$

$$v_{2n} = -\frac{(5-\sqrt{3})v_0}{12} \tag{2.67}$$

Then, the velocity v_1 is given as

$$v_1 = \sqrt{v_{1t}^2 + v_{1n}^2} \tag{2.68}$$

Using the equations (2.63), (2.66), and (2.68), we have

$$v_1 = \sqrt{\frac{1}{4} + \left(\frac{(2\sqrt{3}-1)}{6}\right)^2}\,v_0 = (\sqrt{22-4\sqrt{3}})\frac{v_0}{6}$$

Similarly, the velocity v_1 is given as

$$v_2 = \sqrt{v_{2t}^2 + v_{2n}^2} \tag{2.69}$$

Using the equations (2.64), (2.67), and (2.69), we have

$$v_2 = \sqrt{\frac{3}{4} + \left(\frac{(5 - \sqrt{3})}{12}\right)^2} \, v_0 = (\sqrt{136 - 10\sqrt{3}})\frac{v_0}{12} \text{ Ans.}$$

Case 2. External impulsive force acts on the system ($m_1 + m_2$).
If any object m_2, say, of the system ($m_1 + m_2$) is constrained to move in the horizontal plane, during the collision between m_1 and m_2, another collision takes place between m_2 and ground. In this process, m_2 experiences an upward (normal) impulsive reaction force N' from the ground. This is an external force to the colliding bodies. Since the weights are non-impulsive forces, their impulses during the short time of collision can be neglected. However, the impulse of the impulsive force N', that is, $\int N' dt$ cannot be ignored. Hence, we cannot conserve the momentum of the system ($m_1 + m_2$) in any direction like the previous cases. However, we can conserve the momentum of the system ($m_1 + m_2$) perpendicular to the external impulsive force N', that is, horizontally. We can conserve the momentum of m_1 along the tangent drawn at the point of collision between m_1 and m_2 because no tangential impulsive force acts on the body m_1. However, we cannot conserve the tangential momentum of m_2 because the component of the impulsive N' force along the tangent is not zero. Let us assume that the velocities of m_1 and m_2 just after the collision are v_1 and v_2, respectively, where $\vec{v_1} = \vec{v_{1t}} + \vec{v_{1n}}$ and $\vec{v_2} = \vec{v_{2t}} + \vec{v_{2n}}$.

Conservation of tangential momentum of

$$m_1: \; m_1 \vec{u_{1t}} = m_1 \vec{v_{1t}} \tag{2.70}$$

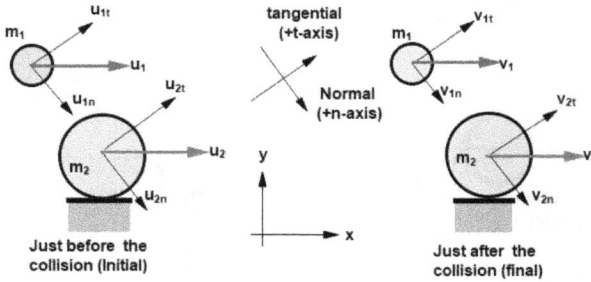

Just before the collision (Initial)   Just after the collision (final)

Conservation of horizontal momentum of the system ($m_1 + m_2$):
Resolving the velocities along horizontal (x-axis), we have

$$m_1 \vec{u_{1x}} + m_2 \vec{u_{2x}} = m_1 \vec{v_{1x}} + m_2 \vec{v_{2x}} \tag{2.71}$$

where $\vec{v_{1x}} = (\vec{v_{1t}})_x (\vec{v_{1n}})_x$ and $\vec{v_{2x}} = \vec{v_2}$ (because m_2 is constrained to move horizontally).

Coefficient of restitution along the normal:

$$-e(\vec{u_{1n}} - \vec{u_{2n}}) = \vec{v_{1n}} - \vec{v_{2n}} \qquad (2.72)$$

Now we have three equations and three unknown quantities (i.e., v_{1t}, v_{1n} and v_2). Let us apply the above procedure in the following example.

Example 23 A small rubber ball strikes a rough horizontal surface with a velocity v_0 at an angle θ with vertical. If the coefficient of restitution and coefficient of friction between the ball and horizontal surface are e and μ, respectively, find the velocity of the ball just after the impact.

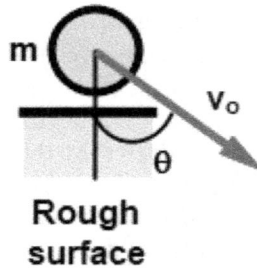

Rough surface

Ans. As the particle collides with the rough horizontal surface, impulsive force come into play in both tangent and normal to the horizontal surface. Let the normal impulse be $\int Ndt$ and horizontal impulse be $\int F\,dt$, where

$$F = \mu N \qquad (2.73)$$

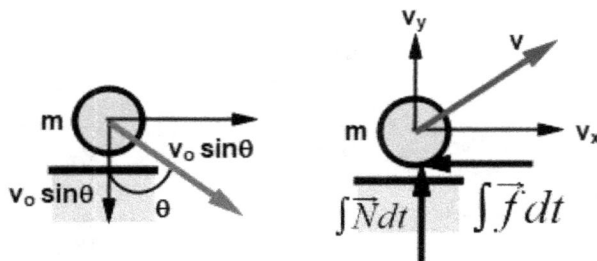

Writing the impulse momentum equation for the particle in x-direction, we have

$$mv_0 \sin\theta - \int F\,dt = mv_x \qquad (2.74)$$

Similarly, the impulse momentum equation in the y-direction gives

$$-mv_0 \cos\theta + \int N\,dt = mv_y \qquad (2.75)$$

The coefficient of restitution along the y-direction (normal) gives

$$v_y = -e(-v_0 \cos \theta)$$

$$\Rightarrow v_y = ev_0 \cos \theta \qquad (2.76)$$

We have four equations for four unknown quantities (i.e., $\int f \, dt$, $\int N \, dt$, v_x, and v_y). Let us solve them. Substituting $f = \mu N$ in equation (2.74) and eliminating $\int N \, dt$ from equations (2.74) and (2.75), we have

$$v_x = v_0[\sin \theta - \mu(1 + e)\cos \theta] \qquad (2.77)$$

From equations (2.76) and (2.77), we have

$$\overrightarrow{v} = \{v_0 \sin \theta - \mu(1 + e)\cos \theta\}\hat{i} + ev_0 \cos \theta \hat{i} \text{ Ans.}$$

Problem 1 A bullet of mass m is fired from a gun with a muzzle velocity u (velocity of bullet relative to gun). The gun is rigidly mounted with a trolley car at an angle of θ as shown in the figure. The mass of the trolley car plus gun is equal to M. If the trolley car moves with a velocity v on a smooth horozontal ground, find the (a) recoil velocity of the trolley car, (b) velocity of the trolley car just after the firing, (c) value of v so that the trolley car will stop just after the firing of the bullet, (d) horizontal range of the bullet relative to ground, (e) horizontal range of the bullet if initially the car was stationary($v = 0$), and (f) horizontal range of the bullet relative to the trolley car.

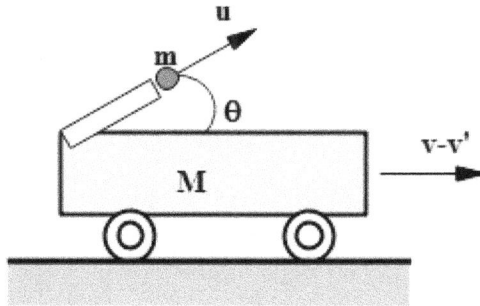

Solution

(a) Let us assume that the trolley car is starting. Let the recoil velocity be v'. Then the horizontal momentum of the system is

$$P_{\text{hor}} = m(u \cos \theta - v') + M(-v') = 0$$

$$\Rightarrow v' = \frac{mu \cos \theta}{M + m}$$

(b) Since the trolley car is initially moving towards the right with a velocity v, the recoil velocity will be subtracted from v to have the final velocity of the trolley car, given as

$$v_f = v - v'$$

$$= v - \frac{mu \cos \theta}{M + m}$$

(c) If $v_f = 0$, $v = \frac{mu \cos \theta}{M + m}$

(d) $R_H = (v_m)_x T$

$$|\vec{v}_{mM_x} + \vec{v}_{M_x}| \, T$$

$$R_H = [u \cos \theta + (v - v')]T$$

$$= \left(u \cos \theta + v - \frac{mu \cos \theta}{M + m} \right) \left(\frac{2u \sin \theta}{g} \right)$$

$$R_H = \left(\frac{Mu \cos \theta}{M + m} + v \right) \left(\frac{2u \sin \theta}{g} \right)$$

(e) If $v = 0$, $R_H = \frac{2Mu^2 \sin \theta \cos \theta}{(M + m)g}$

$$(R_H)_{\text{rel}} = (v_{mM})_x T$$

$$= (u \cos \theta)T$$

$$= u \cos \theta \left(\frac{2u \sin \theta}{g} \right)$$

$$= \frac{2u^2 \sin \theta \cos \theta}{g} \quad \text{Ans.}$$

Problem 2 A trolley car of mass m is at rest on a smooth horizontal surface. There are n numbers of boys on it each of mass m. The boys jump out of the car one by one in one direction with a constant velocity u relative to the trolley car. The total momentum delivered to the trolley car is X. Now, let all the boys jump simultaneously with same relative velocity u. In this case, the total momentum delivered to the trolley car is Y. (a) Find X and Y. (b) In which case will the total momentum delivered be greater?

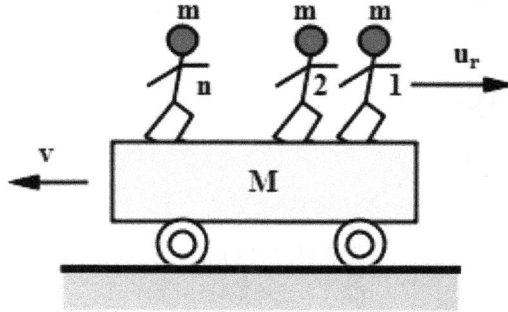

Solution

(a) Let the recoil velocity of the trolley car be v_1. Then the velocity of the boy is $(u_{rel} - v_1)$. The momentum of the system $(M + m)$ is

$$P = m(u_{rel} - v_1) - M'v_1 = 0$$

$$\Rightarrow v_1 = \frac{mu_{rel}}{M' + m}, \text{ where } M' = M + (n - 1)m$$

$$\Rightarrow v = \frac{mu_{rel}}{M + nm}$$

In the second collision of jumping the trolley car gets an additional backward recoil velocity v_2, so following the previous procedure, we have

$$v_2 = \frac{mv_{rel}}{M + (n - 1)m}$$

Just after the nth jumping the recoil velocity of the trolley car is

$$v_n = \frac{mu_{rel}}{M + m}$$

Then, adding all recoil velocities, the final velocity of the trolley car will be

$$v_f = v_1 + v_2 + \ldots\ldots + v_n$$

$$= mu_{rel}\left(\frac{1}{M + nm} + \frac{1}{M + (n - 1)m} + \ldots\frac{1}{M + m}\right)$$

$$= \sum_{n=1}^{n=n} \frac{mu_{rel}}{M + nm} \tag{2.78}$$

If all the boys simultaneously jump with the same relative velocity u_{rel} in the same direction, let the recoil velocity of the trolley car be v'_f. Then, the horizontal momentum of the system $(M + nm)$ is

$$-Mv'_f + nm\left(u_{rel} - v'_f\right) = 0$$

$$\Rightarrow v'_f = \frac{nmu_{rel}}{M + nm} \tag{2.79}$$

We can see that, $v'_f < v_f$.

(b) For two boys, $v'_f = \frac{2mu_{rel}}{M + 2m}$ and $v_f = \frac{mu_{rel}}{M + 2m} + \frac{mu_{rel}}{M + m}$

$$= \frac{(2M + 3m)mu_{rel}}{(M + 2m)(M + m)}$$

Then, $\dfrac{v'_f}{v_f} = \dfrac{\frac{2mu_{rel}}{M + 2m}}{\frac{(2M + 3m)mu_{rel}}{(M + 2m)(M + m)}}$

$$= \frac{2(M + m)}{2M + 3m} < 1$$

Problem 3 Two toy trains of masses M_1 and M_2 are moving side by side with the same velocity v on two parallel smooth tracks. A kid of mass m jumps from the first car to the second with a velocity u relative to the first car with an angle θ with the track. Find the velocities of the cars after the kid jumps.

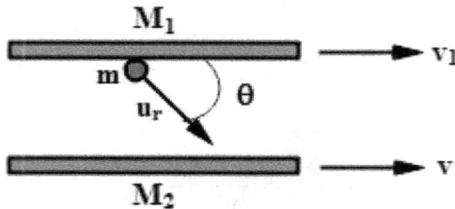

Solution

Let v_1 be the velocity of M_1 just after the the kid jumps.

Just before and after jumping from toy train M_1 we can conserve the linear momentum of $M_1 + m$ along the back to obtain

$$(M_1 + m)v = M_1 v_1 + m(u \cos \theta + v_1)$$

$$\Rightarrow v_1 = \frac{(M_1 + m)v - mu \cos \theta}{M_1 + m}$$

$$= v - \frac{mu \cos \theta}{M_1 + m} \quad \text{Ans.}$$

Just before and after jumping into the train M_2, we can conserve the momentum of $M_2 + m$ along the truck to obtain

$$m(u \cos \theta + v_1) + M_2 v = (M_2 + m)v_2$$

$$\Rightarrow m\left\{u \cos \theta + v - \frac{mu \cos \theta}{M_1 + m}\right\} + M_2 v = (M_2 + m)v_2$$

$$\Rightarrow v_2(M_2 + m) = m\left(\frac{M_1 u \cos \theta}{M_1 + m} + v\right)$$

$$\Rightarrow v_2 = \frac{m}{M_2 + m}\left(\frac{M_1 u \cos \theta}{M_1 + m} + v\right)$$

Problem 4 Two toy trains with one kid in each train have masses M_1 and M_2 and are moving side by side in opposite directions with velocities v_1 and v_2 on two parallel smooth tracks. The kids throw two identical chocolate bags each of mass m towards each other perpendicular to the line of their motion. Find the velocities of the trains after the exchange of the bags.

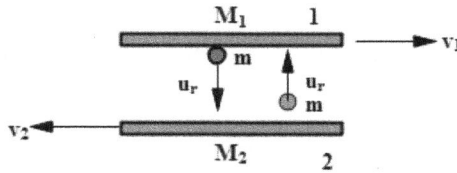

Solution

Exchanging the bags perpendicular to the toy trains does not give any impulsive force to the toy trains along the track. Bag 1 carries a momentum $mv_1\hat{i}$ and bag 2 has a momentum $-mv_2\hat{i}$. So when the bags 1 and 2 are received by the toy trains 2 and 1, respectively, the respective momenta will be vertically added conserving momentum of bag 1 + train 2 just before and after receiving the bag is

$$-M_2 v_2\hat{i} + mv_1\hat{i} = (M_2 + m)v_2$$

$$\Rightarrow \overrightarrow{v}'_2 = \frac{mv_1 - M_2 v_2}{M_2 + m}\hat{i}$$

Similarly conserving momenta of the bag 2 + train 1 just before and after receiving the bag is

$$M_1 v_1\hat{i} - mv_2\hat{i} = (M_1 + m)\overrightarrow{v}'_1$$

$$\Rightarrow \vec{v'}_1 = \frac{M_1 v_1 \hat{i} - m v_2 \hat{i}}{M_1 + m}$$

$$= \left(\frac{M_1 v_1 - m v_2}{M_1 + m} \right) \hat{i} \quad \text{Ans.}$$

Problem 5 A boy of mass m is standing at the end of a boat of mass M. The boy jumps out of the boat with a velocity u relative to the (i) boat and (ii) ground. Find the (a) recoil velocity of the boat and (b) change in KE of the system (boat + boy) just after jumping.

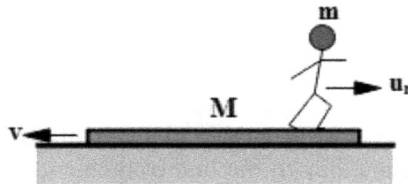

Solution

(a)

(i) Let the recoil velocity of the boat be v. Then, the velocity of the boy is $u - v$ relative to ground. The total momentum of boat + boy system is

$$-Mv + m(u - v) = 0$$

(because every body was at rest initially)

$$\Rightarrow v = \frac{mu}{M + m} \quad \text{Ans.}$$

(ii) If the velocity of the boy is u (relative to ground), the momentum of $M + m$ is

$$-Mv + mu = 0$$

$$\Rightarrow v = \frac{mu}{M} \quad \text{Ans.}$$

(b)

(i) The change in KE of the system is

$$\Delta K = \frac{m}{2} v_m^2 + \frac{Mv^2}{2}$$

$$= \frac{m}{2}(u - v)^2 + \frac{Mv^2}{2}$$

$$= \frac{m}{2}\left(u - \frac{mu}{M + m}\right)^2 + \frac{Mv^2}{2}$$

$$= \frac{m}{2}\left(\frac{Mu}{M + m}\right)^2 + \frac{M}{2}\left(\frac{mu}{M + m}\right)^2$$

$$= \frac{Mmu^2}{2(M + m)}$$

(ii) The change in KE is

$$\Delta K = \frac{1}{2}mv_m^2 + \frac{1}{2}Mv^2$$

$$= \frac{1}{2}mu^2 + \frac{1}{2}M\left(\frac{mu}{M}\right)^2$$

$$= \frac{1}{2}mu^2\left(1 + \frac{m}{M}\right) \text{ Ans.}$$

Problem 6 Two girls A and B each of mass m are standing on a trolley car of mass M, which is moving on a smooth horizontal floor with a velocity v_0. Girl A jumps with a velocity u relative to the trolley car. Then, girl B jumps with a velocity v relative to ground. Find the velocity of the trolley car just after jumping.

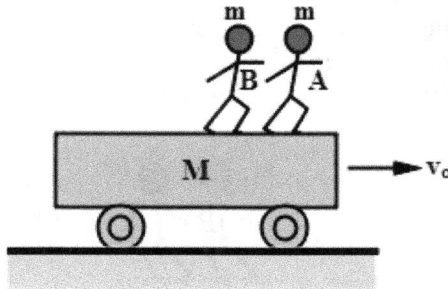

Solution

If girl A jumps out with a velocity u relative to the trolley car, let the velocity of the trolley car be v_1.

Conserving the linear momentum just before and after jumping

$$(M + 2m)v_0 = (M + m)v_1 + m(u + v_1)$$

$$\Rightarrow v_1 = \frac{(M + 2m)v_0 - mu}{M + 2m} \tag{2.80}$$

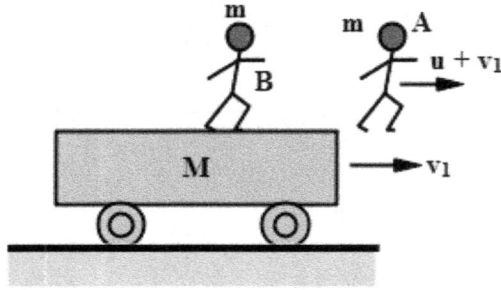

Now girl B jumps with a velocity v relative to ground. Then conserving the momentum of girl B + trolley car M just before and after the jumping, we have

$$(M + m)v_1 = mv + Mv_2$$

$$\Rightarrow v_2 = \left\{ \frac{(M + m)v_1 - mv}{M} \right\} \tag{2.81}$$

Putting v_1 from equations (2.80) in (2.81),

$$v_2 = \frac{(M + m)}{M} \left\{ \frac{(M + 2m)v_0 - mu}{M + 2m} \right\} - \frac{m}{M}v \quad \text{Ans.}$$

Problem 7 Two girls A and B of mass m_1 and m_2 are standing on a trolley car of mass M, which is moving on a smooth horizontal floor with a velocity $\vec{v}_o$. The girls jump simultaneously with velocities u relative to the trolley car and $\vec{v}$ relative to ground. Find the velocity of the trolley car just after jumping.

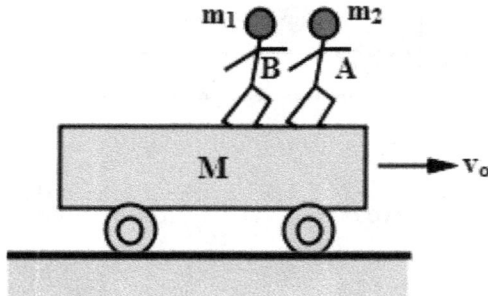

Solution

Let v_1 = velocity of the trolley car just after the jumping of the girls.
The momentum of the system ($A + B$ + trolley car) just before the jumping is

$$\vec{P_i} = (M + m_1 + m_2)\vec{v_0} \tag{2.82}$$

The momentum of the system ($A + B$ + trolley car) just after the jumping is

$$\vec{P_f} = M\vec{v_1} + m_1(\vec{u}+\vec{v_1}) + m_2\vec{v} \tag{2.83}$$

Conserving the momentum just before and after the jumping is

$$M\vec{v_1} + m_1(\vec{u}+\vec{v_1}) + m_2\vec{v} = (M + m_1 + m_2)\vec{v_0}$$

$$\Rightarrow \vec{v_1}(M + m_1) = (M + m_1 + m_2)\vec{v_0} - (m_1\vec{u}+m_2\vec{v})$$

$$\Rightarrow \vec{v_1} = \frac{(M + m_1 + m_2)\vec{v_0} - (m_1\vec{u}+m_2\vec{v})}{(M + m_1)} \quad \text{Ans.}$$

Problem 8 In the previous problem if $\vec{u}$ and $\vec{v}$ are the velocities of the girls relative to the trolley car, (a) find the velocities of the trolley after the jumping of the girls one after the other in the case when (i) girl A jumps first, then B jumps and (ii) girl A jumps first, then B jumps. (b) Are the velocities of the car after the two successive jumping the same?

Solution

The initial momentum is

$$\vec{P_i} = (M + m_1 + m_2)\vec{v_0} \tag{2.84}$$

Let $\vec{v_1}$ = velocity of the trolley car after girl A jumps. As girl A jumps with a velocity $\vec{u}$ relative to the trolley car, the velocity with respect to ground after jumping is $\vec{u}+\vec{v_1}$. Then the momentum of the entire system ($M + m_1 + m_2$) just after jumping of m_1 is

$$\vec{P_f} = (M + m_2)\vec{v_1} + m_1(\vec{u}+\vec{v_1}) \tag{2.85}$$

Since $\vec{P_f} = \vec{P_i}$, we have

$$(M + m_2)\vec{v_1} + m_1(\vec{u}+\vec{v_1}) = (M + m_1 + m_2)\vec{v_0}$$

$$\Rightarrow \vec{v_1} = \vec{v_0} - \frac{m_1\vec{u}}{M + m_1 + m_2} \tag{2.86}$$

Similarly, we can prove that the velocity of the trolley car $\vec{v_2}$ just after girl B jumps is

$$\vec{v_2} = \vec{v_0} - \frac{m_1\vec{u}}{M + m_1 + m_2} - \frac{m_2\vec{v}}{M + m_2} \tag{2.87}$$

If the procedure is reversed, from equation (2.87) interchanging the $\vec{u}$ and $\vec{v}$, we have the velocity of the trolley car just after A jumps as

$$v'_2 = v_0 - \frac{m_2 \vec{v}}{M + m_1 + m_2} - \frac{m_1 \vec{u}}{M + m_1} \tag{2.88}$$

Comparing equations (2.87) and (2.88), we can prove that, in general,

$$\vec{v_2} \neq \vec{v'_2}$$

Hence, if $m_1 = m_2$ and $\vec{u} = \vec{v}$ then $\vec{v_1} = \vec{v'_2}$.

Problem 9 A car of mass m and velocity u collides with a van of mass M moving with a velocity v at an L-shaped corner turning. Find the (a) velocity of the combined mass just after the collision and (b) loss of kinetic energy in collision.

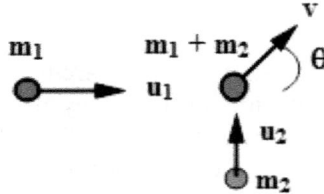

Solution

(a) Just after the inelastic collision let $v =$ common velocity of the combination. The conservation yields

$$m_1 \vec{u_1} + m_2 \vec{u_2} = (m_1 + m_2)\vec{v}$$

$$\Rightarrow \vec{v} = \frac{m_1 \vec{u_1} + m_2 \vec{u_2}}{m_1 + m_2} \tag{2.89}$$

Then, the change in KE of the system is

$$\Delta K = \frac{1}{2}(m_1 + m_2)v^2 - \left(\frac{1}{2}m_1 v_1^2 + \frac{1}{2}m_2 v_2^2\right) \tag{2.90}$$

Putting the value of v from equations (2.89) in (2.90),

$$\Delta K = \frac{1}{2}(m_1 + m_2)\frac{(m_1 \vec{u_1} + m_2 \vec{u_2})^2}{(m_1 + m_2)^2} - \left(\frac{1}{2}m_1 u_1^2 + \frac{1}{2}m_2 u_2^2\right)$$

$$= \frac{m_1^2 u_1^2 + m_2^2 u_2^2 + 2m_1 m_2 \vec{u_1}\vec{u_2}}{2(m_1 + m_2)} - \frac{1}{2}(m_1 u_1^2 + m_2 u_2^2)$$

$$= \frac{1}{2} \left[\frac{m_1 u_1^2 + m_2 u_2^2 + 2m_1 m_2 \overrightarrow{u_1} \overrightarrow{u_2} - (m_1 + m_2)(m_1 u_1^2 + m_2 u_2^2)}{m_1 + m_2} \right]$$

$$= -\frac{1}{2} \left[\frac{-2m_1 m_2 \overrightarrow{u_1} \overrightarrow{u_2} + m_1 m_2 u_1^2 + m_2 u_2 u_2^2}{m_1 + m_2} \right]$$

$$= -\frac{m_1 m_2}{2(m_1 + m_2)} (u_1^2 + u_2^2 - 2\overrightarrow{u_1} \overrightarrow{u_2})$$

$$\Delta K = -\frac{m_1 m_2 |\overrightarrow{u_1} - \overrightarrow{u_2}|^2}{2(m_1 + m_2)}$$

N.B.: This can be obtained by putting $e = 0$ (perfectly inelastic collision) in the formal.

$$\Delta K = \frac{m_1 m_2}{2(m_1 + m_2)} |\overrightarrow{u_1} - \overrightarrow{u_2}|(1 - e^2)$$

(b) Since $\overrightarrow{u_1}$ is perpendicular to $\overrightarrow{u_2}$, $|\overrightarrow{u_1} - \overrightarrow{u_2}|^2 = u_1^2 + u_2^2$, we have

$$\Delta K = \frac{-m_1 m_2}{2(m_1 + m_2)^2} (u_1^2 + u_2^2)$$

Putting $K_i = \frac{1}{2} m_1 u_1^2 + \frac{1}{2} m_2 u_2^2$, we have

$$\eta = \frac{|\Delta K|}{K} = \frac{m_1 m_2 (u_1^2 + u_2^2)}{(m_1 + m_2)^2 (m_1 u_1^2 + m_2 u_2^2)}$$

(c) The velocity of the combination is

$$\overrightarrow{v_1} = \frac{m_1 \overrightarrow{u_1} + m_2 \overrightarrow{v_2}}{m_1 + m_2}$$

$$= \frac{m_1 u_1 \hat{i} + m_2 u_2 \hat{j}}{m_1 + m_2}$$

$$= \frac{m_1 u_1}{m_1 + m_2} \hat{i} + \frac{m_2 u_2}{m_1 + m_2} \hat{j}$$

$$\Rightarrow v_1 = \frac{\sqrt{m_1^2 u_2^2 + m_2^2 u_2^2}}{m_1 + m_2} \quad \text{Ans.}$$

The direction of the velocity is

$$\theta = \tan^{-1} \frac{v_y}{v_x}$$

$$= \tan^{-1} \frac{\left\{ \frac{m_2 u_2}{(m_1 + m_2)} \right\}}{\left\{ \frac{m_1 u_1}{(m_1 + m_2)} \right\}}$$

$$= \tan^{-1} \left(\frac{m_2 u_2}{m_1 u_1} \right) \text{ Ans.}$$

Problem 10 The bob of mass m is released from rest from the given position so that it collides with another hanging bob of mass M after falling from a height h as shown in the figure. If the bobs move as a combined body after the collision and rises a height $h_c = \frac{h}{9}$, find $\frac{m}{M}$.

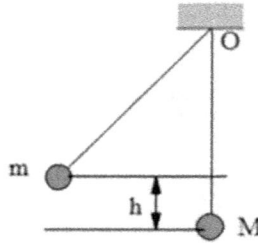

Solution

The velocity of m just before collision is

$$v_0 = \sqrt{2gh}$$

Then the conservation of linear momentum of $(M + m)$ is given as

$$m v_0 = (M + m) v_c$$

$$\Rightarrow v_c = \frac{m v_0}{M + m}$$

Then the height raised by the center of mass is

$$h_c = \frac{v_c^2}{2g}$$

$$= \frac{1}{2g} \left\{ \frac{m v_0}{M + m} \right\}^2$$

$$= \frac{1}{2g} \left(\frac{m}{M + m} \right)^2 v_0^2$$

Putting $\frac{v_0^2}{2g} = h$, we have

$$h_c = \left(\frac{m}{M+m}\right)^2 h \text{ Ans.}$$

If $\frac{h_c}{h} = \eta$

$$m = M\sqrt{\eta} + m\sqrt{\eta}$$

$$\Rightarrow \frac{m}{M} = \frac{\sqrt{\eta}}{1 - \sqrt{\eta}}$$

$$= \frac{\sqrt{\frac{1}{9}}}{1 - \sqrt{\frac{1}{9}}} = \frac{\frac{1}{3}}{\frac{2}{3}} = \frac{1}{2}$$

Problem 11 A block of mass m_1 is kept on another block of mass m_2. The coefficient of friction between the blocks and ground is μ. A ball of mass m collides with the lower block with a velocity v_0 at a height d as shown in the figure. Find (a) the maximum distance moved by the lower block (b) energy lost during the collision.

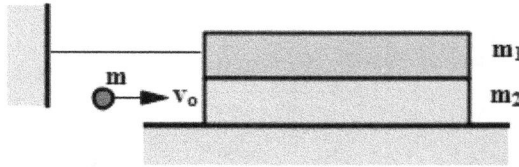

Solution

(a) Just after the collision, let $v =$ velocity of the combination $(m + m_2)$.
   Conserving the momentum

$$mv_0 = (m + m_2)v$$

$$\Rightarrow v = \frac{mv_0}{m + m_2} \quad\quad (2.91)$$

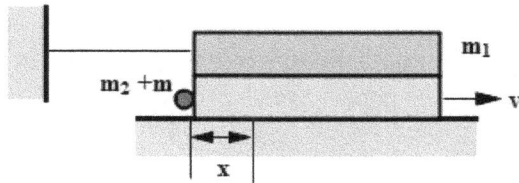

Applying the work–energy theorem

$$W_f = \Delta K$$

$$-\mu_k N_1 x - \mu_k N_2 x = -\frac{1}{2}(m + m_2)v^2$$

We know that

$$N_1 = m_1 g \text{ and } N_2 = (m_1 + m_2 + m)g$$

$$\Rightarrow \mu_k(m_1 + m_2 + m)gx + \mu m_1 gx = \frac{1}{2}(m + m_2)v^2$$

Using equations (2.91) and (2.79)

$$\mu_k(2m_1 + m_2 + m)gx = \frac{1}{2}(m + m_2)\left(\frac{mv_0}{m + m_2}\right)^2 \Rightarrow x = \frac{(m + m_1)}{(m + 2m_1 + m_n)\mu_k g}\left(\frac{mv_0}{m + m_2}\right)^2$$

(b) The fraction of energy loss during the collision is

$$\frac{|\Delta K|}{K} = \left|\frac{\frac{1}{2}(m + m_2)v^2 - \frac{1}{2}mv_0^2}{\frac{1}{2}mv_0^2}\right|$$

$$= 1 - \frac{v^2(m + m_2)}{v_0^2 m} \tag{2.92}$$

Putting equations (2.91) and (2.92)

$$\frac{\Delta K}{K} = 1 - \left[\frac{mv_0}{(m + m_2)}\right]^2\left(\frac{m + m_2}{m}\right)$$

$$= 1 - \frac{m}{m + m_2} = \frac{m_2}{m + m_2} \text{ Ans.}$$

Problem 12 A bullet of mass m hits a wooden block of mass M_1 placed on a smooth surface. It penetrates the block and comes out of it and strikes another block of mass M_2 and sticks to it, as shown in the figure. If the velocity of the first block is η times that of the second block, find the ratio of the velocity of the first block and the initial velocity for the bullet, find the percentage change in the speed of the bullet in its first collision.

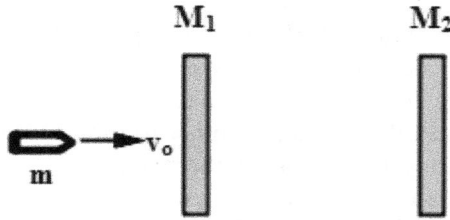

Solution

Conserving linear momentum of $M_1 + m$,

$$mv_0 = M_1 v_2 + mv_1$$

$$\Rightarrow v_2 = \frac{m}{M_1}(v_0 - v_1) \qquad (2.93)$$

Conserving the linear momentum of $M_2 + m$,

$$mv_1 = (M_2 + m)v_3$$

$$\Rightarrow v_3 = \frac{mv_1}{M_2 + m} \qquad (2.94)$$

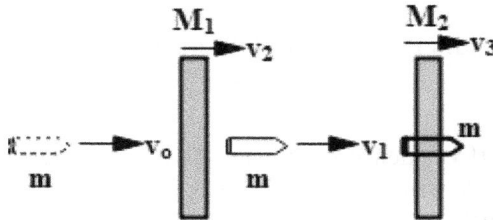

Since the velocity of M_1 is n times the velocity of $M_2 + m$, we can write

$$v_2 = \eta v_3 \qquad (2.95)$$

Using these equations,

$$\frac{m}{M_1}(v_0 - v_1) = \eta \frac{mv_1}{M_2 + m}$$

$$\Rightarrow v_1 \left[\frac{m}{M_1} + \eta \frac{m}{M_2 + m} \right] = \frac{m}{M_1} v_0$$

$$\Rightarrow \frac{v_1}{v_0} = \frac{\dfrac{m}{M_1}}{\dfrac{m}{M_1} + \eta \dfrac{m}{M_2 + m}} \quad \text{Ans.}$$

Problem 13 A hammer of mass M falls freely from a height h onto a vertical iron nail of mass m that stands vertical by penetrating a little bit into a horizontal wooden log. If the collision is inelastic, find the (a) fraction loss of energy of the hammer during the collision, (b) energy transmitted by the hammer to the nail (c) the average resistance offered by the wooden log to the nail if the maximum distance of penetration of the nail into the log $= x$, and (d) the average resistance offered by the wooden log if the time of penetration of the nail into the wooden log $= t$.

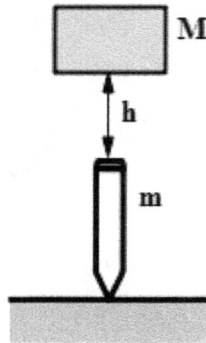

Solution

(a) The block collides inelastically with the nail with a velocity

$$v_0 = \sqrt{2gh} \tag{2.96}$$

Due to perfect inelastic collision ($e = 0$), both the bodies move as a combined block with a velocity u, say just after the collision. Conserving momentum just before and after the collision between M and m, we have

$$(M + m)u = Mv_0$$

$$\Rightarrow v = \frac{Mv_0}{M + m} \tag{2.97}$$

The fraction loss of KE in the collision is

$$\eta = \frac{|\Delta K|}{K} = \frac{\left|\frac{1}{2}(M + m)v^2 - \frac{1}{2}Mv_0^2\right|}{\frac{1}{2}mv_0^2}$$

$$= 1 - \frac{M + m}{M}\left(\frac{v}{v_0}\right)^2 \tag{2.98}$$

Using equations (2.97) and (2.98)

$$\eta = \frac{m}{M + m} \text{ Ans.}$$

(b) Then, the energy transmitted to the nail is

$$K' = (1 - n)K_0 = \left(1 - \frac{m}{M + m}\right)K_0 = \frac{M}{M + m}\left(\frac{M}{2}v_0^2\right)$$

Putting $v_0 = \sqrt{2gh}$ from equation (2.96)

$$K' = \frac{M}{M + m}\left\{\frac{M}{2}(2gh)\right\}$$

$$\Rightarrow K' = \frac{M^2 gh}{M + m} \text{ Ans.}$$

(c) Let $x =$ distance moved by the nail against the reaction force R. Then the work done by R is

$$W_R = -Rx$$

The work done by gravity is

$$W_{gr} = +(M + m)gx$$

The change in KE is

$$\Delta K' = 0 - K'$$

Applying the work–energy theorem,

$W_{total} = W_R + W_{gr} = \Delta K$, we have

$$-Rx - (M + m)gx = -K'$$

$$\Rightarrow -[R - (M + m)g] = -\frac{M^2 gh}{M + m}$$

$$\Rightarrow R = \frac{M^2 gh}{(M + m)x} + (M + m)g \text{ Ans.}$$

(d) If $t =$ time of penetration of the nail with the block, the change in momentum of the system $(M + m)$ is

$$\Delta \overrightarrow{P} = \overrightarrow{P_f} - \overrightarrow{P_i}$$

Since the combined body $(M + m)$ stays at rest finally and $\overrightarrow{P_i} = -(M + m)v$, we have

$$\Delta \overrightarrow{P} -=(M + m)v\hat{j} \tag{2.99}$$

Putting v from equations (2.98) in (2.99),

$$\Delta\vec{P} = (M + m)\frac{Mv_0}{M + m}\hat{j} = Mv_0\hat{j} \tag{2.100}$$

Putting v_0 from equations (2.96) in (2.100),

$$\Delta\vec{P} = M\sqrt{2gh}\hat{j} \tag{2.101}$$

The net impulse, that is, the impulse of the net force $\vec{F} -= \{R - (M + m)g\}\hat{j}$, is

$$\vec{F} = Ft\hat{j} = \{R - (M + m)g\}t\hat{j} \tag{2.102}$$

Using the impulse-momentum equation

$$\vec{I}\left(=\vec{F}t\right) = \Delta\vec{P} \tag{2.103}$$

By substituting $\vec{I}$ and $\Delta\vec{P}$-from equations (2.101) and (2.102), respectively, in equation (2.103), we have

$$\{R - (M + m)g\}t\,\hat{j} = M\sqrt{2gh}\hat{j}$$

$$\Rightarrow R = \frac{M\sqrt{2gh}}{t} + (M + m)g \text{ Ans.}$$

Problem 14 In the following figure, the block of mass m is released from a height h. It has an inelastic collision with the ground. After the inelastic collision the block of mass m continues to move up due to its inertia of motion before it comes down again due to the effect of gravity. Find the time after which the string will be taut again during the downward motion of the block m.

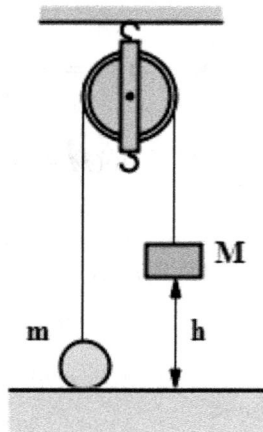

Solution

Let the velocity of the blocks just before the block M hits the ground be v. By conserving energy

$$\Delta K + \Delta U = 0$$

$$\Rightarrow \frac{1}{2}(M + m)v^2 + (mgh - Mgh) = 0$$

$$\Rightarrow v = \sqrt{\frac{2(M - m)gh}{M + m}} \tag{2.104}$$

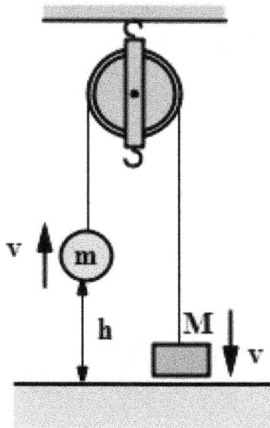

Due to complete inelastic collision, just after the collision, M remains stationary and m moves with an initial velocity v under gravity only until the string gets taut after a time t_1, say, from the collision of M with ground. The time after which the body m returns to its initial point is given as

$$t_1 = \frac{2v}{g} \tag{2.105}$$

Using equations (2.104) and (2.105)

$$t_1 = 2\sqrt{\frac{2(M - m)h}{(M + m)g}} \tag{2.106}$$

The time after which M collides with the ground is

$$t_2 = \frac{v}{a} \tag{2.107}$$

The acceleration of M and m as they move under gravity and string force is

$$a = \frac{Mg - mg}{M + m}$$

$$\Rightarrow a = \frac{M - m}{M + m}g \tag{2.108}$$

Using equations (2.104), (2.107), and (2.108), we have

$$t_2 = \frac{\sqrt{\frac{2(M - m)gh}{M + m}}}{\frac{M - m}{M + m}g}$$

$$\Rightarrow t_2 = \sqrt{\frac{2(M + m)h}{(M - m)g}} \tag{2.109}$$

Then, the time after which again the string will be taut is

$$t = t_1 + t_2 \tag{2.110}$$

Using equations (2.106), (2.109), and (2.110)

$$t = 2\sqrt{\frac{2(M - m)h}{(M + m)g}} + \sqrt{\frac{2(M + m)h}{(M - m)g}}$$

$$= \left(2 + \frac{M + m}{M - m}\right)\sqrt{\frac{2(M - m)h}{(M + m)g}} \quad \text{Ans.}$$

Problem 15 Block A of mass m hangs from a vertical spring of stiffness k. While the block A is in equilibrium, it is struck by another block of mass M with a velocity v_0 as shown in the figure. As a result, the combined block (A + B) moves up until the spring is relaxed. Find the value of v_0.

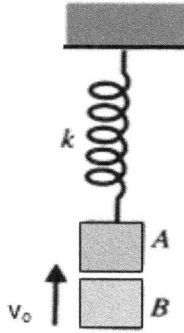

Solution

Let v = velocity of the combined mass $(M + m)$ just after the collision.

As the combined block will stop momentarily when the spring will be undeformed, the change in KE of the combined block $(M + m)$ is

$$\Delta K = -\frac{(M + m)}{2}v^2$$

Let x = initial elongation of the spring, the net work done up to the undeformed position of the spring is

W = work done by spring + work done by gravity.

$$= -\frac{k}{2}\{(0)^2 - x^2\} - (M + m)gx$$

$$= \frac{kx^2}{2} - (M + m)gx$$

Since $W = \Delta K$, according to the work–energy theorem, we have

$$-\frac{(M + m)}{2}v^2 = \frac{kx^2}{2} - (M + m)gx$$

Putting $x = \frac{mg}{k}$, we have

$$\frac{(M + m)}{2}v^2 = -\frac{m^2g^2}{2k} + \frac{(M + m)mg}{k}$$

$\Rightarrow v = \sqrt{\frac{(m + 2M)mg^2}{(M + m)k}}$, where $v = \frac{Mv_0}{M + m}$ obtained by conserving the linear momentum just before and after the collision. Then, we have

$$v_0 = \frac{M + m}{M}v$$

$$\Rightarrow v_0 = \frac{M + m}{M}\left(\sqrt{\frac{(m + 2M)m}{(M + m)k}}\right)g$$

$$= \left(g\sqrt{\frac{(m + 2M)(M + m)m}{M^2 k}} \right) \text{Ans.}$$

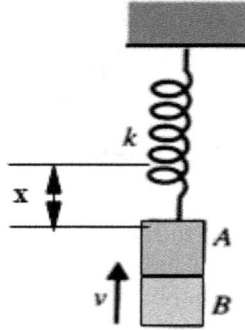

Problem 16 A block of mass m_1 is rigidly connected with the spring of stiffness k. It is released when the spring was compressed by a distance x. If another block of wet mud of mass m_2 sticks to m_1 at its mean position during its return journey, find the (a) maximum deformation of the spring (b) energy lost during the collision.

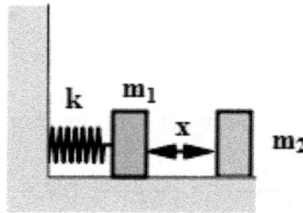

Solution

(a) Let the block m_1 collide with m_2 with a velocity v_0. The conservation of energy yields

$$-\frac{k}{2}x^2 + \frac{1}{2}m_1 v_0^2 = 0$$

$$\Rightarrow v_0 = \sqrt{\frac{k}{m_1}}\, x \tag{2.111}$$

Let $v = $ velocity of the combined mass just after the collision. The conservation of momentum yields

$$(m_1 + m_2)v = m_1 v_0$$

$$\Rightarrow v = \frac{m_1 v_0}{m_1 + m_2} \tag{2.112}$$

Let the maximum elongation of the spring be x_1 and conservation of energy $\Delta U + \Delta K = 0$

$$\Rightarrow \frac{k}{2}x_1^2 - \frac{1}{2}(m_1 + m_2)v^2 = 0$$

$$x_1 = \sqrt{\frac{m_1 + m_2}{k}}\, v \tag{2.113}$$

Using the last three equations,

$$x_1 = \left\{\sqrt{\frac{m_1 + m_2}{k}}\right\}\left\{\frac{m_1}{m_1 + m_2}\right\}\left\{\sqrt{\frac{k}{m_1}}\, x\right\}$$

$$\Rightarrow x_1 = \sqrt{\frac{m_1}{(m_1 + m_2)k}}\, x \text{ Ans.}$$

(b) The energy lost during the collision is

$$|\Delta K| = \frac{1}{2}\frac{m_1 m_2}{m_1 + m_2}v_0^2 \tag{2.114}$$

By using equations (2.111) and (2.114),

$$\Delta K = \frac{1}{2}\frac{m_1 m_2}{m_1 + m_2}\frac{k}{m_1}x^2$$

$$= \frac{m_2 k x^2}{2(m_1 + m_2)} \text{ Ans.}$$

Problem 17 Two bodies of mass $m_1 = 2$ kg moving horizontally with $v_1 = 10$ m s^{-1} collide with another body of mass $m_1 = 3$ kg moving vertically up with a velocity $v_2 = 5$ m s^{-1} at a height $h = 2$ m above the ground. If the bodies stick together just after the collision, find the (a) velocity just after the collision (b) time after which the combined mass reaches the ground (c) horizontal distance traversed by the combined mass (d) maximum height attained by the combined mass (e) speed and (f) angle of striking of the combined mass with ground.

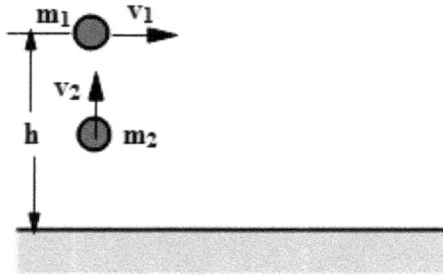

(a) The velocity of the combination just after the collision is

$$\vec{v} = \frac{m_1 v_1 \hat{i} + m_2 v_2 \hat{j}}{m_1 + m_2}$$

$$= \frac{2 \times 10\hat{i} + 3 \times 5\hat{j}}{2 + 3} = 4\hat{i} + 3\hat{j}$$

$$\Rightarrow v = 5 \text{ m s}^{-1} \text{ Ans.}$$

Solution

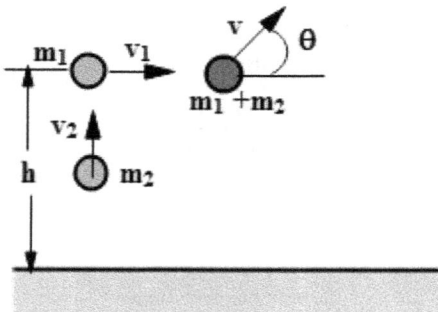

(b) Let it take a time t to reach the ground. Applying

$$\vec{y} = \vec{v}_y + \frac{1}{2}\vec{g}t^2$$

$$-2 = 3t + \frac{1}{2} \times (-10)t^2$$

$$\Rightarrow 5t^2 - 3t - 2 = 0$$

$$\Rightarrow t = \frac{3 \pm \sqrt{19 - 4 \times 5 \times (-2)}}{10}$$

$$= \frac{3 \pm \sqrt{49}}{10} = \frac{7 + 3}{10} = 1 \text{ s Ans.}$$

(c) $R_H = v_x t = 4 \times 1 = 4 \text{ m}$

(d) $h_{max} = h + \frac{v_y^2}{2g} = 2 + \frac{(3)^2}{2 \times 10}$

$$= 2 + 0.45 = 2.45 \text{ m}$$

(e) $v' = \sqrt{v^2 + 2gh}$

$$= \sqrt{(5)^2 + 2 \times 10 \times 2}$$

$$= \sqrt{65} \text{ m s}^{-1}$$

(f) The equal angle $\phi = \tan^{-1} \frac{v'_y}{v'_x}$

$$= \tan^{-1} \frac{\sqrt{u_y^2 + 2gy}}{v_x}$$

$$= \tan^{-1} \frac{\sqrt{(3)^2 + 2 \times 10 \times 2}}{4}$$

$$= \tan^{-1} \frac{7}{4} \text{ with horizontal}$$

Problem 18 Two bodies of masses 2 kg and 1 kg moving with velocities $(2\hat{i} - 3\hat{j} + \hat{k})$ m s^{-1} and $(\hat{i} + 2\hat{j} - 2\hat{k})$ m s^{-1}, respectively, collide inelastically. Find the (a) velocity of the combined mass just after the collision, (b) loss of KE during the collision, and (c) fraction loss of KE during the collision.

Solution

(a) Conserving the momentum just before and after the collision, we have

$$\overrightarrow{v} = \frac{m_1 \overrightarrow{v_1} + m_2 \overrightarrow{v_2}}{m_1 + m_2}$$

$$= \frac{2 \times (2\hat{i} - 3\hat{j} + \hat{k}) + 1(\hat{i} + 2\hat{j} - 2\hat{k})}{2 + 1}$$

$$\overrightarrow{v} = \frac{5\hat{i} - 4\hat{j}}{3}$$

Then, $|\overrightarrow{v}| = \sqrt{\frac{25 + 16}{9}} = \frac{\sqrt{41}}{3} \text{m s}^{-1}$. Ans.

(b) The loss of KE is

$$\Delta K = \frac{1}{2} \frac{m_1 m_2}{m_1 + m_2} |\vec{v_1} - \vec{v_2}|^2$$

$$= \frac{1}{2} \times \frac{2 \times 1}{2 + 1} |(2\hat{i} - 3\hat{j} + \hat{k}) - (\hat{i} + 2\hat{j} - 2\hat{k})|^2$$

$$= \frac{1}{3} |\hat{i} - 5\hat{j} + 3\hat{k}|^2 = \frac{1}{3}(1 + 25 + 9) = \frac{35}{3} \text{ J}$$

(c) The KE of the system $(m_1 + m_2)$ just before the collision is

$$K = K_1 + K_2 = \frac{1}{2}m_1 v_1^2 + \frac{1}{2}m_2 v_2^2$$

$$= \frac{1}{2} \times 2 \times \left|2\hat{i} - 3\hat{j} + \vec{k}\right|^2 + \frac{1}{2} \times 1 |\hat{i} + 2\hat{j} - 2\hat{k}|^2$$

$$= (4 + 9 + 1) + \frac{1}{2}(1 + 4 + 4)$$

$$= 14 + \frac{9}{2} = \frac{37}{2} J$$

Then, $\eta = \frac{\Delta K}{K} = \frac{\frac{35}{3}}{\frac{37}{2}} = \frac{70}{111}$. Ans.

Problem 19 A shell is projected with a velocity v at an angle θ. It explodes at its highest point in to two equal fragments. If one fragment retraces its path, find the (a) velocity of the other fragment just after the explosion (b) distance between the fragments on the ground (c) maximum possible distance between the fragments.
Solution

(a) To trace its path the fragment must recoil back with a velocity $v \cos \theta$. Let the second fragment have velocity v'. Now conserving the linear momentum just before and after the explanation,

$$-\frac{m}{2}v \cos \theta + \frac{m}{2}v' = mv \cos \theta$$

$$v' = 3v \cos \theta \text{ Ans.}$$

(b) Since the time taken by both fragments will be equal to
$$t = \sqrt{\frac{2h}{g}}, \text{ where } h = \frac{v^2 \sin^2 \theta}{2g} \text{ and we have } t = \frac{v \sin \theta}{g}.$$

Then the fragments hit the ground simultaneously at a distance of separation

$$x = v_{rel}t = (v' + v\cos\theta)t$$

$$\Rightarrow x = (3v\cos\theta + v\cos\theta)\frac{v\sin\theta}{g}$$

$$= \frac{4v^2\sin\theta\cos\theta}{g}$$

(c) The maximum possible distance of separation will happen when $\sin^2\theta = 1$ or $\theta = 45°$. Then

$$x\mid_{max} = \frac{2v^2}{g} \text{ Ans.}$$

Problem 20 A bird of mass m flies off the bob of mass M which is hanging by a string. If the string is (a) inextensible (b) extensible, find the change in KE just after the bird flew. Assume that u = velocity of the bird relative to the bob.
Solution

(a) If the string is inextensible the bob will be constrained to move backward with a velocity v, say.
Conserving the horizontal momentum of the system $(M + m)$

$$m(u\cos\theta - v) = Mv$$

$$\Rightarrow v = \frac{mu\cos\theta}{M + m} \tag{2.115}$$

The change in KE of the system is

$$\Delta K = \frac{1}{2}Mu^2 + \frac{1}{2}mu_0^2 \tag{2.116}$$

Using equations (2.115), (2.116)

$$\Delta K = \frac{1}{2}M\left(\frac{mu\cos\theta}{M + m}\right) + \frac{1}{2}m\left(\frac{Mu\cos\theta}{M + m}\right)^2 + \frac{1}{2}mu^2\sin^2\theta$$

$$= \frac{Mmu^2}{2(M + m)^2}(M + m)\cos^2\theta + \frac{1}{2}mu^2\sin^2\theta$$

$$\Rightarrow \frac{Mmu^2 \cos^2 \theta}{2(M+m)} + \frac{1}{2}mu^2 \sin^2 \theta$$

$$= \frac{mu^2}{2}\left[\frac{M}{M+m}\cos^2 \theta + \sin^2 \theta\right]$$

$$= \frac{mu^2}{2}\left[\frac{M+m\sin^2 \theta}{M+m}\right] \text{ Ans.}$$

(b) If the string is extensible, the bob will recoil opposite to $\vec{u}$. Conserving linear momentum

$$m(u - v) + MV = 0$$

$$\Rightarrow v = \frac{mu}{M+m} \text{ Ans.}$$

Then, $\Delta K = \frac{1}{2}m(u - v)^2 + \frac{1}{2}Mv^2$

$$= \frac{1}{2}m\left(u - \frac{mu}{M+m}\right)^2 + \frac{1}{2}M\left(\frac{mu}{M+m}\right)^2$$

$$= \frac{1}{2}\frac{Mmu^2}{M+m} \text{ Ans.}$$

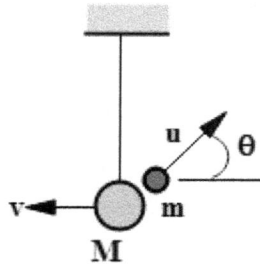

Problem 21 A cannon of mass M fires a shot of mass m with muzzle velocity u. The cannon is connected with the rigid vertical wall by a light spring of stiffness k. Find the (a) recoil velocity of the cannon (b) change in KE in the firing (c) velocity of the cannon just after the firing (d) maximum compression of the spring.

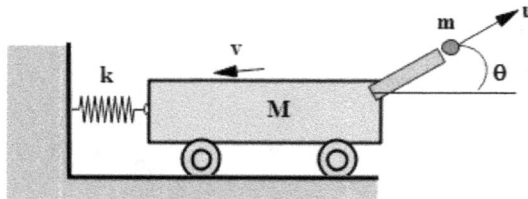

Solution

(a) Let $v =$ recoil velocity of the gun. Then the horizontal velocity of the cannon relative to ground is

$$v_{mx} = u \cos \theta - v$$

The horizontal momentum of the system $(M + m)$ is given as

$$P_x = -Mv + m(u \cos \theta - v) = 0$$

$$\Rightarrow u = \frac{mu \cos \theta}{M + m} \text{ Ans.}$$

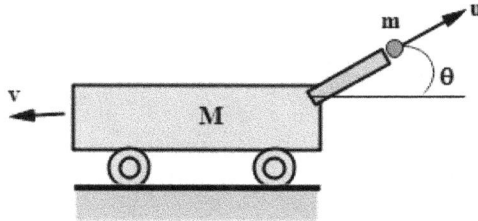

(b) The change in KE of the system $(M + m)$ is

$$\Delta K = \frac{Mv^2}{2} + \frac{1}{2}mv_m^2$$

where $v_m^2 = v_{mx}^2 + v_{my}^2$

$$= (u \cos \theta - u)^2 + (u \sin \theta)^2$$

$$= \left(u \cos \theta - \frac{mu \cos \theta}{M + m} \right)^2 + (u \sin \theta)^2$$

$$\Rightarrow \Delta K = \frac{Mv^2}{2} + \frac{1}{2}m \left\{ \left(\frac{Mu \cos \theta}{M + m} \right)^2 + (u \sin \theta)^2 \right\}$$

$$= \frac{M}{2} \left(\frac{mu \cos \theta}{M + m} \right)^2 + \frac{m}{2} \left(\frac{Mu \cos \theta}{M + m} \right)^2 + \frac{1}{2}mu^2 \sin^2 \theta$$

$$= \frac{Mmu^2 \cos^2 \theta}{2(M + m)^2}(M + m) + \frac{1}{2}mu^2 \sin^2 \theta$$

$$= \frac{Mmu^2 \cos^2 \theta}{2(M + m)} + \frac{1}{2}mu \sin^2 \theta$$

$$= \frac{1}{2}mu^2 \left\{ \frac{M \cos^2 \theta}{M + m} + \sin^2 \theta \right\}$$

$$= \frac{1}{2}mu^2 \left(\frac{M + m \sin^2 \theta}{M + m} \right)$$

$$= \frac{m(M + m \sin^2 \theta)u^2}{2(M + m)} \quad \text{Ans.}$$

(c) The velocity of the cannon is

$$\overrightarrow{v_m} = \overrightarrow{v_{mx}} + \overrightarrow{v_{my}}$$

$$= \frac{Mu \cos \theta}{M + m} \hat{i} + u \sin \theta \hat{j}$$

The angle made by $\overrightarrow{v_m}$ with horizontal is

$$\tan \phi = \frac{v_{my}}{v_{mx}} = \frac{u \sin \theta}{\frac{mu \cos \theta}{M + m}}$$

$$\Rightarrow \tan \phi = \frac{M + m}{M} \tan \theta$$

$$\Rightarrow \phi = \tan^{-1} \left(\frac{M + m}{M} \tan \theta \right)$$

(d) $\frac{k}{2}x^2 = \frac{1}{2}Mv^2$

$$\Rightarrow \frac{kx^2}{2} = \frac{1}{2}M \left(\frac{mu \cos \theta}{M + m} \right)^2$$

$$\Rightarrow x = \sqrt{\frac{M}{k}} \left(\frac{Mu \cos \theta}{M + m} \right) \quad \text{Ans.}$$

Problem 22 Two balls moving along $+$ x-axis experience head-on collision. If $e =$ coefficient of restitution, find the (a) velocity of each ball just after the collision (b) energy loss during the collision (c) impulse acting on the balls.

Solution

(a) Conversing linear momentum of the system $(m_1 + m_2)$

$$m_1 v_1 + m_2 v_2 = m_1 v_1 + m_2 u_2 \tag{2.117}$$

NIF

$$u_1 - u_2 = -e(u_1 - u_2) \tag{2.118}$$

Equations $(2.117) + (2.118) \times m_2$ gives

$$(m_1 + m_2)v_1 = u_1(m_1 - em_2) + m_2u_2(1 + e)$$

$$\Rightarrow v_1 = \frac{m_1 - em_2}{m_1 + m_2}u_1 + \frac{m_2(1 + e)u_2}{m_1 + m_2} \text{ Ans.}$$

Equations $(2.117) - (2.118) \times m_1$

$$m_1v_1 + m_2v_2 = m_1u_1 + m_2u_2$$
$$m_1v_1 - m_1v_2 = -em_1(u_1 - u_2)$$
$$- \quad + \quad +$$

$$v_2(m_1 + m_2) = m_1u_1(1 + e) + (m_2 - em_1)u_2$$

$$v_2 = \frac{m_1u_1(1 + e)}{m_1 + m_2} + \frac{m_2 - em_1}{m_1 + m_2}u_2 \text{ Ans.}$$

(b) During collision the net external force $= 0$. Hence, the KE of the center of mass K_c remains conserved and the change in KE of the system is equal to the change in KE $\Delta K'$ of the system relative to its center of mass.

$$\Rightarrow \Delta K = \Delta K_c + \Delta K' = \Delta K' (\because \Delta K_c = 0)$$

The KE of the two systems relative to center of mass force just before and after the impact are given as

$$K'_i = \frac{1}{2}\frac{m_1m_2(u_1 - u_2)^2}{m_1 + m_2}$$

$$K'_f = \frac{1}{2}\frac{m_1m_2}{m_1 + m_2}(v_1 - v_2)^2$$

Then the change in K' is

$$\Delta K' = K'_f - K'_i$$

$$= \frac{1}{2}\frac{m_1m_2}{m_1 + m_2}\{(v_1 - v_2)^2 - (u_1 - u_2)^2\}$$

Putting $\Delta K' = \Delta K$, $v_1 - v_2 = -e(u_1 - u_2)$, we have

$$\Delta K = -\frac{m_1m_2}{2(m_1 + m_2)}(1 - e^2)(u_1 - u_2)^2$$

In general, for head-on collision

$$\Delta K = -\frac{m_1 m_2}{m_1 + m_2}(1 - e^2)|\overrightarrow{u_1} - \overrightarrow{u_2}|^2 \text{ Ans.}$$

(c) The impulse acting on m_1 is

$$I_1 = m(v_1 - u_1)$$

$$= \left\{\frac{m_1 - em_2}{m_1 + m_2}u_1 + \frac{m_2(1 + e)u_2}{m_1 + m_2} - u_1\right\}$$

$$= m_1\left\{\left(\frac{m_1 - em_2}{m_1 + m_2} - 1\right)u_1 + \frac{m_2(1 + e)u_2}{(m_1 + m_2)}\right\}$$

$$= m_1\left\{\frac{-(e + 1)m_2 u_1}{m_1 + m_2} + \frac{m_2(1 + e)u_2}{m_1 + m_2}\right\}$$

$$= -\frac{m_1 m_2(1 + e)}{m_1 + m_2}(u_1 - u_2)$$

In general,

$$\overrightarrow{I_1} = -\frac{m_1 m_2}{m_1 + m_2}(1 + e)(\overrightarrow{u_1} - \overrightarrow{u_2})$$

and

$$\overrightarrow{I_2} = +\frac{m_1 m_2(1 + e)}{m_1 + m_2}(\overrightarrow{u_1} - \overrightarrow{u_2})$$

Problem 23 In the previous problem if the balls have equal mass and they have velocities $2v_0$ and v_0, find the (a) velocity of the balls just after the collision (b) impulse acting on each ball (c) fraction of energy loss of the system of two balls in the collision.

(a) We know that

$$\overrightarrow{v_1} = \frac{m_1 - em_2}{m_1 + m_2}\overrightarrow{u_1} + \frac{m_2(1 + e)}{m_1 + m_2}\overrightarrow{u_2}$$

$$= \frac{m - \frac{1}{2} \times 2m}{m + 2m}2v_0\hat{i} + \frac{2m\left(1 + \frac{1}{2}\right)}{m + 2m}v_0\hat{i}$$

$$= v_0\hat{i}$$

Similarly $\vec{v_2} = \frac{m_1(1+e)}{m_1+m_2}\vec{u_1} + \frac{m_2-em_1}{m_1+m_2}\vec{u_2}$

$$=\frac{m\left(1+\frac{1}{2}\right)}{m+2m}(2v_0\hat{i}) + \frac{2m - \frac{1}{2} \times m}{m+2m}v_0\hat{i}$$

$$=v_0\hat{i} + \frac{v_0}{2}\hat{i} = \frac{3v_0}{2}\hat{i}$$

(b) $\vec{I_1} = -\frac{m_1m_2}{m_1+m_2}(1+e)\vec{u}$

$$=-\frac{(m)(2m)}{2+2m}\left(1+\frac{1}{2}\right)(2v_0\hat{i} - v_0\hat{i})$$

$$=-mv_0\hat{i} \& \vec{I_2} = mv_0\hat{i}$$

(c) The loss of KE of the system during the collision is

$$|\Delta K| = \frac{1}{2}\frac{m_1m_2}{m_1+m_2}(1-e^2)(\vec{u_1} - \vec{u_2})^2$$

$$=\frac{1}{2}\frac{(m)(2m)}{m+2m}\left(1-\frac{1}{4}\right)(2v_0\hat{i} - v_0\hat{i})^2$$

$$=\frac{1}{2} \times \frac{2}{3}m \times \frac{3}{4}v_0^2 = \frac{mv_0^2}{4}$$

The KE of the system just before the collision is

$$K_i = \frac{1}{2}m_1u_1^2 + \frac{1}{2}m_2u_2^2$$

$$=\frac{1}{2}(m)(2v_0)^2 + \frac{1}{2}(2m)(v_0)^2$$

$$=3mv_0^2$$

Then the fraction of energy lost is

$$\eta = \left|\frac{\Delta K}{K_i}\right| = \frac{\frac{mv_0^2}{4}}{3mv_0^2} = \frac{1}{12} \text{ Ans.}$$

Problem 24 In problem 22, discuss the cases when the ratio of masses of the balls is (a) much lesser than one (b) equal to one (c) much greater than one.

Solution

(a) The final velocity of m_1 is

$$\vec{v_1} = \frac{m_1 - em_2}{m_1 + m_2}\vec{u_1} + \frac{m_2(1 + e)}{m_1 + m_2}\vec{u_2}$$

$$= \frac{1 - e\frac{m_2}{m_1}}{1 + \frac{m_2}{m_1}}\vec{u_1} + \frac{\frac{m_2}{m_1}(1 + e)}{1 + \frac{m_2}{m_1}}\vec{u_2}$$

Since $m_1 >> m_2$ $\frac{m_2}{m_1} \ll 1$. Then we can replace $\frac{m_2}{m_1}$ compared to 1 to obtain.

$$\vec{v_1} = \vec{u_1} + \frac{m_2}{m_1}(1 + e)\vec{u_2}$$

Since $\frac{m_2}{m_1}$ is extremely small, the second turn can be ignored to obtain

$$\vec{v_1} = \vec{u_1}$$

This physically systems that the velocity of the much heavier body practically remain constant.

Similarly $\vec{v_2} = \frac{m_1(1 + e)}{m_1 + m_2}\vec{u_1} + \frac{m_2 em_1}{m_1 + m_2}\vec{u_2}$

$$= \frac{(1 + e)\vec{u_1}}{1 + \frac{m_2}{m_1}} + \frac{\frac{m_2}{m_1} - e}{1 + \frac{m_2}{m_1}}\vec{u_2}$$

Since $\frac{m_2}{m_1} \ll 1$ and $\frac{m_2}{m_1} \ll e$, we have

$$\vec{v_2} \cong (1 + e)\vec{u_1} - e\vec{u_2}$$

This result can be obtained by just applying the NIF setting $v_1 = u_1$ in the formula

$$-e(\vec{u_1} - \vec{u_2}) = \vec{v_1} - \vec{v_2}$$

$$\vec{v_2} = \vec{v_1} + e(\vec{u_1} - \vec{u_2})$$

$$= \vec{u_1} + e\vec{u_1} - e\vec{u_2}$$

$$= \vec{u_1}(1 + e) - e\vec{u_2}$$

(b) If $\frac{m_1}{m_2} = 1$, we have

$$\vec{v_1} = (1 - e)\frac{\vec{u_1}}{2} + \frac{1 + e}{2}\vec{u_2}$$

$$\vec{v_2} = \frac{1+e}{2}\vec{u_1} + \frac{1-e}{2}\vec{u_2}$$

* If $e = 1$ and $\frac{m_1}{m_2} = 1$, $\vec{v_1} = \vec{u_2}$ and $\vec{v_2} = \vec{u_1}$.

This physically signifies that the momenta are exchanged for elastic collision and identical masses.

(c) If $\frac{m_1}{m_2} \ll 1$, ignoring this term in the equations

$$\vec{v_1} = \frac{\frac{m_1}{m_2} - e}{\frac{m_1}{m_2} + 1}\vec{u_1} + \frac{1+e}{1 + \frac{m_1}{m_2}}\vec{u_2}$$

$$\vec{v_2} = \frac{\frac{m_1}{m_2}(1+e)}{1 + \frac{m_1}{m_2}}\vec{u_1} + \frac{1 - e\frac{m_1}{m_2}}{1 + \frac{m_1}{m_2}}\vec{u_2}$$

we have $\vec{v_1} \cong -e\vec{u_1} + (1+e)\vec{u_2}$

So also we have $\vec{v_2} \cong \vec{u_2}$

Alternative method: This physically signifies that the second body's velocity remains practically constant as it is much heavier than the first body. Putting $\vec{v_2} \simeq \vec{u_2}$ in the NIF,
$-e(\vec{u_1} - \vec{u_2}) = \vec{v_1} - \vec{v_2}$ we have

$$v_1 = v_2 - e(\vec{u_1} - \vec{u_2})$$

$$= \vec{u_2} - e\vec{u_1} + e\vec{u_2}$$

$$= -e\vec{u_1} + (e+1)\vec{u_2}$$

Head-on collision (one-dimensional collision):

Problem 25 A ball of mass m is released from rest from a height h on to a surface. The ball collides with the ground many times before coming to rest. The coefficient of restitution is e. Find the average force exerted by the ground on the ball over the time of (a) first collision (b) first downward motion (c) first downward plus first upward motion (d) total up and down motion (e) motion of a cycle considering elastic collision ($e = 1$). You can take the necessary assumptions.

Solution
   (a) Let the ball strike the ground with a velocity v_0 and rebound with a velocity v. Then the change in momentum of the ball is

$$\Delta \vec{P} = m\{v\hat{j} - (-v_0\hat{j})\}$$

$$= m(v + v_0)\hat{j}$$

If Δt = time of contact of the ball with the ground, the average force imparted on the ball due to the ground and the time of impact is $\vec{F}_{av} = \frac{\Delta \vec{p}}{\Delta t}$

$$= \frac{m(v + v_0)\hat{j}}{\Delta t}, \text{ where } v = ev_0$$

$$= \frac{(1 + e)mv_0\hat{j}}{\Delta t}, \text{ where } v_0 = \sqrt{2gh}$$

$$= \frac{(1 + e)m\sqrt{2gh}}{\Delta t}\hat{j} \text{ Ans.}$$

(b) If we average the fore over the time of first downward of fall,

$$\vec{F}'_{av} = \frac{\Delta \vec{P}}{\Delta t'}$$

$$= \frac{(1 + e)m\sqrt{2gh}}{\Delta t'}\hat{j},$$

where $\Delta t' = \frac{v_0}{g} = \sqrt{\frac{2h}{g}}$

$$\Rightarrow F'_{av} = \frac{(1 + e)m\sqrt{2gh}}{\sqrt{\frac{2h}{g}}}$$

$$\Rightarrow F'_{av} = (1 + e)mg\hat{j}$$

(c) If we average the force over the time of free fall for first up and downward motion,

$$\vec{F}_{av_1} = \frac{\Delta \vec{P}}{\Delta t''}$$

$$= \frac{(1 + e)m\sqrt{2gh}}{\Delta t''}\hat{j}$$

where $\Delta t'' = \sqrt{\frac{2h}{g}}(1 + e)$

$$\Rightarrow \vec{F}_{av_1} = mg\hat{j}$$

(d) If we take n-repeated cycles (upward + downward − motion), then the change in momentum will be

$$\Delta \overrightarrow{P} = m(\overrightarrow{v_n} - \overrightarrow{v_0})$$

After the nth collision $v_n = e^n v_0$, we have

$$\Delta \overrightarrow{P} = m(1 + e^n)v_0 \hat{j}$$

The total time will be

$$T = \frac{v_0}{g} + \frac{2ev_0}{g} + \frac{2e^2 v_0}{g} + \ldots + \frac{2e^n v_0}{g}$$

If n is very large C,

$$T = \frac{v_0}{3}[1 + 2(1 + e + e^2 + \ldots)]$$

$$= \frac{v_0}{g}\left[1 + \frac{2e}{1 - e}\right]$$

$$= \frac{v_0}{g}\left(\frac{1 + e}{1 - e}\right)$$

The velocity after a number of collisions will be zero. Then the change in momentum will be

$$\Delta \overrightarrow{P} = mv_0 \hat{j}$$

Hence, the average for any the total time is

$$\overrightarrow{F_{av}} = \frac{\Delta \overrightarrow{P}}{T} = \frac{mv_0 \hat{j}}{\frac{v_0}{g}\left(\frac{1 + e}{1 - e}\right)} = \left(\frac{1 - e}{1 + e}\right)mg$$

(e) If the collision is particularly elastic $\Delta \overrightarrow{P} = 2mv_0 \hat{j}$.

Then the time period of a cycle is

$$\Delta t = \frac{2v_0}{g}$$

Hence, the average fore for any time of each cycle is

$$\overrightarrow{F} = \frac{\Delta \overrightarrow{P}}{\Delta t} = \frac{2mv_0 \hat{j}}{\frac{2v_0}{g}} = mg\hat{j}$$

*N.B.: The gravity force is $\overrightarrow{F} = -mg\hat{j}$.

Problem 26 A ball of mass m collides with the block of mass M with a velocity v_0 when the block hangs from a light spring of stiffness k. If the block moves up till the spring will become relaxed, find the velocity v_0. Assume e = coefficient of restitution of collision.

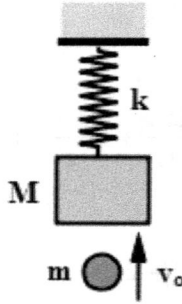

Solution
Conservation of momentum

$$mv_1 + Mv_2 = mv_0 \qquad (2.119)$$

NIF

$$-ev_0 = v_1 - v_2 \qquad (2.120)$$

Solving equations (2.119) and (2.120)

$$m(-ev_0 + v_2) + Mv_2 = mv_0$$

$$\Rightarrow v_2 = \frac{m(1 + e)v_0}{M + m} \qquad (2.121)$$

Applying the work–energy theorem

$$W = \Delta K$$

$$W_{gr} + W_{sp} = \Delta K$$

$$\Rightarrow -Mgx_0 - \frac{K}{2}(0^2 - x_0^2) = \left(0 - \frac{1}{2}Mv_2^2\right)$$

$$\Rightarrow \frac{M}{2}v_2^2 = Mgx_0 - \frac{kx_0^2}{2}$$

$$\Rightarrow v_2 = \sqrt{2gx_0 - \frac{k}{M}x_0^2} \qquad (2.122)$$

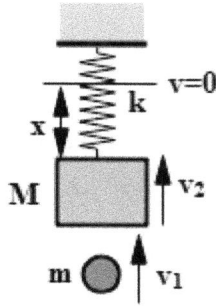

Solving equations (2.121) and (2.122)

$$\frac{m(1 + e)v_0}{M + m} = \sqrt{2gx_0 - \frac{k}{M}x_0^2}$$

Putting

$$x_0 = \frac{Mg}{k}$$

$$\frac{m(1 + e)v_0}{M + m} = \sqrt{2g\frac{Mg}{k} - \frac{K}{M}\left(\frac{Mg}{k}\right)^2}$$

$$= \sqrt{\frac{Mg^2}{k}} = \sqrt{\frac{M}{k}}g$$

$$\Rightarrow v_0 = \frac{(M + mg)\sqrt{\frac{M}{k}}}{m(1 + e)} \quad \text{Ans.}$$

Problem 27 A ball of mass $m = 2$ kg moves towards the right with a velocity of $u = 8$ m s^{-1} towards a vertical wall that is moving towards the left with a velocity of $v = 4$ m s^{-1}. If the ball collides with the wall at a height $h = 5$ m with a coefficient of restitution $e = 0.75$, find the work done by the wall when changing the KE of the ball.

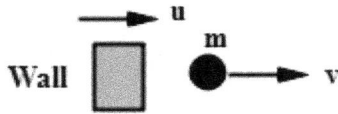

Solution
The heavy moving wall's velocity remains unchanged just after the collision.
Let $v'=$ velocity of the ball just after the collision; the NIF given as

$$u' - v' = -e(u - v)$$

$$\Rightarrow v' = u' + e(u - v)$$

Putting $u' = u$, we have

$$v' = u(1 + e) - ev \qquad (2.123)$$

Then the work done by the wall on the ball is

$$W = \Delta K$$

$$= \frac{1}{2}m(v'^2 - v^2) \qquad (2.124)$$

From equations (2.123) and (2.124)

$$W = \frac{1}{2}m[\{u(1 + e) - ev\}^2 - v^2]$$

$$= \frac{1}{2} \times 2\left[\left\{8\left(1 + \frac{3}{4}\right) - \frac{3}{4} \times 4\right\}^2 - (4)^2\right]$$

$$= 1\left[\left(8 \times \frac{7}{4} - 3\right)^2\right] - 16$$

$$= 121 - 16 = 105 \text{ Joule}$$

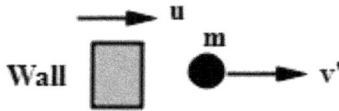

Problem 28 A boy of height h standing in an elevator relates a ball from rest relative to the elevator. The ball strikes the elevator with a coefficient of restriction 'e'. If the elevator moves with constant velocity, find the maximum height attained by the ball relative to the (a) point of impact and (b) elevator's base.

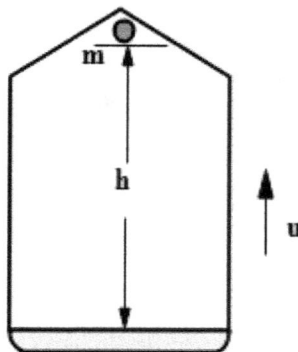

Solution

(a) The velocity of the ball relative to the elevator surface floor is

$$u_{rel} = \sqrt{2gh} \tag{2.125}$$

Then the NIF can be written as

$$-e\overrightarrow{u_{rel}} = \overrightarrow{v_{rel}}$$

$$-e(-u_{rel}) = v' - u$$

where $v' =$ velocity of the ball just after the collision relative to ground.

$$\Rightarrow v' = u + ev_{rel} \tag{2.126}$$

Equations (2.125) and (2.126) give

$$v' = u + e\sqrt{2gh}$$

Then, the maximum height relative to the point of projection is

$$h_{max} = \frac{v'^2}{2g}$$

$$= \frac{\left(u + e\sqrt{2gh}\right)^2}{2g}$$

(b) The maximum height relative to the floor of the elevator is

$$(h_{max})_{rel} = \frac{v_{rel}^2}{2g} = \frac{(eu_{rel})^2}{2g}$$

$$= \frac{e^2}{2g}u_{rel}^2 = \frac{e^2}{2g}(2gh) = e^2h \text{ Ans.}$$

Problem 29 Three identical balls A, B, and C each of mass m are kept on a straight line on a smooth horizontal plane. If the ball A is given a velocity u, collisions take place between the balls and finally the balls move with constant velocities without any collision. Find the final velocities of the balls. Assume that $e = $ coefficient of restitution of collision between A and B and elastic collision between B and C.

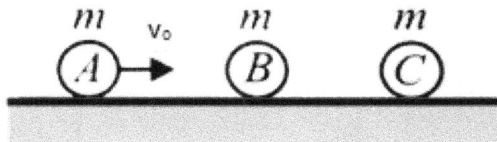

Solution

Ball A collides with ball B:

Conserving momentum

$$mv_0 = mv_1 + mv_2$$

$$\Rightarrow v_1 + v_2 = v_0 \tag{2.127}$$

NIF

$$v_1 - v_2 = -ev_0 \tag{2.128}$$

Solving equations (2.127) and (2.128),

$$v_1 = \frac{1-e}{2}v_0 = \frac{\left(1 - \frac{1}{2}\right)v_0}{2} = \frac{v_0}{4} \text{ and } v_2 = \frac{1+e}{2}v_0 = \frac{1 + \frac{1}{2}}{2}v_0 = \frac{3}{4}v_0$$

Ball B collides with ball C:

Following the above procedure

$$v_2' = \frac{1-e}{2}v_2 = \frac{1-e}{2}\frac{1+e}{2}v_0$$

$$= \frac{1-e^2}{4}v_0 = \frac{1 - \left(\frac{1}{2}\right)^2}{4}v_0 = \frac{3}{16}v_0$$

Similarly, $v_3 = \frac{1+e}{2}v_2 = \frac{1+e}{2}(\frac{1+e}{2}v_0)$

$$= \left(\frac{1+e}{2}\right)^2 v_0 = \frac{\left(1 + \frac{1}{2}\right)^2}{4}v_0 = \frac{9}{16}v_0$$

Now, $v_A = v_1 = \frac{3}{4}v_0$

$$v_B = v_2' = \frac{3}{16}v_0$$

$$v_c = v_3 = \frac{9}{16}v_0$$

Since $v_A > v_B$, A will collide with B again. Let v'_A and v'_B be the velocities just after the collision. Then, conserving momentum,

$$v'_A + v'_B = v_A + v_B$$

$$v'_A - v'_B = -e(v_A - v_B)$$

Solving these two equations, we have

$$v'_A = v_A\left(\frac{1-e}{2}\right) + v_B\left(\frac{1+e}{2}\right)$$

$$v'_B = v_A\left(\frac{1+e}{2}\right) + v_B\left(\frac{1-e}{2}\right)$$

Putting the values of $v_A = \frac{3}{4}v_0$, $v_B = \frac{3}{16}v_0$ and $e = \frac{1}{2}$

$$v'_A = \left(\frac{1}{4}v_0\right)\left(\frac{1-\frac{1}{2}}{2}\right) + \left(\frac{3}{16}v_0\right)\left(\frac{1+\frac{1}{2}}{2}\right)$$

$$= \frac{v_0}{16} + \frac{3}{16}v_0 \times \frac{3}{4} = \frac{v_0}{16}\left(1 + \frac{9}{4}\right) = \frac{13v_0}{64}$$

$$v'_B = v_A\left(\frac{1+e}{2}\right) + v_B\left(\frac{1-e}{2}\right)$$

$$= \left(\frac{1}{4}v_0\right)\left(\frac{1+\frac{1}{2}}{2}\right) + \left(\frac{3}{16}v_0\right)\left(\frac{1-\frac{1}{2}}{2}\right)$$

$$= \frac{1}{4}v_0\left(\frac{3}{4}\right) + \frac{3}{16}v_0 \times \frac{1}{4}$$

$$= \frac{3}{16}v_0\left(1 + \frac{1}{4}\right) = \frac{15}{64}v_0$$

Since $v_c > v'_B > v'_A$, no collision will take place thereafter. So the final velocities of the balls will be $+\frac{13}{64}v_0$, $+\frac{15}{64}v_0$ and $+\frac{9}{16}v_0$, respectively.

Problem 30 A ball of mass m_1 head-on with a stationary ball of mass m_2. If e = coefficient of restriction of the collision, find the (a) friction change in KE (η) of the system ($m_1 + m_2$) (b) draw the graph of η as the function of $\frac{m_1}{m_2}$.

Solution

(a) The change in KE of the system is

$$\Delta K = -\frac{1}{2}\frac{m_1 m_2}{m_1 + m_2} |\vec{u_1} - \vec{u_2}|^2 (1 - e^2)$$

If the second body was at rest just before the collision, putting $u_1 = v_0$ and $u_2 = 0$, we have

$$\Delta K = -\frac{1}{2}\frac{m_1 m_2 v_0^2}{m_1 + m_2}(1 - e^2)$$

where $\frac{m_1 v_0^2}{2} = K_i$ (initial KE of the system).
Then, the friction loss of KE of the system is

$$\eta = \frac{|-\Delta K|}{K} = \frac{m_2}{m_1 + m_2}(1 - e^2) = \frac{1 - e^2}{1 + \frac{m_1}{m_2}}$$

(b) When $\frac{m_1}{m_2} \cong 0$, $\eta = 1 - e^2$.

When $\frac{m_1}{m_2} = 1$, $\eta = \frac{1 - e^2}{2}$.

When $\frac{m_1}{m_2} \rightarrow \infty$, $\eta = 0$.

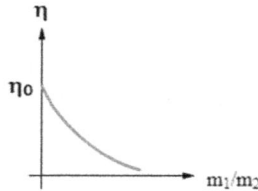

Problem 31 A ball of mass m collides with the block of mass m_1 which is attached with another block of mass m_2 by a light spring of stiffness k. Find the maximum compression of the spring. Assume that $e = $ coefficient of restitution of the collision.

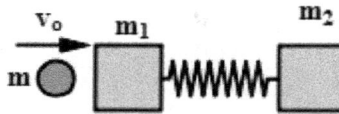

Solution

The first collision takes place between m and m_1 because during the collision the deformation of the spring will be so small that a negligible force is transmitted to m_2 from m_1. So we can conserve the linear momentum of m and m_1.

$$mv + m_1 v_1 = mv_0 \tag{2.129}$$

$$-ev_0 = v - v_1 \tag{2.130}$$

Solving equations (2.129) and (2.130)

$$m(v_1 - ev_0) + m_1 v_1 = mv_0$$

$$\Rightarrow v_1 = \frac{mv_0(1 + e)}{m + m_1} \tag{2.131}$$

Let the maximum compression of the spring be x and that time the velocity of *center of mass* of the system $(m_1 + m_2)$ be v.

Conserving the momentum

$$m_1 v_1 = (m_1 + m_2)v \tag{2.132}$$

Conserving energy, we have

$$\frac{1}{2}(m_1 + m_2)v^2 + \frac{1}{2}kx^2 = \frac{1}{2}m_1 v_1^2 \tag{2.133}$$

Using equations (2.132) and (2.133)

$$\frac{1}{2}(m_1 + m_2)\left(\frac{m_1 v_1}{m_1 + m_2}\right)^2 + \frac{1}{2}kx^2 = \frac{1}{2}m_1 v_1^2$$

$$\Rightarrow \frac{1}{2}kx^2 = \frac{1}{2}\frac{m_1 m_2 v_1^2}{m_1 + m_2}$$

$$\Rightarrow x = \sqrt{\frac{m_1 m_2}{(m_1 + m_2)k}}\, v_1 \tag{2.134}$$

Using equations (2.131) and (2.134)

$$x = \sqrt{\frac{m_1 m_2}{(m_1 + m_2)k}}\left\{\frac{m v_0(1 + e)}{m + m_1}\right\} \text{ Ans.}$$

Problem 32 A pendulum bob of mass m is released from the horizontal position of a taut string. It collides n times each having a coefficient of restitution e. Find the maximum angle made by the string after nth impact with the wall.

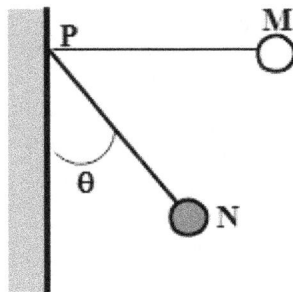

Solution
Let the ball collide with the wall at A with a velocity v_0. Just after n collisions its velocity will be equal to $e^n v_0$. Then, we can conserve the energy between A and the maximum angle θ after collision to obtain

$$mgh = \frac{1}{2}mv_n^2$$

Putting $h = l(1 - \cos \theta) = 2l \sin^2 \frac{\theta}{2}$ and $v_n = ev_0$, we have

$$2g\left(2l \sin^2 \frac{\theta}{2}\right) = (e^n v_0)^2$$

Putting $v_0^2 = 2gl$, we have

$$2 \sin^2 \frac{\theta}{2} = e^{2n}$$

$$\Rightarrow \sin \frac{\theta}{2} = \frac{e^n}{\sqrt{2}}$$

$$\Rightarrow \theta = 2 \sin^{-1}\left(\frac{e^n}{\sqrt{2}}\right) \text{ Ans.}$$

Problem 33 A ball is released from a height h. Find the (a) maximum height attained by the ball after nth collision, (b) total distance covered, and (c) total or net momentum delivered by the ball to the earth, after so many collisions with the ground until it stops. Assume $e =$ coefficient of restitution in each collision.

Solution

(a) The velocity just before the first collision is

$$v = \sqrt{2gh} \tag{2.135}$$

The velocity just after the first collision is

$$v_1 = ev \tag{2.136}$$

Using equations (2.135) and (2.136),

$$v_1 = e\sqrt{2gh} \tag{2.137}$$

Then, the maximum height attained after the first collision is

$$h_1 = \frac{v_1^2}{2g} = \frac{\left(e\sqrt{2gh}\right)^2}{2g}$$

$$\Rightarrow h_1 = e^2 h \tag{2.138}$$

Similarly, the maximum height attained after the second collision is

$$h_2 = e^2 h_1 \tag{2.139}$$

Using equations (2.138) and (2.139),

$$h_2 = e^4 h \tag{2.140}$$

In this way, maximum height attained after nth collision is

$$h_n = e^{2n} h \text{ Ans.}$$

(b) The total distance covered after so many collisions is

$$D = h + 2h_1 + 2h_2 + \ldots.$$

$$= h + 2(h_1 + h_2 + h_3 + \ldots.)$$

$$= h\{2(e^2 h + e^4 h + e^6 h + \ldots.)\}$$

$$= h[1 + 2e^2(1 + e^2 + e^3 + \ldots.)]$$

$$= h\left[1 + 2e^2\left(\frac{1}{1 - e^2}\right)\right]$$

$$D = \frac{1 + e^2}{1 - e^2} h \text{ Ans.}$$

(c) The momentum delivered by earth

$$\Delta \vec{P} = \Delta \vec{P_1} + \Delta \vec{P_2} + \Delta \vec{P_3} + \ldots\ldots$$

where $\Delta \vec{P_1}$, $\Delta \vec{P_2}$, $\Delta \vec{P_3}$ are the momentum delivered by earth (change in momentum of the ball) in the first, second, and third collision, respectively.

$$\Delta \vec{P_1} = m(1 + e)v\hat{j} = m(1 + e)\sqrt{2gh}\,\hat{j}$$

$$\Delta \vec{P_2} = m(1 + e)v_1\hat{j} = me(1 + e)\sqrt{2gh}\,\hat{j}$$

$$\Delta \vec{P_3} = m(1 + e)v_2\hat{j} = me^2(1 + e)\sqrt{2gh}\,\hat{j}$$

$$\Rightarrow \Delta \vec{P} = m(1 + e)\sqrt{2gh}\,(1 + e + e^2 + \ldots.)$$

$$= m(1 + e)\sqrt{2gh}\left(\frac{1}{1 - e}\right)$$

$$= \frac{m(1 + e)}{(1 - e)}\sqrt{2gh} \text{ Ans.}$$

Problem 34 Two spheres are falling freely as shown in the figure. The coefficient of restitution between all collisions is e. (a) If $e = 1/2$, find the velocity of m after being collided with M. (b) If $M = m$ and $e < 1$, find the total time of motion of the balls. (c) If $M \gg m$ and $e < 1$, find the maximum height attained by the body.

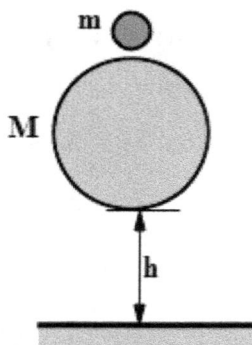

Solution

(a) Let v_0 = velocity of m and M just before collision. Just after colliding with the floor elastically the ball M rebounds upwards with a velocity v_0. Then it collides with the ball m, which has a downward velocity v_0. Let just after the collision v_1 and v_2 be the velocities of M and m, respectively; by conserving the linear momentum, we have:

$$Mv_0 - mv_0 = Mv_1 + mv_2 \tag{2.141}$$

$$\text{NIF} \quad - e\{v_0 - (-v_0)\} = v_1 - v_2$$

$$\Rightarrow 2ev_0 = v_2 - v_1 \tag{2.142}$$

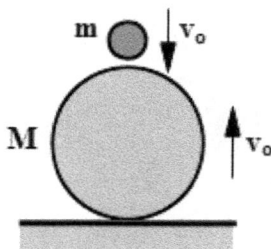

Solving equations (2.141) and (2.142)

$$(M - m)v_0 = Mv_1 + m(2ev_0 + v_1)$$

$$\Rightarrow v_1 = \frac{\{(M - m) - 2me\}v_0}{M + m}$$

$$v_1 = \frac{M - (1 + 2e)m}{M + m}v_0 \tag{2.143}$$

$$v_2 = 2ev_0 + v_1$$

$$= 2ev_0 + \frac{M - (1 + 2e)m}{M + m}v_0$$

$$= v_0 \left[\frac{2eM + 2em + M - m - 2em}{M + m} \right]$$

$$v_2 = v_0 \left[\frac{M(1 + 2e) - m}{M + m} \right] \tag{2.144}$$

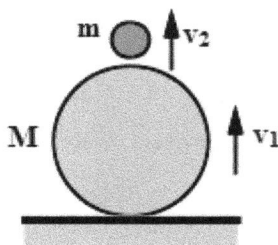

(b) If $m = M$, we have

$$v_1 = \frac{m - (1 + 2e)m}{M + m}v_0 = -ev_0$$

$$v_2 = v_0 \left[\frac{m(1 + 2e) - m}{M + m} \right] = ev_0$$

This means that the ball M moves down with a velocity ev_0 and the ball m moves up with ev_0. Then, the ball M will bounce back with ev_0 after colliding with the floor elastically. Thus, both the balls move upward and then downward with the same velocity at any instant. Finally the initial situation will be repeated with a velocity ev_0 (instead of v_0) because the ball M will collide with the floor with a velocity ev_0 and bounce back with the same speed and then collide with the downward moving ball m. Since their speeds are equal, again they will exchange their momenta in the next cycle. As a result, ball m will move up with a velocity e^2v_0 and M will move down with a velocity e^2v_0. Just after colliding with the floor M will move up with velocity e^2v_0. Again, it will start another cycle with an initial velocity e^2v_0.

After a number of collisions both balls will come to rest. Let the total time of motion of the balls be T.

$$T = \frac{2ev_0}{g} + \frac{2e^2v_0}{g} + \frac{2e^3v_0}{g} + \dots$$

$$\Rightarrow T = \frac{2v_0e}{g}[1 + e + e^2 + e^3 + \dots]$$

$$\Rightarrow T = \frac{2v_0l}{(1 - e)g}$$

After this time $T = \frac{2v_0e}{(1-e)g}$, both the balls will stop bouncing.

(c) If $M \gg m$, just after the collision

$$v_1 \cong v_0 \tag{2.145}$$

Applying NIF, we have

$$-e\{v_0 - (-v_0)\} = v_1 - v_2 \tag{2.146}$$

Using equations (2.145) and (2.146)

$$v_2 = v_0(1 + 2e)$$

Then the maximum height attained by the ball m is given as

$$h_{max} = \frac{v_2^2}{2g} = \frac{\{v_0(1 + 2e)\}^2}{2g}$$

$$\Rightarrow h_{max} = \frac{v_0^2}{2g}(1 + 2e)^2, \text{ where } \frac{v_0^2}{2g} = h$$

$$\Rightarrow h_{max} = (1 + 2e)^2 \times h \text{ Ans.}$$

Problem 35 An elastic ball of mass m moves to and fro between two parallel walls in a gravity-free space. The initial distance of separation between the walls is L and the speed of the ball is v_o. Let us push the left side wall (supposed to be mass-less for the sake of simplicity) slowly to a distance of separation x. Find the average force impressed upon the ball by the walls.

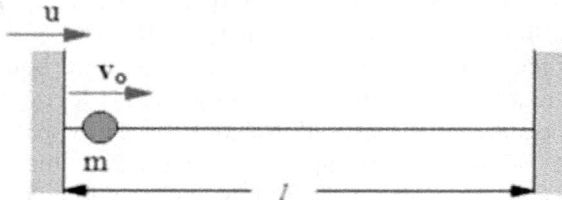

Solution

Applying Newton's impact formula,

$-e(v + u) = u - v'$ where $e = 1$

$$\Rightarrow v' = 2u + v$$

The time taken for the to and fro journey is

$$\Delta t = \frac{x}{v} + \frac{x - dx}{v'} \simeq \frac{x}{v} + \frac{x}{v} \quad (\because u \ll v)$$

The change in momentum of the ball is

$$\Delta \vec{P} \cong 2mv\hat{\imath}$$

Then the average force impressed upon the ball is

$$\vec{F} = \frac{\Delta \vec{P}}{\Delta t} = \frac{2mv}{\frac{x}{v}} = \frac{mv^2}{x}$$

Putting $F = ma = -\frac{mv\, dv}{dx}$, we have

$-\frac{mv\, dv}{dx} = \frac{mu^2}{x}$ (because x decreases)

$$\Rightarrow \frac{du}{v} = -\frac{dx}{x}$$

$$\Rightarrow \int_{v_0}^{v} \frac{dv}{u} = -\int_{\ell}^{x} \frac{dx}{x}$$

$$\Rightarrow \ln \frac{v}{v_0} + \ln \frac{x}{\ell} = 0$$

$$\Rightarrow vx = \text{constant} = v_0 \ell_0$$

$$\Rightarrow v = \frac{v_0 \ell_0}{x} \tag{2.147}$$

Then, the average force impressed upon the wall to move it with a constant velocity is

$$F' = F = \frac{mv^2}{x} \tag{2.148}$$

$$\Rightarrow F_{\text{ext}} = \frac{m}{x} \left(\frac{v_0 \ell}{x} \right)^2$$

$$\Rightarrow F_{\text{ext}} = \frac{mv_0^2 \ell^2}{x^3} \text{ Ans.}$$

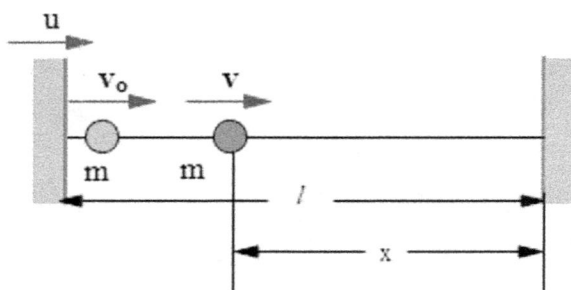

Problem 36 A shell moving with a velocity v_o explodes into two fragments of masses m_1 and m_2. If the energy of explosion is E, find the (a) velocities of the fragments and (b) relative velocity between them just after the explosion.

Solution

(a) The center of mass does not change its momentum or velocity just before and after the explosion because no external force acts on the body during the explosion. Let the fragments m_1 and m_2 have velocities v_1 and v_2 relative to the center of mass whose velocity remains as v_0 towards the $+x$-axis.

Relative to the center of mass frame,

$$m_1 v_1 = m_2 v_2 \tag{2.149}$$

$$\frac{1}{2} \frac{m_1 m_2}{m_1 + m_2} (v_1 + v_2)^2 = E \tag{2.150}$$

Solving equations (2.149) and (2.150),

$$v_1 = \sqrt{\frac{2m_2 E}{m_1(m_1 + m_2)}}$$

$$v_2 = \sqrt{\frac{2m_1 E}{m_2(m_1 + m_2)}}$$

(b) Then the velocity of m_1 just after the explosion relative to grow is

$$v'_1 = v_1 + v_0$$

$$= v_0 + \sqrt{\frac{2m_2 E}{m_1(m_1 + m_2)}}$$

The velocity of m_2 just after the explosion relative to ground is

$$v'_2 = v_0 - v_2$$

$$= v_0 - \sqrt{\frac{2m_1 E}{m_2(m_1 + m_2)}} \quad \text{Ans.}$$

Problem 37 A cricket ball of mass m strikes the ground with a velocity v at an angle θ with vertical. If the coefficient of friction between ground and ball is μ and coefficient of restitution of impact is e, find the (a) angle with respect to vertical, made by the velocity of the ball just after the collision and (b) fraction of energy loss during the collision.

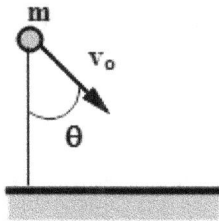

Solution

(a) Along the x-axis, applying $F_x = 0$. So $P_x = C$

$$\Rightarrow mv \sin \phi = mv_0 \sin \theta$$

$$\Rightarrow v \sin \phi = v_0 \sin \theta \qquad (2.151)$$

Along the y-axis, applying the NIF to obtain

$$ev_0 \cos \phi = v \cos \phi \qquad (2.152)$$

From equations (2.151) and (2.152)

$$\frac{v \sin \phi}{v \cos \phi} = \frac{v_0 \sin \theta}{ev_0 \cos \theta}$$

$$\Rightarrow \tan \phi = \frac{\tan \theta}{e}$$

$$\Rightarrow \phi > \theta$$

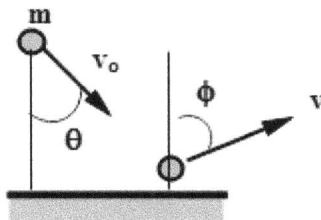

(b) The fraction of energy lost is

$$\eta = \left| \frac{\frac{1}{2}mv^2 - \frac{1}{2}mv_0^2}{\frac{1}{2}mv_0^2} \right| = 1 - \left(\frac{v}{v_0} \right)^2$$

where $v = \sqrt{(ev_0 \cos \theta)^2 + (v_0 \sin \theta)^2}$

$$= v_0 \sqrt{e^2 \cos^2 \theta + \sin^2 \theta}$$

Then, $\eta = 1 - \left(\frac{v}{v_0} \right)^2$

$$= 1 - (e^2 \cos^2 \theta + \sin^2 \theta)$$

$$\eta = (1 - e^2)\cos^2 \theta \text{ Ans.}$$

Problem 38 A cricket ball of mass m is thrown with a horizontal velocity v_0 from a height h above the ground. If the coefficient of friction between ground and ball is μ and coefficient of restitution of impact is e, find the velocity of the ball just after the collision.

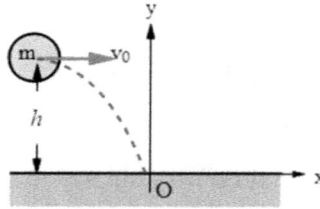

Solution
Newton's impact formula along the y-axis is

$$-ev_y = -v'_y$$

$$\Rightarrow v'_y = ev_y \tag{2.153}$$

Impulse–momentum equations: $\int f_k \, dt = m(v'_x - v_x) \tag{2.154}$

$$\int N \, dt = m(v'_y + v_y) \tag{2.155}$$

Law of kinetic friction: $f_k = \mu N \tag{2.156}$

Using equations (2.154)–(2.156),

$$m\mu(v'_y + v_y) = m(v'_x - v_x)$$

$$v'_x = \mu\left(v_y + v'_y\right) + v_x \tag{2.157}$$

Using equations (2.153) and (2.157),

$$v'_x = \mu\left(v_y + ev_y\right) + v_x$$

$$= \mu v_y(1 + e) + v_x \tag{2.158}$$

Applying $\vec{v} = \vec{u} + \vec{a}\,t$, we have

$$v_x\hat{i} - v_y\hat{j} = v_o\hat{i} + (-g\hat{j})\sqrt{\frac{2h}{g}}$$

$$\Rightarrow v_x = v_0 \text{ and } v_y = g\sqrt{\frac{2h}{g}} = \sqrt{2gh}$$

Putting $v_x = v_0$ and $v_y = \sqrt{2gh}$ in equation (2.158),

$$v'_x = \mu\sqrt{2gh}(1 + e) + v_0 \text{ and } v'_y = ev_y = e\sqrt{2gh}$$

Then, the velocity of the ball just after the collision is

$$\vec{v}' = \vec{v}'_x + \vec{v}'_y = v'_x\hat{i} + v'_y\hat{j}$$

$$= \left\{\mu\sqrt{2gh}(1 + e) + v_0\right\}\hat{i} + e\sqrt{2gh}\hat{j}$$

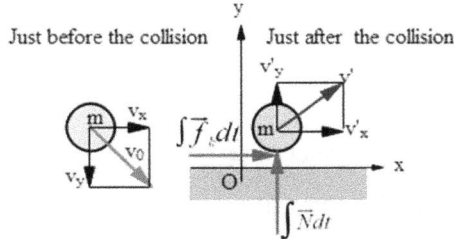

Just before the collision   Just after the collision

Problem 39 A bob of mass m is connected to a fixed point A by an inextensible light string of length l. It is released from rest when the string is horizontal as shown in the figure. The point O is situated at a height $h < l$. If the collision of the ball with the smooth ground and the collision with the ball of mass M placed on the ground are both perfectly inelastic, find the ratio of maximum height attained by the bob, until the bob collides with the ground for the second time and the initial height h of the bob. Put $M = 2m$ and $h = 0.866l$.

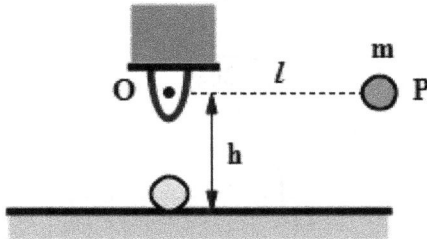

Solution

The velocity with which the bob strikes the ground is given as

$$\frac{1}{2}mv_0^2 = mgh$$

$$\Rightarrow v_0 = \sqrt{2gh}$$

Due to inelastic collision, the bob will not bounce. Its horizontal momentum can be conserved during the first collision. So it will move to the left with a velocity $v = v_0 \cos \theta$. As it collides inelastically with a block of mass M, conserving the linear momentum at 'O',

$$(M + m)v' = mu = mv_0 \cos \theta$$

$$\Rightarrow v' = \frac{mu \cos \theta}{M + m} \tag{2.159}$$

At the same angular position θ of the string in the left side of the point O, due to inelastic collision of the combined block $(M + m)$ with the string, its velocity along the string will be zero just after the collision. So the combined body $(M + m)$ will lose contact with the ground at Q with a velocity

$$v_1 = v' \sin \theta$$

$$= \frac{mu \cos \theta}{M + m} \sin \theta$$

$$v_1 = \frac{mv \sin \theta \cos \theta}{M + m},$$

where $v = v_0 \cos \theta$

Then, the velocity v_1 can be given as

$$v_1 = \frac{mv_0 \sin \theta \cos^2 \theta}{M + m} \tag{2.160}$$

Let, $h' =$ maximum height attained by the body. Then, we can conserve the energy to obtain

$$(M + m)gh - \frac{1}{2}(M + m)v_1^2 = 0$$

$$h' = \frac{v_1^2}{2g} \tag{2.161}$$

Using equations (2.159) and (2.160),

$$h' = \frac{1}{2g}\left(\frac{mv_0 \sin \theta \cos^2 \theta}{M + m}\right)^2$$

Putting $\frac{v_0^2}{2g} = h$, the value of

$$\frac{h'}{h} = \left(\frac{m \sin\theta \cos^2\theta}{M + m}\right)^2$$

Putting $M = 2m$ and $\theta = 30°$,

$$\frac{h'}{h} = \left\{\frac{m\left(\frac{1}{2}\right)\left(\frac{3}{4}\right)}{2m + m}\right\}^2$$

$$= \left(\frac{1}{8}\right)^2 = \frac{1}{64} \text{ Ans.}$$

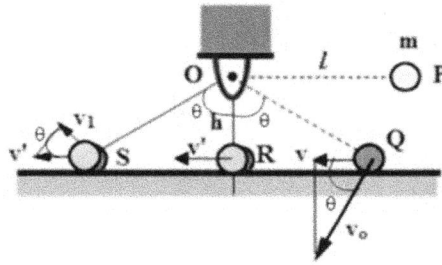

Problem 40 A small sphere of mass m slides from the top along the frictional inclined plane through a height h and horizontal distance x. Then it slides on another inclined plane without bouncing at each junction. Finally, the sphere compresses the spring that is fitted with the vertical wall. Find the maximum compression of the spring.

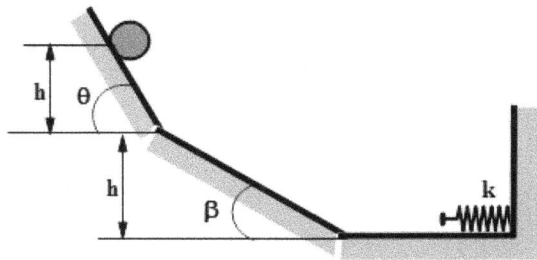

Solution

The component of v_0 perpendicular to AB will be lost and the component of v_0 parallel to AB will be conserved.

Resolving the velocity v_0 along AB, we have

$$v_1 = v_0 \cos(\theta - \beta)$$

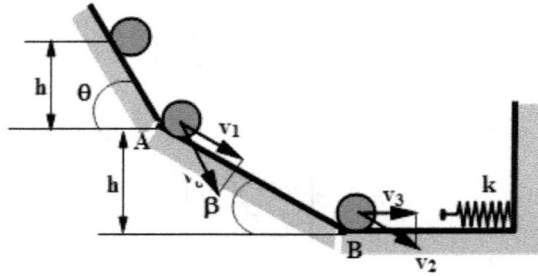

Similarly, the component of v_2 perpendicular to the horizontal surface will be lost and that along the horizontal will remain conserved. Then, the velocity of the ball just after the second impact with the horizontal surface is

$$v_3 = v_2 \cos \beta \tag{2.162}$$

Conserving energy between A and B, we have

$$v_2 = \sqrt{2gh + v_1^2} \tag{2.163}$$

Applying the work–eneregy theorem between the top and A,

$$mgh - \mu mg \cos \theta = \frac{1}{2}mv_0^2,$$

where $\ell \cos \theta = x$

$$\Rightarrow v_0 = \sqrt{2g(h - \mu x)} \tag{2.164}$$

Conserving energy between B and maximum compression of the spring, we have,

$$\frac{1}{2}mv_3^2 = \frac{1}{2}kx^2$$

$$\Rightarrow x = \sqrt{\frac{m}{k}} v_3 \tag{2.165}$$

Putting v_3 from equations (2.162) in (2.165),

$$x = \sqrt{\frac{m}{k}} v_2 \cos \beta \tag{2.166}$$

Putting v_2 from equations (2.163) in (2.166),

$$x = \sqrt{\frac{m}{k}} \left(\sqrt{2gh + v_1^2} \right) \cos \beta$$

$$= \sqrt{\frac{m}{k}(v_1^2 + 2gh)} \cos \beta \tag{2.167}$$

Putting v_1 from equation (2.78) in (2.167),

$$x = \sqrt{\frac{m}{k} \{v_0^2 \cos^2(\theta - \beta) + 2gh\}} \cos \beta \qquad (2.168)$$

Putting v_0 from equations (2.164) in (2.168),

$$x = \left[\sqrt{\frac{m}{k} \{2g(h - \mu x)\cos^2(\theta - \beta) + 2gh\}} \right] \cos \beta$$

$$= \left[\sqrt{\frac{2mg}{k} \{(h - \mu x)\cos^2(\theta - \beta) + h\}} \right] \cos \beta \text{ Ans.}$$

Problem 41 A cylindrical hall has a circular horizontal base. A ball is projected from a point of the wall on the base making an angle θ with the radius of the circular base as shown in the figure. The ball returns to the point of projection after two successive impacts with the wall. If the coefficient of restitution of each impact is e, find the (a) value of $\tan \theta$, (b) speed of striking of the ball at the point of projection, and (c) total time for the round trip. Take $e = 0.5$, $R = 1$ m, and $v_o = 5$ m s^{-1}.

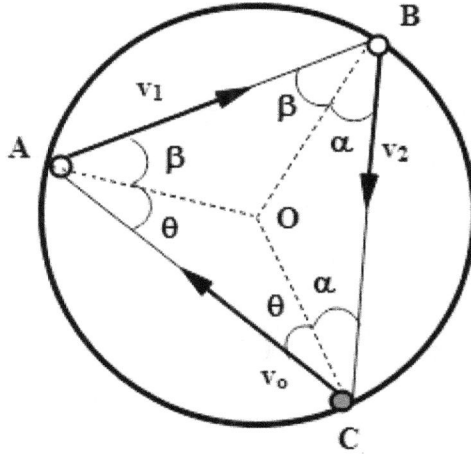

Solution

(a) Using the properties of the triangle, we have $2\theta + 2\beta + 2\alpha = \pi_{\text{rad}}$

$$\theta + \beta + \alpha = \frac{\pi}{2}\text{rad} \qquad (2.169)$$

$$v_0 \sin \theta = v_1 \sin \beta \qquad (2.170)$$

Using Newton's impact formula at A, we have

$$ev_0 \cos \theta = v_1 \cos \beta \qquad (2.171)$$

Using the last two equations, we have

$$\Rightarrow \tan \beta = \frac{\tan \theta}{e} \tag{2.172}$$

Similarly

$$\tan \alpha = \frac{\tan \beta}{e} \tag{2.173}$$

Equations (2.172) and (2.173)

$$\tan \alpha = \frac{\tan v}{e^2} \tag{2.174}$$

$$\theta + \beta = \frac{\pi}{2} - \alpha$$

$$\tan(\theta + \beta) = \tan\left(\frac{\pi}{2} - \alpha\right)$$

$$\Rightarrow \frac{\tan \theta + \tan \beta}{1 - \tan \theta \tan \beta} = \frac{1}{\tan \alpha} \tag{2.175}$$

Putting $\tan \beta = \frac{\tan \theta}{e}$ and $\tan \alpha = \frac{\tan \theta}{e^2}$ from equations (2.172) and (2.174) in (2.175)

$$\frac{\tan \theta + \dfrac{\tan \theta}{e}}{1 - \tan \theta \dfrac{\tan \theta}{e}} = \frac{1}{\dfrac{\tan \theta}{e^2}}$$

$$\tan^2 \theta \left(1 + \frac{1}{e}\right)\left(\frac{1}{e^2}\right) = 1 - \frac{\tan^2 \theta}{e}$$

$$\tan^2 \theta \left[\left(1 + \frac{1}{e}\right)\left(\frac{1}{e^2}\right) + \frac{1}{e}\right] = 1$$

$$\Rightarrow \tan^2 = \frac{1}{\dfrac{1}{e} + \dfrac{1}{e^2} + \dfrac{1}{e^3}} = \frac{e^2 + e^2 + 1}{e^3}$$

$$\Rightarrow \tan \theta = \sqrt{\frac{e^3}{1 + e + e^2}} = \sqrt{\frac{\dfrac{1}{8}}{1 + \dfrac{1}{2} + \dfrac{1}{4}}} = \sqrt{\frac{\dfrac{1}{8}}{\dfrac{4 + 2 + 1}{4}}} = \frac{1}{\sqrt{14}}$$

(b) Since $v_2 \sin \alpha = v_1 \sin \beta = v_0 \sin \theta$

$$v_2 \cos \alpha = ev_1 \cos \beta \text{ and } v_1 \cos \beta = ev_0 \cos \theta$$

we have $v_2 \cos \alpha = e^2 v_0 \cos \theta$.

Then, $(v_2 \sin \alpha)^2 + (v_2 \cos \alpha)^2$

$$= (v_0 \sin \theta)^2 + (e^2 v_0 \cos \theta)^2$$

$$\Rightarrow v_2 = v_0 \sqrt{\sin^2 \theta + e^4 \cos^2 \theta}$$

Similarly $v_1 = (v_1 \sin \theta)^2 + (v_1 \cos \beta)^2$

$$= \sqrt{(v_0 \sin \theta)^2 + (ev_0 \cos \theta)^2}$$

$$= v_0 \sqrt{\sin^2 \theta + e^2 \cos^2 \theta}$$

(c) The time elaspsed by the ball to return to its initial point is

$$T = \frac{2R \cos \theta}{v_0} + \frac{2R \cos \beta}{v_1} + \frac{2R \cos \alpha}{v_2}$$

Putting $v_1 \cos \beta = ev_0 \cos \theta$ and $v_2 \cos \alpha = ev_1 \cos \beta$,

we have $\cos \beta = \frac{ev_0 \cos \theta}{v_1}$ and $\cos \alpha = \frac{e^2 v_0 \cos \theta}{v_2}$

$$\Rightarrow T = \frac{2R \cos \theta}{v_0} + \frac{2 \, \mathrm{Re} \, v_0 \cos \theta}{v_1^2} + \frac{2 \, \mathrm{Re}^2 \, v_0 \cos \theta}{v_2^2}$$

$$\Rightarrow T = \frac{2R \cos \theta}{v_0} \left[1 + \frac{v_0^2}{v_1^2} + \frac{v_0^2}{v_2^2} \right]$$

Putting $v_1^2 = v_0^2 (e \cos^2 \theta + \sin^2 \theta)$, we have

$$T = \frac{2R \cos \theta}{v_0} \left[1 + \frac{1}{e^2 \cos^2 \theta + \sin^2 \theta} + \left(\frac{v_0}{v_2} \right)^2 \right]$$

Since $v_2 \sin \alpha = v_0 \sin \theta$ and $v_2 \cos \alpha = e^2 v_0 \cos \theta$, we have

$$v_2^2 = v_0^2 (\sin^2 \theta + e^4 \cos^2 \theta)$$

Putting $\frac{v_0^2}{v_2^2} = \frac{1}{\sin^2 \theta + e^4 \cos^2 \theta}$, we have

$$T = \frac{2R \cos \theta}{v_0}$$

$$\left[1 + \frac{1}{e^2 \cos^2 \theta + \sin^2 \theta} + \frac{1}{e^4 \cos^2 \theta + \sin^2 \theta}\right] \text{Ans.}$$

Problem 42 A ball is projected from the horizontal ground so that it collides with the vertical wall, which is located at a horizontal distance $x = \frac{R}{4}$ from the point of projection. Then the ball collides with the ground at a horizontal distance x' from the vertical wall. Assume $H = $ the maximum height attained by the ball and $e = $ coefficient of restitution. (a) Find the height h at which the collision takes place on the vertical wall. (b) If u and θ are the speed and angle of projection respectively, prove that $y = \dfrac{xx'\sec^2\theta}{2eu^2} = \dfrac{exx'(ex' + x)\tan\theta}{2(ex + x')^2}$. (c) If the collision with the wall is perfectly elastic and the ball hits the ground directly below the highest position of the ball's trajectory, find the value of h.

Solution

(a) The equation of trajectory of a projectile is given as

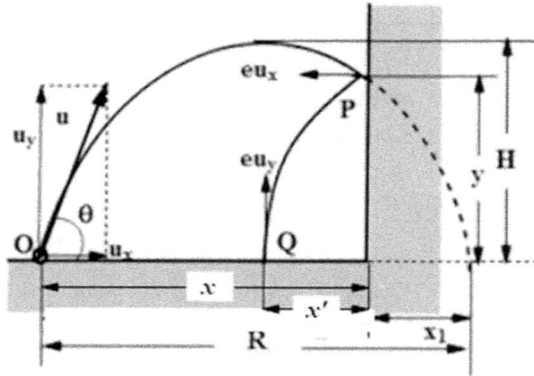

$$y = x \tan \theta - \frac{gx^2}{2v^2 \cos^2 \theta} \tag{2.176}$$

The range of the projectile is

$$R = \frac{v^2 \sin \theta \cos \theta}{g} \tag{2.177}$$

Using equations (2.176) and (2.177)

$$y = x \tan \theta - \frac{gx^2}{\frac{2gR\cos^2\theta}{\sin\theta\cos\theta}}$$

$$= x \tan \theta - x^2 \left(\frac{\tan \theta}{R} \right)$$

$$\Rightarrow y = x \tan \theta \left(1 - \frac{x}{R} \right) \qquad (2.178)$$

Method 1:

$$y = \frac{R}{4} \tan \theta \left(1 - \frac{R/4}{R} \right)$$

$$= \frac{3R}{16} \tan \theta \qquad (2.179)$$

Since $H = \frac{v^2 \sin^2\theta}{2g}$ and $R = \frac{v^2 \sin\theta\cos\theta}{g}$, $\dfrac{H}{R} = \dfrac{\dfrac{v^2\sin^2\theta}{2g}}{\dfrac{2v^2\sin\theta\cos\theta}{g}}$

$$\Rightarrow \frac{H}{R} = \frac{\tan\theta}{4} \qquad (2.180)$$

Using equations (2.179) and (2.180)

$$y = \frac{3}{4} \left(\frac{R}{4H^2} \tan\theta \right) H = \frac{3H}{4} \text{ Ans.}$$

Method 2: We can use the expression $h = \frac{1}{2}gt_1 t_2$

$$= \frac{1}{2}g \left(\frac{3R}{4u\cos\theta} \right) \left(\frac{R}{4\cos\theta} \right)$$

$$h = \frac{3gR^2}{32u^2\cos^2\theta} \qquad (2.181)$$

$$R = \frac{v^2\sin\theta\cos\theta}{g} \qquad (2.182)$$

$$\Rightarrow h = \frac{3g}{32u^2\cos^2\theta} \left(\frac{4u^4\sin^2\theta\cos^2\theta}{g^2} \right)$$

$$= \frac{3u^2\sin^2\theta}{4 \times 2g} = \frac{3}{4} \left(\frac{u^2\sin^2\theta}{2g} \right) = \frac{3}{4}H$$

(b) The time of flight from P to Q = time of flight from P to $R = t_2$, say.

$$\text{We can see that } x + x_1 = R \tag{2.183}$$

Furthermore,

$$\frac{x_1}{x'} = \frac{u \cos \theta t_2}{eu \cos \theta t_2} = \frac{1}{e} \tag{2.184}$$

We know that,

$$y = x \tan \theta \left(1 - \frac{x}{R}\right)$$

$$= x \tan \theta \frac{(R - x)}{R}, \text{ where } R - x = x_1$$

$$= \frac{xx_1 \tan \theta}{R} \tag{2.185}$$

Putting $R = \dfrac{2u^2 \sin \theta \cos \theta}{g}$ and $x_1 = \left(\dfrac{1}{e}\right)x'$ from equations (2.183) in (2.185), we have

$$y = \frac{\dfrac{xx'}{e} \tan \theta}{\left\{\dfrac{2u^2 \sin \theta \cos \theta}{g}\right\}} = \frac{gxx'\sec^2\theta}{2eu^2} \text{ Ans.}$$

The ratio times between O to P and P to Q is The ratio times

$$\frac{t_1}{t_2} = \frac{\dfrac{x}{u \cos \theta}}{\dfrac{x'}{eu \cos \theta}} = \frac{ex'}{x}$$

$$\Rightarrow \frac{t_1}{t_1 + t_2} = \frac{ex'}{ex' + x}$$

$$\Rightarrow t_1 = \frac{ex'}{(ex' + x)}T \tag{2.186}$$

$$t_2 = \frac{xt_1}{ex'} = \frac{xT}{ex' + x} \tag{2.187}$$

As we know,

$$y = \frac{1}{2}gt_1t_2 \tag{2.188}$$

Using equations (2.186), (2.187) and (2.188), we have

$$y = \frac{1}{2}g\left(\frac{ex'}{ex' + x}T\right)\left(\frac{xT}{ex' + x}\right)$$

$$\Rightarrow y = \frac{gexx'}{2(ex' + x)^2}T^2$$

where $T = \frac{2u \sin \theta}{g}$

$$\Rightarrow y = \frac{2exx'u^2 \sin^2 \theta}{g(ex' + x)^2}$$

$$\Rightarrow y = \frac{exx'(ex' + x)\tan \theta}{2(ex' + x)^2}$$

(c) Putting $gT^2 = 8H$, we have

$$y = \frac{4exx'H}{(ex' + x)^2}$$

If the collision is elastic put $e = 1$ and if the ball hits the ground directly below the highest point put $x - x' = 2x'$

$$\Rightarrow x' = \frac{x}{3}.$$

Then, we have

$$y = \frac{4(1)(x), \left(\frac{x}{3}\right)(H)}{\left(\frac{x}{3} + x\right)^2} = \frac{3}{4} H \text{ Ans.}$$

Problem 43 A ball is projected from a point O at an angle α with a velocity u from a horizontal distance x. It collides with the vertical wall at a point P. If OP makes an angle θ with horizontal and $e =$ coefficient of restitution, prove that (a) $\tan \theta = \frac{\tan \alpha}{1 + e}$ (b) $x = \frac{eu^2 \sin 2\theta}{(1 + e)g}$, therefore $d < \frac{eu^2}{(1 + e)g}$.

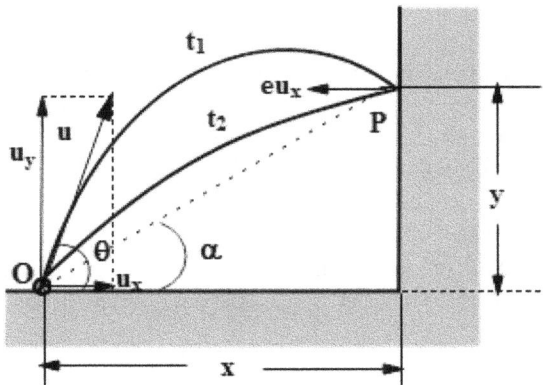

Solution

(a) From the last problem, we know that

$$y = x \tan \theta \left(1 - \frac{x}{R} \right)$$

$$\Rightarrow \frac{y}{x} = \cos \theta \left(1 - \frac{x}{R} \right)$$

$$\Rightarrow \tan \alpha = \tan \theta \left(1 - \frac{x}{R} \right) \qquad (2.189)$$

The sum of the times of flight is $t_1 + t_2 \cong T$

$$\Rightarrow \frac{x}{u \cos \theta} + \frac{x}{eu \cos \theta} = \frac{2u \sin \theta}{g}$$

$$\Rightarrow \left(1 + \frac{1}{e} \right) x = \frac{u^2 \sin \theta \cos \theta}{g}$$

$$\Rightarrow x = \frac{e}{e + 1} R \qquad (2.190)$$

Using equations (2.189) and (2.190),

$$\tan \alpha = \tan \theta \left(1 - \frac{e}{e + 1} \right)$$

$$\Rightarrow \tan \alpha = \frac{\tan \theta}{1 + e} \quad \text{Ans.}$$

(b) Since $x = \frac{e}{e+1} \frac{u^2}{g} \sin 2\theta$ and $\sin 2\theta < 1$

$$\Rightarrow x < \frac{eu^2}{(e + 1)g} \quad \text{Ans.}$$

Problem 44 A ball is projected from a point O from horizontal ground with a velocity u at an angle θ such that it collides at a point A on the vertical wall at its highest position. As a result, the ball bounces back at A, collides with the ground at B, and after the second collision the ball reaches the point of projection as shown in the figure. Find the (a) coefficient of restitution, (b) fraction of total energy loss in the two collisions, (c) time after which the ball reaches the point of projection, (d) speed of the ball just after the second collision, (e) angle of striking of the ball at the point O, and (f) tan Ø, where Ø is the angle of elevation of OP.

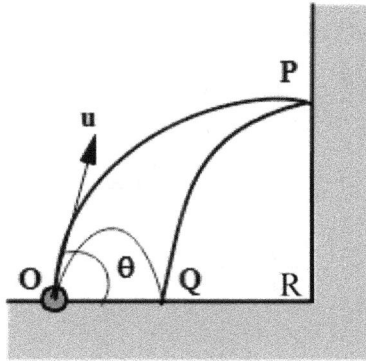

Solution

(a) The time of flight between O and P is

$$T = \frac{u \sin \theta}{g}$$

Furthermore, $OR = OQ + QR$ (2.191)

The distance between O and R $= OR = u_x T = u \cos \theta . \dfrac{u \sin \theta}{g}$

$$= \frac{u^2 \sin \theta \cos \theta}{g} \tag{2.192}$$

The distance between R and Q $= QR = (eu\cos\theta)T = eu\cos\theta\sqrt{\dfrac{2H}{g}}$, where $H = PR$

$$= eu \cos \theta \sqrt{\frac{2}{g}\left(\frac{u^2 \sin^2 \theta}{2g}\right)}$$

$$= \frac{eu^2 \sin \theta \cos \theta}{g} \tag{2.193}$$

The distance between O and Q $= OQ = u_x T' = \left(\dfrac{2eu_y}{g}\right)(eu \cos \theta)$

$$= \frac{2e^2}{g}u^2 \sin \theta \cos \theta \tag{2.194}$$

2-108

Using all the above four equations,

$$\frac{u^2 \sin \theta \cos \theta}{g} = \frac{eu^2 \sin \theta \cos \theta}{g} + \frac{2e^2u^2 \sin \theta \cos \theta}{g}$$

$$\Rightarrow 2e^2 + e - 1 = 0$$

$$\Rightarrow e = \frac{-1 \pm \sqrt{1 + 8}}{4} = \frac{3 - 1}{4} = 0.5 \text{ Ans.}$$

(b) The total loss of energy in the first and second collision is

$$\Delta E = \Delta E_1 + \Delta E_2$$

$$= -\left[\frac{1}{2} mu^2 \cos^2 \theta (1 - e^2) + \frac{1}{2} mu^2 \sin^2 \theta (1 - e^2) \right]$$

$$= -(1 - e^2) \frac{mu^2}{2}$$

Then the partion loss of energy is

$$\eta = -\frac{\Delta E}{KE_0} = (1 - e^2) = 1 - \frac{1}{4} = \frac{3}{4} \text{ Ans.}$$

(c) The time after which the body reaches the point of projection is

$$T = T_{OP} + T_{PQ} + T_{QO}$$

$$\Rightarrow T = \frac{u \sin \theta}{g} + \frac{u \sin \theta}{g} + \frac{2eu \sin \theta}{g}$$

$$= \frac{2u \sin \theta}{g}(1 + e) \text{ Ans.}$$

(d) The velocity of the body just after the second collision is

$$\vec{v} = -eu \cos \theta \hat{i} + eu \sin \theta \hat{j}$$

$$\Rightarrow v = eu = \frac{u}{2}$$

(e) The angle of striking with respect to horizontal is

$$\beta = \tan^{-1} \frac{v_y}{v_x}$$

$$=\tan^{-1}\frac{eu\sin\theta}{eu\cos\theta}$$

$$=\tan^{-1}\tan\theta = \theta \text{ Ans.}$$

(f) The angle of elevation of P is

$$\tan\phi = \frac{PR}{OR}$$

where $PR = H_{max} = \frac{u^2\sin^2\theta}{2g}$ and

$$OR = \frac{u^2\sin\theta\cos\theta}{g}$$

$$\Rightarrow \tan\phi = \frac{\frac{u^2\sin^2\theta}{2g}}{\frac{u^2\sin\theta\cos\theta}{g}} = \frac{\tan\theta}{2} \text{ Ans.}$$

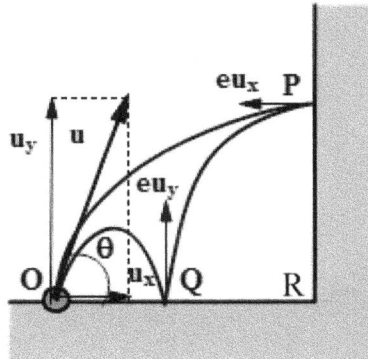

Problem 45 A ball is released from rest from a height h on to a plate that moves with a horizontal velocity u. Assuming μ = coefficient of friction and e = coefficient of restitution, find the (a) velocity of the ball just before second collision (b) time of flight of the ball (c) angle made by the velocity of the ball with horizontal.

Solution

(a) The velocity just before collision is

$$v_0 = \sqrt{2gh} \tag{2.195}$$

The vertical velocity just after the collision is

$$v_y = ev_0 = e\sqrt{2gh} \tag{2.196}$$

The vertical impulse is

$$N\delta t = m(1 + e)v_0$$

$$N\delta t = m(1 + e)\sqrt{2gh} \tag{2.197}$$

The horizontal impulse is

$$m\Delta \vec{v}_x = \mu N\delta t \hat{i} = m\mu(1 + e)\sqrt{2gh}\,\hat{i}$$

$$\Rightarrow \vec{v}_x = \mu(1 + e)\sqrt{2gh}\,\hat{i} \tag{2.198}$$

$$\vec{v}_y = e\sqrt{2gh}\,\hat{j}$$

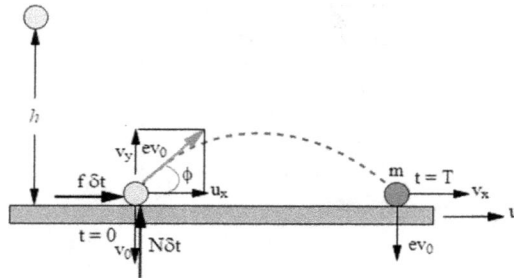

The velocity of the ball just after the collision is

$$\vec{v} = \sqrt{2gh}\,\{\mu(1 + e)\hat{i} + e\hat{j}\}\ \text{Ans.}$$

(b) The time of flight of the ball is

$$T = \frac{2v_y}{g} = \frac{e\sqrt{2gh}}{g} = e\sqrt{\frac{2h}{g}}$$

The (horizontal range) relative to the plate is

$$R = v_{\text{rel}}T$$

$$\Rightarrow R = (v_x - u)T$$

$$= e\{\mu(1 + e) - u\}\sqrt{\frac{2h}{g}} \text{ Ans.}$$

(c) The angle made by the velocity of the ball with horizontal is

$$\phi = \tan^{-1}\frac{v_y}{v_x}$$

$$= \tan^{-1}\frac{\sqrt{2gh}}{\mu(1 + e)\sqrt{2gh}}$$

$$= \tan^{-1}\frac{e}{\mu(1 + e)} \text{ Ans.}$$

Problem 46 A smooth ball of mass m collides with a stationary plank of mass M with a velocity v_0 at an angle θ with vertical. The plank is placed on a smooth horizontal surface. The coefficient of restitution for the collision between the ball and plank is e and the coefficient of kinetic friction between the plank and the ball is μ. Find the (a) velocity of the ball just after the collision (b) angle made by the velocity of the ball with vertical just after the collision (c) the velocity of the plank just after the collision.

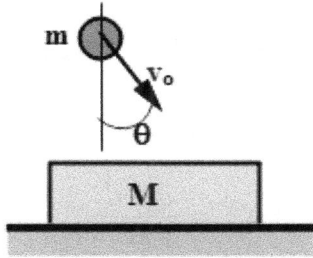

Solution

(a) The vertical impulse on the ball is

$$N\delta t = mv_0(1 + e)\cos\theta \tag{2.199}$$

The horizontal impulse on the ball is

$$-\mu N\delta t = m(v_1 - v_0\sin\theta) \tag{2.200}$$

Using equations (2.199) and (2.200)

$$-\mu m v_0(1 + e)\cos \theta = m(v_1 - v_0 \sin \theta)$$

$$\Rightarrow v_1 = v_0\{\sin \theta - \mu(1 + e)\cos \theta\}$$

$$\Rightarrow v_1 = v_0\{\sin \theta - \mu(1 + e)\cos \theta\} \qquad (2.201)$$

Then the velocity of the ball just after the impact is

$$\vec{v_m} = v_1\hat{i} + ev_0 \cos \theta\hat{j}$$

$$\vec{v_m} = v_0[\{\sin \theta - \mu(1 + e)\cos \theta\}\hat{i} + e \cos \theta]\hat{j} \text{ Ans.}$$

(b) The angle ϕ made by $\vec{v_m}$ with vertical is given as

$$\tan \phi = \frac{v_{mx}}{v_{my}}$$

$$= \frac{\sin \theta - \mu(1 + e)\cos \theta}{e \cos \theta}$$

$$= \frac{\tan \theta - \mu(1 + e)}{e}$$

$$\Rightarrow \phi = \tan^{-1}\left\{\frac{\tan \theta - \mu(1 + e)}{e}\right\}$$

N.B.: If $\mu = 0$, $\phi = \tan^{-1}\frac{\tan \theta}{e}$ as derived earlier.

(c) The horizontal impulse on the plank is

$$\mu N\delta t\hat{i} = M\Delta\vec{v}$$

$$\Rightarrow \Delta\vec{v} = \frac{\mu N\delta t\hat{i}}{M} \qquad (2.202)$$

$$\Rightarrow \vec{v_M} = \frac{\mu m v_0(1 + e)\cos \theta}{M}\hat{i}$$

$$\Rightarrow \vec{v_M} = \left\{\frac{\mu m v_0(1 + e)\cos \theta}{M}\right\}\hat{i} \text{ Ans.}$$

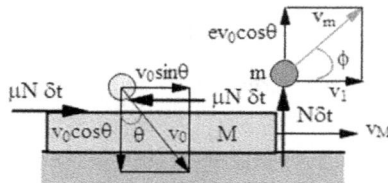

Problem 47 A ball is projected horizontally from the top of a hemispherical smooth surface of radius R with a minimum speed so that the ball escapes the sphere just after the projection. Then the ball collides with a smooth horizontal surface many times before coming to rest. Assume e = coefficient of restitution of each collision. Find the (a) total time of the ball in air, (b) total horizontal distance covered by the ball until it stops bouncing.

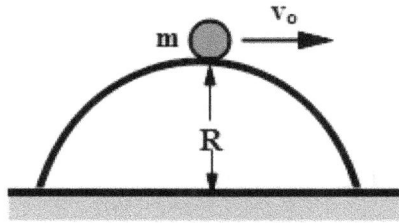

Solution

(a) The net vertical force acting on the ball is

$$mg - N = ma_y$$

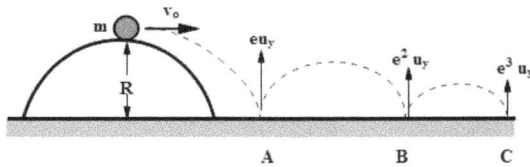

Putting $a_y = \frac{v_0^2}{R}$ for $N = 0$, we have

$$mg = \frac{mv_0^2}{R}$$

$$\Rightarrow v_0 = \sqrt{gR} \tag{2.203}$$

The vertical velocity just before the first collision is

$$v_y = \sqrt{2gR}$$

The vertical velocity just after the first collision is

$$v'_y = ev_y = e\sqrt{2gR}$$

Then the time between the first and second collision is

$$t_1 = \frac{2v'_y}{g} = 2e\sqrt{\frac{2R}{g}}$$

The time between the second and third collision is

$$t_2 = \frac{2v''_y}{g} = \frac{2ev'_y}{g} = 2e^2 \sqrt{\frac{2R}{g}}$$

This collision goes on repeating until the ball stops bouncing after a time
$T = t_0 + t_1 + t_2 + \ldots\ldots$. where $\frac{1}{2}gt_0^2 = R$

$$\Rightarrow T = \sqrt{\frac{2R}{g}} + 2e\sqrt{\frac{2R}{g}} + 2e^2\sqrt{\frac{2R}{g}} + \ldots.$$

$$= \sqrt{\frac{2R}{g}}(1 + 2e + 2e^2 + 2e^3 + \ldots)$$

$$= \sqrt{\frac{2R}{g}}[1 + 2e(1 + e^2 + e^3 + \ldots.)]$$

$$= \sqrt{\frac{2R}{g}}\left[1 + 2e\left(\frac{1}{1-e}\right)\right]$$

$$\Rightarrow T = \frac{1+e}{1-e}\sqrt{\frac{2R}{g}}$$

(b) The total horizontal distance covered by the ball during time T is

$$x = (v_x)T = v_0 T$$

$$x = \frac{(1+e)v_0}{1-e}\sqrt{\frac{2R}{g}} \qquad\qquad (2.204)$$

Using equations (2.203) and (2.204)

$$x = \frac{1+e}{1-e}\sqrt{gR}\cdot\sqrt{\frac{2R}{g}}$$

$$= \sqrt{2}\left(\frac{1+e}{1-e}\right)R \text{ Ans.}$$

Problem 48 Two identical discs each of mass m and radius r are placed in contact on a smooth horizontal surface. Another disc of mass M and radius $R = \eta r$ strikes the discs symmetrically as shown in the figure. Assume that all discs are smooth and e = coefficient of restitution of impacts. (a) For what value of e will the striking disc (i) move forward, (ii) stop, and (iii) move backwards, just after the impact? (b) Find the velocities of the discs just after the impact.

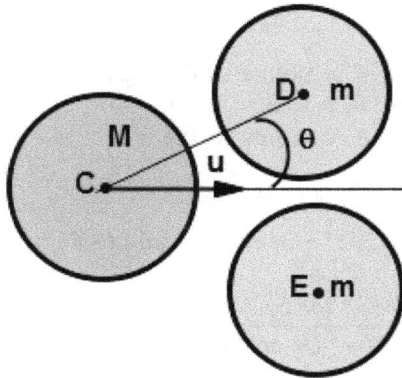

Solution

(a) Conserving horizontal momentum of the system $(M + 2m)$,

$$Mu = 2mv' \cos \theta + Mv \qquad (2.205)$$

NIF along the normal

$$-eu \cos \theta = v \cos \theta - v' \qquad (2.206)$$

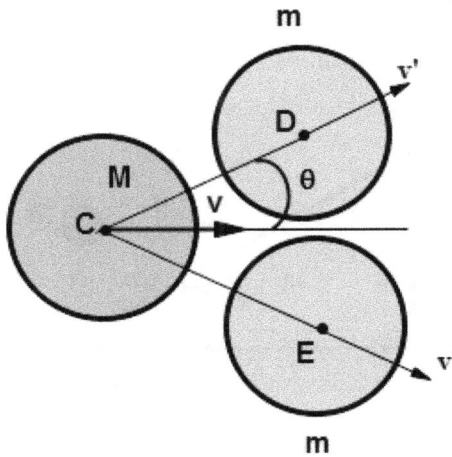

Solving equations (2.205) and (2.206)

$$Mu = 2m(v \cos \theta + eu \cos \theta)\cos \theta + Mv$$

$$\Rightarrow v(M + 2m \cos^2 \theta) = Mu - 2meu \cos^2 \theta$$

$$\Rightarrow v = \frac{(M - 2me \cos^2 \theta)u}{M + 2m \cos^2 \theta} \quad \text{Ans.}$$

If $M > 2me \cos^2 \theta$, v points to the right.
If $M = 2mu \cos^2 \theta$, $v = 0$.
If $M < 2me \cos^2 \theta$, v points to the left.

(b) The velocity of the discs m is

$$v' = v \cos \theta + eu \cos \theta$$

$$= \frac{(M - me \cos^2 \theta)u}{M + 2m \cos^2 \theta} \cdot \cos \theta + eu \cos \theta$$

$$= u \cos \theta \left[\frac{M - me \cos^2 \theta}{M + 2m \cos^2 \theta} + e \right]$$

$$= \left\{ \frac{M(1 + e) + me \cos^2 \theta}{M + 2m \cos^2 \theta} \right\} u \cos \theta \quad \text{Ans.}$$

Problem 49 A ball of mass m collides with the smooth stationary wedge of mass M with a horizontal velocity u, say. As a result, the ball moves vertically up just after the collision. Find the value/s of m/M assuming $\theta = 53°$.

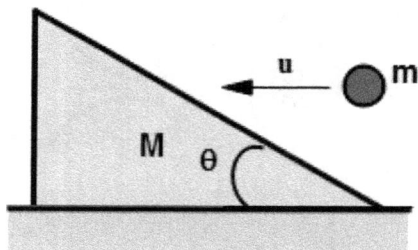

Solution
Conservation of horizontal line as momentum

$$Mv' = mu \tag{2.207}$$

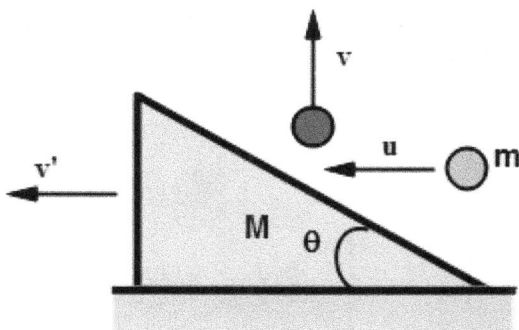

Conserving linear momentum of the ball along the slant, we have

$$mu \cos \theta = mv \sin \theta$$

$$\Rightarrow u \cos \theta = v \sin \theta \tag{2.208}$$

Newton's law of impact (along the y-axis or normal to the slant)

$$-ev_{\text{approach}} = v_{\text{separation}}$$

$$-e\{(-u \cos \theta) - 0\} = v \cos \theta - (-v \sin \theta)$$

$$\Rightarrow eu \cos \theta = v \cos \theta + v' \sin \theta \tag{2.209}$$

$$\Rightarrow e = \frac{v \cos \theta + v' \sin \theta}{u \cos \theta}$$

$$= \frac{v}{u} + \frac{v'}{u} \tan \theta \tag{2.210}$$

Putting $\frac{v}{u} = \cos \theta$ from equation (2.208) and $\frac{v'}{u} = \frac{m}{M}$ from equations (2.207) in (2.210),

$$e = \cos \theta + \frac{m}{M} \tan \theta \leqslant 1$$

$$\Rightarrow \frac{m}{M} \leqslant \frac{(1 - \cos \theta)}{\tan \theta}$$

$$\Rightarrow \frac{m}{M} \leqslant \frac{1 - \cot \theta}{\tan \theta}$$

Putting $\theta = 53°$, $\cos \theta = \frac{3}{4}$ and $\tan \theta = \frac{4}{3}$,

$$\frac{m}{M} \leqslant \frac{1 - \frac{3}{4}}{\frac{4}{3}} = \frac{1}{4} \times \frac{3}{4} = \frac{3}{16}$$

$$\frac{m}{M} \leqslant \frac{3}{16} \text{ Ans.}$$

Problem 50 The ball of mass m collides with the smooth wedge of mass M with a vertical downward velocity v. Find the maximum compression of the spring of stiffness k. Assume e = coefficient of restitution.

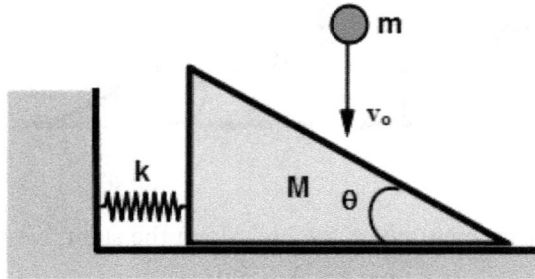

Solution

The momentum of the ball parallel to the slant just before and after the collision remains conserved. So $v_0 \sin \theta$ is the velocity of the ball just after the collision along the slant for the system $(M + m)$.

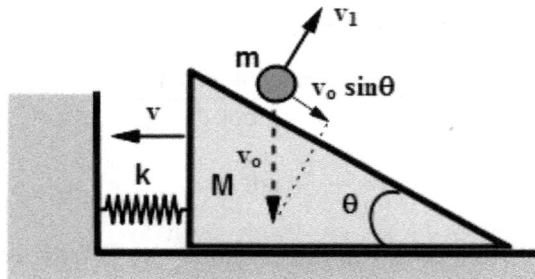

Since, $P_{\text{horizontal}} \Rightarrow 0$ we have

$$MV - m(v_1 \sin \theta + v_0 \sin \theta \cos \theta) = 0$$

$$\Rightarrow Mv = m(v_1 \sin \theta + v_0 \sin \theta \cos \theta)$$

$$\Rightarrow Mv = m \sin \theta (v_1 + v_0 \cos \theta) \tag{2.211}$$

Newton's impact (collision) formula perpendicular to the slant

$$-e(v_0 \cos \theta) = -v \sin \theta + (-v_1)$$

$$\Rightarrow v \sin \theta + v_1 = +ev_0 \cos \theta \tag{2.212}$$

Solving equations (2.211) and (2.212)

$$Mv = m \sin \theta \{v_0 \cos \theta + ev_0 \cos \theta - v \sin \theta\}$$

$$\Rightarrow v = \frac{mv_0 \sin \theta \cos \theta (1 + e)}{M + m \sin^2 \theta} \quad \text{Ans.}$$

Conservation of energy $\Delta U + \Delta K = 0$

$$\Rightarrow \frac{k}{2} x^2 - \frac{M}{2} v^2 = 0$$

$$\Rightarrow x = \sqrt{\frac{M}{k}} v$$

$$= \sqrt{\frac{M}{k}} \frac{mv_0 \sin \theta \cos \theta (1 + e)}{M + m \sin^2 \theta} \quad \text{Ans.}$$

Problem 51 The bob of mass m collides perpendicular to the slant of a smooth wedge of mass M with a velocity v_0 as shown in the figure. Find the (a) velocity of the (i) wedge and (ii) bob and (b) condition for which the bob will stop momentarily just after the collision.

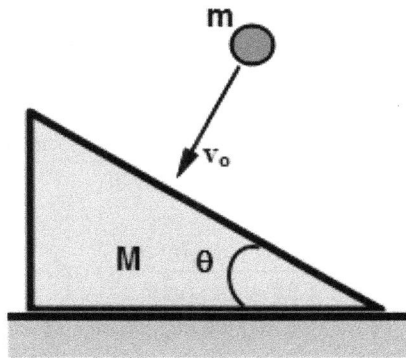

Solution

(a)

(i) Conserving horizontal momentum of the system $(M + m)$,

$$Mv - mv_1 \sin \theta = mv_0 \sin \theta \qquad (2.213)$$

Newton's impact formula along the normal to the slant

$$-ev_0 = -v \sin \theta + (-v_1)$$

$$\Rightarrow ev_0 = v \sin \theta + v_1 \qquad (2.214)$$

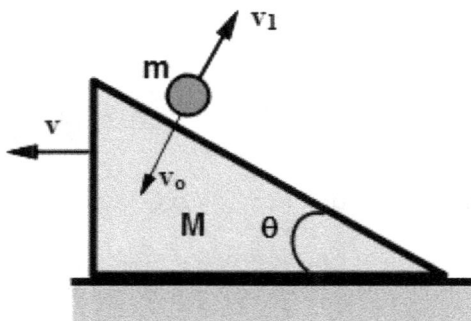

Solving equations (2.213) and (2.214),

$$Mv - m(ev_0 - v \sin \theta)\sin \theta = mv_0 \sin \theta$$

$$v(M + m \sin^2 \theta) = mv_0 \sin \theta(1 + e)$$

$$\Rightarrow v = \frac{mv_0 \sin \theta(1 + e)}{M + m \sin^2 \theta} \text{ Ans.}$$

(ii) Writing Newton's impact formula along the normal to the slant, we have $v_1 = ev_0 - v \sin \theta$

$$= ev_0 - \frac{mv_0 \sin \theta(1 + e)\sin \theta}{M + m \sin^2 \theta}$$

$$= v_0 \left\{ \frac{Me + me \sin^2 \theta - m \sin^2 \theta - me \sin^2 \theta}{M + m \sin^2 \theta} \right\}$$

$$v_1 = \frac{Me - m \sin^2 \theta}{M + m \sin^2 \theta} v_0 \text{ Ans.}$$

(b) The ball will stop if $v_1 = 0$

$$\Rightarrow Me - m \sin^2 \theta = 0$$

$$\Rightarrow e = \frac{m \sin^2 \theta}{M} \text{ Ans.}$$

Problem 52 The bob of mass m collides with the slant of a smooth wedge of mass M with a horizontal velocity v_0 as shown in the figure. Find (a) the velocity of the wedge just after the collision (b) the angle of inclination of the wedge if the bob moves vertically up just after the collision. Put the coefficient of restitution of the collision $= e = \frac{1}{2}$ and $\frac{m}{M} = \frac{1}{4}$.

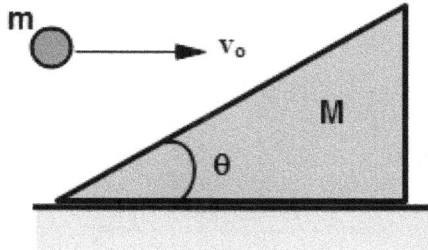

Solution

(a) Since no collision takes place along the slant, conserving the momentum of the ball along the slant just before and after the collision, we have the velocity of the ball $v_0 \cos \theta$ along the slant just after the collision. Let v' and v be the velocities of the ball perpendicular to the slant and the velocity of the wedge just after the collision, respectively.

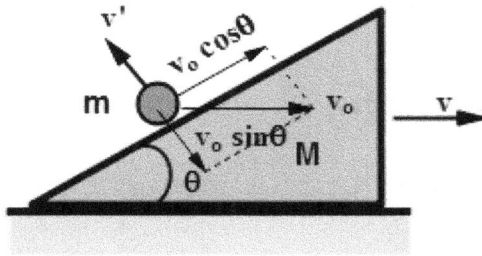

Since the net force $= 0$ along the horizontal on the system $(M + m)$, we have

$$P_{\text{hor}} = mv_0$$

So, $Mv + mv_{\text{horizontal}} = mv_0$

By putting the horizontal component of the velocity of m, we have

$$+Mv + m(v_0 \cos \theta \cos \theta - v' \sin \theta) = mv_0$$

$$\Rightarrow Mv + mv_0 \cos^2 \theta - mv' \sin \theta = mv_0$$

$$\Rightarrow Mv - mv' \sin \theta = mv_0(1 - \cos^2 \theta)$$

$$\Rightarrow Mv - mv' \sin \theta = mv_0 \sin^2 \theta \tag{2.215}$$

Applying the Newton's impact/collision formula along the normal to the slant, we have

$$-ev_{\text{app}} = v_{\text{sep}}$$

$$\Rightarrow -e\{v_0 \sin \theta\} = -v \sin \theta + (-v')$$

$$\Rightarrow ev_0 \sin\theta = v' + v\sin\theta \qquad (2.216)$$

Elininating v' from equations (2.215) by (2.216),

$$Mv = m(ev_0\sin\theta - v\sin\theta)\sin\theta + mv_0\sin^2\theta$$

$$\Rightarrow (M + m\sin^2\theta)v = mv_0\sin^2\theta(1 + e)$$

$$\Rightarrow v = \frac{mv_0\sin^2\theta}{M + m\sin^2\theta}(1 + e) \qquad (2.217)$$

(b) If the ball moves vertically up just after the collision,

$$(v_m)_{\text{horizontal}} = 0$$

$$\Rightarrow v_0\cos\theta.\ \cos\theta = v'\sin\theta$$

$$\Rightarrow v_0\cos^2\theta = (ev_0\sin\theta - v\sin\theta)\sin\theta$$

$$\Rightarrow v\sin^2\theta = (e\sin^2\theta - \cos^2\theta)v_0 \qquad (2.218)$$

Using equations (2.217) and (2.218),

$$\left\{\frac{mv_0\sin^2\theta(1 + e)}{M + \sin^2\theta}\right\}\sin^2\theta = (e\sin^2\theta - \cos^2\theta)v_0$$

$$\Rightarrow \frac{m\sin^4\theta}{M + m\sin^2\theta} + \cos^2\theta = e\sin^2\theta\left\{1 - \frac{m\sin^2\theta}{M + m\sin^2\theta}\right\}$$

$$\Rightarrow \frac{m\sin^2(\sin^2\theta + \cos^2\theta) + M\cos^2\theta}{M + m\sin^2\theta} = \frac{Me\sin^2\theta}{M + m\sin^2\theta}$$

$$\Rightarrow e = \frac{m\sin^2\theta + M\cos^2\theta}{M\sin^2\theta}$$

$$\Rightarrow e = \left(\frac{m}{M} + \cot^2\theta\right)$$

$$\Rightarrow \cot\theta = \sqrt{e - \frac{m}{M}} = \sqrt{\frac{1}{2} - \frac{1}{4}} = \frac{1}{2}$$

Problem 53 A striker is projected with a velocity u from a hole A of a carrom board. If the coefficient of restitution is e, find the (a) velocity (b) time of reaching of the striker at the hole C.

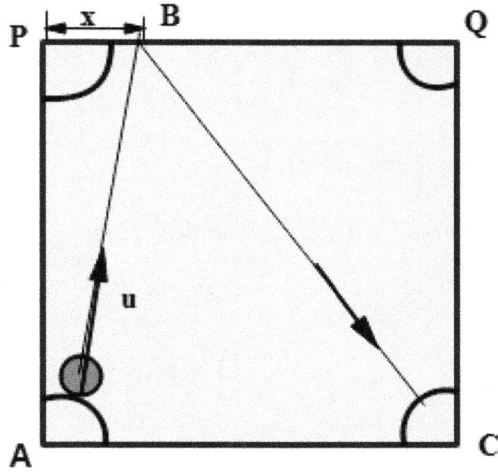

Solution

(a) Newton's impact formula along the y-axis

$$eu \cos \theta = v \cos \phi \tag{2.219}$$

Conservation of momentum along the x-axis

$$u \sin \theta = v \sin \phi \tag{2.220}$$

Using equations (2.219) and (2.220),

$$\tan \phi = \frac{\tan \theta}{e} \tag{2.221}$$

And

$$v = \sqrt{(e^2 \cos^2 \theta + \sin^2 \theta)}\, u \tag{2.222}$$

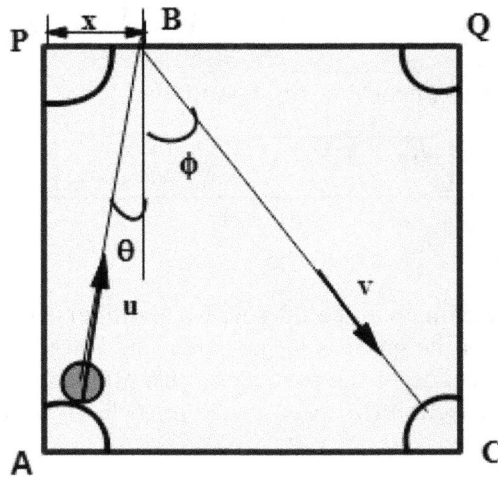

Geometrically,

$$\tan \theta = \frac{x}{l} \tag{2.223}$$

$$\tan \phi = \frac{l - x}{l} \tag{2.224}$$

Using equations (2.221), (2.223) and (2.224),

$$\frac{l - x}{l} = \frac{x}{el}$$

$$\frac{x}{l}\left(\frac{1}{e} + 1\right) = 1$$

$$\Rightarrow \frac{x}{l} = \frac{e}{e + 1} \quad \text{Ans.}$$

(b) The time taken to reach the hole is

$$T = \frac{AB}{u} + \frac{BC}{v}$$

$$= \frac{\sqrt{l^2 + x^2}}{u} + \frac{\sqrt{l^2 + (l - x)^2}}{v}$$

$$= \frac{\sqrt{l^2 + l^2 \frac{e^2}{(e+1)^2}}}{u} + \frac{\sqrt{l^2 + \frac{l^2}{(e+1)^2}}}{v}$$

$$= \frac{l}{u}\left[\sqrt{1 + \left(\frac{e}{e+1}\right)^2} + \frac{u}{v}\sqrt{1 + \left(\frac{1}{e+1}\right)^2}\right]$$

Putting $u/v = 1/\sqrt{e\cos^2\theta + \sin^2\theta}$, we have

$$T = \frac{l\left\{\sqrt{(e + 1)^2 + e^2} + \sqrt{\frac{(e+1)^2 + 1}{e^2\cos^2\theta + \sin^2\theta}}\right\}}{u(e + 1)} \quad \text{Ans.}$$

Problem 54 A particle of mass m_1 collides with a stationary particle of mass m_2 with a velocity u. The ratio of the speed of m_1 just after and before the oblique collision is 4/5. If the angle of deflection of the particle m_1 just after the collision is 37 degrees, find the (a) angle of deflection (b) speed of the particle m_2. Put $m_1/m_2 = 1/2$.

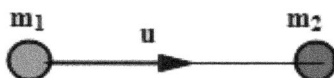

Solution

(a) Conserving momentum along the x-axis is

$$m_1 u = m_1 v_1 \cos \theta_1 + m_2 v_2 \cos \theta_2$$

$$\Rightarrow v_2 \cos \theta_2 = \frac{m_1}{m_2}(u - v_1 \cos \theta_1) \qquad (2.225)$$

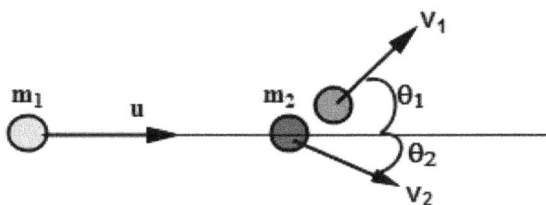

Conserving momentum along the y-axis is

$$m_1 v_1 \sin \theta_1 - m_2 v_2 \sin \theta_2 = 0$$

$$\Rightarrow v_2 \sin \theta_2 = \frac{m_1}{m_2} v_1 \sin \theta_1 \qquad (2.226)$$

From equations (2.225) and (2.226),

$$\tan \theta_2 = \frac{\dfrac{m_1 v_1 \sin \theta_1}{m_2}}{\dfrac{m_1}{m_2}(u - v_1 \cos \theta_1)}$$

$$\theta_2 = \tan^{-1}\left(\frac{v_1 \sin \theta_1}{u - v_1 \cos \theta_1}\right)$$

$$= \tan^{-1}\left(\frac{v_1 \frac{4}{5}}{u - v_1 \frac{3}{5}}\right)$$

$$\Rightarrow \tan^{-1}\left(\frac{3}{5\frac{u}{v_1} - 4}\right)$$

Putting, $\dfrac{u}{v_1} = \dfrac{4}{5}$ we have

$$\theta_2 = \tan^{-1} \frac{4}{5 \times \frac{4}{5} - 3} = 75.9° \text{ Ans.}$$

(b) $v_2 = \sqrt{(v_2 \sin \theta_1)^2 + (v_2 \cos \theta_2)^2}$

$$= \frac{m_1}{m_2} \sqrt{(u - v_1 \cos \theta_1)^2 + (v_1 \sin \theta_1)^2}$$

$$= \frac{m_1}{m_2} \sqrt{u^2 + v_1^2 - 2uv_1 \cos \theta_1}$$

Alternative method:

$$\text{Since } \overrightarrow{P_2} = \overrightarrow{P} - \overrightarrow{P_1}$$

$$P_2 = \sqrt{P^2 + P_1^2 - 2PP_1 \cos \theta_1}$$

$$\Rightarrow m_2 v_2 = \sqrt{m_1^2 u^2 + m_1 v_1^2 - 2m_1 u m_1 v_1 \cos \theta_1}$$

$$\Rightarrow v_2 = \frac{m_1}{m_2} \sqrt{u^2 + v_1^2 - 2uv_1 \cos \theta_1}$$

$$= \frac{1}{2} \sqrt{4^2 + 5^2 - 2 \times 4 \times 5 \times \frac{3}{5}} = 2.06 \text{ m s}^{-1} \text{ Ans.}$$

Problem 55 A cannon of mass M loaded with a shot of mass m is released from rest. It slides down a smooth inclined plane of angle of elevation θ. If the cannon fires the shot with a muzzle velocity u when the cannon has velocity v_0, find the time after which the cannon will stop.

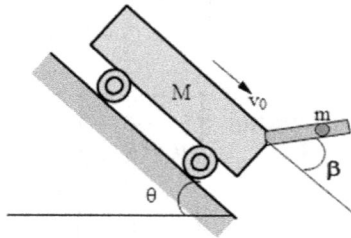

Solution

Let u = velocity of the bullet relative to the gun and v = recoil velocity of the cannon

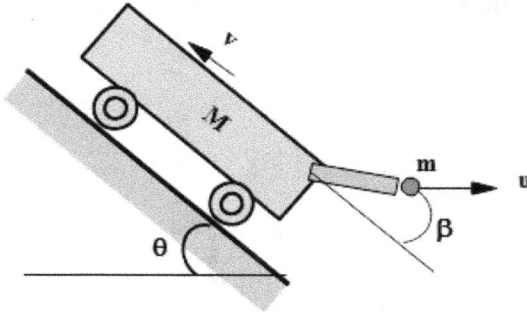

Then,

$$-Mv + m(u \cos \beta - v) = (M + m)v_0$$

$$\text{Or, } v = \frac{mu \cos \beta - (M + m)v_0}{M + m}$$

$$\text{Or, } v = \frac{mu \cos \beta}{M + m} - v_0$$

The time after which the cannon will stop from the instant of firing is

$$\Delta t = \frac{v}{g \sin \theta}$$

$$= \frac{1}{g \sin \theta}\left(\frac{mu \cos \beta}{M + m} - v_0\right)$$

Problem 56 A wet clay pallet collides with block B of mass m' plastically with a velocity u. Block A of mass M is resting on the ground being connected to a hanging block of mass m by an ideal string. Assume that the block leaves the ground due to the impact of the pallet. (a) Find the velocity of the blocks just after the impact. (b) What is the loss of energy in the collision?

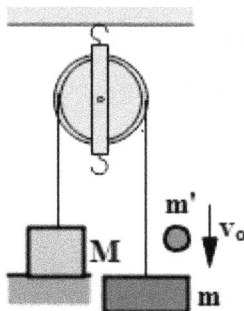

Solution

Let the velocity of the combination block m and m' be v down. Then the block M will move up with velocity v.

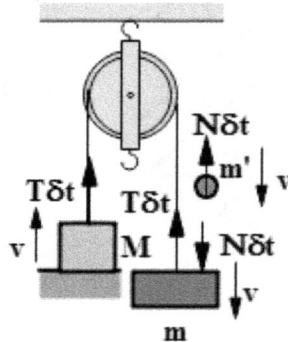

(a) Impulse-momentum equation for

$$m': (-m'g + N)\delta t = m'\{(-v) - (-v_0)\}$$

$$\Rightarrow (N - m'g)\delta t m'(v_0 - v) \tag{2.227}$$

$$m: \{T - (mg - N)\}\delta t = -mv$$

$$\Rightarrow (N + mg)\delta t = mv \tag{2.228}$$

$$M: (-mg + T)\delta t = Mv$$

$$(T - Mg)\delta t = Mv \tag{2.229}$$

Equations $(2.228) + (2.229) - (2.227)$

$$(T - Mg + N + mg - T - N + m'g)\delta t$$

$$= mv + Mv - m'(v_0 - u)$$

$$(m + m' - M)g\delta t = v(M + m + m') - m'v_0$$

Since $\delta t \ll$, we can ignore the left hand term. Then we have

$$v(M + m + m') - m'v_0 = 0$$

$$\Rightarrow v = \frac{m'v_0}{M + m + m'} \quad \text{Ans.}$$

(b) The loss of *KE* during collision is

$$\Delta K = \frac{1}{2}m'v_0^2 - \frac{1}{2}(M + m + m')v^2$$

$$= \frac{1}{2}m'v_0^2 - \frac{1}{2}(M + m + m')\left(\frac{m'v_0}{M + m + m'}\right)^2$$

$$= \frac{1}{2}m'v_0^2 \frac{(M + m)}{M + m + m'}$$

Then, $-\frac{\Delta K}{K} = \frac{M + m}{M + m + m'}$ Ans.

Problem 57 A ball of mass m_1 collides elastically with a stationary ball of mass m_2 so that they deviate relative to the line of motion of the first ball at an angle of θ_1, as shown in the figure.

(a) If the balls are identical, prove that the maximum angle of deviation of the balls is a right angle. (b) Prove that the maximum angle of deflection of m_1 is given as $\sin \theta_m = \frac{m_2}{m_1}$ when $m_2 < m_1$.

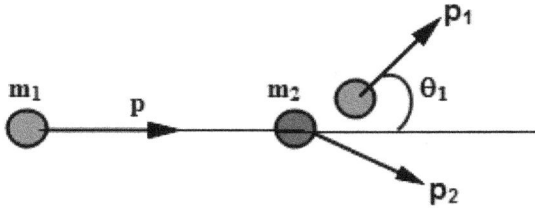

Solution

(a) If $m_1 = m_2 = m$
    The conservation of linear momentum

$$\vec{p} = \vec{p_1} + \vec{p_2}$$

$$\Rightarrow p^2 = p_1^2 + p_2^2 + 2\vec{p_1} \cdot \vec{p_2} \tag{2.230}$$

The conservation of KE in elastic collision is

$$\frac{p^2}{2m} = \frac{p_1^2}{2m} + \frac{p_2^2}{2m}$$

$$\Rightarrow p^2 = p_1^2 + p_2^2 \tag{2.231}$$

Using equations (2.230) and (2.231), we have

$$p_1^2 + p_2^2 = p_1^2 + p_2^2 + 2\vec{p_1} \cdot \vec{p_2}$$

$$\Rightarrow \vec{p_1} \cdot \vec{p_2} = 0$$

$\Rightarrow p_1 \perp \vec{p_2}$ or the angle between $\vec{p_1}$ and $\vec{p_2}$ is 90°.

Then the balls will move perpendicular to each other.

(b) Conservation of linear momentum along the x-axis (line joining the particles)

$$\vec{p_1} + \vec{p_2} = \vec{p}$$

$$\vec{p_2} = \vec{p} - \vec{p_1}$$

$$\Rightarrow p_2^2 = p^2 + p_1^2 - 2pp_1 \cos \theta_1 \tag{2.232}$$

Conservation of kinetic energy in scattering:

$$\frac{p_2^2}{2m_2} + \frac{p_1^2}{2m_1} = \frac{p^2}{2m_1}$$

$$\Rightarrow p_2^2 = \frac{m_2}{m_1}p^2 - \frac{m_2}{m_1}p_1^2 \tag{2.233}$$

Using equations (2.232) and (2.233),

$$p^2 + p_1^2 - 2pp_1 \cos \theta_1 = \frac{m_2}{m_1}p^2 - \frac{m_2}{m_1}p_1^2$$

$$\Rightarrow p_1^2\left(1 + \frac{m_2}{m_2}\right) - 2p_1 p \cos \theta_1 + p^2\left(1 - \frac{m_2}{m_1}\right) = 0$$

For real root of p_1,

$$(-2p \cos \theta_1)^2 \geqslant 4\left(1 + \frac{m_2}{m_1}\right)\left(1 - \frac{m_2}{m_1}\right)p^2$$

$$\Rightarrow \cos^2 \theta_1 \geqslant 1 - \left(\frac{m_2}{m_1}\right)^2$$

$\Rightarrow \sin \theta_1 \leqslant \dfrac{m_2}{m_1}$, where $m_1 > m_2$

If the masses are equal this angle can be equal to or less that a right angle.

Problem 58 A particle of mass m and velocity v_0 collides with a stationary particle of mass M. The particles get scattered by an angle θ in the center of mass reference frame. Find the (a) velocities of m and M in the laboratory reference frame (b) magnitude of fractional change of kinetic energy of m and M.

Solution

(a) Relative to the center of mass frame we can conserve the momentum to zero. So, we can write

$$\left(\overrightarrow{p_{1c}}\right)_i + \left(\overrightarrow{p_{2c}}\right)_i = \left(\overrightarrow{p_{1c}}\right)_f + \left(\overrightarrow{p_{2c}}\right)_f = 0$$

$$\Rightarrow (p_{1c})_i = (p_{2c})_i \text{ and } (p_{1c})_f = (p_{2c})_f$$

$$\Rightarrow p_{1c} = p_{2c} \text{ and } p'_{1c} = p'_{2c}$$

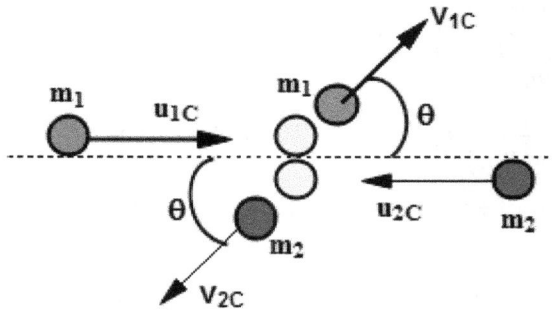

We can also conserve the KE in the center of mass frame because the collision is elastic.

$$\frac{p_{1c}^2}{2m_1} + \frac{p_{2c}^2}{2m_2} = \frac{p'^2_{1c}}{2m_1} + \frac{p'^2_{2c}}{2m}$$

Putting $p_{1c} = p_{2c} = p$ and $p'_{1c} = p'_{2c} = p'$ we have $p^2\left(\frac{1}{2m_1} + \frac{1}{2m_2}\right) = p'^2\left(\frac{1}{2m_1} + \frac{1}{2m_2}\right)$

$$\Rightarrow p = p'$$

$$\Rightarrow p_{1c} = p_{2c} = p'_{1c} = p'_{2c}$$

This means that

$$v_{1c} = u_{1c} \text{ and } v_{2c} = u_{2c} \tag{2.234}$$

In other words, the magnitudes of initial final velocities of m_1 relative to the center of mass frame are equal. Similarly, the magnitude of velocity of m_2 just before and after the collision with respect to the center of mass frame are equal. This is the concept of elastic scattering. If $\theta =$ angle of scattering in the center of mass frame, the velocities of m_1 and m_2 are given as

$$|\vec{v_1}| = |\vec{v_{1c}} + \vec{v_c}|$$

$$= \sqrt{v_{1c}^2 + v_c^2 + 2v_{1c} \cdot v_c \cos\theta}$$

Putting, $v_{1c} = u_{1c}$ we have

$$v_1 = \sqrt{u_{1c}^2 + v_c^2 + 2u_{1c}v_c \cos\theta}$$

Putting $u_{1c} = |\vec{v_0} - \vec{v_c}| = |\vec{v_0} - \frac{m_1\vec{v_0}}{m_1 + m_2}| = \frac{m_2 v_0}{m_1 + m_2}$ and $v_c = \frac{m_1 v_0}{m_1 + m_2}$, we have

$$v_1 = \sqrt{\left(\frac{m_2 v_0}{m_1 + m_2}\right)^2 + \left(\frac{m_1 v_0}{m_1 + m_2}\right)^2 + 2\frac{m_1 m_2 v_0^2}{(m_1 + m_2)^2}\cos\theta} \qquad (2.235)$$

$$= \frac{v_0}{m_1 + m_2}\sqrt{m_1^2 + m_2^2 + 2m_1 m_2 \cos\theta} \text{ Ans.}$$

(b) The fraction of energy lost (delivered) by m_1 in elastic scattering is

$$\frac{\Delta K_1}{K_1} = \frac{v_1^2 - v_0^2}{v_0^2} = 1 - \left(\frac{v_1}{v_0}\right)^2 \qquad (2.236)$$

Using equations (2.235) and (2.236)

$$\frac{\Delta K_1}{K_1} = 1 - \frac{m_1^2 + m_2^2 + 2m_1 m_2 \cos\theta}{(m_1 + m_2)^2}$$

$$= \frac{(m_1 + m_2)^2 - (m_1^2 + m_2^2 + 2m_1 m_2 \cos\theta)}{(m_1 + m_2)^2}$$

$$= \frac{2m_1 m_2(1 - \cos\theta)}{(m_1 + m_2)^2} \text{ Ans.}$$

Chapter 3

Gravitation

3.1 Introduction

In the standard model of particle physics gravitational force is the first among the four fundamental forces of nature: gravitation, electromagnetism, nuclear and weak forces. In fact, 5% of the content of the Universe is comprised of normal matter that is manifested as planets, stars, and galaxies. Newton was the first person to predict the existence of an all-pervading attractive gravitational force between all material bodies. The concept of gravitational force was founded by Newton following Keplar's laws.

3.2 Kepler's laws

The famous astronomer Tycho Brahe was the supervisor of the observatory of Uraniborg established by King Frederick-II of Denmark. Johann Kepler (1571–1630) was the assistant of Brahe, who verified and corrected the astronomical tables of that time to make more accurate astrological calculations for the royal family. Kepler undertook Brahe's work after his sudden death. Brahe was a supporter of the Ptolemaic geo(earth)-centric system whereas Kepler was a supporter of the Copernican system of helio (sun)-centric arrangement of planets. Keplar found the Brahe's astronomical data fitted into a geocentric model, but did not fit into the heliocentric model. Even though Brahe's data worked mostly for the Ptolemaic model, the disagreement of '8 minutes of arc' between the observed and computed positions inspired Kepler to rethink the 'exact nature of the orbits of planets'. As explained earlier, in both Copernican and Ptolemaic models, the motion of the planets was assumed to be circular and epicyclic.

Kepler's 1st law: Keplar repeated the observations of the planetary positions relative to the Sun. He found that, due to the difference in the observed and computed positions of some planets, in general, the motion of the planets could not be totally circular relative to the Sun. He assumed many types of orbits and finally the observed result was fitted into the calculated results when he assumed *elliptical* *orbits*. The theory and observations both concurred on the elliptical orbits of the planets. This is the essence of Kepler's first law, which states that

The planets move around the Sun in elliptical orbits with the Sun as the focus.

Elliptical Orbit

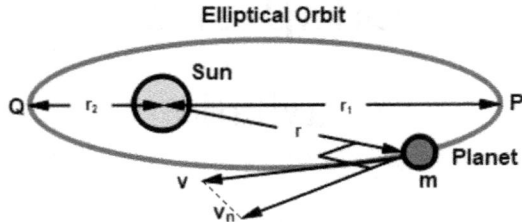

Kepler's 2nd law:
To match the observed positions of the planets, Kepler followed the heliocentric model giving up the idea of uniform circular motion of the planets. He assumed that the speed of the planets varies when they move in elliptical orbits. When the planets are nearest to the Sun, they move with the greatest speeds and vice versa. How do the speeds vary with the radial distance? To match both theory and observation, Kepler found that the speed v varies inversely proportional to the radial distance r of the planet from the Sun. Hence, we can write

$$v \propto \frac{1}{r}$$

Or, $vr = $ constant.
This forms the basis of the 2nd law, which states that

The radius vector of the planets drawn from the Sun sweeps equal areas in equal time interval.

In other words, areal velocity of a planet is constant:
$$\Rightarrow \frac{dA}{dt} = C \text{ (Constant)}$$

N.B.: We can prove that Kepler's 2nd law is, in essence, conservation of angular momentum of the planets. This states that

The angular momentum of a planet is always conserved about the Sun.

Kepler's 3rd law:
After the discovery of the 1st and 2nd laws, Kepler took a long time to compare the time periods of revolution of different planets relative to the Sun. He tabulated the time period against the size of an orbit, which is twice the length of the semimajor axis of the elliptical orbit. He found that the ratio of square of time period of revolution of the planets and the cube of the semimajor axis of the orbit of the corresponding planets is a constant quantity. This is the basis of Kepler's 3rd law, which states that

The square of the periods of revolution of the planets are directly proportional to the cubes of the semimajor axes of their axes of their respective orbits.

$$T^2 \propto R^3,$$

where $R = \frac{1}{2}(r_1 + r_2)$.

3.3 Newton's law of universal gravitation

Kepler's laws are purely kinematical without any notion of the concept of *force*, because at that time no concept of force had been expounded. So, Keplar's laws tell us how the planets are moving, but these laws cannot tell us why the planets are moving. In Keplar's time it was a major subject of research to understand the cause of motion of the planets. While Kepler was deriving his laws, his contemporary Galileo, who was a strong advocate of the Copernican heliocentric model like Keplar, proclaimed that the planets kept on moving in their orbits owing to the inertia of their motion. Newton was born in 1642 and in the same year Galilio died. Newton extended Galileo's work and developed the laws of motion following Galileo's ideas of inertia. These laws are familiar as Newton's 1st law or law of inertia, which states that

An object tends to continue its state of motion or rest which is characterised by the momentum of the object.

According to Kepler's 1st law, the planets move in curved paths (ellipses and circles). Furthermore, the speeds of the planets in elliptical path change according to Kepler's 2nd law. Even though the orbits are elliptical, some planets such as Mars and Earth (closer to the Sun) move in circles (special case of an ellipse of eccentricity equal to one) with uniform speeds, but the direction of velocity (motion) keeps on changing while the planets revolve around the Sun. When the velocities of such planets in circular orbits change, we can say they are accelerating towards the center of their circular orbits. Hence, Newton argued that all planets must experience a certain force to accelerate (change their direction and magnitude of velocities).

What is the nature of the force? How does this force vary with distance from the Sun? Let's see in the following example.

Example 1 (Finding the nature of gravitational force by Newton)
Following Kepler's laws how did Newton prove the inverse square law of gravitational force?
Solution
Let's assume that a planet moves in a circular path around the Sun. According to Kepler's 2nd law, the speed of the planet is constant because its radial distance (semimajor axis) or radius of the orbit remains constant. As studied in uniform circular motion, the planet must accelerate towards the center of the circle. Hence, a radially inward force must act on the planet. As we know, the planet cannot pull itself according to Newton's third law. So, it must be the Sun that pulls the planet towards it (center of the circular orbit), which is known as *centripetal force*. Let F be the force of gravitational attraction due to the Sun on the planet, which is the centripetal force in this case.

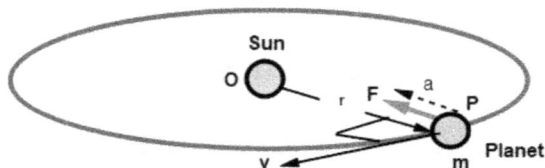

Applying Newton's 2nd law, we have

$$F = ma,$$

where $a = r\omega^2$

$$\Rightarrow F = mr\omega^2 \tag{3.1}$$

According to Kepler's 3rd law

$$\omega^2 r^3 = C \tag{3.2}$$

Putting $\omega^2 = \frac{C}{r^3}$ from equations (3.2) in (3.1), we have

$$F = \frac{mC}{r^2}$$

Since C and m are constants, we have

$$F \propto \frac{1}{r^2} \text{ Ans.}$$

The force of attraction between the Sun and planet is inversely proportional to the square of the distance between the Sun and the planet.

N.B.: In the above example, if $v = \frac{C}{mr}$ according to Kepler's 2nd law, you will get $F = \frac{C^2}{mr^3}$. Do we mean that $F \propto \frac{1}{r^3}$, which contradicts the above result? The answer is No. In general, for any orbit we can write

$$a = \frac{v^2}{r'} = r\omega^2 \text{ and } L = mvr.$$

where $r =$ the distance of the planet from the Sun and $r' =$ radius of curvature of the path. These two quantities will be equal for a circle, but in general, they are different for elliptical paths.

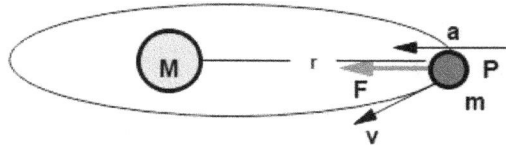

Then the gravitational force is given by Newton's 2nd law as

$$F = ma = m\frac{v^2}{r'}$$

Putting $v = L/mr$ in the last equation, we have

$$F = m\frac{(L/mr)^2}{r'} = \frac{L^2}{mr'r^2}$$

Since $L =$ angular momentum $=$ constant, for a particular radius of curvature at a point in the ellipse, the force acting on the planet obeys the inverse square law of distance between the planet and Sun.

Example 2 (Verification of inverse square law)
After deriving the inverse square law, Newton tried to verify it for circular orbits. He took the case of the Moon orbiting around the Earth in a circular path. How did he verify the inverse square law? How did he prove that the pull of the Earth to a projectile and the pull of Earth to the orbiting moon are of the same origin (nature)?
Solution
As the Moon moves in a circular orbit of radius r, it covers a very small angle 'θ' during one second. Hence, the radial distance y_1 fallen by the Moon is given as

$$y_1 = r - r\cos\theta = r(1 - \cos\theta)$$

$$= 2r\sin^2\frac{\theta}{2} = 2r\left(\frac{\theta}{2}\right)^2$$

$$= \frac{r\theta^2}{2}, \quad \text{where } r\theta = s$$

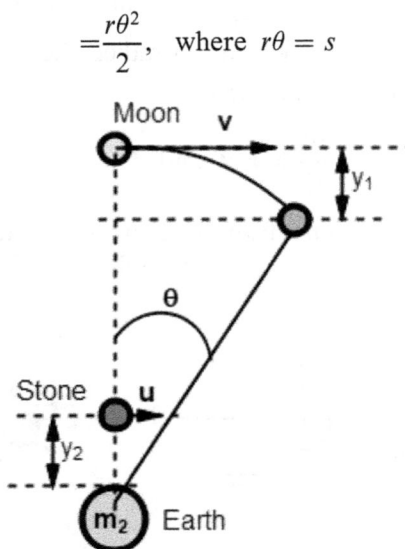

Then, $y_1 = \frac{s^2}{2r}$, where s = arc length = distance covered by the Moon in one second.

The distance s can be given by

$$s = r\omega t = (r)\left(\frac{2\pi}{T}\right)(1) = \frac{2\pi r}{T}$$

Substituting $s = \frac{2\pi r}{T}$ in $y_1 = \frac{s^2}{2r}$,

we have

$$y_1 = \frac{2\pi^2 r}{T^2}$$

Substituting $r = 5.9 \, R_e = 5.8 \times 10^8$ m and $T = 27.3$ days $\cong 2.4 \times 10^6$ s, we have $y_1 \cong 1.3 \times 10^{-3}$ m.

The vertical distance fallen by a horizontally projected body (a stone, say) near the Earth's surface during a time t is

$$y_2 = \frac{1}{2}gt^2$$

The distance fallen due to gravity during one second is

$$y_2 = \frac{1}{2}g(1)^2 = \frac{1}{2} \times 9.8 = 4.9 \text{ m}$$

$$\Rightarrow \frac{y_1}{y_2} = \frac{1.3 \times 10^{-3}}{4.9} \simeq 2.7 \times 10^{-4} \simeq \frac{1}{3700}$$

Since the vertical (radial) distance fallen is directly proportional to acceleration of the particle, we have

$$\frac{y_1}{y_2}\left(= \frac{a_1}{a_2} \right) = \frac{1}{3700} \tag{3.3}$$

Since the ratio of radii of the orbits is

$$\frac{r_2}{r_1} = \frac{R_e}{69 R_e},$$

$$\Rightarrow \left(\frac{r_2}{r_1} \right)^2 \simeq \frac{1}{3600} \tag{3.4}$$

Using equations (3.3) and (3.4),

$$\frac{y_1}{y_2} = \frac{a_1}{a_2} = \left(\frac{r_2}{r_1} \right)^2 \tag{3.5}$$

This signifies that acceleration due to gravity (or gravitational force) is inversely proportional to the square of the distance.

Recapitulating, the force acting on an object thrown above the Earth's surface and the force acting on a satellite (Moon, etc) due to the Earth have the same origin, which is called 'mutual gravitational force of attraction'. This force exists everywhere between any two objects in the Universe.

Thus, Newton verified that the gravitational force between the Earth and a stone, say, is inversely proportional to the square of the radius (distance of separation between the Earth and stone). But how does the inverse square law hold good for the Earth and stone even though the Earth is not a point object? We will see in Example 11 that a spherically symmetrical body such as any planet or star behaves as a point mass and thus obeys inverse square law like the particles.

N.B.: But gravitational forces between any two extended bodies separated by a distance r, say, comparable with their linear dimensions cannot follow 'inverse square law'.

Inverse square law of gravitational force: From the foregoing discussion we can understand that inverse square law is valid for particles and spherically symmetrical bodies. When two bodies are separated through a distance (very large compared to their linear dimensions, i.e., length, breadth, and height), we can treat them as particles. In this sense, all heavenly bodies like stars, planets, and satellites can be treated as particles as they are far apart (millions of light years in many cases).

Hence, Newton explains that each particle in this universe attracts the others. The force of attraction between any two particles must proportionally vary with their masses (according to Newton's 2nd law of motion) and is inversely proportional to the square of the distance between the particles.

According to Newton's 3rd law, each particle experiences equal and opposite force:

$$\overrightarrow{F_1} = -\overrightarrow{F_2}$$

Hence, $\overrightarrow{F_1}$ and $\overrightarrow{F_2}$ can be called an action-reaction pair. The line of action of gravitational forces $\overrightarrow{F_1}$ and $\overrightarrow{F_2}$ passes through the line joining the particles m_1 and m_2.

Since $|\overrightarrow{F_1}| = |\overrightarrow{F_2}| = F_1$ (say), according to Newton's law of gravitation $F \propto m_1 m_2$, so also $F \propto \frac{1}{r^2}$

Then, combining last two expressions, we can write

$$F \propto \frac{m_1 m_2}{r^2}$$

Equalising both sides by a constant 'G', we have

$$F = G\frac{m_1 m_2}{r^2}$$

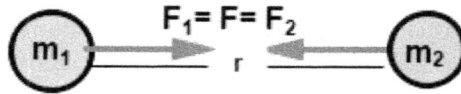

It was experimentally verified that G does not depend on any material medium and the nature of bodies. Hence, G is a universal gravitational constant having the measured value $G = 6.67 \times 10^{-11}\,\mathrm{Nm^2\,kg^{-2}}$.

Thus, the above law is called Newton's law of universal gravitation, which states that

Each particle of the Universe attracts (pulls) any other particle towards it along the line joining team. The gravitational force of attraction between any two particles is directly proportional to the product of their masses and inversely proportional to the square of the distance of separation between the particles. The gravitational force does not depend on the intervening medium. The force acting on m_2 due to m_1 is given as

$$\overrightarrow{F_{21}} = -\overrightarrow{F_{12}} = -\frac{Gm_1 m_2}{r^2}\hat{r}_{21},$$

where $\hat{r}_{21} = \frac{\overrightarrow{r}_{21}}{r}$

3-8

3.4 Inertial and gravitational mass

We have studied that a material particle requires a force to accelerate due to its inertia (or lifelessness). If a horizontal force F pulls an object on a smooth horizontal surface with an acceleration a, the inertial mass of the body is given as the ratio of force and acceleration, according to Newton's 2nd law. This means:

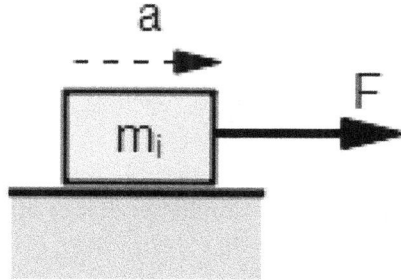

Inertial mass $= m_i = \dfrac{F}{a}$.

If we release the same object in gravity of the Earth, the ratio of gravitational force F_g and the gravitational acceleration $a(= g)$ is called its gravitational mass:

$$\text{Gravitational mass} = m_g = \frac{F_g}{g},$$

where $F_g = m_g g$

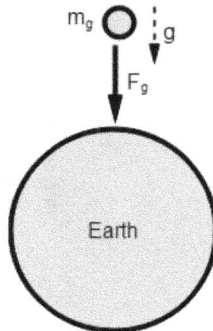

Then, we can state that, owing to the gravitational mass m_g, the body experiences a gravitational pull called gravitational force F_g.

Relation between m_i and m_g: Newton proved that both inertial and gravitational masses are equivalent. 'You cannot logically call them equal because they describe different properties of the same substance'.

Then, using the principle of equivalence, we can write

$$m_i \equiv m_g$$

The above discussion tells us that inertial mass is the property of matter by the virtue of which an object requires a push or pull (force) to accelerate, whereas gravitational mass is the property of matter by the virtue of which it attracts (gravitates) the other bodies; $m_i = \frac{F}{a}$ and m_g can be given as

$$F_g = \frac{G(m_{g_1})(m_{g_2})}{r^2}.$$

N.B.: The inertial mass is a measure of the resistance of an object to change in velocity. The gravitational mass measures the force acting on the object in a gravitational field.

Example 3 Find the time period of a simple pendulum of length l, inertial mass m_i, and gravitational mass m_g.

Solution

At a small angle of inclination θ the net tangential force is

$$F_t = F_g \sin \theta,$$

which accelerates the bob with a tangential acceleration a (say).

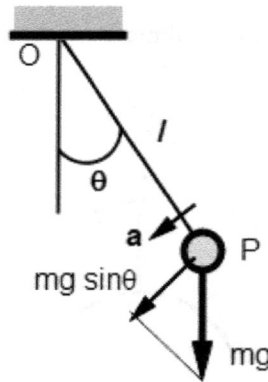

Using Newton's 2nd law,

$$F = F_t = F_g \sin \theta$$

and $m =$ inertial mass $= m_i$, we have

$$a = -\frac{F_g}{m_i} \sin \theta \tag{3.6}$$

For small values of θ,

$$\sin \theta \simeq \theta = \frac{x}{l} \tag{3.7}$$

Using the last two equations, $a = -\frac{F_g x}{m_i l}$

Putting $F_g = m_g g$, we have

$$a = -\frac{m_g g}{m_i l}x$$

Comparing the above equation with

$$a = -\omega^2 x$$

In the chapter Oscillations or Simple Harmonic Motion of the book *Problems and solutions of Oscillations, Waves, Heat and thermodynamics*, we have

$$\omega = \sqrt{\frac{m_g g}{m_i l}}$$

So, the time period of the simple pendulum is

$$T = 2\pi \sqrt{\frac{m_i l}{m_g g}} \text{ Ans.}$$

N.B.: In the foregoing example, we know that $T = 2\pi\sqrt{\frac{l}{g}}$. How can we reconcile it with the above formula? The above example is based on the actual experiment conducted by Newton himself to compare inertial and gravitational mass. He took a hollow bob and put different substances of equal weights. He always found the same time period. Then he concluded that T is independent of $\frac{m_i}{m_g}$. In other words, $m_i = m_g$. However, there is a slight difference between m_i and m_g, which does not affect the calculations in classical physics.

3.5 Gravitational field and field intensity, superposition of gravitational field

Field idea: After the discovery of Newton's law of universal gravitation, people began to wonder how a body can pull another body towards it without touching it? Newton himself could not give any satisfactory explanation for this. Long after Newton, Michael Faraday adopted the field idea. According to this model, everybody has its own field of force. For instance, when we release a stone from any place above the Earth's surface, it accelerates. This means that at any point in the space surrounding the Earth, a force acts on an object. Hence, the space surrounding the Earth is called a region (or field or space) of a gravitational force. A gravitational force is associated with a *gravitational field*.

Gravitational field is the region where a gravitational force acts on an object. Gravitational force field is a vector field because 'force' is a vector. Any particle sets up its own gravitational field in the surrounding space by the virtue of which it pulls the surrounding objects.

Field strength: When you keep a point object of mass m at any point of a gravitational field, it will experience a force $\overrightarrow{F}$, (say). It is experimentally verified that gravitational force due to a system of particles acting on a particle is directly proportional to the mass of the particle.

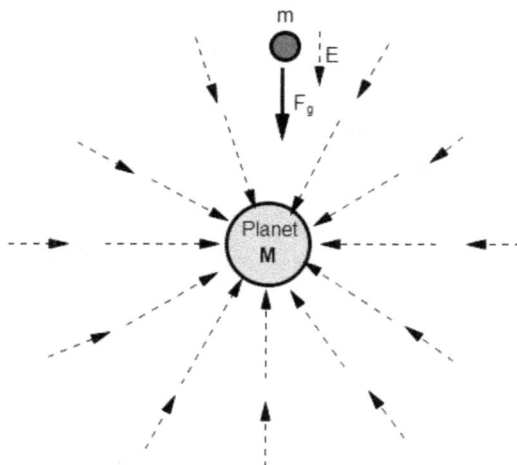

Symbolically,

$$\left|\overrightarrow{F}\right| \propto m$$

Hence, $\frac{|\overrightarrow{F}|}{m}$ must be a constant quantity at any point of a gravitational field. Since the direction of $\overrightarrow{F}$ does not change with mass, we can write

$$\frac{\overrightarrow{F}}{m} = \text{Constant} = \overrightarrow{E}\,(\text{let})$$

We call $\overrightarrow{E}$ the gravitational field strength.

Newton's 2nd law states that $\frac{\overrightarrow{F}}{m} = \overrightarrow{a}$ (acceleration of the particle) whose unit is $\mathrm{m\,s^{-2}}$. Hence, gravitational field strength (or intensity) at any point is equal to the acceleration of a particle placed at that point due to the given gravitational field. However, gravitational field strength $\overrightarrow{E}$ at any point is defined as the force acting on a unit point mass 'm' placed at that point; $\overrightarrow{E} = \frac{\overrightarrow{F}}{m}$; $\overrightarrow{E}$ is a vector quantity pointed in the direction of $\overrightarrow{F}$.

Superposition of $\overrightarrow{E}$: If there are many gravitating systems (objects), each produces its own gravitational field.

Hence, the net force acting on a particle of mass m is given by

$$\overrightarrow{F} = \sum \overrightarrow{F_i},$$

where $\overrightarrow{F_i}$ is the force of gravity due to the ith system.

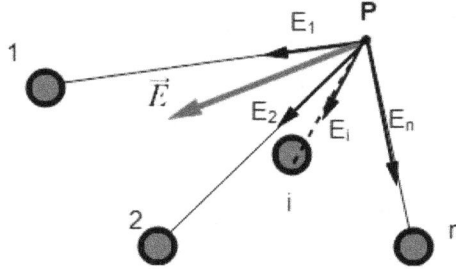

Dividing both sides by mass m of the particle P, we have

$$\frac{\overrightarrow{F}}{m} = \sum \frac{\overrightarrow{F_i}}{m}$$

Substituting $\frac{\overrightarrow{F_i}}{m} = \sum \overrightarrow{E_i}$ (field strength due to the ith system at P), we have

$$\frac{\overrightarrow{F}}{m} = \sum \overrightarrow{E_i},$$

where $\frac{\overrightarrow{F}}{m}$ = total (net) field strength at P = $\overrightarrow{E}$ (say).

Then, the net field is $\overrightarrow{E} = \sum \overrightarrow{E_i}$.

The net gravitational field strength at any point is equal to the vector sum of the field strength of all gravitating systems at that point.

This is known as superposition of gravitational field.

3.6 Calculation of gravitational field intensity

In this section we will find the strength of the gravitational field of different types of mass distribution. For this purpose, we need to use the basic formula

$$\overrightarrow{E} = \sum \frac{\overrightarrow{F}}{m},$$

where $\overrightarrow{F}$ = force acting as a point test mass m placed at the point where we want to find the field intensity.

Point mass: Let's take a M at O (say) and try to find the gravitational field of the mass M at a point P, at a distance r, say, from O. For this, we need to put a test mass m at P. The force acting on m due to M is given as

$$\overrightarrow{F} = -\frac{GMm}{r^2}\hat{r} \qquad (3.8)$$

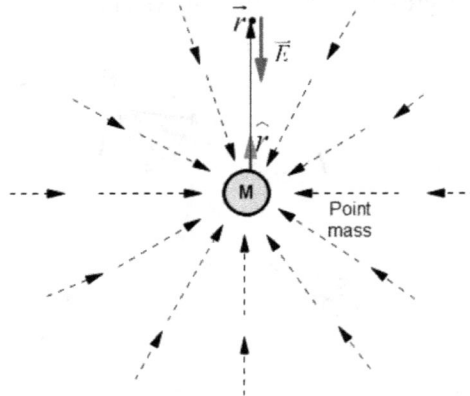

The strength of gravitational field of M at P is given as

$$\overrightarrow{E} = \frac{\overrightarrow{F}}{m} \qquad (3.9)$$

Substituting $\overrightarrow{F}$ from equations (3.8) in (3.9), we have

$$\overrightarrow{E} = \frac{-\frac{GMm}{r^2}\hat{r}}{m}$$

$$\Rightarrow \overrightarrow{E} = \frac{GM}{r^2}\hat{r}$$

The above equation tells us that

The magnitude of strength of gravitational field due to a point mass at point M is directly proportional to the mass M and inversely proportional to the square of distance of separation r between the point mass and the point P where you want to find the field strength. The field strength $\overrightarrow{E}$ points towards the point mass M along the line joining the point mass and the point P under consideration.

Discrete mass distribution: Now you can use expression $\overrightarrow{E} = -\frac{GM}{r^2}\hat{r}$ as a basic formula for field strength due to a point mass. Let's use it in the following example of discrete mass distribution.

Example 4 Two identical particles each of mass M are separated by a distance $2R$. Find the gravitational field strength at P situated at a distance x from O on the perpendicular bisector OP.

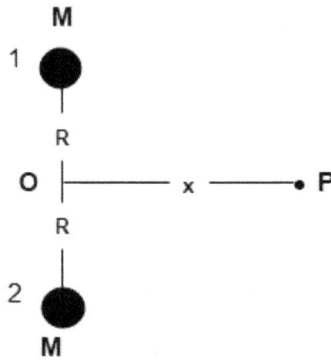

Solution

Since P is situated at equal distance from both particles, the magnitude of $\overrightarrow{E}$, that is, $|\overrightarrow{E}|$ (or E), is the same for both particles, at P. If $\overrightarrow{E_1}$ and $\overrightarrow{E_2}$ are the field strengths due to the particles 1 and 2 at P, respectively, we have

$$E_1 = E_2 = \frac{GM}{r^2} \tag{3.10}$$

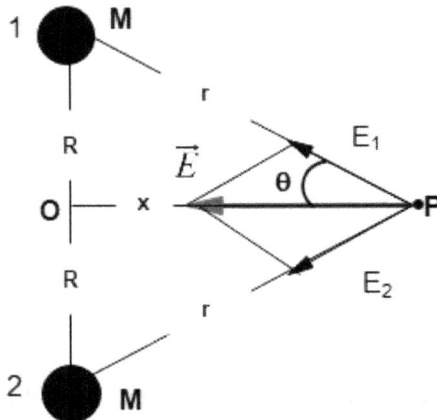

Resolving $\overrightarrow{E_1}$ and $\overrightarrow{E_2}$ along the x- and y-axis, we have

$$E_x = E_1 \cos \theta + E_2 \cos \theta \tag{3.11}$$

$$E_y = E_1 \sin \theta - E_2 \sin \theta \tag{3.12}$$

3-15

Using equations (3.10)–(3.12), the magnitude of total field intensity is

$$E = E_x = \frac{2GM}{r^2} \cos \theta,$$

where $\cos \theta = \frac{x}{r} = \frac{x}{\sqrt{R^2 + x^2}}$ and $r = \sqrt{R^2 + x^2}$

$$\Rightarrow E = \frac{2GMx}{(R^2 + x^2)^{\frac{3}{2}}} \text{ Ans.}$$

N.B.: By putting $dE/dx = 0$, you can find that at $x =$ value of $x = \frac{R}{\sqrt{2}}$ E is maximum, and putting this value, we have maximum value of $E = \frac{4GM}{3\sqrt{3}R^2}$. From the above example let's make the following points:

1. Since $E_y = 0$, the net field strength is directed towards the mid-point O of the line joining the particles. This means that a particle placed at any point on the perpendicular bisector (x-axis) will accelerate towards the origin. Hence, the origin is called 'stable equilibrium position'. The particle will oscillate simple harmonically along the x-axis, about O. In other words, v is maximum and a will be zero at O, which is the condition of stable equilibrium.

2. Since $|\overrightarrow{E}| = \frac{2GMx}{(R^2 + x^2)^{\frac{3}{2}}}$, when $x = 0$, $E = 0$; when $x \to \infty$, $E \to 0$. This means that E attains a maximum value at $x = \frac{R}{\sqrt{2}}$ because $\frac{dE}{dx} = 0$ and $\frac{d^2E}{dx^2} < 0$.

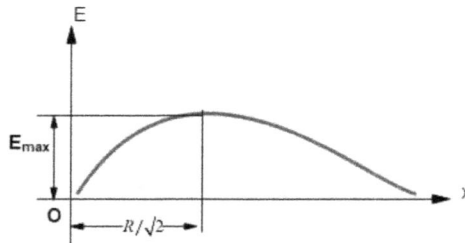

3. Referring to the above equation, the graph $E = f(x)$ is an asymptote. Then using the formula $E = \frac{GM}{r^2}$ for a point mass, let's find the field intensity due to continuous mass distribution.

Continuous mass distribution: In continuous mass distribution (extended object), we cannot distinguish one particle from the others because the particles are distributed continuously. In this case, take an element of mass dm in the extended object. It behaves as a point mass. Then using the basic formula, the field intensity $\overrightarrow{dE}$ due to the point mass at the given point P is

$$d \overrightarrow{E} = -\frac{Gdm}{r^2} \hat{r}$$

Here, you need to check the directions of field intensities due to other elements of the body. If all elementary field intensities have the same direction, we can sum up (integrate) $d\vec{E}$ directly to obtain the net field intensity, which can be given by $\vec{E} = -\hat{r}G \int \frac{dm}{r^2}$.

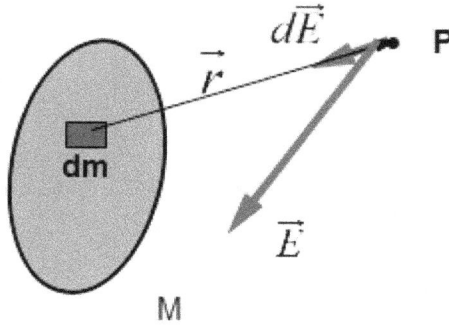

If the directions of field intensities at P are different for different elements of the body, we must resolve dE into components dE_x, dE_y, and dE_z along the x-, y-, and z-axes, respectively. Then integrate the corresponding elementary field intensities to obtain E_x, E_y, and E_z. Hence, the net field intensity of the body at P can be given as

$$\vec{E} = E_x \hat{i} + E_y \hat{j} + E_z \hat{k}$$

In each case, dm must be expressed in terms of r as follows. For linear mass, the linear mass density at any point P can be given as $\lambda = \frac{dm}{dx}$,

$\Rightarrow dm = \lambda \, dx$, where $\lambda = f(x)$

For surface (area) mass distribution, areal mass density is given as

$$\sigma = \frac{dm}{dA}$$

$\Rightarrow dm = \sigma \, dA$, where $\sigma = f(r)$

Similarly, for volume mass distribution volume mass density is given as

$$\rho = \frac{dm}{dV}$$

$\Rightarrow dm = \rho dV$, where $\rho = f(r)$

On a case-to-case basis, we can express dm in terms of r (or any linear dimension).

Let's summarize the above explanation:

To find $\vec{E}$ for an extended object:

1. Take an elementary mass dm

2. Find the field intensity $d\vec{E}$ due to dm

$$d\vec{E} = -\frac{Gdm}{r^2}\hat{r}$$

3. Resolve $d\overrightarrow{E}$ along the x-, y-, and z-axes to find dE_x, dE_y, and dE_z, respectively.
4. Express $dm = \lambda\, dx$ (for linear mass distribution)
   $= \sigma dA$ (areal mass distribution)
   $= \rho dV$ (volume mass distribution)
5. Use the given relations of mass densities λ, σ, and ρ as the function of (linear dimensions)

$$\lambda = f(x), \quad \sigma = f(r) \text{ and } \rho = f(r)$$

6. Integrate dE_x, dE_y, and dE_z to obtain E_x, E_y, and E_z, respectively.
7. Finally, the net field intensity is given as $\overrightarrow{E} = E_x\hat{i} + E_y\hat{j} + E_z\hat{k}$.

Now you can use the above ideas to find field intensities of linear, surface, and volume mass distribution through the following examples.

Linear mass distribution: Under this type, we have wires (straight and bent). Let's take the example of a straight thin wire.

Example 5 Find an expression for field intensity due to the finite straight wire of mass M and length l, at the point P.

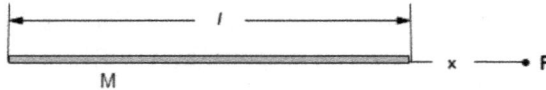

M

Solution

Let's take an elementary mass dm at a distance x from the point P. The field intensity of dm at P is

$$d\overrightarrow{E} = -\frac{Gdm}{x^2}\hat{i} \tag{3.13}$$

where

$$dm = \lambda\, dx \tag{3.14}$$

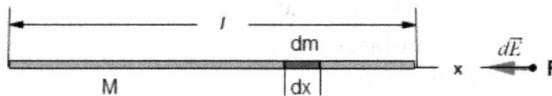

Since the field intensities due to each element of the straight wire have the same direction (to left), we can directly integrate dE.

Then, we have

$$\vec{E} = \int d\vec{E}$$

(3.15)

Using equations (3.13)–(3.15), we have

$$\vec{E} = -G\hat{i} \int \frac{\lambda \, dx}{x^2}$$

Since the wire is uniform, λ is constant. Hence, you can take λ out of the integral to obtain

$$\vec{E} = -G\lambda\hat{i} \int \frac{dx}{x^2}$$

As the wire extends from $x = x_0$ to $x = x_0 + l$, we have

$$\vec{E} = -G\lambda\hat{i} \int_{x_0}^{x_0 + l} \frac{dx}{x^2} = -G\lambda\hat{i} \left[-\frac{1}{x} \right]_{x_0}^{x_0 + l}$$

$$= -G\lambda\hat{i} \left(\frac{1}{x_0} - \frac{1}{x_0 + l} \right) = -\frac{G\lambda l}{x_0(x_0 + l)}$$

$$= -\frac{GM}{x_0(x_0 + l)}\hat{i} \text{ Ans.}$$

In the above example, we found E at a familiar position. Let's derive a general expression of E due to finite wire in the following example.

Example 6 Find the gravitational field strength due to a uniform thin wire AB of linear mass density λ at a point P which is situated at a perpendicular distance R from the wire. The point P subtends internal angles θ_1 and θ_2 at the ends of the wire as shown in the figure.

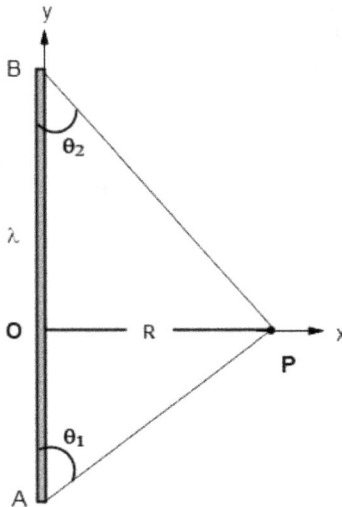

Solution

Drop perpendicular from P onto the wire. The point of intersection O of the perpendicular OX and the wire is assumed as the origin. Now let's take an element dm of the wire at distance y from the origin. If the elementary length of dm is dy,

$$dm = \lambda \, dy$$

Then, the field intensity of dm at P is

$$dE = \frac{G \, dm}{r^2}$$

Substituting $dm = \lambda \, dy$, we have

$$dE = G\lambda \frac{dy}{r^2}$$

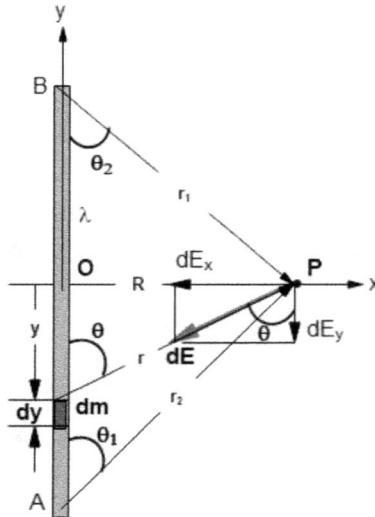

Since the direction of $\overrightarrow{dE}$ differs from element to element, we need to resolve dE along the x- and y-axes to obtain

$$dE_x = dE \sin \theta \text{ and } dE_y = dE \cos \theta$$

Then, the net field intensities in the x- and y-directions are given as

$$E_x = \int dE \sin \theta \text{ and } E_y = \int dE \cos \theta,$$

where $dE = G\lambda \frac{dy}{r^2}$, $\sin \theta = \frac{R}{r}$ and $\cos \theta = \frac{y}{r}$

Then we obtain the following two equations:

$$E_x = G\lambda x \int \frac{dy}{r^3} \text{ and } E_y = G\lambda \int \frac{y \, dy}{r^3}$$

Let's now evaluate the integrals by putting

$$y = R \cot \theta, \quad r = x \operatorname{cosec} \theta$$

and $dy = -R \operatorname{cosec}^2 \theta d\theta$
Then, we have

$$E_R = -\frac{G\lambda}{R} \int_{\theta_1}^{\pi - \theta_2} \sin \theta \, d\theta,$$

where the upper limit of the angle is given as $\theta = \pi - \theta_2$ because $\theta =$ angle between $+y$ direction and BP = position vector of point P relative to the end B,
This gives

$$E_x = \frac{G\lambda}{R}(\cos \theta_1 + \cos \theta_2)$$

Similarly, we have

$$E_y = -\frac{G\lambda}{R} \int_{\theta_1}^{(\pi - \theta_2)} \cos \theta \, d\theta$$

This gives

$$E_y = -\frac{G\lambda}{R}(\sin \theta_2 - \sin \theta_1)$$

N.B.: Student task
1. At any point on the perpendicular bisector of the wire at a distance R, by putting $\theta_1 = \theta_2$ in
   $E_x = \frac{G\lambda}{R}(\cos \theta_1 + \cos \theta_2)$, we have
   $E = E_x = \frac{2G\lambda}{x} \cos \theta$, where $x = R$ and
   $\cos \theta = \frac{l/2}{\sqrt{R^2 + (l/2)^2}}$. Putting $\lambda l = M$, we have
   $$E = \frac{2GM}{R^2\sqrt{4R^2 + l^2}}$$

2. For infinite long straight wire, putting $\theta_1 = \theta_2 = 0$ in the expression
   $E_x = \frac{G\lambda}{R}(\cos \theta_1 + \cos \theta_2)$, we have
   $$E_x = \frac{2G\lambda}{R} \leftarrow \text{ and } E_y = 0$$

3. In general, if the coordinates of A, B, and P are known from the given set, the corresponding angles θ_1 and θ_2 are given as $\sin \theta_1 = \frac{R}{\sqrt{R^2 + y^2}}$,
   $\cos \theta_1 = \frac{y}{\sqrt{R^2 + y^2}}$, $\sin \theta_2 = \frac{R}{\sqrt{R^2 + (l - y)}}$, and $\cos \theta_2 = \frac{l - y}{\sqrt{R^2 + (l - y)^2}}$.

Then substituting the values of θ_1 and θ_2, we can find the E_x and E_y. The directions of E_x is always directed towards the wire AB along OP; the direction of E_y is positive (↑) when $\sin\theta_2 > \sin\theta_1$ (or $\theta_2 > \theta_1$) and vice versa. When $\theta_1 = \theta_2$ in the case of perpendicular bisector, $E_y = 0$.

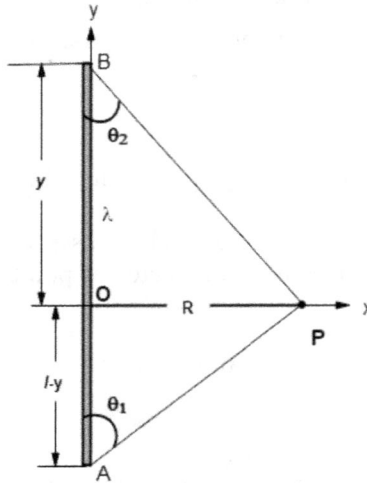

Let's now bend the wire in the form of a circular arc and try to find the field intensity at the center of curvature of the circle in the following example.

Example 7 Find the gravitational field intensity of a thin circular arc having linear mass density λ at the center C of the curvature. The arc subtends an angle β at the center C.
   Solution
   Let's take an element of mass dm at an angle θ. The field intensity of dm at C is
$dE = \frac{G\,dm}{R^2}$, where $dm = \lambda\,dl$

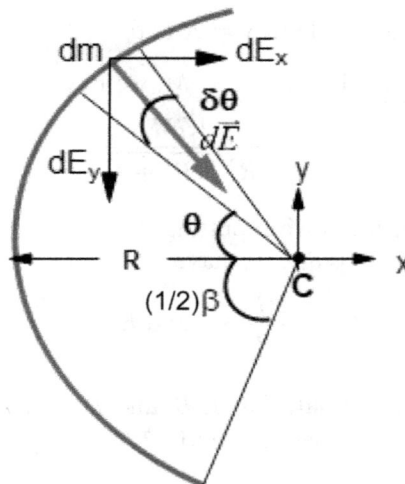

Substituting $dl = R\,d\theta$, we have

$$dE = \frac{G\lambda}{R}d\theta \tag{3.16}$$

Since the angle θ of $d\overrightarrow{E}$ changes, we need to resolve $d\overrightarrow{E}$ in the x- and y-axes to obtain

$dE_x = dE\cos\theta$ and $dE_y = dE\sin\theta$

As the arc is symmetrical about the x-axis, the strengths of the fields of the upper and lower halves of the arc get cancelled. Hence, the net field intensity is directed along the x-axis (towards left):

$$\text{Then } E_{net} = E_x \int dE_x,$$

where

$$dE_x = dE\cos\theta$$

$$\Rightarrow E_{net} = \int dE\cos\theta \tag{3.17}$$

Using equations (3.16) and (3.17), we have

$$E_{net} = \frac{G\lambda}{R}\int_{-\beta/2}^{\beta/2}\cos\theta\,d\theta$$

$$= \frac{2G\lambda}{R}\int_{0}^{\beta/2}\cos\theta\,d\theta$$

$$\Rightarrow E_{net} = \frac{2G\lambda}{R}\frac{\sin\frac{\beta}{2}}{\beta}$$

N.B.: 1. Using the formula $E_{net} = \frac{2G\lambda}{R}\frac{\sin\frac{\beta}{2}}{\beta}$, find the field strength (intensity) of a semicircular wire of mass m and length l as

$$E = \frac{2Gm}{l^2}$$

2. In the above example, if you write $E = \int dE$, where $dE = \frac{G\lambda}{R}d\theta$ you will get $E = \frac{G\lambda\beta}{R}$, which will lead to a wrong result. Since the directions of $d\overrightarrow{E}$ are different for different elements of the arc, you cannot write $|\overrightarrow{E}| = \int |d\overrightarrow{E}|$. Hence, we need to resolve $d\overrightarrow{E}$ into components and then integrate them.

3. In the formula $E = \frac{4G}{R\beta}\sin\frac{\beta}{2}$, if we put $\beta = 0$, we will get $E = 0$. This means that the field strength due to a circular wire at its center is zero.

Let's now find the field strength at the axial points of a circular wire in the following example.

Example 8 Find the expression for the field intensity at any point on the axis of a circular wire of mass m and radius R.

Solution

The field due to the element of mass dm is

$$dE = \frac{G\,dm}{r^2} \tag{3.18}$$

The axial component of dE is

$$dE_x = dE \cos \theta$$

Integrating the axial component, we have

$$E_x = \int dE_x = \int dE \cos \theta \tag{3.19}$$

Using equations (3.18) and (3.19), we have $E_x = G \int \frac{dm}{r^2} \cos \theta$.

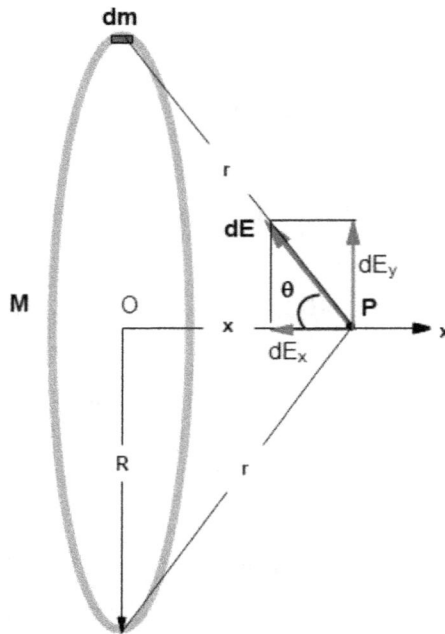

Since each element has the same distance r from P, we can take the constant quantities $\cos \theta$ and r out of the integral.

Then, we have

$$E_x = \frac{G \cos \theta}{r^2} \int dm \tag{3.20}$$

Since we integrate the field of all elements, we can write $\int dm = m$, where $m = $ mass of the ring.

Substituting $\int dm = m$ in equation (3.20), we have

$$\cos \theta = \frac{x}{\sqrt{R^2 + x^2}} \text{ and } r = \sqrt{R^2 + x^2}$$

Then, we have

$$E_x = \frac{Gmx}{(R^2 + x^2)^{3/2}}$$

The radial component of dE is

$$dE_r = dE \sin \theta$$

Since the ring is symmetrical about the axis, the net radial field intensity is zero. Hence, the net field is axially directed towards the center of the circular ring, whose magnitude is given as

$E_{net} = E_x = \frac{Gmx}{(R^2 + x^2)^{3/2}}$ Ans.

N.B.: If we project a particle of mass m along the axis, with a velocity v_0, the motion is periodic as it accelerates towards the center of the circular ring.

Since $E_{net} = 0$ when $x = 0$ and $x \to \infty$.

When $E \to \infty$, the graph of E vs. x is an asymptote. E is maximum when $x = \frac{R}{\sqrt{2}}$.

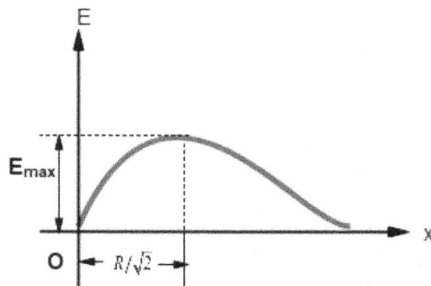

Surface mass distribution: Using the formula for field intensity of a ring, we can find the field intensity at the axial points of a disc because the disc is a combination of thin concentric rings.

Example 9 Derive an expression of gravitational field strength of a uniform disc of radius R, surface mass density σ, at an axial point situated at a distance x from the center of the disc.

Solution

Let's take a thin ring of radius r and thickness dr. The mass of the ring can be given as

$$m_{\text{ring}} = \sigma dA,$$

where dA = area of the strip (shaded portion) of the ring $= 2\pi r dr$ and σ = surface mass density.

Hence,

$$m_{\text{ring}} = 2\pi \sigma r \, dr \tag{3.21}$$

As we derived in the previous example, the field due to the ring is

$$E_{\text{ring}} = \frac{G m_{\text{ring}} x}{(r^2 + x^2)^{3/2}} \tag{3.22}$$

Substituting m_{ring} from equations (3.21) in (3.22), we have

$$E_{\text{ring}} = 2\pi \sigma G x \frac{r \, dr}{(r^2 + x^2)^{3/2}} \tag{3.23}$$

Since the disc is a combination of concentric rings of radii ranging from $r = 0$ to $r = R$ and field due to each ring has the same direction (towards O) the field due to the disc can be given by

$$E_{\text{disc}} = \int E_{\text{ring}} \tag{3.24}$$

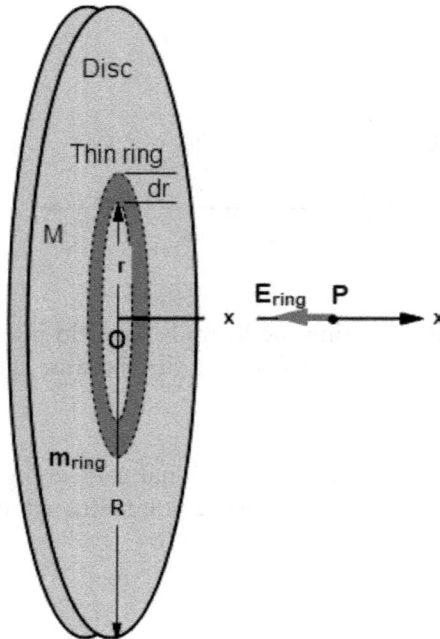

Substituting E_{ring} from equations (3.23) in (3.24), we have

$$E_{\text{disc}} = 2\pi\sigma G x \int_0^R \frac{r\,dr}{(r^2 + x^2)^{3/2}} \Rightarrow E_{\text{disc}} = 2\pi\sigma G\left(1 - \frac{x}{\sqrt{R^2 + x^2}}\right) \text{ Ans.}$$

N.B.: 1. By using the above expression, the field strength due to a uniform annular disc of mass M, inner radius R, and outer radius $2R$, at a distance R from the center of the disc, on its axis can be given as

$$E = \frac{2GM}{3R^2}\left(\frac{1}{\sqrt{2}} - \frac{1}{\sqrt{5}}\right)$$

2. In the foregoing example, we cannot take σ out of the integral if it is a function of radial distance r, that is, non-uniform surface mass distribution.

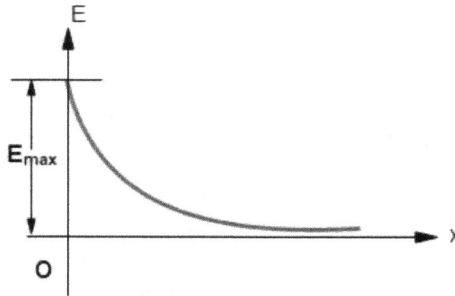

3. In the function $E = 2\pi\sigma G(1 - \frac{x}{\sqrt{R^2 + x^2}})$ if we put $x = 0$ we get $E = 2\pi\sigma G$. If $x \to 0$, $E \to 0$. This means that E is uniform (nearly) at very close points of a finite disc.

4. If we put $R \to \infty$, we will get $E = 2\pi\sigma G$. This tells us that $E = $ constant at any point in space, due to a large sheet.

Let's now take a closed surface to find the field intensity.

Example 10 Derive an expression for intensity due to a thin sphere of mass m and radius R.

Solution

Let's take a point P inside the spherical shell. Let the elementary patches of areas dA_1 and dA_2 at both sides of the point P subtend a solid angle $d\Omega$ at P.

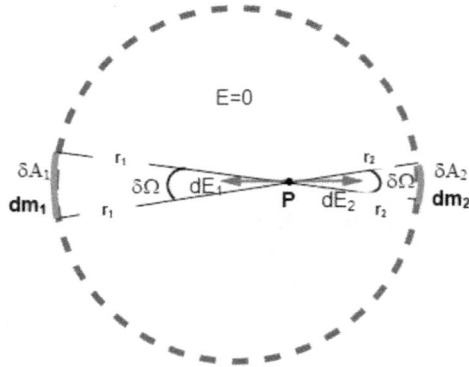

The masses of the elementary patches are $dm_1 = \sigma dA_1$ and $dm_2 = \sigma dA_2$. The fields due to the elements at P are

$$dE_1 = \frac{G dm_1}{r_1^2} = \frac{G \sigma dA_1}{r_1^2} \text{ and}$$

$$dE_2 = \frac{G dm_2}{r_2^2} = \frac{G \sigma dA_2}{r_2^2}$$

Since $d\vec{E_1}$ and $d\vec{E_2}$ oppose each other, the net field is

$$dE = |dE_1 - dE_2| = \left| \frac{G \sigma dA_1}{r_1^2} - \frac{G \sigma dA_2}{r_2^2} \right| = G\sigma \left| \frac{dA_1}{r_1^2} - \frac{dA_2}{r_2^2} \right| \qquad (3.25)$$

Since $\theta_1 = \theta_2 (= \theta$, say) by the properties of tangents at the ends of a chord, where θ_1 and θ_2 are the angles between the surface elements dA_1 and dA_2 and the corresponding perpendiculars to the chords, we can write

$$dA'_1 = dA_1 \cos \theta_1 = dA_1 \cos \theta \text{ and } dA'_2 = dA_2 \cos \theta_2 = dA_2 \cos \theta.$$

where dA' and dA'_2 are the areas perpendicular to the dotted line passing through P. Then, substituting $dA_1 = \frac{dA'_1}{\cos \theta}$ and $dA_2 = \frac{dA'_2}{\cos \theta}$ in equation (3.25), we have

$$dE = \frac{G\sigma}{\cos \theta} \left(\frac{dA'_1}{r_1^2} - \frac{dA'_2}{r_2^2} \right)$$

We know that the opposite solid angles are equal, given as $\frac{dA'_1}{r_1^2} = \frac{dA'_2}{r_2^2} = d\Omega$ (solid angle) according to solid geometry. Using the last two equations, we have
$dE = 0$
This means that the net gravitational field intensity inside a thin spherical shell of uniform mass distribution is zero.

In other words,

The gravitational field strength inside a thin uniform spherical shell is zero.

The field strength at any point outside the spherical shell situated at a radial distance r can be calculated as follows.

Take a thin ring at an angular position θ. The mass of the ring is

$$m_r = \sigma dA,$$

where $dA =$ area of the strip of the ring, as shown in the following figure, given by

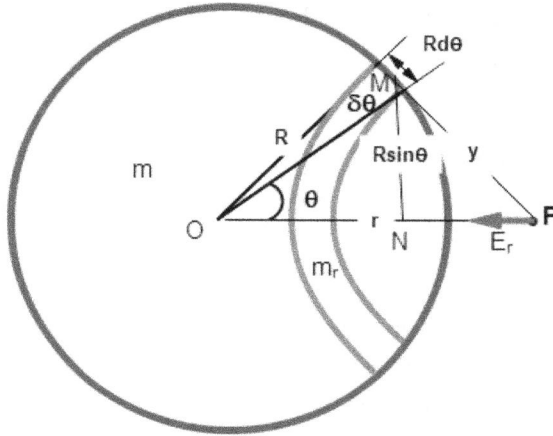

$$dA = 2\pi(R \sin \theta)(Rd\theta) = 2\pi R^2 \sin \theta d\theta$$

This gives

$$m_r = 2\pi\sigma R^2 \sin \theta d\theta \tag{3.26}$$

Then, the field due to the ring at P is

$$E_r = \frac{Gm_r x}{y^3} \tag{3.27}$$

In the $\triangle OMP$, we have

$$y^2 = r^2 + R^2 - 2rR \cos \theta \tag{3.28}$$

Differentiating equation (3.28), we have

$$y \, dy = rR \sin \theta d\theta \tag{3.29}$$

In $\triangle MNP x = r - R \cos \theta$, where $R \cos \theta = \frac{-y^2 + R^2 + r^2}{2r}$ from equation (3.28)

This gives $x = \frac{r^2 - R^2 + y^2}{2r}$.

Substituting m_r from equation (3.26), $\sin \theta d\theta$ from equation (3.29), and x from equation (3.30) in equation (3.27) we obtain

$$E_r = \frac{\pi\sigma GR}{r^2}\left(\frac{r^2 - R^2}{y^2} + 1\right)dy$$

3-29

Finally, integrating E_r we obtain

$$E_{\text{sph}} = \int E_r$$

Substituting E_r, we have

$$E_{\text{sph}} = \frac{\pi \sigma GR}{r^2} \int_{r-R}^{r+r} \left(\frac{r^2 - R^2}{y^2} + 1 \right) dy$$

After evaluating the integration, we have

$$E_{\text{sph}} = \frac{4 \pi G \sigma R^2}{r^2},$$

where $4 \pi R^2 = m$ (mass of the shell)

$$\Rightarrow E_{\text{sph}} = \frac{Gm}{r^2}.$$

N.B.: We note the following points from the foregoing discussions:
1. $E = 0$; $r < R$; field inside the shell is zero.
2. $E = \frac{Gm}{r^2}$; $r \geqslant R$; field varies obeying inverse square law, outside the shell.

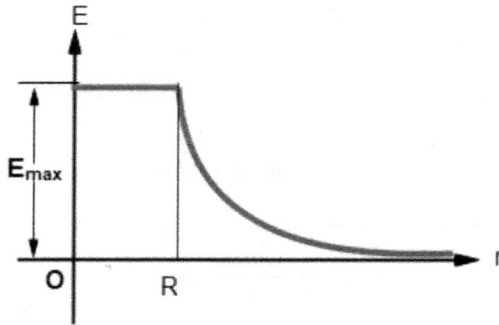

3. The $E-r$ graph tells us that E decreases from $\frac{Gm}{R^2}$ to zero obeying inverse square law.
4. Since $E = \frac{Gm}{r^2}$ for any outside point of a symmetrical spherical mass distribution, it behaves as a point mass. However, for any other mass distribution (linear, areal, and other non-uniform distribution of mass), we cannot substitute the object by a point mass.
5. For any inside point P of the spherical shell, we have

$$E_{\text{sph}} = \frac{\pi \sigma GR}{r^2} \int_{R-r}^{R+r} \left(\frac{r^2 - R^2}{y^2} + 1 \right) = 0$$

Volume mass distribution: Let's now use the expression of field intensity due to a thin spherical shell to find the field due to a solid sphere.

Example 11 Derive an expression for the field strength due to a uniform solid sphere of mass m and radius R.

Solution

Let's take a point P at a radial distance r from the center O of the sphere. Draw a sphere of radius r passing through P having its center at O, shown as a shaded region in the following figure.

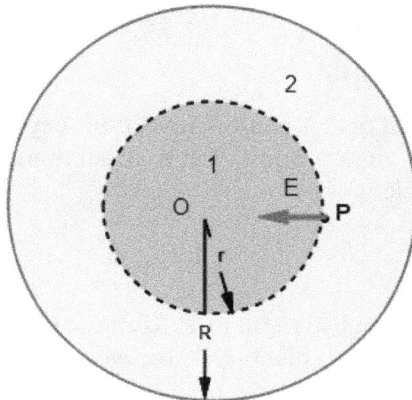

Since the point P is inside of the hollow sphere 2, the field at P due to the hollow sphere is zero. Then, the field at P solely due to the hollow sphere is zero. Then, the field at P is solely due to the shaded sphere 1. Since the point P is situated just outside of the uniform sphere 2, using the concept of spherical symmetry, the field intensity due to the sphere at P can be given by

$$E = \frac{Gm_1}{r^2},$$

where m = mass of the sphere which can be given as

$$m = \frac{4}{3}\pi r^3 \rho \text{ and } \rho = \text{ volume mass density.}$$

Then, we have

$$E = \frac{4}{3}G\pi\rho r; \, r \leqslant R \text{ Ans.}$$

N.B.: If we consider the point outside or the surface of the sphere, following the concept of spherical symmetry, we have

$E = \frac{GM}{r^2}; \, r \geqslant R$, where M = total mass of the sphere.

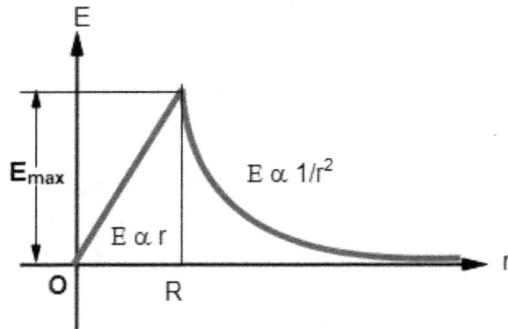

3.7 Work done by gravity

We know that gravitational field is conservative as it obeys inverse square law. If we release a particle of mass m at a point P in a gravitational field, the gravitational force acting on the particle is

$$\overrightarrow{F_{gr}} = m\overrightarrow{E},$$

where $\overrightarrow{E}$ = gravitational field strength at P. If the particle moves in any arbitrary curve, for an elementary displacement $d\overrightarrow{s}$, the work done by gravity is

$$dW_{gr} = \overrightarrow{F_{gr}} . d\overrightarrow{s},$$

where $\overrightarrow{F_{gr}} = m\overrightarrow{E}$.

This gives

$$dW_{gr} = m\overrightarrow{E} . d\overrightarrow{s}$$

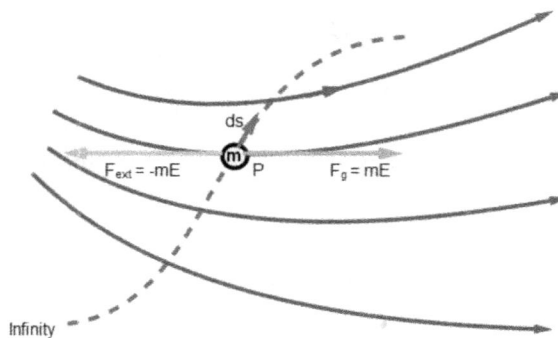

When the particle undergoes displacement from point 1 to point 2, integrating the elementary work done, the total work done by gravity is

$$W_{\text{gr}} = \int dW_{\text{gr}},$$

where $dW_{\text{gr}} = m \overrightarrow{E} . d \overrightarrow{s}$

$$\Rightarrow W_{\text{gr}} = \int_1^2 m \overrightarrow{E} . d \overrightarrow{s}$$

$$\Rightarrow W_{\text{gr}} = m \int_1^2 \overrightarrow{E} . d \overrightarrow{s}$$

Example 12 Find the work done by gravity when a particle of mass m is displaced from the center of the Earth through a diametrical chute to the (a) surface of the Earth and (b) infinity. Assume $R =$ radius of the Earth and other conditions are ideal.

Solution

(a) As we have derived earlier,

$$E = \frac{GM}{R^3} r; r \leqslant R$$

Hence, the work done by gravity is

$$W_{\text{gr}} = m \int_0^R \overrightarrow{E} . d \overrightarrow{s},$$

where $\overrightarrow{E} = -\frac{GM}{R^3}\hat{r}$ and $d \overrightarrow{s} = dr \, \hat{r}$.

$$\Rightarrow W_{\text{gr}} = -\frac{GMm}{R^3} \int_0^R r \, dr$$

After evaluating the integral, we have

$$W_{\text{gr}} = -\frac{GMm}{2R} = -\frac{1}{2}\left(\frac{GM}{R^2}\right) mR,$$

Putting $\frac{GM}{R^2} = g$, we have

$$W_{\text{gr}} = -\frac{mgR}{2} \text{ Ans.}$$

(b) As we have derived earlier,

$$E = \frac{GM}{r^2}; r \leqslant R$$

The work done by the external agent in slowly removing the particle from surface to infinity is

$$W'_{\text{gr}} = -\int_R^\infty E\, dr = -\int_R^\infty \frac{GMm}{r^2}\, dr$$

$$= -\frac{GMm}{R} = -\frac{GMmR}{R^2} = -mgR$$

Adding the last two works done, the work done by the external agent in slowly removing the particle from center to infinity is

$$W = -\frac{1}{2}mgR - mgR = -\frac{3}{2}mgR$$

3.8 Gravitational potential energy between two particles

For the sake of simplicity, let's consider a system of two particles of masses m_1 and m_2 separated by a distance r. The expression of potential energy of interaction between the two point masses m_1 and m_2 can be given as

$$U = -\frac{Gm_1m_2}{r}$$

The above expression tells us that the gravitational potential energy of interaction between two point masses (particles) is directly proportional to the product of their masses and inversely proportional to the distance of separation between the particles. Gravitational potential energy between two particles is always negative because the gravitational force is attractive, whereas electrostatic potential energy between the charge particles can be positive if the electrostatic force of interaction between the charge particles is repulsive and vice versa.

We know that the work done by gravity in bringing the particles from infinity to the given configuration is given as

$$W_{\text{gr}} = \frac{Gm_1m_2}{r}$$

In other words, the work done by the external agent in slowly bringing the particles from infinity to the given configuration can be given by

$$W_{\text{ext}} = -\frac{Gm_1m_2}{r}$$

Hence, the potential energy is given as

$$U = -W_{gr} = W_{ext}$$

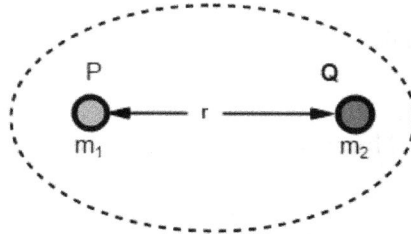

Closed/isolated system of two partices

The above discussion states that the gravitational potential energy of interaction between two particles can be defined as the negative work done by their mutual gravitational forces or the minimum work done by an external agent in bringing (assembling) the particles from infinity to the given configuration.

3.9 Gravitational potential

Let's take an extended object that is generally a system of particles. Then bring a particle of mass m slowly from infinity (a far distance) to a point P in any arbitrary path. In this process, you have to fight against the gravitational force F_{gr} acting on the particle due to the other objects. To move the particle slowly, with zero acceleration, you have to push the particle with a force $\overrightarrow{F_{ext}}$ such that this will nullify the gravitational force $\overrightarrow{F_{gr}}$.

Since $\overrightarrow{F_{ext}} + \overrightarrow{F_{gr}} = m\overrightarrow{a} = 0$, we have

$$\overrightarrow{F_{ext}} = -\overrightarrow{F_{gr}} \qquad (3.30)$$

While the particle undergoes an elementary displacement $d\overrightarrow{s}$, the elementary work done by the external agent is

$$dW_{ext} = \overrightarrow{F_{ext}} \cdot d\overrightarrow{s} \qquad (3.31)$$

Substituting $\overrightarrow{F_{ext}} = -\overrightarrow{F_{gr}}$ from equation (3.30) in (3.31), we have

$$dW_{ext} = -\overrightarrow{F_{gr}} \cdot d\overrightarrow{s}$$

Integrating the elementary works, the external work done in bringing the particle from infinity to the point P is

$$W_{ext} = \int dW_{ext} = -\int_{\infty}^{P} \overrightarrow{F_{gr}} \cdot d\overrightarrow{s}, \qquad (3.32)$$

where $\overrightarrow{F_{\text{gr}}} = m\,\overrightarrow{E}\,;\overrightarrow{E} = $ gravitational field intensity.

$$\Rightarrow W_{\text{ext}} = -m\int_{\infty}^{P}\overrightarrow{E}.d\,\overrightarrow{s}$$

Since the gravitational field is conservative and central, $m\int_{\infty}^{P}\overrightarrow{E}.d\,\overrightarrow{s}$, that is, the total work done by gravity between infinity and the point P does not depend on the path followed by the particle.

Definition of potential: This means that the minimum work done W_{ext} (in shifting the particle from infinity) will be different for different masses of the particles. However, in each case, the ratio of W_{ext} and mass m of the particle, that is, $\frac{W_{\text{ext}}}{m}$, is a constant quantity for the point P.

Hence, $\frac{W_{\text{ext}}}{m} = -\int_{\infty}^{P}\overrightarrow{E}.d\,\overrightarrow{s}$.

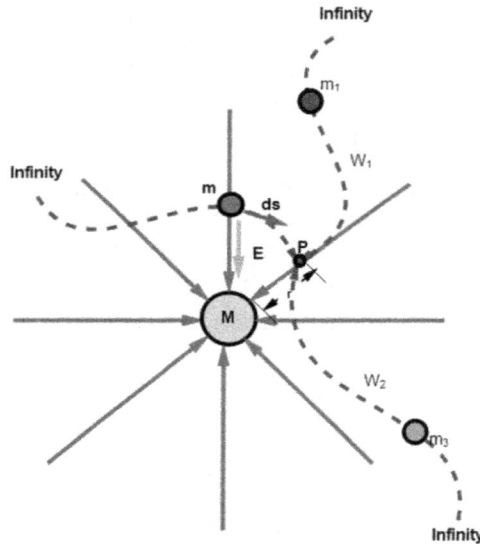

The above expression tells us that the work done by an external agent in slowly shifting any particle of mass 1 kg from infinity to any point in the gravitational field of an object is a 'unique quantity' that can be termed as 'potential' of the field at that point because it shows the stamina of the field. The potential of a point P can be given by

$$V = -\int_{\infty}^{P}\overrightarrow{E}.d\,\overrightarrow{s}$$

Physical significance of potential: If an external agent does more work in shifting a particle of mass 1 kg from infinity to any point, we can say that the potential of the point is more. Hence, it is related with the work done by (or stamina of) the external agent or the gravitational field. Thus, similar to the field strength $\overrightarrow{E}$, potential V at is a scalar; $\overrightarrow{E}$ is the force per unit mass and V is the work done per unit mass. The unit of gravitational potential is joule kg^{-1} and it is a negative quantity, whereas electrostatic potential can be positive and negative. In some books you may find that electrostatic potential can be positive and negative. In other books you may find the potential being defined as 'potential energy per unit mass'. If you follow this definition, you must be more careful to understand the meaning of 'potential energy per unit mass'. Here, the word potential energy means the potential energy of interaction between the particle and the object (or system), but not the total potential energy of interaction between all particles of the object.

N.B.: Every point of a gravitational field is characterised by field strength (or intensity) $\overrightarrow{E}$ and potential V:

$$\overrightarrow{E} = \frac{\overrightarrow{F}}{m}$$

$$V = \frac{W_{\text{ext}}(\infty \to P)}{m} = -\int_{\infty}^{P} \overrightarrow{E}.d\,\overrightarrow{s}$$

Different work will be done in shifting the particles of different masses to a particular point P. However, the ratio of work done and the mass is always a constant quantity for a point that is known as potential of a field at that point.

Different particles will experience different amount of force at the point P. However, the ratio of force and mass is always a constant quantity for that point, which is known as the strength or intensity of the gravitational field.

3.10 Calculation of gravitational potential

Discrete mass distribution: We will find gravitational potential of discrete and continuous mass distribution as we did for gravitational field intensity. The basic formula, that is, $V = -\frac{Gm}{r}$, gives us the potential due to a point mass. For a group of particles, we can find the potential due to each and then add them algebraically to obtain the total potential, at any point. Hence, the potential at P due to discrete mass distribution of n particles of masses $m_1, m_2, \ldots m_n$ situated at distances $r_1, r_2, \ldots r_n$ from the point P can be given as

$$V = V_1 + V_2 + \ldots + V_n,$$

where $V_1 = -\frac{Gm_1}{r_1}, V_2 = -\frac{Gm_2}{r_2}\ V_n = -\frac{Gm_n}{r_n}$.

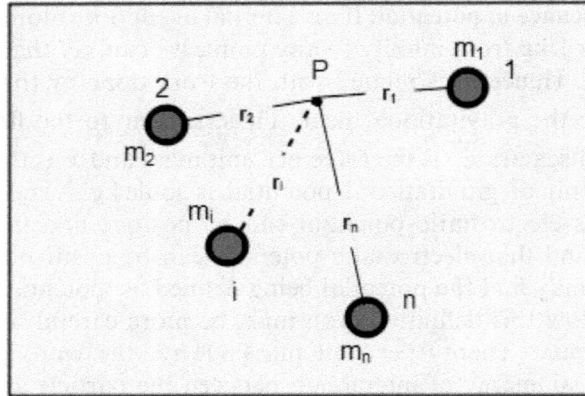

$$\Rightarrow V = -G\left(\frac{m_1}{r_1} + \frac{m_2}{r_2} + \ldots\ldots + \frac{m_n}{r_n}\right)$$

In a nutshell, we can write the above expression as

$$V = -G\sum_{i=1}^{i=n}\frac{m_i}{r_i}$$

Let's use the above formula in the following example.

Example 13 Find the gravitational potential of a system of two particles of masses m_1 and m_2 separated by a distance r, at a point where the field intensity of the system is zero.

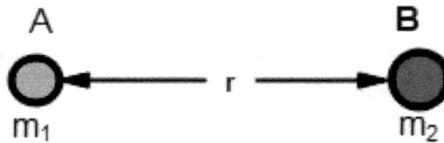

Solution
Let the field at P be zero.
Then, we have

$$\overrightarrow{E_{net}} = \overrightarrow{E_1} + \overrightarrow{E_2} = 0$$
$$\Rightarrow E_1 = E_2$$

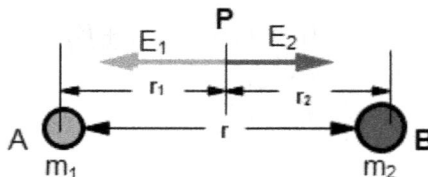

$$\Rightarrow \frac{Gm_1}{r_1^2} = \frac{Gm_2}{r_2^2}$$

$$\Rightarrow \frac{r_1}{r_2} = \left(\frac{m_1}{m_2}\right)^{1/2} \tag{3.33}$$

As per the given system

$$r_1 + r_2 = r \tag{3.34}$$

Using equations (3.33) and (3.34), we have

$$r_1 = \frac{r\sqrt{m_1}}{\sqrt{m_1} + \sqrt{m_2}} \tag{3.35}$$

$$r_2 = \frac{r\sqrt{m_2}}{\sqrt{m_1} + \sqrt{m_2}} \tag{3.36}$$

The total potential at P is

$$V = -G\left(\frac{m_1}{r_1} + \frac{m_2}{r_2}\right) \tag{3.37}$$

Putting the values of r_1 and r_2 from equations (3.35) and (3.36) in equation (3.37), we have $V = -\frac{G(\sqrt{m_1} + \sqrt{m_2})^2}{r}$ Ans.

Let's now try to develop a general formula of the potential of a continuous mass distribution.

Continuous mass distribution: For any extended object, first of all we consider an element of mass dm. Then, the potential due to the extended object is

$$dV = -G\frac{dm}{r}$$

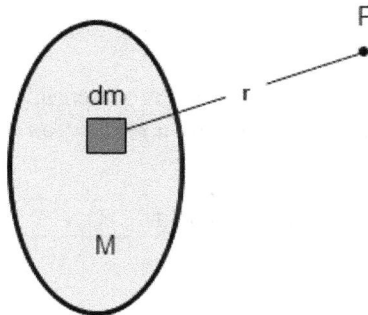

The total potential due to the extended object is

$$V = \int dV = -G \int \frac{dm}{r},$$

where $dm = \lambda dx$ for linear mass distribution;
$= \sigma \, dA$ for surface mass distribution; and
$= \rho \, dV$ for volume mass distribution.

Example 14 Find the gravitational potential at P due to the uniform thin rod of linear mass density λ and length l, as shown in the figure.

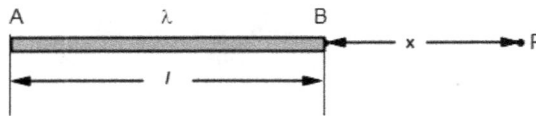

Solution
Potential due to the element dm at P is, $dV = -G\frac{dm}{r}$, where $dm = \lambda \, dr$

$$\Rightarrow dV = -G\lambda \frac{dr}{r}$$

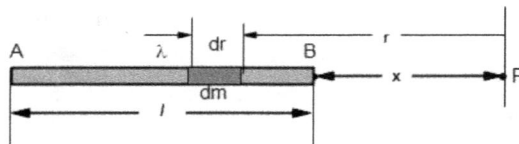

Integrating the elementary potentials, total potential due to the rod is

$$V = \int dV = -G\lambda \int_{x}^{l+x} \frac{dr}{r}$$

$$V = -G\lambda \ln \frac{l+x}{x} \quad \text{Ans.}$$

N.B.: Following the above procedure, the potential due to the rod at a point situated at its perpendicular bisector having a separation x from the rod can be given as

$$V = \frac{-2G\lambda}{x} \ln\left(\frac{R + \sqrt{R^2 + x^2}}{x} \right)$$

Example 15 Find the gravitational potential of a thin circular arc of mass m and radius of curvature R, at the center of curvature.

Solution

The potential due to the element dm at C is $dV = -G\frac{dm}{R}$.

Then, the total potential due to the arc is

$$V = \int dV = -\int_0^m G\frac{dm}{R}$$

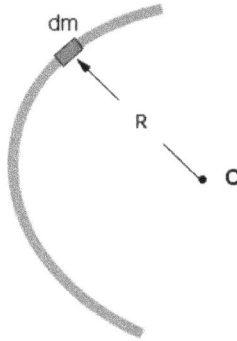

Since G and R are constant, taking them out of the integral, we have

$$V = -\frac{G}{R}\int_0^m dm$$

$$\Rightarrow V = -\frac{Gm}{R} \text{ Ans.}$$

N.B.: Since each element of a ring is equidistance from the center of the ring, the potential at the center of a ring of mass m and radius R can be given as $V = -\frac{Gm}{R}$.

Example 16 Find the gravitational potential of a thin circular ring of mass m and radius R, at a point P on the axis of the ring located at a distance x from the center the ring.

Solution

The potential of the element dm at P is given by

$$dV = -\frac{Gdm}{r}$$

Since each element dm of the ring is situated at a distance r from the axial point P, the potential of the ring is given by

$$V = \int dV = -\int_0^m G\frac{dm}{r}$$

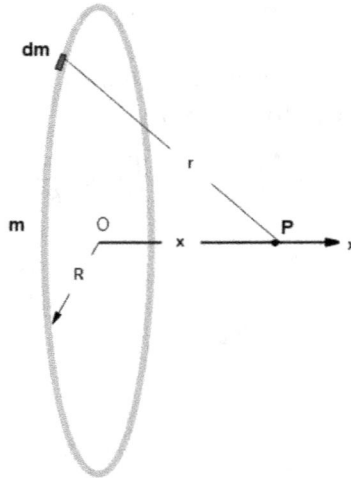

$$\Rightarrow V = -\frac{Gm}{r}\left(\text{but not, } -\frac{Gm}{R}\right),$$

where $r = \sqrt{R^2 + x^2}$.

$$\Rightarrow V = -\frac{Gm_{\text{ring}}}{\sqrt{R_{\text{ring}}^2 + x^2}} = -\frac{Gm}{\sqrt{R^2 + x^2}}$$

At $x = 0$, $V = \frac{Gm}{R}$; If $x \to \infty$, $V \to 0$

This means that the potential varies with x, as shown in the following figure.

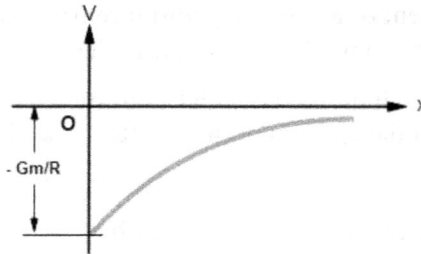

Example 17 Using the formula of the gravitational potential obtained for the ring in the previous example, derive an expression for gravitational potential at the axial point that is situated at a distance x from the center of a uniform disc of surface mass density σ and radius R.

Solution

Let's take a thin ring of radius r and thickness dr. Then, the mass of the ring is $m_{ring} = \sigma(2\pi r \, dr)$, as explained earlier.

As derived in the previous example the potential due to the ring is

$$V_{ring} = -\frac{Gm_{ring}}{\sqrt{R_{ring}^2 + x^2}}$$

where $m_{ring} = 2\pi\sigma r \, dr$ and $R_{ring} = r \Rightarrow V_{ring} = 2\pi\sigma G \frac{r \, dr}{\sqrt{r^2 + x^2}}$.

Integrating V_{ring}, we have

$V_{disc} = \int V_{ring}$

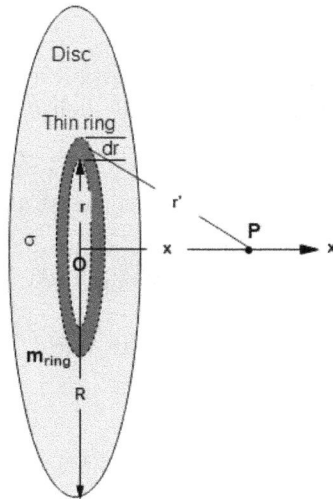

$$= -2\pi\sigma G \int_0^R \frac{r \, dr}{\sqrt{r^2 + x^2}}$$

Let $\sqrt{r^2 + x^2} = t$; then, $r \, dr = t \, dt$ when $r = 0$, $t = x$; when $r = R$,

$$t = \sqrt{R^2 + x^2}$$

$$\Rightarrow V_{disc} = -\pi\sigma G \int_R^{\sqrt{R^2 + x^2}} \frac{t \, dt}{t}$$

After evaluating the integral, we have

$$V_{disc} = -\pi\sigma G\left(\sqrt{R^2 + x^2} - x\right) \text{ Ans.}$$

N.B.: 1. At $x = 0$, $V = -\pi G\sigma R$ and at $x \to \infty$, $V \to 0$.
2. If $R \to \infty$, $V \to \infty$, V is infinite for a large sheet.

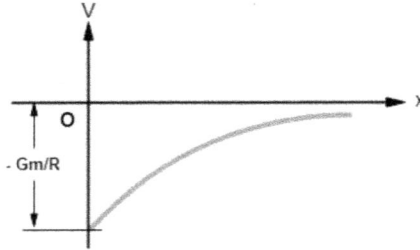

Example 18 Derive an expression for the gravitational potential at a point having a radial distance r, outside of a thin spherical shell of surface mass density σ and radius R.

Solution

Take a thin ring (portion between two red arcs) at an angular position θ. As explained earlier, the mass of the ring is

$$m_{\text{ring}} = \sigma(2\pi R^2 \sin\theta \, d\theta)$$

$$\Rightarrow m_{\text{ring}} = 2\pi\sigma R^2 \sin\theta \, d\theta \tag{3.38}$$

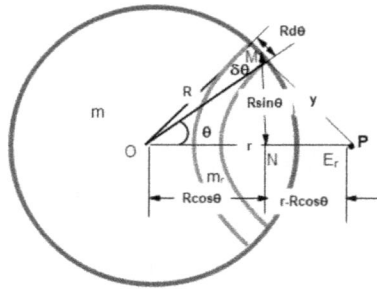

The potential due to the ring at P is

$$V_{\text{ring}} = -\frac{Gm_{\text{ring}}}{y} \tag{3.39}$$

Using the last two equations,

$$V_{\text{ring}} = -\frac{2\pi\sigma R^2 G \sin\theta \, d\theta}{y} \tag{3.40}$$

In the triangle, as shown in the last figure, applying the cosine rule, we have

$$R^2 + r^2 - 2Rr \cos \theta = y^2$$

Differentiating both sides

$$2Rr \sin \theta \, d\theta = 2y \, dy$$

$$\Rightarrow \sin \theta \, d\theta = \frac{y \, dy}{Rr} \tag{3.41}$$

Substituting $\sin \theta \, d\theta$ from equations (3.41) in (3.40), we have

$$V_{\text{ring}} = -\frac{2\pi\sigma GR}{r} dy \tag{3.42}$$

The thin spherical shell is the combination of coaxial rings of angular positions θ ranging from $0°$ to $180°$. Then, the value of y must vary from $(r - R)$ to $(R + r)$. Integrating V_{ring} we have

$$V_{\text{sphere}} = \int V_{\text{ring}} \tag{3.43}$$

Using the last two equations

$$V_{\text{sphere}} = -\frac{2\pi\sigma GR}{r} \int_{r-R}^{r+R} dy$$

$$V_{\text{sphere}} = -\frac{4\pi\sigma GR^2}{r} = -\frac{Gm}{r}(\because 4\pi\sigma R^2 = m) \text{ Ans.}$$

To find the potential inside the spherical shell, we take the point P inside the shell. Then, the lower limit becomes $R - r$ and the upper limit of the integral remains the same $(= R + r)$.

So, the aforementioned integral can be written as

$$V_{\text{sphere}} = -\frac{2\pi\sigma GR}{r} \int_{R-r}^{R+r} dy.$$

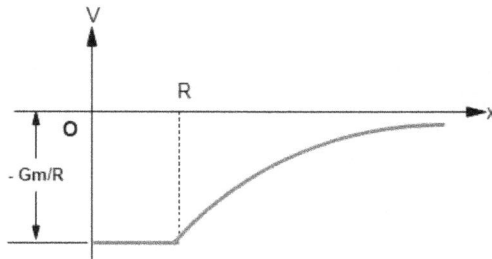

Substituting $\sigma(4\pi R^2) = m$ (mass of the spherical shell), we have

$$V = -\frac{Gm}{r}; r \geqslant R$$

$$= -\frac{Gm}{r}; r \leqslant R$$

This means that the potential remains constant, that is, $-\frac{Gm}{R}$, and its magnitude $|V|$ decreases hyperbolically to zero (or V increases from $-\frac{Gm}{R}$ to zero) when $r \to \infty$ (we go radially away).

Example 19 Find the gravitational potential of a uniform solid sphere of radius R and volume mass density ρ at any point having a radial distance r.
  Solution
  We divide the sphere into parts: sphere 1 (shaded portion) of radius r containing the point P and the hollow sphere 2 of inner radius r and outer radius R. If we can find potential due to each sphere at P and then sum up, we will get the potential due to the given sphere at P. Let's proceed with this idea.
  Let the potentials due to sphere 1 and hollow sphere 2 at P be V_1 and V_2, respectively. Since the sphere 1 is composed of thin spherical shells of radii ranging from 0 to r and the point P is situated outside of each thin sphere (spherical shells), the potential due to 1 at P is equal to the sum of potentials due to each thin shell. Taking a thin spherical shell of mass dm, the potential at P is

$$dV = -G\frac{dm}{r}$$

Integrating dV, the total potential due to the sphere of radius r at P is

$$V_1 = \int dV = -\frac{G}{r}\int_0^{m'} dm = -\frac{Gm'}{r},$$

where $m' =$ mass of the sphere $1 = \frac{4}{3}\pi r^3 \rho$

$$\Rightarrow V_1 = -\frac{4\pi G \rho r^2}{3} \tag{3.44}$$

Let's now calculate V_2. For this let's take a spherical shell of mass dm and radius x. Since the point P lies inside the shell, its potential at P is given as
$dV = -G\frac{dm}{x}$ as discussed earlier.
Integrating dV, the potential due to the hollow sphere 2 at P is
$V_2 = \int dV = -\int_r^R G\frac{dm}{x}$, where $dm = \rho 4\pi x^2 dx$

$$\Rightarrow V_2 = -4\pi \rho G \int_r^R x\, dx$$

$$\Rightarrow V_2 = -2\pi\rho G(R^2 - r^2) \tag{3.45}$$

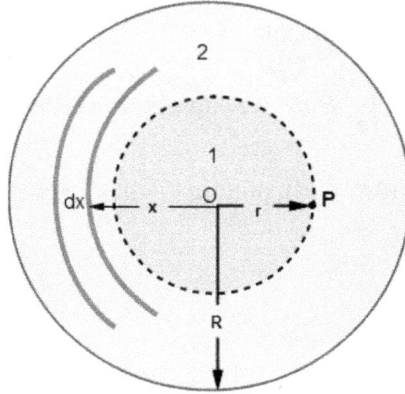

Adding V_1 and V_2, we have the total potential at P given by

$$V = V_1 + V_2 \tag{3.46}$$

Substituting V_1 from equations (3.44), (3.45) in (3.36), we have

$$V = \frac{4\pi G\rho}{6}(3R^2 - r^2); \; r \leqslant R \; \text{Ans.}$$

As discussed earlier, for any outside point ($r > R$), we have

$$V = -\frac{Gm'}{r},$$

where $m' =$ mass of the total sphere. Because we have taken the point P outside of the given sphere

$$m' = = \frac{4}{3}\pi R^3 \rho$$

Using last two equations,

$$V = -\frac{4\pi R^3 \rho G}{3r}; \; r \geqslant R \; \text{Ans.}$$

Alternative method:
We can also use the following expression to get the same result.

$$V = -\int_\infty^r E \, dr$$

to obtain the same result.

As derived earlier,

$$E = \frac{GM}{R^3}r; \, r \leqslant R \quad \text{and} \quad E = \frac{GM}{r^2}; \, (r \text{ is greater than or equal to } R)$$

Putting these values in the expression

$$V = -\int_{\infty}^{R} E \, dr - \int_{R}^{r} E \, dr, \text{ we can obtain the following expression}$$

N.B.: 1. Since $V = -\frac{4\pi G\rho}{6}(3R^2 - r^2)$, putting $\frac{4\pi R^3}{3}\rho = m$, we have $V = -\frac{Gm}{2R^3}(3R^2 - r^2)$, for $r \leqslant R$.

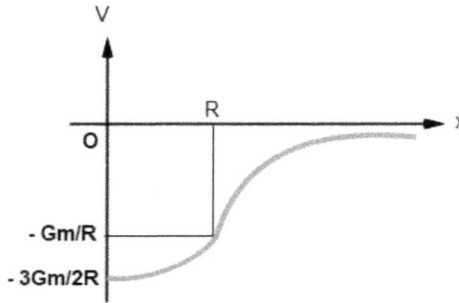

This means that potential increases from $-\frac{3Gm}{2R}$ to $-\frac{Gm}{R}$ (or its magnitude decreases from $\frac{3Gm}{2R}$ to $\frac{Gm}{R}$) parabolically from center to the surface of the sphere.

2. As the sphere possesses spherical symmetry, it behaves as a point mass for the points outside the sphere. Hence, potential $V = -\frac{Gm}{r}$ for $r \geqslant R$. This means that potential varies hyperbolically outside the sphere like a thin uniform spherical shell.

3.11 Gravitational potential energy of a group of particles

We have derived the expression of gravitational potential energy by using the concept of work done and the work–energy theorem for a two-particle system. Let's apply the work–energy theorem for a many-particle system to develop a general formula for potential energy of interaction of a group of particles. As discussed earlier, potential energy of a two-particle system of masses m_1 and m_2 is equal to the minimum work done by an external agent (or negative of work done by gravity) in assembling the system (particles) from infinity to the given configuration. Let's assume that we have n particles of mass $m_1, m_2, m_3, ..., m_n$. If we bring m_1 slowly from infinity and place it at position (point) 1, we do not have to perform any work because no external field is present.

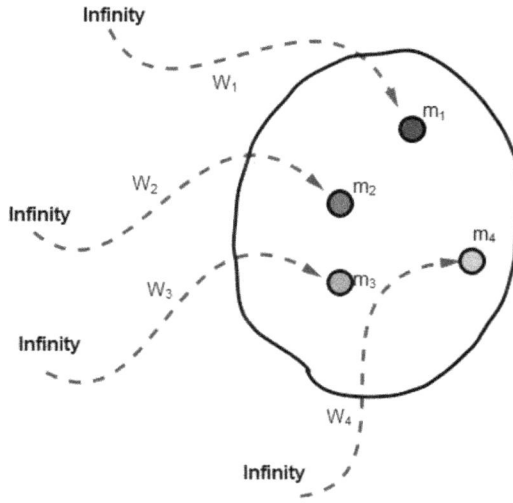

$$W_1 = 0 \tag{3.47}$$

Now bring the particle m_2 slowly from infinity and place it at the point 2. In this process the work done against the gravity of m_1 is

$$W_2 = -\frac{Gm_1m_2}{r_{21}} \tag{3.48}$$

Similarly when you bring the particle m_3 slowly from infinity to the point 3, you have to do work $-\frac{Gm_1m_2}{r_{31}}$ against the gravity of m_1, $-\frac{Gm_2m_3}{r_{32}}$ against the gravity of m_2. Hence, the total work done in shifting m_3 is

$$W_3 = -\frac{Gm_1m_3}{r_{31}} - \frac{Gm_2m_3}{r_{32}} \tag{3.49}$$

Likewise, the total external work done in bringing the particle m_4 slowly from infinity to the point 4 is

$$W_4 = -\frac{Gm_1m_4}{r_{41}} - \frac{Gm_2m_4}{r_{42}} - \frac{Gm_3m_4}{r_{43}} \tag{3.50}$$

Adding equations (3.47)–(3.50), we have the total work

$$W = -\frac{Gm_1m_2}{r_{21}} - \frac{Gm_1m_3}{r_{31}} - \frac{Gm_2m_3}{r_{32}} - \frac{Gm_1m_4}{r_{41}} - \frac{Gm_2m_4}{r_{42}} - \frac{Gm_3m_4}{r_{43}} \tag{3.51}$$

Let's now do the same thing but in reverse order. First of all, you bring m_4 from infinity and place it at the point 4. The work done is zero because no field was there. Hence,

$$W'_1 = 0 \tag{3.52}$$

Then bring m_3 slowly from infinity and place it at point 3. The external work done is

$$W'_2 = -\frac{Gm_3m_4}{r_{34}} \tag{3.53}$$

Similarly total work done in bringing m_2 from infinity to the point 2

$$W'_3 = -\frac{Gm_2m_4}{r_{24}} - \frac{Gm_2m_3}{r_{23}} \tag{3.54}$$

The external work done in bringing m_1 from infinity to the point 1

$$W'_4 = -\frac{Gm_1m_4}{r_{14}} - \frac{Gm_1m_3}{r_{13}} - \frac{Gm_1m_2}{r_{12}} \tag{3.55}$$

Adding equations (3.52)–(3.55), we have

$$W' = -\frac{Gm_3m_4}{r_{34}} - \frac{Gm_2m_4}{r_{24}} - \frac{Gm_2m_3}{r_{23}}$$

$$-\frac{Gm_1m_4}{r_{14}} - \frac{Gm_1m_3}{r_{13}} - \frac{Gm_1m_2}{r_{12}} \tag{3.56}$$

In both cases we have exactly the same configurations of the particles. Hence, the external work done must be the same in both cases. Then

$$W + W' = 2W \tag{3.57}$$

Substituting W from equation (3.51) and W' from equations (3.56) in (3.57) and grouping the terms, we have

$$2W = m_1\left(-\frac{Gm_2}{r_{12}} - \frac{Gm_3}{r_{13}} - \frac{Gm_4}{r_{14}}\right) + m_2\left(-\frac{Gm_1}{r_{21}} - \frac{Gm_3}{r_{23}} - \frac{Gm_4}{r_{24}}\right)$$

$$+ m_3\left(-\frac{Gm_1}{r_{31}} - \frac{Gm_2}{r_{32}} - \frac{Gm_4}{r_{34}}\right) + m_4\left(-\frac{Gm_1}{r_{41}} - \frac{Gm_2}{r_{42}} - \frac{Gm_3}{r_{43}}\right) \tag{3.58}$$

Substituting $-\frac{Gm_2}{r_{12}} - \frac{Gm_3}{r_{13}} - \frac{Gm_4}{r_{14}} = V_1$ (potential at 1 due to all particles except m_1), $-\frac{Gm_1}{r_{21}} - \frac{Gm_3}{r_{23}} - \frac{Gm_4}{r_{24}} = V_2$, $-\frac{Gm_1}{r_{31}} - \frac{Gm_3}{r_{32}} - \frac{Gm_4}{r_{34}} = V_3$, and $-\frac{Gm_1}{r_{41}} - \frac{Gm_3}{r_{42}} - \frac{Gm_4}{r_{43}} = V_4$ in equation (3.58), we have

$$2W = m_1V_1 + m_2V_2 + m_3V_3 + m_4V_4$$

Following the above procedure for n particles, we have

$$2W = m_1V_1 + m_2V_2 + \cdots + m_nV_n$$

In a nutshell, we can write

$$2W = \sum_{i=1}^{i=n} m_i V_i$$

where V_i = potential at ith point due to all particles except m_1.

$$\Rightarrow W = W_{\text{ext}} = \frac{1}{2}\sum_{i=1}^{i=n} m_i V_i$$

As we know,
$W_{\text{ext}} = U - U_\infty$ and $U_\infty = 0$ (because $F = 0$, at infinity). Then, we have

$$U = \frac{1}{2}\sum_{i=1}^{i=n} m_i V_i$$

Example 20 Eight identical particles each of mass m are situated at the vertices of a cube of side l. Find the gravitational potential energy of interaction of the system of eight particles.
Solution
In a cube we have eight corners having equal masses. Hence, $V_1 = V_2 = \cdots = V_8$ and $m_1 = m_2 = \cdots = m_8 = m$.

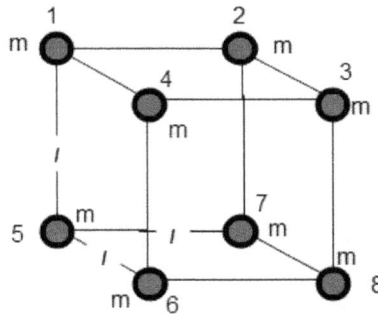

Then using the expression

$$U = \frac{1}{2}\sum m_i V_i$$

we have

$$U = \left(\frac{1}{2}m_1 V_1\right)(8),$$

where $m_1 = m$ and V_1 = potential at point 1.

Then, we have

$$U = 4mV_1,\tag{3.59}$$

where

$$V_1 = -G\frac{m}{l}\left(\frac{3}{1} + \frac{3}{\sqrt{2}} + \frac{1}{\sqrt{3}}\right),$$

$$U = -\frac{12Gm^2}{l}\left(1 + \frac{1}{\sqrt{2}} + \frac{1}{3\sqrt{3}}\right) \text{ Ans.}$$

N.B.: Using the formula $U = \frac{1}{2}\sum m_i V_i$, the gravitational potential energy of interaction of the system of six identical particles, each of mass m, placed at the vertices of a regular hexagon of side l, can be given as

$$U = -\frac{6Gm^2}{l}\left(1 + \frac{1}{\sqrt{3}} + \frac{1}{2}\right)$$

Potential energy due to continuous mass distribution: Let's take the expression $U = \frac{1}{2}\sum m_i V_i$ for discrete mass distribution. In continuous mass distribution we cannot distinguish one particle from the others. Hence, we write $m_1 = dm$ and $V_1 = V$; then replacing the sign '$\sum$' by internal sign '$\int$' in the above formula, we have

$$U = \frac{1}{2}\int V \, dm$$

where V = potential at the point P and $dm = \lambda \, dx$ (for linear mass distribution);
$= \sigma \, dA$ (for surface mass distribution); and
$= \rho \, dV$ (for volume mass distribution).

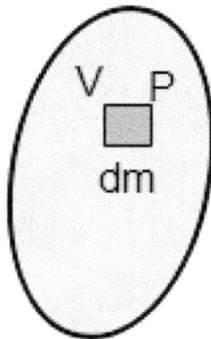

Remember that the above formula is presented without any proof, which is beyond the scope of this book. Let's use the above formula.

Example 21 Find the gravitational potential energy possessed by a thin spherical shell of mass m and radius R.

Solution

Let's take an elementary mass dm at P. The potential at P is

$$V = -\frac{GM}{R}$$

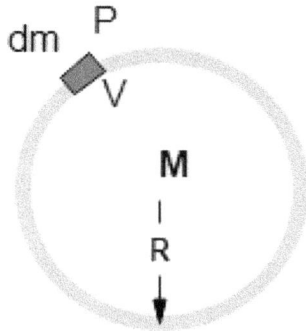

Substituting $V = \frac{GM}{R}$ in the formula,
$U = \frac{1}{2} \int V \, dm$, we have

$$U = \frac{1}{2} \int_0^M \left(-\frac{GM}{R} \right) dm$$

Since $\frac{GM}{R}$ is constant, take it out of the integral to obtain

$$U = -\frac{GM}{2R} \int_0^M dm$$

$$\Rightarrow U = -\frac{GM^2}{2R} \text{ Ans.}$$

N.B.: In the previous example, if you write $dm = dM$ and do not take it out of the integral, you will get the wrong result as follows:

$$U = -\frac{1}{2} \int_0^M \frac{GM}{R} \cdot dM = -\frac{GM^2}{4R}$$

3.12 Earth's gravitational field

We have talked about the gravitational force between two particles of the universe. We found the field strength, potential energy, work done, and potential due to any arbitrary object. In this section, let's talk about the gravity of our planet Earth. According to Newton's law of gravitation, Earth pulls each object. For the sake of simplicity let's assume that Earth is a perfect sphere having uniform density. Due to its spherical symmetry, it behaves as a point mass to an object outside the Earth. Then the gravitational force acting on a point mass m at a distance r (outside the Earth) from the center of the Earth is

$$\overrightarrow{F_g} = -\frac{GMm}{r^2}\hat{r},$$

where M = mass of the Earth.

Hence, the strength of Earth's gravity at a radial distance r is

$$\overrightarrow{E} = \frac{\overrightarrow{F_g}}{m} = -\frac{GM}{r^2}\hat{r}$$

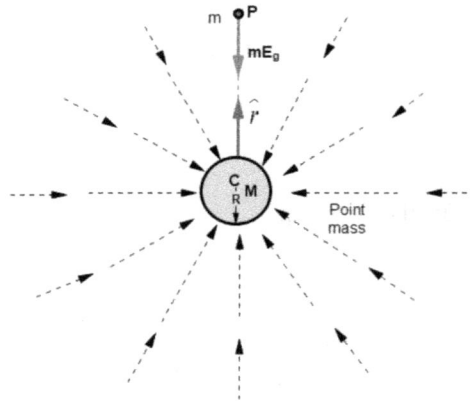

The above expression states that the Earth's gravitational field strength (intensity) at a point on or above the Earth is directly proportional to the mass of the Earth and inversely proportional to the square of the distance of the point from the center of the Earth. The strength of Earth's gravity is directed towards the center of the Earth.

If no other force acts on the particle of mass m placed in Earth's gravitational field, the net force is equal to the gravitational force $\overrightarrow{F_g}$. According to Newton's 2nd law, the acceleration of the particle is given by

$$\overrightarrow{a} = \frac{\overrightarrow{F_g}}{m}, \text{ where } \frac{\overrightarrow{F_g}}{m} = \overrightarrow{E}.$$

Hence, the strength of Earth's gravitational field at any point is equal to the acceleration of a particle subjected to Earth's gravity at that point. In other words,

$\overrightarrow{E}$ can be called 'acceleration due to gravity' of the Earth denoted as $\overrightarrow{g}$. If you put $r = R$ in the expression $g = \frac{GM}{r^2}$, we have $g = \frac{GM}{R^2}$. Substituting $M = 6 \times 10^{24}$ kg and $R = 6400$ km, we have $g = 9.8$ m s^{-2} ($= g_0$, say); we call it acceleration due to gravity at (or near) Earth's surface.

Example 22 Assuming the Earth as a homogeneous sphere, draw the graph of gravitational field and potential versus radial distance.

Solution

Since Earth is assumed as uniform and possesses spherical symmetry, the field strength g and potential due to Earth can be given as

$$g = \frac{GM}{R^3}r; \; r \leqslant R$$

$$= \frac{GM}{r^2}; \; r \geqslant R$$

For any inside point, the potential is

$$V = -\frac{GM}{2R}(3R^2 - r^2); \; r \leqslant R$$

For any outside point, the potential is

$$V = -\frac{GM}{r}; \; r \geqslant R$$

We have plotted the graphs as shown in the following figure. We can see that both graphs $g = f(r)$ and $V = f(r)$ are continuous.

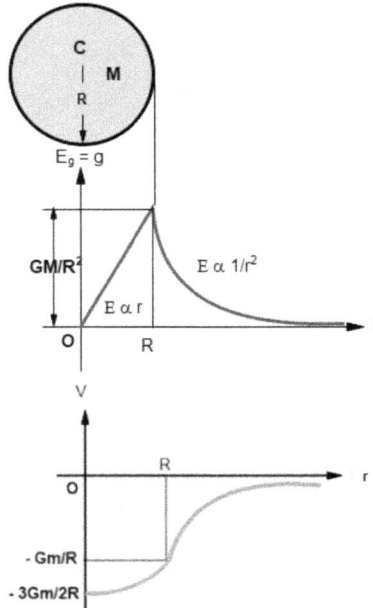

In the above figures (graphs), the magnitude of gravitational field intensity is taken which is always positive; but, the gravitational potential is always negative because Newtonian gravitational force is attractive.

3.13 Variation of $|\vec{g}_{\text{eff}}|$ and apparent weight

The magnitude of g will vary with radial distance (i.e., height and depth). Its 'apparent value' is affected by the rotation of the Earth. The mass distribution and shape of the Earth also affect the value of g. Let's discuss one by one.

Due to height: At any point at a height h, $r = (R + h)$. The value of g at P is $g = \frac{GM}{r^2}$, where $r = (R + h)$

$$\Rightarrow g = \frac{GM}{(R + h)^2}$$

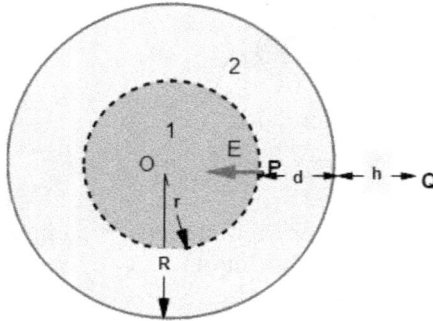

Dividing and multiplying by R^2, we have

$$g = \frac{GM}{R^2} \frac{R^2}{(R + h)^2}$$

Substituting $\frac{GM}{R^2} = g_0$, we have

$$g = \frac{g_0}{\left(1 + \frac{h}{R}\right)^2}$$

The above expression can be written as

$$g = g_0\left(1 + \frac{h}{R}\right)^{-2}$$

Since $(1 + x)^n \cong 1 + nx$ for $x \ll 1$, we have

$g \cong g_0(1 - \frac{2h}{R})$, for small $h(\ll R)$.

$\Rightarrow g = g_0(1 - \frac{2h}{R})$ is valid if $h \ll R$.

The above equation tells us that:

For a small value of height $h(\ll R)$, the acceleration due to gravity g decreases linearly with h. However, in a strict sense g decreases with h obeying the inverse square law.

Example 23 At what altitude does the weight of a body decrease by 0.01%?

Solution

Let the weight decrease from W_0 to W at a height h.

Since $(1 - \frac{W}{W_0}) = 0.01\% = 1 \times 10^{-4}$, putting $\frac{W}{W_0} = \frac{g}{g_0} = 1 - \frac{2h}{R}$, we have

$h = \frac{R}{2} \times 10^{-4}$, where $R = 6400$.

This gives $h = 320$ m Ans.

N.B.: We have the following alternate method to obtain the expression:

$$\frac{\delta g}{g} = -\frac{2h}{R}$$

Since $g = \frac{GM}{r^2}$, taking logarithm of both sides, we have in $g = $ In GM–2In r

Taking differentials of both sides, we have

$\frac{\delta g}{g} = -2\frac{\delta r}{r}$

If we put $\delta r = h$, we obtain $(\frac{\delta g}{g})_{r=R} = -\frac{2h}{r}$.

N.B.: The above formula is an approximation formula valid up to the variation of 10% (approximately). However, the formula $g = \frac{g_0}{(1 + \frac{h}{R})^2}$ is an exact expression valid for any value of h.

Due to the depth: Let's take a point P at a depth h. Then $r = R - d$.

As derived earlier

$$g = \frac{GM}{R^3}r \text{ for } r \leqslant R,$$

where

$$\frac{GM}{r^2} = g_0$$

This gives $g = \frac{g_0 r}{R}$, where $r = R - d$

$$\text{Then } g = g_0\left(1 - \frac{d}{R}\right)$$

The above expression tell us that g varies (decreases) linearly with depth from g_0 to zero when we go from the surface to the center of the Earth.

Example 24 At what height will the magnitude of g be equal to that at a depth $d(=\frac{R}{2})$, where R = radius of the Earth?

Solution

The value of g at the depth d is given by

$$g_1 = g_0\left(1 - \frac{d}{R}\right) \tag{3.60}$$

The value of g at the depth h is given by

$$g_2 = \frac{g_0}{\left(1 + \frac{h}{R}\right)^2}$$

$$\simeq g_0\left(1 - \frac{2h}{R}\right)(\because h \ll R) \tag{3.61}$$

For the equal weights

$$g_1 = g_2 \tag{3.62}$$

This gives $h/R = 1/2$ Ans.

Effect of rotation of the Earth and latitude on the apparent weight: Let's take an object of mass m at a latitude θ. Since the object is stationary relative to the rotating Earth, it experiences a centrifugal force $F_{CF} = mr\omega^2\rightarrow$ in addition to the tangential contact force $F_t\nwarrow$ (friction, tension, or any other constraint forces), normal contact force $N\nearrow$, and gravitational force $F_g\swarrow$ due to the Earth.

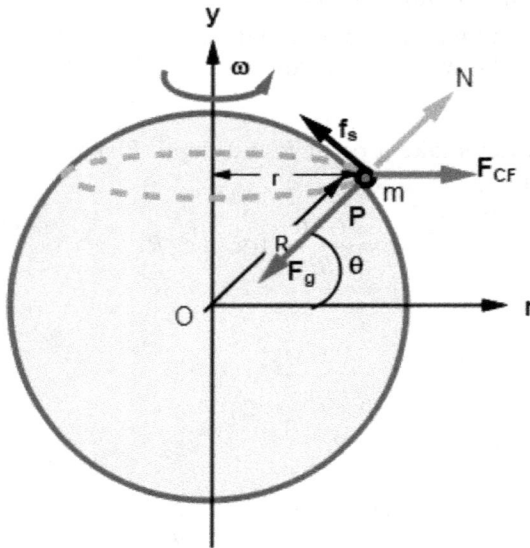

Using Newton's 2nd law in radial direction, we have

$$\sum F_r = N + F_{CF} \cos \theta - F_g = ma_r,$$

where $a_r = 0$, $F_{CF} = mr\omega^2$ and $F_g = \frac{GMm}{R^2}$.

Then, we have

$$N = \frac{GMm}{R^2} - mr\omega^2 \cos \theta$$

Substituting $\frac{GM}{R^2} = g_0$ (acceleration due to gravity at the equatorial plane), $r = R \cos \theta$ (because the object moves in a circle of radius r, but not R, due to the rotating Earth), we have

$$N = m(g_0 - R\omega^2 \cos^2 \theta)$$

Dividing both sides by m, we have

$$\frac{N}{m}(=g_{\text{eff}}, \text{say}) = g_0 - R\omega^2 \cos^2 \theta$$

We can call N the 'apparent weight' of the object. The normal reaction N is less than the real weight mg_0 by a factor $R\omega^2 \cos^2 \theta$ due to the rotation (angular velocity ω) and latitude θ of Earth.

N.B.: 1. When $\theta = 0$(at the equator), $N = m(g_0 - R\omega^2)$.

This means that the apparent weight is minimum at the equator. In other words, the effect of rotation is maximum at equatorial points.

2. When $\theta = 90°$ (at the poles), $N = mg_0$.

This means that the apparent weight is maximum, that is, equal to the weight of the object at the poles. In other words, there is no effect of rotation of Earth at the poles.

3. Since $g_{\text{eff}}(=\frac{N}{m}) = g_0 - R\omega^2 \cos^2 \theta$ and $g_0 = \frac{GM}{R^2}$, the value of g does not vary with rotation because $g = \frac{GM}{r^2}$, rather it is the value of $\frac{N}{m}$ or g_{eff} that varies with rotation and latitude.

Shape of the Earth: Generally we assume that the Earth is spherical for deducing the expressions of field, potential, energy etc. However, the Earth is not a perfect sphere, rather it is spheroidal. Due to the rotation of the Earth, the mass of the Earth centrifuges more towards the equator. Consequently the Earth is bulged at the equator. Hence, the equatorial radius is more than the polar radius by 21 km approximately; $R_e - R_p \simeq 21$ km. Since $g \propto \frac{1}{r^2}$, the value of g increases from equator to pole.

Non-uniformity of the Earth: We derived all the formulae assuming that the density of Earth is uniform. In fact, Earth has a non-uniform distribution of mass due to the presence of solid, liquid, and gas inside Earth's crust in the form of natural gases, minerals, oils, etc. Hence, the non-uniform distribution of mass affects the ideal variation of $|\vec{g}|$.

3.14 The motion of planets

Angular momentum of a planet: Kepler tells us that the planets move around the Sun in elliptical orbits. Newton's law of universal gravitation tells us that the Sun attracts the planets along the line of their separation.

Elliptical Orbit

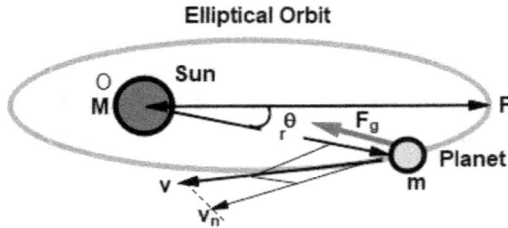

The gravitational pull of the Sun on the planet is

$$\overrightarrow{F_g} = -\frac{GMm}{r^2}\hat{r}$$

Since $\overrightarrow{r}$ and $\overrightarrow{F}$ are collinear, the torque of the force F about the Sun is zero. Symbolically, the torque due to gravity about the Sun $= |\overrightarrow{\tau_{gr}}|_s = |\overrightarrow{r} \times \overrightarrow{F_g}|$ $= rF \sin 180° = 0$.

Since the torque acting on the planet due to the Sun is

$$\overrightarrow{\tau} = \frac{d\overrightarrow{L}}{dt} = 0,$$

the angular momentum of the planet is

$$\overrightarrow{L} = m\overrightarrow{r} \times \overrightarrow{v} = \text{constant}$$

This tells us that:

The torque produced by gravitational force acting on the planet due to the Sun about the Sun is zero as the gravitational force passes through the center of the Sun. Hence, the angular momentum of the planet about the Sun remains constant. Please note that since the Sun is too massive, the center of mass of the Sun–planet system lies practically inside the Sun.

Example 25 Let the angular velocity of a planet relative to the Sun be $\overrightarrow{\omega}$. If its moment of inertia about the Sun is I, (a) find the expression for the angular momentum of the planet relative to the Sun, using the basic formula $\overrightarrow{L} = m\overrightarrow{r} \times \overrightarrow{v}$, and (b) prove that areal velocity of the planet is constant.

Solution

(a) The angular momentum of a planet about the Sun is

$$\vec{L} = m\vec{r} \times \vec{v} = mv_n\hat{k}$$

where v_n= component of the velocity $\vec{v}$ perpendicular to $\vec{r}$ as shown in the figure.

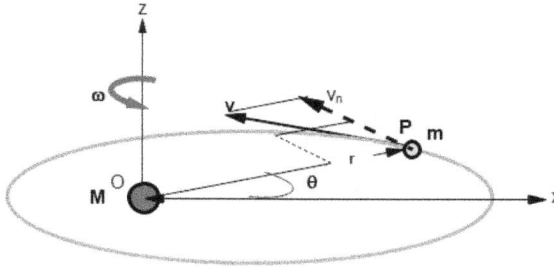

Then, we have

$$\vec{L} -= mv_\perp r\hat{k} \tag{3.63}$$

Substituting $v_\perp = r\omega$ in equation (3.63), we have

$$\vec{L} = mr^2\omega\hat{k} \tag{3.64}$$

Writing $\omega\hat{k} = \vec{\omega}$ and substituting $mr^2 = I$ (moment of inertia of the planet about Sun) in equation (3.64), we have

$$\vec{L} = I\vec{\omega}$$

(b) Put $\omega = \frac{d\theta}{dt}$ in the expression $L = mr^2\omega$ to obtain $L = mr^2\frac{d\theta}{dt}$. Then substitute $\frac{1}{2}r^2\frac{d\theta}{dt} = \frac{dA}{dt}$ to have $\frac{dA}{dt} = \frac{L}{2m}$. Since $L = C$, we have $\frac{dA}{dt} = $ constant.

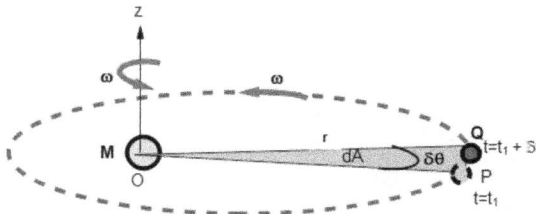

Kepler's 2nd law is the direct consequence of law of conservation of angular momentum about the Sun. The center of mass of the Sun–planet system lies very close to center of the Sun.

Energy of a planet: As explained earlier, gravitational field is a conservative field because the gravitational force is a central force because it obeys the inverse square law of distance. The gravitational potential energy between two particles of masses m_1 and m_2 separated by a distance r can be given by

$$U = -\frac{Gm_1m_2}{r},$$

where $m_1 = M$ and $m_2 = m$.

Then, the potential of the Sun–planet system is

$$U = -\frac{GMm}{r} \tag{3.65}$$

Since the Sun is practically very massive when compared to the planet, the kinetic energy of the Sun–planet system is mostly contributed by the planet. Hence, the kinetic energy of the Sun–planet system is given as

$$K = \frac{1}{2}mv^2 \tag{3.66}$$

Adding equations (3.65) and (3.66) the total mechanical energy of the Sun–planet system is

$$E = U + K = -\frac{GMm}{r} + \frac{1}{2}mv^2$$

Roughly, you can call E 'the total mechanical energy of the planet', but in a strict sense, it is the total mechanical energy of the Sun–planet system.

Example 26 A planet of mass m revolves around the Sun of mass M in an elliptical orbit. The minimum and maximum distances of the planet from the Sun are r_1 and r_2, respectively. Find the (a) velocities of the planet when it is at the minimum and maximum distances from the Sun and (b) total angular momentum of the Sun–planet system.
  Solution

(a) Let's assume that the speeds of the planet at the nearest and farthest points are v_1 and v_2, respectively.

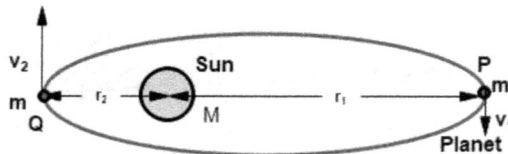

$L = C$: conserving the angular momentum of the planet between point 1 and 2, we have

$$L = mv_1 r_1 = mv_2 r_2 \tag{3.67}$$

$$\Rightarrow v_1 r_1 = v_2 r_2 \tag{3.68}$$

$E = C$: conserving total mechanical energy of the Sun–planet system between the portions 1 and 2, we have

$$E = \frac{1}{2}mv_1^2 - \frac{GMm}{r_1} = \frac{1}{2}mv_2^2 - \frac{GMm}{r_2}$$

$$\Rightarrow v_1^2 - v_2^2 = 2GM\left(\frac{1}{r_1} - \frac{1}{r_2}\right) \tag{3.69}$$

Solving equations (3.68) and (3.69), we have

$$v_1 = \sqrt{\frac{2GMr_2}{(r_1 + r_2)r_1}} \quad \text{and} \quad v_2 = \sqrt{\frac{2GMr_1}{(r_1 + r_2)r_2}} \quad \text{Ans.}$$

(b) Putting the obtained values of either v_1 or v_2 in equation (3.67), the total angular momentum is

$$L = m\sqrt{\frac{2GMr_1 r_2}{(r_1 + r_2)}} \quad \text{Ans.}$$

Since $E = $ constant, when KE increases, potential energy decreases and vice versa so that the sum of these two energies remains constant. When the planet is speeding up from its minimum speed, gravity is doing positive work and when the planet slows down attaining a maximum speed, gravity does negative work. As a whole, in a round trip, the net work done by gravity is zero.

3.15 Motion of planets and satellites in circular orbit

Let a planet (or satellite) of mass m move around the Sun (or planet) with a uniform speed v in a circular orbit of radius r. Let the mass and radii of the central body planet (say) be M and R, respectively.

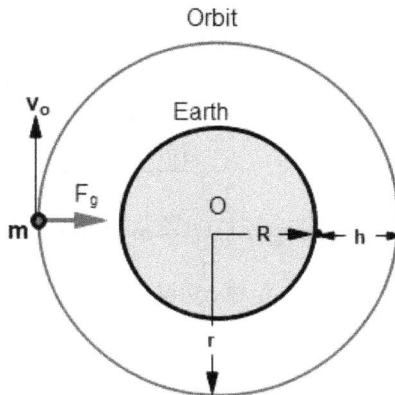

Orbital speed: Applying Newton's 2nd law on the satellite m, we have

$$F_g = ma,$$

where $F_g = \frac{GMm}{r^2}$ and $a = \frac{v^2}{r}$.

This gives

$$v = \sqrt{\frac{GM}{r}}$$

This is called orbital speed and does not depend on the mass of the satellite (or any revolving body).

Angular speed: The angular speed of the satellite is

$$\omega = \frac{v}{r}, \quad \text{where } v = \sqrt{\frac{GM}{r}}.$$

This gives $\omega = \sqrt{\frac{GM}{r^3}}$.

Period of revolution: The time period of revolution in uniform circular motion is given as

$$T = \frac{2\pi}{\omega}, \quad \text{where } \omega = \sqrt{\frac{GM}{r^3}}.$$

Then, we have

$$T = 2\pi\sqrt{\frac{r^3}{GM}}$$

Hence, the square of the time period is directly proportional to the cube of radius of the circular orbits.

Angular momentum: The angular momentum of the satellite relative to the planet is

$$L = mvr, \quad \text{where } v = \sqrt{\frac{GM}{r}}.$$

Then, we have

$$L = m\sqrt{GMr}$$

Gravitational potential energy: The gravitational potential energy between M and m, that is, the planet–satellite system (or Sun–planet system), can be given as

$$U = -\frac{GMm}{r}$$

Kinetic energy: The kinetic energy of $M + m$ system is given as

$$K = \frac{1}{2}mv^2,$$

where

$$v = \sqrt{\frac{GMm}{2r}}$$

Total mechanical energy: Summing up kinetic and potential energies of the $M + m$ system, we have

$$E = U + K,$$

where $U = -\frac{GMm}{r}$ and $K = \frac{GMm}{2r}$

$$\Rightarrow E = -\frac{GMm}{r} + \frac{GMm}{2r}$$

$$\Rightarrow E = -\frac{GMm}{2r}$$

Energy graphs: The expressions for kinetic and potential energy are

$$K = \frac{GMm}{2r} = |E| = \frac{|U|}{2}$$

This tells us that all energies vary with r hyperbolically. Hence, U–r, K–r, and E–r graphs are rectangular hyperbolas. K is always positive, whereas U and E are negative.

If a satellite has an orbit of radius r, its total energy must be negative. If $E \geqslant 0$, the satellite will no longer be a bounded particle. It will be a free particle.

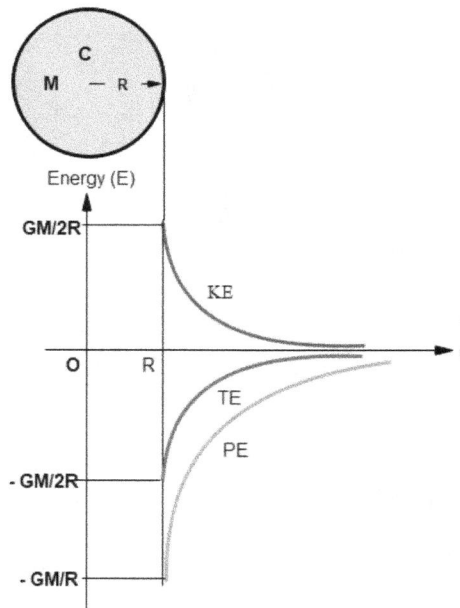

We note the following points from the foregoing discussion:

1. Orbital speed $v \propto \dfrac{1}{\sqrt{r}}$
2. Angular velocity $\omega \propto \dfrac{1}{r^{3/2}}$
3. Time period $T \propto r^{3/2}$
4. Angular momentum $L \propto \sqrt{r}$
5. Potential energy $|U| \propto \dfrac{1}{r}$
6. Kinetic energy $K \propto \dfrac{1}{r}$
7. Total energy $|E| \propto \dfrac{1}{r}$
8. $|U| = 2\,|E| = 2K$
9. Total energy of $M + m$ system is negative so as to make the system bounded.

Example 27 A satellite of mass m is lifted from the Earth's surface to a height h and put in the circular orbit. If the mass and radius of the Earth are M and R, respectively, find the total work done by the external agent in this process.

Solution

The work done by the external agent in shifting from point 1 to 2 is

$$W_1 = \Delta U = U_2 - U_1,$$

where

$U_1 = -\dfrac{GMm}{R}$ and $U_2 = -\dfrac{GMm}{R+h}$

Then

$$W_1 = \frac{GMmh}{R(R + h)} \tag{3.70}$$

By putting the satellite in the circular orbit of radius $r = R + h$, work done by the external agent is

$$W_2 = \Delta K = K$$

where the KE of the satellite is given by

$$K = \frac{GMm}{2(R + h)}$$

Then, we have

$$W_2 = \frac{GMm}{2(R + h)} \tag{3.71}$$

Adding equations (3.70) and (3.71) the total work done by the external agent is

$$W = W_1 + W_2 = \frac{GMm}{R + h}\left(\frac{h}{R} + \frac{1}{2}\right) \text{ Ans.}$$

N.B.:

1. In the above example, the PE, KE, and TE are given as follows:

$$W = \frac{GMm}{R + h}$$

$$U = -\frac{GMm}{R + h},$$

$$E = -\frac{GMm}{2(R + h)}.$$

2. Since $\frac{W_1}{W_2} = \frac{h}{2R}$, when h increases $\frac{W_1}{W_2}$ increases.

3.16 Escape velocity

Case 1 Particle is projected from an isolated system to infinity: Suppose a particle of mass m is placed at any point P in the gravitational field of a system of particles (it may be discrete or continuous). You are asked to find the minimum velocity that should be given to the particle so that it will not come back to the system A. In this case, in the gravitational potential energy U of interaction between all possible pairs, each pair must contain the particle m which is given as

$$U = -Gm\sum\frac{m_i}{r_i},$$

where m_i and r_i are the mass of ith particle of the system A and distance between m and m_i.

Let's project the particle with a velocity v. Now the particle has kinetic energy

$$K = \frac{1}{2}mv^2$$

Then we add the kinetic energy $K(=\frac{1}{2}mv^2)$ with the potential energy U to obtain the total mechanical energy E_1 of the particle.

The total mechanical energy is given as

$$E_1 = U_1 + K_1 = -GM\sum\frac{m_i}{r_i} + \frac{1}{2}mv^2 \tag{3.72}$$

After getting the total mechanical energy of the particle at initial (given) point P, we have to equate this with the total mechanical energy at the final point Q, say. If there are not other gravitating bodies except the given system of particles, the particle P must go to infinity in order not to return. Hence, the point Q is at infinity. As the distance between each element (or particle) of the system and infinity is infinitely large, the gravitational potential energy between the particle and system is zero at final position of the particle. This means, $U_2 = 0$. For a minimum speed v of projection, the final speed, that is, speed at infinity, is zero. Hence, $K_2 = 0$. Then,

$$E_2 = U_2 + K_2 = 0 \tag{3.73}$$

According to the conservation of mechanical energy, we have

$$E_2 = E_1 \qquad (3.74)$$

Then using equations (3.72)–(3.74), we have

$$v = \sqrt{2G\sum \frac{m_i}{r_i}}$$

The minimum velocity of projection so that the particle will escape from the system is called escape velocity and is denoted by v_e. If the speed of projection $v \geqslant v_e$, the particle will never return back to the initial position.

If the system is comprised of continuously distributed particles, you can find the potential V of the system at P. Then find the potential energy $U = mV$, where $V = -G\int \frac{dm}{r}$.

Then find

$$E = \frac{1}{2}mv^2 + mV$$

Finally equating it to zero (E at infinity), we have
$v = \sqrt{-2V}$, where V is a negative quantity.

The above equation is also valid for a discrete particle system.

Case 2 Particle is projected from a system A to another system: Sometimes we have two systems A and B. A particle of mass m is projected from A so that it will not return to A again. This means that the particle will escape from C. In this case, let's see how to find the v_{escape}.

First of all we can find the potential energy possessed by the particle m at its initial position due to both systems A and C. We can get it if we find the total potential of the systems A and B at the point P. Let it be V_1.

Then,

$$U_1 = mV_1$$

Now add its kinetic energy
$K_1 = \frac{1}{2}mv_1^2$ with U_1 to obtain total energy

$$E_1 = \frac{1}{2}mv_1^2 + mV_1 \qquad (3.75)$$

As the particle is projected from system A to system B, you cannot take the final position of the particle at infinity. Then, what is the final position of the particle? You may think that it should be at some point Q on system B. Well, let's analyse. Let's take the point Q as the final point of the particle m. We can find the sum of the potential of A and B at Q be V_2. Then the gravitational potential energy possessed by the particle at Q is $U_2 = mV_2$. Now we have to find the final kinetic energy of the particle, that is, KE of the particle at Q, but we can find it directly. Hence, we need to think more carefully to find a point where we can find the KE of the particle. So, how do we find the final point, say, C? It must be somewhere in between P and Q because the particle moves in a straight line from P to Q. To find the exact location of C, we need to analyse the effect of gravitational field of A and B on the particle. At P, the system A pulls the particle towards the left and system B pulls the particle towards the right. Since A is nearer to P, we expect the gravitational field of A to dominate the gravitational field of B. Hence, there must be point C in between P and Q, where the gravitational effects of both A and B nullify each other. In other words, the net force acting on particle at C will be zero. This means that the force acting on the particle points towards A until it crosses the point C. As a result, the particle goes on slowing down from the point P to C. If it just passes through point C with negligible or zero velocity, the dominating gravity of system B at the right-hand side of the point C will pull the particle towards B. Hence, the desired (final) point C must be a force-free point. At C, no net force acts on the particle. In other words, gravitational field intensity is zero at C. After equating the field strengths of A and B, find the location of the point C. Then find the total potential V' of the system $(A + B)$ at C and then find the final potential energy $U_2 = mV_C$. Setting $v = 0$ at C, we have $K = 0$.

Hence, we have $E_2 = mV_C$.

Now equating E_1 and E_2, we have $mV_C = mV_1 + \frac{1}{2}mv^2$.

This gives $v = \sqrt{2(V_C - V_1)}$.

More generally, escape velocity is the minimum velocity required for a particle to escape from system A to system B, given as

$$v = \sqrt{2(V_C - V_1)},$$

where $V_1 =$ potential at point A and $V_C =$ potential at C where field strength is zero. For an isolated system, C is at infinity; $V_C = 0$. Hence, $v = \sqrt{-2V_1}$.

Example 28 Find the escape velocity of a particle that is situated at a radial distance r above the Earth's surface. Assume M as mass of the Earth.

Solution

At the point P the gravitational potential energy of m is

$$U_1 = -\frac{GMm}{r}$$

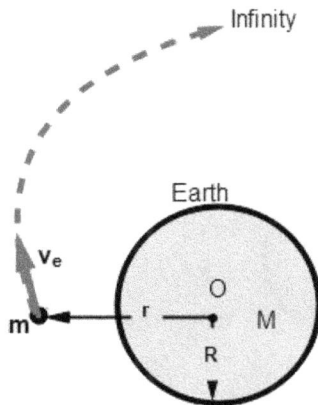

If we give a velocity v, the KE of the particle is

$$K_1 = \frac{1}{2}mv^2$$

Then, the total mechanical energy is

$$E_1 = U_1 + K_1 = -\frac{GMm}{r} + \frac{1}{2}mv^2 \tag{3.76}$$

Since the particle escapes to infinity, for minimum v the total mechanical energy at infinity is

$$E_2 = U_2 + K_2 = 0 \tag{3.77}$$

Equating E_1 and E_2, and simplifying the factors we have

$$v = \sqrt{\frac{2GM}{r}}$$

Since the escape velocity is

$$v_{\text{escape}} = \sqrt{\frac{2GM}{r}} \left(\text{but not always} \sqrt{\frac{2GM}{R}} \right)$$

and the orbital velocity is

$$v_{\text{orbital}} = \sqrt{\frac{GM}{r}},$$

we can write $v_{\text{escape}} = \sqrt{2}\,v_{\text{orbital}}$ for a given point orbit.

N.B.: If we project the particle from the center of the Earth along a smooth diametrical chute, the escape speed will be $\sqrt{\frac{3GM}{R}}$.

Example 29 Two planets A and B of masses $2M$ and $\frac{M}{2}$, respectively, each of radius R are separated by a distance $4R$. A body is thrown from point 1 of planet A such that it goes to (a) the planet B and (b) infinity. Find the escape speed of the body in each case.

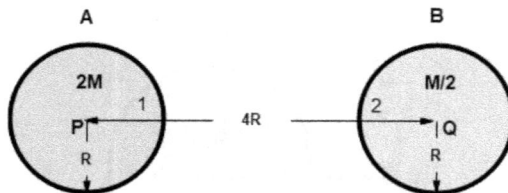

Solution

(a) Referring to the aforementioned theory, we can directly use the formula

$$v = \sqrt{2(V_C - V_1)}, \tag{3.78}$$

where V_1 and V_C are the potential at the points 1 and C, respectively.

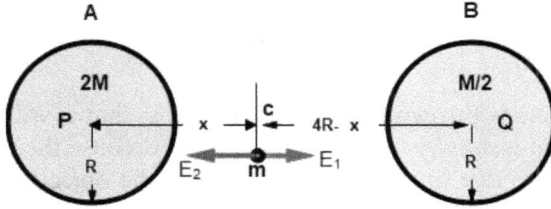

The potential V_1 is given as

$$V_1 = -\frac{G(2M)}{R} - \frac{G\left(\frac{M}{2}\right)}{3R}$$

$$\Rightarrow V_1 = -\frac{13GM}{6R} \tag{3.79}$$

To find the critical (null or zero-force) point C, we can equate the field of both planets A and B at C. So, we can write

$$E_A = E_B, \tag{3.80}$$

where the field of

$$E_A = \frac{G(2M)}{x^2} \tag{3.81}$$

$$E_B = \frac{G\left(\frac{M}{2}\right)}{(4R - x)^2} \tag{3.82}$$

Using the last three equations, we have

$$x = \frac{8R}{3} \tag{3.83}$$

Then, the potential at C is

$$V_C = -\frac{G(2M)}{x} - \frac{G\left(\frac{M}{2}\right)}{4R - x} \tag{3.84}$$

Using the last two equations, we have

$$V_C = -\frac{9GM}{8R} \tag{3.85}$$

Finally putting V_1 from equation (3.79) and V_C from equation (3.98) in equation (3.78) and simplifying the factors, we have

$$v = \frac{5}{2}\sqrt{\frac{GM}{3R}} \quad \text{Ans.}$$

(b) If the particle is thrown from the planet A so that it will go to infinity, the critical point is *infinity*. So, putting $V_C = 0$ (because the potential at infinity is always zero due to infinite distance from the objects) in equation (3.78), the escape speed is

$$v = \sqrt{2(V_C - V_1)} = \sqrt{2(-V_1)} \tag{3.86}$$

Using equations (3.79) and (3.86), we have

$$v_e = \sqrt{\frac{13GM}{3R}} \quad \text{Ans.}$$

3.17 Orbital velocity and nature of orbits of a satellite

When we project a body at any point P above any planet, Earth, say, its total energy can be given as

$$E = -\frac{GMm}{r} + \frac{1}{2}mv^2$$

For a particular position $\vec{r}$ of the satellite relative to the planet,

$$U = -\frac{GMm}{r}$$

This means that E can be varied by imparting different velocities to the body. We know that the body moves with a velocity

$$v = \sqrt{\frac{GM}{r}}$$

to stay in a circular orbit. Its corresponding total energy is given as

$$E_0 = -\frac{GMm}{2r}$$

If $v < \sqrt{\frac{GM}{r}} (=v_0)$, the body falls on the Earth following parabolic (projectile) path.

3-72

If $v > \sqrt{\frac{GM}{r}}\,(=v_0)$ and $v < \sqrt{\frac{2GM}{r}}\,(=v_e)$, the body will revolve in an elliptical orbit about the planet. In other words, $-\frac{GMm}{2r} < E < 0$.

If $v > v_e(=\sqrt{\frac{2GM}{r}})$, the body escapes from the planet following a hyperbolic path. Then, the orbit will be an unbound orbit. In this case, $E > 0$.

If $v = v_e$, the body escapes from the planet following a parabolic path. Then, the orbit will be an unbound orbit because $E = 0$.

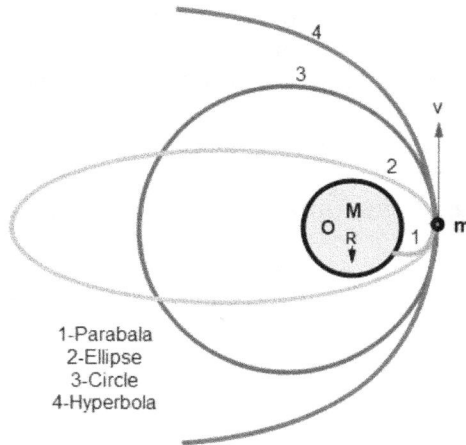

1-Parabala
2-Ellipse
3-Circle
4-Hyperbola

We note the following points from the above discussions:

1. If $0 < v < v_0$, orbit is bound and parabolic; $E < -\frac{GMm}{2r}$.
2. If $v = v_0$, orbit is bound and circular; $E = -\frac{GMm}{2r}$.
3. If $v_0 < v < v_e$, orbit is bound and elliptical; $0 > E > -\frac{GMm}{2r}$.
4. If $v = v_e$, orbit is unbound and parabolic; $E = 0$.
5. If $v > v_e$, orbit is unbound and hyperbolic; $E > 0$.
6. When $E < 0$, the orbit is called bound (or the particle is a bounded one), and when $E \geqslant 0$, the orbit is unbound (or the particle is a free particle).

3.18 Weightlessness

As you know, the weight of a body of mass m at any point is given as $W = mg$, where g = gravitational acceleration due to the Earth at that point. Hence, the word 'weightlessness' does not mean that the weight is lost or reduced. Since g cannot be zero in Earth's gravity, the weight of the body can never be zero in Earth's gravitational field. Then the 'weightlessness' bears a completely different meaning, which is explained as follows.

Let's imagine that you are standing on a weighing machine kept on an elevator. When the elevator moves with a downward acceleration a, the normal reaction between you and the weighing machine (spring balance, say) is given by

$$N = m(g - a),$$

where m = your mass.

This tells us that the spring balance reads less than your weight. Remember that the spring balance records 'N' but not necessarily the weight 'mg'. If the elevator is allowed to fall freely, $a = g$.

Then

$$N = m(g - g) = 0$$

This means that the spring balance will show no reading at all. Physically you feel as if you are floating in space having no weight. This happens because you have lost contact with the surface of the weighing machine. Since there is no pressing force, the deformation (compression) of the spring is zero and hence the spring balance pointer records a zero reading.

1. Any 'freely falling' body possesses 'weightlessness' by losing contact with surrounding objects.
2. The word 'weightlessness' means zero contact forces but not zero weight of the object.
3. The linear dimensions of the freely falling body must be negligible compared to the radius of the Earth so that no significant variation of 'g' occurs in the body.

Example 30

(a) Explain the fact that both an orbiting satellite and the astronaut experience weightlessness.

(b) Can a bucket of water whirled in a vertical plane be called 'freely falling'? If yes, does it possess weightlessness? Explain.

Solution

(a) Since the size of the satellite is negligible compared to the radius of the Earth, the Earth pulls both astronaut and satellite independently with the same acceleration.

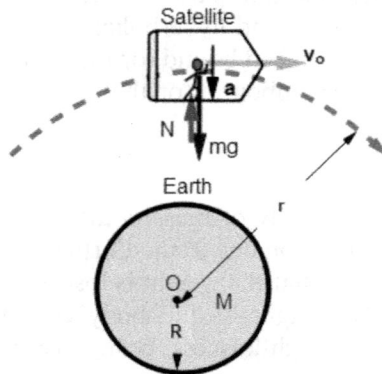

This satisfies the condition that the normal reaction between the satellite and astronaut is zero. Hence, the astronaut and satellite experience weightlessness.

(b) Generally, no, because $\vec{a} \neq \vec{g}$. If $v = \sqrt{Rg}$, at the highest point, where R = radius of the circular path, $\vec{a} = \vec{g}$. Then the water falls freely experiencing weightlessness.

Since no other forces are acting on the satellite and astronaut except the gravitational force due to the Earth, disregarding the other forces, we can say that both objects are in a state of 'free fall'. The words 'free fall' may mean that the object should approach the Earth's surface under Earth's gravity reducing the distance of separation between object and the Earth, but the satellite does not seem to reduce its distance from the Earth as it moves in a circular orbit. Then, how can you say that the satellite is in a free fall? To answer this question, let's resort to the idea given by Newton himself as follows:

If there were no gravity, the satellite would have moved in a straight line AP obeying Newton's 1st law, but due to the effect of Earth's gravity the satellite moves in a circle, falling through a radial distance PQ. In this way we can say that the satellite falls freely under the gravity of the Earth.

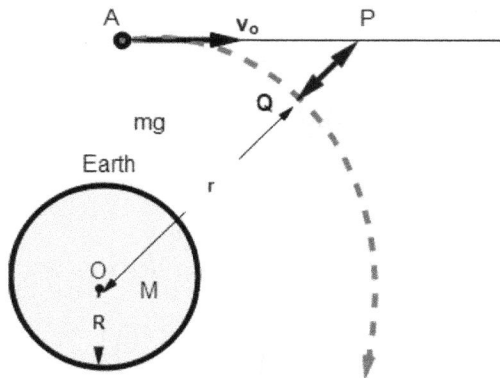

3.19 Earth as an inertial reference frame

We have discussed the effect of rotation (spin) of the Earth on the apparent weight of an object on Earth's surface, ignoring its orbital motion. Let's now ignore the spin of the Earth and take its orbital motion into account. As we know, the Earth revolves around the Sun in an elliptical orbit. For the sake of simplicity let's assume that the Earth revolves in a circular orbit of radius r.

The gravitational field strength of the Sun at the center O of the Earth is

$$E_O = \frac{GM_s}{r^2}$$

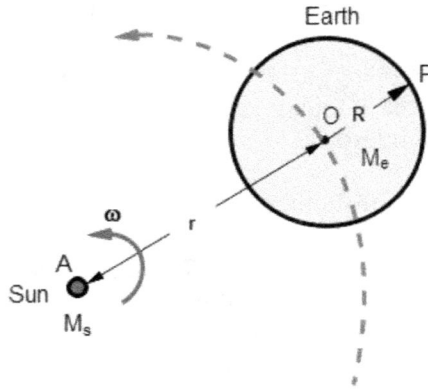

If we take a point P on the Earth's surface, the gravitational field strength of the Sun at P is

$$E_P = \frac{GM_s}{(r + R)^2}$$

The difference in field intensities at O and P is

$$\Delta E = E_O - E_P$$

$$= \frac{GM_s}{r^2} - \frac{GM_s}{(r + r)^2}$$

$$= \frac{GM_s}{r^2}\left[1 - \left(1 + \frac{R}{r}\right)^{-2}\right]$$

$$= \frac{GM_s}{r^2}\left[1 - \left(1 - \frac{2R}{r}\cdots\right)\right]$$

The higher powers of $(\frac{R}{r})$ are neglected as $\frac{R}{r} \ll 1$.

$$\Rightarrow \Delta E = 2GM_s\frac{R}{r^3}$$

Since $R \ll r$, we have $\Delta E \cong 0$.

This means that the variation of E due to the Sun on the Earth's surface is negligible for most of the practical experiments. In other words, almost all points of the Earth accelerate towards the Sun practically with the same acceleration. Hence, the effect of orbital motion of the Earth on the calculation of weight of an object is also negligible for most practical cases.

If we compare ΔE with $g = 9.8$ m s^{-2}, we have

$$\frac{\Delta E}{g} = \frac{2GM_s \frac{R}{r^3}}{\frac{GM_e}{R^2}} = 2\left(\frac{R}{r}\right)^3 \frac{M_s}{M_e} = 10^{-6}$$

This tells us that the fractional change (error) in weight of the body due to orbital motion of the Earth is one part of a million, that is, too small for any gross calculation. However, for very small forces due to two gravitating bodies, this effect can also be taken into account. Thus, the weight of an object does not change significantly due to the centripetal acceleration of the Earth. Hence, the Earth can be practically treated as inertial reference frame. Similarly, the effect of other heavenly bodies on the Earth can be neglected.

N.B.:

1. Since the square of period of revolution $T(=3.15 \times 10^7$s$)$ of Earth around the Sun is much greater than the average distance $r(=1.5 \times 10^{11}$ m$)$ between Sun and Earth, the centripetal acceleration $a(=\frac{4\pi^2 r}{T^2})$ is negligible and small compared to $g(=9.8$ ms$^{-2})$. Hence, the orbital motion of the Earth has a negligible effect on the apparent weight of an object on the Earth's surface.

2. Since $r \gg R$ (radius of the Earth), almost all portions of the Earth accelerate towards the Sun with the same acceleration (rate) approximately. Thus, the percentage error of weight measurement of an object at any two places of the Earth is negligible due to the orbital motion of the Earth. Hence, the Earth can be treated as an inertial reference frame.

Field and force

Problem 1 An isolated straight heavy rod AB of mass m is kept in a gravity-free space. Find the ratio of gravitational fields at the points P and Q. Assume that $AR = RB = a$.

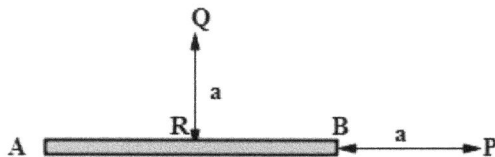

Solution
Referring to Example 5, the value of gravitational field intensity at P is given as

$$\Rightarrow E_P = G\lambda\left(\frac{1}{a} - \frac{1}{x}\right) \tag{3.87}$$

Putting $y = a$ and $x = 2a$ in equation (3.87), we have

$$\Rightarrow E_P = G\lambda\left(\frac{1}{a} - \frac{1}{3a}\right) = \frac{2G\lambda}{3a}$$

Putting $\lambda = m/2a$, we have

$$\Rightarrow E_P = \frac{Gm}{3a^2} \tag{3.88}$$

The gravitational field intensity at Q is

$$E_Q = \frac{G\lambda}{y}(\cos\theta + \cos\theta)$$

$$= \frac{2G\lambda}{y}\cos\theta$$

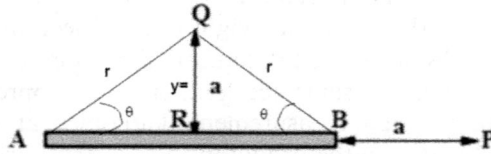

$$= \frac{2G\lambda}{y}\left\{\frac{\left(\frac{\ell}{2}\right)}{r}\right\}$$

$$= \frac{G(\ell\lambda)}{yr}$$

$$= \frac{Gm}{y\sqrt{\frac{\ell^2}{4} + y^2}}(\because \lambda = ml)$$

$$E_Q = \frac{2Gm}{y\sqrt{\ell^2 + 4y^2}} \tag{3.89}$$

Putting $y = a$ and $l = 2a$ in equation (3.89), we have

$$E_Q = \frac{2Gm}{a\sqrt{4a^2 + 4a^2}}$$

$$\Rightarrow E_Q = \frac{Gm}{\sqrt{2}\,a^2} \tag{3.90}$$

From equations (3.88) and (3.90), we have

Then, $\frac{E_P}{E_Q} = \frac{\sqrt{2}}{3}$ Ans.

Problem 2 Two parallel semi-infinite rods are separated by a distance $2R$, where $R =$ radius of the semicircular rod joining the straight rods as shown in the figure. Find the gravitational field intensity at the center O of the semicircle.

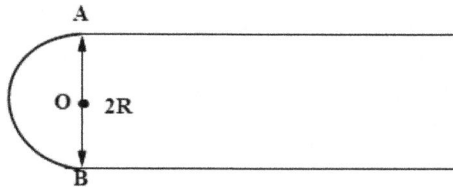

Solution

Let's consider the x-axis parallel to the wires 1 and 3 fixing the origin at O. Then the field due to the wires 1 and 2 along the y-axis will cancel each other.

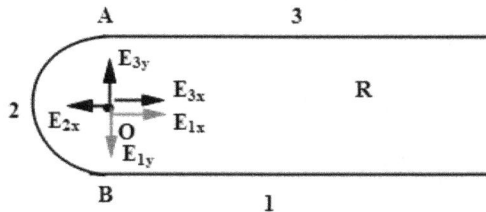

The net field due to wires 1 and 3 is

$$E = E_{1x} + E_{3x} = 2E_{1x}(\because E_{1x} = E_{3x}),$$

where

$$E_{1x} = \frac{G\lambda}{R}\,|\sin 90 - \sin 0| = \frac{G\lambda}{R}$$

Then, we have

$$\overrightarrow{E} = \frac{2G\lambda}{R}\hat{i}$$

The field due to the semicircle is

$$\overrightarrow{E'} = -\frac{G\lambda}{R}\frac{\sin\frac{\pi}{2}}{\frac{\pi}{2}}\hat{i} = -\frac{2G\lambda}{\pi R}\hat{i}$$

Then, the net field due to the given arrangement is

$$\overrightarrow{E_{net}} = \overrightarrow{E} + \overrightarrow{E'}$$

$$= \frac{2G\lambda}{R}\hat{i} - \frac{2G\lambda}{\pi R}\hat{i}$$

$$= \frac{2G\lambda}{R}\left(1 - \frac{1}{\pi}\right)\hat{i}$$

Problem 3 A uniform sphere and a uniform ring each of mass M and radius R are kept such that the distance between their centers is equal to a. Find the ratio of the fields contributed by the ring and sphere at the (a) mid-point P of the line joining their centers and (b) point Q situated at a distance $R/2$ from the center of the sphere towards the center of the ring. Put $a = 2R$.

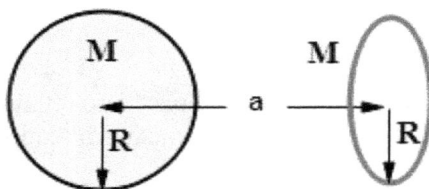

Solution

(a) Let E_1 and E_2 be the fields due to the sphere and ring at a point P, respectively.

The field due to the sphere 1 at P is

$$\overrightarrow{E_1} = -\frac{GM}{x^2}\hat{i} \tag{3.91}$$

The field due to ring 2 at P is

$$\overrightarrow{E_2} = +\frac{GM(a-x)}{\{R^2 + (a-x)^2\}^{\frac{3}{2}}}\hat{i} \tag{3.92}$$

As per the given condition, we have

$$\eta E_1 = E_2 \tag{3.93}$$

Using equations (3.91), (3.92) and (3.93)

$$\eta\frac{GM}{x^2} = \frac{GM(a-x)}{\{R^2 + (a-x)^2\}^{\frac{3}{2}}}$$

$$\Rightarrow \eta = \frac{(a-x)x^2}{\{R^2 + (a-x)^2\}^{\frac{3}{2}}}$$

where $x = \frac{a}{2} = R$

Putting $a = 2R$ and $x = R$,

we have $\eta = \frac{1}{2\sqrt{2}}$.

(b) For the point Q, $x < R$. Then, we have $\eta = \frac{E_2}{E_1}$, where $E_1 = \frac{GM}{R^3}x$ and

$$E_2 = \frac{GM(a-x)}{\{R^2 + (a-x)^2\}^{\frac{3}{2}}}$$

$$\Rightarrow \eta = \frac{R^3(a-x)}{x\{R^2 + (a-x)^2\}^{\frac{3}{2}}}$$

$$= \frac{R^3\left(2R - \frac{R}{2}\right)}{(R/2)\left\{R^2 + \left(2R - \frac{R}{2}\right)^2\right\}^{\frac{3}{2}}}$$

$$= \frac{3}{\left(\frac{13}{4}\right)^{\frac{3}{2}}} = \frac{24}{13^{\frac{3}{2}}}$$

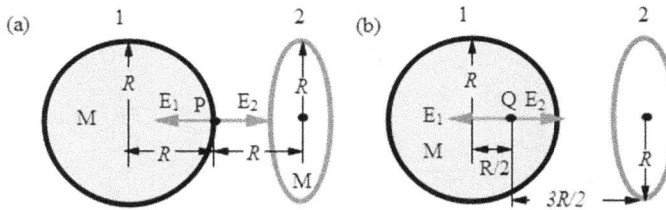

(a)   1   2   (b)   1   2

Problem 4 A uniform rod of linear mass density λ is aligned along the axis of a uniform ring of mass M and radius R. Find the force acting on the ring due to the rod if the rod is (a) finite having length l and (b) semi-infinite.

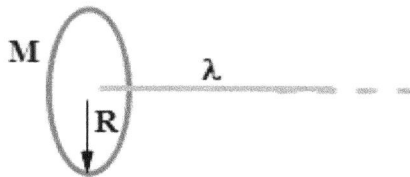

Solution

(a) The gravitational field of the ring at a distance x is

$$E = \frac{GMx}{(R^2 + x^2)^{\frac{3}{2}}} \tag{3.94}$$

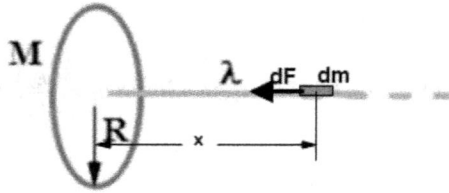

Then the force acting on an element dm of the rod is

$$dF = Edm,$$

where $dm = \lambda dx$

$$\Rightarrow dF = E\lambda dx \qquad (3.95)$$

The net force acting on the rod is
$F = \int dF = \int E\lambda dx$
Since $\lambda = \text{constant}$,

$$F = \lambda \int_0^\ell Edx \qquad (3.96)$$

Using equations (3.94) and (3.96),

$$F = \lambda \int_0^\ell \frac{GMx\, dx}{(R^2 + x^2)^{\frac{3}{2}}}$$

$$= \frac{Gm\lambda}{2} \int_0^\ell \frac{2x\, dx}{(R^2 + x^2)^{\frac{3}{2}}} \qquad (3.97)$$

Let $R^2 + x^2 = u^2$.
Then, $2x\, dx = 2u\, du$.
So, the integration will be

$$I = \int \frac{2x\, dx}{(R^2 + x^2)^{\frac{3}{2}}}$$

$$= \int \frac{2u\, du}{(u^2)^{\frac{3}{2}}}$$

$$= 2\int \frac{du}{u^2} = 2\left(\frac{u^{-2+1}}{-2+1}\right)$$

$$= \frac{-2}{u} = \frac{-2}{\sqrt{R^2 + x^2}} \qquad (3.98)$$

Using the result of the integration from equations (3.98) in (3.97),

$$F = \frac{GM\lambda}{2}\left(\frac{-2}{\sqrt{R^2 + x^2}}\right)\Bigg|_0^\ell$$

3-82

$$= -GM\lambda \left(\frac{1}{\sqrt{R^2 + \ell^2}} - \frac{1}{R} \right)$$

$$= \frac{GM\lambda \left(\sqrt{R^2 + \ell^2} - R \right)}{R\sqrt{R^2 + \ell^2}} \tag{3.99}$$

Putting $\lambda\ell = m$, we have a rightward force

$$F = \frac{GMm\left(\sqrt{R^2 + \ell^2} - R \right)}{R\ell\sqrt{R^2 + \ell^2}} \text{Ans.}$$

(b) If the rod is very long (semi-infinite) ($l \to \infty$), we have

$$F = \lim_{\ell \to \infty} GM\lambda \left(\frac{1}{R} - \frac{1}{\sqrt{R^2 + \ell^2}} \right)$$

$$= \frac{GM\lambda}{R} \text{ towards the right. Ans.}$$

Problem 5 Two perpendicular rods each of mass m and length $L = 2a$ are kept on a horizontal surface. Find the gravitational field intensity at a point P as shown in the figure.

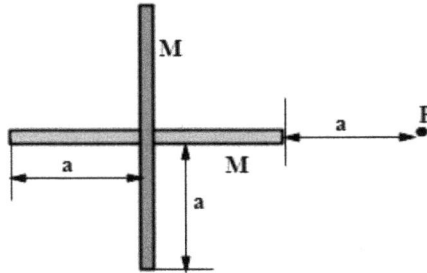

Solution

Referring to Problem 1, the gravitational field due to an element dm of the rod 1 at P is given as

$$dE_1 = \frac{G\,dm}{x^2}$$

The net field at P is

$$E_1 = \int dE_1 = \int G\frac{dm}{x^2},$$

where $dm = \frac{m}{\ell}dx$

$$\Rightarrow E_1 = \frac{Gm}{\ell} \left| \int_a^{3a} \frac{dx}{x^2} \right|$$

$$= \frac{Gm}{\ell} \left| \frac{1}{x} \right|_a^{3a}$$

$$\frac{Gm}{\ell}\left|\frac{1}{a} - \frac{1}{3a}\right| = \frac{2Gm}{3\ell a}$$

$$\Rightarrow E_1 = -\frac{2Gm}{3\ell a}\hat{i} \tag{3.100}$$

The gravitational field intensity due to rod 2 at P

$$E_2 = \frac{G\lambda}{2a}(\cos\theta + \cos\theta),$$

where $\lambda = \frac{M}{\ell}$ and $\cos\theta = \frac{a}{\sqrt{a^2 + (2a)^2}}$

$$\Rightarrow E_2 = \frac{GM}{2\ell a}\frac{a}{\sqrt{a^2 + 4a^2}}$$

$$\Rightarrow \overrightarrow{E_2} = -\frac{GM}{2\sqrt{5}\ell a}\hat{i} \tag{3.101}$$

The net field at P is

$$\overrightarrow{E} = \overrightarrow{E_1} + \overrightarrow{E_2} \tag{3.102}$$

Using equations (3.100)–(3.102),
$\overrightarrow{E} = -\frac{GM}{3\ell a}\hat{i} - \frac{GM}{2\sqrt{5}\ell a}\hat{i}$, where $\ell = 2a$

$$\Rightarrow \overrightarrow{E} = -\frac{GM}{2a^2}\left(\frac{1}{3} + \frac{1}{2\sqrt{5}}\right)\hat{i} \text{ Ans.}$$

Problem 6 A uniform rod is kept horizontal. The gravitational field at a point O is E'. Let the rod be bent into the shape of an arc subtending an arc θ at the center O of curvature. Now, the gravitational field at O will be E. (a) Find the value of E/E'. (b) Evaluate the value of E/E' for semicircular arc. Assume $AB = BC$.

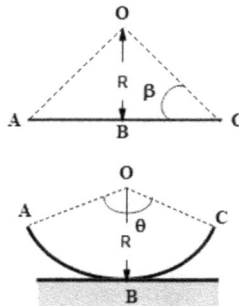

Solution

(a) The gravitational field due to the arc at O is

$$E = \frac{2G\lambda}{R}\frac{\sin\frac{\theta}{2}}{\theta} \tag{3.103}$$

If the arc is made a straight rod, its length $\ell = R\theta$, then the new gravitational field at P due to the rod is

$$E' = \frac{G\lambda}{R}(2\cos\phi),$$

where $\cos\phi = \dfrac{\frac{\ell}{2}}{\sqrt{\frac{\ell^2}{4}+R^2}} = \dfrac{\ell}{\sqrt{\ell^2+4R^2}}$

$$\Rightarrow E' = \frac{2G\lambda\ell}{R\sqrt{\ell^2+4R^2}}, \text{ where } \ell = R\theta$$

$$\Rightarrow E' = \frac{2G\lambda\theta}{R\sqrt{\theta^2+4}} \tag{3.104}$$

Using equations (3.103) and (3.104), the ratio of fields is

$$\frac{E}{E'} = \frac{\frac{2G\lambda\sin\frac{\theta}{2}}{R\theta}}{\frac{2G\lambda\theta}{R\sqrt{\theta^2+4}}}$$

$$= \frac{\sqrt{\theta^2+4}\,\sin\frac{\theta}{2}}{\theta^2} \text{ Ans.}$$

(b) For a semicircle $\theta = \pi$, then we have

$$\frac{T}{T'} = \frac{\sqrt{\pi^2+4}}{\pi^2} \text{ Ans.}$$

Problem 7 A uniform rod BC of length a is kept perpendicular to an infinite rod of linear mass density λ as shown in the figure. Find the gravitational force of interaction between the rods.

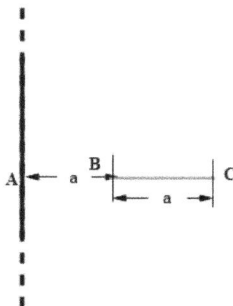

Solution

The force acting on an element dm of the rod due to the infinite rod is

$$dF = (dm)E,$$

where

$$E = \frac{2G\lambda}{x} \text{ and } dm = \frac{m}{\ell}dx$$

$$\Rightarrow dF = \frac{2G\lambda m}{\ell}\frac{dx}{x}$$

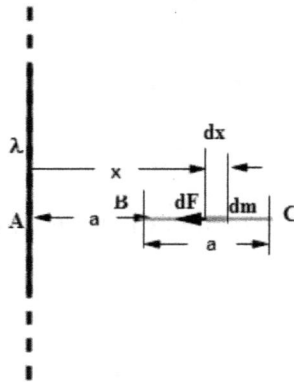

Integrating, the total force acting on the rod is

$$F = \int dF$$

$$= \frac{2G\lambda m}{\ell}\int_{x=a}^{x=2a}\frac{dx}{x}$$

$$= \frac{2G\lambda m}{\ell}\ln x\Big|_{a}^{2a}$$

$$= \frac{2G\lambda m}{\ell}\ln 2 \text{ Ans.}$$

Problem 8 A uniform rod (red line) of linear mass density λ is lying along the tunnel dug inside a hypothetical stationary planet (Earth, say) of mass M as shown in the figure. Find the (a) net gravitational force acting on the rod and (b) normal reaction offered by the chute on the rod.

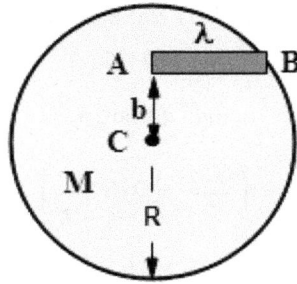

Solution

(a) The component of force acting along the chute on an element is

$$dF_x = dF \cos \theta$$

$$= (dm)E \cos \theta$$

$$= (\lambda dx)\left(\frac{GM}{R^3}r\right)\left(\frac{x}{r}\right)$$

$$= \frac{G\lambda M}{R^3}x \, dx$$

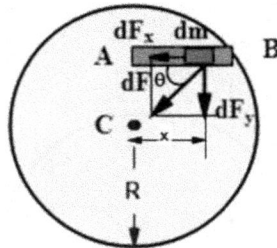

Then, the net force acting on the body is

$$F_x = \frac{G\lambda M}{R^3} \int_0^{\sqrt{R^2 - b^2}} x \, dx$$

$$= \frac{G\lambda M}{R^3}\frac{x^2}{2}\Big|_0^{\sqrt{R^2 - b^2}}$$

$$\Rightarrow F_x = \frac{G\lambda M}{2R^3}(R^2 - b^2) \text{ Ans.}$$

(b) The normal reaction offered by the element on the chute is

$$dN = dF \sin \theta$$

$$= (E\ dm)\sin \theta$$

$$= \left(\frac{GM}{R^3}r\right)(\lambda dx)\left(\frac{b}{r}\right)$$

$$= \frac{GM\lambda b}{R^3}dx$$

Then the total normal reaction is

$$N = \int dN$$

$$= \frac{GM\lambda b}{R^3}\int_0^{\sqrt{R^2 - b^2}} dx$$

$$= \frac{GM\lambda b\sqrt{R^2 - b^2}}{R^3}\ \text{Ans.}$$

Problem 9 Imagine an isolated uniform spherical planet of radius R and density ρ. A spherical cavity is formed inside the sphere. The distance between the centers of the sphere and cavity is $AB = r$.

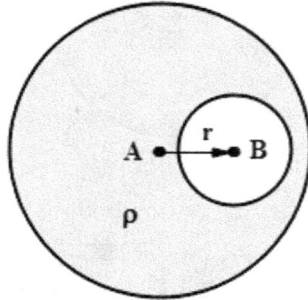

(a) Find the gravitational field E inside the cavity.
(b) If we release an object from rest from the farthest point P of the cavity, find the time after which it reaches the nearest point Q of the cavity. Please assume that PQ and AB are collinear $= 2b$.

Solution

(a) The cavity is imagined as a superposition of positive and negative volume mass density ρ and $-\rho$, respectively.

Then the given sphere with cavity is equivalent to the complete sphere of radius R with positive mass density $+\rho$ and the sphere of radius r with negative density $-\rho$. Let's take a point P inside the cavity. The field at P is the superposition of fields due to spheres of densities $+\rho$ and $-\rho$.

$$\overrightarrow{E} = \overrightarrow{E_1} + \overrightarrow{E_2}$$

$$= -\frac{4\pi\rho G}{3}\overrightarrow{r_1} + \left(-\frac{4\pi(-\rho)G}{3}\overrightarrow{r_2}\right)$$

$$\Rightarrow \overrightarrow{E} = -\frac{4\pi\rho G}{3}(\overrightarrow{r_1} - \overrightarrow{r_2}) \tag{3.105}$$

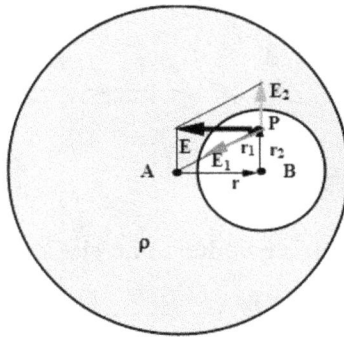

Since $\overrightarrow{r} + \overrightarrow{r_2} = \overrightarrow{r_1}$, we have

$$\overrightarrow{r_1} - \overrightarrow{r_2} = \overrightarrow{r} \tag{3.106}$$

Then, using equations (3.105) and (3.106)

$$\overrightarrow{E} = -\frac{4\pi\rho G}{3}\overrightarrow{r} \quad \text{Ans.}$$

This means that the net field inside the cavity points parallel to the $-\overrightarrow{r}$-vector as shown in the diagram.

(b) If we release an object, the acceleration of the object is

$$\overrightarrow{a} = \frac{F}{m} = \frac{m\overrightarrow{E}}{m} = \overrightarrow{E}$$

$$= -\frac{4\pi\rho G}{3}\overrightarrow{r}$$

Then, the time taken to reach the other end of the diameter of the cavity is

$$t = \sqrt{\frac{2(2b)}{a}} = 2\sqrt{\frac{b}{a}}, \text{ where } a = \frac{4\pi\rho Gr}{3}$$

$$\Rightarrow t = 2\sqrt{\frac{b}{\frac{4\pi\rho Gr}{3}}}$$

$$\Rightarrow t = \sqrt{\frac{3b}{\pi\rho Gr}} \ \text{Ans.}$$

Problem 10 A particle A of mass m is kept at a distance a from the end O of a rod whose density varies with x as

$$\lambda = \lambda_0\left(1 + \frac{x}{a}\right)$$

Find the gravitational force acting on the particle due to the rod.

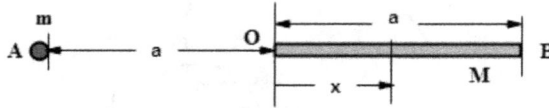

Solution
The force acting on the particle m due to an element dM is

$$dF = \frac{G(dM)m}{(a + x)^2}$$

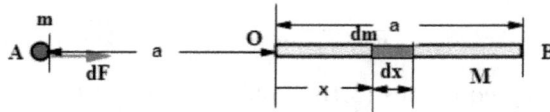

The total force acting on particle A due to the rod is

$$F = \int dF$$

$$= Gm\int \frac{dM}{(a + x)^2}$$

$$= Gm\int \frac{\lambda\,dx}{(a + x)^2}$$

$$= Gm\int \lambda_0\frac{\left(1 + \frac{x}{a}\right)dx}{(a + x)^2}$$

3-90

$$= \frac{Gm\lambda_0}{a} \left(\int_0^a \frac{dx}{x+a} \right)$$

$$= \frac{Gm\lambda_0}{a} \mid \ln(x+a) \mid_0^a$$

$$\Rightarrow F = \frac{Gm\lambda_0}{a} \ln 2 \tag{3.107}$$

The value of λ_0 can be calculated by using the expression

$$M = \int dM = \int \lambda dx,$$

where $\lambda = \lambda_0(1 + \frac{x}{a})$. Then, we have

$$M = \lambda_o \int_0^a (1 + x/a)dx$$

$$\Rightarrow M = 3\lambda_o a/2 \tag{3.108}$$

Putting the value of λ_0 from equations (3.108) in (3.107), we have
$\Rightarrow F = (2/3)\frac{GMm}{a^2} \ln 2$. Ans.

Problem 11 A circular portion of radius r is removed from a uniform infinite thin sheet of surface mass density σ.
(a) Find the gravitational field E at a point P on the axis of the circle.

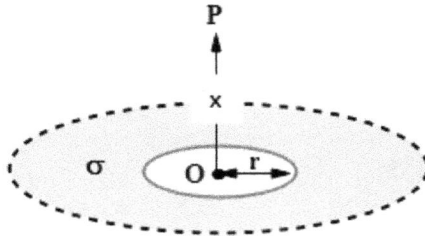

(b) Draw the graph of E versus x.
(c) If a particle placed at the point O is slightly displaced along the axis OP and released, describe the nature of motion of the particle.

Solution
The given figure is equivalent to the sum of an infinite plate of surface mass density σ and a circular sheet of radius with a surface mass density-σ.
(a) Then the field due to the given object at P is

$$\overrightarrow{E} = \overrightarrow{E_1} + \overrightarrow{E_2} \tag{3.109}$$

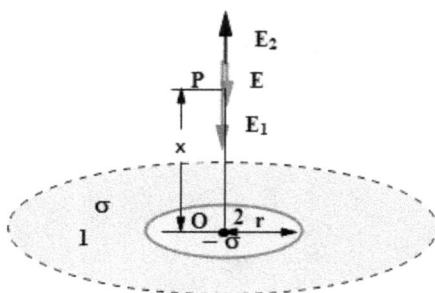

For infinite plate

$$\vec{E_1} = -2\pi\sigma G\hat{j} \tag{3.110}$$

For disc

$$E_2 = 2\pi\sigma G\left(1 - \frac{x}{\sqrt{r^2 + x^2}}\hat{j}\right) \tag{3.111}$$

By using equations (3.109), (3.110) and (3.111), we have

$$\vec{E} = 2\pi\sigma G\left[1 - \frac{x}{\sqrt{r^2 + x^2}} - 1\right]\hat{j}$$

$$\Rightarrow \vec{E} = -2\pi\sigma G\frac{x}{\sqrt{r^2 + x^2}}\hat{j} \text{ Ans.}$$

(b) When $x \to 0$, we have $E \to 0$. When $x \to \infty$, again $E \to 2\pi\sigma G$.

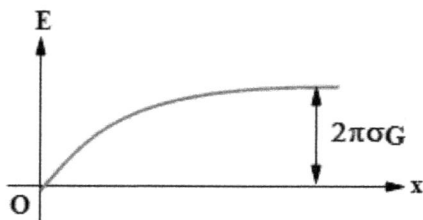

(c) This means that a particle of mass m located at a distance x from the center O experiences an attraction force

$$F = -mE$$

$$\Rightarrow F = -\left(2\pi\sigma G\frac{x}{\sqrt{r^2 + x^2}}\right)$$

$$= -2\pi Gm\sigma\frac{x}{\sqrt{r^2 + x^2}}$$

For small x, $r^2 + x^2 \simeq r^2$

$$\Rightarrow F = -2\pi Gm\sigma\frac{x}{r}$$

Comparing the last expression with the general expression

$$F = -kx$$

the effective spring constant is

$$k = 2\pi Gm\frac{\sigma}{r}$$

Then, the frequency of small oscillation is

$$\omega = \sqrt{\frac{k}{m}} = \sqrt{\frac{2\pi G\sigma}{r}}$$

Then, the motion of the particle will be simple harmonic. Ans.

Potential, potential energy, work–energy theorem, and energy conservation theorem

Problem 12 A body of mass m is falling freely from a height $h = R$ onto an isolated planet of mass M and radius R. As a result, the body moves in a straight line along the line joining the center of the planet and body. A diametrical chute is made inside the planet so that the body can pass through it. Find the time taken by the body to reach the surface of the planet.
  Solution
  Method 1 (force method)
  At any distance of separation r, the net force acting on the particle is

$$F = \frac{GMm}{r^2}$$

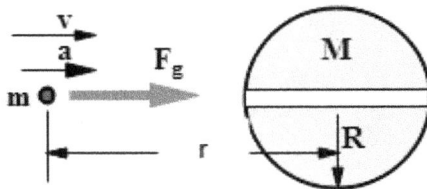

The acceleration of the particle is

$$a = \frac{F}{m} = \frac{GM}{r^2}$$

Since $a = -\frac{v\,dv}{dr}$, we have

$$-\frac{v\,dv}{dr} = \frac{GM}{r^2}$$

$$\Rightarrow v\,dv = -GM\frac{dr}{r^2}$$

Integrating both sides

$$\int_0^v v\,dv = -GM \int_{2R}^r \frac{dr}{r^2}$$

$$\Rightarrow \frac{v^2}{2} = GM\left(\frac{1}{r} - \frac{1}{2R}\right)$$

$$\Rightarrow v = \sqrt{\frac{GM}{R}\frac{(2R - r)}{r}} \tag{3.112}$$

Method 2 (energy method)
Conserving energy between the initial and final (when the particle is just above the surface of the planet), we have

$$\frac{mv^2}{2} - \frac{GMm}{r} = -\frac{GMm}{2R}$$

$$\Rightarrow v = \sqrt{\frac{GM}{R}\frac{(2R - r)}{r}}$$

This is the same expression as we obtained earlier given as equation (3.112). Now we can write

$$v = \frac{dr}{dt} = \sqrt{\frac{GM}{R}}\,\sqrt{(2R - r)}$$

$$\Rightarrow \frac{\sqrt{r}\,dr}{\sqrt{2R - r}} = \sqrt{\frac{GM}{R}}\,dt$$

$$\Rightarrow \int_{2R}^R \frac{dr\,\sqrt{r}}{\sqrt{2R - r}} = \sqrt{\frac{GM}{R}} \int_0^t dt$$

Put $r = 2R \sin^2 \theta$
Then, $dr = 2R(2 \sin \theta \cos \theta d\theta)$

Putting $r = 2R$ and R, we have $\theta = \frac{\pi}{2}$ and $\frac{\pi}{4}$, respectively. After substituting in the above integration, we have

$$t = \sqrt{\frac{R}{GM}} \int_{\frac{\pi}{2}}^{\frac{\pi}{4}} \sqrt{\frac{2R \sin^2 \theta}{2R - 2R \sin^2 \theta}} . 4R \sin \theta \cos \theta d\theta$$

$$= \sqrt{\frac{R}{GM}} 4R \int_{\frac{\pi}{2}}^{\frac{\pi}{4}} \sin^2 \theta \, d\theta$$

$$= \sqrt[2]{\frac{R^3}{GM}} \left| \int_{\frac{\pi}{2}}^{\frac{\pi}{4}} (1 - \cos 2\theta) d\theta \right|$$

$$= \sqrt[2]{\frac{R^3}{GM}} \left| \left(\theta - \frac{1}{2} \sin 2\theta \right) \right|_{\frac{\pi}{2}}^{\frac{\pi}{4}}$$

$$= \sqrt[2]{\frac{R^3}{GM}} \left| -\frac{\pi}{4} - \frac{1}{2} \left(\sin \frac{\pi}{2} - \sin \pi \right) \right|$$

$$= 2 \left(\frac{\pi}{4} + \frac{1}{2} \right) \sqrt{\frac{R^3}{GM}}$$

$$= \left(\frac{\pi}{2} + 1 \right) \sqrt{\frac{R^3}{GM}} \quad \text{Ans.}$$

Problem 13 Let's imagine the Earth as an isolated uniform, spherical, stationary planet having a diametrical chute. A body of mass m starts entering into the chute at the end B of the chute with a velocity v_o. After what time will it pass through the center C of Earth?

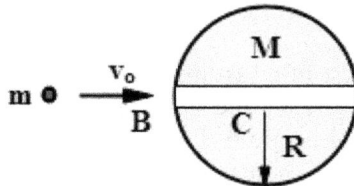

Solution

The velocity of the body at the center C of Earth is maximum, which is given as

$$\frac{1}{2} m v_m^2 + \left(-\frac{3GMm}{2R} \right) = \frac{1}{2} m v_0^2 + \left(-\frac{GMm}{R} \right)$$

$$\Rightarrow v_m = \sqrt{v_0^2 + \frac{GM}{R}} \qquad (3.113)$$

Since the motion of the body inside the Earth is given as

$$a = -\frac{GM}{R^3}r \qquad (3.114)$$

its equation of motion is

$$r = A \sin \omega t,$$

where $A =$ amplitude of oscillation and $\omega =$ angular frequency of oscillation

$$\Rightarrow v = \frac{dr}{dt} = A\omega \cos \omega t, \qquad (3.115)$$

where $A\omega = v_{\max} = v_m$

$$\Rightarrow v = v_m \cos \omega t \qquad (3.116)$$

So, the time interval between $v = v_o$ (at the position B) and $v = v_m$ (at the mean position C) is given as

$t = \frac{1}{\omega} \cos^{-1} \frac{v}{v_m}$, where $\omega = \sqrt{\frac{GM}{R^3}}$

$$\Rightarrow t = \sqrt{\frac{R^3}{GM}} \cos^{-1} \frac{v}{v_m} \qquad (3.117)$$

Putting $v = v_0$ and $v_m = \sqrt{v_0^2 + \frac{GM}{R}}$ from equations (3.113) in (3.117), we have

$$t = \sqrt{\frac{R^3}{GM}} \cos^{-1}\left\{\frac{v_0}{\sqrt{v_0^2 + \frac{GM}{R}}}\right\} \text{ Ans.}$$

Problem 14 A satellite is first lifted from the Earth's surface to a height h. In this process the work done is W_1. Then the satellite is given an adequate speed so as to put t in a circular orbit at that height h. The extra work done is W_2, say. Find the (a) total work done and (b) ratio of these two works done as the function of height h and analyse the result.

Solution

(a) By lifting from P to Q the work done by the external agent against gravity is

$$W_1 = \Delta U_{gr} = U_Q - U_P$$

$$= \left(-\frac{GMm}{R+h} \right) - \left(-\frac{GMm}{R} \right)$$

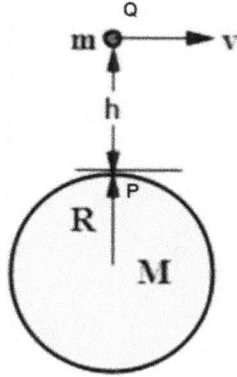

$$\Rightarrow W_1 = \frac{GMmh}{R(R+h)} \tag{3.118}$$

By imparting an orbital velocity v, the additional work done is

$W_2 = \Delta K = \frac{1}{2}mv^2$, where $v = \sqrt{\frac{GM}{r}}$

$\Rightarrow W_2 = \frac{GMm}{2r}$, where $r = R + h$

$$\Rightarrow W_2 = \frac{GMm}{2(R+h)} \tag{3.119}$$

Then the total work done is

$$W = W_1 + W_2 = \frac{GMm}{R+h}\left(\frac{h}{R} + \frac{1}{2} \right) \text{Ans.}$$

(b) The ratio of work done is

$$\frac{W_1}{W_2} = \frac{GMmh}{R(R_1h)} \times \frac{1(R+h)}{GMm} = \frac{2h}{R}; \ R \leqslant h \leqslant \infty)$$

When $h \to \infty$, $\frac{W_1}{W_2}$ increases from 2 to infinite.

Problem 15 An annular ring of inner radius a and outer radius $2a$ is placed horizontally. A particle of mass m is released from the point P on the axis of the ring. Find the (a) gravitational potential at P (b) velocity with which the particle will pass through the center of the ring. Assume a uniform surface mass density σ of the ring.

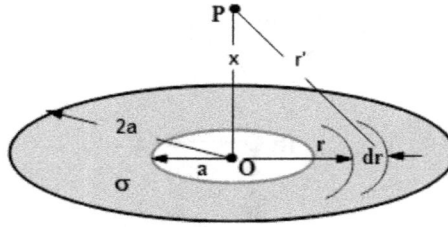

Solution

Method 1

(a) Let's take a concentric ring of thickness dr and radius r. Its gravitational potential at P is

$dV_P = -\frac{G\,dm}{r}$, where $dm = \sigma(2\pi\,r\,dr)$ and $r' = \sqrt{r^2 + x^2}$

$$\Rightarrow dV_P = 2\pi G\sigma \frac{r\,dr}{\sqrt{r^2 + x^2}}$$

Then the potential due to the angular disc at P is

$$V_P = \int d\,V_P = -2\pi G\sigma \int_{r=a}^{r=2a} \frac{r\,dr}{\sqrt{r^2 + x^2}}$$

$$= -2\pi G\sigma\left(\sqrt{r^2 + x^2}\right)\Big|_{r=a}^{2=2a}$$

$$= -2\pi G\sigma\left(\sqrt{4a^2 + x^2} - \sqrt{a^2 + x^2}\right) \tag{3.120}$$

Putting $x = a$, at P

$$V_P = -2\pi G\sigma a(\sqrt{5} - \sqrt{2})\ \text{Ans.}$$

Using equation (3.120) the potential at O is obtained by putting $x = 0$

$$\Rightarrow V_O = -2\pi G\sigma\left(\sqrt{4a^2 + 0} - \sqrt{a^2 + 0}\right)$$

$$= -2\pi G\sigma a$$

Method 2

The given annular disc is equivalent to the sum of a disc of radius $2a$ of surface mass density σ and a smaller disc of radius a with a surface mass density $-\sigma$. In other words, the annular disc is equivalent to the full disc of radius $2a$ from which a concentric disc of radius a is deducted. Then the potential due to the annular disc is

$$V = V_1 - V_2, \tag{3.121}$$

where V_1 and V_2 are the potentials due to the discs of radii $2a$ and a, respectively, at the axial point P. We can write

$$V_1 = 2\pi\sigma G\left(\sqrt{4a^2 + x^2} - x\right) \tag{3.122}$$

$$V_2 = -2\pi\sigma G\left(\sqrt{a^2 + x^2} - x\right) \tag{3.123}$$

Using the last three equations

$$V = -2\pi\sigma G\left(\sqrt{4a^2 + x^2} - \sqrt{a^2 + x^2}\right)$$

Putting $x = a$, we can get the same answer.

(b) Conserving energy between O and P we have

$$K_P + U_P = K_O + U_O$$

$$0 + mV_P = \frac{1}{2}mv^2 + mV_O$$

$$\Rightarrow v = \sqrt{2(V_P - V_O)}$$

Putting the obtained values of V_P and V_O

$$v = \sqrt{2[\{-2\pi G\sigma a(\sqrt{5} - \sqrt{2})\} - \{-2\pi G\sigma a\}]}$$

$$= \sqrt{4\pi G\sigma a(-\sqrt{5} + \sqrt{2} + 1)} \text{ Ans.}$$

Problem 16

Consider a hypothetical planet of mass M and radius R having a uniform volume mass density ρ. A diametrical chute is made inside the planet. If a body of mass m is approaching from infinity towards the planet, find the (a) velocity at P and O of the body and (b) the ratio of the velocities.

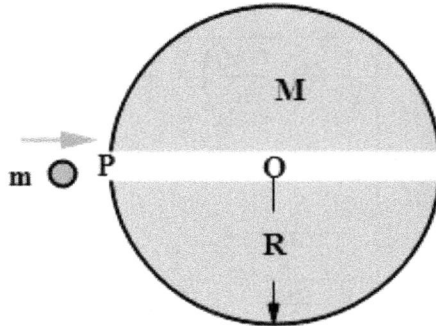

Solution

(a) The potential function of the sphere is

$$V = -\frac{GM}{2R^3}(3R^2 - r^2)$$

Putting $r = R$, the potential at P is

$$V_P = -\frac{GM}{2R^3}(3R^2 - R^2)$$

$$= -\frac{GM}{R}$$

Putting $r = 0$, the potential at Q is

$$V_Q = -\frac{GM}{2R^3}(3R^2 - 0)$$

$$= -\frac{3GM}{2R}$$

Conserving energy between,

$$U_O + K_O = U_P + K_P = U_\infty + K_\infty$$

$$\Rightarrow +mV_O + mv_0^2/2 = -mV_P + \frac{1}{2}mv_P^2$$

$$= 0 + 0 = 0$$

$$\Rightarrow v_O = \sqrt{-2V_O} = \sqrt{-2\left(\frac{-3GM}{2R}\right)}$$

$$= \sqrt{\frac{3GM}{R}}$$

(b) Similarly, $v_P = \sqrt{-2V_P}$

$$= \sqrt{-2\left(-\frac{GM}{R}\right)} = \sqrt{\frac{2GM}{R}}$$

(c) $\frac{v_P}{v_O} = \frac{\sqrt{-2V_P}}{\sqrt{-2V_O}} = \sqrt{\frac{V_P}{V_O}}$

$$= \sqrt{\frac{-\frac{GM}{R}}{-\frac{3GM}{2R}}} = \frac{1}{\sqrt{3/2}} \quad \text{Ans.}$$

Conservation of energy and linear momentum

Problem 17 Two spherical planets of masses m_1 and m_2 are released from rest when they are separated by a distance x. Find the (a) relative velocity between the planets just before they meet and (b) if x tends to infinity ($x \to \infty$) find the individual velocities when they will meet each other. Assume that $R =$ radius of each planet

Solution

(a) Just before the meeting, let $v_{\text{rel}} =$ relative velocity between the planets.

Then, the change in KE of the system = KE at the time of meeting-initial KE (= zero as they are released from rest from a distance of separation x). This is given as

$$\Delta K = \frac{1}{2} \frac{m_1 m_2}{m_1 + m_2} v_{\text{rel}}^2 \tag{3.124}$$

The change in gravitational PE of the system is

$$\Delta U = \left(-\frac{Gm_1 m_2}{R + R} \right) - \left(-\frac{Gm_1 m_2}{x} \right)$$

$$\Rightarrow \Delta U = -Gm_1 m_2 \left(\frac{1}{2R} - \frac{1}{x} \right) \tag{3.125}$$

Conserving energy,

$$\Delta K + \Delta U = 0 \tag{3.126}$$

Using equations (3.124), (3.125) and (3.126), we have

$$\frac{m_1 m_2 v_{\text{rel}}^2}{2(m_1 + m_2)} - Gm_1 m_2 \left(\frac{1}{2R} - \frac{1}{x} \right) = 0$$

$$\Rightarrow v_{\text{rel}}^2 = \frac{G(m_1 + m_2)(x - 2R)}{Rx}$$

$$\Rightarrow v_{\text{rel}} = \sqrt{\frac{G(m_1 + m_2)}{R} \left(1 - \frac{2R}{x} \right)} \text{ Ans.}$$

(b) Putting $x \to \infty$, we have

$$v_{\text{rel}} = \sqrt{\frac{G(m_1 + m_2)}{R}} \tag{3.127}$$

If v_1 and v_2 are the velocity of m_1 and m_2, respectively,

$$v_1 + v_2 = v_{\text{rel}} \tag{3.128}$$

Conserving linear momentum,

$$m_1 v_1 = m_2 v_2 \tag{3.129}$$

This gives,

$$\frac{v_1}{v_2} = \frac{m_2}{m_1}$$

$$\Rightarrow \frac{v_1}{v_1 + v_2} = \frac{m_2}{m_1 + m_2} \tag{3.130}$$

Using equations (3.128) and (3.130),

$$v_1 = \frac{m_2}{m_1 + m_2} v_{\text{rel}} \tag{3.131}$$

Using equations (3.127) and (3.131),

$$v_1 = \frac{m_2}{m_1 + m_2} \sqrt{\frac{G(m_1 + m_2)}{R}}$$

$$= \sqrt{\frac{G m_2^2}{(m_1 + m_2)R}} \quad \text{Ans.}$$

Using equation (3.129),

$$v_2 = \frac{m_1 v_1}{m_2}$$

$$= \frac{m_1}{m_2} \sqrt{\frac{G m_2^2}{(m_1 + m_2)R}}$$

$$= \sqrt{\frac{G m_1^2}{(m_1 + m_2)R}} \quad \text{Ans.}$$

Alternative method:
The energy conservation yields

$$\Delta K + \Delta U = 0$$

$$\Delta K = -\Delta U$$

$$\Rightarrow \frac{1}{2} m_1 v_1^2 + \frac{1}{2} m_2 v_2^2 = -\left(-\frac{G m_1 m}{2R} \right) \tag{3.132}$$

The momentum conservation yields

$$m_1 v_1 = m_2 v_2 \tag{3.133}$$

Solving equations (3.132) and (3.133),

$$\frac{1}{2}m_1v_1^2 + \frac{1}{2}m_2\left(\frac{m_1v_1}{m_2}\right)^2 = \frac{Gm_1m}{2R} \Rightarrow v_1 = \sqrt{\frac{Gm_2^2}{(m_1+m_2)R}} \quad \text{Ans.}$$

Using equation (3.132),

$$v_2 = \frac{m_1v_2}{m_2} = \frac{m_1}{m_2}\sqrt{\frac{Gm_2^2}{(m_1+m_2)R}}$$

$$= \sqrt{\frac{Gm_1^2}{(m_1+m_2)R}} \quad \text{Ans.}$$

The relative velocity is

$$v_{\text{rel}} = v_1 + v_2$$

$$= \sqrt{\frac{Gm_1^2}{(m_1+m_2)R}} + \sqrt{\frac{Gm_2^2}{(m_1+m_2)R}}$$

$$= \sqrt{\frac{G}{(m_1+m_2)R}}(m_1+m_2)$$

$$= \sqrt{\frac{G(m_1+m_2)}{R}} \quad \text{Ans.}$$

Problem 18 A superman of mass m jumps out of an isolated planet of mass m with a velocity (a) v_r with respect to planet and (b) v with respect to space. Find the maximum height attained by the superman relative to the planet.

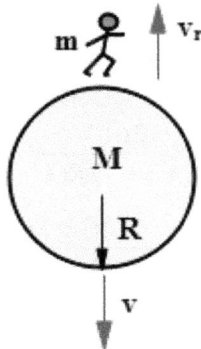

Solution

(a) If v_r = velocity of the superman relative to the free planet, the total momentum of the system $(M + m)$ is given as

$$0 = m(v_r - v) + (Mv)$$

$$\Rightarrow v = \frac{Mv_r}{M + m} \tag{3.134}$$

The KE of the system is

$$K_1 = \frac{1}{2}m(v_r - v)^2 + \frac{1}{2}Mv^2 \tag{3.135}$$

Using equations (3.134) and (3.135),

$$K_1 = \frac{1}{2}m\left(v_r - \frac{mv_r}{M + m}\right)^2 + \frac{1}{2}M\left(\frac{mv_r}{M + m}\right)^2$$

$$\Rightarrow K_1 = \frac{Mmv_r^2}{2(M + m)} \tag{3.136}$$

Let h = maximum height attained by the superman.
By conserving energy between the given position 1 and maximum separated position 2, we have

$$K_1 + U_1 = K_2 + U_2 \tag{3.137}$$

Using equations (3.136) and (3.137),

$$\frac{Mmv_r^2}{2(M + m)} + \left(-\frac{GMm}{R}\right) = 0 + \left(-\frac{GMm}{R + h}\right)$$

$$\Rightarrow \frac{v_r^2}{2(M + m)} - \frac{G}{R} = -\frac{G}{R + h}$$

$$\frac{G}{R + h} = \frac{G}{R} - \frac{v_r^2}{2(M + m)}$$

$$\Rightarrow \frac{G}{R + h} = \frac{2G(M + m) - Rv_r^2}{2R(M + m)}$$

$$\Rightarrow R + h = \frac{2GR(M + m)}{2G(M + m) - Rv_r^2}$$

$$\Rightarrow h = R\left[\frac{2G(M + m)}{2G(M + m) - Rv_r^2} - 1\right] \text{ Ans.}$$

(b) If v' = recoil velocity of the planet

$$Mv' = mv$$

$$\Rightarrow v' = \frac{m}{M}v \tag{3.138}$$

Conserving energy,

$$\frac{1}{2}Mv'^2 + \frac{1}{2}mv^2 = GMm\left(\frac{1}{R} - \frac{1}{R+h}\right) \tag{3.139}$$

Using equations (3.138) and (3.139),

$$\frac{1}{2}M\left(\frac{m}{M}v\right)^2 + \frac{1}{2}mv^2 = GMm\left(\frac{1}{R} - \frac{1}{R+h}\right)$$

$$\Rightarrow \frac{1}{2}mv^2\left(1 + \frac{m}{M}\right) = GMm\left(\frac{1}{R} - \frac{1}{R+h}\right)$$

$$\Rightarrow \frac{1}{R+h} = \frac{1}{R} - \frac{v^2(M+m)}{2GM}$$

$$\Rightarrow \frac{1}{R+h} = \frac{2GM - v^2R(M+m)}{2GMR}$$

$$\Rightarrow h = \frac{2GMR}{2GM - v^2R(M+m)} - R \text{ Ans.}$$

Problem 19 Initially two planets of mass M and m are kept separated at a distance r_o. The planet of mass m is given a velocity v_o such that it moves away from the other planet along the line joining them. Find the maximum distance of separation between the planets.

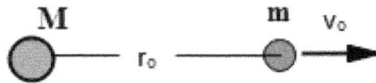

Solution

Let $r = $ maximum distance between the planets. At this distance, instantaneously both the bodies move with the same velocity. Conserving the linear momentum between the initial given position and maximum separation, we have

$$mv_0 = (m + M)v$$

$$\text{So, } v = \frac{Mv_0}{M + m} \tag{3.140}$$

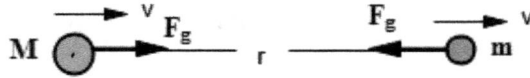

Conserving energy between the aforementioned positions,

$$\frac{1}{2}mv_0^2 - \frac{GMm}{r_o} = -\frac{GMm}{r} + \frac{1}{2}(M+m)v^2. \tag{3.141}$$

Putting 'v' from equation (3.140) in equation (3.141), we have

$$\frac{1}{2}mv_0^2 - \frac{GMm}{r_0} = -\frac{GMm}{r} + \frac{(M+m)}{2}\left(\frac{mv_0}{M+m}\right)^2$$

$$\Rightarrow \frac{Mm}{2(M+m)}v_0^2 = GMm\left(\frac{1}{r_0} - \frac{1}{r}\right)$$

$$\Rightarrow \frac{v_0^2}{2G(M+m)} = \frac{r-r_0}{r\,r_0}$$

$$\Rightarrow \frac{(r-r_0)}{r} = \frac{r_0 v_0^2}{2G(M+m)}$$

$$\text{Or, } 1 - \frac{r_0}{r} = \frac{r_0 v_0^2}{2G(M+m)}$$

$$\text{Or, } r = \frac{r_0}{1 - \dfrac{r_0 v_0^2}{2G(M+m)}} \quad \text{Ans.}$$

Orbital speed

Problem 20 Two planets A and B having mass m_1 and m_2, respectively, are revolving in such a way that they will be able to maintain a constant distance r between them. Find the (a) angular velocity, (b) time period of revolution of the planets, and (c) total energy of the system of two planets.

Solution

Since the bodies maintain a constant distance of separation r they must revolve around the center of mass C of the system $A + B$ or $(m_1 + m_2)$.

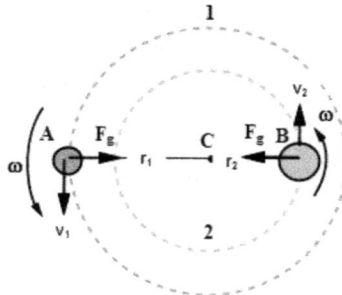

This can be proved by applying Newton's 2nd law on each planet as follows:

(a) Since the planets are attracted by their mutual gravitational force F, the centripetal force acting on each planet is given as

$$F = m_1 r_1 \omega^2 = m_2 r_2 \omega^2, \tag{3.142}$$

where $F = \frac{Gm_1 m_2}{r^2}$

$$\Rightarrow m_1 r_1 = m_2 r_2 \tag{3.143}$$

This proves that the planets must revolve around their fixed center of mass.

For maintaining the constant separation r,

$$r_1 + r_2 = r \tag{3.144}$$

Solving equations (3.142) and (3.144), the radii of the circular orbits of the planets are given as

$$r_1 = \frac{m_2 r}{m_1 + m_2} \tag{3.145}$$

$$r_2 = \frac{m_1 r}{m_1 + m_2} \tag{3.146}$$

The equation (3.142) can be written as

$$\Rightarrow \frac{Gm_1 m_2}{r^2} = m_1 r_1 \omega^2,$$

where $r_1 = \frac{m_2 r}{m_1 + m_2}$

$$\Rightarrow \frac{Gm_1 m_2}{r^2} = \frac{m_1 m_2 r \omega^2}{(m_1 + m_2)}$$

$$\Rightarrow \omega = \sqrt{\frac{G(m_1 + m_2)}{r^3}} \quad \text{Ans.}$$

(b) To maintain the constant separation both planets must have equal time periods of revolution, which can be given as

$$T = \frac{2\pi}{\omega} = 2\pi \sqrt{\frac{r^3}{G(m_1 + m_2)}} \quad \text{Ans.}$$

(c) The total mechanical energy of the two-planet system is

$$E = -\frac{Gm_1 m_2}{r} + \frac{1}{2}\frac{m_1 m_2}{m_1 + m_2} r^2 \omega^2$$

Putting the obtained value of ω, we have

$$= -\frac{Gm_1m_2}{r} + \frac{1}{2}\frac{m_1m_2r^2}{m_1 + m_2}\left\{\frac{G(m_1 + m_2)}{r^3}\right\}$$

$$= -\frac{Gm_1m_2}{2r} \text{ Ans.}$$

Problem 21 A hypothetical planet is comprised of an infinite rod of linear mass density λ that passes through a uniform solid sphere of mass m and radius r as shown in the figure. Find the (a) orbital velocity and (b) period of revolution of a satellite of mass m orbiting very close to the planet about the infinite rod.

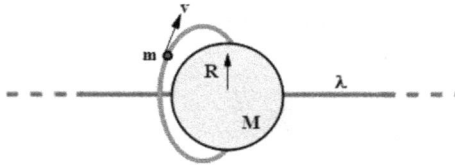

Solution

(a) The gravitational field strength due to the rod at the orbit is

$$E_1 = \frac{2\lambda G}{R} \tag{3.147}$$

The gravitational field strength due to the sphere at P the orbit is

$$E_2 = \frac{GM}{R^2} \tag{3.148}$$

The field strength at the orbit is

$$E = E_1 + E_2$$

$$\Rightarrow E = \frac{G}{R}\left(2\lambda + \frac{M}{R}\right) \tag{3.149}$$

Then, the centripetal force acting on the satellite is

$$mE = \frac{mv^2}{R}$$

$$\Rightarrow v = \sqrt{ER} \tag{3.150}$$

Using equations (3.149) and (3.150), we have

$$v = \sqrt{G\left(2\lambda + \frac{M}{R}\right)} \quad \text{Ans.}$$

(b) The period of revolution of the satellite is

$$T = \frac{2\pi R}{v} = \frac{2\pi R}{\sqrt{G\left(2\lambda + \frac{M}{R}\right)}} \quad \text{Ans.}$$

N.B.: If the spherical planet is absent, the orbital velocity will be

$$v = \sqrt{2G\lambda}$$

So the orbital velocity is independent of the radius of the orbit.

Problem 22 Three identical planets each of mass m are located at the vertices of an equilateral triangle of side l. They are given with equal speed v such that they will always move in a circle about the center of mass of the system of three planets that coincides with the center of the circle. Find the (a) value of v and (b) total energy of the revolving planets.

Solution

(a) The given condition is that the planets must maintain the same distance from each other. For this to happen, each planet should be given a velocity perpendicular to the line joining it and the center of mass of the system of three planets. In other words, all planets must revolve in the same sense with the same angular velocity. Let v = orbital velocity of each planet.

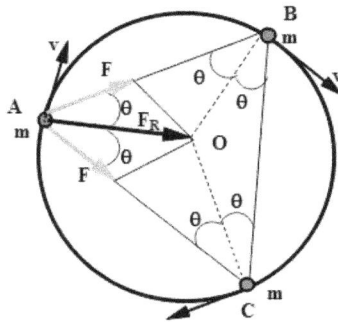

Then, the kinetic energy of the system is

$$K = \frac{1}{2}mv^2 \times 3 = \frac{3}{2}mv^2$$

The potential energy of the system is

$$U = -\frac{Gm^2}{\ell} \times 3 = -\frac{3Gm^2}{\ell}$$

Then, the total energy of the system is

$$E = K + U$$

$$\Rightarrow E = \frac{3}{2}mv^2 - \frac{3Gm^2}{\ell} \tag{3.151}$$

Since each planet revolves around the center of mass O of the system, the radius of the circular path is

$$r = \frac{\ell}{2}\sec 30° = \frac{\ell}{2} \times \frac{2}{\sqrt{3}}$$

$$\Rightarrow r = \frac{\ell}{\sqrt{3}}$$

If $F =$ mutual gravitational force between any two planets, the net force acting on each planet is

$$F_{\text{net}} = F \cos 30^0 + F \cos 30^o = \sqrt{3}F,$$

where $F = \frac{Gm^2}{\ell^2}$

Since the net gravitational force acting on each planet is directed towards the center of the circular orbit, the centripetal force is equal to the net gravitational force. Applying Newton's 2nd law on each planet, we have

$$\Rightarrow F_{CP} = \frac{\sqrt{3}\,Gm}{\ell^2} = \frac{mv^2}{r}, \text{ where } r = \frac{\ell}{\sqrt{3}}$$

$$\Rightarrow \frac{\sqrt{3}\,Gm}{\ell^2} = \frac{mv^2}{\frac{\ell}{\sqrt{3}}}$$

$$\Rightarrow v = \sqrt{\frac{Gm}{\ell}} \tag{3.152}$$

(b) Putting v from equations (3.152) in (3.151), we have

$$E = \frac{3}{2}m\left(\sqrt{\frac{Gm}{\ell}}\right)^2 - \frac{3Gm^2}{\ell} = -\frac{3Gm^2}{2\ell} \text{ Ans.}$$

Problem 23

   (a) Consider a hypothetical galaxy of mass M and radius R having a uniform density of stars in its spherical volume. Find the variation of the orbital velocity of the stars relative to the center of the galaxy.

   (b) Consider a point P at a distance $r = R/2$ and another point Q at a distance $r = 2R$. Compare the escape velocities at these two points. Assume $\rho = $ average density of the galaxy.

Solution

   (a) At any radial distance r from the galactic center C, we can write for $r \leqslant R$, $\rho = \rho_0 = $ constant.

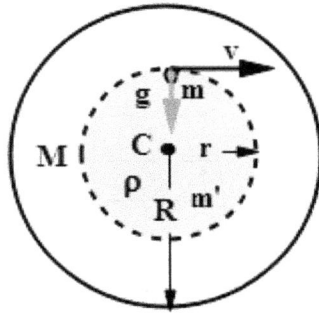

The flux of field intensity $\vec{g}$ is

$$\oint \vec{g} \cdot d\vec{A} = 4\pi m G$$

Taking the spherical Gaussian surface,

$$\Rightarrow g 4\pi r^2 = 4\pi G \left(\rho_0 \frac{4}{3} \pi r^3 \right)$$

$$\Rightarrow g = \frac{4\pi \rho_0 G}{3} r; \quad r \leqslant R. \tag{3.153}$$

Then, the orbital velocity at the radial distance r is given as

$$v = \sqrt{gr} \tag{3.154}$$

Using the last two equations, we have

$$v = \sqrt{\frac{4\pi \rho_0 G}{3} r \cdot r}$$

$$\Rightarrow v = \sqrt{\frac{4\pi \rho_0 G}{3}} r$$

This means that the orbital velocity varies linearly with r inside the galaxy.

$$\Rightarrow v \propto r \text{ for } r \leqslant R$$

For outside the galaxy, $r \geqslant R$, $\rho = 0$.
Then the flux of $\vec{g}$ is

$$\oint \vec{g} \cdot d\vec{A} = 4\pi Gm'$$

$$\Rightarrow g 4\pi r^2 = 4\pi G \frac{4}{3}\pi \rho_0 R^3$$

$$\Rightarrow g = \frac{4\pi \rho_0 G R^3}{3r^2}$$

Then the orbital velocity is

$$v = \sqrt{gr}$$

$$= \sqrt{\frac{4\pi \rho_0 G R^3}{3r^2} \cdot r}$$

$$= \sqrt{\frac{4\pi \rho_0 G R^3}{3r}}$$

$$\Rightarrow v \propto \frac{1}{\sqrt{r}} \text{ for } r \geqslant R$$

This means that outside the galaxy, the orbital velocity is proportional to the reciprocal of the square root of the radial distance.

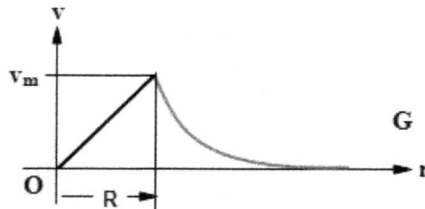

(b) The gravitational potential of the galaxy is given by the formula:

$$V = -\frac{4\pi G\rho}{6}(3R^2 - r^2); \; r \leqslant R$$

$$= -\frac{4\pi G R^3}{3r}; \; r \geqslant R$$

3-112

Then by putting $r = \frac{R}{2}$

$$V_P = \frac{4\pi G\rho}{6}\left\{3R^2 - \left(\frac{R}{2}\right)^2\right\}$$

$$= -\frac{4\pi G\rho}{6}\left(\frac{11R^2}{4}\right)$$

$$\Rightarrow V_P = \frac{-11\pi G\rho R^2}{6} \tag{3.155}$$

Similarly, by putting $r = 2R$, we have

$$V_Q = -\frac{4\pi G R^3\rho}{3(2R)} = -\frac{2\pi\rho G R^2}{3} \tag{3.156}$$

Using the last two equations, we have

$$\frac{V_P}{V_Q} = \frac{11 \times 3}{6 \times 2} = \frac{11}{4} \tag{3.157}$$

The escape velocity is

$$v_e = \sqrt{2(-V)} \tag{3.158}$$

So, the ratio of the escape velocities at p and Q is

$$\frac{(v_e)_P}{(v_e)_Q} = \sqrt{\frac{V_P}{V_Q}} = \sqrt{\frac{11}{4}} \text{ Ans.}$$

Conservation of angular momentum and energy

Problem 24 A satellite of mass m is revolving around a fixed planet of mass M in an orbit of radius r_1. If it gets an impulse, it will move to the higher orbit of radius r_2. Find the (a) initial velocity of the satellite just after receiving the impulse, (b) extra velocity received by the satellite, and (c) impulse received by the satellite.

Solution

(a) Let v_1 = velocity of projection of the satellite at the point A of the orbit 1 and v_2 = velocity of the satellite at the point B of the orbit 2, conserving angular momentum, we have

$$mv_1 r_1 = mv_2 r_2$$

$$\text{Or, } v_2 = \frac{v_1 r_1}{r_2} \tag{3.159}$$

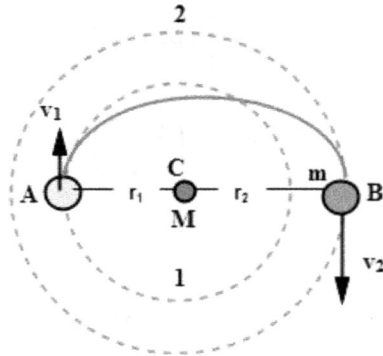

Conserving the energy,

$$\frac{1}{2}mv_1^2 - \frac{GMm}{r_1} = \frac{1}{2}mv_2^2 - \frac{GMm}{r_2} \qquad (3.160)$$

Using equations (3.159) and (3.160),

$$\frac{1}{2}mv_1^2\left\{1 - \left(\frac{v_2}{v_1}\right)^2\right\} = GMm\left(\frac{r_2 - r_1}{r_1 r_2}\right),$$

where $\frac{v_2}{v_1} = \frac{r_1}{r_2}$

$$\Rightarrow v_1^2\left\{1 - \left(\frac{r_1}{r_2}\right)^2\right\} = 2GM\left(\frac{r_2 - r_1}{r_1 r_2}\right)$$

$$\Rightarrow v_1^2\frac{r_2^2 - r_1^2}{r_2^2} = \frac{2GM(r_2 - r_1)}{r_1 r_2}$$

$$\Rightarrow v_1 = \sqrt{\frac{2GMr_2}{r_1(r_1 + r_2)}} \quad \text{Ans.}$$

(b) The initial orbital velocity is

$$(v_1)_0 = \sqrt{\frac{GM}{r_1}}$$

Then the additional velocity imparted to the satellite is

$$v_1 - v_{1_0} = \sqrt{\frac{2GM_{r_2}}{r_1(r_1 + r_2)}} - \sqrt{\frac{GM}{r_1}}$$

$$\Delta \vec{v} = \sqrt{\frac{GM}{r_1}}\left[\sqrt{\frac{2r_2}{r_1 + r_2}} - 1\right] \quad \text{Ans.}$$

3-114

(c) The impulse applied is

$$m\Delta v = m\sqrt{\frac{GM}{r_1}}\left[\sqrt{\frac{2r_2}{r_1+2r_2}} - 1\right] \text{ Ans.}$$

Problem 25 A body of mass m is projected horizontally from a tall mountain with a velocity v_o, which is greater than the orbital velocity and less than the escape velocity. The velocity of projection at P is given as $v_0 = \sqrt{\frac{\eta GM}{R}}$, where $v_{\mathrm{orb}} = \sqrt{\frac{GM}{R}}$ and η is greater than one. If the maximum height attained by the body is $h = R$, find the value of η.

Solution

Under the given condition the body will move in an elliptical path sweeping the Earth at the point of projection P. Following an elliptical path, the body will attain a maximum height h at Q. Let $v = $ velocity of the body at Q.

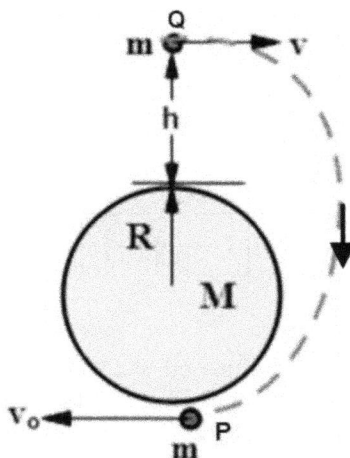

Conserving the angular momentum of the body at P and Q,

$$mv_0 R = mv(R + h)$$

$$\Rightarrow v = \frac{v_0 R}{R + h} \tag{3.161}$$

Conserving energy of the body at P and Q,

$$\frac{1}{2}mv_0^2 - \frac{GMm}{R} = \frac{1}{2}mv^2 - \frac{GMm}{(R + h)}$$

$$\Rightarrow \frac{GMm}{R+h} = -\frac{1}{2}m(v_0^2 - v^2) + \frac{GMm}{R}$$

$$\Rightarrow \frac{GM}{R+h} = -\frac{1}{2}(v_0^2 - v^2) + \frac{GM}{R} \tag{3.162}$$

Using equations (3.161) and (3.162),

$$\frac{1}{2}\left\{v_0^2 - \left(\frac{v_0 R}{R+n}\right)^2\right\} = GM\left(\frac{1}{R} - \frac{1}{R+n}\right)$$

$$\frac{1}{2}v_0^2\left\{1 - \frac{R^2}{(R+h)^2}\right\} = \frac{GMh}{R(R+h)}$$

Putting $v_0 = \sqrt{\frac{\eta GM}{R}}$, we have

$$\frac{1}{2}\left(\eta \frac{GM}{R}\right)\left\{1 - \frac{R^2}{(R+h)^2}\right\} = \frac{GMh}{R(R+h)}$$

$$\Rightarrow \eta = \frac{2h}{(R+h)\left[1 - \left(\frac{R}{R+h}\right)^2\right]}$$

If $h = R$, we get

$$\eta = \frac{2(R)}{(R+R)\left[1 - \left(\frac{R}{R+R}\right)^2\right]} = \frac{4}{3} \text{ Ans.}$$

Escape velocity

Problem 26 The planet Saturn is comprised of a spherical portion of mass M, ring of mass m, and radius r as shown in the figure. Find the escape velocity v_e of a body from the position as shown in the figure. Put $r = 2R$ and $M = 2m$.

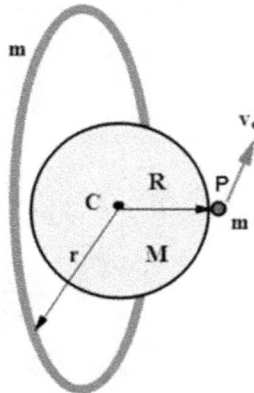

Solution

The potential due to the ring at the surface point P of the planet is

$V_1 = -\frac{Gm}{r}$, where $r = \sqrt{R^2 + 4R^2} = \sqrt{5}\,R$

$$\Rightarrow V_1 = -\frac{Gm}{\sqrt{5}\,R} \tag{3.163a}$$

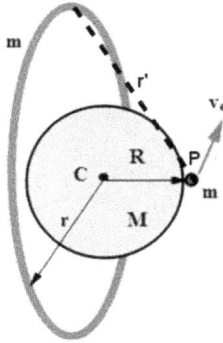

The potential of the sphere at its surface point P is

$$V_2 = -\frac{GM}{R} = -\frac{G(2m)}{R} \tag{3.163b}$$

The total potential at P is

$$V_P = V_1 + V_2 = \frac{-Gm}{R}\left(\frac{1}{\sqrt{5}} + 2\right) \tag{3.164}$$

Conserving the energy between the point of projection and infinity

$$\frac{1}{2}mv_e^2 + mV_P = 0 \tag{3.165}$$

Using the equations (3.164) and (3.165)

$$v_e = \sqrt{-2\left(-\frac{Gm}{R}\right)\left(2 + \frac{1}{\sqrt{5}}\right)}$$

$$= \sqrt{2\left(2 + \frac{1}{\sqrt{5}}\right)\frac{Gm}{R}} \quad \text{Ans.}$$

Problem 27 Let's imagine that a uniform spherical planet is floating in the sky. Assume that $\rho =$ average density and $R =$ radius of the planet. Let's form a spherical cavity of radius $R/2$ that passes through the center O of the planet. Find the minimum velocity v_o of a body of mass m that should be projected radially from the center A of the cavity so that it will never return.

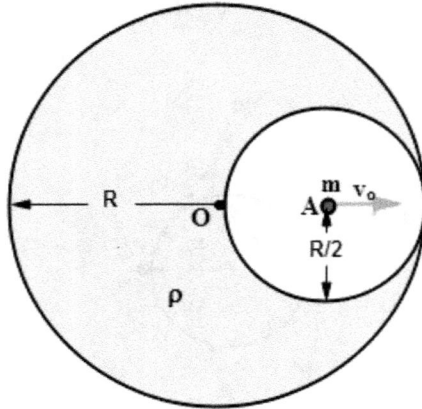

Solution

The net potential at A is

$$V = V_1 + V_2 \tag{3.166}$$

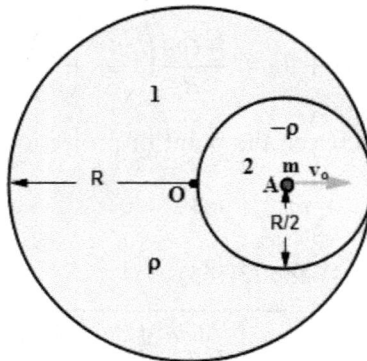

The potential due to the full bigger sphere is

$V_1 = \frac{4G\pi\rho}{6}(3R^2 - r^2)$, where $r = R$

$$\Rightarrow V_1 = -\frac{2G\pi\rho}{3}\left(3R^2 - \frac{R^2}{4}\right)$$

The potential due to the inner sphere is

$$= -\frac{11\pi G\rho R^2}{6} \tag{3.167}$$

$$V_2 = -\frac{2G\pi(-\rho)}{3}\left(\frac{3}{4}R^2 - r^2\right), \text{ where } r = 0$$

$$\Rightarrow V_2 = \frac{1}{2}\pi G\rho R^2 \tag{3.168}$$

So the net potential at A is

$$V = V_1 + V_2 = \pi\rho GR^2\left(\frac{1}{2} - \frac{11}{6}\right) \tag{3.169}$$

Then, the escape velocity is given as

$$v_e = \sqrt{2(-V)} \tag{3.170}$$

Taking inifinity as the final portion using equations (3.169) and (3.170), we have

$$v_e = \sqrt{-2\left\{\pi\rho GR^2\left(\frac{1}{2} - \frac{11}{6}\right)\right\}}$$

$$= \left(\sqrt{\frac{8}{3}G\pi\rho}\right)R \text{ Ans.}$$

Allied problems

Problem 28 Let's assume that Earth is an isolated uniform, spherical, stationary planet. If M, R, and ρ are mass, radius, and density of Earth, respectively, find the (a) variation of gravitational pressure inside Earth as the function of the radial distance r and (b) gravitational pressure at the center of Earth.
   Solution

   (a) The pressure difference dP along the radial direction is given by equating the gravitational force to pressure force on an element dm, as $\delta F = (dP)(\delta A) = (\delta m)E$.
      The $-$ve sign is due to the fact that pressure decreases radially away as r increases.

$$\Rightarrow -dP = \frac{\delta m}{\delta A}\left(\frac{GM}{R^3}r\right)$$

$$\Rightarrow -dP = \frac{\rho(\delta A)(dr)}{\delta A}\left(\frac{GM}{R^3}r\right)$$

$$\Rightarrow -dP = \rho \frac{GM}{R^3} r d r$$

$$\Rightarrow -\int_0^P dP = \frac{GM\rho}{R^3} \int_R^r r d r$$

$$\Rightarrow P = -\frac{GM\rho}{R^3} \left(\frac{r^2 - R^2}{2} \right),$$

where $\rho = \frac{M}{\frac{4}{3}\pi R^3} = \frac{3M}{4\pi R^3}$

$$\Rightarrow P = -\frac{GM}{2R^3} \left(\frac{3M}{4\pi R^3} \right) (R^2 - r^2)$$

$$\Rightarrow P = \frac{3GM^2}{8\pi R^6} (R^2 - r^2)$$

$$= \frac{3GM^2}{8\pi R^4} \left\{ 1 - \left(\frac{r}{R} \right)^2 \right\}$$

(b) At the center of Earth, the pressure is maximum (where $r = 0$). Then the maximum pressure is given as.

$$P_{max} = \frac{3GM^2}{8\pi R^4}$$

$$= \frac{3G^2M^2}{8\pi R^4} \cdot \frac{R^3}{G}$$

$$\Rightarrow P_{max} = \frac{3g^2}{G}, \text{ where } g = \frac{GM}{R^2}$$

Putting $G = 6.67 \times 10^{-11} \, \text{m}^3 \, \text{kg}^{-1} \, \text{s}^{-2}$,

we have $P_{max} = 1.73 \times 10^{11} \, \text{Pa}$. Ans.

Problem 29 An Earth satellite moving in a circular orbit of radius r experiences a viscous drag due to cosmic dust, which can be given as

$$F = -\alpha v^2,$$

where α is a positive constant and v is the speed of the satellite. As a result, the satellite slowly releases energy and collapses onto the Earth following a spiral path. Find the time after which the satellite will reach Earth's surface.

Solution

The total energy of the satellite at any radial distance r is

$$E = -\frac{GMm}{2r}$$

If the path is assumed to be nearly circular, then the rate of decrease in total energy is

$$\frac{dE}{dt} = +\frac{GMm}{2r^2}\frac{dr}{dt}, \tag{3.174}$$

where $\frac{dr}{dt}=$ rate of change of radial distance.

According to the work–energy theorem, work done by the non-conservative (friction) is

$$W = \frac{dE}{dt}$$

$$-F \cdot v = \frac{dE}{dt} \tag{3.175}$$

Using equations (3.174) and (3.175),

$$\frac{GMm}{2r^2}\frac{dr}{dt} = -Fv, \text{ where } F = \alpha v^2$$

$$\Rightarrow +\frac{GMm}{2r^2}\frac{dr}{dt} = -\alpha v^2 \cdot v$$

$$\Rightarrow -\frac{GMm}{2r^2}\frac{dr}{dt} = \alpha v^3 \tag{3.176}$$

Putting $v^2 \simeq \frac{GM}{r}$, we have

$$-\frac{GMm}{2r^2}\frac{dr}{dt} = \alpha\frac{GM}{r}\sqrt{\frac{GM}{r}}$$

$$\Rightarrow -\frac{dr}{2\sqrt{r}} = \alpha\sqrt{GM}\ dt$$

Intigrating both sides

$$-m\int_r^R \frac{dr}{\sqrt{r}} = \alpha\sqrt{GM}\int_0^t dt$$

$$\Rightarrow -\sqrt{r}\,|_r^R = \frac{\alpha\sqrt{GM}}{m}t$$

$$\Rightarrow -(\sqrt{R} - \sqrt{r}) = \frac{\alpha\sqrt{GM}}{m}t$$

$$\Rightarrow t = \frac{\sqrt{r} - \sqrt{R}}{m\alpha\sqrt{GM}} = \frac{\sqrt{r} - \sqrt{R}}{m\alpha\sqrt{\frac{GM}{R^2}}\sqrt{R}},$$

where $\frac{GM}{R^2} = g$

$$\Rightarrow t = \frac{1}{m\alpha\sqrt{g}}\left(\sqrt{\frac{r}{R}} - 1\right) \text{Ans.}$$

IOP Publishing

Problems and Solutions in Many-Particle Systems

Pradeep Kumar Sharma

Chapter 4

Fluid statics

4.1 Introduction

In fluid statics, a fluid does not move relative to the container. Fluid is that which can flow; so there is a small or negligible shearing stress between the layers in a fluid. In other words, a fluid is a non-rigid system of particles. In this chapter we will see how to use the formula $\vec{F} = m\vec{a_C}$ for a segment of a fluid (liquid or gas) to define pressure. From the concept of *pressure*, we will develop the expressions for hydro-static force, hydrostatic torque, buoyant force, etc.

After studying this chapter you will understand the mechanism of many hydro-statical phenomena such as why dams are wider at their bases, how hydraulic brakes work, and how to measure the pressures in dams, tanks of gas pipelines etc. You will also understand why ships float but a cube of iron sinks, why a hydrogen balloon accelerates up, etc, in this chapter.

4.2 Fluids and solids

States of matter: Generally, matter has three forms (states): solid, liquid, and gas. We will not discuss the fourth state of matter called 'plasma' in this chapter.

Solid: Practically, we can see that any solid, a piece of stone, say, can stay together for years without changing its shape and size. In other words, a solid has a definite shape and size.

Liquid: A liquid completely filled in a container can assume the shape of the container. If we pour the liquid into different containers, it will attain different shapes. But the volume or size of the liquid remains the same. So, a liquid can have a definite size but different shapes. Gravity plays a significant role in this. However, in free space, a liquid drop assumes a spherical shape because of its surface tension. We will discuss this in chapter 6 'Properties of matter'.

Gas: A gas can expand spreading all around. It can fill a closed container of any shape and size. So, a gas does not have a definite shape and size.

doi:10.1088/978-0-7503-6447-8ch4

Recapitulating, a solid has both definite shape and size; a liquid has a definite size but no definite shape; and a gas has neither a definite size nor a definite shape.

Reaction forces in matter: Matter responds to an external force by either accelerating or by deforming its shape or size. Let's interpret it by considering the variation of interatomic force F with the distance of separation r of two neighbouring atoms.

We can see that at $r = r_0 = 10^{-10}$ m, $F = 0$. At this distance of interatomic separation, no force acts on the atoms. At $r < r_0$, force F becomes positive, which signifies that a repulsion force acts on the atoms. At $r > r_0$, the force becomes negative. This physically signifies an attraction force between the atoms. If $r \rightarrow \infty$, the interatomic force of attraction decreases to zero. When $r \rightarrow 0$, the interatomic repulsion becomes incredibly large. With the above information let's try to explain the behaviour of solid, liquid, and gas when we apply a force on them.

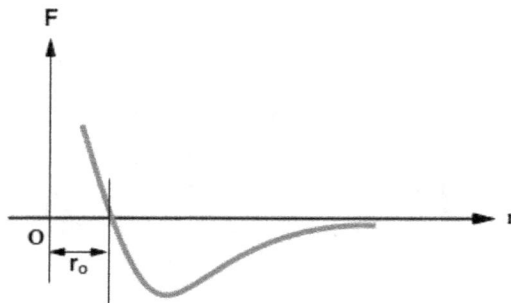

Solid: In solid state, the atoms are strongly attached to neighbouring atoms maintaining an equilibrium (mean) distance $r_0 = 10^{-10}$ m. This means that each atom oscillates about its mean position by small amplitudes. Hence, the molecules and atoms in a solid have strong spring- like interatomic forces. This strong interatomic force in a solid is essentially the cause of a repeating three-dimensional structure, which is known as a lattice structure, as shown in the following figure.

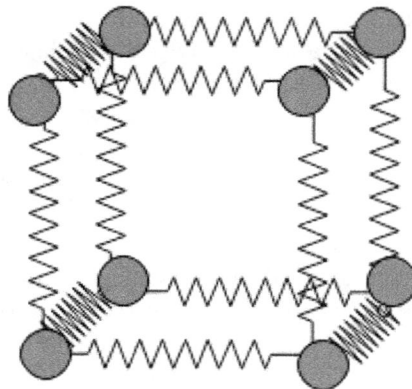

If we compress a solid, the interatomic distance r will be less than r_0. This causes a strong repulsive force between the atoms. This appears as a gross manifestation of a compressive reaction force. Thus, a solid can respond to a compressive or pushing external force.

When you pull a solid, the interatomic separation increases ($r > r_0$). As the force becomes negative in the graph, we can imagine it as a strong attraction force between the atoms. As all atoms start pulling the other atoms, a tensile reaction or tension appears inside the solid.

If you try to twist a solid by applying a torque, a shearing force will appear in the solid. This can also increase the intermolecular separation. As a result, an appropriate shearing force or torque in response to the applied shearing force or torque results.

Since a solid can resist (or withstand) the applied tensile and compressive forces with a little deformation compared to its dimensions, we can say that a solid maintains a definite volume (size). Since a solid can withstand a shearing force, it does not change its shape appreciably. Hence, a solid possesses a definite shape.

Recapitulating, a solid can resist external forces (tensile, compressive, and shearing) by exerting equal and opposite reaction forces on the external agent. A solid can provide a reaction force in any direction.

Liquid: In liquids, molecules are closely spaced at a relatively greater distance compared to that in solids. This causes a weaker intermolecular force.

When we compress a liquid, intermolecular distance will be less than r_0. This gives rise to a strong intermolecular repulsive force. Hence, a liquid can resist a compressive force like solids without any appreciable deformation.

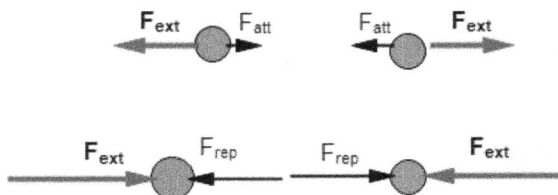

As a simple experiment, you trap a drop of soap water between your fingers; when you slowly separate your fingers more and more, you can see a thin soap film trapped between the fingers which is also being stretched. This means, according to the F-r graph, the large increase in the intermolecular distance r between the molecules causes a significant decrease in the intermolecular force of attraction F. Hence, a liquid can produce a little reaction force to an applied tensile force. This manifests as *surface tension*. Due to the small value of surface tension of water (72 dyne cm^{-1}) we can easily blow a soap bubble.

Moreover, it is not difficult to wave our hands or walk in a swimming pool since the weaker intermolecular attraction cannot noticeably withstand the applied shearing force. This means the shape of the liquid changes as one liquid layer slides on the other, in response to a shearing force.

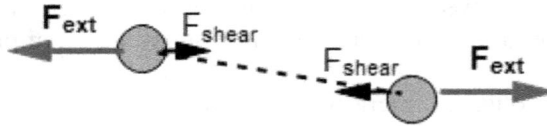

Recapitulating, a liquid can resist a compressive force perfectly and maintains a definite size (volume). A liquid can withstand neither a tensile force nor a shearing force. Due to the weak resistance to the shearing force, the shape of the liquid assumes the shape of the containers.

Gas: In gases, the average spacing between molecules is greatest. Hence, the intermolecular force between the gas molecules is very small. So, gases cannot withstand a shearing stress with negligible deformation. However, pulling or pushing a piston of a gas jar can deform the gas just like an elastic body. Thus, the trapped gases can withstand the tensile and compressive forces although it does not have a define shape and size.

Definition of fluid: Applying a force can displace a fluid. For instance, by switching a fan we can experience a flow of air; a rotating blade can churn a liquid. So, one layer of a liquid (or gas) can slide over the other layers by the application of a small shearing force. In other words, liquids and gases can flow under the action of a shearing force. Due to the flowable character of a liquid or gas, they are known as 'fluids'.

Classification of fluids

Viscous fluid: We now understand why a fluid (liquid or gas) offers little resistance to an external shearing force. This appears in the form of 'viscosity' of fluid. For instance, when you continuously rotate your hand in bucket of water, the water will rotate in response to the shearing force given by your hand in the water. When you remove your hand, the water will stop after some time. When a train moves, the air touching the train moves in the direction of motion of the train in response to the shearing force given by the train to the air. We will discuss the effect of viscosity of liquid in chapter 6 'Properties of matter'.

From the above discussion we understand that a fluid has an internal friction that is called viscous force. These fluids are called 'viscous or real fluid', such as honey, mobil, crude oils, etc.

A fluid is said to be viscous when it exerts a viscous force to the motion of a body in it. Viscosity affects the relative motion of a fluid with respect to the surface in contact, but it does not affect the equilibrium of a fluid.

Perfect fluid: As explained earlier, a fluid, a liquid, say, needs a small shearing force in order to change its shape. On the other hand, a liquid can withstand any amount of a compressive force. This means that a shearing or viscous force in a liquid is negligible compared to the compressive reaction force exerted by the liquid. Then, we can neglect the viscous force of a liquid in comparison with its compressive reaction force to call it an 'ideal or perfect liquid'. So, any layer of a perfect liquid can slide over its other layers without any friction.

Since the compression of a liquid is negligible compared to its total volume, we can ignore the compression of a liquid to make it an ideal or perfect liquid. So, ideal liquids are incompressible. A perfect fluid is not capable of giving any resistance to an external shearing force.

4.3 Definition of fluid statics

Every segment of a fluid has a large number of molecules moving randomly. The velocity of each molecule changes from time to time. For a stationary fluid, the average velocity of each molecule is zero. When a fluid moves relative to the container, it is said to be flowing. The center of mass of each molecule acquires some extra velocity in the direction of flow by the action of a shearing force or any other external force acting on the liquid. Since each molecule experiences a number of forces due to other molecules, it is impossible to describe the motion of the fluid, molecule-wise. As a fluid is a system of particles (molecules), we need to use the mechanics of a system of particles for a fluid. We will apply the basic laws of mechanics such as Newton's laws, conservation of momentum and energy, etc, for a fluid.

The branch of mechanics that deals with the momentum, energy, force, pressure, etc, of a fluid is known as 'mechanics of fluid'. When each element of the fluid is at rest relative to the container, the mechanics of the fluid is called 'fluid statics' and the mechanics that deals with the motion of fluid relative to the container is known as 'fluid dynamics'.

Let's discuss the properties of a fluid at rest.

Interconversion of states of matter: We have explained the difference between solid, liquid, and gas by the concept of intermolecular force. However, the distinction between these three states of matter is incomplete without mentioning the terms *temperature* and *pressure* of the matter. For instance, when you pull a metal through a thin hole (die), it draws into thin wires. This property is called ductility of metals. This is possible because the layers of the metal flow by the application of large shearing (tangential) force in the process of pulling it through a die (hole). We can make thin sheets of metals by hammering them. This happens due to the flowability of metals under increased shearing force. Rocks also can flow in glaciers due to a large shearing stress inside the Earth's crust.

On the other hand, heating a metal weakens the intermolecular (atomic) force. As a result, metal melts into a liquid. Further increase in temperature converts the liquid metal into a gaseous or vapour state. So, the states of matter are changeable. They are interconvertible by the application of suitable pressure (shearing force) and supplying heat energy (increase in temperature).

Increase in pressure converts gases to liquids and liquids to solids. Increase in temperature converts solids to liquids and liquids to gases. Increase in external shearing stress can cause the solids to flow like liquids.

4.4 Density

Average density: If we take 1 litre of water from one place, it weighs 1 kg. If we bring two litres of water from another place, its mass will be 2 kg. This tells us that different quantities (masses) of water occupy different volumes. Hence, neither the mass nor the volume can be treated as a property (a constant quantity) for a liquid. When we take their ratio, that is, $\frac{\text{mass}}{\text{volume}}$ (mass per unit volume), we will get a constant quantity 1 kg per liter. This constant quantity can be treated as a property of water, which is known as 'average density' of water.

In general, the average density ρ of a liquid is given as the ratio of mass of the liquid and the volume occupied by the liquid.

$$\rho_{ac} = \frac{m}{V}$$

If we apply a huge pressure of $3000 \, \text{N m}^{-2}$ on the water surface in a jar by a piston, we can compress it by a very little amount. As the volume decreases, the density increases by a very small factor 15×10^{-8}. This tells us that a liquid is approximately incompressible. In other words, average density of an ideal liquid is constant. However, pressure and temperature can change the average density in a strict sense.

Absolute density: If you press an object (a piece of matter) by a huge force, molecules and atoms will come closer. Then, the volume of the object will decrease considerably but it is not the minimum possible volume! As we know, the majority of space occupied by atoms and molecules is vacuum. Most of the mass of an atom is concentrated in the nucleus whose volume is 10^{-12} times smaller than the volume of the atom. This means that if we remove the electrons (which share negligible mass to the atom), the volume of the matter will be negligibly smaller than its original volume. This state of matter is called a 'plasma state'. Due to the minimum possible volume, a piece of matter in a plasma state has incredibly large density. This is known as 'absolute density'. For instance, a neutron star consisting of clusters of neutrons (nuclei) has a density of $0.5 \times 10^{12} \, \text{kg m}^{-3}$!

Absolute density is defined as the ratio of mass of a substance and the minimum possible volume occupied by it (volume of the matter consisting of the nuclei only).

4.5 Specific gravity (relative density)

Definition: It is convenient to express the average density of a liquid as a unitless and dimensionless quantity by dividing its density with the density of a standard substance (water, say). Then, the relative density (RD) is given as

$$RD = \frac{\rho_L}{\rho_w},$$

where ρ_L = density of liquid at any temperature θ and ρ_w = density of water at that temperature ($\theta > 4 \, °C$).

The above equation tells us that *RD* is a unitless and dimensionless number that tells us how many times a liquid is denser than water, at a given temperature.

As we know, generally a liquid expands with rise in temperature. Hence, the density of liquid decreases with rise in temperature because mass of the liquid does not change. This will be discussed in a chapter of *Problems and Solutions of Waves and Oscillations and Heat and Thermodynamics*.

In this chapter we will talk about the density of a liquid at a given temperature.

Calculation of specific gravity of a mixture

(a) Volumes and specific gravities are given:

Let V_1, V_2, ..., V_n be the volumes and S_1, S_2, ..., S_n be the specific gravities of the given substances constituting the mixture.

The mass of the mixture is

$$m = \rho_1 V_1 + \rho_2 V_2 + \cdots + \rho_n V_n$$

Substituting $\rho_1 = S_1 \rho_w$, $\rho_2 = S_2 \rho_w$ and so on, we have

$$m = \rho_w (S_1 V_1 + S_2 V_2 + \cdots + S_n V_n)$$

The volume of the mixture is

$$V = V_1 + V_2 + \cdots + V_n$$

Then, the density of the mixture is

$$\rho = \frac{m}{V}$$

Substituting m and V, we have

$$\rho = \frac{(S_1 V_1 + S_2 V_2 + \cdots + S_n V_n)\rho_w}{V_1 + V_2 + \cdots + V_n}$$

Finally, substituting $\frac{\rho}{\rho_w} = S$ (specific gravity of the mixture), we have

$$S = \frac{V_1 S_1 + V_2 S_2 + \cdots + S_n V_n}{V_1 + V_2 + \cdots + V_n}$$

In a nutshell, we have
$S = \frac{\sum V_i S_i}{\sum V_i}$, where $i = 1, 2, ..., n$.

(b) Masses and specific gravities are given:

Let m_1, m_2, ..., m_n be the masses and S_1, S_2, ..., S_n be the specific gravities of the substances constituting the mixture.

The total mass of the mixture is $m = m_1 + m_2 + \cdots + m_n$.

The volume of the mixture is

$$V = V_1 + V_2 + \cdots + V_n,$$

where $V_1 = \frac{m_1}{\rho_w S_1}$, $V_2 = \frac{m_2}{\rho_w S_2}$, ..., $V_n = \frac{m_n}{\rho_w S_n}$

This gives

$$V = \frac{1}{\rho_w}\left(\frac{m_1}{S_1} + \frac{m_2}{S_2} + \cdots + \frac{m_n}{S_n}\right)$$

Then, the density of the mixture is

$$\rho = \frac{m}{V}$$

Substituting m and V, we have

$$\rho = \frac{(m_1 + m_2 + \cdots + m_n)\rho_n}{\left(\frac{m_1}{S_1} + \frac{m_2}{S_2} + \cdots + \frac{m_n}{S_n}\right)}$$

Substituting $\frac{\rho}{\rho_w} = S$ (specific gravity of the mixture), we have

$$S = \frac{m_1 + m_2 + \cdots + m_n}{\frac{m_1}{S_1} + \frac{m_2}{S_2} + \cdots + \frac{m_n}{S_n}}$$

In a nutshell, we have
$S = \frac{\sum m_i}{\sum \frac{m_i}{S_i}}$, where $i = 1, 2, \ldots, n$.

Example 1 When equal volumes of two liquids are mixed, the specific gravity of a mixture is 4; when equal weights of the same liquids are mixed, the specific gravity of the mixture is 3. What are the specific gravities of two liquids?
Solution
Let S_1 and S_2 be the specific gravities of two liquids. Since the volumes of liquids are equal, $V_1 = V_2 (=V$ say).
Substituting $S = 4$, $V_1 = V_2 = V$ in the expression $S = \frac{V_1 S_1 + V_2 S_2}{V_1 + V_2}$, we have

$$4 = \frac{V S_1 + V S_2}{V + V}$$

This gives

$$S_1 + S_2 = 8 \tag{4.1}$$

In the second case, since the masses of liquids are equal, $m_1 = m_2 (=m$ say).
Substituting $S = 3$, $m_1 = m_2 = m$ in the expression
$S = \frac{m_1 + m_2}{\frac{m_1}{S_1} + \frac{m_2}{S_2}}$, we have

$$3 = \frac{m + m}{\frac{m}{S_1} + \frac{m}{S_2}}$$

This gives

$$\frac{S_1 + S_2}{S_1 S_2} = \frac{2}{3} \tag{4.2}$$

Solving equations (4.1) and (4.2), we have $S_1 = 6$ and $S_2 = 2$. Ans.

N.B.: You can show that the specific gravity of a mixture of n liquids is greater, when equal volumes are taken compared to that when equal weights are taken. Assume no chemical reaction and change in volume of the liquids.

4.6 Hydrostatic force

Definition: If you put your hand in a bucket of water, do you feel any shearing force unless you move your hand? This tells us that a fluid does not exert a shearing (tangential) force on a surface in contact if it is stationary relative to the fluid. Then, the fluids (liquids and gases) can exert forces normal to the contacting surfaces. Let's took at the following illustrations to support this fact.

While standing in a swimming pool we feel as if water is pressing our body from all sides. Atmosphere presses our bodies from all directions. In all cases we do not feel any shear on our skins.

If you try to stop the water leakage from a tank by pressing your finger on the hole, you can feel as if the water pushes your finger out of the hole.

N.B.: From the above explanation, we have the following points:

1. Hydrostatic force is a normal contact force dF_n (or normal reaction) of the liquid on the surface of the container or any surface dA imagined inside the liquid.

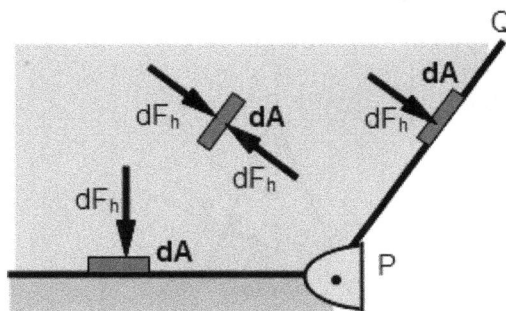

2. The hydrostatic force acts in the direction of outward normal given by the unit vector $\hat{n}$. (The outward normal is the normal drawn to the surface outside the liquid in this case. If the liquid is in the left side, outward normal will be in the right side and vice versa.)

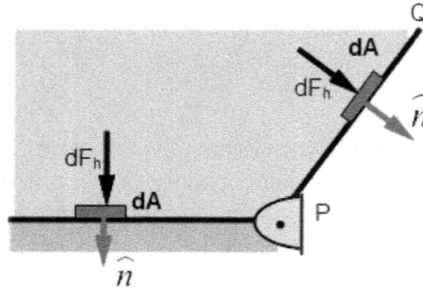

Example 2 We have three vessels of different shapes filled with water. Let's take a small patch of area dA in each container. In each case, what will be the direction of hydrostatic force?

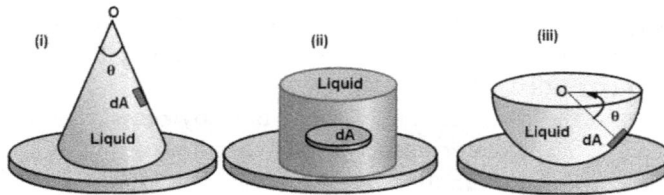

Solution

   (i) Since the hydrostatic force points in the direction of outward normal drawn to the surface in contact, water pushes the red patch at an angle of $\theta/2$ with horizontal.

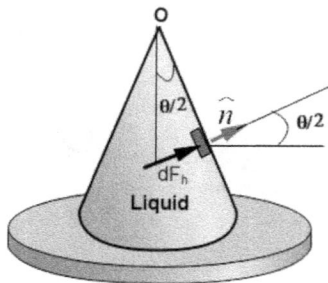

   (ii) In this case, water pushes the given upper surface of the circular plate vertically down because the outward normal is pointing downwards. Ans.

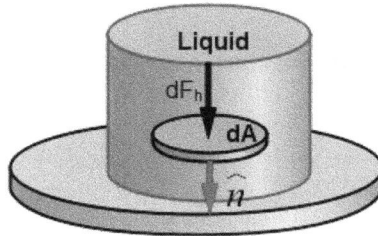

(iii) The water pushes the red patch radially out at an angle of $\theta = 30°$ with horizontal.

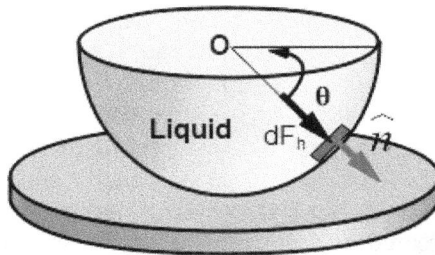

In all cases the hydrostatic force dF is perpendicular to the surface area dA and points in the direction of outward normal $\hat{n}$.

Micro-interpretation of hydrostatic force: When we look at the microscopic level, each liquid molecule collides elastically with the surface on the container (or any surface placed inside it). During each elastic collision, the momentum of the liquid changes normal to the surface. As a consequence, the average force exerted by the liquid molecules is directed along the outward normal to the surface in contact. That is why the hydrostatic force is a 'normal reaction'.

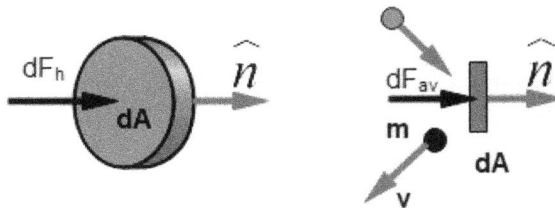

Hydrostatic force is a normal contact force caused by the exchange of momenta of liquid molecules perpendicular to the surface in contact.

4.7 Hydrostatic pressure

Why do we need the quantity 'pressure'? In the last section we learned that the hydrostatic force arises from the collision of liquid molecules with the surface in contact. Each liquid molecule pushes the surface normally reversing its momentum normal to the surface. So, the hydrostatic force is distributed over the area of the given surface rather than acting at any specified point of the surface.

As a surface can be flat or curved, a flat surface can have different size (magnitudes) and orientations (directions) and a curved surface can have different shape and size. For this purpose, we can take an elementary (very small) patch. As the hydrostatic force dF always acts perpendicular to the surface area, its orientation will be the same for a flat surface and varies from point to point on a curved surface.

For the sake of simplicity, to keep the orientation same, let's first consider a flat surface. For a flat surface, each element has the same angle of orientation. This means liquid pushes each element of a flat surface in the same direction along outward normal. In other words, hydrostatic force dF acts on each element (segment) of a flat surface in the same direction.

If we take a smaller surface, the hydrostatic force (thrust) of the liquid is smaller; if we take a bigger surface, more hydrostatic force will act on it. It is practically more difficult to press your finger against a bigger hole than a smaller hole on the wall of a tank of water, to prevent the water leakage. Thus, neither hydrostatic force nor area of the surface can be treated as a constant quantity. Hence, we need the ratio of hydrostatic force dF and area dA to define a pressure as an intensity of force in order to characterise a liquid as its property. It is comparable to the density, which is defined as the ratio of mass and volume, given as

$$\frac{m}{V} = \rho$$

Average pressure: Let's assume that the liquid presses the finite patch of area ΔA with a force ΔF. On an average, each unit area feels the thrust of liquid, which is equal to total force divided by total area, that is, $\frac{\Delta F}{\Delta A}$. This is known as average surface density of force which is more familiar as 'average pressure' denoted by the symbol P_{av}

$$P_{av} = \frac{\Delta F}{\Delta A}$$

Pressure is a scalar quantity because it is defined as a ratio of magnitude of force (acting perpendicular to a flat surface of area A) and the area A. The quantity *pressure* is used as a scale to measure the hydrostatic force.

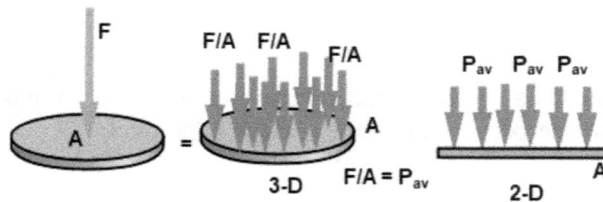

Average or uniform pressure distribution

However, P_{av} cannot give you the actual amount of force acting on each unit area. Hence, it is just a mathematical quantity giving a statistics of average distribution of force on a given area. It is similar to the average marks of your class (total marks divided by total students) giving us statistics of intellectual strength (merit) of your class.

When you have the average pressure P_{av} over any area A, the force acting on that area can be vectorially given as

$$\Delta \overrightarrow{F} = P_{av} \Delta A \hat{n}$$

where $\hat{n} =$ unit vector directed along outward normal to the surface.

Outward normal

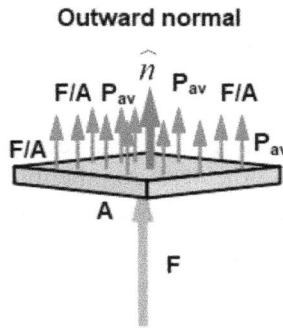

For instance, a vertical gate in a dam experiences a side thrust F due to water. If its area of contact with water is A, the average pressure is given as

$$P_{av} = F / A$$

Hydrostatic pressure at a point: Generally, we are interested in 'pressure at a point' because we experience (feel) it practically. We will present the physical significance of pressure at a point shortly. Pressure at a point can be denoted as P (without the suffix 'av' at the bottom of the letter P). To find the pressure at a point, we need to reduce the finite area to zero; mathematically we write that $\Delta \overrightarrow{A} \rightarrow 0$ (or $d\overrightarrow{A}$); $d\overrightarrow{A}$ means an elementary (point-like) area. Let the hydrostatic force acting on it equal to $d\overrightarrow{F}$. Dividing $|d\overrightarrow{F}|$ with $|d\overrightarrow{A}|$, we have the pressure P at a point. Since pressure at a point is a pressure averaged over a vanishingly small area ($\Delta \overrightarrow{A} \rightarrow 0$), we can write

$$P = \lim_{A \to 0} P_{av} = \lim_{A \to 0} \frac{F}{A}$$

When $A \to 0$, we have $F \to 0$; however, the ratio of two vanishingly small quantities is *finite*.

This means $\lim_{A \to 0} \frac{F}{A}$ can be written as $\frac{dF}{dA}$ using the notations of calculus. In this way, we can define the pressure at a point as

$$P = \frac{dF}{dA}$$

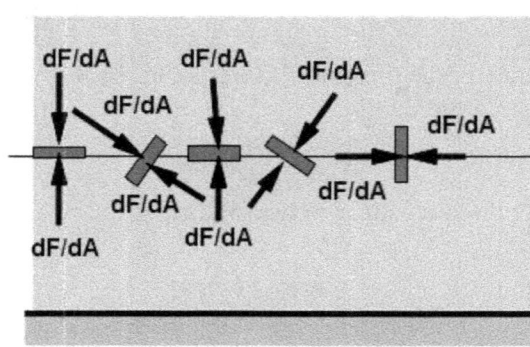

Pressure at a point of a liquid at rest is a characteristic or property of the liquid. It is a point (or microscopic) function. It shows the intensity of force at that point.

When we have pressure P at a point (on an elementary area $d\vec{A}$), the hydrostatic force on the elementary area can be vectorially given as $d\vec{F} = Pd\vec{A}$.

This formula is equally valid for both flat and curved surfaces, because an elementary area (patch) behaves as a flat surface. This formula works at any point of the liquid.

Uniform and non-uniform pressure: When the pressure at any point of a flat surface is the same, we say that the distribution of hydrostatic force is uniform. In other words, if the liquid pushes all elements of a surface with equal force, the pressure is said to be uniform. In this case, average pressure over any portion of the surface is a constant quantity.

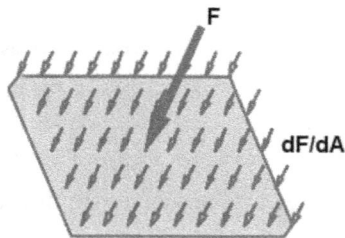

Total Area =A
Total Force =F(Big arrow)

F/A=dF/dA=P (Small arrows)

If pressure varies from point to point, we call it non-uniform pressure. This means that the liquid pushes different elements of the surface with different magnitudes of force.

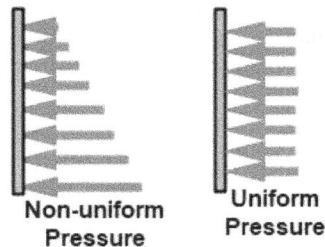

Non-uniform
Pressure

Uniform
Pressure

Pressure is the same in any direction: We know that a liquid molecule has equal probability of exchanging its momentum in all directions, because a molecule has no preferred direction of its random motion. So, a liquid can exert a force of the same magnitude on a small (elementary) flat area when it is rotated about a point and kept at any orientation in space. We have a conceptual device consisting of a piston fitted with a spring in an evacuated tube. The same pressure is felt by the small light piston in all orientations of the device, which can be experienced by observing an equal compression of the spring.

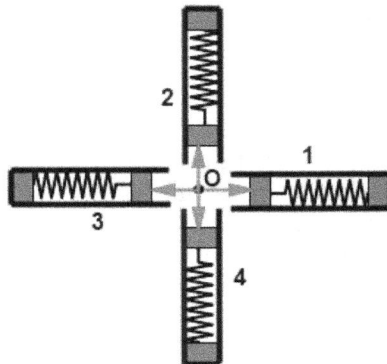

The pressure at a point is the same in all directions in a stationary liquid. Pressure is directional but not specified. So, in a strict sense, it is a *tensor*.

Physical significance of pressure at a point: If you sink in a lake, you feel as if your body is pressed by the liquid (water) from all sides normal to the skin of your body. You feel the pressure on each and every point of your skin.

Sharp objects (needle, say) hurt you more than the blunt objects when they are pressed against your skin by equal forces. This is because of the greater pressure of sharp objects due to the smaller surface area of contact.

N.B.: We feel the pressure as the intensity of force at different points of our bodies. It pinches more if the pressure is more. If the thrust at a point is more painful, the intensity of force, that is, the pressure is more at that point. In this way, pressure is defined as a scalar quantity used as a scale to measure the distributed forces.

2. Pressure at a point of a liquid is a property (or characteristic) of the liquid.

3. In a stationary fluid, the force per unit area (pressure) across any imaginary surface is normal to the surface for its all orientations. At any position (or depth) of the elementary surface, pressure due to the liquid is constant for all orientations.

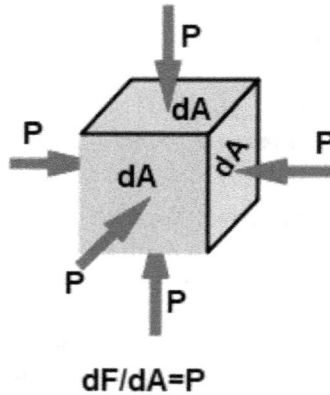

$$dF/dA=P$$

4.8 Pressure in a non-accelerating liquid

A non-accelerating liquid means the container is either at rest or moving with constant velocity. Neither the container nor the liquid should rotate. Furthermore, the liquid should not move relative to the container to obey the definition of 'hydrostatics'.

For the sake of simplicity let's consider a stationary liquid. As the stationary liquid is in equilibrium, the net force acting on any segment of the liquid is zero. Let's use this idea to find the pressure difference between any two points 1 and 2 (say) inside the liquid.

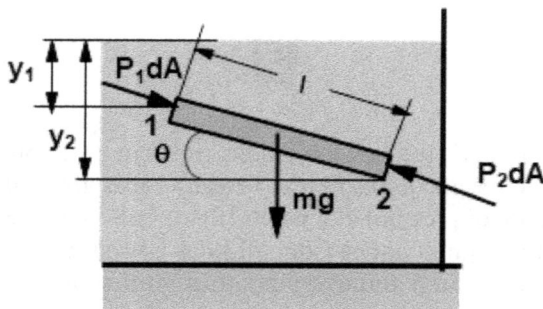

Let's assume that the depths (vertical distance from the free surface of liquid) of the points 1 and 2 are y_1 and y_2, respectively. Imagine a thin cylinder (tube) of liquid between the points 1 and 2. If the area of cross-section of the cylinder is δA, the forces acting on the cylinder along its length are hydrostatic forces $P_1 \delta A \searrow$, $P_2 \delta A \nwarrow$, and $\delta mg \sin \theta \searrow$.

Since the tube of liquid is in equilibrium, its acceleration is zero.

So, the net force acting on the cylinder is

$$F_{\text{net}}(=P_1\delta A + \delta mg \sin \theta - P_2\delta A) = 0$$

This gives

$$P_2 - P_1 = \frac{\delta mg}{\delta A} \sin \theta,$$

where $\delta m = \rho(\delta A)l$.

Then, we have

$$P_2 - P_1 = \rho lg \sin \theta$$

Substituting $l \sin \theta = y_2 - y_1$, we have

$$P_2 - P_1 = \rho g(y_2 - y_1)$$

N.B.: From the given equation we conclude the following points for non-accelerating liquid:

1. Hydrostatic pressure in a non-accelerating liquid arises due to the effect of gravity.
2. When the points are situated at the same depths, $(y_1 = y_2)$, we have $P_2 = P_1$. This means pressure does not vary in the horizontal direction for a non-accelerating liquid. In other words, pressures at two points at the same depths are the same for a stationary (or non-accelerating) liquid.
3. When two points lie on the same vertical line

$$\text{We have} \quad P_2 - P_1 = \rho g y,$$

where $y(=y_2 - y_1)$ is the distance between two points. However, in general, y is the vertical distance between two points (not horizontal distance).
4. Substituting $y_1 = 0$, $P_1 = P_0$ (atmospheric pressure), $y_2 = y$, $P_2 = P$, we have

$$P = P_0 + \rho g y$$

This tells us that hydrostatic pressure increases linearly with the depth.

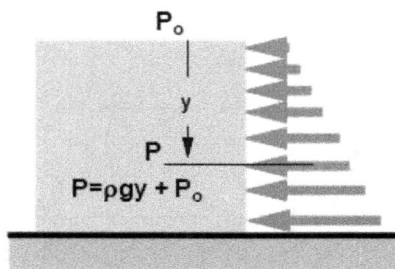

Example 3 We derived the expression $P_2 - P_1 = \rho g(y_2 - y_1)$ to find the pressure difference between two points 1 and 2 in a liquid, which can be directly connected by a straight line 1–2 drawn inside the liquid in a rectangular vessel. If we consider a liquid in a tube of arbitrary shape and size (a vertical circular tube of radius R completely filled with a liquid of density ρ, say) we can no longer connect the points 1 and 2 by a straight line. In this case, how is the above formula valid to find the pressure difference between these points?

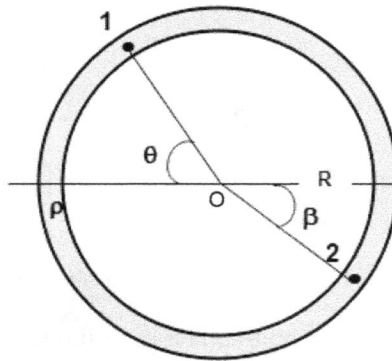

Solution

Let's draw a series of vertical and horizontal lines from point 1 to point 2. This means the points are connected by the vertical and horizontal lines. We learned that pressure does not change horizontally in a non-accelerating liquid. Since difference in pressure is directly proportional to the vertical distance AC, according to the expression

$$P_2 - P_1 = \rho g(\text{AC}),$$

where $\text{AC} = R \sin \theta + R \cos \beta = R(\sin \theta + \cos \beta)$.

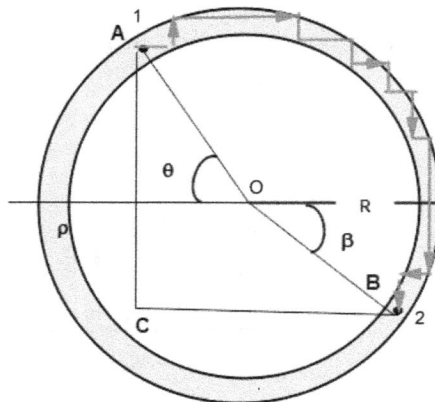

$$\Rightarrow P_2 - P_1 = \rho g R(\sin \theta + \cos \beta) \text{ Ans.}$$

N.B.:

1. The pressure difference between any two points 1 and 2 in a non-accelerating liquid filled in an open or closed tube of arbitrary shape and size is given as

$$P_2 = P_1 + \rho g y,$$

where y is the vertical distance between two points. The point at a greater depth has more pressure.

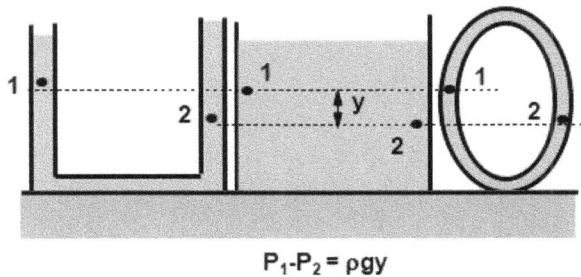

$$P_1\text{-}P_2 = \rho g y$$

2. Hence, levels of liquid in the vessels maintain the same height. Hydrostatic pressure does not depend on the shape and size of the vessels. It depends only on the vertical distance y.

3. The free surface of a non-accelerating liquid always remains horizontal because each element of the free surface of the liquid is at the same atmospheric pressure. So, they must be at the same height h.

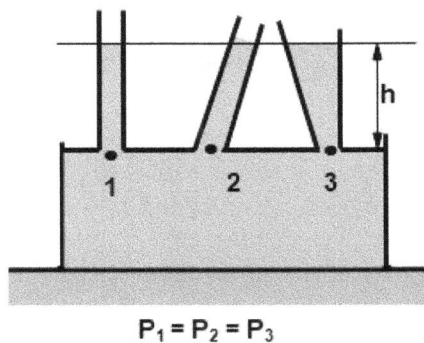

$$P_1 = P_2 = P_3$$

4. For a homogeneous liquid, hydrostatic pressure increases with depth; $P \propto y$. So, the dams are made thicker at their bases.

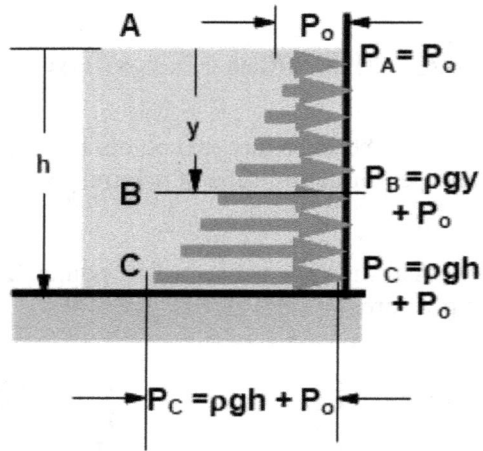

5. For any non-accelerating liquid, hydrostatic pressure is uniform at the same depth.

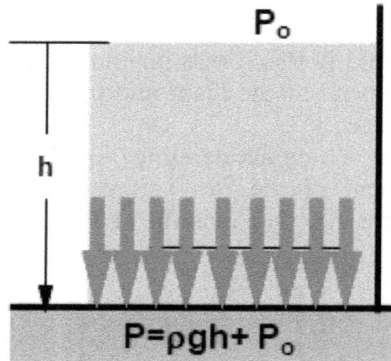

Example 4 A cubical vessel of side $AB = b$ filled with a liquid of density ρ is kept at rest on an inclined plane of angle of inclination θ. Find the pressure difference across the base of the vessel between A and B in terms of b.

Solution

The pressures at A and B are

$$P_A = P_0 + \rho g y_1$$

$$P_B = P_0 + \rho g y_2$$

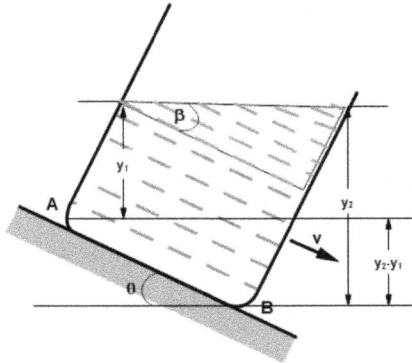

Then, we have

$$P_B - P_A = P_0 + \rho g y_2 - \left(P_0 + \rho g y_1\right)$$

$$= \rho g(y_2 - y_1),$$

where $y_2 - y_1 = AB \sin \theta = b \sin \theta$.

Then, we have $P_B - P_A = \rho g b \sin \theta$ Ans.

N.B.: In the above example, if the length of the liquid column perpendicular to the inclined plane at the back and front are given as h_1 and h_2, respectively, and if you write

$$P_B - P_A = \rho g(h_1 - h_2),$$

it will lead to the wrong answer because in the formula, $P = P_0 + \rho g y$, 'y' is the vertical distance of the points A and B from the free surface (horizontal) of the non-accelerating liquid. You can see that $y_1 = h_1 \cos \theta$ and $y_2 = h_2 \cos \theta$; where h_1 and h_2 are the distances (lengths) of the liquid column perpendicular to the inclined plane at the left and right sides of the vessel.

Pressure gradient: If we take the variation of density of fluid (for instance, the density of atmosphere decreases with altitude) into account, we cannot use the last formula. So, we have to introduce the expression of *pressure gradient* at a point in a liquid.

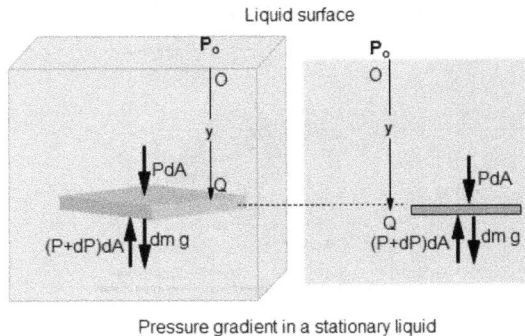

Pressure gradient in a stationary liquid

For this, let's consider a thin horizontal slice of liquid at a depth y. If the thickness of the slice is dy and its area is dA, its weight is given as $dmg = \rho(Ady)g$.

If the pressure at the top and bottom of the slice are P and $P + dP$, respectively, the hydrostatic force PdA pushes the slice down and the hydrostatic force $(P + dP)A$ pushes the slice up.

Since the slice is in equilibrium under the action of the forces $dmg\downarrow$, $PdA\downarrow$, and $(P + dP)dA\uparrow$, the net force acting on the slice is zero. This means

$$F_{\text{net}} = -dmg - PdA + (P + dP)dA = 0$$

This gives $dP = \frac{dm}{dA}g$,

where $\frac{dm}{dA} = \rho dy$ (since $dm = \rho Ady$, we have $\frac{dm}{dA} = \rho dy$).

Hence, $\frac{dP}{dy} = \rho g$.

N.B.: From the above expression we conclude the following points for non-accelerating liquid:

1. The pressure gradient at a point in a non-accelerating liquid is equal to the product of density of liquid and acceleration due to gravity at the point.
2. If the density of a fluid is assumed to be constant (in the case of an ideal liquid), the pressure gradient is a constant quantity for a height h for which we can take g as a constant if $h \ll R$ (radius of Earth).
3. If pressure gradient is given (or is found as the slope of P-y graph), the density of the liquid can be given as

$$\rho = \frac{\frac{dP}{dy}}{g} = \frac{\text{Slope of } P-y \text{ graph}}{\text{Acceleration due to gravity}} = \frac{\tan\theta}{g}$$

4. When ρ and g are constants, using $\frac{dP}{dy} = \rho g$ the hydrostatic pressure (including atmosphere) at a depth y is given as

$$P = \rho g y + P_0,$$

where $P_0 =$ atmospheric pressure.

This means pressure increases linearly with depth for a homogeneous non-accelerating liquid.

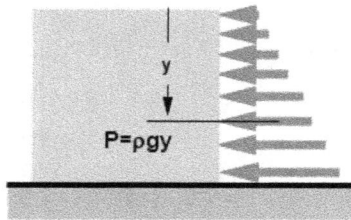

5. When $\rho = f(y)$ and $g = C$, using the expression

$$\frac{dP}{dy} = \rho g,$$

the hydrostatic pressure at a depth y is given as

$$P = g \int_0^y \rho \, dy + P_0$$

6. Pressure gradient $\frac{dP}{dy}$ is positive $(+\rho g)$ when we take y as depth (from the top of the liquid) and negative $(-\rho g)$ when we take y as height (from ground level).
7. In a non-accelerating homogeneous liquid, pressures at the same horizontal level are the same to equilibrate the horizontal elementary tube. This means

$$F_{\text{net}} = P_1 \delta A - P_2 \delta A = 0$$

Then, $P_1 = P_2$

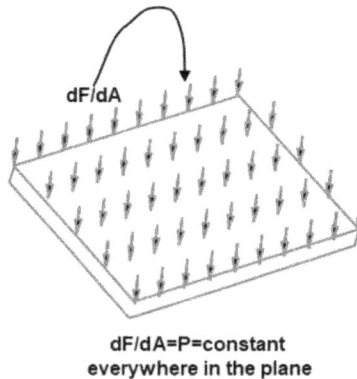

dF/dA

dF/dA=P=constant
everywhere in the plane

In other words, when pressures are the same at all the points of the horizontal plane, the corresponding densities must be equal.

Example 5 We know that the pressure of the atmosphere (layers of gases of thickness of 360 km from the Earth's surface) is P_0 at the Earth's surface; $P_0 = 1.013 \times 10^5 \text{ N m}^{-2}$. Taking the variation of densities of gases into account and ignoring the variation of temperature, derive an expression for the atmospheric pressure at a height h from the Earth's surface. Assume that, $M =$ average molecular mass of all gases of the atmosphere and $T =$ temperature of the atmosphere.
Solution
Assuming a thin horizontal strip of thickness dh at a height h, we have

$$\frac{dP}{dh} = -\rho g \tag{4.3}$$

According to general gas law for an ideal gas of mass m, we have

$$PV = \frac{m}{M}RT$$

This gives the density of the gas as

$$\rho\left(=\frac{m}{V}\right) = \frac{PM}{RT} \tag{4.4}$$

Substituting ρ from equations (4.4) in (4.3), we have

$$\frac{dP}{dh} = -\frac{PMg}{RT}$$

This gives

$$\frac{dP}{P} = -\frac{Mg}{RT}dh$$

If pressure changes (decreases) from P_0 to P when we move from the Earth's surface to a height h, then we have

$$\int_{P_0}^{P} \frac{dP}{P} = -\frac{Mg}{RT} \int_0^h dh$$

$$\Rightarrow [\ln P]_{P_0}^{P} = -\frac{Mgh}{RT}$$

$$\Rightarrow P = P_0 e^{-\frac{Mgh}{RT}} \text{ Ans.}$$

Example 6 In the foregoing example, find the height of the center of mass of a thin vertical cylinder of atmosphere (assuming that the height of the atmosphere is infinite).

Solution

The height of the center of mass in a continuous mass distribution is given as

$$h_c = \frac{\int_0^\infty h\rho\, dh}{\int \rho\, dh} \tag{4.5}$$

As obtained in the last problem,

$$\rho = \frac{PM}{RT} \tag{4.6}$$

Using the last two equations, we have

$$h_c = \frac{\int_0^\infty h\rho\, dh}{\int_0^\infty \rho\, dh} = \frac{\int_0^\infty h\frac{PM}{RT}dh}{\int_0^\infty \frac{PM}{RT}dh}$$

Taking the constant M/RT out of the integral in the numerator and denominator and cancelling them out, we have

$$\Rightarrow h_c = \frac{\int_0^\infty hP\,dh}{\int_0^\infty P\,dh} \tag{4.7}$$

With reference to the result of the last problem, the atmospheric pressure at any altitude h is

$$P = P_0 e^{-\frac{Mgh}{RT}} \tag{4.8}$$

Using the last two equations, we have

$$\Rightarrow h_c = \frac{\int_0^\infty h\left(P_0 e^{-\frac{Mgh}{RT}}\right)dh}{\int_0^\infty P_0 e^{-\frac{Mgh}{RT}}\,dh} = \frac{\int_0^\infty h e^{-\frac{Mgh}{RT}}\,dh}{\int_0^\infty e^{-\frac{Mgh}{RT}}\,dh}$$

After evaluating the integration and simplifying the factors, we have

$$h_c = \frac{RT}{Mg} \text{ Ans.}$$

N.B.:

1. The variation of pressure with altitude h given by $P = P_0 e^{-\frac{Mgh}{RT}}$ is valid based on the following assumptions:
    i. The constituent gases of the atmosphere are ideal.
    ii. The variation of temperature (decrease in temperature) with height is neglected.
    iii. The variation of acceleration due to gravity with height is neglected.
2. The presence of gravity increases the rate of collision of the fluid molecules if we come down to the bottom of a fluid. This in turn increases the pressure.
3. If the temperature varies linearly as

$$T = T_1 + ky,$$

we have

$$P = P_1\left(\frac{T_1}{T_1 + ky}\right)^{\frac{1}{kR}}$$

4.9 Pressure due to many non-accelerating liquid layers

Until now we have talked about the pressure due to one liquid. Let's now consider two immiscible liquids of densities ρ_1 and ρ_2 placed in a container. Now a question arises, how do they remain inside the container? What will the shape of their interface be? Will it be horizontal or vertical or inclined or curved? Let's discuss.

Interface of two immiscible liquids: Let's take a horizontal thin tube of cross-section δA, consisting of these two liquids. If the pressures at the points 1 and 2 are P_1 and P_2, respectively, the hydrostatic forces at these points are $P_1 \delta A \rightarrow$ and $P_2 \delta A \leftarrow$, respectively. Under the action of these horizontal forces, the tube is in equilibrium in horizontal. Equating these forces, we have

$$P_1 \delta A = P_2 \delta A$$

This gives

$$P_1 = P_2,$$

where $P_1 = \rho_1 g y$ and $P_2 = \rho_2 g y$.

Then, we have

$$\rho_1 = \rho_2$$

This means at any two points in the same horizontal line the densities of these liquids must be the same. In other words, the same liquid must be present at all points of the same depth.

Then, the interface of two non-accelerating immiscible liquids is horizontal.

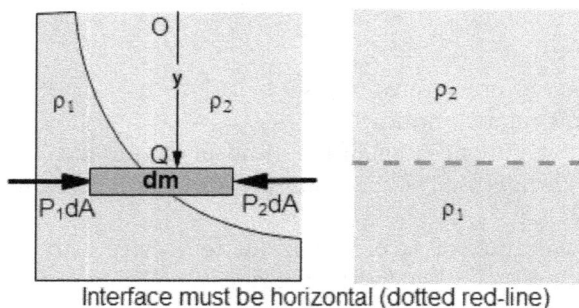

Interface must be horizontal (dotted red-line)

Pressure due to multiple liquids: Let's assume that n number of immiscible liquids are kept in a container one over the other. Generally, lighter liquids stay above the heavier liquids because heavier liquids settle down. Otherwise, you have to provide thin separating polyethene membrane between the liquids.

Let the densities and heights of the liquids above the point under consideration be $(\rho_1, h_1), (\rho_2, h_2), \ldots, (\rho_n, h_n)$, respectively. Let's now take a thin vertical cylinder containing all liquids above the given point. Since the thin cylinder is in equilibrium under the downward atmospheric force $P_0 \delta A$, weights of the corresponding liquid columns, i.e., $\delta m_1 g, \delta m_2 g, \ldots, \delta m_n g$, respectively, and upward hydrostatic force $P \delta A$ at the given point (P = hydrostatic pressure at that point), the sum of all the above forces is zero.

$$F_{\text{net}} = -\sum \delta m_1 g - P_0 \delta A + P \delta A$$

Since $\delta m_i = \rho_i h_i \delta A$, by substituting $F_{net} = 0$, we have

$$P = P_0 + g\sum \rho_i h_i$$

If you consider the hydrostatic forces only, we have

$$P_{hydro} = g\sum \rho_i h_i; \ \ i = 1, 2, \cdots, n.$$

The given expression tells us that total hydrostatic pressure at a point is equal to sum of hydrostatic pressure of each liquid:

$$P = \sum P_i, \ \text{where} \ \ P_i = \sum \rho_i g h_i$$

2. When a tube contains n number of immiscible liquids, pressure at the lowest point is

$$P = \sum \rho_i g h_i$$

where h_i = height (but not necessarily the lengths of the liquids columns).

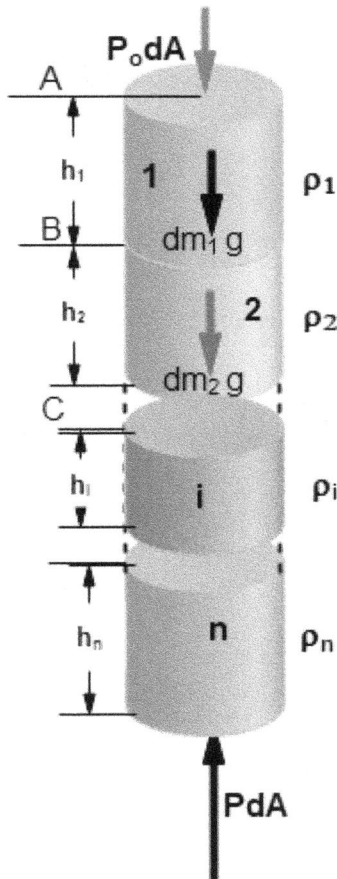

3. The pressure P does not depend on the shape and size of the tube.

Example 7 (Manometer) A U-tube containing mercury is connected to a spherical tank A of gas (or liquid). We need to find the pressure inside the tank observing from the difference in mercury columns in the U-tube.

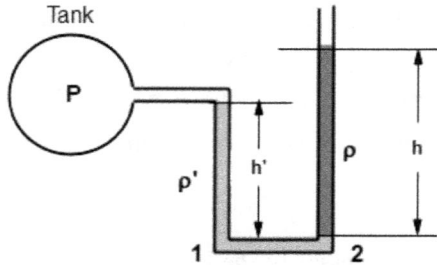

Solution

Let the pressure inside the tank be P. Applying Pascal's law, the pressure at point 1 is

$$P_1 = P + \rho'gh' \tag{4.9}$$

Similarly, the pressure at 2 is

$$P_2 = P_o + \rho gh \tag{4.10}$$

Since the portion of tube 1–2 is horizontal,

$$P_1 = P_2 \tag{4.11}$$

Substituting P_1 from equation (4.9) and P_2 from equations (4.10) in (4.11), we have

$$P + \rho'gh' = P_0 + \rho gh$$

$$\Rightarrow P_1 + \rho'gh' = P_0 + \rho g(y_2 - y_1)$$

Substituting $y_2 - y_1 = h$. we have

$$P = P_0 + \rho gh - \rho'gh' \text{ Ans.}$$

N.B.:

1. If ρ' is much lesser than ρ in the case of gases, we have $P \cong P_0 + \rho gh$.
2. Manometer is a pressure measuring device. The formula used here is given as

$$P = P_0 + g\sum \rho_i h_i,$$

where P_0 = atmospheric pressure and, ρ_i and h_i are the density and height of ith fluid column, respectively.

Example 8 In the foregoing example, we connect another spherical tank B to the open end of the U-tube; observing the heights h_1, h_2 of the columns of fluids of densities ρ_1 and ρ_2 of C and B, respectively. Find the difference in pressures, $P_A - P_B$. This is what we call 'differential manometer' which gives the difference in pressures of the fluids in the tanks.

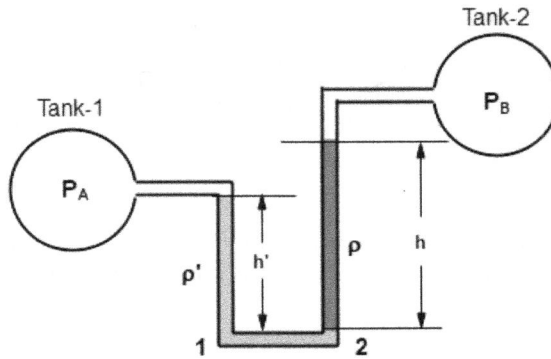

Solution

Since the gases inside the tanks press the fluid inside the tubes to maintain the given lengths, applying Pascal's law, the pressure at point 1 is

$$P_1 = P_A + \rho'gh' \tag{4.12}$$

Similarly, the pressure at point 2 is

$$P_2 = P_B + \rho gy \tag{4.13}$$

Since the portion of tube 1–2 is horizontal and non-accelerating (rest),

$$P_1 = P_2 \tag{4.14}$$

Substituting P_1 from equation (4.12) and P_2 from equations (4.13) in (4.14), and simplifying the factors, we have $P_A - P_B = (\rho h - \rho'h')g$ Ans.

Example 9 A circular tube of uniform cross-section is filled with two liquids of densities ρ_1 and ρ_2 such that half of each liquid occupies a quarter of the tube. If the line joining the free surfaces of the liquids and center O of the circle makes an angle θ, find the value of θ.

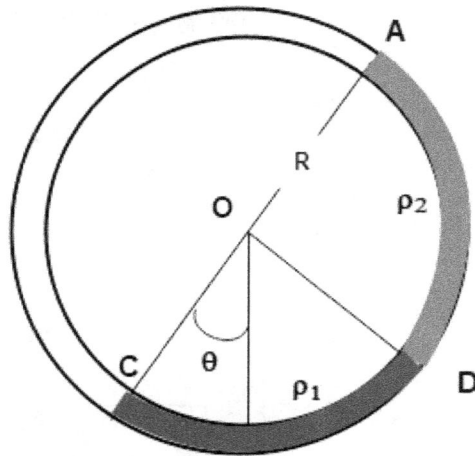

Solution

Let's find the pressure at the lowest point B. Since the liquid of density ρ_2 and height of liquid column h'_2 is there, in the right-hand side of the point B, the pressure at B is

$$P_1 = \rho_1 g h_1 \tag{4.15}$$

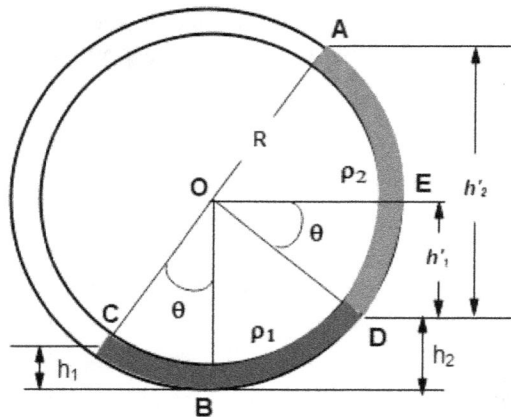

Since two liquid columns of heights h_1 and h_2 and densities ρ_1 and ρ_2 are situated above the point B, on the left-hand side, the pressure at B is

$$P_2 = \rho_1 g h_2 + \rho_2 g h'_2 \tag{4.16}$$

Equating P_1 and P_2 from equations (4.15) and (4.16), we have

$$\rho_1 h_2 + \rho_2 h'_2 = \rho_1 h_1 \tag{4.17}$$

Substituting $h'_2 = R \sin \theta + R \cos \theta$,
$h_2 = R(1 - \cos \theta)$ and $h_1 = R(1 - \sin \theta)$ in equation (4.17), we have

$$\rho_1 R(1 - \cos \theta) + \rho_2 R(\sin \theta + \cos \theta) = \rho_1 R(1 - \sin \theta)$$

$$\Rightarrow \frac{\cos \theta + \sin \theta}{\cos \theta - \sin \theta} = \frac{\rho_1}{\rho_2}$$

$$\Rightarrow \tan \theta = \frac{\rho_1 - \rho_2}{\rho_1 + \rho_2} \text{ Ans.}$$

Example 10 Explain the principle of a barometer and find the length of the barometric mercury column.

Solution

First of all, take a glass tube of 1m long, say, closed at one of its ends. Pour mercury into the tube until it is completely filled. Bring a tub partially filled with mercury. Then invert the glass tube and slowly insert it into the mercury tub. You can see that the level of mercury in the glass tube is slowly decreasing. At the steady state, the mercury column remains stationary at a height h, say, measured from the level of mercury in the tub. We can find h by using Newton's 2nd law.

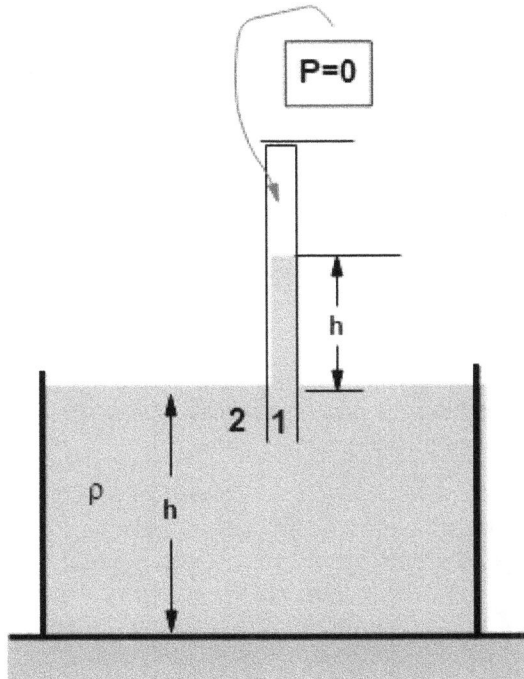

The forces acting on the column of mercury in the tube are (i) weight of the mercury column and (ii) upward thrust by the liquid at point 1, which is equal to $P_{atm} A$. This means that under the action of the forces $P_{atm} A \uparrow$ and $mg \downarrow$, the mercury column is in equilibrium.

Using Newton's 2nd law, we have

$$F_{net}(=-mg + P_{atm} A) = ma,$$

where $a = 0$ (for equilibrium).

Substituting $m = A\rho h$, $\rho = 13.6 \times 10^3$ kg m^{-3} and $P_{atm} = 1.01 \times 10^5$ N m^{-2}, we have $h \cong 76$ cm.

The space above the free surface of mercury in the tube is vacuum. In fact, there is a little vapour of mercury present in that free space due to the evaporation of mercury, but it does not give a considerable pressure. In this way, the vapour pressure of mercury in the free space (or physical vacuum) can be neglected; $P = 0$, in the free space.

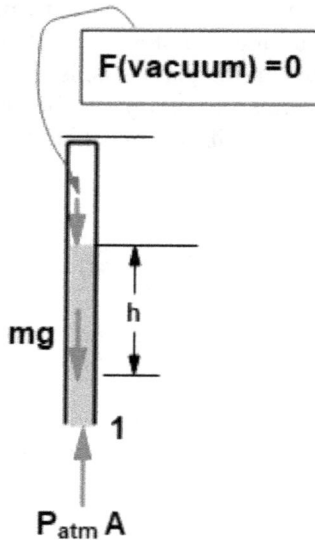

N.B.: If you go on pushing the glass tube more and more into the tub of mercury, the height of the mercury column remains constant.

Example 11 A triangular tube of uniform cross-section has three liquids of densities ρ_1, ρ_2 and ρ_3. Each liquid column has length l which is equal to length of sides of the equilateral triangle. (a) Find the length x of the liquid of density ρ_1 in the horizontal limb of the tube, if the triangular tube is kept in the vertical plane. (b) If $\rho_3 = \frac{\rho_1 + \rho_2}{2}$, find x.

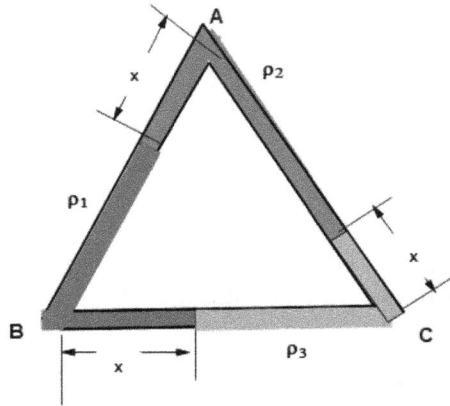

Solution

(a) Let's consider two points 1 and 2 in the horizontal limb. Pressure at 1 is

$$P_1 = \rho_1 g h_1 + \rho_2 g h_2$$

Pressure at 2 is

$$P_2 = \rho_2 g h'_2 + \rho_3 g h_3$$

Since $P_1 = P_2$ for non-accelerating liquid, we have

$$\rho_1 h_1 + \rho_2 h_2 = \rho_2 h'_2 + \rho_3 h_3$$

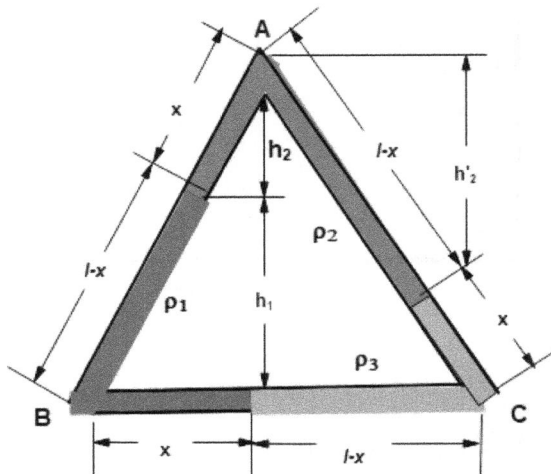

Substituting $h_1 = h'_3 = (l - x)\sin 60°$ and $h_2 = h_3 = x \sin 60°$, we have

$$x = \frac{(\rho_2 - \rho_1)l}{2\rho_2 - (\rho_1 + \rho_3)} \quad \text{Ans.}$$

(b) Putting $\rho_3 = \frac{\rho_1 + \rho_2}{2}$, we have

$x = \frac{l}{3}$ after simplification. This shows that the result does not depend on the individual values of the densities.

4.10 Hydrostatic pressure in a vertically accelerating liquid

Upward acceleration: Let's fill a cubical vessel with a liquid of density ρ. If we accelerate the vessel vertically up, how do we find the pressure at a point situated at a depth y (measured from the free surface of the liquid)?

Consider a thin vertical liquid column of cross-sectional area δA assuming the atmospheric pressure P_0 at the top of the liquid column. Referring to the free-body diagram, we have the following forces acting liquid column.

(i) Atmospheric force $P_0 \delta A$ downward
(ii) Hydrostatic force $P \delta A$ upward
(iii) Weight δmg of the liquid column.

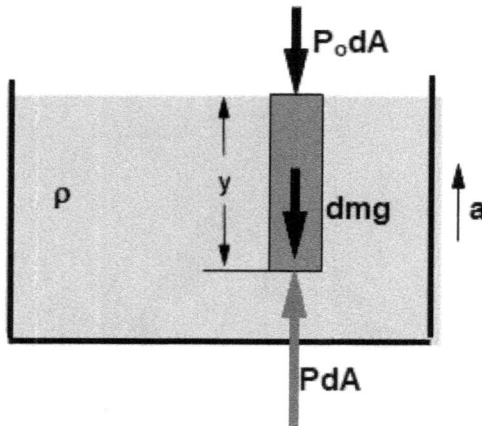

Net force acting on the liquid column is

$$F = P\delta A - P_0 \delta A - \delta mg$$

This net force F force accelerates the liquid column up with an acceleration a. Applying Newton's 2nd law, we have

$$F = p\delta A - P_0 \delta A - \delta mg = \delta ma$$

This gives

$$P = P_0 + \frac{\delta m}{\delta A}(g + a)$$

Substituting $\delta m = \rho \delta Ay$, we have

$$P = P_0 + \rho(g + a)y$$

From the above expression we understand that the total pressure at a depth y is the sum of the following three pressures:

P_0 = Pressure due to atmosphere

ρgy = Pressure due to gravity

ρay = Pressure due to acceleration a.

Then, the total hydrostatic pressure at O is

$$P_{\text{hydro}} = \rho(g + a)y$$

This means that the pressure increases linearly with depth y in the vertically downward direction. However, pressure does not vary in horizontal.

Downward acceleration: From the previous discussion, we have

$$P_{\text{hydro}} = \rho(g + a)y$$

If the vessel accelerates down, substituting $a = -a$, we have

$$P_{\text{hydro}} = \rho(g - a)y$$

Now we have following cases:

Case 1: $(a < g)$

When $a < g$, we have

$$P_{\text{hydro}} = \rho(g - a)y > 0$$

This means pressure varies with y as

$$P = P_0 + \rho(g - a)y,$$

which is greater than P_0.

Hence, pressure increases linearly in vertically downward direction, with the depth y.

Case 2: $(a = g)$

When $a = g$, we have $P_{\text{hydro}} = 0$.

This means the hydrostatic pressure at any point inside the liquid is zero. Hence, pressure at each point of the liquid is equal to atmospheric pressure P_0; $P = P_0$. This is a freely falling case which tells that the hydrostatic pressure inside a freely falling liquid is zero. Hence, pressure remains constant in any direction in a freely falling liquid.

If a hole is made on the container (or the container is open), the pressure inside the liquid is equal to atmospheric pressure.

Case: $(a > g)$

When $a > g$, we have

$$P = P_0 - \rho(a - g)h$$

Since $a > g$, $P < P_0$.

This means pressure decreases linearly with the depth y, but how can the pressure inside a liquid be lesser than the atmospheric pressure? Let's see in the following example.

Example 12 A cubical vessel is completely filled with water. If it is sealed and made to accelerate down with an acceleration $a = 2g$, find the ratio of the pressure at the top and bottom of the vessel.

Solution

Let's take a thin vertical column of liquid. As the liquid moves with a downward acceleration greater than g, the upper surface of the vessel must push the liquid with certain force dN, say. When this force favours the weight dmg of the liquid column, the net force is

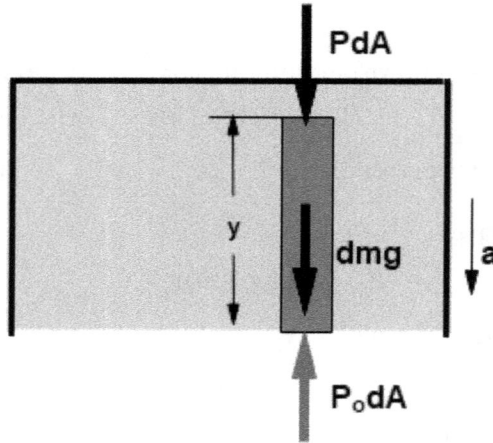

$$F_{\text{net}} = \delta N + \delta mg - P_0 \delta A$$

Since $F_{\text{net}} = \delta ma$, we have

$$\delta N + \delta mg - P_0 \delta A = \delta ma$$

Then, we have

$$\delta N = P_0 \delta A + \delta m(a - g)$$

Then, the pressure at the top of the vessel is

$$P = \frac{\delta N}{\delta A} = P_0 + \frac{\delta m}{\delta A}(a - g),$$

where $\delta m = \rho(\delta A)h$.

This gives us

$$P = P_0 + \rho(a - g)h$$

In general, the pressure at a vertical distance y from the free surface of the liquid (bottom of the vessel) is given as

$$P = P_0 + \rho(a - g)y$$

Then, the pressure at the top of the vessel can be given as

$$P_{\text{top}} = P_0 + \rho(2g - g)h$$

$$P_{\text{top}} = P_0 + \rho g h$$

The pressure at the bottom of the vessel is

$$P_{\text{bottom}} = P_0$$

Then, the ratio of the pressures is

$$P_{\text{top}}/P_{\text{bottom}} = (P_0 + \rho g h)/P_0 \text{ Ans.}$$

N.B.: The above equation tells us that pressure increases vertically upward linearly with the vertical distance y measured from the free surface of the liquid.

Recapitulating, when the vessel accelerates up or down with an acceleration a, pressure at a perpendicular distance y from the free surface of the liquid can be given as

$$P = P_0 + \rho \, |g \pm a| \, y$$

Take $g + a$, when the vessel accelerates up; $|g - a|$ when the vessel accelerates down. We can see that in all cases of vertical acceleration, the pressure increases linearly with the vertical distance y from the free surface of the liquid.

Remember that, pressure increases vertically downwards when the liquid accelerates up or down with $a < g$ and upward with $a > g$. However, the pressure decreases vertically downwards when the closed liquid accelerates down with $a > g$.

4.11 Hydrostatic pressure in a horizontally accelerating liquid

If we move the vessel with a horizontal acceleration a, how does the pressure of the liquid inside the vessel vary?

Consider two points 1 and 2 in a thin horizontal liquid column of length x and area of cross-section δA. Referring to free-body diagram, the net horizontal force acting on the liquid column is

$$F_{\text{net}} = P_1 \delta A - P_2 \delta A$$

Since this force accelerates the liquid column with an acceleration a, applying Newton's 2nd law, we have

$$P_1 \delta A - P_2 \delta A = \delta m a,$$

where $\delta m = \rho(\delta A)x$. This gives $P_2 = P_1 - \rho a x$.

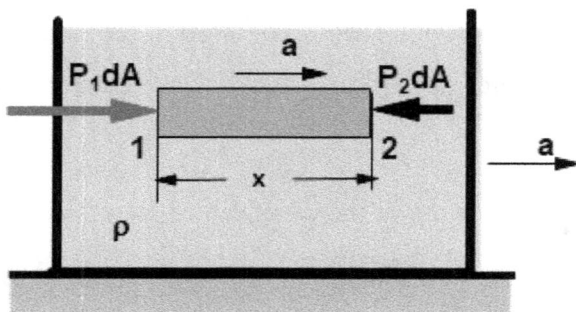

The given equation tells us that pressure decreases linearly with horizontal distance measured in the direction of acceleration of the liquid. Hence, pressures at different points at the same depth or height will not be equal for a horizontally accelerating liquid.

Let's now see if there is any variation of pressure in vertical line due to acceleration of the vessel. For this, referring to free-body diagram in previous section, we have

$$F_{net} = P_1 \delta A + \delta m g - P_2 \delta A = 0 \left(\text{because } a_y = 0\right)$$

This gives

$$P_2 = P_1 + \frac{\delta m}{\delta A} g,$$

where $\frac{\delta m}{\delta A} = \rho y$.

Then,

$$P_2 = P_1 + \rho g y$$

The above equation tells us that there is no effect of horizontal acceleration of liquid on the variation of pressure in the vertical line. The pressure increases linearly with depth as usual due to the effect of gravity.

Example 13 A rectangular closed trough of length l and height h is completely filled with a liquid of density ρ. It moves with an acceleration a. If P_0 is the atmospheric pressure at the hole made at the corner A of the trough, find the (a) expression for pressure at the point C and (b) pressure at D, E, and F.

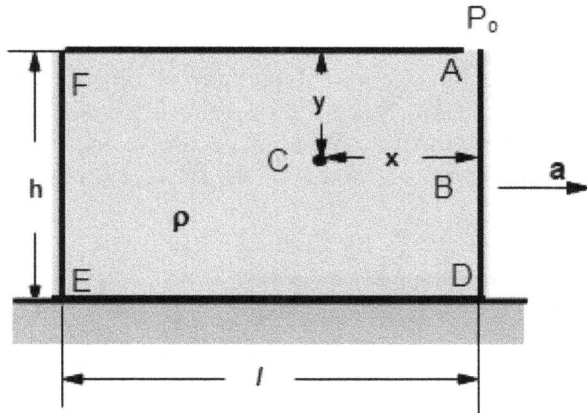

Solution

(a) Directly using the formula

$P_B = P_A + \rho gy$, where $P_A = P_0$ (given), we have

$$P_B = P_0 + \rho gy \qquad\qquad (4.18)$$

Similarly using the formula

$$P_B = P_C - \rho ax \qquad\qquad (4.19)$$

Equating P_B from equations (4.18) and (4.19), we have

$$P_C = P_0 + \rho ax + \rho gy \text{ Ans.}$$

(b) Using the last formula, we have

$$P_D = P_0 + \rho gh$$

$$P_E = P_A + \rho al + \rho gh = P_0 + \rho al + \rho gh$$

$$P_F = P_A + \rho al = P_0 + \rho al$$

N.B.: The pressure at any point inside a horizontally accelerating liquid is given as

$$P(=P_C) = P_0 + \rho gx + \rho gy,$$

where $P_0 = $ Atmospheric pressure

$\rho ax = $ Pressure due to the acceleration

$\rho gy = $ Pressure due to gravity.

Since pressure is a scalar quantity, we can add all the above three pressures to get the total pressure P. Hence, the pressure at any point or pressure difference between two points in a horizontally accelerating liquid does not depend on the path we follow. It solely depends upon the horizontal distance x and vertical distance y between the points.

Equi-pressure lines: We have seen that pressure at different points at the same horizontal line will be different. Since the pressure increases vertically downward and decreases in the direction of acceleration of the vessel, you can expect equal pressure between any two points A and B in the straight (dotted red) line AB.

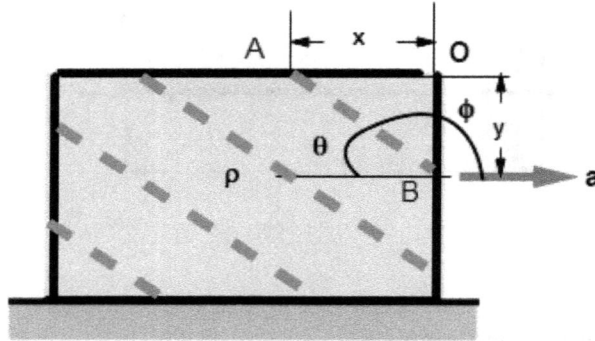

**Dotted red lines are
equi-pressure lines**

As we know,

$$P_A = P_O + \rho a x \text{ and } P_B = P_O + \rho g y,$$

Then, we have $P_B - P_A = \rho(gy - ax)$
Substituting $P_B = P_A$, we have

$$y = \frac{a}{g}x$$

Since $\frac{y}{x} = \tan \theta$, we have

$$\theta = \tan^{-1}\frac{a}{g}$$

This equation tells us that the equi-pressure lines are the parallel straight lines making an obtuse angle $\phi = (180 - \tan^{-1}\frac{a}{g})$ with the direction of acceleration of the liquid. The angle ϕ does not depend on the density of liquid.

Example 14 Referring to the last problem, let's now open the cap of the vessel and assume that the height of the vessel is enough so that the liquid will not come out of the vessel. Find the
(a) Angle of orientation of the equi-pressure lines.
(b) Pressures at M, N, and Q.

Solution

(a) Let's assume that the free surface of the liquid gets inclined as shown as dotted red-lines. Let's take an element of mass dm at any point inside the liquid. Let this element remain on the equi-pressure line AB. The element experiences a normal contact force δN due to the surrounding liquid and its weight dmg. Resolving δN horizontally and vertically, we have $\delta N \sin\theta$ and $\delta N \cos\theta$ respectively. $\delta N \sin\theta$ pushes the element of mass dm with a horizontal acceleration a. Using Newton's 2nd law, we have

$$\delta N \sin\theta = \delta ma \tag{4.20}$$

Since δm does not accelerate in vertical, we have net force
$F_{net} = \delta mg - \delta N \cos\theta = 0$ ($\delta N \cos\theta$ nullifies the gravity)
This gives

$$\delta N \cos\theta = \delta mg \tag{4.21}$$

Eliminating δN from equation (4.20) by using equation (4.21), we have

$$\tan\theta = \frac{a}{g} \quad \text{Ans.}$$

(b) Since θ is a constant, the surface of the liquid is flat making an angle $\phi(=180-\theta)$ with the direction of its acceleration. This means the angle orientation of the equi-pressure lines will not change.

From the geometry of the surface of liquid, we understand that the right half of the surface comes down and the left half of the surface rises up.

Hence, we have

$$h_1 = h - \frac{l}{2}\tan\theta$$

$$\text{and } h_2 = h + \frac{l}{2}\tan\theta$$

The pressure at M is

$$P_M = P_0 + \rho g h_1$$

Substituting $h_1 = h - \frac{l}{2}\tan\theta$, we have

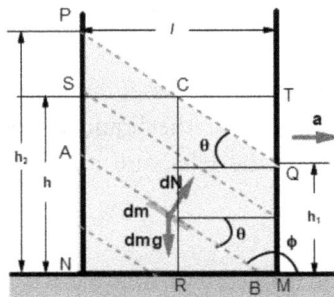

**Dotted red lines are
equi-pressure lines**

$$P_M = P_0 + \rho g h - \frac{\rho l g}{2} \tan \theta,$$

where $g \tan \theta = a$

This gives

$$P_M = P_0 + \rho g h - \frac{\rho l a}{2}$$

The pressure at Q is

$$P_Q = P_M + \rho \frac{l}{2} a$$

Substituting
$P_M = P_0 + \rho g h - \frac{\rho l a}{2}$, we have

$$P_Q = P_0 + \rho g h$$

The pressure at N is

$$P_N = P_Q + \rho a \frac{l}{2},$$

where $P_Q = P_0 + \rho g h$
Then, we have

$$P_N = P_0 + \rho g h + \rho a \frac{l}{2}$$

You can also directly obtain P_N by substituting

$$h_2 = h + \frac{l}{2} \tan \theta, \text{ in the expression}$$

$$P_N = P_0 + \rho g h_1, \text{ where } \tan \theta = \frac{a}{g} \text{ Ans.}$$

N.B.: On opening the cap of the vessel the pressure at any point in the right-hand side of the plane CQ of the liquid decreases by $\rho a \frac{l}{2}$. Similarly, the pressure at any point in the left-hand side of the plane CQ of the liquid increases by $\rho a \frac{l}{2}$.

Example 15 A vertical U-tube is filled with a liquid with a liquid of density ρ up to a height $b/2$. It accelerates slowly from rest to a constant acceleration $a = \frac{g}{2}$.

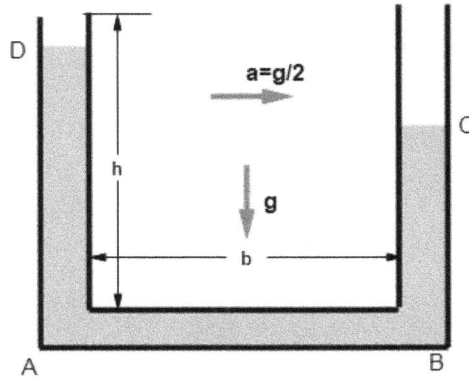

Find the
(a) Difference in heights of the liquid column in the limbs.
(b) Pressure difference between the points A and B.
(c) Lengths of the liquid column in each vertical limb.

Solution

(a) The pressures at A and B are as follows:

$$P_A = P_0 + \rho g(AD) \tag{4.22}$$

$$P_B = P_0 + \rho g(BC) \tag{4.23}$$

$$P_A = P_B + \rho ba \tag{4.24}$$

Substituting P_A and P_B from equations (4.22) to (4.24), we have

$$P_o + \rho g(AD) = P_0 + \rho g(BC) + \rho ba$$

$$\Rightarrow \rho g(AD - BC)) = \rho ba$$

$$\Rightarrow (AD - BC) = ba/g = b(g/2)/g = b/2. \text{ Ans.}$$

(b) From equation (4.24), the pressure difference between A and B is given as
$$P_A - P_B = \rho ba,$$
where $a = g/2$.
    Then, we have
$$P_A - P_B = \rho g b/2$$

(c) The height of the liquid column in the right hand limb is

$$h_1 = h - \frac{b}{2} \tan \theta$$

$$= \frac{b}{2} - \frac{ba}{2g} = \frac{b}{4} (\because a = g/2)$$

The height of the liquid column in the left hand limb is

$$h_2 = h + \frac{b}{2} \tan \theta$$

$$= \frac{b}{2} + \frac{ba}{2g} = \frac{3b}{4} (\because a = g/2)$$

Example 16 (Rotating liquid) A horizontal thin tube completely filled with a liquid of density ρ rotates about a vertical axis passing through one of its ends, with an angular velocity ω. Two vertical thin tubes are fitted with the horizontal tube to measure the pressure difference between A and B. Find the

   (a) Pressure gradient at a distance x from the axis of rotation.
   (b) Pressure at the point which is located at a distance x from the axis of rotation, assuming P_0 as the pressure of the liquid at the axis of rotation.

Solution

   (a) Let's consider a thin vertical strip of liquid of thickness dx at a distance x from the axis of rotation. Let the pressures at right- and left-hand sides of the strip be $P + dP$ and P, respectively. Referring to free-body diagram, the net force

$$\delta F = (P + \delta P) - P dA = (\delta P) dA$$

This force accelerates the strip with an acceleration a towards the axis of rotation.

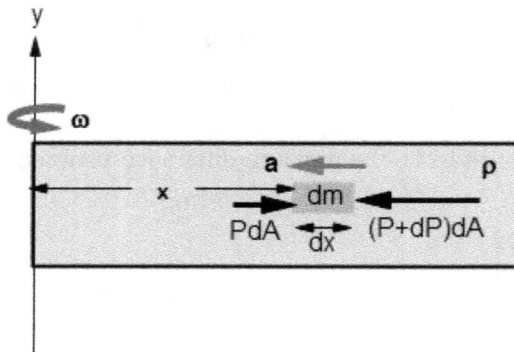

Applying Newton's 2nd law, we have

$$\delta F = \delta m. \, a$$

(b) Substituting $\delta F = (\delta P)A$ and $\delta m = A\rho\delta x$, we have

$$\delta P. \, A = (A\rho\delta x)a, \quad \text{where } a = x\omega^2$$

this gives

$$\frac{\delta P}{\delta x} = \rho\omega^2 x$$

(c) From the above expression, we have

$$\delta P = \rho\omega^2 x\delta x$$

Since pressure increases from P_0 to P when we move from axis of rotation by a distance x, we have

$$\int_{P_0}^{P} \delta P = \rho\omega^2 \int_0^x x\delta x$$

This gives

$$P = P_0 + \frac{\rho\omega^2 x^2}{2} \quad \text{Ans.}$$

Example 17 A cylindrical tank filled with a viscous liquid rotates with an angular velocity ω. What does the free surface of liquid look like?

Solution

Let's consider an element of liquid surface at a distance x from the axis of rotation. The forces acting on the element are

    (i) Reaction force δN

    (ii) Gravity force δmg.

The net horizontal force acting on the element is

$$\delta N \sin\theta = \delta ma_r,$$

where $a_r = x\omega^2$.

This gives

$$\delta N \sin\theta = \delta mx\omega^2 \tag{4.25}$$

The net vertical force is

$$\delta N \cos\theta - \delta mg = 0$$

This gives

$$\delta N \cos\theta = \delta mg \tag{4.26}$$

Using equations (4.25) and (4.26), we have

$$\tan \theta = \frac{x\omega^2}{g}$$

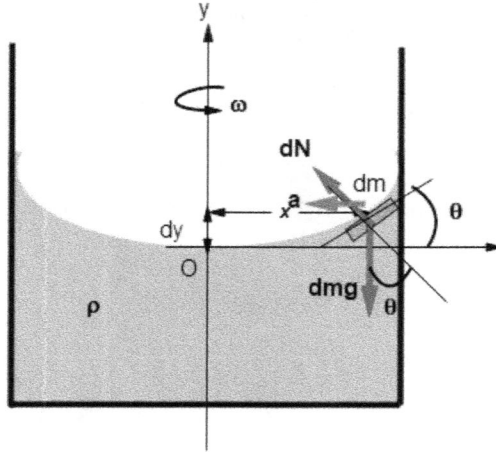

Substituting the slope of the curve

$$\tan \theta = \frac{dy}{dx}, \text{ we have}$$

$$\frac{dy}{dx} = \frac{x\omega^2}{g}$$

This gives

$$dy = \frac{\omega^2}{g} x \, dx$$

Integrating, we have

$$y = \frac{\omega^2}{g} \int_0^x x \, dx$$

This gives

$$y = \frac{\omega^2 x^2}{2g} \text{ Ans.}$$

N.B.: The free surface of a rotating liquid is parabolic. This means the equi-pressure surfaces are paraboloids. Let's summarize following points for rotating liquids.

4-46

We have three cases of rotating liquid as follows:
  (i) Vessel cap closed (forced surface).
  (ii) Vessel top is open (free surface).
  (iii) Vessel is in the form of a tube of any shape and size (for the sake of simplicity, we have a U-tube).

  (a) In each of the above cases, the equi-pressure surfaces are paraboloid given as

$$y = \frac{\omega^2 x^2}{2}$$

  (b) The pressure increases radially outward in the same horizontal line which can be given as

$$P_2 = P_1 + \frac{\rho \omega^2 x^2}{2}$$

  (c) The pressure in the liquid increases linearly with vertical distance (depth) y, given as

$$P_2 = P_1 + \rho g y.$$

  (d) If the liquid surface is made free, it assumes a shape of a paraboloid. For the liquid in the tubes, the free liquid surfaces in the limbs must be the surface of the same parabola.

Example 18 A vertical U-tube containing a liquid of density ρ rotates about the vertical axis with a constant angular velocity $\omega = \sqrt{\frac{g}{l}}$. Find the difference in heights of the liquid columns in the vertical limbs.

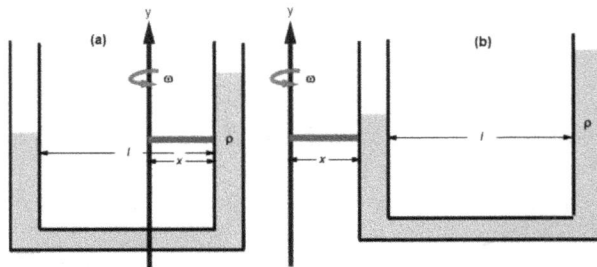

Solution

(a) As derived earlier, the free surfaces of liquid in each limb belong to the equipressure surfaces given as

$$y = \frac{\omega^2 x^2}{2}$$

So, the difference in heights of the liquid columns in the vertical limbs of a U-tube is

$$\Delta y = \frac{\omega^2 \{x^2 - (l - x)^2\}}{2g}$$

$$\Rightarrow \Delta y = \frac{\omega^2}{2g}(2x - l)l$$

Putting $\omega = \sqrt{\frac{g}{l}}$, we have

$$\Delta y = \frac{1}{2l}(2x - l)l = x - l/2$$

(b) The difference in heights of the liquid columns in the vertical limbs of a U-tube is

$$\Delta y = \frac{\omega^2 \{(l + x)^2 - x^2\}}{2g}$$

$$\Rightarrow \Delta y = \frac{\omega^2}{2g}(2x + l)l$$

Putting $\omega = \sqrt{\frac{g}{l}}$, we have

$$\Delta y = \frac{1}{2l}(2x + l)l = x + l/2$$

Example 19 A vertical U-tube has two liquids 1 and 2. The heights of liquid columns in both limbs are h and $2h$ and the length of each liquid column in the horizontal limb is equal to h.

(a) If the tube is stationary and density of the liquid 1 is 2ρ, find the density of the liquid 2.

(b) If we accelerate the tube towards right until the heights of liquid columns will be same, find the acceleration of the tube.

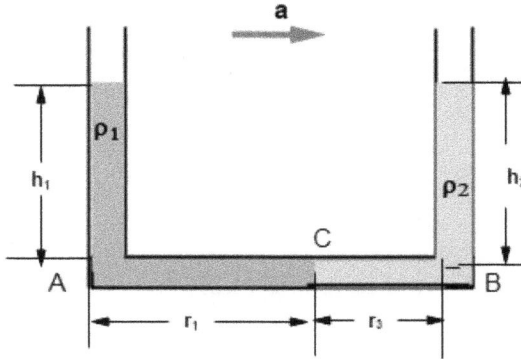

Solution

(a) First of all, we take two points A and B at the bottom of the left and right vertical limbs. The pressure at A is

$$P_A = P_0 + (2\rho)gh$$

The pressure at B is

$$P_B = P_0 + \rho_2 g(2h),$$

Since $P_A = P_B$ for stationary liquid, we have

$$P_0 + 2\rho gh = P_0 + \rho_2 g(2h)$$

This gives $\rho_2 = \rho$ Ans.

(b) If the levels of liquids in both limbs are equal, conserving the total volume of the liquids, the height of each liquid column will be equal to $h_1 = h_2 = \frac{3h}{2}$.

The lengths of liquids in the horizontal limbs are $r_1 = \frac{h}{2}$ and $r_2 = \frac{3h}{2}$.
The pressure at the interface 'C' of the liquids is
$P_C = P_A - (2\rho)a(\frac{h}{2})$, where

$$P_A = P_0 + (2\rho)g\frac{3h}{2}$$

This gives

$$P_C = P_0 + (2\rho)g\left(\frac{3h}{2}\right) - (2\rho)a\left(\frac{h}{2}\right)$$

$$= P_0 + 3\rho gh - \rho ah \tag{4.27}$$

P_C can also be given as

$$P_C = P_B + \rho a\left(\frac{3h}{2}\right),$$

where

$$P_B = P_0 + \rho g\left(\frac{3h}{2}\right)$$

This gives

$$P_C = P_0 + \frac{3}{2}\rho g h + \frac{3}{2}\rho a h \qquad (4.28)$$

Eliminating P_3 from equations (4.27) and (4.28), we have

$$a = \frac{3}{5}g \ \text{Ans.}$$

Example 20 If we rotate the tube about a vertical axis passing through right-hand side limb such that two liquid levels attain the same height. Find the angular speed of rotation.

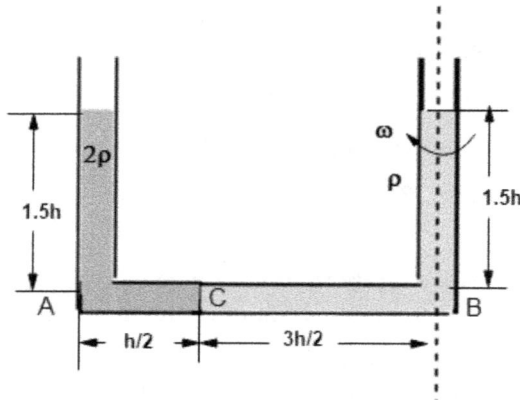

Solution
The pressure at A is

$$P_A = P_0 + (2\rho)g(1.5h) \qquad (4.29)$$

The pressure at B is

$$P_B = P_0 + \rho g(1.5h) \qquad (4.30)$$

As derived earlier, the pressure increases parabolically. So the pressure at C is $P_C = P_B + \frac{\rho \omega^2 x^2}{2}$, where $x = 3h/2$

$$\Rightarrow P_C = P_B + \frac{\rho \omega^2 (1.5h)^2}{2} \qquad (4.31)$$

Similarly, the pressure at A is

$$P_A = P_C + \frac{2\rho\omega^2\{(2h)^2 - (1.5h)^2\}}{2} \tag{4.32}$$

From equations (4.29) and (4.30),

$$P_A - P_B = 1.5\rho gh \tag{4.33}$$

From equations (4.31) and (4.32),

$$P_A - P_B = (2.5/4)\rho\omega^2h^2 \tag{4.34}$$

From equations (4.33) and (4.34), we have

$$(23/8)\rho\omega^2h^2 = 1.5\rho gh$$

$$\Rightarrow \omega = \sqrt{\frac{12g}{23h}} \text{ Ans.}$$

4.12 Hydrostatic force on a flat surface

Let's consider a vertical gate of surface area A. We need to find the thrust of the liquid on the gate. For this, we take a thin horizontal strip at a depth y. So, the pressure P will be same at all points of the strip. Then, the hydrostatic force acting on the strip is

$dF = P\,dA$, where $P = \rho gy$

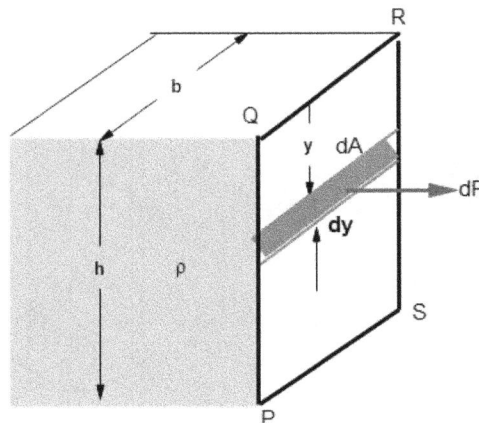

Then, we have

$$dF = \rho gy dA$$

Since dF has the same direction at all points of the patch, integrating dF, the total hydrostatic force (side thrust of liquid) is given as

$$F = \int dF$$

Substituting $dF = \rho g y dA$, we have

$$F = \int \rho g y dA$$

Since ρg is a constant, taking it out of the integral, we have

$$F = \rho g \int y \, dA$$

Substituting

$$\int y \, dA = y_C A,$$

where y_C = centroid of the patch of area A, we have obtained an important general expression

$$F = \rho g y_C A.$$

Example 21 Liquid of density ρ pushes the vertical plate of a gate towards the right and to the base plate of the gate downwards. The dimensions of the base and vertical plates are, $(b \times c)$ and $(b \times h)$ respectively. Find the hydrostatic force on (a) vertical plate and (b) base.

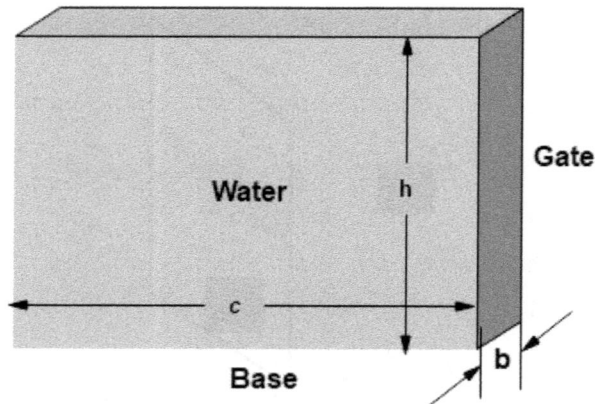

Solution

(a) The hydrostatic force acting on the vertical gate is

$$F = \rho g y_C A,$$

where $y_C = \frac{h}{2}$ and $A = bh$.
This gives

$$F = \frac{1}{2}\rho g b h^2 \text{ Ans.}$$

(b) The hydrostatic force acting on the base is

$$F_{\text{base}} = \rho g (y_C)_{\text{base}} A_{\text{base}},$$

where $(y_C)_{\text{base}} = h$ and $A_{\text{base}} = bc$
This gives

$$F_{\text{base}} = \rho g b c h \text{ Ans.}$$

N.B.: If the patch is rectangular (or symmetrical about horizontal axis), we have
$dA = b\, dy$
Then

$$F = \int P\, dA = b \int_0^h P\, dy,$$

where $P = \rho g y$.
This gives $F = \rho g \frac{h^2}{2} b$.
More generally, we can write

$$F = P_{av} A,$$

where $P_{av} = \frac{\int_0^h P\, dy}{h}$ and $A = bh$.
In hydrostatics of single liquid

$$P_{av} = \frac{1}{2}\left(P_{\text{top}} + P_{\text{bottom}}\right)$$

But, be careful! the above expression is valid only for rectangular or symmetrical flat area like a circular patch.

However, you can use expression $F = \rho g y_C A$ (but not always $P_{av}.\ A$ for any arbitrary flat surface).

Let's take an example of multiple liquid layers and find the hydrostatic force on a flat rectangular surface using the average pressure method.

Example 22 Two immiscible liquids of densities ρ and 2ρ and thickness (heights) h and $2h$, respectively, push a vertical plate. Find the total side thrust given by the liquids on the vertical plate.
Solution
Let the hydrostatic forces acting on the rectangular patches of areas A_1 and A_2 of the gate be F_1 and F_2.

Then, the total hydrostatic force on the gate is $F = F_1 + F_2$, where $F_1 = P_{av_1} A_1$ and $F_2 = P_{av_2} A_2$; P_{av_1} and P_{av_2} are the average pressures on the areas A_1 and A_2, respectively.
Then, $F = P_{av_1} A_1 + P_{av_2} A_2$

Substituting $P_{av_1} = \frac{P_A + P_B}{2}$, $A_1 = bh_1 = bh$

$$P_{av_2} = \frac{P_B + P_C}{2} \text{ and } A_2 = bh_2 = b(2h),$$

we have

$$F = \frac{(P_A + P_B)}{2} bh + \frac{(P_B + P_C)}{2} 2bh$$

Substituting $P_A = 0$, $P_B = \rho g h$ and

$$P_C = (\rho_1 h_1 + \rho_2 h_2)g = [\rho h + (2\rho)(2h)]g = 5\rho g h,$$

we have

$$F = \frac{13}{2} \rho g h^2 b \text{ Ans.}$$

N.B.: We should remember that the formula $F = \sum P_{av_i} A_i$ is not valid for any arbitrary shape and size of the area of the patch such as triangular area. In that case, we need to derive a general expression for the vertical patch.

Hydrostatic force due to many liquid layers: We have n number of liquid layers; densities and heights of the liquids are, (ρ_1, h_1), (ρ_n, h_2), ..., (ρ_n, h_n), respectively. Let's assume a vertical patch inside the nth liquid. If the patch has an arbitrary shape and size (of area A) how to find the hydrostatic force acting on the patch, Let's see.

Let's take a thin horizontal strip of area dA at a distance y from the top of nth liquid. The hydrostatic force dF on the thin strip is

$$dF = PdA \tag{4.35}$$

where P is the hydrostatic pressure on the thin strip. We know that the pressure due to multiple liquids is

$$P = \rho_1 g h_1 + \rho_2 g h_2 + \cdots + \rho_n g y \tag{4.36}$$

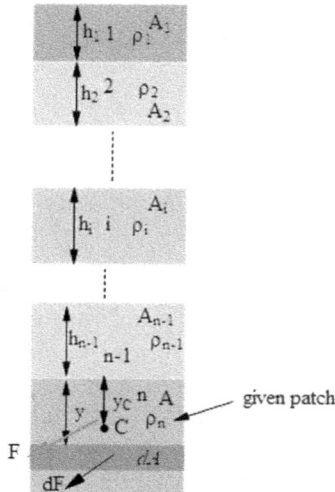

The side hydrostatic force F acts on the given patch of area A due to all liquids

Substituting P from the equations (4.36) in (4.35),
We have

$$dF = \left(\rho_1 g h_1 + \rho_2 g h_2 + \cdots + \rho_n g y\right) dA$$

Then, the total hydrostatic force on the strip is $F = \int dF$

$$= \int \left(\rho_1 g h_1 + \rho_2 g h_2 + \cdots + \rho_n g y\right) dA$$

$$= \sum_{i=1}^{i=n-1} \left(\rho_i g h_i\right) A + \rho_n g \int y \, dA$$

$$F = \left(\sum_{i=1}^{i=n-1} \rho_i g h_i + \rho_n y_C\right) g A,$$

where y_C = vertical distance of the centroid of the given area (patch) from the top of nth liquid.

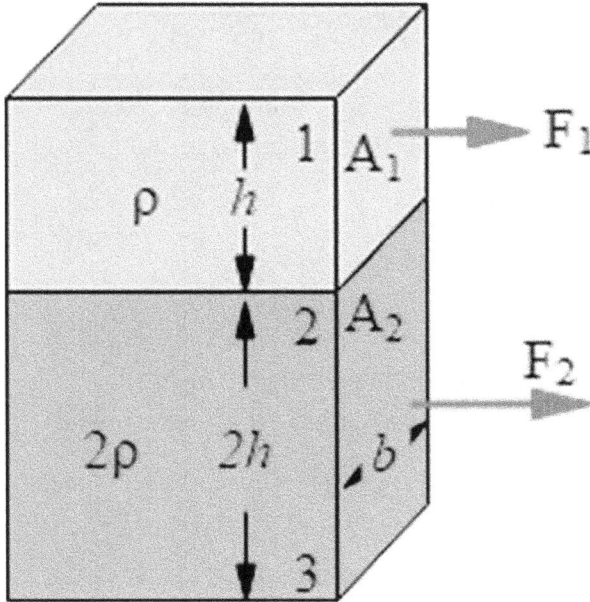

Example 23 A composite plate consists of a triangular area (1) and rectangular area (2) as shown in the figure. The plate experiences force due to liquids of densities ρ_1 and ρ_2. Find the hydrostatic forces on the (a) area 1, (b) area 2, and (c) total composite area of the plate. Assume $\rho_1 = \rho$ and $\rho_2 = 2\rho$.

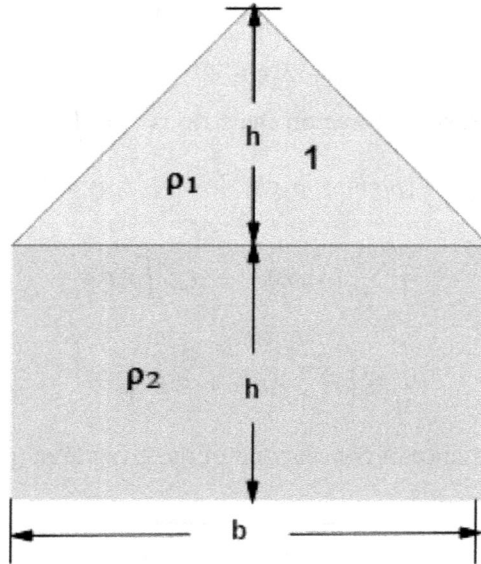

Solution

(a) The hydrostatic force on the triangular area is

$$F_1 = \rho_1 g y_{C_1} A_1, \quad \text{where } \rho_1 = \rho, \ y_{C_1} = \frac{2h}{3} \ \text{and} \ A_1 = \frac{1}{2} b \cdot h$$

This gives

$$F_1 = \frac{\rho g h^2 b}{3} \ \text{Ans.}$$

(b) The hydrostatic force on the rectangular area is

$$F_2 = \left(\rho_1 g h_1 + \rho_2 g y_{C_2} \right) A_2$$

Substituting $\rho_1 = \rho, h_1 = h$, $\rho_2 = 2\rho$, $y_{C_2} = \frac{h}{2}$ and $A_2 = bh$, we have

$$F_2 = 2\rho g h^2 b \ \text{Ans.}$$

(c) The total hydrostatic force on the composite patch is

$$F = F_1 + F_2 = \frac{\rho}{3} g h^2 b + 2\rho g h^2 b = \frac{7}{3} \rho g h^2 b \ \text{Ans.}$$

Example 24 In the last example, if we invert the composite area, find the total hydrostatic force on the composite area.

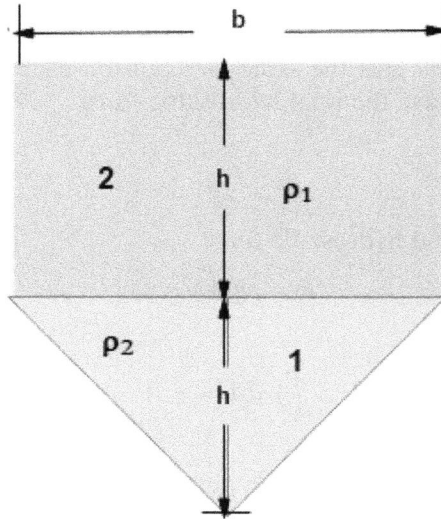

Solution

The hydrostatic force on surface 2 is

$$F'_1 = \rho_1 g y_{C_1} A_2 = (\rho)g(h/2)(bh) = \rho g b h^2/2$$

The hydrostatic force on surface 1 is

$$F''_2 = \rho_1 g h A_1 + \rho_2 g y_{C_2} A_2$$

$$=(\rho)(g)h(bh) + (2\rho)(g)\left(\frac{h}{3}\right)(bh/2) = 4\rho g b h^2/3$$

Hence, the total hydrostatic force on surface 2 due to liquid 1 and 2 is

$$F_2 = F'_2 + F''_2 = 11\rho g h^2 b/6 \text{ Ans.}$$

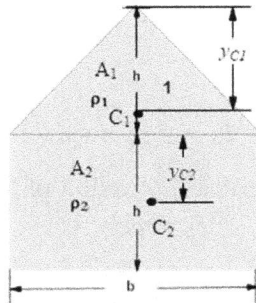

Hydrostatic force on an inclined plate: Now we can incline the vertical gate and see how the hydrostatic force changes its magnitude and direction acting on the area A.

Let's consider an elementary horizontal strip of area dA on the plane of the gate at a depth y. The hydrostatic force on the area dA is

$$dF = \rho g y dA$$

Since hydrostatic force has the same direction on each strip, summing up the elementary forces, we have the total hydrostatic force
$F = \int dF$, where

$$dF = \rho g y \, dA$$

Then we have the total hydrostatic force

$$F = \rho g \int y \, dA,$$

where

$$\int y \, dA = y_C A$$

This gives

$$F = \rho g y_C A$$

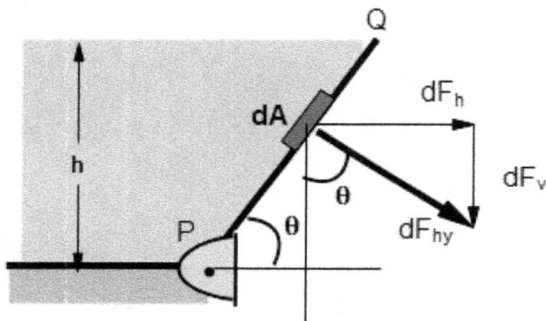

This means that the total hydrostatic force does not change its magnitude, whereas its direction changes by an angle $\pi/2 - \theta$.

Resolving the net force F in horizontal, the horizontal thrust is

$$F_h = F \sin \theta,$$

where $F = \rho g y_C A$.
Then, we have

$$F_h = \rho g y_C A \sin \theta$$

Substituting $A \sin \theta = A_v$ (vertical projection of A)

$$F_h = \rho g y_C A_v$$
$$= \rho g y_C A \sin \theta$$

Similarly, resolving F vertically, we have the vertical thrust
$F_v = F \cos \theta$, where

$$F = \rho g y_C A$$

This gives the vertical thrust as

$$F_v = \rho g y_C A \cos \theta,$$

where $A \cos \theta = A_h$ (horizontal projection of A).
  So, $F_v = \rho g y_C A_h$

$$= \rho g y_C A \cos \theta$$

The above expressions tell us that when we change the orientation of the patch inside the liquid, the horizontal and vertical thrust change. However, the total thrust remains the same if we the position of centroid of the patch does not move.

Example 25 A thin disc of radius R is pivoted about a horizontal axis AB. The axis of rotation of the disc lies at a depth $2R$ from the free surfaces of a liquid of density ρ. Initially the disc is vertical. If we rotate the disc about the axis AB by an acute angle ϕ, find the (a) horizontal, (b) vertical thrust, and (c) net thrust exerted by one side the liquid on the disc.

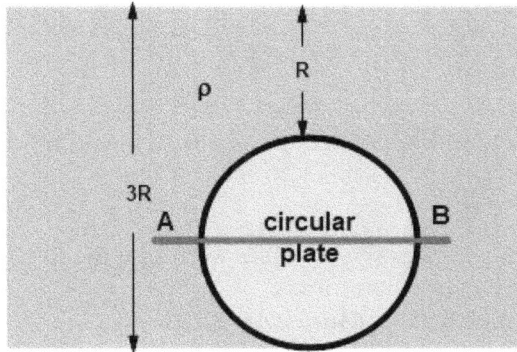

Solution

(a) By rotating the disc by an angle ϕ, the normal to the plane of the disc will make an angle

$$\theta = 90 - \phi$$

The horizontal thrust at any angular position ϕ with respect to vertical is

$$F_h = \rho g y_C A \cos \theta,$$

$$\Rightarrow F_h = \rho g y_C A \cos(90 - \phi)$$

Substituting $y_C = 2R$ and $A = \pi R^2$,
we have

$$F_h = 2\pi \rho g R^3 \sin \phi \text{ Ans.}$$

(b) The vertical thrust at the angular position ϕ is

$$F_v = \rho g y_C A \sin\theta,$$

where $y_C = 2R$, $A = \pi R^2$ and $\theta = 90 - \phi$.
Then, we have

$$F_v = 2\pi\rho g R^3 \cos\phi \text{ Ans.}$$

(c) The net hydrostatic thrust is

$$F_{\text{net}} = \rho g y_C A = \rho g (2R)(\pi R^2) = 2\pi\rho g R^3$$

N.B.: Do not try to write

$$A_h = \pi(R\sin\phi)^2 = A\sin^2\phi$$

and

$$A_v = \pi(R\cos\phi)^2 = A\cos^2\phi,$$

where $A = \pi R^2$.

This is because the projections of the disc are not circular. The correct expressions are $A_h = A\sin\phi$ and $A_v = A\cos\phi$.

Example 26 In the previous example, while we rotate the disc, find the change in
(a) Magnitude of hydrostatic force.
(b) Magnitude of change in hydrostatic force.

Solution

(a) On rotating the disc the net hydrostatic force $\vec{F}$ changes its direction by an angle ϕ, whereas its magnitude is constant, which is given as

$$F = 2\pi\rho g R^3$$

Hence, the change in magnitude of hydrostatic force is zero.
(b) $\Delta F = 0$ and

$$\left|\Delta\vec{F}\right| = \left|\vec{F_2} - \vec{F_1}\right| = 2F\sin\frac{\phi}{2},$$

where

$$F = 2\pi\rho g R^3$$

then, we have the magnitude of the change in hydrostatic force is

$$\left|\Delta\vec{F}\right| = 4\pi\rho g R^3 \sin\frac{\phi}{2} \text{ Ans.}$$

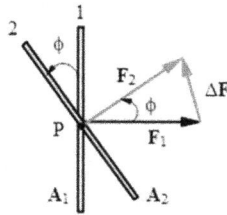

Example 27 A rectangular plate is kept vertical, partly projected outside (above) the liquid of density ρ, as shown in the figure. If the plate is rotated clockwise by an angle θ about horizontal axis passing through its bottom. Find the (a) hydrostatic force, (b) horizontal, and (c) vertical thrust acting on the plate. Assume $b =$ breadth of the plate.

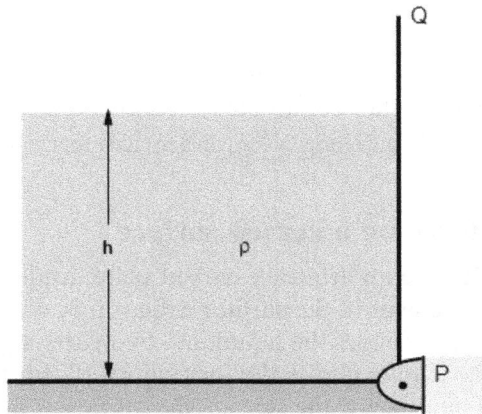

Solution

After rotating the plate through an angle θ, the level of liquid remains constant. Hence, the y-coordinate of the centroid of spattered area of the rectangular plate remains the same, which is given as

$$y_C = \frac{h}{2}$$

However, the area A of contact with the liquid increases from A_0 to $\frac{A_0}{\cos\theta}$, where $A_0 = hb$

Hence, the hydrostatic force is

$$F = \rho g y_C A,$$

where $y_C = \frac{h}{2}$ and $A = \frac{A_0}{\cos\theta}$.

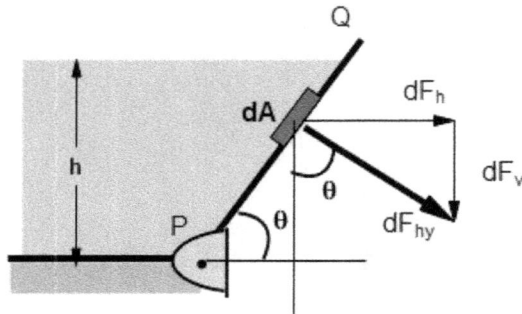

This gives

$$F = \rho g \frac{hA_0}{2 \cos \theta}$$

Substituting $A_0 = hb$, we have

$$F = \frac{\rho g h^2 b}{2} \sec\theta \text{ Ans.}$$

N.B.: Since the original area of constant A_v is the projection of new area $A = \frac{A_0}{\cos \theta}$ and y_C remains the same, the horizontal thrust remains constant unlike the previous example. But, do you think that $A = \frac{A_0}{\cos \theta}$ for any arbitrary plate?

4.13 Hydrostatic force on a curved surface

Horizontal thrust: Let's take an arbitrary curved plate inside a liquid of density ρ. The liquid pushes each element of the surface either up or down that depends upon the nature curve. The plate divides the liquid into two parts; liquid above and liquid below the surface. The liquid above the surface exerts horizontal force F_h and vertical force F_v. The liquid below the surface can also exert horizontal and vertical forces on the plate. Hence, the net hydrostatic force exerted on the plate by the surrounding liquid is zero (if the thickness of the surface is negligible). But, in general, the net horizontal hydrostatic force will be always zero for a non-accelerating liquid. By ignoring the thickness of the surface, it just makes the things easier to consider the hydrostatic force on either concave (or convex) surface. Keeping the above idea in mind, Let's find the horizontal thrust of liquid on the curved surface on either side. For the sake of simplicity, let's assume that the liquid is present in the left side of the curved surface.

Let's take an elementary patch of area dA at a depth y on the plate. The hydrostatic force acting on the element is

$$\overrightarrow{dF} = Pd\overrightarrow{A} ,$$

where $P = \rho g y$.
This gives

$$dF = \rho g y dA$$

Since $d\vec{F}$ changes its direction from point to point on the curved surface, we cannot just integrate (sum) the elementary force $d\vec{F}$ to obtain the total hydrostatic force. We need to resolve the force $d\vec{F}$ into two components, horizontal component dF_h and vertical component dF_v. Resolving the force $d\vec{F}$ horizontally, we have

$$dF_h = \rho g y dA \sin \theta,$$

where $dA \sin \theta = dA_v =$ vertical projection (or component of the elementary area dA). This gives $dF_h = \rho g y dA_v$.

The total horizontal thrust on the curved surface is

$$F_h = \int dF_h = \rho g \int y dA_v$$

Since

$$\frac{\int y dA_v}{A_v} = y_C,$$

where $y_C =$ depth of the centroid of the vertical projection of the curved surface, we have

$$Flh = \rho g y_C A_v,$$

where $\rho g y_C (=P_C) =$ hydrostatic pressure at the centroid of A_v.

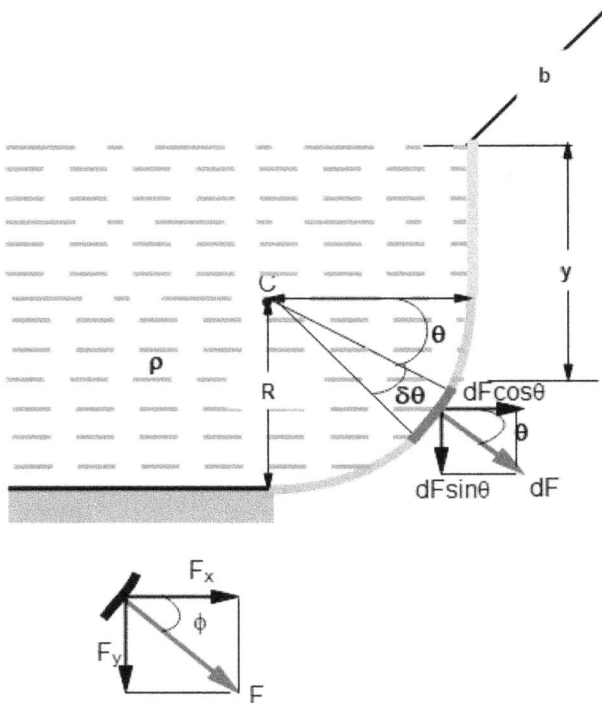

From the above formula we can understand that the horizontal thrust of liquid on a curved surface is equal to the product of vertical component or projection A_v of the curved surface and pressure of the liquid at the centroid of A_v. It is given as

$$F_h = P_C A_v, \text{ where } P_C = \rho g y_C.$$

Example 28 A cylinder of radius R and length R is kept in front of a liquid of height R and density ρ. Find the (a) horizontal, (b) vertical, and (c) total thrust acting on the cylinder.

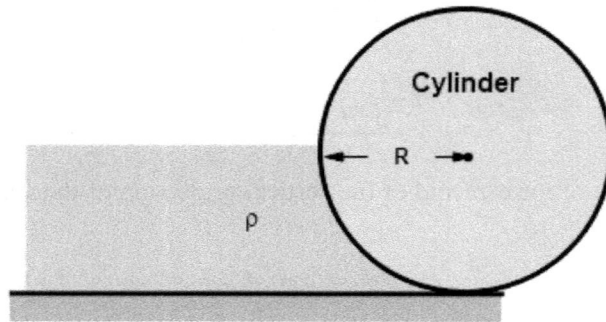

Solution

(a) The horizontal thrust on the curved surface of the cylinder is given as

$$F_h = \rho g y_C A_v,$$

where $y_C = \frac{R}{2}$, $A_v = R^2 = $ vertical projection.
This gives $F_h = \frac{1}{2}\rho g R^3$ Ans.

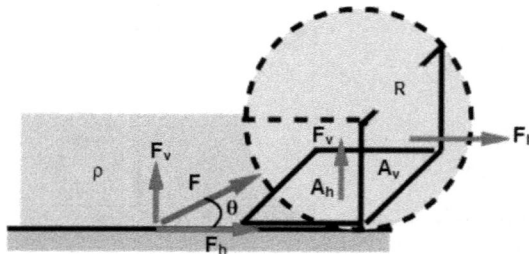

(b) The vertical thrust on the curved surface of the cylinder is given as

$$F_h = \rho g y_C A_h,$$

where $y_C = R$, $A_h = R^2 = $ horizontal projection.
This gives $F_v = \rho g R^3$.
(c) The net thrust is

$$F = \sqrt{F_h^2 + F_v^2} = \sqrt{5}\,\rho g R^3/2$$

The angle made by the net force with horizontal is

$$\theta = \tan^{-1}(F_v/F_h) = \tan^{-1}2$$

Example 29 Prove that the net horizontal thrust on a partially or fully immersed body is zero.
Solution
Any immerged body can be imagines as the parallel array of thin horizontal tubes. Let's consider an elementary tube at a depth y whose ends have area dA_1 and dA_2. Since dA_1 is vertical the horizontal thrust acting on dA_1 is given as

$$dF_1 = P_1 dA_v,$$

where $P_1 = \rho g y$ and $dA_1 = dA_v$.
This gives

$$dF_1 = \rho g y dA_v$$

Taking the other end of area dA_2 of the elementary tube, the horizontal thrust on it is

$$dF_2 = P_2 dA_v$$

and the vertical component of
$dA_2 = dA_v$.
Putting $P_2 = \rho g y$, we have

$$dF_2 = \rho g y dA_v \qquad (4.37)$$

From equations (4.1) and (4.37), we can see that the net horizontal force acting on the elementary horizontal tube is

$$dF = dF_1 - dF_2 = 0$$

Since we have assumed the body as a bundle of horizontal thin tubes and each tube tube experiences zero horizontal thrust, we can conclude that the net horizontal force acting on the immerged (totally or fully) is zero.

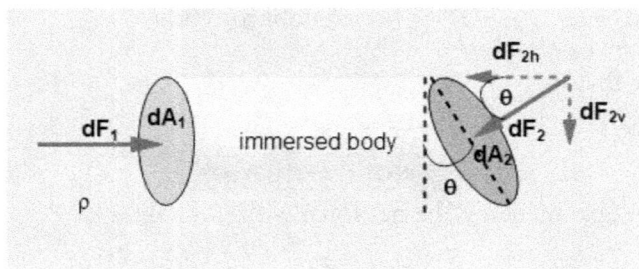

In other words, the net horizontal hydrostatic thrust on a totally or partially immersed body is zero.

If the liquid exerts zero side thrust (force) of an immersed body, then, we should expect that the net hydrostatic force acting on the body must be vertical. Let's see whether the net hydrostatic force is vertically upward or downward.

Vertical thrust on a curved surface: Let's now resolve the force dF vertically to obtain the vertical thrust dF_v on the elementary area dA

$$dF_v = dF \sin \theta,$$

where $dF = \rho gy dA$

This gives

$$dF_v = \rho gy dA \sin \theta$$

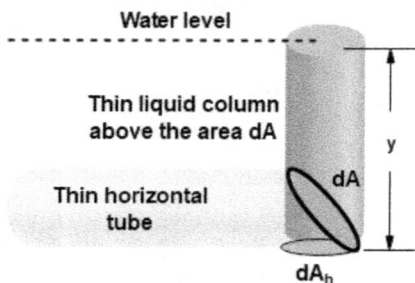

Substituting $dA \sin \theta = dA_h$ (= horizontal projection of dA), we have

$$dF_v = \rho gy dA_h$$

Since $y dA_h = dV$ (volume of thin vertical liquid column of height y above the elementary area), we have $dF_v = \rho g dV$.

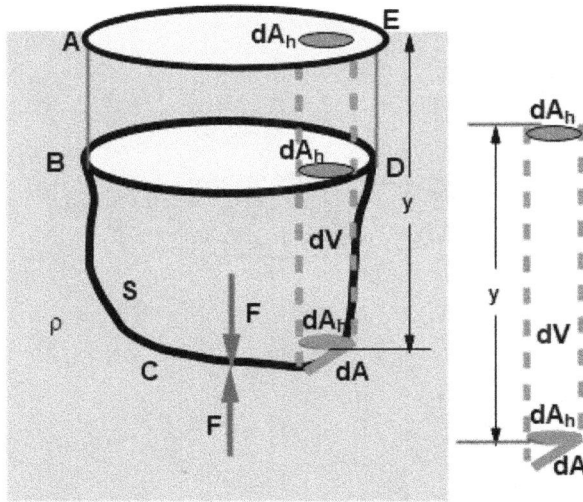

Volume ABCDE is generated by the
curved surface S with the liquid surface

Summing up (integrating) dF_v, the vertical thrust on the curved surface S (BCD) is given as

$$F_v = \int dF_v = \int \rho g y dA_h = \rho g \int y dA_h,$$

where $\int y dA_h = V$ = volume generated by the curved surface S with the surface of liquid.

Then, we have

$$F_v = \rho g V$$

If the surface S is closed, the volume V will be equal to the volume of the immersed portion of the body. If the closed surface is fully immersed, $V =$ the volume enclosed by the surface. Then, we have the net hydrostatic force which is vertically upward can be called buoyant force, given as

$$F_v = F_b = \rho g V,$$

where $V =$ volume of the immerged portion of the body. Then, the above equation tells us that;

The vertical thrust acting on a curved surface is equal to the weight of the volume of portion of the body immerged in the liquid, that is, volume of the liquid displaced by the body. That is what is known as *Archimedes' Principle*.

Example 30 A closed vessel of an arbitrary shape and a circular cap of area A is kept in a liquid of density ρ as shown in the figure such that the cap is located at a depth of h. Find the vertical thrust of liquid acting on the curved surface of the vessel.

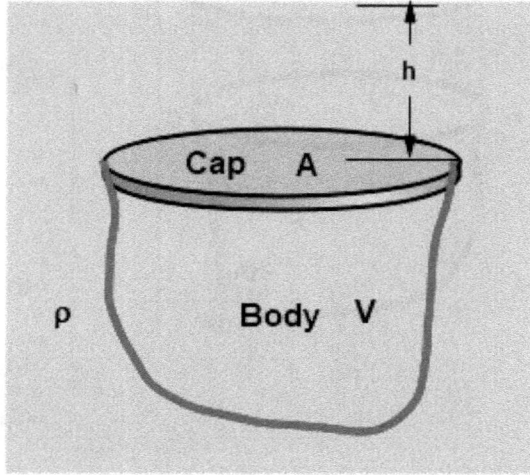

Solution

Vertical (downward) thrust of liquid on the cap is

$$F_1 = PA = (\rho g h)(\pi R^2) = \pi \rho g h R^2, \qquad (4.38)$$

The vertical thrust of liquid on the curved surface is F_2, say.

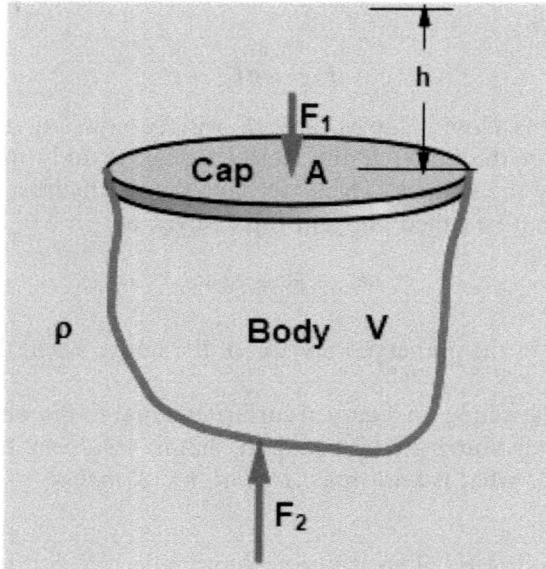

We know that the net vertical force on the sphere is

$$F_b = V\rho g, \tag{4.39}$$

where V = volume of the body (given).

The net hydrostatic force = buoyant force

$$= F_b = F_2 - F_1 = V\rho g \tag{4.40}$$

Using equations (4.39) and (4.40),

$$F_2 = F_1 + V\rho g \tag{4.41}$$

Using equations (4.38) and (4.41),

$$F_2 = \pi\rho g h R^2 + V\rho g \text{ Ans.}$$

Example 31 A sphere of radius R is kept in a liquid of density ρ as shown in the figure. Find the hydrostatic force located at a depth of h. Find the vertical acting on the (a) lower hemisphere, (b) upper hemisphere, and (c) complete sphere.
Solution

(a) Let's divide the sphere into two parts, upper hemisphere AHB and lower hemisphere ACB.

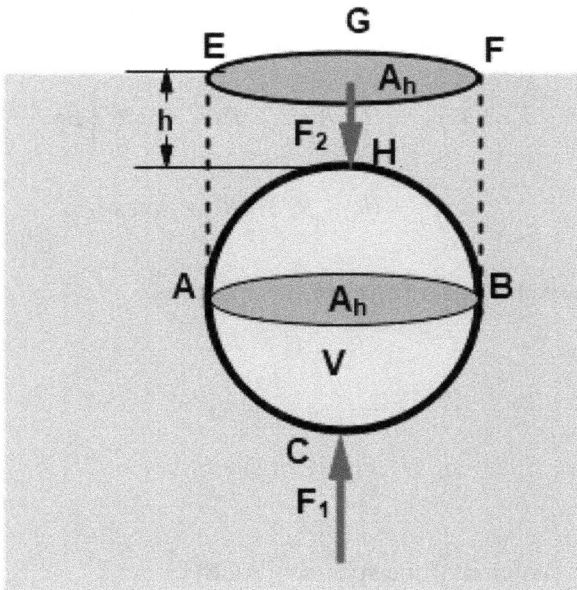

Vertical thrust of liquid on ACB is

$$F_1 = V_1 \rho g \qquad (4.42)$$

where V_1 = volume of the liquid column EACBF.
= Volume of the cylinder EABF + volume of the hemisphere ACB

$$\Rightarrow V_1 = \pi R^2 (h + R) + \frac{2}{3} \pi R^3 \qquad (4.43)$$

Using last two equations, we have

$$F_1 = V_1 \rho g = \left\{ \pi R^2 (h + R) + \frac{2}{3} \pi R^3 \right\} \rho g$$

$$\Rightarrow F_1 = (h + 5R/3) \pi R^2 \rho g \text{ Ans.}$$

(b) Similalrly, the vertical thrust of liquid on AHB is

$$F_2 = V_2 \rho g,$$

where V_2 = volume of the portion EAHBF
= Volume of the cylinder EABF − volume of the hemisphere AHB

$$\Rightarrow V_1 = \pi R^2 (h + R) - \frac{2}{3} \pi R^3 \qquad (4.44)$$

Using last two equations, we have

$$F_1 = V_1 \rho g = \left\{ \pi R^2 (h + R) - \frac{2}{3} \pi R^3 \right\} \rho g$$

$$\Rightarrow F_1 = (h + R/3) \pi R^2 \rho g \text{ Ans.}$$

(c) Hence, the net vertical force on the sphere is

$$F_b = F_1 - F_2$$

$$= V_1 \rho g - V_2 \rho g$$

$$= (V_1 - V_2) \rho g$$

$$= V \rho g,$$

where V = volume of the sphere = ACBH

Putting $V = \frac{4}{3}\pi R^3$, we have

$$F_b = \frac{4}{3}\pi R^3 \rho g$$

Example 32 A solid hemispherical vessel of radius R is immersed in a liquid of density ρ as shown in the figure. Find the (a) net hydrostatic force acting on the vessel, (b) vertical thrust of liquid on the curved surface of the vessel, and (c) net force acting on the vessel. Put $h = R$.

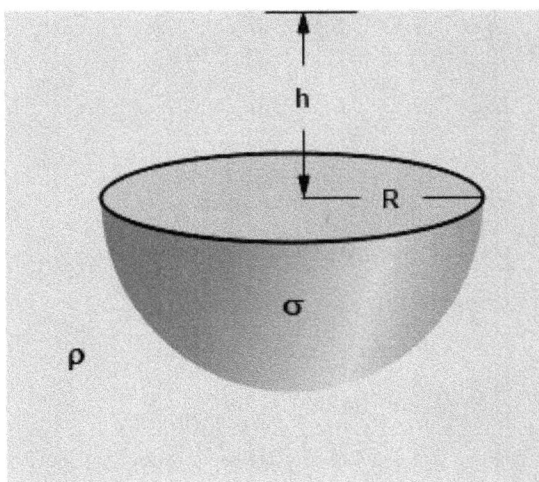

Solution

(a) Let F_1 be the upward force acting on the curved surface of the hemisphere. If the liquid presses the top down with a force F_2, say, the net the net hydrostatic force = buoyant force is given as

$$F_b = F_1 - F_2 = V\rho g \tag{4.45}$$

Putting $V = \frac{2}{3}\pi R^3$, we have

$$F_b = \frac{2}{3}\pi R^3 \rho g \text{ (up)}$$

(b) From equation (4.45),

$$F_1 = F_2 + V\rho g \tag{4.46}$$

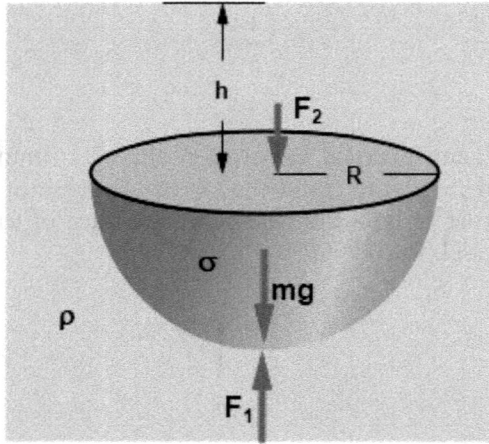

The liquid presses the top down with a force

$$F_2 = PA = (\rho gh)(\pi R^2) = \pi \rho gh R^2,$$ (4.47)

Using equations (4.46) and (4.47), we have

$$F_1 = \pi \rho gh R^2 + V\rho g$$

Putting $V = \frac{2}{3}\pi R^3$, we have

$$F_1 = \pi \rho gh R^2 + \frac{2}{3}\pi R^3 \rho g, \text{ where } h = R.$$

Then, we have

$$F_1 = \frac{5}{3}\pi R^3 \rho g \text{ Ans.}$$

The net force acting on the vessel is

$$F = F_b - F_{gr} = V\rho g - mg$$

Putting $m = V\sigma$, we have

$$F = V\rho g - V\sigma g = Vg(\rho - \sigma) \text{ (up)}$$

If $\rho > \sigma$, the net force is up; If $\rho < \sigma$, the net force is down; If $\rho = \sigma$, the net force is zero. Ans.

Example 33 A solid hemisphere of radius R is made to just sink in a liquid of density ρ. Find the (a) vertical thrust on the curved surface (b) vertical thrust on the flat surface (c) side thrust on the hemisphere (d) total hydrostatic force (= buoyant force) acting on the hemisphere.

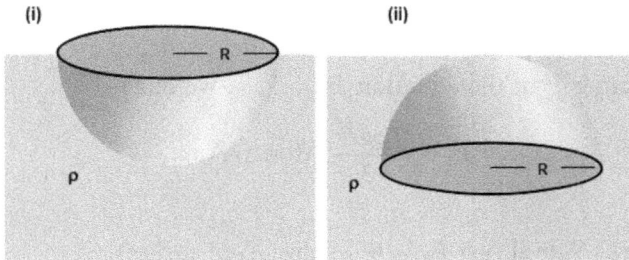

(i)   (ii)

Solution

(a)

(i) Vertical thrust of the liquid is equal to weight of the liquid column above the curved (spherical) surface

$$F_v = V\rho g,$$

where V = volume generated by the curves surface with liquid surface = volume of hemisphere

$$= \frac{2}{3}\pi R^3$$

(i)   (ii)

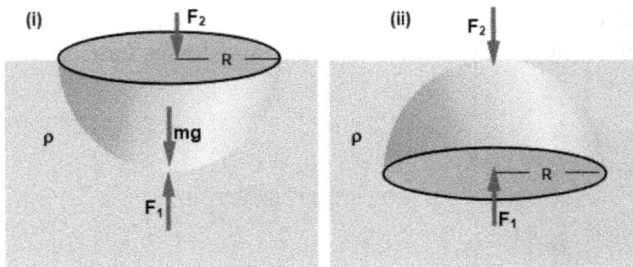

Substituting V in the equation $F_v = V\rho g$, we have

$$F_1 = \frac{2\pi\rho g R^3}{3} \text{ (down) Ans.}$$

(ii) Vertical thrust of the liquid on the curved surface = weight of the liquid column above the curved (spherical) surface
$F_2 = V\rho g$, where V = volume generated by the curved portion = volume of the cylinder − volume of hemisphere

$$=(\pi R^2)R - \frac{2}{3}\pi R^3$$

$$=\frac{\pi R^3}{3}$$

(b) Substituting V in the equation $F_2 = V\rho g$, we have

$$F_2 = \frac{\pi \rho g R^3}{3} \text{ (down) Ans.}$$

(c) (i) and (ii) Side thrust $F_h = 0$ (as discussed earlier) Ans.
(d) (i) and (ii) The total hydrostatic force is

$$F_b = V\rho g,$$

where $V =$ volume of the hemisphere. Then, we have

$$F_b = \frac{2\pi}{3}R^3\rho g \uparrow \text{ Ans.}$$

Example 34 A tortoise just sinks in water of density ρ. The tortoise is assumed as a hemisphere of radius R. Find the:
   (a) Horizontal thrust
   (b) Vertical thrust
   (c) Total hydrostatic force
   (d) Angle of orientation of total hydrostatic force acting on the tortoise
   (e) Ratio of vertical thrusts on the upper and lower half of the tortoise.

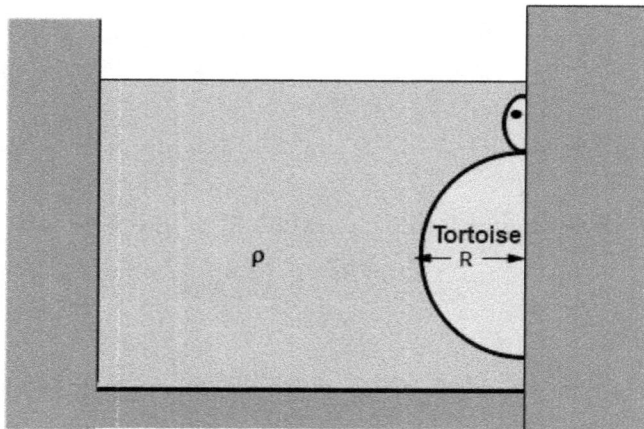

Solution

(a) Let the horizontal and vertical thrusts on the tortoise be F_h and F_v, respectively.

We know that

$$F_h = \rho g y_C A_v,$$

where $y_C = R$ and $A_v = \pi R^2$

This gives $F_h = \rho g \pi R^3$ (right). Ans.

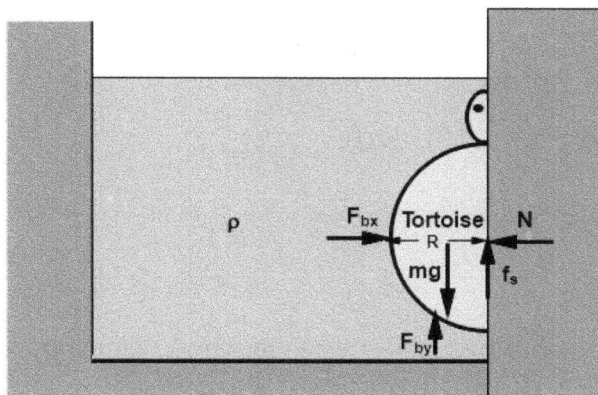

(b) Similarly using the formula

$F_v = \rho V g$, where V = volume of the tortoise $= \frac{2}{3} \pi R^3$, we have

$$F_v = \frac{2}{3} \rho g \pi R^3 \text{ (up) Ans.}$$

(c) Hence, the net hydrostatic force on the tortoise is

$$F = \sqrt{F_h^2 + F_v^2} = \sqrt{(\rho g \pi R^3) + \left(\frac{2}{3}\rho g \pi R^3\right)^2} = \sqrt{\frac{13}{3}} \rho g \pi R^3 \text{ Ans.}$$

(d) The angle of orientation of the force F is

$$\phi = \tan^{-1}\frac{F_h}{F_v} = \tan^{-1}\frac{\rho g \pi R^3}{\frac{2}{3}\rho g \pi R^3} = \tan^{-1}\frac{3}{2} \text{ Ans.}$$

(e) The downward vertical thrusts on the upper half of the tortoise = weight of the liquid column above the top half curved (spherical) surface:

$$F_{\text{top}} = V \rho g,$$

where V = volume generated by the top half with the water surface =

$$=\frac{1}{2}\left\{(\pi R^2)R - \frac{2}{3}\pi R^3\right\} = \frac{1}{6}\pi R^3$$

Substituting V in the equation $F_2 = V\rho g$, we have

$$F_{top} = \frac{\pi \rho g R^3}{6} \text{ (down)}$$

The upward vertical thrusts on the bottom half of the tortoise = weight of the liquid column above the bottom half curved (spherical) surface;

$$F_{top} = V\rho g,$$

where V = volume generated by the bottom half with the water surface =

$$=\frac{1}{2}\left\{(\pi R^2)R + \frac{2}{3}\pi R^3\right\} = \frac{5}{6}\pi R^3$$

Substituting V in the equation $F_2 = V\rho g$, we have

$$F_{top} = \frac{5\pi \rho g R^3}{6} \text{ (up)}$$

Then, ratio of vertical thrusts on the upper and lower half of the tortoise is equal to 1:5. Ans.

Hydrostatic paradox

Let's take three different vessels of the same base area A (say) and fill the vessels with a liquid (water, say) up to a height h, say. Ignore the weight of the vessels. If I ask you which liquid exerts more force on the ground, you can promptly say vessel 3, because it contains the maximum amount of water as it has the maximum volume, then it comes to vessel 2; vessel 1 pushes the ground with the least force because it contains the least amount of water.

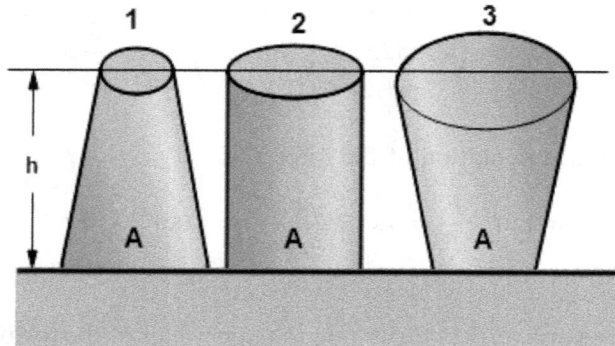

My next question is which liquid exerts more force on the base of the vessels? The answer is 'all three vessels'. It is because both pressure P at the base and the base area remain the same in all cases. This seems to be paradoxical but actually this happens. What is the cause of this paradox? Let's see.

The curved surface of vessel 1 presses the liquid down with a force F_v favouring the minimum weight of the liquid. The curved surface of vessel 2 presses the liquid horizontally radially inwards. So the vertical component F_v is zero. In vessel 3, the curved surface lifts the liquid up, nullifying the maximum weight of the liquid partly. Thus, it causes the same force at the base compared to the other vessels.

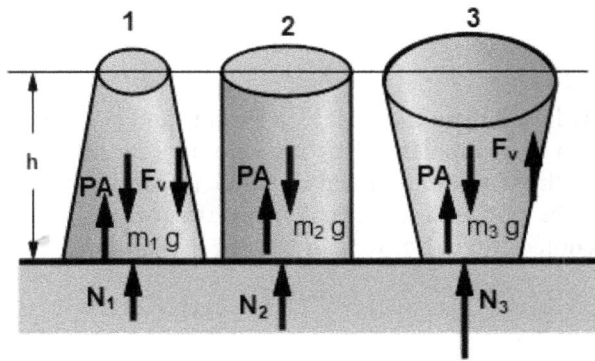

We conclude that

$$N_3 > N_2 > N_1,$$

where N_1, N_2, and N_3 are the forces of liquids acting on the vessels (or ground), whereas the pressing forces of the liquid to the base of the vessels are equal:

$$F_1 = F_2 = F_3 = PA,$$

where $P = \rho g h$; F_1, F_2, and F_3 are the hydrostatic forces acting at the bases of the vessels.

4.14 Archimedes' principle

Explanation 1

Let's take a thin shell completely filled with water. If we immerse it fully in water, because the shell is extremely thin, the pressure difference across the thickness of the shell is negligible. Hence, each element of the pot is pressed by the surrounding water (liquid) with equal and opposite forces. As a consequence, the net hydrostatic force on the pot is zero. This means the downward force acting on the pot due to the liquid inside the pot is equal to the upward force exerted by the liquid outside the pot.

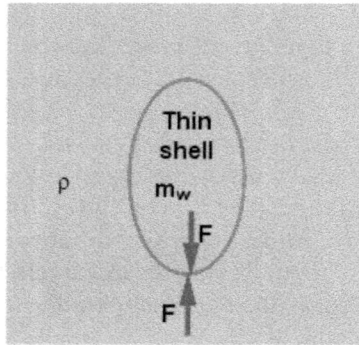

Water is present inside and
outside the shell

If we remove the liquid from the shell by a syringe and close the hole, now the net hydrostatic force on it is unbalanced by an amount $F = mg$, where $m =$ mass of water inside the shell. After the removal of water from the shell, the outside water continues to push the pot with a vertically upward force F. That is what we call 'buoyant force' denoted by F_b. So, we can write $F_b = F$ and $F = mg$. Then, we have

$$F_b = mg,$$

where $m = \rho V$, and $V =$ volume of water removed or displaced by the shell. Then we can say that

$$F_b = mg = V\rho g$$

This result also holds good for a partial immersion followed by the same argument.

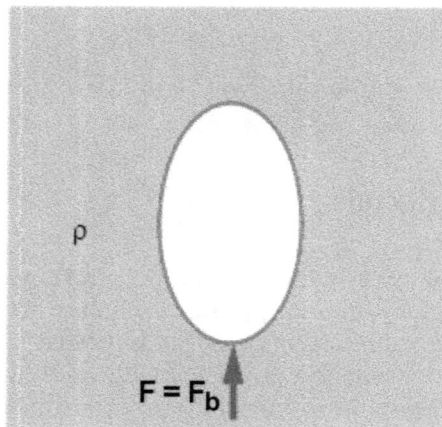

Water inside the shell
is removed

The above equation tells us that whenever a body is partially or totally immersed in a liquid (or fluid), the net hydrostatic force acting on the body is called *buoyant force,* which is equal to the weight of the liquid displaced by the body. This is known as *Archimedes' principle.*

Explanation 2

Take two identical wooden logs of uniform cross-section A, say. The right side log is immersed totally and the left side log is immersed partially as shown in the figures.

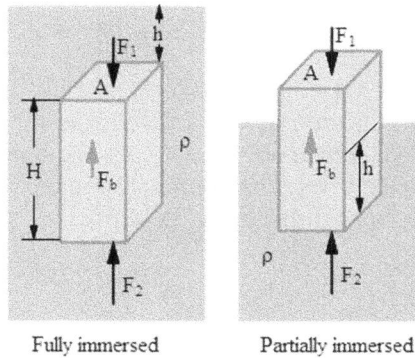

Fully immersed   Partially immersed

We learned that the net side thrust is zero in both cases. So, the net hydrostatic force is vertical, which is known as buoyant force and can be given as

$$F_b = F_2 - F_1, \tag{4.48}$$

where F_1 and F_2 are the total hydrostatic force acting on the top and bottom faces of the logs.

In the case of totally immersed

$$F_1 = P_1 A = (P_{atm} + \rho g h)A \tag{4.49}$$

$$F_2 = P_2 A = \{P_{atm} + \rho g(H + h)\}A \tag{4.50}$$

Using the last three equations, we have

$$F_b = \rho g H A = \rho g V, \tag{4.51}$$

where V = volume of the body.

In the case of partial immersion,

$$F_1 = P_1 A = P_{atm} A \tag{4.52}$$

$$F_2 = P_2 A = (P_{atm} + \rho g h)A \tag{4.53}$$

Using the last three equations (4.48), (4.52), and (4.53), we have

$$F_b = \rho g h A = \rho g V',$$ (4.54)

where $V = Ah$ = volume of the immersed portion of the body.

In the equations (4.51) and (4.54), the volume V and V' can be termed as a common factor as 'volume of the immersed portion of the body'. This is equal to the volume of the liquid displaced by the body due to its immersion. Let's denote this volume as V_d. So, the general formula for buoyant force is given as

$$F_b = \rho g V_d$$

You may drop the subscript 'd' and simply denote V = volume of the liquid displaced.

Center of buoyancy: When a body is partially or fully immersed, the volume of liquid displaced is equal to the volume of portion of the solid immersed as shown by the dotted lines. For partial immersion, the volume of liquid displaced is less than the volume of the body. However, for complete immersion, the volume of liquid displaced is equal to the volume of the body. In both cases, we need to locate volume V_d of the liquid displaced.

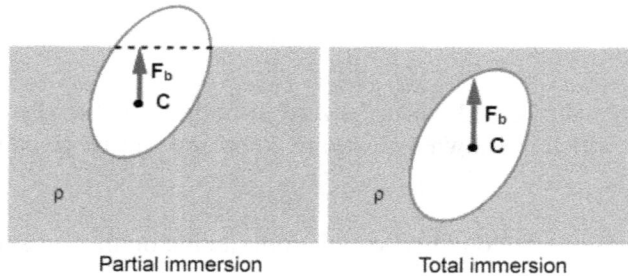

Partial immersion   Total immersion

Then, we can find the 'centroid' of this volume assuming the presence of liquid. At that centroid 'C' of the volume (but not center of mass of the bodies) of the displaced liquid, the buoyant force F_b is assumed to be acting. This is known as 'the center of buoyancy'.

Center of buoyancy is the centroid of the region from which the liquid is displaced. In other words, it is the center of mass of the displaced liquid. This means that F_b acts vertically up at the center of buoyancy but not necessarily the center of mass of the body.

4.15 Application of archimedes' principle

Example 35 Explain the origin of buoyant force due to a non-accelerating liquid (or fluid) acting on a body immersed partially or totally in the fluid.

Solution

When a body is immersed completely or partially in a liquid, the hydrostatic force dF always acts normal to the surface. Its magnitude at deeper points is greater than that at the points nearer to the liquid surface. This happens because pressure increases linearly with depth. Then, the lower portion of the body experiences greater upward hydrostatic thrust and the upper portion of the body experiences less downward thrust of liquid. As there is no net side thrust due to the stationary liquid, the net hydrostatic force, that is, buoyant force, acts vertically upwards.

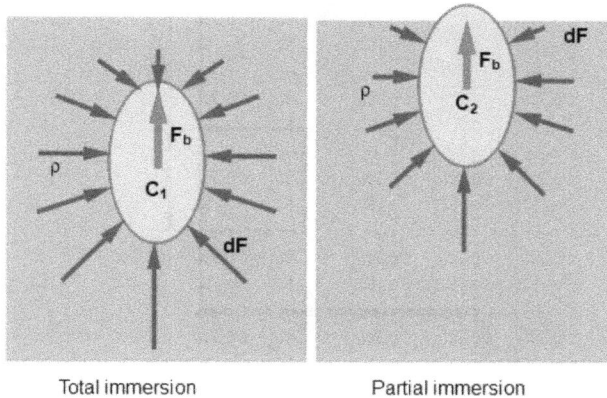

Total immersion   Partial immersion

Buoyant force points vertically up if the fluid does not accelerate. It happens because the hydrostatic pressure increases linearly with depth.

Example 36 If we weld the vertical face of a cube of side l in a water tank, the density of water is ρ. Find the hydrostatic force acting on the cube.

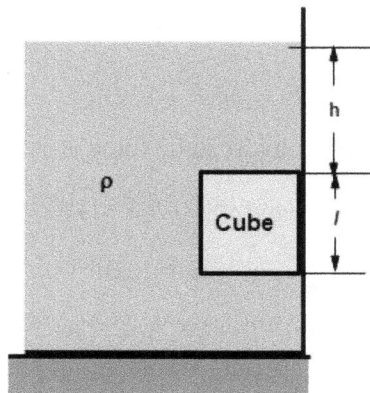

Solution
If we weld the vertical face of a cube in a water tank, all vertical faces are not exposed to the liquid. The liquid does not surround the body. Hence, we cannot call

it an 'immersion'. So, directly we cannot apply the idea of 'buoyant force'. However, the net hydrostatic force acting on the body is

$$\overrightarrow{F} = F_1\hat{i} + (F_3 - F_2)\hat{j},$$

where $F_3 - F_2 = F_b$.

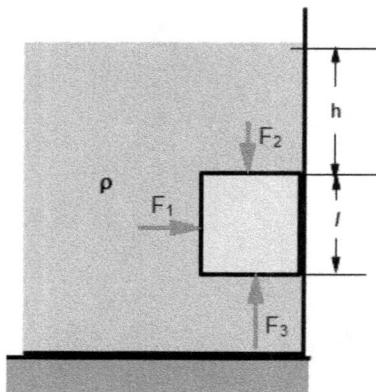

$$\Rightarrow \overrightarrow{F} = F_1\hat{i} + F_b\hat{j}, \tag{4.55}$$

where

$$F_b = \rho g V_d = \rho g l^3 \tag{4.56}$$

The side thrust F_1 can be given as

$$F_1 = \rho g y_c A,$$

where $y_c = (h + l/2)$ and $A = l^2$

$$F_1 = \rho g(h + l/2)l^2$$

Using equation (4.55), the net hydrostatic force is

$$\overrightarrow{F} = \rho g(h + l/2)l^2\hat{i} + \rho g l^3\hat{j}$$

$$\Rightarrow \overrightarrow{F} = \rho g l^2\{(h + l/2)\hat{i} + l\hat{j}\}$$

N.B.: This is not equal to buoyant force $F_b = V\rho g$ when the body is completely immersed.

Example 37 If we weld the vertical face of a cube of side l in a water tank, the density of water is ρ. Find the hydrostatic force acting on the cube.

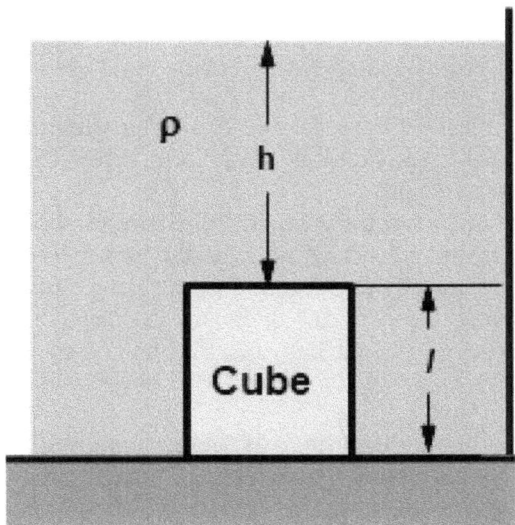

In this case, the base of the body is welded (or pressed) so that the liquid cannot exert force at the base. Then, the total hydrostatic force is

$$\vec{F} = (F_1 - F_3)\hat{i} - F_2\hat{j}$$

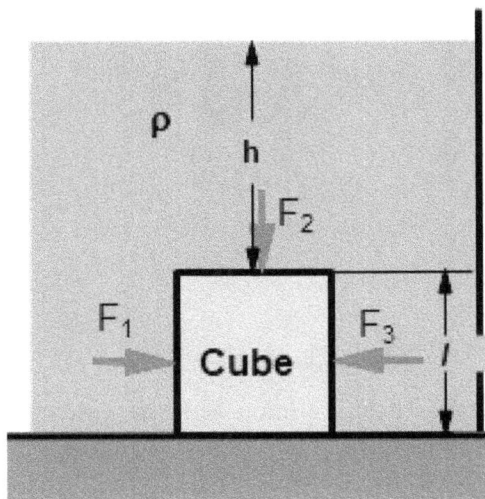

Since the net side thrust is zero,

$$F_1 = F_3, \,,$$

Then, we have

$$\vec{F} = -F_2\hat{j}\,,$$

where $F_2 = \rho g y_c A = \rho g h l^2$,

$$\Rightarrow \overrightarrow{F} = -F_2 \hat{j} = -\rho g h l^2 \hat{j} \text{ (down)}$$

N.B.: This is not equal to buoyant force $F_b = V \rho g$ when the body is completely immersed. It means that if the liquid is not present below the base of the cube, we cannot call it an immersed body.

An immersed body must touch the liquid at its base (or bottom). The liquid must surround the body. The net hydrostatic force on the immersed body is called buoyant force, which is directed vertically up for a non-accelerating liquid. If the body does not come under the case of immersion, we should not use the idea of buoyant force. However, the net hydrostatic force acting on the body can be found by using the first principle.

Example 38 A body of mass m density σ is totally immersed in a fluid of density ρ. Find the (a) net force acting on the body and (b) acceleration of the body.
   Solution

   (a) The forces acting on the body are (i) buoyant force F_b and (ii) weight mg. Then, the net force acting on the body is
       $F = F_b - mg$, where $F_b = V \rho g$

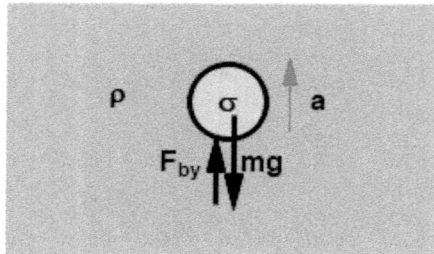

   This gives
$$F = V \rho g - mg$$

   Since the body is totally immersed, the volume of liquid displaced is equal to the volume of the body. Substituting $V = \frac{m}{\sigma}$, where $\sigma = $ density of the body, we have

$$F = mg\left(\frac{\rho}{\sigma} - 1\right) \text{ Ans.}$$

   (b) If the body moves with an acceleration a, applying Newton's 2nd law, we have
       $a = \frac{F}{m}$, where $F = mg(\frac{\rho}{\sigma} - 1)$

This gives

$$a = g\left(\frac{\rho}{\sigma} - 1\right) \text{(up) Ans.}$$

N.B.: When
  (i) $\rho > \sigma$, the acceleration is $a = (\frac{\rho}{\sigma} - 1)g\uparrow$.
  (ii) When $\rho = \sigma$, the acceleration is zero.
  (iii) When $\rho < \sigma$, the acceleration is

$$a = \left(1 - \frac{\rho}{\sigma}\right)g\downarrow$$

N.B.: If the density of the body is greater than the density of the surrounding fluid, it accelerates down. When their densities are equal, the body moves with constant velocity inside the liquid. If the body is less dense than fluid, it accelerates up. In all the cases, no other forces must act other than gravity and the buoyant forces.

Example 39 A rubber ball of density $\sigma(<\rho)$ is released from rest from a height h above a swimming pool. If the density of water is ρ and the ball does not touch the bottom of the swimming pool, find the (a) maximum distance covered by the ball inside the water during its descent and (b) total time of periodic motion for one cycle. Neglect friction between ball, water, and air.

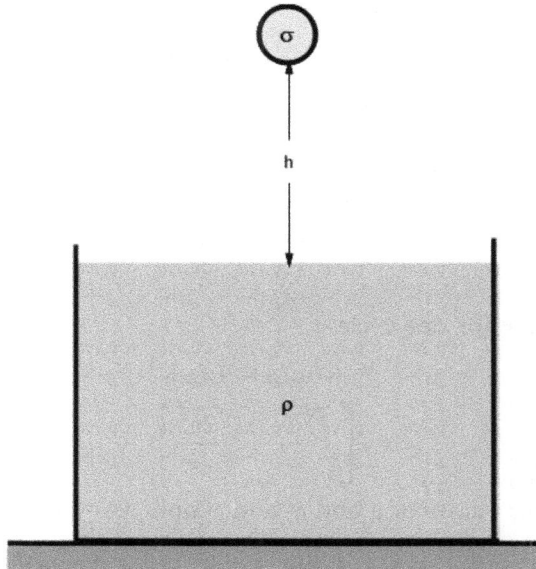

Solution

(a) Let the ball accelerate up inside the water with an acceleration a. Applying $v^2 = u^2 + 2as$ for the ball from A to B and from B to C, we have the following equations:

$$v_B^2 = v_A^2 + 2gh \tag{4.57}$$

$$v_C^2 = v_B^2 + 2(-a)h' \tag{4.58}$$

Substituting $v_A = 0$ as the ball is released from rest from A and $v_C = 0$ as it remains stationary momentarily at C, we have

$$v_B^2 = 2gh = 2ah'$$

This gives

$$h' = \frac{g}{a}h,$$

where $a = g(\frac{\rho}{\sigma} - 1)$.

Then, we have

$$h' = \frac{h}{\left(\frac{\rho}{\sigma} - 1\right)} \quad \text{Ans.}$$

Alternative method:

Applying WE theorem from A to C, we have

$$W_{\text{gr}} + W_{\text{buoyant}} = \Delta K,$$

where $W_{\text{gr}} = mg(h + h')$,

$$W_{\text{buoyant}} = -F_b h' \text{ and}$$

$$\Delta K = K_C - K_A = 0$$

After substituting
$F_b = mg(\frac{\rho}{\sigma} - 1)$, we have

$$h' = \left(\frac{\rho}{\sigma} - 1\right)h$$

(b) The total time for one cycle is

$$T = 2[t_{\text{ascent}} + t_{\text{descent}}]$$

$$= 2\left[\sqrt{\frac{2h}{g}} + \sqrt{\frac{2h'}{a}}\right]$$

Putting the values of a and h' and simplifying the factors, we have

$$T = 2\sqrt{\frac{2h}{g}}\left(1 + \sqrt{\frac{\sigma}{\rho - \sigma}}\right)$$

Example 40 A hemispherical iron vessel of mass m, radius R, and density σ is floating in liquid of density ρ.

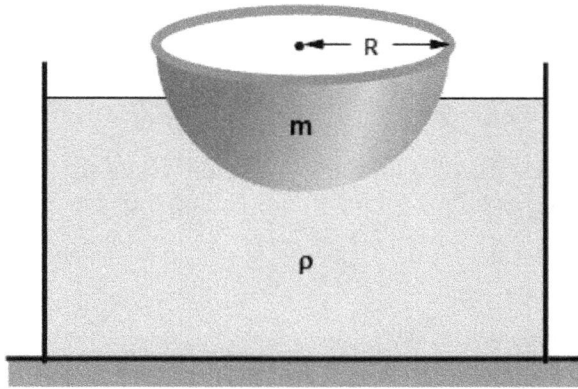

Find the:
 (a) Volume of liquid displaced
 (b) Acceleration of the vessel, when it is made to immerse completely by pushing it into the liquid
 (c) Greatest mass required to keep inside the body (bowl) so as to float water
 (d) Minimum force needed to equilibrate the body when it sinks completely.

Solution

 (a) Since the body floats, its acceleration is zero. This means

$$F_{net} = F_b - mg = 0$$

Then,

$$F_b = mg$$

where $m =$ mass of the body (mass of the liquid displaced).
 Substituting

$$F_b = V\rho g,$$

we have $V = \frac{m}{\rho}$ Ans.

 (b) If the vessel is pushed down by applying a downward force on it, it will sink completely. In this stage, water will enter into the vessel and fill it completely. After a complete immersion of the vessel, the buoyant force will be less than the weight of the vessel and the net downward force can be given as

$$F = mg - F_b = mg - V\rho g$$

Putting $V = m/\rho$ = volume of liquid displaced for a totally immersed body, we have

$$F = mg - F_b = mg(1 - \rho/\sigma)$$

Since ρ/σ is less than one, the acceleration of the vessel is given as

$$a = g(1 - \rho/\sigma) \uparrow \text{ (up) Ans.}$$

(c) Let the greatest mass be M loaded on to the vessel so that it will be about to sink. For this to happen the amount of liquid displaced is

$$V = \frac{2\pi R^3}{3}$$

So, the buoyant force is

$$F_b = V\rho g = \frac{2\pi \rho g R^3}{3}$$

This will be just equal to the weight of the body plus that of the loaded mass M. So, we can write

$$\frac{2\pi \rho g R^3}{3} = (M + m)g$$

$$\Rightarrow M = \frac{2\pi \rho R^3}{3} - m \text{ Ans.}$$

(d) The minimum downward force required to sink is

$$F = Mg = \left(\frac{2\pi \rho R^3}{3} - m\right)g \text{ Ans.}$$

N.B.: Difference between immersion and flotation: For a body to float, one condition is that the density of the body is less than that of the liquid. But you should not falsely assume that when the density of a body is greater than the density of surrounding liquid, the body sinks completely. To disprove this statement, in the above example, the density of steel vessel is greater than the density of the liquid; yet the vessel floats. So massive ships float on water. For a body to float, the necessary and sufficient condition is that the buoyant force must be equal to the weight of the body. In other words, for a floating body, the mass of the liquid displaced is equal to the mass of the body. So, 'flotation' is different from 'partial immersion'. Any object ($\sigma \geqslant \leqslant \rho$) can be made fully or partially immersed by the application of external forces. But when a body floats in a non-accelerating liquid the buoyant force balances the gravity. When we push a floating body down (into) the liquid, the volume of liquid displaced increases; so, the buoyant force will be greater than the weight of the body. Eventually, the net force is up and the body will accelerate up if the body is released. Thus, any floating body oscillates when it is either pushed or pulled up or down and released.

If a body floats in a non-accelerating liquid, the buoyant force will be equal to the weight of the body, $F_b = mg$; if the liquid accelerates and the body floats, $F_b \gtrless mg$ because the net force acting on the body is not zero.

However, the net force acting on a completely immersed body can be zero if $\rho = \sigma$; non-zero if $\rho \neq \sigma$; $F_b \gtrless mg$.

In other words, the net force acting on a floating body in a non-accelerating liquid is zero; the mass of the liquid displaced is equal to the mass of the body whereas the net force acting on a totally immersed body may be zero (or non-zero) because the mass of the liquid displaced need not be equal to the mass of the body.

However, buoyant force acting on the bodies is always equal to the weight of the liquid displaced by the body totally or partially immersed. Hence, partial immersion, complete immersion, flotation, amd sinking of a body should be properly understood.

Example 41 A vertical cylinder of length l, cross-section A, and density σ is pushed into liquid of density ρ through a distance x and released.

    (a) Find the:

          (i) Net force acting on the cylinder

          (ii) Equilibrium distance x, if $\sigma < \rho$

    (b) Analyse the motion of the cylinder when $\rho = \sigma$.

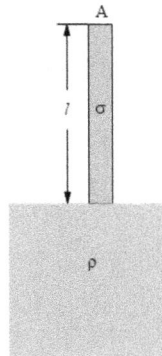

Solution

    (a)

          (i) The forces acting on the body are F_b and mg.
              The net force acting on it is $F_{net} = F_b - mg$,
              where $F_b = V\rho g$; V = volume of liquid displaced.

This gives

$$F_{net} = V\rho g - mg \tag{4.59}$$

When the cylinder is pushed by a distance x into the liquid, the volume of liquid displaced is $V = Ax$.
Substituting $V = Ax$ in equation (4.59), we have

$$F_{net} = (A\rho x - m)g \tag{4.60}$$

Substituting $m = Al\sigma$ in equation (4.60), we have
$F_{net} = (\rho x - \sigma l)Ag$ Ans.
(ii) If the body floats (remains in equilibrium) substitute $F_{net} = 0$ to obtain $x = \frac{\sigma}{\rho}l$ Ans.

(b) If $\frac{\sigma}{\rho} = 1$, the depth of penetration is

$$x = \frac{\sigma}{\rho}l = l.$$

At this depth it will remain as it is if released from rest.

The speed of the cylinder increases until it completely sinks; then it moves with a constant velocity.
For any depth of penetration $x(<l)$,
If $\frac{x}{l} = \frac{\sigma}{\rho}$, $a = 0$ (the body remains in equilibrium)
If $\frac{x}{l} > \frac{\sigma}{\rho}$, (the body accelerates up)
If $\frac{x}{l} < \frac{\sigma}{\rho}$, (the body accelerates down) Ans.

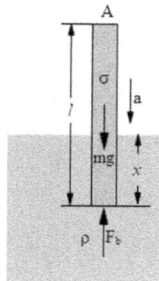

Example 42 A uniform vertical cylinder (as described in the last example) is released from rest when it just touches the liquid surface of a deep lake.

(a) Find the acceleration of the cylinder as the function of distance x moved in the liquid during its descent.

(b) What is the velocity of the cylinder as the function of x?

(c) Calculate the maximum distance x when $\rho > \sigma$.

(d) Plot the velocity of the cylinder versus distance x, if $\frac{\rho}{\sigma} = 1$.

Solution

(a) Let's assume that the cylinder accelerates down. Referring to the last example,

$$F_{net} = -mg + F_b = m(-a)$$

Substituting $F_b = A\rho g x$ and $m = Al\sigma$ we have $a = g(1 - \frac{\rho x}{\sigma l})$ Ans.

(b) Using the kinematical expression $v\,dv = a\,dx$ and substituting $a = g(1 - \frac{\rho x}{\sigma l})$, we have $v\,dv = g(1 - \frac{\rho x}{\sigma l})dx$.

Since the cylinder speeds up from zero to v during the displacement x, we have

$$\int_0^v v\,dv = g\int_0^x \left(1 - \frac{\rho x}{\sigma l}\right)dx$$

Then, the velocity of the cylinder is

$$v = \sqrt{2gx\left(1 - \frac{\rho x}{2\sigma l}\right)} \text{ Ans.}$$

(c) Substituting $v = 0$, we have

$$x_{max} = \frac{2\sigma l}{\rho} = 2x_0,$$

where $x_0 = \frac{\sigma l}{\rho}$ Ans.

(d) Putting $x = l$, we have

$$v = \sqrt{2gx\left(1 - \frac{\rho x}{2\sigma l}\right)} = \sqrt{2gl\left(1 - \frac{\rho l}{2\sigma l}\right)}$$

Putting $\frac{\sigma}{\rho} = 1$, we have

$$v = \sqrt{gl} = v_{max}$$

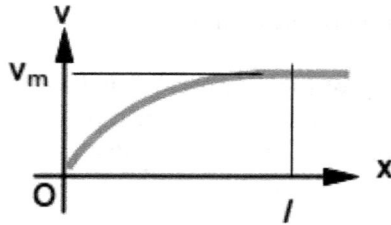

The cylinder will move with the same velocity with zero acceleration until it touches the bottom of the lake.

Example 43 A cylinder of radius R, height $H = 2R$, and density σ has a conical cut at its bottom. The top of the cylinder is kept at depth $h = R$ from the liquid surface. If the density of liquid is ρ, (a) find the hydrostatic force acting on the curved (conical) surface of the cylinder. (b) If we hold the body how much minimum force do we have to apply?

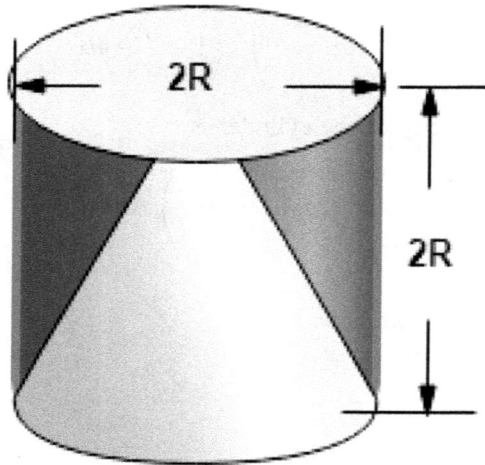

Solution

Let's assume that the liquid presses the top and bottom of the cylinder with forces F_1 and F_2 respectively. Since the net side thrust is zero, the net hydrostatic force, that is, the buoyant force, is

$$F_b = F_2 - F_1,$$

where $F_b = V\rho g$.

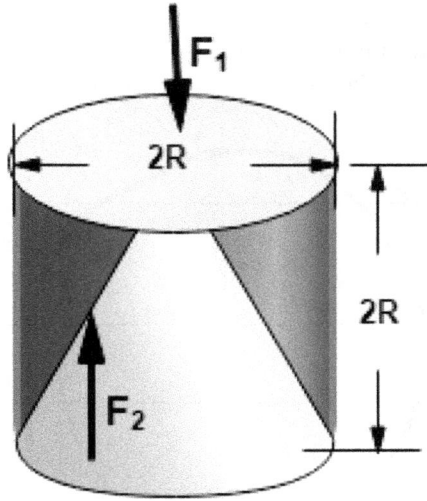

This gives

$$F_2 = V\rho g + F_1 \tag{4.62}$$

The thrust of liquid on the top of the body is

$$F_1 = P_1 A = (\rho g h)(\pi R^2) \tag{4.63}$$

Subtracting the volume of the removed cone from the cylinder, the volume of the body is

$$V = \pi R^2 H - \frac{1}{3}\pi R^2 H = \frac{2}{3}\pi R^2 H \tag{4.64}$$

Using the last three equations, we have

$$F_2 = \left(\frac{2}{3}\pi R^2 H\right)\rho g + \pi R^2 \rho g h$$

$$\Rightarrow F_2 = \pi R^2 \rho g\left(h + \frac{2}{3}H\right)$$

Putting $h = R$ and $H = 2R$, we have

$$F_2 = \pi R^2 \rho g\left\{R + \frac{2}{3}(2R)\right\} = \frac{7}{3}\pi R^3 \rho g \text{ Ans.}$$

Example 44 Derive an expression for net hydrostatic force acting on a body immersed in a composite liquid consisting of n layers of immiscible liquids. The densities and thickness of the liquid layers are $\rho_1, \rho_2, \ldots, \rho_n$ and $h_1, h_2, \ldots, h_n$, respectively.

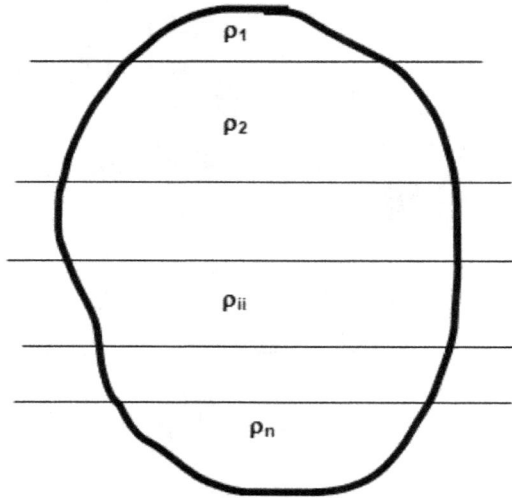

Solution

When a body is immersed in the layers of many liquids, each liquid exerts its pressure at the bottom of the body *via* other liquids obeying Pascal's law.

All liquids present above the lowest point of the body contribute their buoyant (net hydrostatic) forces and push the body up even though they do not touch every part of the body.

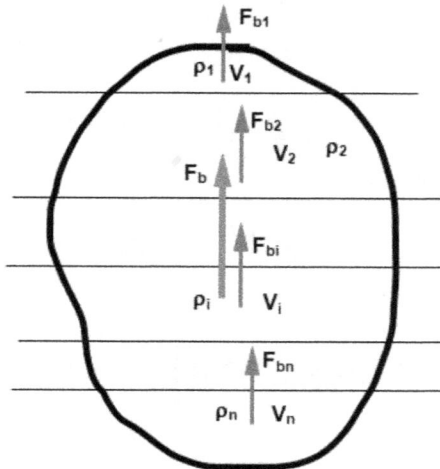

Let's assume that V_1, V_2, ..., V_n are the volumes of displaced liquids of densities ρ_1, ρ_2, ..., ρ_n, respectively. Then, the net buoyant force acting on the liquid is equal to the sum of buoyant force $F_{b_1} = \rho_1 V_1 g$, $F_{b_2} = \rho_2 V_2 g$, $F_{b_n} = \rho_n V_n g$ exerted by the liquids.

Hence, $F_b = F_{b_1} + F_{b_2} + \cdots + F_{b_n}$.

Substituting $F_{b_1} = \rho_1 V_1 g$, $F_{b_2} = \rho_2 V_2 g$, ..., $F_{b_n} = \rho_n v_n g$, we have

$$F_b = \rho_1 V_1 g + \rho_2 V_2 g + \cdots + \rho_n V_n g$$

In a nutshell, $F_n = \sum_{i=1}^{i=n} (\rho_i V_i) g$ Ans.

Example 45 A cube is floating in two liquids 1 and 2 of densities ρ_1 and ρ_2. If the height of the liquid columns are x and $l - x$, find:

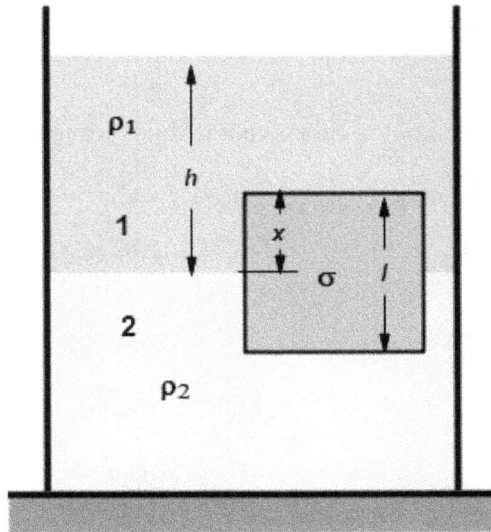

(a) the value of x.
(b) the buoyant force acting on the cube by the bottom liquid.

Solution
Method 1

(a) The cube is at rest and the net force acting on it is zero. Referring to the free-body diagram, we have three forces acting on it. There are two upward buoyant forces F_{b1} and F_{b2} due to two liquids 1 and 2 and the weight mg of the cube.

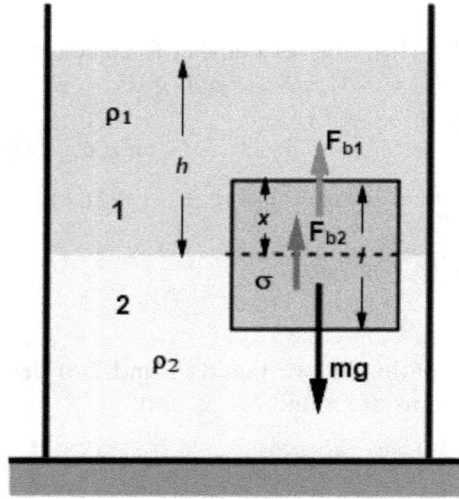

For equilibrium of the cube, the total buoyant force is equal to the weight of the cube.

$$F_b = \rho_1 V_1 g + \rho_2 V_2 g = mg$$

Putting $V_1 = Ax$ and $V_2 = (l-x)A$ and $m = Al\sigma$, we have

$$\rho_1 Axg + \rho_2(l - x)Ag = Al\sigma g$$

$$\Rightarrow \rho_1 x + \rho_2(l - x) = l\sigma$$

$$\Rightarrow x(\rho_1 - \rho_2) = l(\sigma - \rho_2)$$

$$\Rightarrow x = \frac{(\sigma - \rho_2)l}{(\rho_1 - \rho_2)} \text{ Ans.}$$

(b) Liquid 2 touches the base of the cube and can exert a force $F_2 = P_2 A$, where P_2 = pressure contribution due to liquid 2 at the bottom of the cube = $P_2 = \rho_2 g(l - x)$. So, we have

$$F_2 = P_2 A = \rho_2 g(l - x)A$$

Putting $A = l^2$, we have

$$F_2 = P_2 A = \rho_2 g \left[l - \frac{(\sigma - \rho_2)l}{(\rho_1 - \rho_2)} \right] l^2$$

So, we have

$$F_2 = \frac{\rho_1 g(\rho_1 - \sigma)l^3}{(\rho_1 - \rho_2)}$$

Method 2

The cube is in equilibrium under the action of one downward hydrostatic force F_{h1} acting at the top and one upward hydrostatic force F_{h2} acting at the bottom of the cube and the weight mg of the cube.

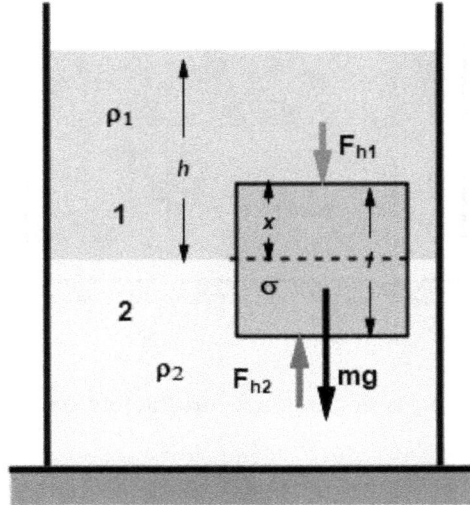

For equilibrium of the cube, the total hydrostatic force is equal to the weight of the cube.

$$F_h = F_{h2} - F_{h1} = mg,$$

where $F_{h1} = \rho_1(h - x)gA$ and

$$F_{h2} = \rho_1 hgA + \rho_2(l - x)gA \text{ and } m = Al\sigma.$$

Then, we have

$$\Rightarrow \rho_1 hgA + \rho_2(l - x)gA - \rho_1(h - x)gA = \sigma Alg$$

$$\Rightarrow \rho_1 h + \rho_2(l - x) - \rho_1(h - x) = \sigma l$$

$$\Rightarrow \rho_1 x + \rho_2(l - x) = l\sigma$$

$$\Rightarrow x(\rho_1 - \rho_2) = l(\sigma - \rho_2)$$

$$\Rightarrow x = \frac{(\sigma - \rho_2)l}{(\rho_1 - \rho_2)} \text{ Ans.}$$

Example 46 A sphere of mass m and density σ is kept on a block of mass M floating in a liquid of density ρ. If the sphere m is dropped into the liquid, what happens to the height of liquid level and depth of penetration of block M?

Discuss the cases when the density of the sphere m (i) is greater than, (ii) lesser than, and (iii) equal to the density of liquid.

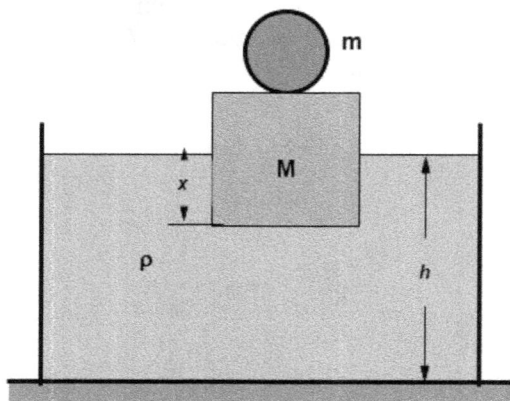

Solution

Since the mass $(M + m)$ is floating, net force acting on it is zero.

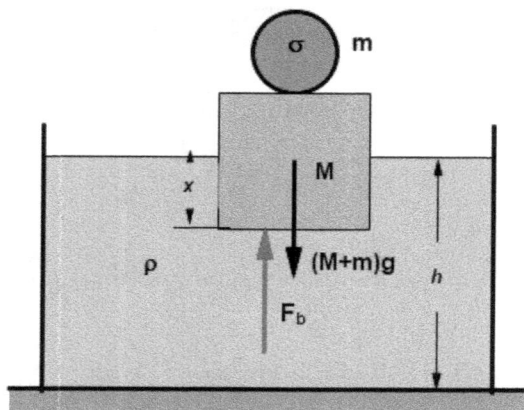

Referring to the free-body diagram, we have
$F_{net} = (M + m)g - F_b = 0$, where $F_b = V_1 \rho g$
This gives the volume of liquid displaced V_1 as

$$V_1 = \frac{M + m}{\rho} \tag{4.65}$$

When block m is dropped, it sinks because the density of the sphere is greater than that of the liquid. Then the volume of liquid displaced by block m is

$$V'_2 = \frac{m}{\sigma}$$

As the block M floats, the volume of liquid displaced by it is

$$V''_2 = \frac{M}{\rho}$$

because $F'_b(=V''_2\rho g) = Mg$

Hence, the total volume of liquid displaced is

$$V_2 = V'_2 + V''_2 = \frac{M}{\rho} + \frac{m}{\sigma} \tag{4.66}$$

Since $V_2 < V_1$ when compared from equations (4.65) and (4.66), we conclude that the height of the liquid level decreases. As block m is dropped, the depth of penetration of block M decreases. Ans.

When block m is dropped, for $\sigma \geqslant \rho$, it floats. Then the volume of liquid displaced by the sphere m is

$$V''_2 = \frac{m}{\rho}$$

Hence, the total volume of liquid displaced is

$$V_2 = V'_2 + V''_2 = \frac{M}{\rho} + \frac{m}{\rho} = \frac{M+m}{\rho} \tag{4.67}$$

Since $V_2 = V_1$ when compared from equations (4.65) and (4.67), the overall height of the liquid remains unchanged.

4.16 Buoyant force in accelerating liquid

When a liquid accelerates, pressure decreases in the direction of acceleration with a gradient

$$\frac{dP}{dx} = \rho a_0$$

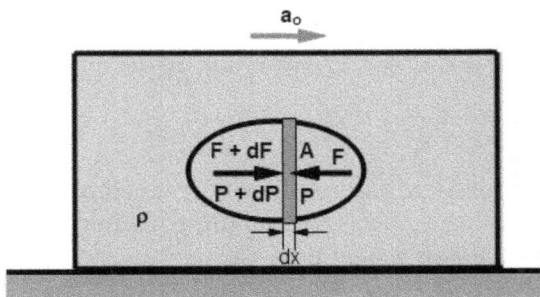

Let's keep a body inside the accelerating liquid. The component of hydrostatic force on the thin strip of length dx of the body having area dA is

$$dF = (dP)A,$$

where $dP = \rho a_0 dx$

$$\Rightarrow dF = \rho a_0 A dx$$

Summing up all elementary forces in the direction of acceleration $\vec{a_0}$, the net hydrostatic force (buoyant force) is

$$F_b = \int dF,$$

where $dF = \rho a_0 A dx$.
Then,

$$F_b = \int \rho a_0 A dx$$

If the density ρ of the liquid and its acceleration $\vec{a_0}$ is constant, we have

$$F_b = \rho a_0 \int A dx$$

Substituting $\int A dx = V$ (volume of the body = volume of liquid displaced), we have

$$\vec{F_b} = V \rho \vec{a_0}$$

The above expression tells us that the net hydrostatic force, that is, buoyant force, is F_b due to an accelerating liquid acting on a totally immersed body points in the direction of acceleration of the liquid. This buoyant force is equal to the product of volume of liquid displaced, density of liquid, and acceleration of liquid. In other words, that the buoyant force $\vec{F_b}$ due to an accelerating liquid acting on a body fully immersed in it is equal to the product of mass m of the liquid displaced and acceleration a_o of the liquid; $\vec{F_b} = m\vec{a_o}$.

However, the net force acting on the body is equal to the product of mass of the body and acceleration of the body given as $\vec{F}_{net} = m\vec{a}$.

We know that another buoyant force acts on the body due to the effect of gravity, which can be given as

$$\vec{F}'_b = -V\rho \vec{g}$$

Since $\vec{g}$ points vertically downward, $\vec{F}'_b(=-V\rho\vec{g})$ is directed vertically up.

Now let's combine the two effects, acceleration of liquid and gravitational effect. Due to acceleration of liquid, we have

$$\vec{F_b} = V\rho \vec{a_o}$$

Due to gravity, we have

$$\vec{F'}_b = -V\rho\vec{g}$$

The combined effect of both gives rise to a net buoyant force

$$\vec{F_b} = \vec{F_b} + \vec{F'}_b;$$

Substituting $\vec{F}_{-b} = V\rho\vec{a_o}$ and $\vec{F'}_b = -V\rho\vec{g}$.
Then, we have

$$\vec{F_b} = V\rho(\vec{a_o} - \vec{g})$$

The above expression tells us that whenever a body is totally immersed in an accelerating liquid, the combined effect of 'gravitational acceleration $\vec{g}$ and acceleration $\vec{a}$ of the liquid' gives a buoyant force $\vec{F_b}$, which is equal to the product of $V\rho$ (mass of liquid displaced) and $(\vec{a_o} - \vec{g})$. This means $\vec{F_b}$ points in the direction of $(\vec{a_o} - \vec{g}) = {}^{\prime}\vec{g}_{\text{eff}}{}^{\prime}$.

Example 47 Does the depth of penetration of a cube floating in a liquid depend on its acceleration if the liquid accelerates vertically?
Solution
The forces acting on the cubical block are $F_{\text{hydro}}\uparrow$ and $mg\downarrow$.

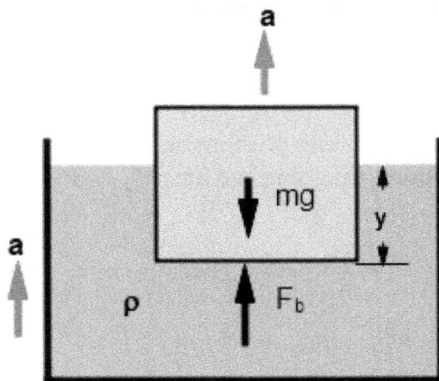

The net force acting on the body (cube) is
$F = F_{\text{hydro}} - mg$, where

$$F_{\text{hydro}} = \int PdA = \rho(g + a)yA$$

This gives $F = A\rho(g + a)y - mg$.
If this force accelerates the block with an acceleration a, we have

$$F = A\rho(g + a)y - mg = ma$$

This gives

$$y = \frac{m}{A\rho} \text{ Ans.}$$

So, the depth of immersion remains constant and does not depend on the acceleration of the liquid.

Example 48 A ball of density σ is released from rest from the center of an accelerating trough completely filled with a liquid of density ρ. If the trough moves with a (a) horizontal and (b) vertical acceleration $\vec{a_0}$, find the acceleration of the ball at the given instant.

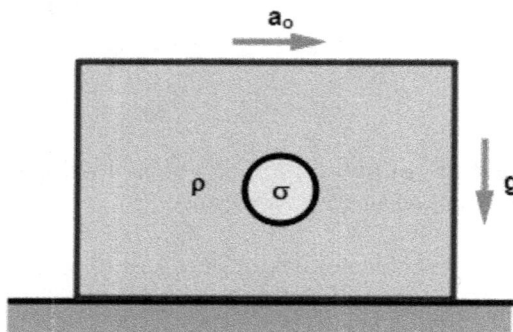

Solution
Let's assume that the ball moves with an acceleration $\vec{a}$.
Referring to the free-body diagram, we have $F_x = V\rho a_0 = ma_x$.
This gives

$$a_x = \frac{V\rho a_0}{m} = \frac{\rho}{\sigma}a_0 (\text{because } m = \sigma V)$$

Similarly, $F_y = V\rho g - mg = ma_y$, where $m = V\sigma$.
This gives

$$a_y = \left(\frac{\rho}{\sigma} - 1\right)g$$

Then the net acceleration of the body is $\vec{a} = a_x\hat{i} + a_y\hat{j}$.
Substituting $a_x = \frac{\rho}{\sigma}a_0$ and $a_y = (\frac{\rho}{\sigma} - 1)g$, we have

$$\vec{a} = \frac{\rho}{\sigma}a_0\hat{i} + \left(\frac{\rho}{\sigma} - 1\right)g\hat{j} \text{ Ans.}$$

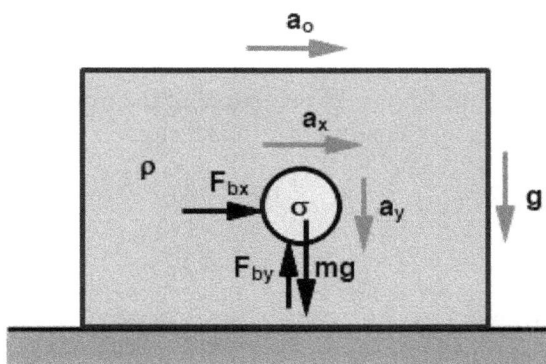

Alternative method:

You can also obtain the answer by directly writing $\vec{F} = m\vec{a}$ and substituting

$$F = \vec{F_b} + \vec{F_b}' + m\vec{g}$$

$$= V\rho a_0 \hat{i} + V\rho g \hat{j} - mg\hat{j} \text{ and } m = \sigma V$$

Example 49 A rubber ball being tied with a light inextensible string is resting inside a stationary sealed trough completely filled with a liquid. The tension in the string connecting the ball and base of the trough is T_0. When the trough moves with an (a) upward and (b) downward acceleration a_0, find the tension in the string.

Solution

(a) The forces acting on the body are buoyant force due to gravity, that is, $F_b = V\rho g\uparrow$, tension $T\downarrow$, buoyant force due to acceleration, that is, $F'_b = V\rho a\uparrow$, and weight $mg\downarrow$. Then, the net force acting on the body is

$$\vec{F} = \vec{F_b} + m\vec{g} + \vec{F'}_b + \vec{T}$$

$$= V\rho g - mg + V\rho a - T$$

Substituting $\vec{F} = m\vec{a}$, where $\vec{a} = a_0$ (as there is no relative acceleration between the body and liquid trough, because the string is taut with a tension T), we have

$$ma_0 = V\rho g - mg + V\rho a_0 - T$$

This gives

$$T = (g + a_0)(V\rho - m), \tag{4.68}$$

When $a_0 = 0$, putting $T = T_0$, we have

$$T_0 = g(V\rho - m) \tag{4.69}$$

Eliminating $(V\rho - m)$ from equations (4.68) and (4.69), we have $T = T_0(1 + \frac{a_0}{g})$ Ans.

(b) Following the aforementioned procedure, if the trough accelerates vertically down, the tension in the string can be given as

$$T = T_0\left(1 - \frac{a_0}{g}\right)$$

Example 50 A rubber ball being tied with a light inextensible string is resting inside a stationary sealed trough completely filled with a liquid. The tension in the string connecting the ball and base of the trough is T_0. When the trough moves with a horizontal acceleration a, find the (a) tension in the string and (b) angle made by the string with vertical.

Solution

The forces acting on the body are buoyant force due to gravity, that is, $F_b = V\rho g\uparrow$, tension $T\downarrow$, buoyant force due to acceleration, that is, $F_{bx} = V\rho a\rightarrow$, and weight $mg\downarrow$. Referring to the free-body diagram, then, the net force acting on the body is

$$\vec{F} = \vec{F_{bx}} + m\vec{g} + \vec{F_{by}} + \vec{T}$$

$$= (V\rho g - mg)\hat{j} + V\rho a\hat{i} - \vec{T}$$

Substituting $\vec{F} = m\vec{a}$, where $\vec{a} = a_0\hat{i}$ (as there is no relative acceleration between the body and liquid trough, because the string is taut with a tension T), we have

$$(V\rho g - mg)\hat{j} + V\rho a_o\hat{i} - \vec{T} = ma_0\hat{i}$$

This gives

$$(V\rho g - mg)\hat{j} + V\rho a_o\hat{i} - \vec{T} = ma_o\hat{i} \tag{4.70}$$

$$T = (V\rho - m)\sqrt{g^2 + a_o{}^2} \tag{4.71}$$

When $a_o = 0$, putting $T = T_0$, we have

$$T_0 = g(V\rho - m) \tag{4.72}$$

Eliminating $(V\rho - m)$ from equations (4.71) and (4.72), we have $T = T_0\left(\frac{\sqrt{g^2 + a_o{}^2}}{g}\right)$ Ans.

(b) Writing force equations in horizontal and vertical, we have the following two equations:

$$V\rho a_o + T \sin \theta = m a_o$$

$$T \sin \theta = (m - V\rho)a_o \qquad (4.73)$$

$$V\rho g - mg + T \cos \theta = 0$$

$$T \cos \theta = (m - V\rho)g \qquad (4.74)$$

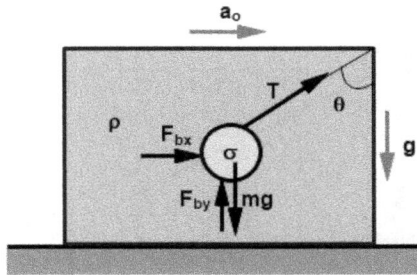

Using equations (4.73) and (4.74), we have

$$\tan \theta = a_o/g$$

$$\Rightarrow \theta = \tan^{-1} a_o/g$$

N.B.: Then, we can conclude that the angle of inclination of the string does not depend on the presence of the liquid.

Example 51 A uniform rod of length l, density σ is pivoted smoothly at the point P below the water level at a depth h, as shown in figure. If the density of water is ρ and the rod remains stationary, find the (a) angle made by the rod with horizontal, (b) length of the rod lying outside water, and (c) reaction at the pivot.

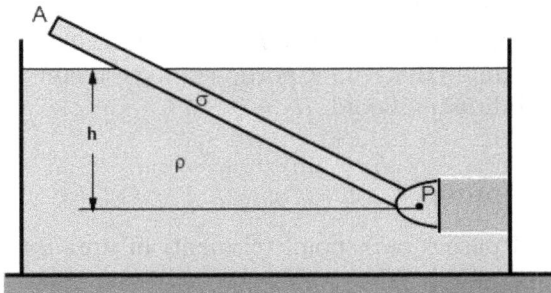

Solution

The forces acting on the rod are:

   (i) Weight $mg\downarrow$
   (ii) Buoyant force $F_b\uparrow$
   (iii) Assumed reaction forces $R_x\rightarrow$ and $R_y\uparrow$ at the pivot

The net torque acting on the rod is zero because it is in rotational equilibrium ($\alpha = 0$).

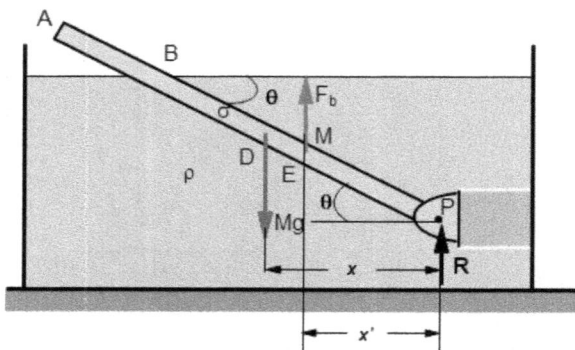

Taking the torques of all forces about the pivot O, we have

$$\tau_0 = F_b x' - mgx = 0, \text{ where } x = \frac{l}{2}\cos\theta \text{ and } x' = \frac{h}{2}\cos ec\theta \cos\theta$$

This gives

$$F_b h \cos ec\theta \cos\theta = mgl \cos\theta$$

Substituting $F_b = V\rho g = \{A(h \cosec \theta)\}\rho g$, $m = (Al\sigma)$, we have

$$\theta = \sin^{-1}\frac{h}{l}\sqrt{\frac{\rho}{\sigma}}$$

The rod is in both translational and rotational equilibrium; $a = 0$ and $\alpha = 0$. This means $F = 0$ and $\tau = 0$.

You can observe that the buoyant force F_b acts at the center of buoyancy, that is, the geometrical center (mid-point) of the portion of the rod immersed in liquid. Since there is no horizontal thrust of liquid, $R_x = 0$.

4.17 Hydrostatic torque

As we learned, liquid pushes each point (element) of area $\overrightarrow{dA}$ of any surface in contact normally with a force $\overrightarrow{dF} = P\overrightarrow{dA}$, that every force has its rotational effect, that is, torque. The torque of $d\vec{F}$ about a fixed-point O (on the surface, say) is $d\vec{\tau} = \vec{r} \times d\vec{F}$.

Then, the net torque $\vec{\tau}$ of liquid about the point O is equal to the summation of elementary torques.

$$\vec{\tau} = \int d\vec{\tau},$$

where $d\vec{\tau} = \vec{r} \times d\vec{F}$.

This gives

$$\vec{\tau} = \int \vec{r} \times d\vec{F},$$

where $d\vec{F} = Pd\vec{A}$

Then, we have $\vec{\tau} = \int \vec{r} \times Pd\vec{A}$.

Center of force: Let's assume that the total hydrostatic force acting on the surfaces is F, which is given as

$$\vec{F} = \int P\, d\vec{A}$$

If we choose a point C such that the net hydrostatic force $\vec{F}$ supposed to be acting at C can produce a torque

$$\vec{\tau} = \int \vec{r} \times \vec{F} \text{ about } O$$

If this torque is equal to the net hydrostatic torque, we call the point C, 'center of force'. The position vector $\vec{r_C}$ of 'center of force' can be given as $\int \vec{r_C} \times \vec{F} = \int \vec{r_C} \times d\vec{F}$.

Let's apply the above idea in the following examples.

Example 52 An L-shaped gate of dimensions as shown in the figure is in equilibrium under the action of hydrostatic forces. (a) Find the torque produced by the liquid of density ρ due to its side thrust about the point P passing through the zz' axis. (b) Find the location of center of force on the gate.

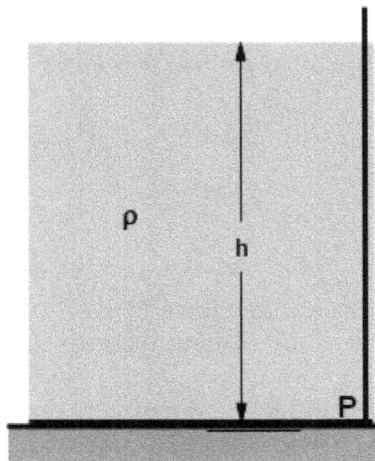

Solution

(a) The hydrostatic force acting on the thin horizontal strip of thickness dy at a depth y is

$$dF = PdA,$$

where $P = \rho gy$ and $dA = ldy$.
This gives $dF = \rho gly\,dy$.

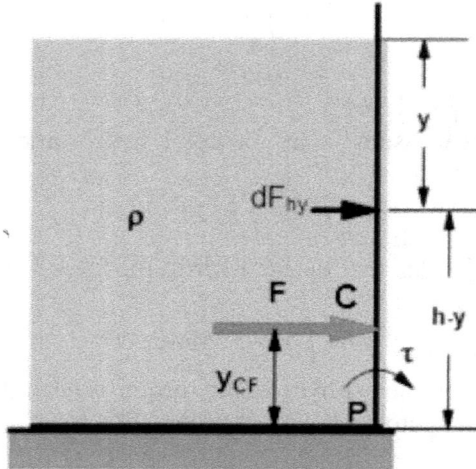

This elementary force produces a torque

$$d\tau = (h - y)dF$$

about the pivot P ($z - z'$ axis of rotation).
Substituting $dF = \rho gly\,dy$, we have

$$d\tau = (h - y)\rho gly\,dy$$

Integrating $d\vec{\tau}$, the total torque $\vec{\tau}$ is given as

$$\tau = \int d\vec{\tau} = \int_0^h \rho gl(h - y)y\,dy$$

This gives

$$\tau = \frac{\rho glh^3}{6} \quad \text{Ans.}$$

(b) The net horizontal hydrostatic force $F = \frac{\rho glh^2}{2}$ (as found earlier).
The torque produced by F about the pivot O is

$$\tau' = \frac{\rho glh^2}{2}y_C$$

Equating τ' with $\tau = \frac{\rho g l h^3}{6}$, we have

$$y_C = \frac{h}{3} \text{ Ans.}$$

N.B.: If the surface is flat and rectangular, for a single liquid, the center of force is situated at a height of $\frac{h}{3}$ from the pivot, where h is the height of liquid column above the pivot. Remember that the above result is not valid for any arbitrary flat surface. In general, 'center of force' is given by the expression $y_C = \frac{\int y \, dF}{\int dF}$ for flat surfaces.

Example 53 A hemicylindrical gate of radius R and length l pivots about its axis. Find the:

(a) Torque offered by the liquid about the axis
(b) Horizontal thrust
(c) Vertical thrust of liquid on the gate
(d) Net hydrostatic force if we allow the liquid to be present at other side of the gate.

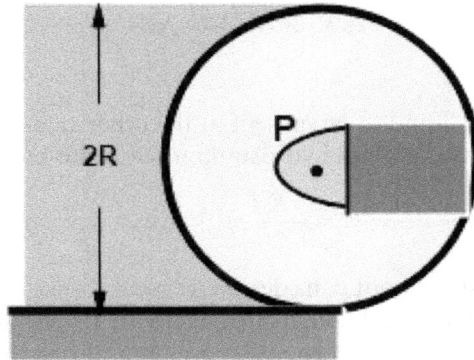

Solution

(a) Since hydrostatic force dF acting at each element at each element passes through the center (pivot P), it cannot produce a torque about P.

(b) As discussed earlier, the net horizontal thrust is

$$F_x = \rho g y_C A,$$

where $y_C = R$ and $A = (2R)(l)$

$$\Rightarrow F_x = 2\rho g R^2 l$$

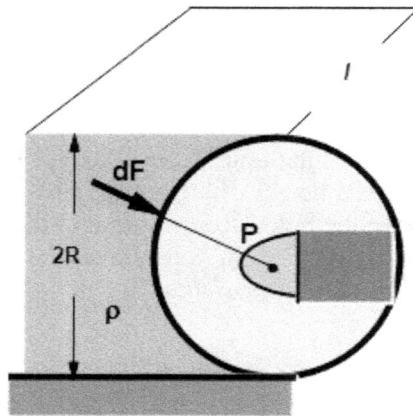

(c) The vertical thrust is

$$F_y(=F_b) = V\rho g,$$

where V = volume of liquid displaced $= \frac{\pi}{2} R^2 l$

$$\Rightarrow F_y = \frac{\pi \rho R^2 g l}{2} \text{ Ans.}$$

(d) If we allow the liquid to be present at the other side of the gate, the net side thrust is zero. So, the net hydrostatic force is the buoyant force

$$F = \frac{\pi R^2 l}{2} \rho g \uparrow \text{ Ans.}$$

N.B.: You should not consider the vertical thrust F_b at the center of mass of the hemicylinder and take its torque about P. This will be wrong because you need to consider the torque of horizontal thrust that will counteract the torque due to F_b about the same point P.

4.18 Pascal's law and its application

Let's consider a tube completely filled with a liquid. At the end A of the tube, let's push the light piston by a force F, say. Then, the pressure at A will be given as

$$P_A = \frac{F}{A} + P_{atm}$$

Hence, pressure at B is given as

$$P_B = P_A + \rho g h$$

Let's now push the piston a bit harder by an impulsive force δF. This increases the pressure at A by

$$\delta P_A \left(= \frac{\delta F}{A} \right)$$

The pressure difference travels along the tube as a pressure wave. After passing through B, it gets reflected between the boundaries several times. Finally, the waves die out and the liquid attains a steady (equilibrium) state.

Since $P_B = P_A + \rho g h$, differentiating both sides, we have $\delta P_B = \delta P_A$ (because h is constant).

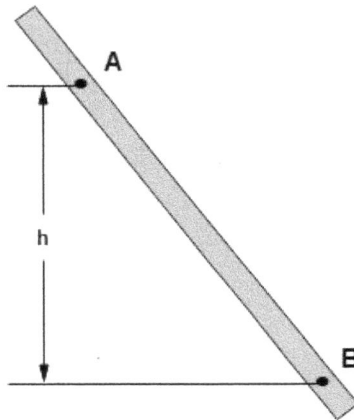

This means that after a steady state is attained, same change in pressure is felt at B. This mathematical result was experimented and practically verified by 'Pascal'. Hence, the above idea is known as 'Pascal's law', which states that whenever a change in pressure occurs at any point of a fluid, at steady state the same change in pressure is felt everywhere inside the liquid.

In other words, whenever you try to change the pressure at any point of a fluid, the same change in pressure is transmitted in all directions inside the liquid.

Let's use the above idea in the following example.

Example 54 (Hydraulic brake) Explain the 'hydraulic brake'.
  Solution
  To understand the hydraulic brake, let's take two vertical tubes of cross-sections A_1 and $A_2(A_1 < < A_2)$. Connect the tubes by a horizontal tube. Fill the tube with a liquid. Place two smooth pistons on the free surfaces of liquid in the vertical limbs. Let the pistons remain in equilibrium. Let's now, apply a force F_1 on the piston 1. This increases the pressure at 1 by $\Delta P_1(=\frac{F_1}{A_1})$. This pressure difference is transmitted undiminished to piston 2, This means the pressure difference ΔP_2 at 2 is $\Delta P_1(=\frac{F_1}{A_1})$. This pushes piston 2 with an upward force $F_2 = (\Delta P_2)A_2$, where $\Delta P_2 = \Delta P_1 = \frac{F_1}{A_1}$.

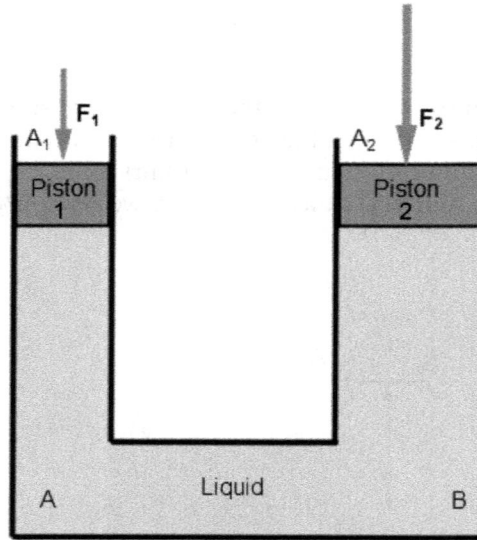

This gives us

$$F_2 = \left(\frac{A_2}{A_1}\right)F_1$$

If you want to equilibrate piston 2, we need to push it down with a force F_2. Since $A_2 \gg A_1$, we have $F_2 \gg F_1$.

This means that by giving a small force F_1 at 1 we get a large force F_2 at 2. The output force F_2 is $(\frac{A_2}{A_1})$ times greater than the input force F_1, Hence, you can term $(\frac{A_2}{A_1})$ as the 'magnification factor'. The drivers in trains, cranes, and all heavy automobiles apply a small force by pressing the piston while applying the brakes and thus generate huge torques due to large output forces.

The main idea of a hydraulic brake is to obtain a large force by giving a small force so that equal pressure is transmitted from the input end to output end. $F_{(output)} = (\frac{A_{output}}{A_{inpit}})F_{input}$.

N.B.: The valve in a tube has less cross-section than the radius so as to increase the output pressure by the greater magnification factor.

Problem 1 A solid sphere of radius R is just immersed in a liquid of density ρ. Find the (a) hydrostatic force acting on the upper and lower hemisphere and (b) ratio of these two forces.

Solution

(a) Let F_1 and F_2 be the hydrostatic forces acting on the top and bottom hemisphere, respectively. The net hydrostatic force is

$F_b = F_2 - F_1 = V\rho g$, where $V = \frac{4}{3}\pi R^3 \rho$.

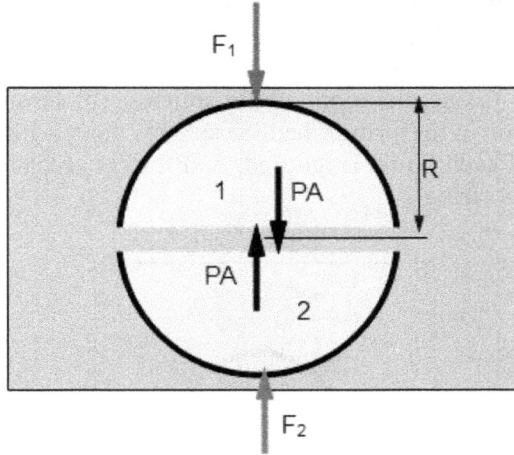

$$\Rightarrow F_2 - F_1 = \frac{3}{2}\pi R^3 \rho g \tag{4.75}$$

For the upper hemisphere,

$$PA - F_1 = F_{b_1}$$

$$PA - F_1 = V_1 \rho g$$

$$\Rightarrow F_1 = PA - V_1 \rho g, \text{ where } P = \rho g R$$

$$\Rightarrow F_1 = (\rho g R)\pi R^2 - \left(\frac{2}{3}\pi R^3\right)\rho g$$

$$\Rightarrow F_1 = \frac{1}{3}\rho \pi R^3 g$$

Using equation (4.75) putting the obtained value of F_1, we have

$$F_2 = F_1 + \frac{4}{3}\pi R^3 \rho g$$

$$= \frac{1}{3}\pi \rho R^3 g + \frac{4}{3}\pi \rho R^3 g$$

$$\Rightarrow F_2 = \frac{5}{3}\pi R^3 \rho g \text{ Ans.}$$

(b) Then the ratio of forces is $\frac{F_2}{F_1} = \frac{5}{1}$ Ans.

Problem 2 Two identical spheres A and B each of radius R and density σ are placed in a trough of two immiscible liquids as shown in the figure. The spheres are connected by an inextensible string. Assume that the string does not get slacken when the sphere B is given a downward velocity v. (i) If we consider the viscosity of the liquids, find the (a) acceleration of the spheres, (b) terminal velocity of the spheres, and (c) tension in the string when the spheres move with a terminal velocity. (ii) If the viscosity of each liquid is ignored, find the (a) acceleration of the spheres and (b) tension in the string.

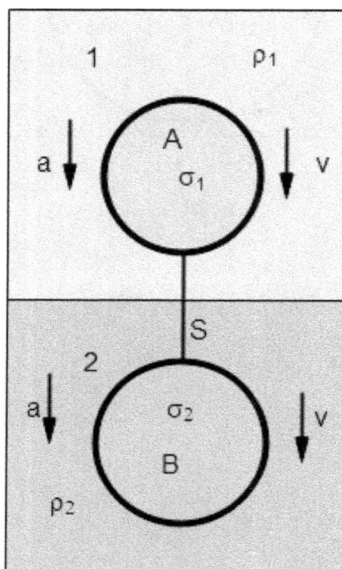

Solution

(i)

(a) Referring to the free-body diagram of m_1:

$$T - f_{v_1} - F_{b_1} + m_1 g = m_1 a \tag{4.76}$$

$$f_{v_1} = 6\pi \eta_1 v \tag{4.77}$$

$$F_{b_1} = V \rho_1 g = \frac{m_1}{\sigma} \rho_1 g \tag{4.78}$$

Using the last three equations:

$$T - 6\pi \eta_1 \eta v - \frac{m_1}{\sigma} \rho_1 g + m_1 g = m_1 a$$

$$\Rightarrow T - 6\pi \eta_1 \eta v + m_1 g \left(1 - \frac{\rho_1}{\sigma_1} \right) = m_1 a \tag{4.79}$$

Similarly, for m_2:

$$m_2g - T - F_{b_2} - f_{v_2} = m_2 a \tag{4.80}$$

$$F_{b_2} = \frac{m_2}{\sigma_2}\rho_2 g \tag{4.81}$$

$$f_{v_2} = 6\pi\eta_2 r_2 v \tag{4.82}$$

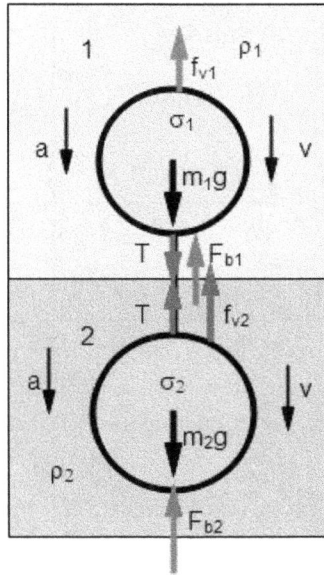

Using the last three equations:

$$m_2 g\left(1 - \frac{\rho_2}{\sigma_2}\right) - T - 6\pi\eta_2 r_2 v = m_2 a \tag{4.83}$$

Solving equations (4.79) and (4.83):

$$m_1 g\left(1 - \frac{\rho_1}{\sigma_1}\right) + m_2 g\left(1 - \frac{\rho_2}{\sigma_2}\right)$$

$$-6\pi\left(\eta_1 r_1 + \eta_2 r_2\right)v = (m_1 + m_2)a$$

Or,

$$\Rightarrow a = \frac{\left[m_1 g\left(1 - \frac{\rho_1}{\sigma_1}\right) + m_2 g\left(1 - \frac{\rho_2}{\sigma_2}\right) - 6\pi\left(\eta_1 r_1 + \eta_2 r_2\right)v\right]}{m_1 + m_2} \quad \text{Ans.}$$

(b) If the bodies move with terminal velocity v,
putting $a = 0$:

$$v = v_t = \frac{m_1 g\left(1 - \frac{\rho_1}{\sigma_1}\right) + mg\left(1 - \frac{\rho_1}{\sigma_2}\right)}{6\pi\left(\eta_1 r_1 + \eta_2 r_2\right)} \quad \text{Ans.}$$

(c) Putting the obtained value of $v = v_t$ and $a = 0$ in equation (4.79), we have:

$$T = 6\pi\eta_1 r_1 v_t - m_1 g\left(1 - \frac{\rho_1}{\sigma_1}\right)$$

$$= 6\pi\eta_1 r_1 \frac{m_1 g\left(1 - \frac{\rho_1}{\sigma_1}\right) + m_2 g\left(1 - \frac{\rho_2}{\sigma_2}\right)}{6\pi\left(\eta_1 r_1 + \eta_2 r_2\right)} - m_1 g\left(1 - \frac{\rho_2}{\sigma_1}\right)$$

$$= \frac{\eta_1 r_1 m_2\left(1 - \frac{\rho_2}{\sigma_2}\right) - \eta_2 r_2 m_1\left(1 - \frac{\rho_1}{\sigma_1}\right)}{\eta_1 r_1 + \eta_2 r_2} g \quad \text{Ans.}$$

(ii)

(a) If viscosity $\simeq 0$, the equations (4.79) and (4.83) become:

$$m_1 g\left(1 - \frac{\rho_1}{\sigma_1}\right) + T = m_1 a \tag{4.84}$$

$$m_2 g\left(1 - \frac{\rho_2}{\sigma_2}\right) - T = m_2 a \tag{4.85}$$

Solving the last two equations:

$$a = \frac{m_1\left(1 - \frac{\rho_1}{\sigma_1}\right) + m_2\left(1 - \frac{\rho_1}{\sigma_2}\right)}{m_1 + m_2} g$$

(b) $\left(\frac{1}{m_1} + \frac{1}{m_2}\right)T = g\left(1 - \frac{\rho_2}{\sigma_2}\right) - g\left(1 - \frac{\rho_1}{\sigma_1}\right)$

$$\Rightarrow T = \frac{m_1 m_2}{m_1 + m_2}\left(\frac{\rho_1}{\sigma_1} - \frac{\rho_2}{\sigma_2}\right) g \quad \text{Ans.}$$

Problem 3 A uniform rod of mass M, length l, and density σ fitted with a particle of mass m is immersed partially in the liquid of density ρ. Find the depth of immersion of the rod for (a) rotational and (b) translational equilibrium of the rod.

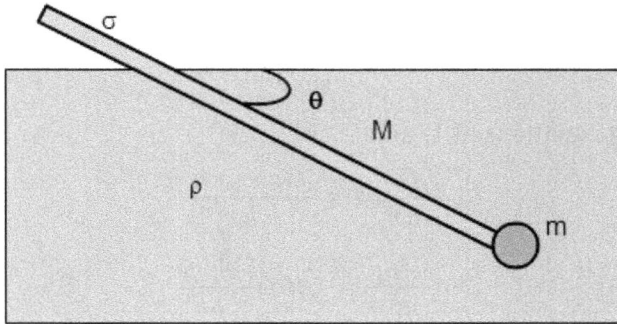

Solution

(a) Let E = center of buoyancy and D = center of mass of the rod. For equilibrium, $\tau = 0$.

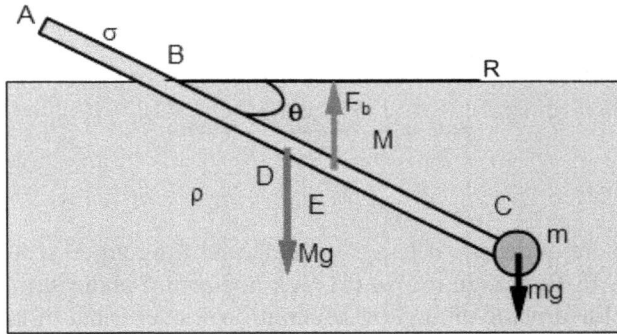

$$\Rightarrow Mgx_1 - F_b x_2 = 0$$

$$\Rightarrow Mg(DC \cos \theta) - F_b EC \cos \theta = 0$$

$$\Rightarrow F_b = \frac{Mg(PC)}{PD}, \text{ where } DC = \frac{\ell}{2} \text{ and } EC = \frac{h}{2 \sin \theta}, \text{ where } h = RC$$

$$\Rightarrow F_b = Mg \frac{\frac{\ell}{2}}{\frac{h}{2 \sin \theta}}$$

$$\Rightarrow F_b = \frac{Mgh \sin \theta}{h}$$

By the Archimedes principle,

$$F_b = V\rho g = (Ax')\rho g, \text{ where } x' = \frac{h}{\sin \theta}$$

$$\Rightarrow F_b = A\frac{h}{\sin\theta}\rho g \tag{4.86}$$

Using equations (4.1) and (4.86),

$$\frac{Ah}{\sin\theta}\rho g = \frac{Mgb\sin\theta}{h}$$

$$\Rightarrow \frac{Ah\rho}{\sin\theta} = (A\ell\sigma)\frac{l\sin\theta}{h}$$

$$\Rightarrow h = \sqrt{\frac{\sigma}{\rho}}\, l\sin\theta$$

(b) For the transitional equilibrium, $a = 0$:

$$F_b = (M + m)g$$

$$\Rightarrow \frac{Ah}{\sin\theta}\rho g = (M + m)g$$

$$\Rightarrow h = \frac{(M + m)\sin\theta}{A\rho} \quad \text{Ans.}$$

Problem 4 A sphere of density σ hangs in an accelerating liquid trough. The density of the liquid is ρ. If the trough moves (a) on a horizontal plane with an acceleration a, (b) down a inclined plane of angle of inclination θ, coefficient of kinetic friction μ, and (c) on a smooth inclined plane of angle of inclination θ, find the tension in the string.

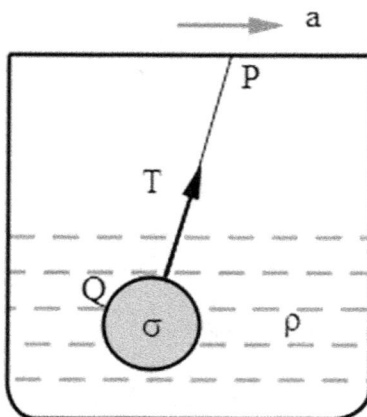

Solution

(a) The buoyant force due to gravity is

$$\vec{F}_{b_1} = -V\rho\vec{g} \tag{4.87}$$

The buoyant force due to accelerating liquid is

$$\vec{F}_{b_2} = V\rho\vec{a} \tag{4.88}$$

The weight of the body is

$$\vec{W} = m\vec{g} \tag{4.89}$$

Then the tension of the string is given as

$$\vec{F}_{net} = \left(\vec{T} + \vec{W} + \vec{F}_{b_1} + \vec{F}_b\right) = m\vec{a}$$

$$\Rightarrow \vec{T} = m\vec{a} - \vec{W} - \left(\vec{F}_{b_1} + \vec{F}_{b_2}\right) \tag{4.90}$$

Using the last four equations, we have

$$\vec{T} = m\vec{a} - m\vec{g} - \{(-V\rho\vec{g}) + V\rho\vec{a}\}$$

$$\Rightarrow \vec{T} = m(\vec{a} - \vec{g}) - V\rho(\vec{a} - \vec{g})$$

$$= (\vec{a} - \vec{g})(m - V\rho), \text{ where } V = \frac{m}{\sigma}$$

$$\Rightarrow \vec{T} = \left(1 - \frac{\rho}{\sigma}\right)m(\vec{a} - \vec{g}) \text{ Ans.}$$

(b) Since the system $(M + m)$ slides down the slant with an acceleration a,

$$F_x = (M + m)a$$

$$(M + m)g \sin\theta - f_k = (M + m)a \tag{4.91}$$

$$f_k = \mu N \tag{4.92}$$

$$N - (M + m)g \cos\theta = 0 \tag{4.93}$$

Using the last three equations,

$$\vec{a} = g(\sin\theta - \mu\cos\theta)\hat{i} \tag{4.94}$$

Let us recast the equation:

$$\vec{T} = \left(1 - \frac{\rho}{\sigma}\right)m(\vec{a} - \vec{g})$$

4-119

In component form, we have

$$\vec{T} = m\left(1 - \frac{\rho}{\sigma}\right)[g(\sin\theta - \mu\cos\theta)\hat{i} - g(\sin\theta\hat{i} + \cos\theta\hat{j})]$$

$$\Rightarrow \vec{T} = mg\left(1 - \frac{\rho}{\sigma}\right)(-\mu\hat{i} + \hat{j})\cos\theta$$

Then, we have

$$T = mg\left(1 - \frac{\rho}{\sigma}\right)\sqrt{1 + \mu^2}\cos\theta \text{ Ans.}$$

The angle made by the string with y-axis (perpendicular to the slant) is given as

$$\beta = \tan^{-1}\frac{T_x}{T_y}$$

$$\Rightarrow \beta = \tan^{-1}\mu, \text{ where } \mu = \tan\phi$$
$$\Rightarrow \beta = \phi \text{ Ans.}$$

(c) If the surface is smooth $\beta = 0$, this means the string will remain perpendicular to the slant.

Then, the tension in the string will be

$$T = mg\left(1 - \frac{\rho}{\sigma}\right)\cos\theta \text{ Ans.}$$

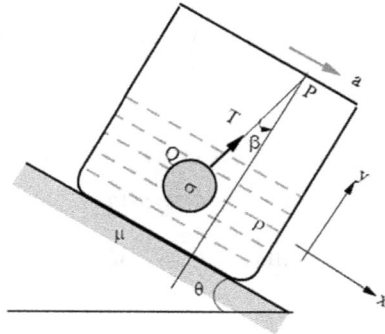

Problem 5 A conical vessel of height h and radius of the base R is filled with a liquid of mass m. A cardboard of mass m' is placed on the vessel. The vessel is inverted slowly by putting the palm on the cardboard. If we remove the palm and fix the inverted vessel for demonstration with a ceiling, the cardboard remains in equilibrium. Find the (a) contact force between the cardboard and vessel and (b) maximum value of h so as to equilibrate the cardboard which is assumed to be light.

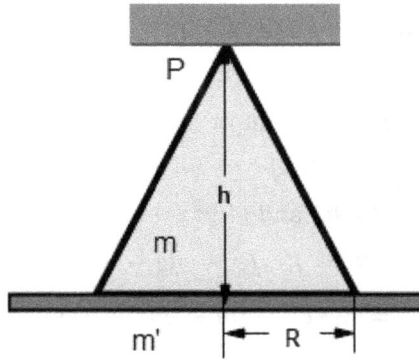

Solution

(a) The pressure of liquid on the plate is

$$P = \rho g h \tag{4.95}$$

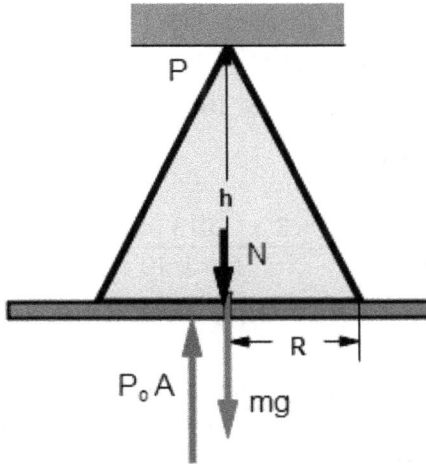

The downward hydrostatic force acting on the plates is

$$N = P.\,A = \rho g h (\pi R^2)$$

$$\Rightarrow N = \pi \rho g h R^2 \tag{4.96}$$

If the downward contact force offered by the vessel onto the plate is N', for equilibrium of the plate

$$mg + N' + N = \text{Atmospheric force} = P_0 A$$

$$\Rightarrow N' = P_0 A - mg - N \tag{4.97}$$

Using equations (4.96) and (4.97),

$$N' = P_0 A - mg - \pi \rho g h R^2 \tag{4.98}$$

Since $\frac{\pi R^2 h \rho g}{3} = m'g$ = weight of the liquid,

$$\pi R^2 h \rho g = 3m'g \tag{4.99}$$

Using equations (4.98) and (4.99),

$$N' = P_0 \pi R^2 - mg - 3m'g$$

$$= P_0 \pi R^2 - (m + 3m')g \text{ Ans.}$$

(b) If the plate is massless, putting $m = 0$, when $N' = 0$

$$P_0 \pi R^2 = m'g$$

$$\Rightarrow P_0 \pi R^2 = \frac{\rho \pi R^2 h}{3} g$$

$$\Rightarrow h = \frac{3 P_0}{\rho g}$$

Putting $P_0 = 1.013 \times 10^5$ N m^{-2}

$$\rho = 1 \times 10^3 \text{ kg m}^{-3}$$

$g \cong 10$ m s^{-2}, we have

$$h \cong \frac{3 \times 1.013 \times 10^5}{10^3 \times 10}$$

$$\simeq 3.013 \times 10 = 30.13 \text{ m Ans.}$$

Problem 6 A hemispherical beaker contains a liquid of density ρ. The radius of the base is R and the height of the liquid column in the thin vertical limb of the beaker is h. As the base of the beaker is absent, the water inside the beaker is in equilibrium by the weight of the beaker. Find the (a) contact force between the ground and beaker and (b) minimum weight of the beaker to equilibrate it.

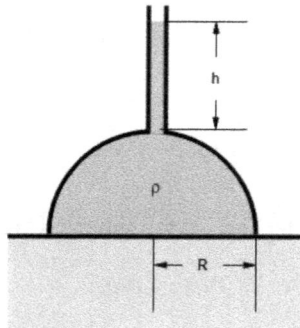

Solution

(a) The following forces are acting on the liquid:

(i) Weight mg

(ii) Vertically downward force F offered by the curved surface of the vessel

(iii) Normal reaction offered by the horizontal surface.

Since the liquid is at rest,

$$F_{\text{net}} = 0$$

$$\Rightarrow mg + F = N, \tag{4.100}$$

where $N = PA = (\rho g h + \rho g R)\pi R^2$

$$= \rho g(h + R)\pi R^2 \tag{4.101}$$

Then, $F = N - mg = \pi \rho g R(h + R) - mg,$

where $m = \frac{2}{3}\pi R^3 \rho + ah\rho$

$$\Rightarrow F = \rho g\left(\pi R^3 - \frac{2}{3}\pi R^3 + \pi R^2 h\right)$$

$$\Rightarrow F = \pi \rho g R^2\left(\frac{R}{3} + h\right) \text{Ans.}$$

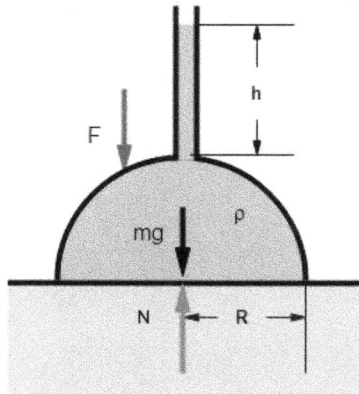

(b) Since the vessel is in equilibrium, its minimum weight will be equal to the upward force F exerted by the liquid.

$$\Rightarrow m'g = \pi\rho g R^2\left(\frac{R}{3} + h\right)$$

$$\Rightarrow m' = \pi\rho R^2\left(\frac{2}{3} + h\right) \text{Ans.}$$

Problem 7

(a) A solid cone of radius R, height h, and density σ is floating in a liquid of density ρ. Find the depth of immersion h of the cone. (b) If we invert the cone and then place it gently on the liquid, find the depth of its immersion. (c) Find the force acting on the curved surface of the cone in case (b).

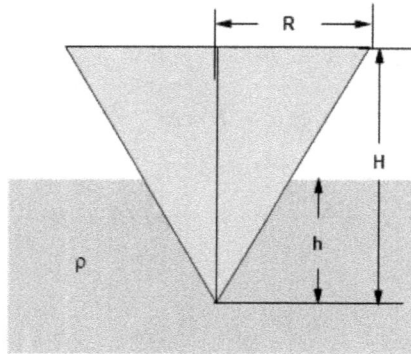

Solution

For equilibrium of the cone,

$$F_b = mg$$

$\Rightarrow V\rho g = mg$, where V = volume of liquid displaced.

$$\Rightarrow \frac{\pi r^2 h}{3}\rho g = \frac{\pi R^2 \sigma}{3}g$$

$$\Rightarrow r^2 h\rho = R^2\sigma \tag{4.102}$$

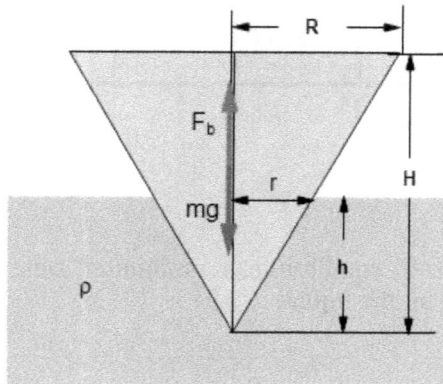

Since $\frac{r}{R} = \frac{h}{H}$,

$$\Rightarrow \left(\frac{h}{H}\right)^2 R^2 h \rho = R^2 \sigma$$

$$\Rightarrow h = \left(\frac{\sigma}{\rho}\right)^{\frac{1}{3}} H \text{ Ans.}$$

(b) As the cone is floating buoyant force is
$F_b = mg$

$$\Rightarrow V' \rho g = mg$$

$$\Rightarrow V' = \frac{m}{\rho}, \tag{4.103}$$

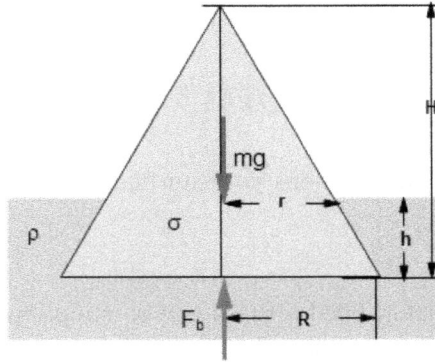

where $V' = \frac{\pi H R^2}{3} - \frac{\pi h r^2}{3}$

$$= \frac{\pi R^2 H}{3}\left(1 - \frac{h r^2}{H R^2}\right)$$

Since $\frac{r}{R} = \frac{h}{H}$, we have

$$V' = \frac{\pi R^2 H}{3}\left(1 - \frac{h^3}{H^3}\right) \tag{4.104}$$

Putting $m = \frac{\pi R^2}{3}\sigma H$ and the value of V' from equations (4.104) in (4.103),

$$\frac{\pi R^2 H}{3}\left(1 - \frac{h^3}{H^3}\right) = \frac{\pi R^2 \sigma H}{3\rho}$$

$$\Rightarrow 1 - \frac{h^3}{H^3} = \frac{\sigma}{\rho}$$

$$\Rightarrow h = \left(1 - \frac{\sigma}{\rho}\right)^{\frac{1}{3}} H \text{ Ans.}$$

(c) Let $F_y =$ force acting on the curved surface of the cone

$$F_y + mg = N$$

$$F_y = N - mg \qquad (4.105)$$

Furthermore, $N = (P_{bottom})A_{bottom} = (\rho g h)\pi R^2$.
Putting the obtained value of h,

$$N = \rho g \left(1 - \frac{\sigma}{\rho}\right)^{\frac{1}{3}} H \pi R^2$$

$$\Rightarrow N = \pi \rho g H R^2 \left(1 - \frac{\sigma}{\rho}\right)^{\frac{1}{3}} \qquad (4.106)$$

Using the last two equations you can find the force acting on the curved surface. Ans.

Problem 8 A tube completely filled with water is rotating about the vertical axis with an angular velocity ω. A small cube of length r and density σ is released in the liquid inside the tube so that it does not touch the surface of the tube. Find the acceleration of the cube.

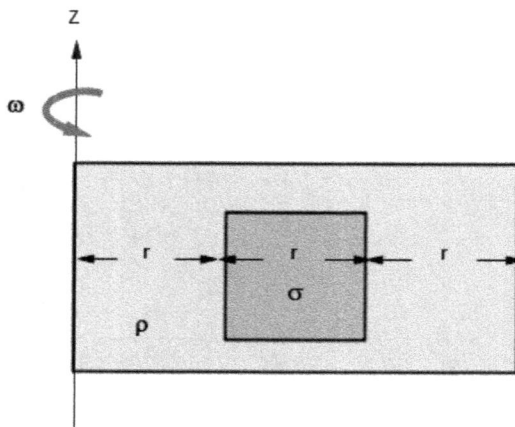

Solution

The horizontal pressure difference across the cube is

$$P_2 - P_1 = \rho \omega^2 \int_r^{2r} r\, dr$$

$$\Rightarrow P_2 - P_1 = \rho \omega^2 (4r^2 - r^2)/2$$

$$\Rightarrow P_2 - P_1 = \frac{3\rho^2 \omega^2 r^2}{2} \tag{4.107}$$

Then the net horizontal force acting on the cube is

$$F_x = (P_2 - P_1)$$

$$= \frac{3\rho \omega^2 r^2 A}{2} \text{ (in ward)}$$

The horizontal acceleration of the cube is $a_x = \frac{F_x}{m} = \frac{3\rho \omega^2 r^2 A}{2m}$, where $m = A\sigma r$

$$\Rightarrow \vec{a_x} = -\frac{3\rho \omega^2 r}{\sigma}\hat{i} \text{ radially inward.}$$

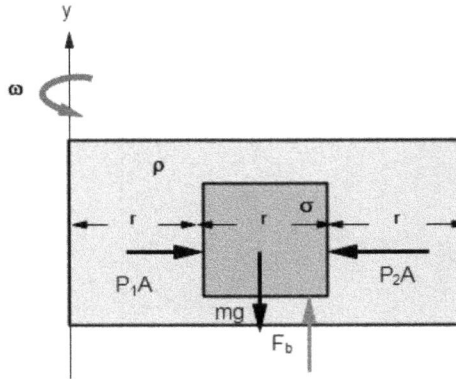

The net vertical force is

$$F_y = mg - F_b$$

$$= mg - V\rho g$$

$$= mg - \left(\frac{m}{\sigma}\right)\rho g$$

$$= -mg\left(1 - \frac{\rho}{\sigma}\right) \tag{4.108}$$

Then, the vertical acceleration is

$$a_y = \frac{F_y}{m}$$

$$\Rightarrow \overrightarrow{a_y} = -g\left(1 - \frac{\rho}{\sigma}\right)\hat{j}$$

Hence, the net acceleration is

$$\overrightarrow{a} = \overrightarrow{a_x} + \overrightarrow{a_y}$$

$$= -\left\{\frac{3\rho\omega^2 r}{\sigma}\hat{i} + g\left(1 - \frac{\rho}{\sigma}\right)\hat{j}\right\} \text{ Ans.}$$

Problem 9 Two monometers containing two different liquids are inserted into the rotating test tube having been filled completely with another liquid. The levels of total liquid column in both the monometers are equal as shown in the figure. Find the angular velocity ω of the rotating tube.

Solution

Let P_A= pressure at the axis A. Then, the pressure at B is

$$P_B = P_A + \frac{\rho\omega^2}{2}r_1^2 \tag{4.109}$$

Similarly, the pressure at C is

$$P_C = P_A + \frac{\rho}{2}(r_1 + r_2)^2\omega^2 \tag{4.110}$$

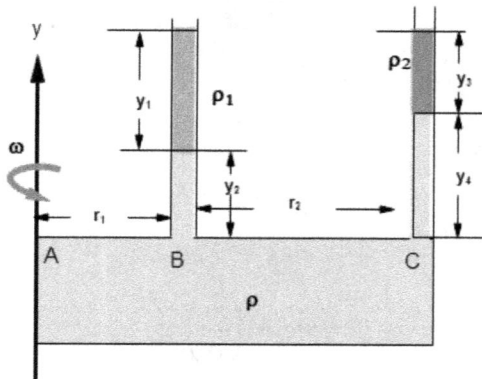

By applying Pascal's law,

$$P_B = P_0 + \rho_1 g y_1 + \rho g y_2 \tag{4.111}$$

$$P_C = P_0 + \rho_2 g y_3 + \rho g y_4 \tag{4.112}$$

Equations (4.109) and (4.110)

$$P_C - P_B = \frac{\rho \omega^2}{2}\{(r_1 + r_2)^2 - r_1^2\} \tag{4.113}$$

Equations (4.111) and (4.112)

$$P_C - P_B = g(\rho_2 y_3 - \rho_1 y_1) + \rho g(y_4 - y_2) \tag{4.114}$$

Using equations (4.113) and (4.114)

$$\rho \frac{\omega^2}{2}\{(r_1 + r_2)^2 - r_1^2\}$$

$$= g(\rho_2 y_3 - \rho_1 y_1) + \rho g(y_4 - y_2)$$

$$\Rightarrow \omega = \sqrt{\frac{2g}{\rho}\left[\frac{(\rho_2 y_3 - \rho_1 y_1) + \rho(y_4 - y_2)}{(r_1 + r_2)^2 - r_1^2}\right]}$$

Problem 10 A cubical vessel is filled with water up to a height h. If it is moved with a constant horizontal acceleration a, find the (a) nature of liquid meniscus, (b) heights of liquid in front and back surface of the vessel, and (c) pressures at the bottom of the vessel at front, mid-point and back of the base.

Solution

(a) If θ = angle of inclination of the free surface of the liquid with horizontal, referring to Example 14,

$$\tan \theta = \frac{dy}{dx} = -\frac{a}{y}$$

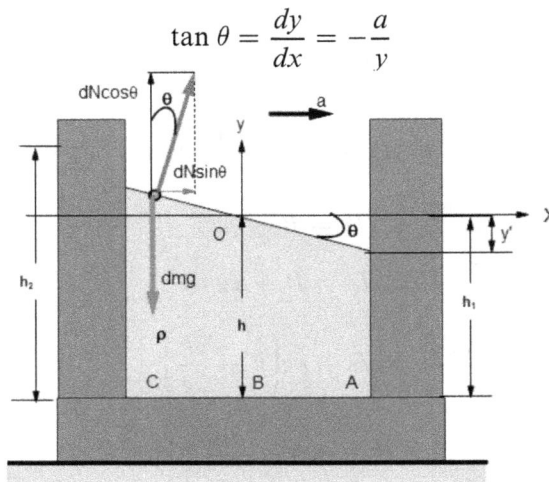

$$\Rightarrow dy = -\frac{a}{g}dx$$

$$\Rightarrow \int_0^y dy = -\frac{a}{g}x$$

$$\Rightarrow y = -\frac{a}{g}x$$

This is an equation of a straight line. So the free surface is an inclined flat surface.

(b) Then, $h_1 = h - y$

$h - \frac{a}{g}x$, where $x = \frac{\ell}{2}$

$$\Rightarrow h_1 = h - \frac{a\ell}{2g} \quad \text{Ans.}$$

Similarly, $h_2 = h + y$

$$= h + \frac{a}{g}x$$

$$= h + \frac{a}{g}\frac{\ell}{2}$$

$$\Rightarrow h_2 = h + \frac{a\ell}{2g} \quad \text{Ans.}$$

(c) Then, pressures at A, B, and C are

$$P_A = P_0 + \rho g h_1$$

$$= P_0 + \rho g\left(h - \frac{a\ell}{2g}\right)$$

$$= P_0 + \rho g h - \frac{\rho a\ell}{2} \quad \text{Ans.}$$

$$P_B = P_0 + \rho g h$$

$$P_C = P_0 + \rho g h_2$$

$$= P_0 + \rho g\left(h + \frac{a\ell}{2g}\right)$$

$$= P_0 + \rho g h + \frac{\rho a\ell}{2} \quad \text{Ans.}$$

Problem 11 A vertical U-tube having two immiscible liquids of densities ρ_1 and ρ_2 is rotating about the vertical axis with an angular velocity ω. If $\rho_1/\rho_2 = 2$, $h_1 = 0.5$ m, $h_2 = 1/3$ m, $r_1 = 0.3$ m, $r_2 = 1/2$ m, and $r_3 = 0.2$ m, find the value of ω.

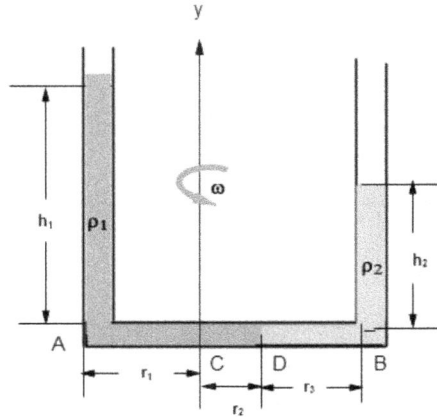

Solution
Applying Pascal's law, we have

$$P_A = P_0 + \rho_1 g h_1 \tag{4.115}$$

$$P_B = P_0 + \rho_2 g h_2 \tag{4.116}$$

Let the pressure at C be P_C.
Since $\frac{dP}{dr} = \rho \omega r^2$, $P = P_C + \frac{\rho \omega^2 r^2}{2}$
So the pressure at A is

$$P_A = P_C + \frac{1}{2}\rho_1 \omega^2 r_1^2 \tag{4.117}$$

$$P_D = P_C + \frac{1}{2}\rho_1 \omega^2 r_2^2 \tag{4.118}$$

Since $\int_{P_D}^{P_B} dP = \rho \omega^2 \int_{r_2}^{(r_2 + r_3)} r\, dr$

$$\Rightarrow P_B = P_D + \frac{1}{2}\rho_2 \omega^2 \{(r_2 + r_3)^2 - r_2^2\} \tag{4.119}$$

Using equations (4.118) and (4.119),

$$P_B = P_C + \frac{1}{2}\rho_1 \omega^2 r_2^2 + \frac{1}{2}\rho_2 \omega^2 \{(r_2 + r_3)^2 - r_2^2\} \tag{4.120}$$

Equations (4.115) and (4.116),

$$P_A - P_B = (\rho_1 h_1 - \rho_2 h_2)g \tag{4.121}$$

Equations (4.118), (4.119), (4.120) and (4.121),

$$\frac{1}{2}\omega^2\left[\rho_1(r_2^2 - r_1^2) + \rho_2\{(r_2 + r_3)^2 - r_2^2\}\right] = (\rho_1 h_1 - \rho_2 h_2)g$$

$$\Rightarrow \omega = \sqrt{\frac{2(\rho_1 h_1 - \rho_2 h_2)g}{\rho_1(r_2^2 - r_1^2) + \rho_2\{(r_2 + r_3)^2 - r_2^2\}}}$$

$$= \sqrt{\frac{2\left(2 \times \frac{1}{2} - 1 \times \frac{1}{3}\right)10}{2(50^2 - 30^2) + 1\{(50 + 20)^2 - 50^2\}10^{-4}}}$$

$$= \sqrt{\frac{2\left(\frac{2}{3}\right) \times 10 \times 10^4}{2 \times 1600 + 20 \times 120}}$$

$$= \sqrt{\frac{4000}{(32 + 24)(3)}}$$

$$= \sqrt{\frac{4000}{3 \times 68}}\, \text{rads}^{-1}$$

$$= \sqrt{\frac{1000}{51}}\, \text{rad s}^{-1}$$

Problem 12 A uniform rod of length l is pivoted smoothly at its top end P, which is located at a height h above the liquid level. If the densities of the liquid and rod are ρ and σ, respectively, find the (a) length of the rod immersed in the liquid and (b) angle made by the rod with horizontal.

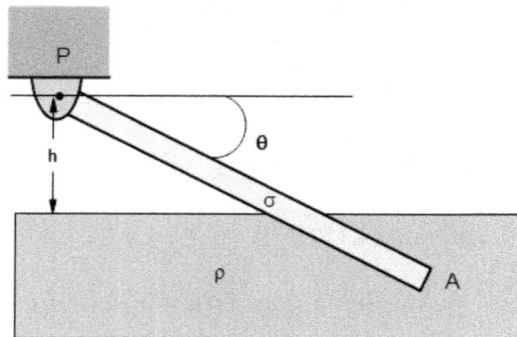

Solution

(a) The net torque due to buoyant force F_b and mg about P is zero as the rod is m equilibrium.

So, $\tau_{\text{buoy}} = \tau_{\text{gr}}$

$$\Rightarrow F_b\left(\ell - \frac{x}{2}\right)\cos\theta = mg\frac{\ell}{2}\cos\theta$$

$$\Rightarrow F_b\left(\ell - \frac{x}{2}\right) = mg\frac{\ell}{2} \tag{4.122}$$

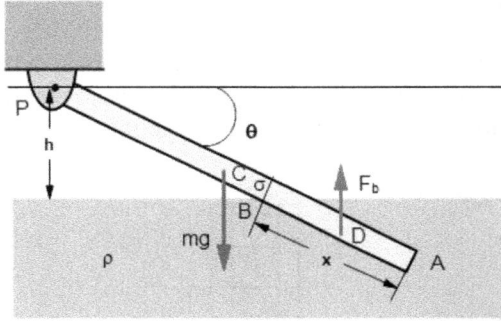

The buoyant force is

$$F_b = \rho g x A \tag{4.123}$$

Putting $m = (A\sigma\ell)$, we have

$$A\rho g x\left(\ell - \frac{x}{2}\right) = (A\sigma\ell)g\frac{\ell}{2}$$

$$\Rightarrow 2\rho x\left(\ell - \frac{x}{2}\right) = \sigma\ell^2$$

$$\Rightarrow \rho x(2\ell - x) = \sigma\ell^2$$

$$\Rightarrow \rho x^2 - 2\rho\ell x + \sigma\ell^2 = 0$$

$$\Rightarrow x = \frac{2\rho\ell \pm \sqrt{4\rho^2\ell^2 - 4\sigma\rho\ell^2}}{2\rho}$$

$$= \ell\left[1 \pm \sqrt{1 - \frac{\sigma}{\rho}}\right]$$

Since $x < \ell$, we have

$$x = \ell\left(1 - \sqrt{1 - \frac{\sigma}{\rho}}\right) \text{ Ans.}$$

(b) The angle of inclination of the rod with horizontal is
$\theta = \sin^{-1}\left(\frac{h}{\ell - x}\right)$, where $\ell - x = \ell\sqrt{1 - \frac{\sigma}{\rho}}$

$$\Rightarrow \theta = \sin^{-1}\frac{h}{\ell\left(1 - \frac{\sigma}{\rho}\right)} \text{ Ans.}$$

Problem 13 A piston of mass M, say, having a hollow axial cylindrical rod of radius r is placed on a liquid of density ρ. The height of the liquid inside the axial tube of the piston is h. If the radius of the piston is R, find the mass of the piston with its axle.

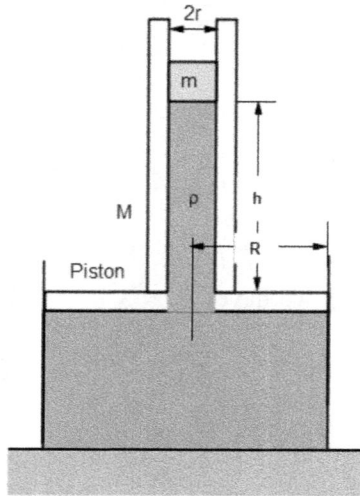

Solution
The pressure at Q is

$$P_Q = \frac{mg + \pi r^2 \rho h g}{\pi r^2} + P_0$$

$$\Rightarrow P_Q = \frac{mg}{\pi r^2} + \rho g h + P_0 \tag{4.124}$$

The upward force acting on the piston is

$$F_{\text{up}} = P_Q A = P_Q. \, \pi(R^2 - r^2) \qquad (4.125)$$

The downward force acting on the piston is

$$F_{\text{down}} = Mg + P_0 A, \qquad (4.126)$$

where $A = \pi(R^2 - r^2)$. The net force acting on the piston is

$$F_{\text{up}} + F_{\text{down}} = 0 \qquad (4.127)$$

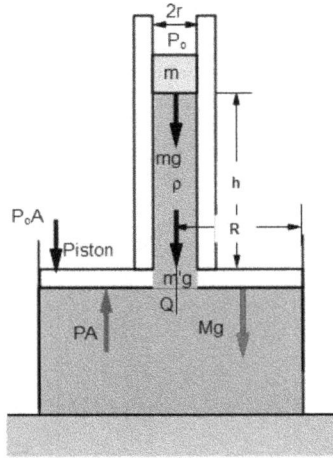

Using equations (4.125)–(4.127),

$$P_Q \pi(R^2 - r^2) = Mg + P_0 A$$

$$\Rightarrow P_Q = \frac{Mg}{\pi(R^2 - r^2)} + P_0 \qquad (4.128)$$

Using equations (4.124) and (4.128), $\frac{mg}{\pi r^2} + \rho g h + P_0 = \frac{Mg}{\pi(R^2 - r^2)} + P_0$

$$\Rightarrow M = \pi \left[\frac{m}{\pi r^2 + \rho h} \right] (R^2 - r^2) \text{ Ans.}$$

Problem 14 An inclined planer (rectangular) gate is smoothly pivoted at its axis P. If a force F acts at the top of the gate perpendicular to the plane of the gate, it remains in equilibrium under the thrust of the liquid of density ρ. Find the value of F.

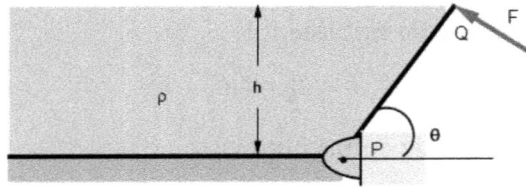

Solution

(a) The hydrostatic pressure at a depth y is

$$P = \rho g y$$

The hydrostatic force acting on the strip is

$$dF = P \, dA = (\rho g y)(b \, dy)$$

$$\Rightarrow dF = \rho g b \, y dy \tag{4.129}$$

Then, the torque acting on the gate is

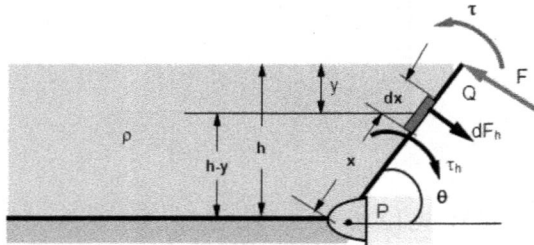

$$\tau = \int x \, dF \tag{4.130}$$

Using equations (4.129) and (4.130),

$$\tau_{hy} = \int x \, \rho g b \, y \, dy$$

$$= \rho g b \int y \, x \, dy, \text{ where } x = (h - y)\text{cosec } \theta$$

$$\Rightarrow \tau_{hy} = \rho g b \text{ cosec } \theta \int_0^h y(h - y) \, dy$$

$$=\rho g b \, \mathrm{cosec} \, \theta \left(\frac{y^2}{2} h - \frac{y^3}{3} \right) \Big|_0^h$$

$$=\rho g b \, \mathrm{cosec} \, \theta \left(\frac{h^3}{2} - \frac{h^3}{3} \right)$$

$$=\rho \frac{g b h^3}{6} \, \mathrm{cosec} \, \theta \text{ Ans.}$$

The net torque about the pivot O is
$\tau_P = F\ell - \tau_{hy}$, where

$$\tau_P = I\alpha = 0$$

$$\Rightarrow F\ell - \frac{\rho g h^3 b \, \mathrm{cosec} \, \theta}{6}$$

$$\Rightarrow F = \frac{\rho g h^3 b \, \mathrm{cosec} \, \theta}{6\ell} \text{ Ans.}$$

Problem 15 A horizontal light and rigid bar AB is smoothly pivoted at P. A bob of mass m, say, hangs at B. To counterbalance the weight of the bob, a block of mass M and density σ hangs at A and floats on a liquid of density ρ. If $h =$ depth of immersion of the block, find the value of m.

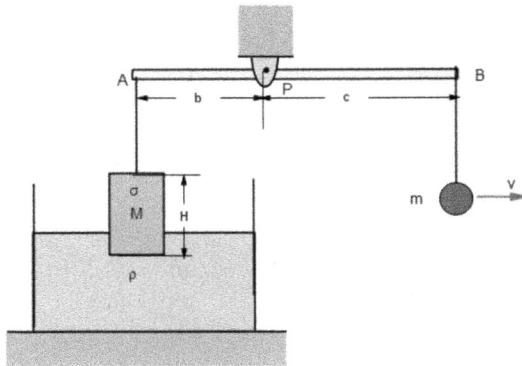

Solution

Referring to the free-body diagram:

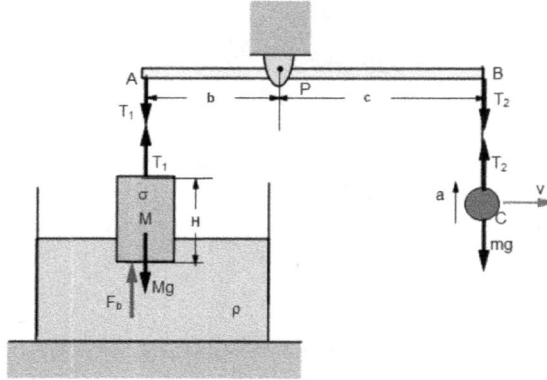

Force equation:

$$F_b + T_1 = Mg \tag{4.131}$$

$$F_b = A\rho gh \tag{4.132}$$

Torque equation:

$$\tau_P = 0$$

$$T_1 b = T_2 c \tag{4.133}$$

Force equation on m:

$$T_2 - mg = \frac{mv^2}{\ell}$$

$$\Rightarrow T_2 = m\left(g + \frac{v^2}{\ell}\right) \tag{4.134}$$

Using equations (4.131) and (4.132),

$$T_1 = mg - A\rho gh$$

$$= A\sigma gH - A\rho gh$$

$$\Rightarrow T_1 = Ag(\sigma H - \rho h) \tag{4.135}$$

Using equations (4.133)–(4.135),

$$Ag(\sigma H - \rho h)b = m\left(g + \frac{v^2}{\ell}\right)c$$

$$m = \frac{Ag(\sigma H - \rho h)b}{\left(g + \frac{v^2}{l}\right)c} \quad \text{Ans.}$$

Problem 16 A fixed quarter-cylindrical gate of radius R is used to restrict the liquid of density ρ in the reservoir from flowing. Find the magnitude and direction of the thrust of the liquid on the gate.

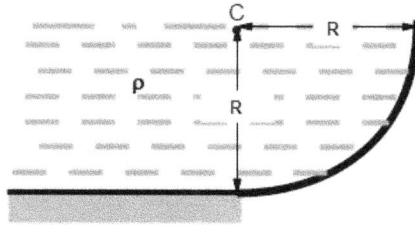

Solution
The force acting on the elementary strip is

$$\delta F = P\delta A$$

$$= \rho g y(R\, d\theta)(\ell)$$

$$= \rho g R\ell\, y d\theta, \text{ where } y = R\sin\theta$$

$$\Rightarrow \delta F = \rho g R^2 \ell \sin\theta d\theta \tag{4.136}$$

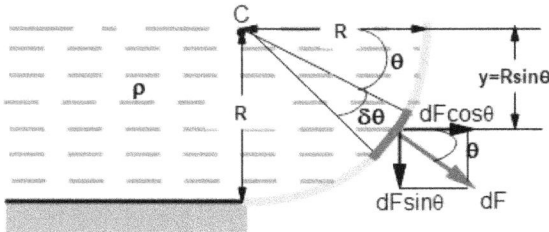

The net horizontal thrust of water on the gate is

$$F_x = \int dF_x = \int dF \cos\theta \tag{4.137}$$

Using equations (4.136) and (4.137),

$$F_x = \rho g R^2 \ell \int_0^{\frac{\pi}{2}} \sin\theta \cos\theta d\theta$$

$$= \frac{\rho g R^2 \ell}{2} \int_0^{\frac{\pi}{2}} \sin 2\theta d\theta$$

$$= \frac{\rho g R^2 \ell}{2} \left[-\frac{1}{2} \cos 2\theta \right]_0^{\frac{\pi}{2}}$$

$$= \frac{\rho g R^2 \ell}{2} \text{ Ans.}$$

The vertical thrust of water on the gate is

$$F_y = \int dF_y = \int dF \sin \theta \qquad (4.138)$$

Using equations (4.136) and (4.138),

$$F_y = \int_0^{\frac{\pi}{2}} \rho g R^2 \ell \sin^2 \theta$$

$$= \rho g R^2 \ell \int_0^{\frac{\pi}{2}} \sin^2 \theta$$

$$= \frac{\rho g R^2 \ell}{2} \left(\because \int_0^{\frac{\pi}{2}} \sin^2 \theta d\theta = \frac{1}{2} \right) \text{ Ans.}$$

Then the net force is

$$F = \sqrt{F_x^2 + F_y^2}$$

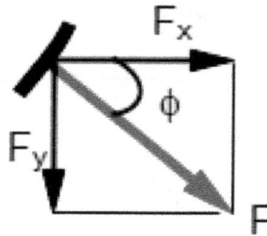

$$= \sqrt{\frac{(\rho g R^2 \ell)^2}{2} + \frac{(\rho g R^2 \ell)^2}{2}} = \sqrt{2} \rho g R^2 \ell$$

$$\phi = \tan^{-1} \frac{F_y}{F_x} = \tan^{-1} 1 = \frac{\pi}{4} \text{rad Ans.}$$

Problem 17 A liquid vessel moves down a smooth inclined plane of angle of inclination θ. Find the angle β made by the liquid surface with the inclined plane if the (a) acceleration of the vessel is a down the plane, (b) the inclined plane is smooth, and (c) there is a coefficient of kinetic friction μ between the inclined plane and the vessel.

Solution

(a) Taking an element dm at the surface of liquid and writing the force equation on it, we have

$$\delta F_x = \delta ma$$

$$\Rightarrow \delta mg \sin\theta + \delta N \sin\beta = \delta ma$$

$$\Rightarrow \delta N \sin\beta = \delta m(a - g\sin\theta) \tag{4.139}$$

Force equation on dm normal to the slant:

$$\delta F_y = 0$$

$$\Rightarrow \delta N \cos\beta = \delta mg \cos\theta \tag{4.140}$$

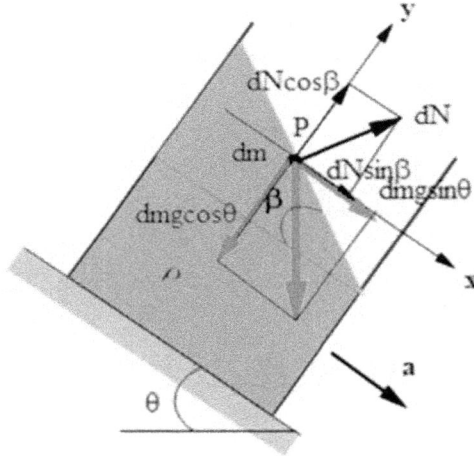

Equations $(4.139) \div (4.140)$

$$\frac{\delta N \sin\beta}{\delta N \cos\beta} = \frac{\delta m(a - g\sin\theta)}{\delta mg \cos\theta}$$

$$\Rightarrow \tan\beta = \frac{a - g\sin\theta}{g\cos\theta}$$

$$\Rightarrow \beta = \tan^{-1}\left\{\frac{a - g\sin\theta}{g\cos\theta}\right\}$$

(b) If the surface is smooth $a = g\sin\theta$. Then $\beta = 0$ and the liquid level will be parallel to the slant.

(c) If $a = g(\sin\theta - \mu\cos\theta)$, we get

$$\beta = \tan^{-1}\left(-\frac{\mu g\cos\theta}{g\cos\theta}\right) = \tan^{-1}(-\mu)$$

$$\Rightarrow \beta = -\phi$$

If $a = 0$ or $v = c$, we have

$$\beta = \tan^{-1}(-\tan\theta)$$
$$\Rightarrow \beta = -\theta$$

This means that the liquid level will be horizontal.

Problem 18 Two immiscible liquids 1 (top) and 2 (bottom) of density ρ and 2ρ and height h and $2h$, respectively, are obstructed by a gate PQ, which is pivoted at P. By applying external force or torque the gate remains vertical. Find the (a) net hydrostatic torque acting on the gate about P and (b) center of hydrostatic force.

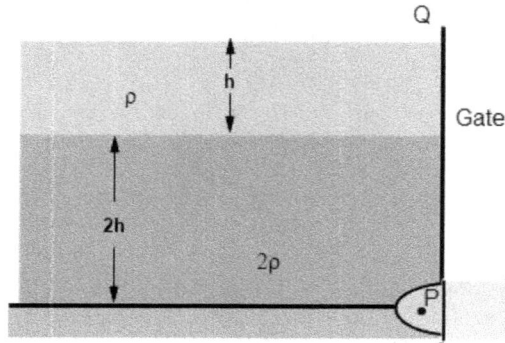

Solution

(a) The horizontal hydrostatic force acting on the elementary strip of area dA is

$$dF = PdA \tag{4.141}$$

This force produces a torque about the pivot P, which is given as

$$d\tau = ydF$$

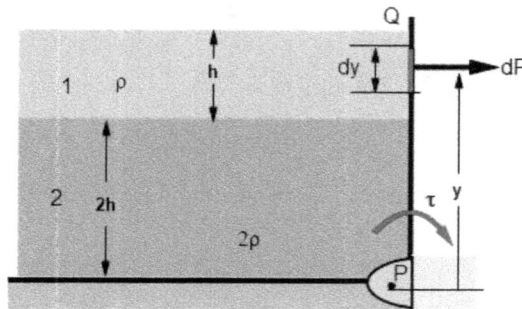

4-142

The net hydrostatic torque is

$$\tau = \int d\tau = \int y dF \tag{4.142}$$

By using equations (4.141) and (4.142)

$$\tau = \int y P dA$$

$$\Rightarrow \tau = \int_{2h}^{3h} y P_1 dA + \int_0^{2h} y P_2 dA \tag{4.143}$$

$$P_1 = \rho g(3h - y) \text{ for liquid 1} \tag{4.144}$$

$$P_2 = \rho g h + (2\rho)g(2h - y) \text{ for liquid 2}$$

$$= \rho g(5h - 2y) \tag{4.145}$$

Using equations (4.143)–(4.145)

$$\tau = \int_{2h}^{3h} y \cdot \rho g(3h - y)b dy + \int_0^{2h} \rho g y(5h - 2y) \, b dy$$

$$= \rho g b \left[\int_{2h}^{3h} y(3h - y) dy + \int_0^{2h} (5h - 2y) y dy \right]$$

$$= \rho g b \left[\left(\frac{3hy^2}{2} - \frac{y^3}{3} \right)_{2h}^{3h} + \left(\frac{5hy^2}{2} - \frac{2y^3}{3} \right)_0^{2h} \right]$$

$$= \rho g b \left[\frac{3}{2}(9h^2 - 4h^2)h - \frac{(27h^3 - 8h^3)}{3} + \frac{5h}{2}(4h^2) - \frac{2}{3} \times 8h^3 \right]$$

$$= \rho g b h^3 \left[\frac{15}{2} - \frac{19}{3} + 10 + \frac{16}{3} \right]^{-2}$$

$$= \frac{33 \rho g b h^3}{2} \text{ Ans.}$$

(b) The position of center of force is given as

$$y_c = \frac{\tau}{F},$$

where $F = F_1 + F_2 = Pav_1A_1 + Pav_2A_2 = \left(\dfrac{\rho gh}{2} + 0\right)(hb) + \dfrac{\rho gh + \rho gh + (2\rho)g(2h)}{2}\{(2h)b\}$

$$= \rho gh^2 b\left(\dfrac{1}{2} + 6\right) = \dfrac{13}{2}\rho gh^2 b$$

Putting the obtained values of F and τ, we have

$$y_c = \dfrac{\frac{33\rho gh^3 b}{2}}{\frac{13}{2}\rho gh^2 b}$$

$= \dfrac{33h}{13}$ from the pivot Ans.

Problem 19 A liquid column of height h is obstructed by a vertical flat gate PQ, which is smoothly pivoted at the lowest point P. The gate is kept in equilibrium by applying a horizontal thrust F at a height of $\dfrac{2h}{3}$. Find the (a) force F and (b) horizontal reaction at the pivot.

Solution

(a) Let $R = $ reaction force acting at the pivot P. The hydrostatic force acting on the plate is y.

$$F_h = P_{av}A$$

$$= \left(\dfrac{\rho gh + 0}{2}\right)(bh)$$

$$= \dfrac{\rho gbh^2}{2} \qquad (4.146)$$

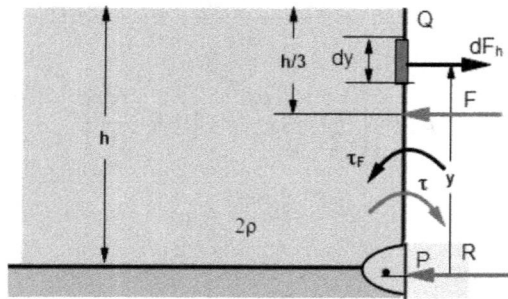

The net horizontal force acting on the plate is

$$F_h - F - R = 0 (\because a = 0)$$

$$\Rightarrow R = F_h - F \qquad (4.147)$$

Equating the torques of the liquid and the applied force F about P,
$\int y' dF_h = Fh''$, where $dF_h = PdA = \rho gy dA$

$$\Rightarrow \int_0^h (h - y)(\rho gy)(bdy) = F\left(\frac{2h}{3}\right)$$

$$\Rightarrow \rho gb \int_0^h y(h - y)dy = 2h\frac{F}{3}$$

$$\Rightarrow \rho gb \left[h\frac{y^2}{2} - \frac{y^3}{3} \right]_0^h = 2F\frac{h}{3}$$

$$\rho gb\frac{h^3}{6} = 2F\frac{h}{3}$$

$$\Rightarrow F = \frac{\rho gbh^2}{4} \text{ Ans.}$$

(b) Putting the value of F in equation (4.147)

$$R = F_h - \frac{\rho gbh^2}{4} \tag{4.148}$$

Using equations (4.146) and (4.148)

$$R = \frac{\rho gh^2 b}{2} - \frac{\rho gbh^2}{4}$$

$$= \frac{\rho gbh^2}{4} \text{ Ans.}$$

Problem 20 Two immiscible liquids 1 (top) and 2 (bottom) of density ρ and 2ρ and height h and $2h$, respectively, are obstructed by a gate PQ, which is pivoted at P. By applying external force or torque the gate remains vertical. Find the hydrostatic force acting on the (a) gate and (b) base. Assume b and l are the breadth and length of the gate.
Solution

(a) The angular hydrostatic pressure on A_1 is
$P_{av_1} = \frac{P_A + P_B}{2}$, where $P_A = 0$ and $P_B = \rho gh$

$$\Rightarrow P_{av_1} = \frac{\rho gh}{2} \tag{4.149}$$

The average hydrostatic pressure on A_2 is

$Pav_2 = \frac{P_B + P_C}{2}$, where $P_B = \rho g h$ and $P_C = \rho g h + (2\rho)g(2h)$

$$\Rightarrow Pav_2 = \frac{\rho g h + \rho g h + 4\rho g h}{2}$$

$$\Rightarrow Pav_2 = 3\rho g h \tag{4.150}$$

Then, the net force acting on the plate is

$$F = F_1 + F_2$$

$$= Pav_1 A_1 + Pav_2 A_2 \tag{4.151}$$

where $A_1 = hb$ and $A_2 = (2h)b$.
Using equations (4.149)–(4.151)

$$F = \left(\frac{\rho g h}{2}\right)(bh) + (3\rho g h)(2hb)$$

$$\Rightarrow F = \rho g b h^2 \left(\frac{1}{2} + 6\right)$$

$$= \frac{13}{2}\rho g b h^2 \text{ Ans.}$$

Alternative method:
The hydrostatic force is

$$F = \int P \, dA$$

$= \int P_1 \, dA + \int P_2 \, dA$, where $dA = b \, dy$, $P_1 = \rho g y$ and $P_2 = \rho g h = (2\rho)g y$

$$\Rightarrow F = \rho g b \int_0^h y \, dy + \int_0^{2h} (\rho g h + 2\rho g y) \, b \, dy$$

$$= \rho g b \frac{y^2}{2}\Big|_0^h + \rho g h b \int_0^{2h} dy + 2\rho g b \frac{y^2}{2}\Big|_0^{2h}$$

$$= \rho g b \frac{h^2}{2} + \rho g b (2h^2) + \rho g b (4h^2)$$

$$= \rho g b h^2 \left(\frac{1}{2} + 6\right) = \frac{13\rho g b h^2}{2} \text{ Ans.}$$

(b) The hydrostatic pressure at the base is $P = \rho g h + (2\rho)g(2h) = 5\rho g h$.
The area of the base is

$$A = b\ell$$

$\Rightarrow$ The hydrostatic force on the base is

$$F = PA = 5\rho g b h \ell \text{ Ans.}$$

Problem 21 Consider two triangular halves PQS (section 1) and RQS (section 2) of a vertical rectangular gate PQRS that experiences a hydrostatic side thrust. Find the (a) hydrostatic thrust acting on sections 1 and 2 and (b) ratio of these thrusts (forces).

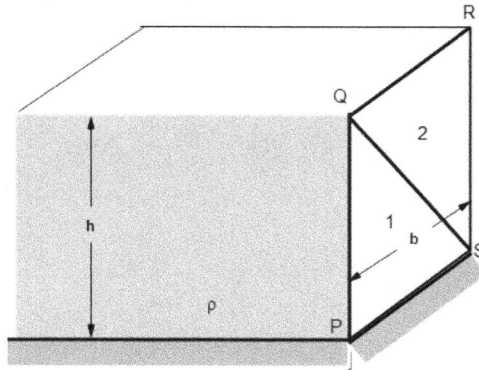

Solution

(a) The center of area or centroid of triangle 1 is

$$y_1 = \frac{2h}{3}$$

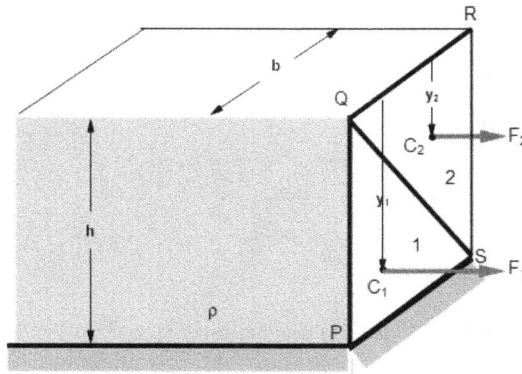

Then, the hydrostatic force acting on area 1 is

$$F_1 = \rho g y_1 A_1$$

$$= \rho g \left(\frac{2h}{3} \right) \left(\frac{1}{2} bh \right)$$

$$= \frac{\rho g b h^2}{3} \text{ Ans.}$$

The center of area of sector (triangle) 2

$$y_2 = \frac{h}{3}$$

Then, the hydrostatic force acting on area 2 is

$$F_2 = \rho g y_2 A_2$$

$$= \rho g \left(\frac{h}{3} \right) \left(\frac{1}{2} bh \right) = \frac{\rho g h^2 b}{6} \text{ Ans.}$$

(b) The ratio of forces is

$$\frac{F_1}{F_2} = \frac{bh^2}{3} \times \frac{6}{h^2 b} = 2 \text{ Ans.}$$

Problem 22 A smooth vertical U-tube moves with an acceleration a containing two immiscible liquids of densities ρ and 3ρ. The cross-sections of the left- and right-hand limbs of the tube are A_1 and A_2, respectively. Find the value of h.

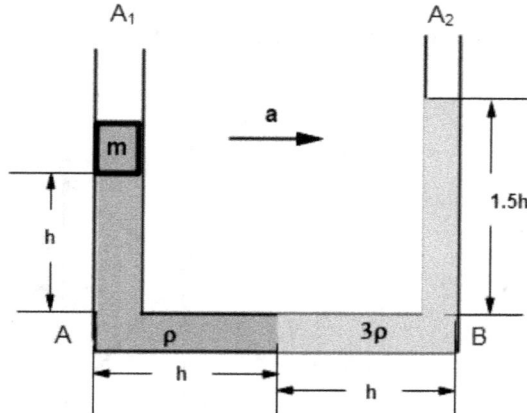

Let's start from the point A of the right side limb and end at the point T of the left side limb. Using Pascal's law,

$$P_Q = P_A = \rho g(1.5) = P_0 + 1.5\rho gh \tag{4.152}$$

$$P_S = P_T + \frac{mg}{A_1} + \rho gh$$

$$= P_0 + \frac{mg}{A_1} + \rho gh \tag{4.153}$$

Since the U-tube is accelerating

$$P_S - P_Q = (2\rho)(h)a + \rho h(a)$$

$$\Rightarrow P_S - P_Q = 3\rho ha \tag{4.154}$$

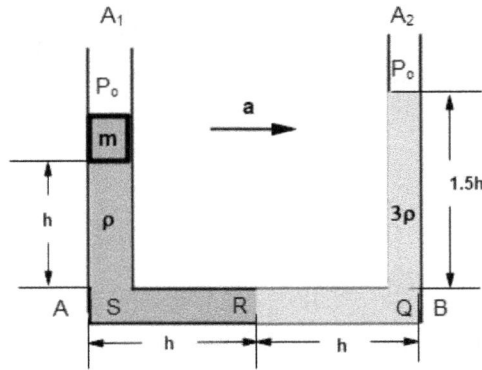

Using equations (4.152)–(4.154),

$$P_0 + \frac{mg}{A_1} + \rho gh - \left(P_0 + \frac{3}{2}\rho gh\right) = 3\rho ha$$

$$\Rightarrow \frac{mg}{A_1} - \frac{\rho gh}{2} = 3\rho ha$$

$$\Rightarrow \rho h\left(3a + \frac{g}{2}\right) = \frac{mg}{A_1}$$

$$\Rightarrow h = \frac{mg}{\rho A_1\left(3a + \frac{g}{2}\right)} \quad \text{Ans.}$$

Problem 23 A vertical L-shaped tube contains a liquid of density ρ. A piston of mass m is loaded on to the liquid. A spring of stiffness k is attached to the other end of the tube. A smooth block of mass m is connected with the spring. Find the angular frequency of small oscillation f of the block.

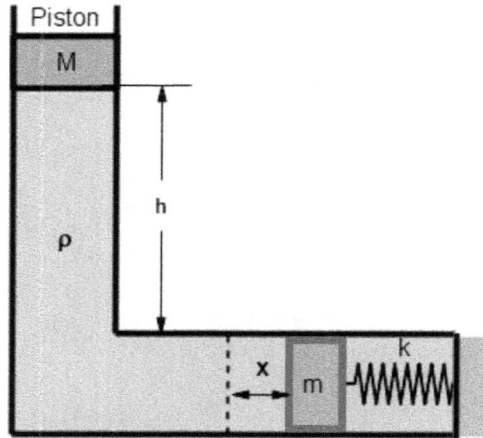

Solution

Force equation on m:

$$(P_0 + \rho gh)A = kx_0 + P_0 A$$

$$\Rightarrow kx_0 = \rho ghA \tag{4.155}$$

Let's displace the piston towards the left by a distance x. Then the net force acting on the piston is

$$F_{\text{net}} = ma$$

$$\Rightarrow k(x + x_0) - \rho g(h - x)A = ma \tag{4.156}$$

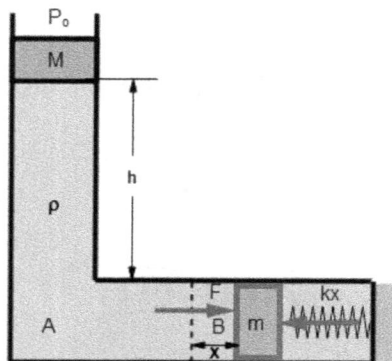

$$\Rightarrow (k + \rho gA)x = ma \ (\because kx_0 - \rho ghA \text{ from equation (4.155))}$$

$$\Rightarrow a = \frac{k + \rho g A}{m} x \qquad (4.157)$$

Comparing equation (4.157) with $a = \omega^2 x$, we have

$$\omega = \sqrt{\frac{k + \rho g A}{m}} \text{ Ans.}$$

Problem 24 A vertical U-tube of different cross-sections contains mercury of height h_o in both the limbs. An immiscible liquid (water, say) is slowly poured until the length of the water column will be equal to h. Find the shift x' of the mercury in the right side limb. Assume that the ratio of radii of the left side and right side limbs is η and the levels of liquids are equal in both sides.

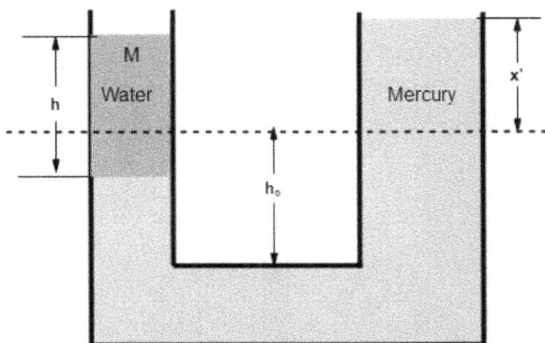

Solution

Let x = displacement of water column in the left side limb and x'= displacement of mercury column in the right side limb.

$$\Rightarrow \pi r^2 x = \pi (\eta r)^2 x'$$

$$\Rightarrow x' = \frac{x}{\eta^2} \qquad (4.158)$$

Pressure equalisation at A and B.

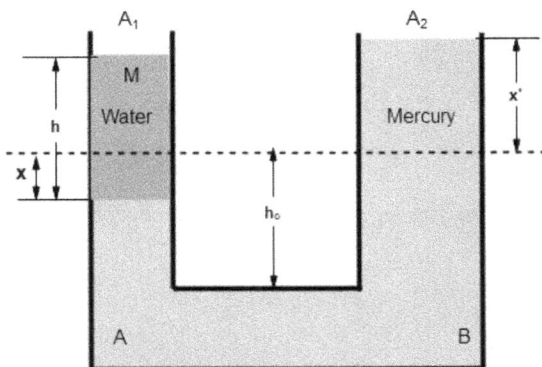

$$P_A = P_B$$

$$\Rightarrow \rho_w gh + \rho_m g(h_0 - x) + P_0 = \rho_m g(h_0 + x') + P_0$$

$$\Rightarrow \rho_w h = \rho_m (x + x') \qquad (4.159)$$

Using equations (4.158) and (4.159),

$$\rho_w h = \rho_m (x'\eta^2 + x')$$

$$\Rightarrow x' = \frac{\rho_w}{(\eta^2 + 1)\rho_m} \quad \text{Ans.}$$

Problem 25 A vertical cylindrical rod of length L and density is released from rest when its bottom end just touches the water surface. The height of the water column is equal to h in a large container. Find the velocity of the rod with which it sinks completely.

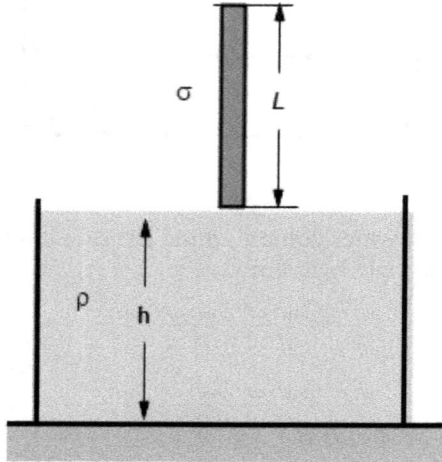

Solution
The rod touches the water surface with a velocity

$$v_0 = \sqrt{2gh} \qquad (4.160)$$

At the depth of penetration y, the buoyant force force is

$$F_b = A\rho gy$$

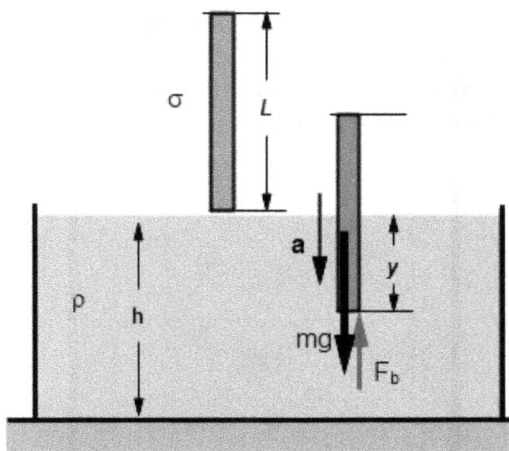

The net force acting on the rod is

$$F = mg - F_b$$

$$\Rightarrow ma = mg - A\rho gy$$

$$\Rightarrow a = g - \frac{A\rho gy}{m} = g - \frac{A\rho gy}{A\sigma L}$$

$$\Rightarrow a = g\left(1 - \frac{\rho}{\sigma L}y\right)$$

$$\Rightarrow \frac{v\,dv}{dy} = g\left(1 - \frac{\rho}{\sigma L}y\right)$$

$$\Rightarrow v\,dv = g\left(1 - \frac{\rho}{\sigma L}y\right)dy$$

$$\Rightarrow \int_0^v v\,dv = g\int_0^L \left(1 - \frac{\rho y}{\sigma L}\right)dy$$

$$\Rightarrow \frac{v^2}{2} = g\left(L - \frac{\rho L^2}{2\sigma L}\right) = gL\left(1 - \frac{\rho}{2\sigma}\right)$$

$$\Rightarrow v = \sqrt{2gL\left(1 - \frac{\rho}{2\sigma}\right)} \text{ Ans.}$$

Problem 26 Two immiscible liquids of densities 3ρ and 2ρ are placed in a container. The height of the liquid columns in the container are $2h$ and h, respectively. A small rubber ball of mass m and density σ is released from rest from the bottom of the container. Disregarding the viscosity of the liquids, find the (a) velocity of the rubber ball at each interface, (b) total time of ascent of the ball, and (c) maximum height attained by the ball. Assume that $\rho = \sigma$.

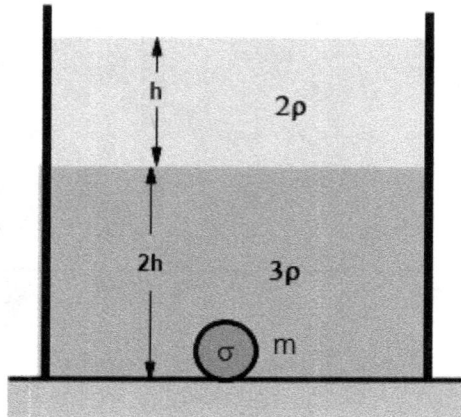

Solution

(a) The velocity at the interface B is

$$v_1 = \sqrt{2a_1(2h)} = \sqrt{4a_1h}$$

where $a_1 = g(1 - \frac{3\sigma}{\sigma}) = -2g$

This means that $\overrightarrow{a_1}$ is directed up

$$\Rightarrow v_1 = \sqrt{4 \times (2g)h}$$

$$v_1 = 2\sqrt{2gh} \text{ Ans.}$$

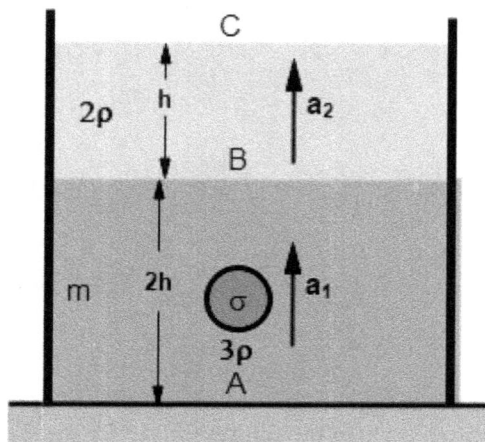

The velocity at the interface C is

$$v_2 = \sqrt{v_1^2 + 2a_2h} \; a_2 = g\left(1 - \frac{\sigma}{2\sigma}\right),$$

where

$$\Rightarrow v_2 = \sqrt{2gh + 2\left(\frac{g}{2}\right)h}$$

$$= \sqrt{9gh} = 3\sqrt{gh} \; \text{Ans.}$$

(b) The total time to go to the highest position is

$$t = t_1 + t_2 + t_3$$

$$= \frac{v_1 - 0}{a_1} + \frac{v_2 - v_1}{a_2} + \frac{v_2 - 0}{g}$$

$$= \frac{v_1}{a_1} + \frac{v_2 - v_1}{a_2} + \frac{v_2}{g}$$

$$= \frac{2\sqrt{gh}}{2g} + \frac{3\sqrt{gh} - 2\sqrt{2gh}}{\frac{g}{2}} + \frac{3\sqrt{gh}}{g}$$

$$= \sqrt{\frac{h}{g}} \{9 - 3\sqrt{2}\} \; \text{Ans.}$$

(c) The maximum height attained is

$$h_{\max} = 2h + h + \frac{v_2^2}{2g}$$

$$= 3h + \frac{9gh}{2g}$$

$$= 7.5h \; \text{Ans.}$$

IOP Publishing

Problems and Solutions in Many-Particle Systems

Pradeep Kumar Sharma

Chapter 5

Fluid dynamics

5.1 Definition of hydrodynamics

An element of fluid has millions of molecules. So, an element of fluid may mean a 'system or cluster of molecules of fluid'. We know that a molecule of any material is the smallest particle having all the properties of that material. In atomic stage, it loses all properties of matter. On the other hand, the property of a liquid (or liquid state) such as pressure, etc, cannot be explained by a single molecule; it depends on the presence of other molecules. The intermolecular distance, that is, the distance of separation between the molecules, plays a significant role in defining such properties. An element of fluid can be referred to as a fluid particle, which is not just a molecule of a fluid. In a broader sense, a fluid particle can be defined as an element of fluid, which consists of several fluid molecules. The size of an element of fluid may be mathematically infinitely small, but physically it is in the order of micro- or nano-meter scale so that you can call it a fluid particle (parcel). The key concept is that the fluid element must reflect the property of the fluid to define the streamline motion.

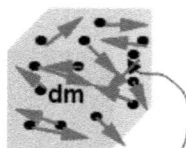

Liquid molecules in the element *dm* move randomly

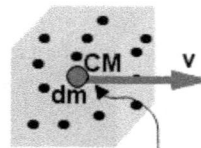

The CM of the element *dm* defines the liquid motion

As we defined, any element of fluid has many atoms and molecules inside it, each moving relative to others; rotating, translating, and vibrating. So, the motion of a fluid

doi:10.1088/978-0-7503-6447-8ch5

does not refer to the internal motion of electrons, protons, atoms, molecules, etc. Each molecule collides with other molecules and follows a *zig-zag* path, which is known as 'Brownian motion'. Can you call it 'motion of fluid'? No. Then, what is meant by 'motion of fluid'? It may mean the motion of the center of mass of an element of the fluid. This may be a bit confusing. To understand this let's move a completely filled water bottle. In this case, the center of mass of water moves, but you cannot call it motion of water in the sense we intend to discuss in this chapter. This is because the liquid (water) does not move relative to the container (bottle) even though its center of mass is moving. From the above discussion we understand that whenever any liquid (or gas) moves relative to the container, we call it fluid motion (or fluid dynamics). Thus, hydrodynamics is defined as the study of motion of fluid relative to the surface in contact.

For instance, air flow (relative to the Earth), motion of water (water flow) in rivers and channels, and flow of liquids and gases in pipelines are familiar examples of flow (motion) of fluids. We call this chapter 'hydrodynamics' because we deal with the dynamics (motion) of (hydro) water-like bodies called fluids.

Example 1 Which of the following belong to hydrodynamics?
   (A) Water in a moving tanker
   (B) Steady state rotating water buckets
   (C) Moving car in stationary air
   (D) Water leakage from a bottle
   (E) Rain fall.

Solution
In (A) and (B), in steady state, water does not move relative to the container. Hence, these examples belong to hydrostatics. However, the rotating liquid in a fixed container can be an example of hydrodynamics even though we can apply the concept of hydrostatic force in this case. In (C), air moves relative to the surface of the car, although it is at rest relative to the ground. So, relative to the car it is flow of air, but relative to the ground, it belongs to hydrostatics.

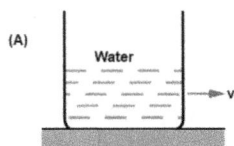

(A)
Water

Tanker moves but water does not
move relative to the tanker

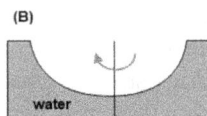

(B)
water

Beaker is stationary but water
rotates relative to the beaker

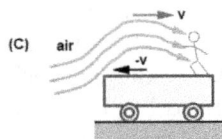

(C) air

The boy feels an air flow in opposite
direction of motion of the trolley car

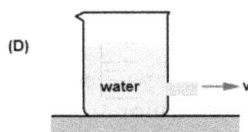

(D)
water

Beaker is stationary but in water
leakage, water moves out relative
to the beaker

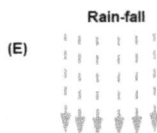

(E)
Rain-fall

Falling of water drops
is discontinuous, so it
is not fluid flow

In (D), water level falls (moves) down relative to the bottle and the steam of water falls freely after escaping from the bottle. Hence, (C) and (D) can be treated as the examples of hydrodynamics. The rain-fall in (E) is not hydrodynamics as it is discontinuous.

The key idea to define hydro or fluid dynamics is that the fluid must move relative to the container. For instance, earth is a container for flow of water in rivers, waterfalls and flow of air in atmosphere.

5.2 General characteristics of fluid flow

Every small segment (element) of fluid is a system of particles (fluid molecules). The internal motion of the molecules (relative to *center of mass*) gives us internal kinetic energy KE_{int} and internal (intrinsic) angular momentum (or spin angular momentum). The intermolecular force gives us the internal potential energy of the system denoted by U_{int}. The sum of K_{int} and U_{int} gives the total internal energy. The change in internal energy in a fluid defines its internal 'pressure'. The average distance of separation between two neighbouring molecules defines the 'density' of fluid. An element of a liquid is subjected to various external forces such as gravity, electrostatic, and magnetostatic forces in addition to internal friction (shearing force). As we know, the internal forces cannot accelerate the *center of mass*. So, the external forces are responsible for the acceleration by changing the velocity of *center of mass* of the elements of liquids (or gases). Hence, the 'velocity of *center of mass*' of an element of liquid can define the motion of a fluid.

From the above discussion, it is relevant that
1. The motion of a liquid can be characterised by (i) density of liquid, (ii) pressure of liquid, and (iii) velocity (of *center of mass*) of liquid elements.
2. Density defines the compressibility of fluids.
3. Pressure defines the internal energy of fluids.
4. Velocity of *center of mass* defines the translational kinetic energy (of *center of mass*) of fluids.

Using the ideas of pressure, velocity, and density of liquid at a point, let's categorize the following types of motion of a liquid (or gas).

Streamline (steady) or turbulent flow: As mentioned above, each element of a fluid is characterized by the point functions such as pressure, velocity, and density.

When we observe the leaves of a tree in a gentle breeze, they do not wiggle; forming a stationary pattern. A leaf seems to orient in the same direction, which signifies that the pressure and velocity of the wind do not change at a fixed point. In other words, each leaf experiences a definite pressure and velocity of the wind.

Another example is motion of water from a tap when the tap is just opened. You can see a constant (fixed) pattern of water jet falling from the tap. Since the pattern (shape and size of the falling liquid) does not change, we conclude that pressure and velocity of the liquid remain constant (steady) at any point. This type of motion is called 'steady flow'.

There are several examples of fluid motion, where pressure and velocity of fluid change from time to time at a point. This makes the flow turbulent (or unsteady). The familiar examples are water waves at the seashore, a cyclone, flooded river, and motion of water after complete opening of a water tap. We will talk about streamline motion in later sections. In steady flow, no collision takes place between two liquid elements (particles). Hence, mechanical energy is conserved. On the contrary, mechanical energy is lost in the form of heat, sound, etc, due to the collision between the particles of liquid in turbulent motion.

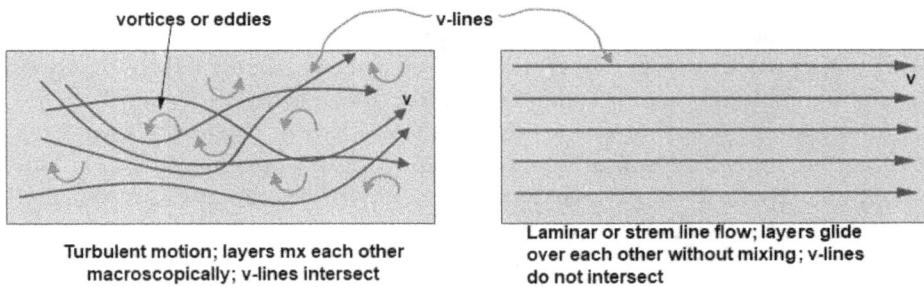

Turbulent motion; layers mx each other macroscopically; v-lines intersect

Laminar or strem line flow; layers glide over each other without mixing; v-lines do not intersect

Smooth and viscous: As we know, a real liquid possesses a little friction between its layers as it gives resistance to the external shearing force applied on the liquid. Consequently, all liquid layers cannot move with equal velocities. For instance, the velocity of water is maximum at the surface and is minimum at the base (bottom) of a river or channel. This is because of more frictional force between any two layers of water at the deeper levels. This phenomenon is known as 'viscosity'. We will discuss this in a separate section at the end of this chapter.

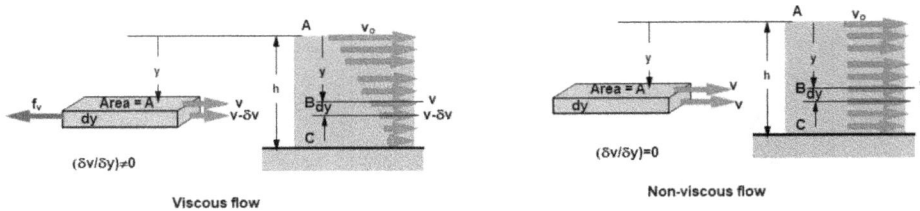

Viscous flow

Non-viscous flow

Thus, the flow of a real liquid is viscous. However, ideally all particles of a liquid move with the same velocity in a vertical plane if we neglect its viscosity. Some liquids are much less viscous. We can call them practically non-viscous liquids such as kerosene, water, and other thinner (rarer) liquids.

Compressible and incompressible liquids: As discussed in hydrostatics, gases can be compressed much more easily than the liquids when compressive forces are applied. The compression in a liquid is much smaller than the original dimensions of the liquid. Hence, we can neglect the compression of a liquid. Then, we can call the flow of liquids an incompressible flow practically. However, the compression of a gas can also be neglected when a shearing force acts on it. This means when a flat object (a car, aeroplane, etc) moves in air (or any gas) with low speeds, the flow (motion) of air relative to the object can be regraded as (nearly) incompressible. In incompressible flow, density of gas remains constant at all points. If the objects move rapidly with speeds more than the speed of sound, say, the surrounding layers of air get compressed and the flow of air may be treated as compressible.

Rotational and irrotational: When you put a tiny flywheel (peddle) inside the flowing water in a power channel or river, it will rotate when the axis of the flywheel is kept horizontal and perpendicular to the direction of flow; the flywheel spins with maximum angular speed. However, the flywheel does not rotate when we keep its axis along the line of motion of liquid or vertical. This experiment tells us that the flow of liquid is rotational. Whirlpools (vortices or eddies) in air and water are the examples of rotational flow. Observing the circular motion of bulk of water you may think that rotational flow means circular motion of a liquid but that is not correct. In the given example of flow of water in a power channel, even though the water moves horizontally, the flow of liquid is rotational because it rotates the flywheel. However, sometimes liquids seem to rotate, but its flow particles move with the same angular velocity. If you put a tiny peddle in the bucket, it will not spin (rotate). That means the flow of water is irrotational. Then what factor decides whether the flow of fluid is rotational or irrotational? Actually it is the 'spin' of the elements of water that makes the flow rotational flow. When the center of mass or an element of water moves, the particles of water must possess a non-zero angular momentum about the center of mass, which is known as 'spin angular momentum'. If an element of a fluid rotates (or spins as its center of mass moves), the fluid flow is said to be rotational.

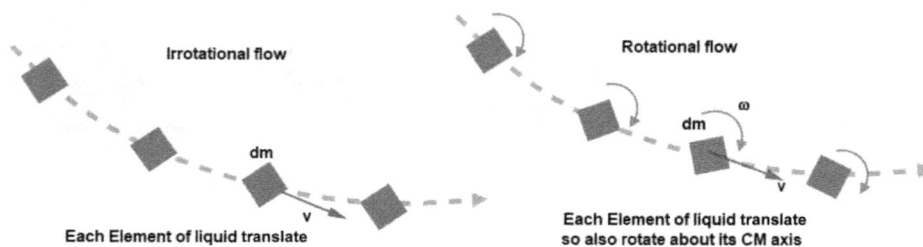

Irrotational flow

Rotational flow

dm

dm

ω

v

v

Each Element of liquid translate

Each Element of liquid translate so also rotate about its CM axis

After realising the above characteristics of motion of a fluid, let's characterise the motion of an ideal fluid.

5.3 Streamline motion

When we observe a gentle flow of water in a stream, river, or channel it seems as if water is steady; the pattern of liquid remains stationary. This means all liquid particles (center of mass of all elements of liquid) move with the same velocity at a given point (spot). The path traced by the *center of mass* of an element of liquid is called the streamline or line of force. In streamline motion liquid elements do not collide in order to possess a definite velocity and pressure at a point. Hence, two lines of forces will never exist in a streamline motion. In other words, we can conserve mechanical energy of each element of liquid. We will discuss it as Bernoulli's equation in a later section. In streamline motion, we can think of any segment of liquid as a bundle of streamlines or lines of force. This is called a 'tube of flow'. For instance, when water flows in a smooth tube (pipe), the tube of flow of water assumes the shape of the pipe. For a streamline flow, the patterns of streamlines do not change with time. Hence, you can also call it steady flow. Needless to say, in steady flow, the liquid (or gas) can be treated as nearly incompressible.

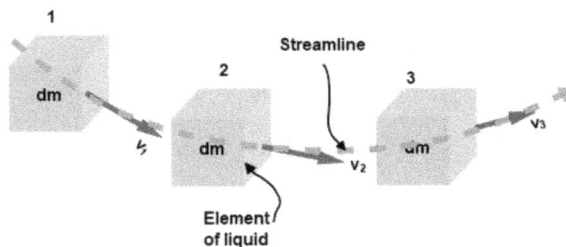

In streamline motion of a fluid we have the following assumptions:

1. The fluid must be incompressible; V (volume) $= C$ (constant).
2. The motion of the fluid must be steady; $\vec{v} = C$ and U (internal energy) $= C$, at any point.
3. The fluid motion is irrotational ($\vec{L}_{spin} = 0$).
4. The motion of fluid must be non-viscous ($f = 0$).

If the above four conditions are met we can conserve the mechanical energy of each particle (element) of liquid. In other words, the mechanical energy of a tube of liquid remains constant. So, the liquid layers must move parallel without intermixing. Furthermore, the velocity of liquid element does not change with time at a given point in a streamline flow.

Irrotational flow

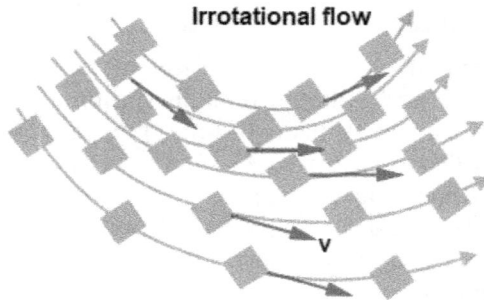

Each element of liquid translates

5.4 Equation of continuity

Let's consider a streamline flow of a liquid in a tube. For the sake of simplicity, let's push the liquid by a piston. If the piston has velocity v, it moves through an elementary distance dx during an elementary time dt. Then, we have $dx = v\, dt$.

During the time dt, the volume of liquid displaced by the piston is $dV = A dx$, where A = area of the piston.

Substituting $dx = v\, dt$

we have $dV = Av\, dt$

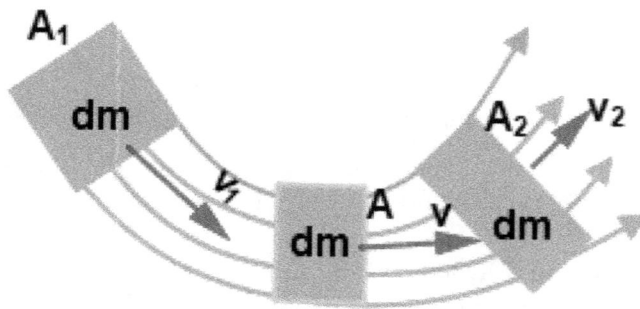

The volume rate of flow of liquid is given as $R = \frac{dV}{dt} = vA$.

Let's assume that the liquid is incompressible and there is no body to vomit (supply) or drink (suck) the liquid inside the tube. Then, an equal amount of liquid will pass through any cross-section of the tube of flow. Hence, we can write

$$R = vA = \text{constant}$$

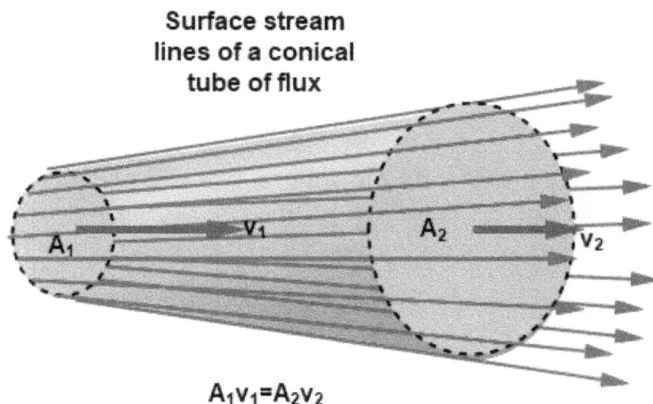

**Surface stream
lines of a conical
tube of flux**

$A_1v_1 = A_2v_2$

From the above equation, we have the following points:
1. The volume rate of flow of an incompressible liquid in streamline motion is constant in a tube of flow, which is equal to the product of velocity of liquid and the area of cross-section of the tube of flow. At any two points in a tube of flow $A_1v_1 = A_2v_2$.
2. For viscous flow, the equation of continuity also holds good.
3. For rotational flow, the motion is not streamline. Hence, the equation of continuity does not hold good.

Example 2 An ideal liquid in a tube of uniform cross-sectional area $A_1 = 0.001$ m^2 flows with a speed $v_1 = 2$ m s^{-1}. At the junction C, the streamlines of the liquid divide into two parts; part of the liquid flux (liquid) moves in a tube of uniform cross-sectional area $A_2 = 0.02$ m^2 with a velocity $v_2 = 3$ m s^{-1}. Find the velocity of liquid in the other branch of uniform cross-sectional area $A_3 = 0.03$ m^2.

Solution
If you look at the junction C, there is no source, no sink. Then, the net incoming flux (rate) = the net outgoing flux.

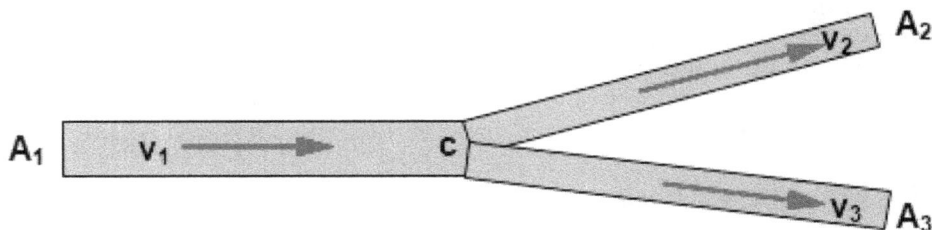

$$R_1 = R_2 + R_3,$$

where $R_1 = A_1 v_1$, $R_2 = A_2 v_2$, and $R_3 = A_3 v_3$.

Then $A_1 v_1 = A_2 v_2 + A_3 v_3$

$$v_3 = \frac{A_1 v_1 - A_2 v_2}{A_3}$$

$$= \frac{(0.01)(2) - (0.02)(3)}{(0.03)} \text{m s}^{-1} = -\frac{4}{3}\text{m s}^{-1} \text{ Ans.}$$

The negative sign signifies that the direction of flow of liquid is just opposite to the assumed direction of v_3.

N.B.:

For a junction (without any source and sink), the algebraic sum of rates (or currents) of flow is zero.

The incoming rate is assumed as negative and the outgoing rate is assumed as positive. You can assume it in a reverse way because it is just a convention.

Then, $\sum R_i = 0$.

In other words, $\sum R_{\text{incoming}} = \sum R_{\text{outgoing}}$; $R_i = +A_i v_i$ (outgoing) and $-A_i v_i$ (incoming).

5.5 Flux of $\overrightarrow{v}$-field (ϕ_v)

Meaning of flux: Flux is a word used in various ways in science and engineering bearing different meanings; mass flux, volume flux, energy flux, momentum flux, E-flux, B-flux, J-flux, v-flux, ..., flux of iron, etc. Here, we use the word 'flux' as 'rate of flow'. If it is rate of flow of mass, it is called mass flux; if it is rate of flow of volume, we call it volume flux.

This means the word 'flux' of anything roughly represents the 'time rate of flow of that thing', denoted by ϕ.

Volume flux: Then the volume flux can be given as $\phi(=R) = \frac{dV}{dt}$.

Substituting the volume flow rate $R = A_n v$ for an incompressible liquid, we have $\phi = v A_n$,

where $A_n =$ cross-sectional area perpendicular to the direction of flow.

Physical significance of ϕ: When the rate of flow is more, we can say that the flux of liquid is more. We can increase the flux in three different ways: by increasing the speed v keeping the tube cross-section constant, by increasing the size of the tube (cross-section A) with a constant flow velocity, or by increasing both A and v.

This means $\phi \propto A_n$ and $\phi \propto v$.

More ϕ means more rapid flow of liquid. Accordingly, you need to imagine more streamlines or lines of flux (or force) of liquid. This means, $\phi \propto$ number of streamlines.

In other words $\phi \equiv$ number of lines of forces (streamlines).

The above expression relates the physical quantity 'flux' to the geometrical (mathematical) quantity 'number of lines of force'.

Flux passing through a flat surface: If you ask 'how much liquid flows in the tube?', it has no precise meaning. If you ask 'what is the rate of flow through any surface in the tube?', it makes better sense. For instance, let's imagine a flat surface perpendicular to the flow. The rate of flow (flux) passing through this surface is given as $R(=\phi) = vA$.

If we rotate the plane (surface) by an angle θ, we can see that less liquid passes through the surface. Then the flux of liquid crossing the surface is given as $\phi = v_n A$, where $v_n (= v\cos\theta)$ is the component of velocity of liquid particles perpendicular to the surface. This gives $\phi = vA\cos\theta$.

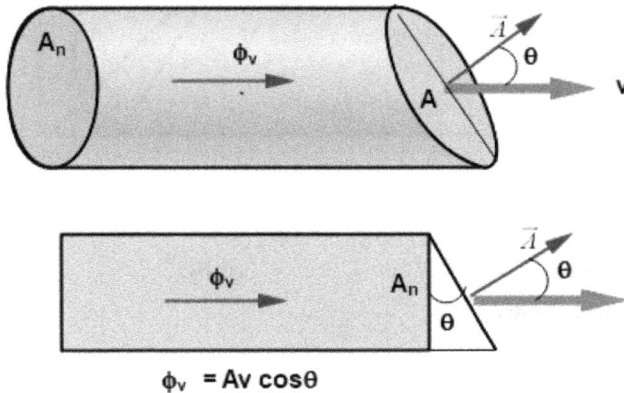

$$\phi_v = Av\cos\theta$$

Substituting $A\cos\theta = A_n$, where A_n is the projection of the surface perpendicular to the direction of flow, we have $\phi = vA_n$.

Summarizing the above facts, we can conclude that:

1. The flux of a liquid passing through a flat surface area A with a velocity v is given as $\phi = vA\cos\theta = \overrightarrow{v}.\overrightarrow{A}$, where θ=angle between $\overrightarrow{v}$ and $\overrightarrow{A}$ and $\overrightarrow{A}$ is the area vector directed along outward normal.

2. Since $\cos\theta$ can be positive, negative, and zero, we can have positive, negative, and zero flux. Hence, flux is an 'algebraic scalar quantity'. When ϕ is +ve, it represents an outward flux; if ϕ is −ve, it represents inward flux. Outward and inward direction is decided by the location of the observer.

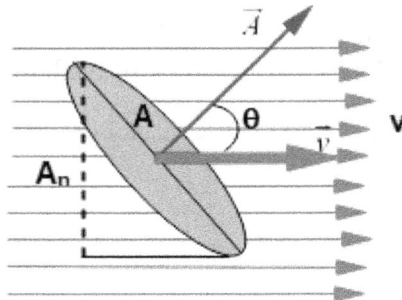

Example 3 A non-viscous liquid flows through a tube of uniform cross-sectional area A. Consider two flat surfaces A_1 and A_2, and find the ratio of fluxes passing through the surfaces.

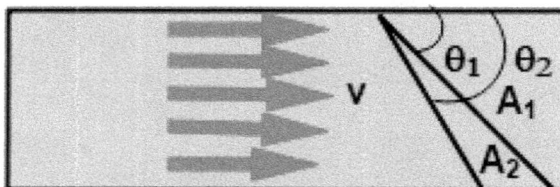

Solution

The flux ϕ_1 passing through A_1 is $\phi_1 = vA_1 \cos \theta_1$.

The flux ϕ_2 passing through A_2 is $\phi_2 = vA_2 \cos \theta_2$.

Then the ratio of fluxes $\frac{\phi_1}{\phi_2} = \frac{vA_1 \cos \theta_1}{vA_2 \cos \theta_2} = \frac{A_1 \cos \theta_1}{A_2 \cos \theta_2}$.

Since $A_1 \cos \theta_1 = A_2 \cos \theta_2 = A$,

we have $\frac{\phi_1}{\phi_2} = 1:1$ Ans.

Alternative method:

Since $\phi_v = vA \cos \theta = vA_n$ ($\because A \cos \theta = A_n$); so, $\frac{\phi_1}{\phi_2} = 1:1$. Ans.

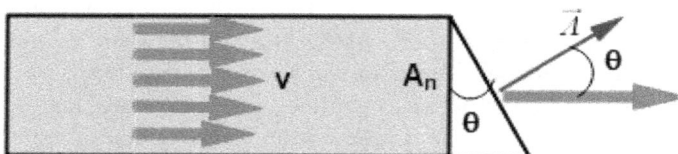

$$\phi_v = Av \cos\theta$$

Example 4 In the above example, if you consider two arbitrary flat areas A_1 and A_2 inside the tube, find the ratio of fluxes passing through the areas.

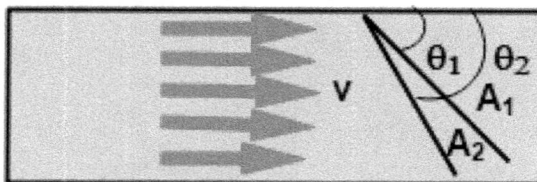

Solution

The flux ϕ_1 passing through A_1 is $\phi_1 = vA_1 \cos(90 - \theta_1) = A_1 v \sin \theta_1$.

The flux ϕ_2 passing through A_2 is $\phi_2 = vA_2 \cos(90 - \theta_2) = A_2 v \sin \theta_2$.

Then the ratio of fluxes is

$$\frac{\phi_1}{\phi_2} = \frac{vA_1 \sin \theta_1}{vA_2 \sin \theta_2} = \frac{A_1 \sin \theta_1}{A_2 \sin \theta_2}$$

The flux passing through a given area A is equal to the flux passing through the projection of the area perpendicular to the flow velocity $\phi = vA_n = v_n A$.

Flux passing through a curved surface: If the surface is curved, take an elementary area $d\vec{A}$. The flux passing through the elementary area is

$$d\phi = \vec{v}.d\vec{A} = vdA \cos \theta.$$

Then the total flux is $\phi = \int d\phi$,

$$\text{where } d\phi = \vec{v}.d\vec{A}$$

This gives $\phi = \int \vec{v}.d\vec{A}$.

The above expression tells us that the surface integral of velocity gives the volume flux passing through the curved surface.

Flux passing through a closed surface: If you want to find the total flux passing through a closed surface, consider an elementary patch (area) on the surface. Find the flux $d\phi$ passing through the elementary area. The total passing through the closed surface is $\phi = \oint d\phi$. Then, substitute the value of $d\phi = \int \vec{v}.d\vec{A}$, where $\oint$ is called a closed integral. This tells us that the closed surface integral of velocity gives us the total flux passing through the closed surface.

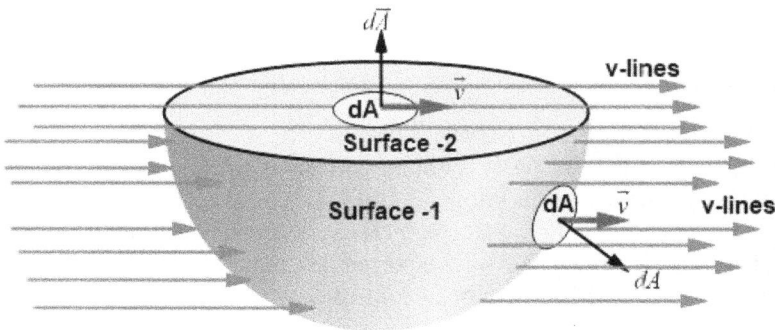

Algebraic nature of flux: When the net flux ϕ passing through a closed surface is positive, we must think of a 'source' of generating the flux. This signifies that the outward flux is greater than the inward flux. In other words, the amount of liquid going out of the closed surface is greater than the amount of liquid coming into (entering) the closed surface. Similarly if the flux ϕ is negative the net flux is inward. This means the inward flux is greater than the outward flux. In that case, somebody must be there to suck the liquid. This is known as 'sink'. If ϕ is zero, then the inward and outward fluxes are equal. This means there is no source and sink in the region enclosed by the curved surface.

1. In a closed surface, if $\phi > 0$, net flux is outward. Hence, there is a *source* in the region bounded by closed surface.
2. If $\phi < 0$, net flux is inward. Hence, there is a *sink* in the region bounded by a closed surface.
3. If $\phi = 0$. there is no *source* and no *sink*.

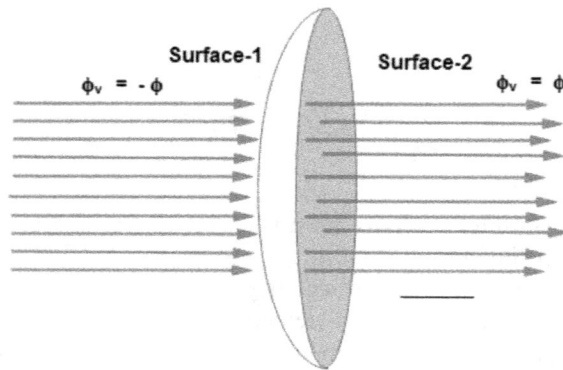

Surface-1 (curved) has inward flux $-\varnothing$ and the surface-2 (flat) has outward flux $+\varnothing$. So total flux passing through the closed surface (surface-1 plus surface-2) is zero. This is due to the algebraic nature of flux.

5.6 Flux density

Let's consider a flow of a non-viscous liquid in a tube of uniform cross-section. We have explained that the flux passing through a plane area A is $\phi = vA_n$.

Average flux density: As you know, for different planes of same orientation, A_n is different. Hence, the flux passing through different areas will be different in the tube. However, if you divide the flux ϕ by the perpendicular projection of area A_n, we have a constant quantity, that is, velocity of liquid.

$$\frac{\phi}{A_n} = v$$

Since the flux ϕ is divided by area A_n, we call the term '$\frac{\phi}{A_n}$' 'surface (or area) density of flux' or simply 'flux density' averaged over the area. Since flux density is equal to 'velocity', it is defined as a vector quantity.

When $A_n = 1$, we have $v = \phi$

This means that the flux density of $\vec{v}$ field is numerically equal to the flux (or number of lines of force) crossing through a unit area perpendicularly.

Flux density at a point: When the streamlines are curved and the liquid flows in a tube of non-uniform cross-section, we consider a small area ΔA_n perpendicular to a bundle of streamlines. Then find the flux $\Delta \phi$ passing through this area. This will give you an average flux density as discussed earlier. If we reduce the area ΔA_n to a point, the flux $\Delta \phi$ will tend to decrease. Then, the ratio of $\frac{\Delta \phi}{\Delta A_n}$ when $\Delta A_n \rightarrow 0$ will give us the flux density at a point, which can be stated as $\lim\limits_{\Delta A_n \rightarrow 0} \frac{\Delta \phi}{\Delta A_n} = \frac{d\phi}{dA_n} = v$.

Physical significance of flux density: As you know, the flow of liquid is associated with moving liquid particles. Hence, a moving liquid is characterised by velocity $\vec{v}$. This means a moving liquid is a field of velocity vector (or $\vec{v}$-field). In this way, the streamlines can also be termed as v-flux lines of v-lines. Hence, the density of v-flux at any point is defined as velocity of liquid at that point. When we say flux density is more, this means that v is more. Hence, the lines of forces are more crowded. We will use the above ideas of flux and flux density in electromagnetism to define the electric flux density $\vec{D}$ and magnetic flux density $\vec{B}$.

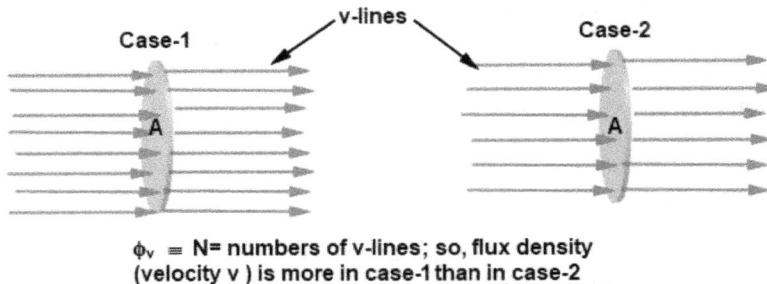

$\phi_v \equiv N=$ numbers of v-lines; so, flux density
(velocity v) is more in case-1 than in case-2

V-lines or streamlines are more crowded in case-1 than in case-2; so the flux density of v-flux lines, that is, velocity v is greater in case-1 than in case-2. Physically, the tube of liquid moves faster through area A in case-1 than in case-2.

We note the following points from the above discussion.

The average flux (velocity) of a moving liquid (v-field) is defined as the ratio of flux and perpendicular area. The flux density is defined as vector quantity. If the streamlines or v-lines are more crowded in any region, the flux density (velocity) is higher in that region (or point) and vice versa. The average flux density can also be defined as the number of lines of force or streamlines or v-lines passing through a unit area perpendicularly.

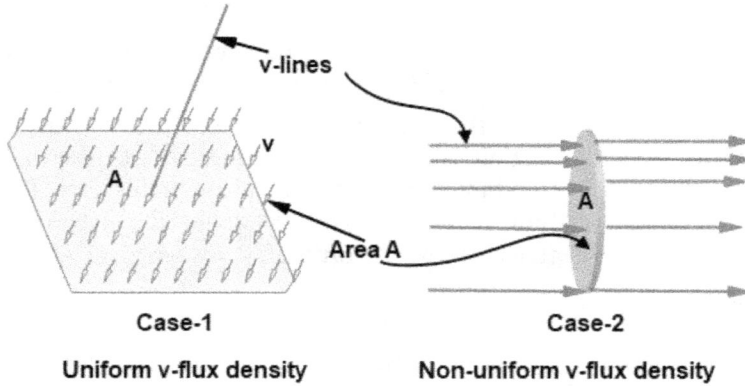

Case-1   Case-2

Uniform v-flux density   **Non-uniform v-flux density**

You can see two types of flux density: uniform and non-uniform, which convey different physical meaning in physics.

5.7 Equation of state of fluid motion

Forces acting on an element: Let's consider an elementary cube of liquid of mass δm. The forces acting on the element in the x-direction (x-direction should not be confused with horizontal because x-axis can have any direction other than horizontal) are:

(i) Reaction forces (hydrostatic forces) due to surrounding liquid $F_{1_x} = P_{1_x}\, \delta A$ and $F_{2_x} = P_{2_x}\, \delta A$

(ii) x-component of gravity force and other conservative forces like electrostatic and magnetic forces F_{con_x}

(iii) x-component of shearing (viscous) force F_{vis_x}

Then the net force acting on the element in the x-direction is

$$F_x = F_{1_x} - F_{2_x} + F_{com_x} - F_{vis}$$

Dividing both sides by the volume δV of the element, we have

$$\frac{F_x}{\delta V} = \frac{F_{1_x} - F_{2_x}}{\delta V} + \frac{F_{con_x}}{\delta V} - \frac{F_{vis}}{\delta V} \tag{5.1}$$

In the above equation, we have four terms. Let's name them one by one.

The first term $\frac{F_x}{\delta V}$ = net external force per unit volume $= (f_x)_{ext}$, say.

The second term $\frac{F_{1_x} - F_{2_x}}{\delta V} = \frac{P_{1_x} A - P_{2_x} A}{A\delta x}$

$$= -\frac{(P_{2_x} - P_{1_x})}{\delta x} = -\frac{\delta P}{\delta x} = -\frac{\partial P}{\partial x}, \text{ say}$$

We call it the pressure gradient.

The third term is $\frac{F_{con_x}}{\delta V} = \frac{1}{\delta V}\left(-\frac{\partial U}{\partial x}\right) = -\frac{\partial U_v}{\partial x}$, say

where U_v = potential energy density.

Hence, you can call $\frac{F_{con}}{\delta V}$ the potential energy density gradient ($=f_{con}$, say).

The last term $\frac{F_{con}}{\delta V}$ can be called viscous force per unit volume ($=f_{vis}$, say).

Substituting the values of all terms in equation (5.1) we have

$$f_x = \frac{\partial P}{\partial x} - \frac{\partial U_v}{\partial x} - (f_{vis})_x$$

where $f_x = \frac{\delta ma}{\delta V} = \rho a_x$ (according to Newton's 2nd law).

This gives

$$\rho a_x = -\frac{\partial P}{\partial x} - \frac{\partial U_v}{\partial x} - (f_{vis})_x \tag{5.2}$$

Similarly, we have

$$\rho a_y = -\frac{\partial P}{\partial y} - \frac{\partial U_v}{\partial y} - (f_{vis})_y \tag{5.3}$$

$$\rho a_z = -\frac{\partial P}{\partial z} - \frac{\partial U_z}{\partial z} - (f_{vis})_z \tag{5.4}$$

You can write $-\frac{\partial U_v}{\partial x}$ instead of f_{con_x}. Similarly $-\frac{\partial U_v}{\partial y} = f_{con_y}$ and $-\frac{\partial U_v}{\partial z} = f_{con_z}$.

Acceleration: If many particles pass through a point with different velocities, we can say that v is a time varying function. When a particle passes through different points with different velocities, we say that v is a space-dependent function. Combining the above two effects we can say that as a liquid moves (flows) in a tube or any channel, not only its velocity may change from time to time at a particular spot, but also it may vary from point to point inside the liquid. In that case, the flow can be called 'unsteady'. Thus, in general you may consider the velocity v as space and time varying.

Symbolically $v = f(x, y, z, t)$

This means that the acceleration of a particle of a liquid can be given as $\vec{a} = \frac{d\vec{v}}{dt}$,

where $v = f(x, y, z, t)$

Then, we have $\vec{a} = \frac{\partial \vec{v}}{\partial t} + v_x \frac{\partial v}{\partial x}\hat{i} + v_y \frac{\partial v}{\partial y}\hat{j} + v_z \frac{\partial v}{\partial z}\hat{k}$.

Equation of state for streamline flow: As we know, the velocity v does not vary with time at a particular place in the case of streamline motion. Then $\frac{\partial v}{\partial t}$ (but not $\frac{d\vec{v}}{dt}$) $= 0$.

Then we have

$$a_x = v_x \frac{\partial v}{\partial x} \tag{5.5}$$

$$a_y = v_y \frac{\partial v}{\partial y} \tag{5.6}$$

$$a_z = v_z \frac{\partial v}{\partial z} \tag{5.7}$$

Substituting a_x, a_y, and a_z from equations (5.5)–(5.7) in (5.2)–(5.4), we have

$$\rho v_x \frac{\partial v}{\partial x} = -\frac{\partial P}{\partial x} + (f_{con})_x - (f_{vis})_x \tag{5.8}$$

$$\rho v_y \frac{\partial v}{\partial x} = -\frac{\partial P}{\partial x} + (f_{con})_y - (f_{vis})_y \tag{5.9}$$

$$\rho v_z \frac{\partial v}{\partial x} = -\frac{\partial P}{\partial x} + (f_{con})_z - (f_{vis})_z \tag{5.10}$$

Equation of state for an ideal non-viscous liquid flow: For an ideal liquid we neglect viscous forces

$$(f_{vis})_x = 0, \ (f_{vis})_y = 0 \ \text{and} \ (f_{vis})_z = 0$$

For the sake of simplicity, let's assume that only the gravitational force acts on the liquid.

Then $(f_{con})_x = \frac{\delta m g_x}{\delta V} = \rho g_x$

$$(f_{con})_y = \frac{\delta m g_y}{\delta V} = \rho g_y$$

$$\text{and} \ (f_{con})_z = \frac{\delta m g_z}{\delta V} = \rho g_z$$

Substituting the values of f_{vis} and f_{con} in equations (5.8)–(5.10), we have

$$\rho v_x \frac{\partial v}{\partial x} = -\frac{\partial P}{\partial x} + \rho g_x \tag{5.11}$$

$$\rho v_y \frac{\partial v}{\partial y} = -\frac{\partial P}{\partial y} + \rho g_y \tag{5.12}$$

$$\rho v_z \frac{\partial v}{\partial z} = -\frac{\partial P}{\partial z} + \rho g_z \tag{5.13}$$

5.8 Bernoulli's theorem

In this section we need to find the energy associated with the liquid. First of all take an elementary segment (element) of a liquid. Then find the work done by x-, y-, and z-components of forces acting on it. Since the work is defined as a way (process) of energy transfer, this will give us the total energy possessed by that element. Finally dividing the total energy δE by the volume δV of the element, we can get total energy density $E_v = \frac{\delta E}{\delta V}$. We can find '$E_v$' by two methods: (i) force method and (ii) work–energy method.

Force method: In this method, first of all we find the volume density net force on the element $\vec{f} = \frac{\delta m\vec{a}}{\delta V} = \rho\vec{a}$.

Then the work done by f in displacing the *center of mass* of the element through an elementary displacement $d\,\vec{r}$ is

$$dW(=\vec{f}\,.d\,\vec{r}) = (dm\,\vec{a}).\ d\,\vec{r}$$

Substituting $\vec{f} = -\left\{\frac{\partial P}{\partial x}\hat{i} + \frac{\partial P}{\partial y}\hat{j} + \frac{\partial P}{\partial z}\hat{k} + \rho(g_x\hat{i} + g_y\hat{j} + g_z\hat{k}\right\}$

$$d\,\vec{r} = dx\hat{i} + dy\hat{j} + dz\hat{k} \text{ and } \vec{a} = v_x\frac{\partial v_x}{\partial x}\hat{i} + v_y\frac{\partial v_y}{\partial y}\hat{j} + v_z\frac{\partial v_z}{\partial z}\hat{k}$$

we have $-(\partial P_x + \partial P_y + \partial P_z) - \rho g_x\,dx - \rho g_y\,dy - \rho g_z\,dz = \rho d\left\{\frac{(v_x^2 + v_y^2 + v_z^2)}{2}\right\}.$

If you take the x-axis horizontally and y-axis vertically we have $g_x = 0$, $g_z = 0$, and $g_y = g$.

Then, substituting $v_x^2 + v_y^2 + v_z^2 = v^2$; v = speed of *center of mass* of the element), $\partial P_x + \partial P_y + \partial P_z = dP$ (total elementary change in pressure), and $g_x = g_z = 0$ and $g_y = g$, in the last expression, we have $-dP = \rho g dy = d(\rho\frac{v^2}{2})$. This gives $d(P + \frac{\rho v^2}{2} + \rho g y) = 0$. Hence, $P + \frac{\rho v^2}{2} + \rho g y = $ constant.

Work–energy method: Let's consider the flow of a liquid in a tube as streamline, non-viscous, steady, and irrotational. This means that pressure and velocity at any point in the liquid remain constant.

Let's now define the system as the segment BC of liquid flowing in the tube at any instant $t = t_1$. On this system we need to apply the work–energy theorem. For this, we need to know the forces acting on the system.

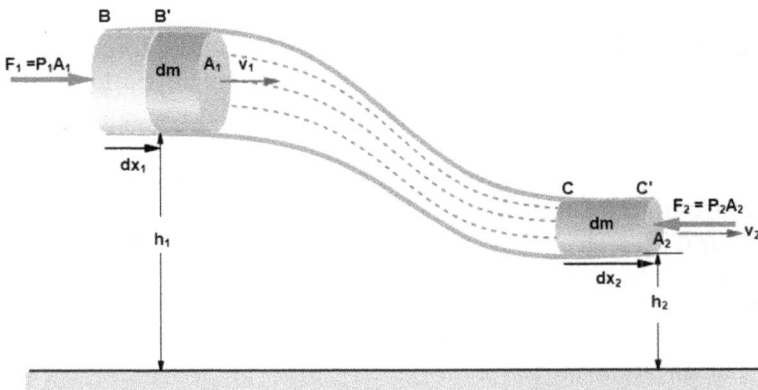

As an extended liquid along the long tube, liquid is present at both the left- and right-hand side of the system (segment BC of liquid). Referring to the free-body diagram, you can see that the liquid in the left-hand side of the face (end) B of the system pushes the system (BC) with a force F_1 towards the right. Similarly, the liquid in the right-hand side of the end C of the system pushes the system (BC) towards the left with a force F_2. You know that F_1 and F_2 are known as hydrostatic forces. Apart from this, gravity acts on (each element of) the system.

Thus, the system moves under the hydrostatic pressure forces (F_1 and F_2), gravity, and other constraint reaction forces offered by the wall of the tube. The internal forces acting on each liquid element by the surrounding liquid will cancel out as a whole for the system BC, leaving us with the external forces only, which can change the momentum and energy of the segment or system BC.

Let's assume that the end B goes to B' moving through an elementary distance dx_1, during elementary time dt. During this time, F_1 pushes the end B in the direction of its motion (displacement dx of B). Hence, F_1 performs a positive work given as

$$dW_1 = F_1\, dx_1$$

At the same time, the force F_2 pushes the end C against its motion (displacement dx_2 of C). Hence, the work done by F_2 is negative during the time dt, which is given as

$$dW_2 = -F_2\, dx_2.$$

Then, the total work done by the hydrostatic forces F_1 and F_2 during time dt is

$$dW = dW_1 + dW_2.$$

Substituting

$$dW_1 = F_1 dx_1 \text{ and } dW_2 = -F_2 dx_2,$$

we have

$$dW = F_1 dx_1 - F_2 dx_2 \qquad (5.14)$$

The work–energy theorem tells us that whenever the external forces perform a non-zero work on a system, the total energy of the system changes. Let's now find the change in 'total energy' of the system. For this purpose, let's follow the logic as given below.

As the liquid flows continuously, let's fix our eyes on 'our system BC'. During time dt, B moves to B', C moves to C'. This means that the system changes its shape from BC to $B'C'$; its size (mass and volume) remains constant. Let's divide the system BC (at $t = t_1$) and $B'C'$ (at $t = t_1 + dt$) into two parts: (i) the portion $B'C$ common to BC and $B'C'$ (different shapes of the system) and (ii) the remaining portion of the system (which has portions BB' (at $t = t_1$) and CC' (at $t = t_1 + dt$) having an elementary mass 'dm').

Since the flow is ideal, pressure and velocity of the liquid do not change with time. Hence, the total mechanical energy of the common portion $B'C$ remains constant during the time dt; let this energy be E'. If we add the energy of the elementary portion of mass dm, it will give the total energy of the system.

Following the above logic, let's find the total energy of the system. The total energy of the system at $t = t_1$ is

$$E_1 = E' + U_1 + K_1.$$

where $U_1 =$ gravitational PE of $(BB')dm$ (at $t = t_1$) $= dm\, gh_1$.

$K_1 =$ kinetic energy of $(BB')dm$ (at $t = t_1$) $= \frac{1}{2}dmv_1^2$

Then we have

$$E_1 = E' + dm\, gh_1 + \frac{dm}{2}v_1^2 \tag{5.15}$$

Similarly, the total energy of the system at $t = t_1 + \delta t$ is

$$E_2 = E' + U_2 + K_2,$$

where $U_2 =$ potential energy of $(CC')\, dm$ (at $t = t_1$) $= dmgh_2$

$K_2 =$ kinetic energy of $(CC')\, dm$ (at $t = t_1 + dt$) $= \frac{1}{2}dm\, v_2^2$

Then we have

$$E_2 = E' + dm\, gh_2 + \frac{dm}{2}v_2^2 \tag{5.16}$$

Using equations (5.15) and (5.16), the change in mechanical energy of the system during time δt is $dE = E_2 - E_1$.

Substituting E_1 and E_2, we have

$$dE = \frac{1}{2}dm(v_2^2 - v_1^2) + dmg(h_2 - h_1) \tag{5.17}$$

Substituting dW from equation (5.14) and dE from equation (5.17) in the work–energy theorem $dW = dE$.

We have $F_1 dx_1 - F_2 dx_2 = dmg(h_2 - h_1) + \frac{1}{2}dm(v_2^2 - v_1^2)$.

Then dividing both sides by volume dV of the elementary mass dm, we have

$$\frac{F_1 dx_1}{dV} - \frac{F_2 dx_2}{dV} = \frac{dm}{dV}g(h_2 - h_1) + \frac{1}{2}\frac{dm}{dV}(v_2^2 - v_1^2)$$

Substituting $dV = A_1 dx_1 = A_2 dx_2$ and $\frac{dm}{dV} = \rho$, we have

$$\frac{F_1}{A_1} - \frac{F_2}{A_2} = \rho g(h_2 - h_1) + \frac{1}{2}\rho(v_2^2 - v_1^2)$$

Finally, substituting $\frac{F_1}{A_1} = P_1$ (hydrostatic pressure at B) and $\frac{F_2}{A_2} = P_2$ (hydrostatic pressure at C), we have

$$P_1 - P_2 = \rho g(h_2 - h_1) + \frac{\rho}{2}(v_2^2 - v_1^2)$$

Finally, we have

$$P_1 + \rho gh_1 + \frac{\rho v_1^2}{2} = P_2 + \rho gh_2 + \frac{\rho v_2^2}{2}$$

Energy densities and conservation of energy: In the above equation, we have three terms. The first term is pressure P. Since this is equal to the work done by hydrostatic pressure per unit volume, we can term it 'pressure energy density' based on the fact that work done is a process of energy transfer.

The second term is $\rho g h$, which is equal to gravitational energy per unit volume. In other words, it is called 'gravitational potential energy density'. The third term is $\frac{\rho v^2}{2}$, which is equal to kinetic energy of liquid per unit volume. Hence, we call it kinetic energy density.

From the above expression it is relevant that the sum of the pressure energy density P, kinetic energy density $\frac{\rho v^2}{2}$, and gravitational potential energy density $\rho g h$ at any point of a streamline flow is a constant quantity. In other words, the density of total energy of liquid at any point of an ideal streamline flow is constant. This is in accordance with the conservation of energy; symbolically,

$$P + \rho g h + \frac{1}{2}\rho v^2 = C.$$

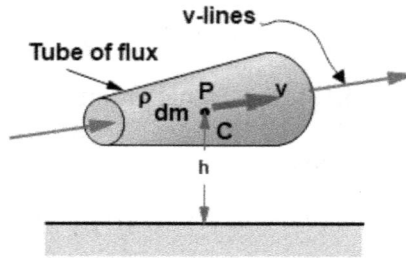

This is what we call 'Bernoulli's theorem'. If any other conservative forces are there, we can just add the potential energy density of the respective conservative force fields.

If any point C of a streamline flow, if pressure is P, height is h, and velocity is v, for a density ρ at the given point, the Bernoulli's equation is

$$P + \frac{\rho v^2}{2} + \rho g h = \text{constant}.$$

Then, we can say that

The sum of pressure energy, kinetic energy, and gravitational energy of fluid per unit volume at any point of an ideal streamline flow remains constant.

This can be given in a different way if we divide both sides by a constant 'ρg'. Then we have

$$\frac{P}{\rho g} + h + \frac{v^2}{2g} = \text{constant}$$

Since each term gives the dimension of a height (head), we call them 'heads'. The first term '$\frac{P}{\rho g}$' is called 'pressure head'. The second term 'h' is called 'gravitational head' and the third term '$\frac{v^2}{2g}$' is called 'velocity head'.

Then, we can say that

The sum of pressure head, gravitational head, and velocity head of an ideal streamline motion remains constant.

5.9 Applications of Bernoulli's theorem and equation of continuity

Now we have two equations; (i) Continuity equation, which obeys the conservation of mass (and volume for incompressible liquids) and (ii) Bernoulli's equation, in accordance with the conservation of energy (or work–energy theorem). We can use the above two equations to solve the problems of dynamics of an ideal liquid as follows.

Let's take a thin tube of flux. Choose any two points 1 and 2. Let the cross-sectional areas of the tube at points 1 and 2 be A_1 and A_2, respectively. The heights of the points from horizontal ground be h_1 and h_2, respectively. Let's assume the pressures at 1 and 2 are P_1 and P_2, respectively. Furthermore, assume that the liquid flows through the points with velocities. Furthermore, assume that the liquid flows through the points with velocities v_1 and v_2, respectively. We can write the following equations at points 1 and 2 as follows:

$$\text{Equation of continuity Rate of flow} = \phi = A_1 v_1 = A_2 v_2 \tag{5.18}$$

$$\text{Bernoulli's equation } P_1 + \frac{1}{2}\rho v_1^2 + \rho g h_1 = P_2 + \frac{1}{2}\rho v_2^2 + \rho g h_2 \tag{5.19}$$

Here, we have two equations, but six unknown quantities such as v_1, v_2, P_1, P_2, h_1, and h_2. If any four unknown quantities are given directly or indirectly in the problem, we can solve for the other two 'unknown quantities of the problem'.

Let's apply Bernoulli's equation and equation of continuity in the following examples. So, you remember the following steps.

Consider a thin tube of flux (or any tube of flux of given areas). Locate the points at which we want to find the flow rate (by writing continuity equation) and energy density (by writing Bernoulli's equation). Then solve the two equations for two unknown quantities.

Example 5 (Bernoulli's effect in horizontal flow) An ideal liquid flows in a truncated horizontal pipe of cross-sectional areas A_1 and A_2 at points 1 and 2 as shown in the figure below. If the pressures of liquid at 1 and 2 are P_1 and P_2, respectively, find the (a) velocity of liquid at point 2 and (b) the rate of flow.

Solution

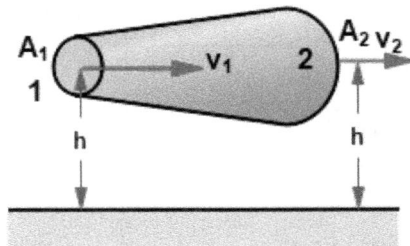

Putting $h_1 = h_2$ in the Bernoulli's equation, we have

$$P_1 + \frac{\rho v_1^2}{2} = P_2 + \frac{\rho v_2^2}{2} \qquad (5.20)$$

The equation of continuity gives

$$A_1 v_1 = A_2 v_2 \qquad (5.21)$$

Substituting $v_1 = \frac{A_2 v_2}{A_1}$ from equations (5.21) in (5.20), we have

$$\frac{\rho}{2}\left\{ v_2^2 - \left(\frac{A_2}{A_1} v_2\right)^2 \right\} = P_1 - P_2$$

This gives $v_2 = \sqrt{\dfrac{2(P_1 - P_2)}{\rho(1 - \frac{A_2^2}{A_1^2})}}$ Ans.

(b) Putting the obtained value of v_2 in equation (5.21), we have

$$R = A_2 v_2 = A_1 A_2 \sqrt{\frac{2(P_1 - P_2)}{\rho(A_1^2 - A_2^2)}} \text{ Ans.}$$

Since $P + \rho\frac{v^2}{2} =$ constant, P is more when v is less and vice versa. Remember that P is the 'static pressure' and $\frac{\rho v^2}{2}$ is the *dynamic pressure*. The sum of two pressures is a constant when the liquid flows horizontally. In other words, the pressure of the fluid is lower where the fluid moves faster. This is known as 'Bernoulli's effect'.

Example 6 (Venturi meter) Describe the working principle of a venturi meter with relevant expressions.

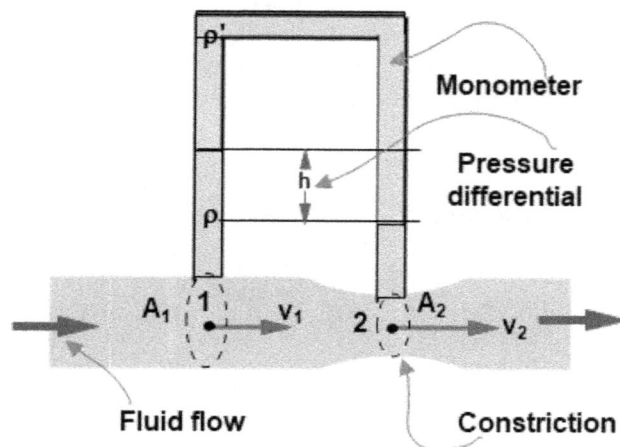

Solution

This is a device used to measure the speed of a liquid in a horizontal pipe. Let the pressures at points 1 and 2 of the tube be P_1 and P_2 and the corresponding speeds of liquid be v_1 and v_2, respectively. By applying Bernoulli's equation, we have

$$P_1 - P_2 = \frac{1}{2}\rho(v_2^2 - v_1^2) \tag{5.22}$$

The cross-sectional area of the tube at points 1 and 2 of the tube are given as A_1 and A_2.

Then applying equation of continuity at 1 and 2, we have

$$A_1 v_1 = A_2 v_2 \tag{5.23}$$

Solving these equations velocities of liquid at 1 and 2 can be given as

$$v_1 = A_1 \sqrt{\frac{2(P_1 - P_2)}{\rho(A_1^2 - A_2^2)}} \tag{5.24}$$

$$v_2 = A_2 \sqrt{\frac{2(P_1 - P_2)}{\rho(A_1^2 - A_2^2)}} \tag{5.25}$$

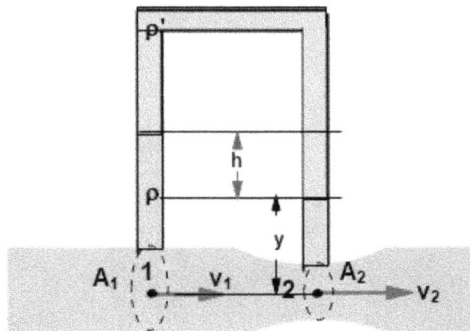

Pressure at 1 greater than that at 2
because velocity at 2 is greater than that at 1.

Let's connect a vertical U-tube filled with a liquid of density ρ' (that is what we call a manometer). The ends of the manometer limbs are connected to the regions where cross-sectional areas A_1 and A_2 of the pipe. We can record the difference of height h in the manometer. The pressure difference between 2 and 1 is

$$P_1 - P_2 = \rho g(h + y) - (\rho'gh + \rho gy) = (\rho - \rho')gh \tag{5.26}$$

Finally, using the last three equations, we have

$$v_1 = A_1 \sqrt{\frac{2(\rho' - \rho)gh}{\rho(A_1^2 - A_2^2)}} \tag{5.27}$$

$$v_2 = A_2 \sqrt{\frac{2(\rho' - \rho)gh}{\rho(A_1^2 - A_2^2)}} \tag{5.28}$$

We can note that since $A_2 < A_1$ we have $v_2 > v_1$. From equation of continuity if we apply $v_2 > v_1$, we can get $P_1 > P_2$. Hence, the liquid is accelerated in the direction of net force (decreasing pressure).

Example 7 (Spray or atomizer) Describe the working principle of an atomiser.
   Solution
   Take an L-shaped tube having opening at A, B, and C. The vertical limb is immersed in the liquid. Blow the air at the end A by mouth or any rubber pump. Air flows through the tube AB. Take a point 1 inside the vertical tube BC and point 2 at the end of horizontal tube AB. At 1, there is no air flow; $v_1 = 0$. At point 2, the speed of air is v_2, say.
   Since the points are not far apart, $h \cong 0$;so $\rho gh \simeq 0$.
   Then

$$P_1 = P_2 + \frac{1}{2}\rho v_2^2$$

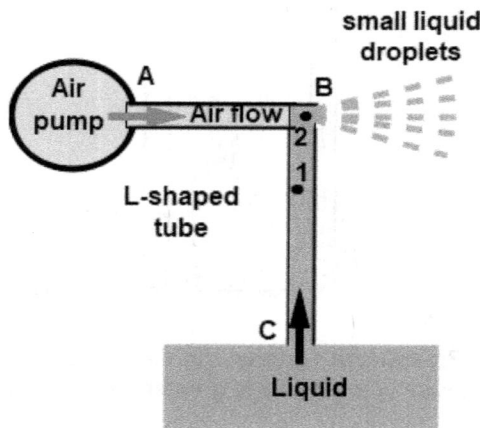

This implies that the pressure at 1 will be greater than pressure at 2. As a consequence, the liquid will be pushed up and escape through the end B as tiny droplets by the forced air. That is what we call a spray or an atomizer.

Example 8 (Narrowing water jet) A stream of liquid jet comes out of a water tap of radius R with a downward velocity v_0. Find the (a) radius of the tube of liquid at a vertical distance y from the water tap and (b) the thrust exerted by the liquid in the process of collision if the liquid jet strikes the ground after falling a distance h and the liquid does not bounce just after collision.

Solution

(a) Here, the tube of flux has length y. We have chosen points 1 and 2 in the tube of flux to apply the foregoing two equations. Since both points are subjected to atmosphere, we have $P_1 = P_2 = P_{\text{atm}}$.

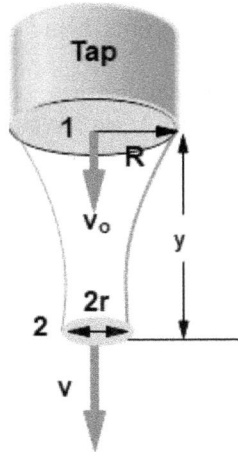

Then, Bernoulli's equation can be given as

$$P_{\text{atm}} + \frac{\rho v_1^2}{2} + \rho g h_1 = P_{\text{atm}} + \frac{\rho v_2^2}{2} + \rho g h_2$$

Substituting $h_1 = y$, $h_2 = 0$ (point 2 is the reference), $v_1 = v_0$, and $v_2 = v$, we have

$$v = \sqrt{v_0^2 + 2gy} \tag{5.29}$$

The equation of continuity gives

$$A_1 v_1 = A_2 v_2$$

where $A_1 = \pi R^2$, $v_1 = v_0$, $A_2 = \pi r^2$ and $v_2 = \sqrt{v_0^2 + 2gy}$.

This gives $r = \dfrac{R\sqrt{v_0}}{(v_0^2 + 2gy)^{1/4}}$ Ans.

(b) The thrust exerted by the liquid is

$$F_{\text{imp}} = v_{\text{rel}}(dm/dt),$$

where $v_2 = \sqrt{v_0^2 + 2gy}$ and $dm/dt = (\pi R^2)\rho v_0$

$$\Rightarrow F_{\text{imp}} = \pi R^2 \left(\sqrt{v_0^2 + 2gh} \right)$$

N.B.: As the pressure does not vary between two points of a tube of flux, $\frac{\rho v^2}{2} + \rho g h = $ constant. This means v is more where height h is less and vice versa.

Example 9 (Siphon) A liquid moves in a smooth tube of uniform cross-section. Find the pressure P_2 at point 2 of the tube if the pressure at point 1 of the extended tube is P_1 and the vertical distance between points 1 and 2 is h.

Solution

Since the cross-section of the tube is uniform, $A_1 = A_2 = A$. Then, applying the equation of continuity at points 1 and 2, we have

$$v_1 = v_2 \tag{5.30}$$

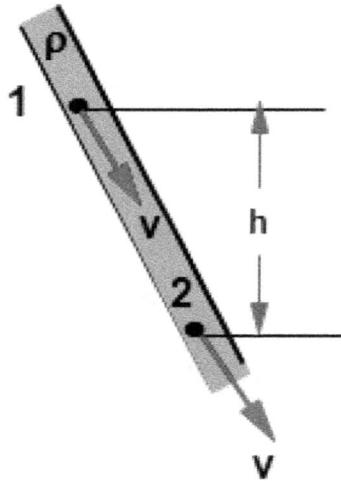

Substituting $v_1 = v_2(=v)$ in Bernoulli's equation, we have

$$P_1 + \frac{1}{2}\rho v^2 + \rho g h_1 = P_2 + \frac{1}{2}\rho v^2 + \rho g h_2$$

This gives $P_1 + \rho g h_1 = P_2 + \rho g h_2$.

Since $h_1 = h_2 = h$

we have $P_2 = P_1 + \rho g h$ Ans.

N.B.: You will get the same expression if the tube (i) is not straight and (ii) has non-uniform cross-section but at points 1 and 2 areas of cross-section are equal. When a liquid flows in a tube of uniform cross-section (may have any arbitrary shape), the speed of liquid remains constant obeying the equation of continuity. In that case, Bernoulli's equation assumes a form $P + \rho g h = c$. This means pressure at lower points will be more than that at the higher points of the liquid by a factor $\rho g h$, where h = vertical distance (but not necessarily the length of the tube) between the two points 1 and 2. However, when the tube is straight and vertical, $l = h$. In general, $P_2 = P_1 + \rho g h$. This is the principle of a siphon.

Example 10 (Motion of liquid in an inclined tapering tube) A liquid (water, say) moves down from an extended tube of non-uniform cross-section. The areas of cross-sections at point 1 and 2 are A_1 and $A_2 = \frac{A_1}{2}$ and the corresponding heights are h_1 and h_2, respectively. If the liquid moves at point 2 with a velocity of 1 m s^{-1}, assuming $h_1 = 5$ m, $h_2 = 2$ m, $\rho = 10^3$ kg m^{-3}, and $g = 9.8$ m s^{-2} find the (a) difference in pressure energy densities between points 1 and 2 and (b) (i) work done per unit volume by gravitational force, (ii) difference in kinetic energy density at points 1 and 2, and (iii) difference in pressure heads.

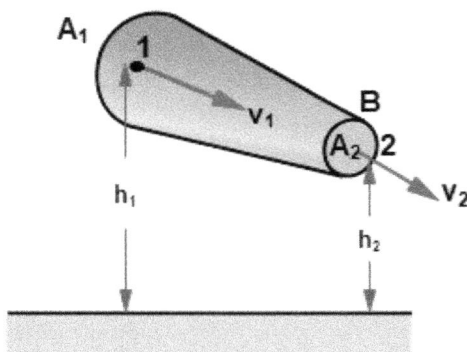

Solution
(a) Substituting $A_1 = 2A_2$ and $v_2 = 1$ m s^{-1} in the equation of continuity

$$A_1 v_1 = A_2 v_2,$$

we have

$$v_1 = \frac{A_2}{A_1} v_2 = \frac{1}{2} \text{ m s}^{-1}$$

Now substitute $v_1 = \frac{1}{2}$ m s^{-1}, $v_2 = 1$ m s^{-1}, $h_1 = 5$ m, and $h_2 = 2$ m in Bernoulli's equation

$$P_1 + \frac{\rho v_1^2}{2} + \rho g h_1 = P_2 + \frac{\rho v_2^2}{2} + \rho g h_2,$$

we have $P_1 + \frac{1}{2}(10^3)\left(\frac{1}{2}\right)^2 + (10^3)(9.8)(5) = P_2 + \frac{1}{2}(10^3)(1)^2 + (10^3)(9.8)(2)$

This gives $P_1 - P_2 = -290\,25$ J m^{-3} Ans.

(b) (i) The work done per unit volume by gravitational force is $u_{gr} = \rho g h = (10^3)(9.8)(5 - 2) = 2.94 \times 10^4$ J m^{-3}.

(ii) The difference in kinetic energy density at points 1 and 2 is

$$\Delta k_v = \frac{\rho v_2^2}{2} - \frac{\rho v_1^2}{2} = \frac{\rho}{2}(v_2^2 - v_1^2)$$

$$= \frac{10^3}{2}(v_2^2 - v_1^2) = \frac{10^3}{2}\{(1)^2 - (0.5)^2\} = \frac{3 \times 10^3}{8} = 3.75 \times 10^2 \text{ J m}^{-3}. \text{ Ans.}$$

(iii) The difference in pressure at points 1 and 2 is

$$(P_1 - P_2)/\rho g = -290\ 25/\{(10^3)(9.8)\} \cong 2.9 \text{ m}. \text{ Ans.}$$

The negative work done by the hydrostatic pressure signifies that since the pressure at 2 is greater than the pressure at 1 ($P_2 > P_1$), the pressure force does net negative work by pushing liquid upward. Gravity is doing positive work in pulling the liquid down. Then, the sum of density of work done (work done per unit volume) by gravity and pressure force is positive. Hence, kinetic energy density increases by 3.75×10^2 J m^{-3}.

Example 11 (velocity of efflux) Explain the concept of velocity of efflux taking the example of a cylindrical (or cubical) vessel filled with an ideal liquid. The liquid comes out of a hole made at a depth y. The areas of the base and hole are A_1 and A_2, respectively. Find the velocity of efflux.

Solution

Choose two points 1 and 2 at the free surface of liquid and at just outside the hole, respectively.

Since points 1 and 2 are exposed to atmosphere $P_1 = P_2 = P_{atm}$, we get a mathematical advantage. However, you can take any other point inside the tube instead of taking it at the free surface of the liquid. The equation of continuity gives

$$A_1 v_1 = A_2 v_2 \tag{5.31}$$

Bernoulli's equation gives

$$P_{atm} + \frac{\rho v_1^2}{2} + \rho g h_1 = P_{atm} + \frac{\rho v_2^2}{2} + \rho g h_2$$

Substituting $h_1 = y$, $h_2 = 0$ (reference point 2), we have

$$v_2 = \sqrt{v_1^2 + 2gy} \tag{5.32}$$

Substituting $v_1 = \frac{A_2}{A_1} v_2$ from equation (5.31) in equation (5.32)

we have $v_2 = \sqrt{\dfrac{2gy}{1 - (\frac{A_2}{A_1})^2}}$ Ans.

Case 1: If area of the hole is much less than area of the base, the velocity of efflux is given as $v \cong \sqrt{2gy}$, which is the same as free fall velocity, but it is an approximated result.

Case 2: When $A_1 = A_2$, the liquid falls with gradually decreasing area of cross-section losing contact with the wall of the tube.

Case 3: If area of the hole is comparable to the area of the base, the velocity of efflux is given more strictly as

$$v(=v_2) = \sqrt{\frac{2gy}{1 - \left(\frac{A_2}{A_1}\right)^2}}$$

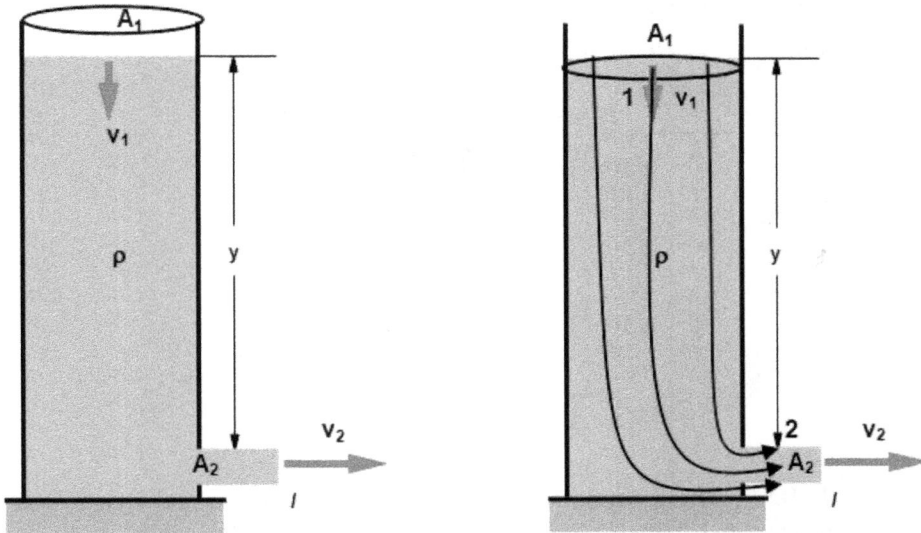

N.B.: The velocity of efflux is the velocity of the escaping liquid relative to the container (but not necessarily relative to the ground when the container moves).

Example 12 (Horizontal range of the escaping liquid) A small hole is made at a depth h and height h' from the base of a cylindrical vessel which is completely filled with a non-viscous liquid. Put $h + h' = H$. Find the (a) horizontal range and (b) speed of striking of the first liquid particle coming out of the vessel with the ground (c) maximum range of the liquid particle.
Solution

(a) For a small hole the range of the liquid particle is given as

$$x = vt, \text{ where } v = \sqrt{2gh}$$

The time of flight of the water is

$$t = \sqrt{\frac{2h'}{g}}, \text{ where } h' = H - h.$$

Then we have $x = 2\sqrt{hh'}$ Ans.

(b) The speed of striking of the liquid is $v' = \sqrt{v^2 + 2gh'}$

Substituting $v = \sqrt{2gh}$, we have $v' = \sqrt{2g(h + h')}$

(c) Putting $h' = H - h$ in $x = 2\sqrt{hh'}$, we have $x = 2\sqrt{h(H - h)}$.

For the range to be maximum, dx/dh or $dx^2/dh = 0$;

$\frac{dx^2}{dh} = 2\frac{d\{h(H - h)\}}{dh} = 0$; so, we have $h = H/2$.

Putting $h = H/2$ in $x = 2\sqrt{h(H - h)}$, we have $x_{max} = H$.

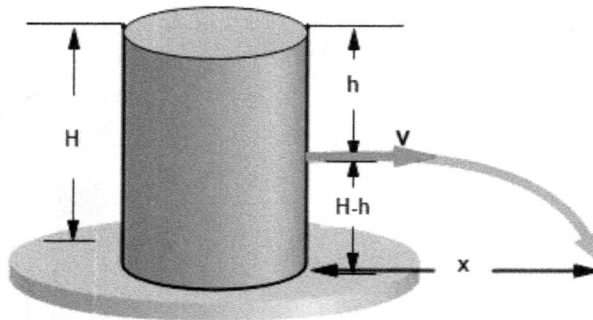

N.B.: 1. The formula $x = 2\sqrt{hh'}$ tells us that if we make two holes at equal vertical distance from the top and bottom of the liquid, both liquid jets will strike the same spot (but not simultaneously) assuming h = constant.

2. If $h = \frac{H}{2}$, the range is maximum and $R_{max} = H$. This result is based on the assumption that area of the hole is negligible compared to the base area of the vessel. In a strict sense, $R_{max} < H$ and $h > \frac{H}{2}$ for maximum range if the correction factor $\frac{1}{\sqrt{1 - (\frac{A_{hole}}{A_{base}})^2}}$ is taken into account.

Example 12 (Time of emptying the vessel) A cylindrical vessel of base area A_1 is filled with a liquid up to height H. A hole of area A_2 is made at the bottom. Find (a) the height of the liquid column inside the vessel as function of time and (b) time after which (i) the first half of the liquid, (ii) second half of the liquid, and (iii) total liquid escapes from the vessel.

Solution

(a) The velocity of efflux at any height y of liquid column is $v = \sqrt{2gy}$. Then, the rate of flow of liquid out of the vessel is $R = A_2 v = A_2 \sqrt{2gy}$.

At the same rate, the mass (and volume) of liquid decreases in the vessel. Hence, $-\frac{dV}{dt} = R$.

Substituting the volume of liquid $V = A_1 y$ and $R = A_2 \sqrt{2gy}$, we have

$$-A_1 \frac{dy}{dt} = A_2 \sqrt{2gy}$$

This gives $\quad - A_1 \frac{dy}{\sqrt{y}} = A_2 \sqrt{2g}\, dt$

If the height of liquid level decreases from H to y during time t, we have

$$-A_1 \int_H^y \frac{dy}{\sqrt{y}} = A_12\sqrt{2g} \int_0^t dt$$

This gives $y = (\sqrt{H} - \frac{A_2 t}{A_1} \sqrt{\frac{g}{2}})^2$ Ans.

(b)

    (i) For the first half of liquid to empty, put $h = H/2$ in the obtained expression to have

$$t_1 = (\sqrt{2} - 1)\frac{A_1}{A_2}\sqrt{\frac{H}{g}}$$

    (ii) For the last half of liquid to empty, put $y = H/2$ in the obtained expression to have

$$t_2 = \frac{A_1}{A_2}\sqrt{\frac{H}{g}}$$

    (iii) For the total liquid to empty, put $y = 0$ in the obtained expression to have

$$T = \frac{A_1}{A_2}\sqrt{\frac{2H}{g}}$$

N.B.:

1. The height of liquid column varies (decreases) parabolically with time.
2. If you want to find 'time of emptying the vessel', put $y = 0$.
3. As the liquid level drops, speed of efflux decreases. Hence, the rate of flow (or rate of decrease in mass, volume, and height) decreases. This means that the liquid level drops at a lower rate with time.

4. If you pour the liquid into the vessel at a rate Q, say, the above equation becomes $Q - A_1\frac{dy}{dt} = A_2\sqrt{2gy}$. When $Q > A_2\sqrt{2gy}$, level of liquid increases; if $Q = A_2\sqrt{2gy}$, level of liquid remains constant; and if $Q < A_2\sqrt{2gy}$, level of liquid decreases.

Example 13 (Impact force of liquid) A cylindrical vessel of mass M and area of the base A_1 is filled with a liquid of density ρ up to a height h. The liquid goes out of the vessel through a hole of area A_2 made at a distance y from the free surface of liquid. Find the (a) thrust of the escaping liquid on the vessel as the function of y and (b) acceleration of the vessel at the given instant if the hole is small.

Solution

(a) The outgoing liquid kicks the vessel back with a force known as impact force as explained in chapter 1 'System of particles'. The impact force of liquid is given by

$$F_{imp} = v_{rel}\frac{dm}{dt},$$

where v_{rel} = velocity of outgoing (escaping) liquid relative to the vessel (= velocity of efflux).

Substituting $v_{rel} = v$ (= velocity of efflux)

$$\text{and } \frac{dm}{dt} = A_2\rho v$$

we have

$$F_{imp} = A_2\rho v^2 \tag{5.33}$$

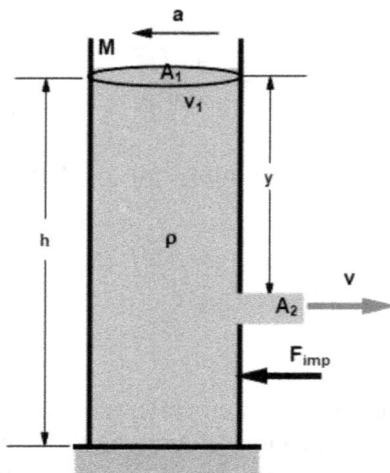

Substituting $v = \dfrac{\sqrt{2gy}}{\sqrt{1 - (\frac{A_2}{A_1})^2}}$ in equation (5.33), we have

$$F_{\text{imp}} = \frac{2A_2\rho gy}{\left(1 - \frac{A_2^2}{A_1^2}\right)} \quad \text{Ans.}$$

(b) If the hole is much smaller than the base of the vessel $\dfrac{A_2^2}{A_1^2} \cong 0$; so, the impact force is

$$F_{\text{imp}} = 2A_2\rho gy$$

Then the acceleration of the vessel is

$$a = F_{\text{imp}}/m = \frac{2A_2\rho gy}{m}, \text{ where the total mass is } m = A_1\rho h + M$$

$$\Rightarrow a = \frac{1A_2\rho gy}{A_1\rho h + M}$$

N.B.:

1. If the vessel is smooth and mass of the vessel is negligible and the hole is made at the lowest position of the curved surface, substituting $M = 0$ and $y_0 = 0$, the acceleration of the vessel can be given as $a = \dfrac{2A_2}{A_1}g$, which is a constant.

2. If you want to find the impact force offered by the liquid at the time of striking the ground, apply the formula $F = v_{\text{rel}}\dfrac{dm}{dt}$, but be careful to write $v_{\text{rel}} = \sqrt{v^2 + 2gy_0}$ (but not v), because this is the velocity of liquid with which it strikes the ground assuming that the liquid does not bounce just after the collision.

Example 14 (Pitot tube) What is a pitot tube? Explain the working principle of a pitot tube.

Solution

A pitot tube is used to measure the speed of a flowing gas. There are many openings (small holes) on the pitot tube. You take any hole and mark it as point 1. The air (or any gas) has velocity v_1 at point 1 as it passes just above the hole.

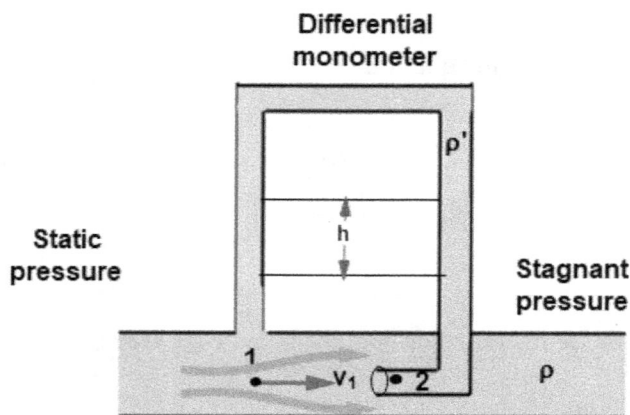

Differential monometer

Static pressure

ρ'

h

Stagnant pressure

1

v_1

2

ρ

You can see that the end of the manometer tube is directly subjected to air pressure. If we take a point 2 in the horizontal limb of the manometer, there we have no motion of air. This means if you choose the point 2, $v_2 = 0$. Since the difference in heights at points 1 and 2 is very small, write Bernoulli's equation at 1 and 2 as

$$P_1 + \frac{1}{2}\rho v_1^2 = P_2 + \frac{1}{2}\rho v_2^2,$$

where $v_1 = v$ and $v_2 = 0$.

This gives

$$P_1 + \frac{1}{2}\rho v^2 = P_2 \qquad (5.34)$$

Substituting $P_2 - P_1 = \rho'gh$ and observing from the manometer in equation (5.34), we obtain

$$v = \sqrt{\frac{2gh\rho'}{\rho}}$$

A pitot tube is used as an air-speed indicator.

Example 15 (Dynamic lift) Describe the 'principle of aerodynamic lift'. Derive the expression for aerodynamic lift on any object of arbitrary shape and size.

Solution

Let's project an air foil. Air flows relative to the surface of the foil forming a streamline flow. The foil divides the streamlines into two parts moving with different speeds v_1 and v_2 at top and bottom (points 1 and 2, respectively) relative to the foil. The streamlines become more crowded above than below the foil. In general, when a body is thrown in a fluid, the difference in speeds of fluid at opposite sides (top and bottom, etc) of the body depends on the shape, size, and spin of the body. Furthermore, the texture of the surface also plays a significant role to make the speed of the fluid different at different points of contact of the body. Here, we emphasize the principle of 'dynamic lift' by applying Bernoulli's equation

$$P + \frac{\rho v^2}{2} + \rho g h = c$$

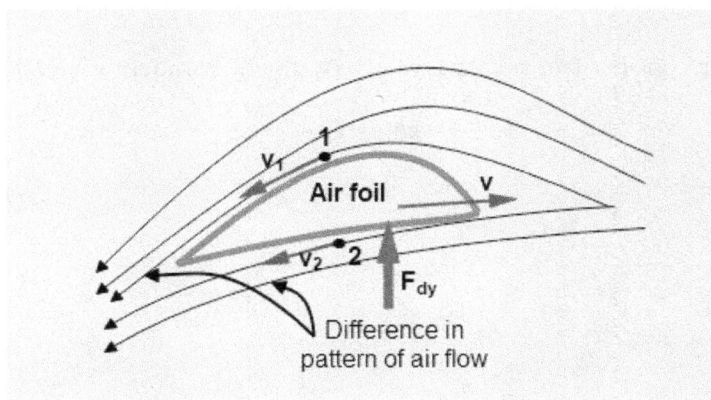

between two points 1 and 2 in a tube of flux at a small vertical distance of separation.

$$P_1 + \frac{\rho v_1^2}{2} + \rho g h_1 = P_2 + \frac{\rho v_2^2}{2} + \rho g h_2$$

Since the distance between points 1 and 2 is very small compared to the height of the center of mass of the body from the ground, h_1 and h_2 are nearly equal. So, we can write

$$P_1 + \frac{\rho v_1^2}{2} = P_2 + \frac{\rho v_2^2}{2}$$

If $v_1 < v_2$, we have $P_2 > P_1$. Then, the pressure difference is

$$P_2 - P_1 = \frac{\rho}{2}(v_1^2 - v_2^2) \tag{5.35}$$

This pushes the body from higher pressure to lower pressure with a force

$$F = (P_2 - P_1)A, \tag{5.36}$$

where A = effective area.

Substituting $P_2 - P_1 = \frac{\rho}{2}(v_1^2 - v_2^2)$ from equations (5.35) in (5.36), we have

$$F = \frac{\rho A}{2}(v_1^2 - v_2^2)$$

If we put $\frac{v_1 + v_2}{2} = u$ (average velocity of fluid) and $v_1 - v_2(=u)$, that is, the difference in velocities at points 1 and 2, we have $F = \rho A u v$. Since this force comes from the motion of the fluid, it is called hydro (or aero) dynamic force denoted as F_{dy}.

Example 16 (Thrust on a rocket) A rocket ejects the fuel (hot gases) of density ρ from its chamber, which is maintained at a pressure P_1, say. The atmospheric pressure is P_0. If the base area of the cylindrical chamber is A_1 and the area of the hole through which hot gases escape from the rocket is A_2, find (a) velocity of efflux of the fuel and (b) thrust exerted on the rocket given by the escaping fuel.

Solution

(a) Let's consider two points 1 and 2. Applying Bernoulli's equation, we have

$$P_1 + \frac{1}{2}\rho v_1^2 + \rho g h_1 = P_2 + \frac{1}{2}\rho v_2^2 + \rho g h_2$$

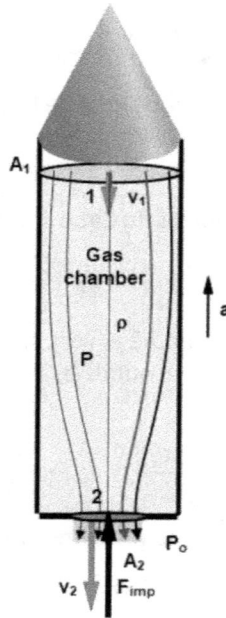

Since $\rho g(h_1 - h_2) = \rho g h$ is negligible compared to $P_1 = P$ (pressure inside the chamber) and $P_2 = P_0$ (atmospheric pressure), we have

$$P + \frac{1}{2}\rho v_1^2 = P_0 + \frac{1}{2}\rho v_2^2 \tag{5.37}$$

Equation of continuity gives

$$A_1 v_1 = A_2 v_2 \tag{5.38}$$

Since $v_1 = \frac{A_2}{A_1}v_2$ and $A_2 \ll A_1$, we can ignore v_1 when compared to v_2.

Then the equation (5.37) gives

$$v_2 = \sqrt{\frac{2(P - P_0)}{\rho}} \text{ Ans.}$$

(b) The impact force of the ejected gas is given as

$$F = v_{\text{rel}} \frac{dm}{dt},$$

where $v_{\text{rel}} = v_2 = \sqrt{\frac{2(P - P_0)}{\rho}}$ and $\frac{dm}{dt} = A_2 \rho v_2 = A_2 \rho \sqrt{\frac{2(P - P_0)}{\rho}}$

Then we have $F = 2A_2(P_1 - P_0)$ Ans.

N.B.: By mistake do not be tempted to write $F = (P_1 - P_2)A$, where A = effective area = A_2 or A_1 or $(A_1 - A_0)$ or $\frac{A_1 + A_0}{2}$. Remember that the thrust on the rocket is not an aerodynamic force, rather it is an impact force given as

$$F_{\text{imp}} = v_{\text{rel}} \frac{dm}{dt} = \rho A_2 v_2^2$$

where as $F_{\text{dynamic}} = (P_1 - P_2)A_{\text{eff}}$,

$$\text{where } P_1 - P_2 \cong \frac{1}{2}\rho v^2$$

Distinguish the forces $F_{\text{imp}} = \rho A_2 v_2^2$, $F_{\text{dynamic}} \cong \rho A_{\text{eff}} v_2^2$, $F_{\text{static}} = \rho g h A_{\text{eff}}$, and $F_{\text{buoyant}} = V \rho g$.

Example 17 (Magnus effect) The direction of motion of a spinning ball moving in air is unpredictable. It is evident in cricket, baseball, football, tennis ball, and other ball games. It is quite spectular to watch the batsman being baffled by a spinning ball. This means that a spinning ball must experience some extra force in addition to other common forces like gravity, viscous force, and buoyant force, which causes the direction of the ball to deviate from its predicted path. This special force is responsible for deflecting the ball from its predicted path. This is an aerodynamic force and the effect is called 'magnus effect'. Explain the origin of this force. If the spin of the ball is given as $\omega = \frac{v}{2R}$, where v = velocity of center of mass of the ball, find the aerodynamic lift. Discuss the effect when there is no spin.

Solution

The central concept is the Bernoulli's equation between two points when the vertical distance between these points does not contribute significant gravitational potential energy density ($\rho g h$) when compared to kinetic energy densities $\frac{\rho v^2}{2}$. In other words, $\rho g h \ll \frac{\rho v^2}{2}$. Then, the Bernoulli's equation can be reduced to $P + \rho v^2 = c$.

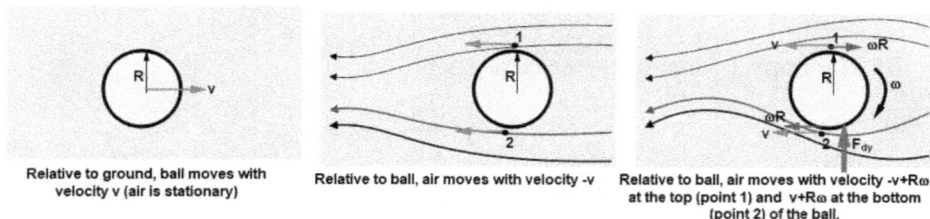

Relative to ground, ball moves with velocity v (air is stationary)

Relative to ball, air moves with velocity -v

Relative to ball, air moves with velocity -v+Rω at the top (point 1) and v+Rω at the bottom (point 2) of the ball.

Let's apply this equation for two points 1 and 2 of a stream of air when viewed from a spinning ball. If the ball moves with a velocity v, air sweeps the top and bottom of the ball with velocities $-v$ (to the left). Due to the spinning with an angular velocity ω, the ball rotates a thin layer of air touching it with the same angular velocity. As the ball pushes the air with an extra velocity $R\omega$, the net velocity of two points 1 (top) and 2 (bottom) of the flux of air sweeping relative to the center of the ball are given as $v_1 = v - R\omega$ and $v_2 = v + R\omega$, respectively. As the speed of air at the top is less than that at the bottom, the pressure at the top must be less than the pressure at the bottom. In other words, $P_1 < P_2$. Hence, the net dynamic force $F_{dy} = F$, say, acting on the ball is directed upward as shown in the figure (extreme right).

The aerodynamic force acting on a spinning ball is caused by the difference in pressure across the ball. The pressure difference is mainly caused by the difference in speeds of streams of air relative to the ball, but not because of the pressure $\rho g h$, which is much less than the dynamic pressure.

Problem 1 A liquid of density ρ flows along a horizontal pipe having an abrupt change in cross-section from A_1 to A_2. As the velocity increases from v_1 to v_2, compute the change in the pressure difference across the junction of the abrupt change in cross-section.

Solution

Let the mass Δm change its velocity from v_1 to v_2 while moving from region 1 to region 2. Then the change in momentum of the liquid is

$$\Delta p = \Delta m(V_2 - V_1)$$

The rate of change in momentum is

$$\frac{\Delta p}{\Delta t} = \frac{\Delta m}{\Delta t}(v_2 - v_1),$$

where $\frac{\Delta m}{\Delta t} = A_1 \rho v_1$.

So, the force impressed on the liquid is

$$F_2 = A_1 \rho v_1(v_2 - v_1)$$

The pressure difference is

$$\frac{F_2}{A_1} = (\Delta P) = \rho v_1(v_2 - v_1) \tag{5.39}$$

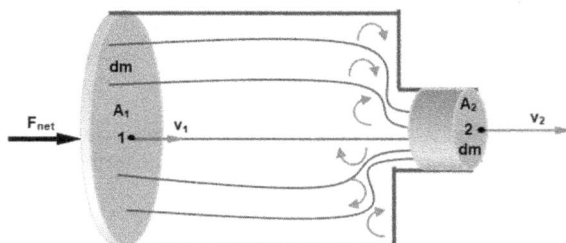

If we disregard the energy loss, the Bernoulli's equation between any two points 1 and 2 yields

$$P_1 + \frac{1}{2}\rho v_1^2 = P_2 + \frac{1}{2}\rho v_2^2 \Rightarrow P_1 - P_2 = \Delta P' = \frac{1}{2}\rho(v_2^2 - v_1^2) \tag{5.40}$$

The pressure difference in the last two cases differs by

$$(\Delta P)_{\text{extra}} = \Delta P - \Delta P'$$

$$= \rho v_1(v_2 - v_1) - \frac{1}{2}\rho(v_2^2 - v_1^2)$$

$$= \frac{1}{2}\rho(v_2 - v_1)\{2v_1 - (v_1 + v_2)\}$$

$$= -\frac{1}{2}\rho(v_2 - v_1)^2$$

This difference of excess pressure is due to the sudden (abrupt) change in cross-section of the tube.

Problem 2 A ball of radius R moves to right with a velocity v. If it spins with a clockwise angular velocity ω, find the aerodynamic force exerted on the ball.

Solution

As the ball is moving to right, the air flows to the left relative to the ball with the speed v. As the ball spins, it imparts an additional velocity ωR to the air flow; at the top it opposes and at the bottom it favours the air flow. Then, the velocity of the air flow relative to the ball at the top point 1 and bottom point 2 is given as

$$v_P = v + \omega R \tag{5.41}$$

$$v_Q = v - \omega R \tag{5.42}$$

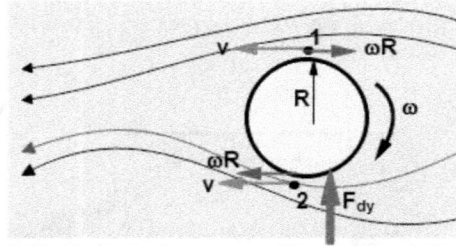

**Relative to ball, air moves with velocity -v+Rω
at the top (point 1) and v+Rω at the bottom
(point 2) of the ball.**

Applying Bernoulli's equation at 1 and 2, we have

$$P_1 + \frac{1}{2}\rho v_P^2 = P_2 + \frac{1}{2}\rho v_Q^2 \tag{5.43}$$

We neglect the diameter (height) $2R$ between 1 and 2 in comparison with the height of the *center of mass* of the ball relative to the ground.

Using the last equation, we have

$$P_2 - P_1 = \frac{1}{2}\rho(v_1^2 - v_2^2) \tag{5.44}$$

Putting values of v_1 and v_2 from equations (5.41), (5.42), respectively, in equation (5.44), we have

$$P_2 - P_1 = \frac{1}{2}\rho\{(v + R\omega)^2 - (v - R\omega)^2\}$$

$$\Delta P = \frac{1}{2}\rho\{4vR\omega\} = 2\rho vR\omega$$

Then the vertical upward aerodynamic force acting on the ball is

$$F = (\Delta P)A_{\text{eff}}$$

$$=(2\rho vR\omega)(\pi R^2)$$

$$=2\pi\rho v\omega R^3 \text{ Ans.}$$

Problem 3 A cylindrical vessel completely filled with a liquid of density ρ has a tap of length l fitted at a depth h from the water surface. The combined mass is m. The area of cross-section of the tap is a, which is η times less than the area of cross-section of the vessel. Find the (a) impact force of outgoing water and (b) height h so that the vessel will topple just after opening the tap.

Solution

(a) The general formula for the velocity of efflux is

$$v = \sqrt{\frac{2gh}{1 - \frac{a^2}{A^2}}} \qquad (5.44)$$

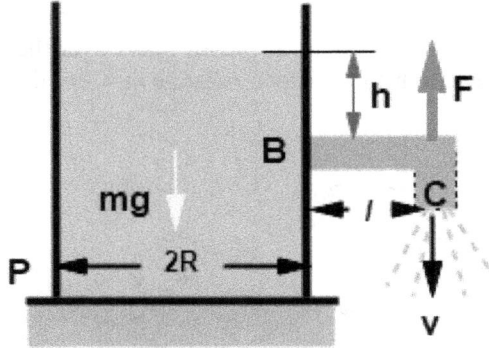

The upward reaction force of water is

$$F = v\frac{dm}{dt}$$

$$\Rightarrow F = v(a\rho v) = a\rho v^2 \qquad (5.45)$$

Using equations (5.44) and (5.45)

$F = a\rho\frac{2gh}{1 - \frac{a^2}{A^2}}$, where $\frac{a}{A} = \eta$

$$\Rightarrow F = \frac{2a\rho gh}{1 - \eta^2} \text{ Ans.}$$

(b) For toppling the vessel about P, the torque produced by F about P is

$$\overrightarrow{\tau_F} = (\ell + 2R)F\hat{k}$$

$$= (\ell + 2R)\frac{2a\rho gh}{1 - \eta^2}\hat{k} \text{ Ans.}$$

(c) For toppling the torque of the reaction force of water must be greater than the gravitational torque about P.

$$\overrightarrow{\tau_F} \geqslant \overrightarrow{\tau_{gr}}$$

$$\Rightarrow (\ell + 2R) \left\{ \frac{2a\rho g h}{1 - \eta^2} \right\} \geq mgR$$

$$\Rightarrow h \geq \frac{mR(1 - \eta^2)}{2(\ell + 2R)a\rho} \quad \text{Ans.}$$

Problem 4 A house has two identical roof sheets each of area A. The angle of inclination between the sheets is θ. The sheets are welded together and just placed on the house without any nut-bolt system. If the air flows over the sheets with a velocity v, find the value of v so as to lift the sheets. Assume $m =$ combined mass of the sheets and $\rho =$ density of air.

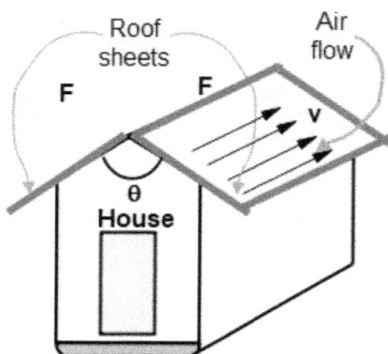

Solution

Let v be the velocity of air just above the roof. Since the velocity of air just below the roof (inside the house) is negligible, the pressure inside the house is greater than that outside the house by

$$\Delta P = \frac{1}{2}\rho(v_{\text{out}}^2 - v_{\text{in}}^2), \quad \text{where } v_{\text{in}} \approx 0$$

$$\Rightarrow \Delta P = \frac{1}{2}\rho v^2 \qquad (5.46)$$

The hydrodynamic force F acting on each sheet of the roof is

$$F = (\Delta P)A \qquad (5.47)$$

By using equations (5.46) and (5.47)

$$F = \frac{1}{2}\rho v^2 A \qquad (5.48)$$

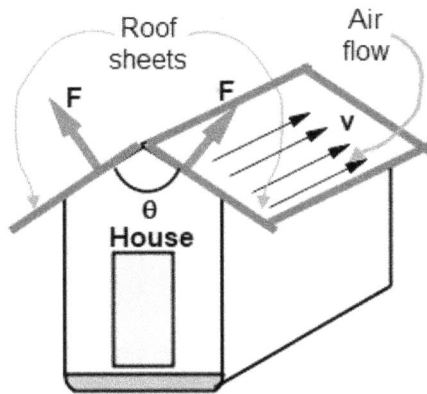

Since there are two sheets on the roof, each experiencing force F, the net upward force acting on the roof is

$$F_y = 2F \cos\left(\frac{180 - \theta}{2}\right)$$

$$\Rightarrow F_y = 2F \sin\frac{\theta}{2} \qquad (5.49)$$

To lift the roof the net hydrodynamic lift must be greater or equal to the weight of the roof. This means that

$$F_y \geqslant mg \qquad (5.50)$$

Putting the value of F_y from equation (5.49) in expression (5.50), we have

$$2F \sin\frac{\theta}{2} \geqslant mg \qquad (5.51)$$

Putting the value of F from equation (5.48) in expression (5.51), we have

$$2\left(\frac{1}{2}\rho v^2 A\right)\sin\frac{\theta}{2} \geqslant mg$$

$$\Rightarrow v \geqslant \sqrt{\frac{mg}{\rho A}} \operatorname{cosec}\frac{\theta}{2} \quad \text{Ans.}$$

Problem 5 A cylindrical vessel filled with water up to a height $H = 4h$. If two small holes 1 and 2 are made on the same vertical line on the vessel as shown in the figure and if the steams of water intersect at a point $P(x, h)$, (a) find the range R_1, (b) prove that the range of a liquid jet can be given as $R = 2\sqrt{(\text{depth})(\text{height})}$, (c) find the value of h', (d) determine the ratio of range of the liquid jets on the ground, and (e) find the ratio of velocity of striking of the water jets at the ground.

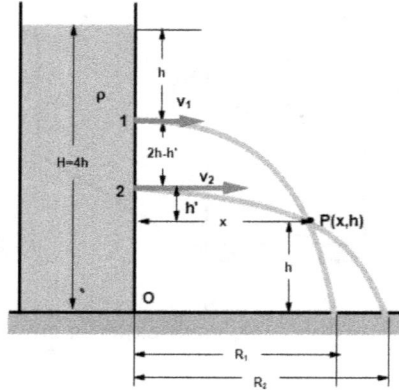

Solution

(a)

(i) The velocities at points 1 and 2 are

$$v_1 = \sqrt{2gh} \tag{5.52}$$

The time of flight of the upper liquid jet is

$$t_1 = \sqrt{\frac{2(H - h)}{g}} \tag{5.53}$$

Then, $R_1 = v_1 t_1$

$$= \sqrt{2gh} \cdot \sqrt{\frac{2(H - h)}{g}} = 2\sqrt{h(H - h)}. \text{ Ans.}$$

(b) Putting h = depth and $H - h$ = height of the hole, the horizontal range of any liquid jet can be given as

$$R = 2\sqrt{(\text{depth})(\text{height})}. \text{ Ans.}$$

(c) Using the last expression, we have

$$R = x = 2\sqrt{(\text{depth})(\text{height})}$$

By equation x for the liquid jets 1 and 2, we have
(Depth)(height) = constant

$$d_1 h_1 = d_2 h_2$$

Putting height and depths of both the holes,

$$(h)(2h) = (3h - h')h'$$

Solving the last equation, we have $h' = h$. Ans.
(d) The range of the first liquid jet is

$$R_1 = 2\sqrt{h(3h)} = 2\sqrt{3}\,h \tag{5.54}$$

The range of the second liquid jet (stream) can be given as
$R_2 = 2\sqrt{(3h - h')(h + h')}$, where $h = h'$. Then we have

$$R_2 = 4h \tag{5.55}$$

Then the ratio of the ranges of the water jets on the ground is

$$\frac{R_1}{R_2} = \frac{2\sqrt{3}\,h}{4h} = \frac{\sqrt{3}}{2} \text{ Ans.}$$

(e) The ratio of velocity of striking of the water jets at the ground is

$$\frac{v'_1}{v'_2} = \frac{\sqrt{v_1^2 + 2g(2h)}}{\sqrt{v_2^2 + 2gh}}$$

$$= \frac{\sqrt{2gh + 4gh}}{\sqrt{4gh + 2gh}} = 1:1 \text{ Ans.}$$

N.B.: As the liquid jets intersect at the point P, for both the jets the horizontal distance x is the same. Hence, the vertical distance of the top hole from the top surface of water must be equal to the vertical distance of the bottom hole from the base of the vessel (water surface). Thus, the holes must be equidistant from top and bottom, respectively. The general formula for the range of a liquid jet is $R = 2\sqrt{(\text{depth})(\text{height})}$.

This means that P will be located on the ground and the ranges will be equal for both jets. In other words, both water jets strike at the same spot on the ground.

Problem 6 A U-shaped siphon of area of cross-section a is immersed in a non-viscous liquid kept in a cylindrical tank of base area A as shown in the figure. Find the (a) velocity of efflux of the liquid and (b) maximum value of the velocity of efflux.

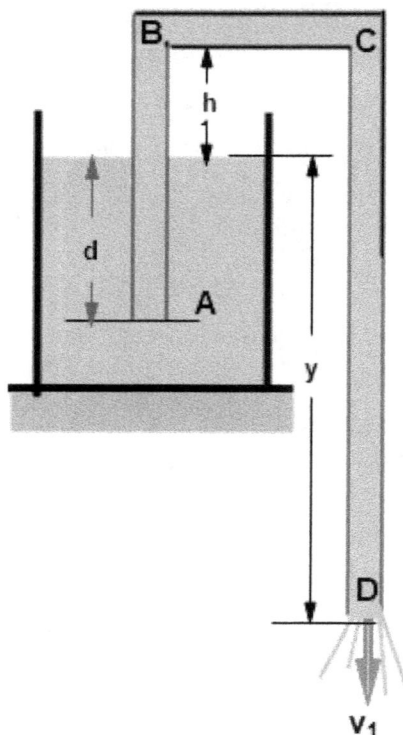

Solution

(a) The equation of continuity between A and D:

$$v'A = va \tag{5.56}$$

Bernoulli's equation between A and D:

$$P_A + \frac{1}{2}\rho v'^2 = P_D + \frac{1}{2}\rho v^2 - \rho g h,$$

where

$$P_A = P_o \tag{5.57}$$

Pascal's law:

$$P_D = P_0 + \rho g y \tag{5.58}$$

Using equations (5.56)–(5.58),

$$P_o + \frac{1}{2}\rho v'^2 = P_o + \rho g y + \frac{1}{2}\rho v^2$$

$$\Rightarrow \rho(v^2 - v'^2) = 2gy \tag{5.59}$$

Using equations (5.56) and (5.59),

$$\Rightarrow v^2 \left(1 - \frac{a^2}{A^2}\right) = 2gy$$

$$\Rightarrow v = \sqrt{\frac{2gy}{1 - \frac{a^2}{A^2}}}$$

If $a \ll A$, $v \cong \sqrt{2gy}$

(b) Applying Bernoulli's equation between B and Q, we have

$$P_0 + \frac{1}{2}\rho v'^2 = P_B + \rho gy + \frac{1}{2}\rho v^2,$$

where $v' = \frac{a}{A}v$

$$\Rightarrow P_0 - P_B - \rho gy = \frac{1}{2}\rho v^2 \left(1 - \frac{a^2}{A^2}\right)$$

$$\Rightarrow \left(1 - \frac{a^2}{A^2}\right)v^2 = \frac{2(P_0 - P_B)}{\rho} - 2gy$$

Since $v \geqslant 0$, $\frac{2(P_0 - P_B)}{\rho} - 2gy \geqslant 0$

$$\Rightarrow y \leqslant \frac{(P_0 - P_B)}{\rho g}$$

So, $y_{max} = \frac{P_0}{\rho g}$ (by putting $P_B = 0$)

If $y < y_{max}$, we have

$$v_{max} = \sqrt{2\left(\frac{P_0}{\rho} - gy\right)}$$

Problem 7 A liquid is filled in a cylindrical vessel. The vessel is made to rotate about its vertical axis with a constant angular velocity ω. The height of the liquid column in the vessel is maintained constant. A hole is made at a depth y from the surface of liquid. In steady state, find the (a) velocity of efflux and (b) velocity of the liquid with which it will hit the ground. Put $\omega = 2\sqrt{\frac{g}{R}}$.

Solution

(a) The pressure gradient at a radial distance r is

$$\frac{dP}{dr} = \rho \omega^2 r$$

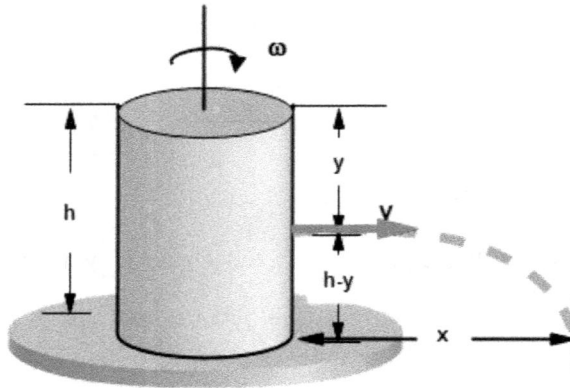

Separating the variables and integrating both sides,

$$\int_{P_1}^{P} dP = \rho\omega^2 \int_0^R r \, dr$$

$$\Rightarrow P - P_1 = \frac{\rho\omega^2 R^2}{2}$$

$$\Rightarrow P = P_1 + \frac{\rho\omega^2 R^2}{2} \tag{5.60}$$

The pressure at 1 is

$$P_1 = \rho g y + P_0 \tag{5.61}$$

By using equations (5.60) and (5.61),

$$P = P_0 + \rho g y + \frac{\rho\omega^2 R^2}{2} \tag{5.62}$$

Using Bernoulli's equation between points A and B,

$$P_A + \frac{1}{2}\rho v_A^2 = P_B + \frac{1}{2}\rho v_B^2$$

Since $v_A \ll v_B$ because the hole is very small compared to the cross-section of the vessel, we have

$$P_A = P_B + \frac{1}{2}\rho v_B^2,$$

where $P_A = P$, $v_B = v$ and $P_B = P_0$

$$\Rightarrow P = P_0 + \frac{1}{2}\rho v^2$$

$$\Rightarrow v = \sqrt{\frac{2(P - P_0)}{\rho}} \tag{5.63}$$

Using equations (5.62) and (5.63),

$$v = \sqrt{\frac{2\left(\rho g y + \frac{\rho \omega^2 R^2}{2}\right)}{\rho}}$$

$$= \sqrt{2\left(g y + \frac{\omega^2 R^2}{2}\right)} \quad \text{Ans.}$$

(b) The liquid will hit the ground with a velocity

$$v' = \sqrt{v^2 + 2g(h - y)}$$

Putting the obtained value of v, we have

$$v' = \sqrt{2gh + \frac{\omega^2 R^2}{2}}$$

If $\omega = 2\sqrt{\frac{g}{R}}$, we have $v' = \sqrt{2gh + \frac{4g}{R} \cdot \frac{R^2}{2}}$

$$= \sqrt{2g(R + h)} \quad \text{Ans.}$$

Problem 8 A cylindrical vessel of moment of inertia I is fitted with two identical horizontal tubes each of mass m and length $l = R$ and area of cross-section a. A liquid of density ρ is filled up to a height H, and this height is maintained constant. A hole is made at a depth $h = H/2$. Find the (a) torque due to the horizontal thrust of the escaping liquid on the vessel about the axis yy' of rotation and (b) angular acceleration of the vessel. Neglect the mass of water flowing through the tubes.

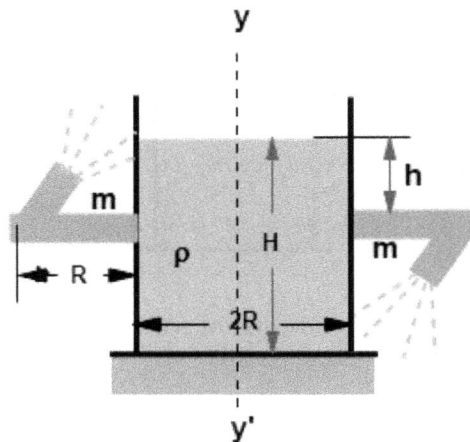

Solution

(a) The velocity of efflux is

$$v_{\text{rel}} = \sqrt{2g\frac{H}{2}}$$

$$\Rightarrow v_{\text{rel}} = \sqrt{gH} \tag{5.64}$$

Then the reaction force of the liquid jet is

$$F = v_{\text{rel}}\frac{dm}{dt} = v_{\text{rel}}a\rho v_{\text{rel}} = a\rho v_{\text{rel}}^2 \tag{5.65}$$

Using equations (5.64) and (5.65),

$$F = a\rho gH \tag{5.66}$$

The torque exerted by each reaction force about the axis of rotation is

$$\tau = F(2R) + F(2R)$$

$$\Rightarrow \tau = 4FR \tag{5.67}$$

Using equations (5.66) and (5.67),

$$\tau = 4a\rho gHR \text{ Ans.}$$

(b) The moment of inertia (M.I.) of the system about the vertical axis of rotation is

$$I = I_{\text{vessel}} + I_{\text{rods}} + I_{\text{liquid}}$$

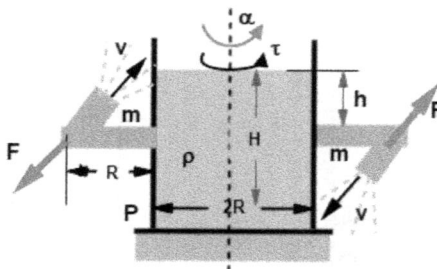

Since the vessel is smooth, the liquid will not rotate. Then, $I = I_{\text{vessel}} + I_{\text{rods}}$. The combined M.I. of the rods is

$$I_{\text{rods}} = 2\int_R^{2R} \left(\frac{m}{R}dx\right)x^2dx$$

$$=2\frac{m}{R}\frac{x^3}{3}\bigg|_R^{2R} = \frac{2mR^2}{3}(8-1) = \frac{14mR^2}{3}$$

Then, $I_{\text{total}} = I_{\text{rods}} + I_{\text{vessel}}$

$$\Rightarrow I_{\text{total}} = \frac{14mR^2}{3} + I$$

Putting I_{total} and τ in the equation

$$\tau = I_{\text{tot}}\alpha, \text{ we have}$$

$$\alpha = \frac{\tau}{I_{\text{tot}}} = \frac{4a\rho gRH}{I + \frac{14mR^2}{3}}$$

$$\Rightarrow \alpha = \frac{12a\rho gH}{3I + 14mR^2} \quad \text{Ans.}$$

Problem 9 A cylindrical vessel is filled with a liquid of density ρ filled up to a height y and this height is maintained constant. A small hole of area a is made at the bottom of the vessel. The escaping liquid collides a smooth block of mass m. If the water falls dead just after the collision, find the (a) velocity of the block as the function of time and (b) terminal velocity of the block. Neglect the mass of water flowing through the tubes.

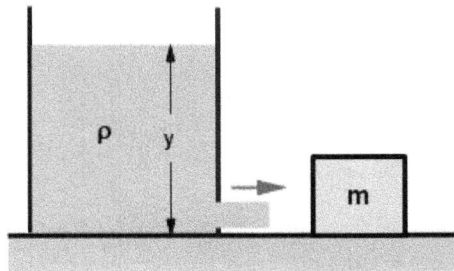

Solution

(a) The velocity of efflux for a fixed height y of the liquid in the vessel is a fixed quantity, which is given as

$$v' = \sqrt{2gy}$$

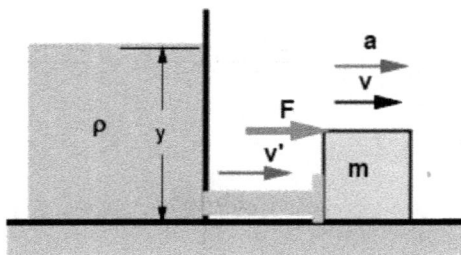

The volume rate of liquid escaping is

$$\frac{dV}{dt} = av' = a\sqrt{2gy}$$

Let the velocity of the block be v. Then the relative velocity between the liquid jet and the block is

$$v_{\text{rel}} = v' - v = a\sqrt{2gy} - v$$

The force imparted on the block is

$$F = \frac{dm}{dt}v_{\text{rel}},$$

where $\frac{dm}{dt} = a\rho v'$ and $v_{\text{rel}} = a\sqrt{2gy} - v$

$$\Rightarrow F = a\rho v'\left(a\sqrt{2gy} - v\right) \text{ where } v' = \sqrt{2gy}$$

$$\Rightarrow F = a\rho\left[(2gy) - \sqrt{2gy} \cdot v\right]$$

$$\Rightarrow m\frac{dv}{dt} = \frac{a\rho\sqrt{2gy}}{m}\left[\sqrt{2gy} - v\right]$$

$$\Rightarrow \int_0^v \frac{dv}{\sqrt{2gy} - v} = \frac{a\rho\sqrt{2gy}}{m}\bigg|_0^t dt$$

$$\Rightarrow -\ln\left(\sqrt{2gy} - v\right)\bigg|_0^v = \frac{a\rho\sqrt{2gy}}{m}t$$

$$\Rightarrow \ln\left(\frac{\sqrt{2gy} - v}{\sqrt{2gy}}\right) = -\frac{a\rho\sqrt{2gy}}{m}t$$

$$\Rightarrow \ln\left(\frac{\sqrt{2gy} - v}{\sqrt{2gy}}\right) = -\frac{a\rho\sqrt{2gy}}{m}t$$

$$\Rightarrow \left(1 - \frac{v}{\sqrt{2gy}}\right) = e^{-\frac{a\rho\sqrt{2gy}\,t}{m}}$$

$$\Rightarrow v = \sqrt{2gy}\left(1 - e^{-\frac{a\rho\sqrt{2gy}}{m}t}\right) \text{ Ans.}$$

(b) As the block accelerates from rest, it speeds up. As a result, the relative velocity between the water jet and the block decreases to zero after a long time where the block experiences a zero force due to the water. Thereafter, the block will move with a constant velocity of magnitude $\sqrt{2gy}$ Ans.

Problem 10 A small hole is made at the bottom of a hemispherical vessel radius R completely filled with water. Find the time after which the total water will escape from the vessel. Assume that $a =$ area of the hole.
  Solution
  The efflux velocity for any height y of the water column in the vessel is

$$v = \sqrt{2gy} \tag{5.68}$$

The rate of decrease in volume is

$$\frac{dv}{dt} = av$$

$$\Rightarrow \frac{(\pi r^2 dy)}{dt} = av, \text{ where } r^2 = R^2 - x^2$$

$$\Rightarrow \frac{\pi(R^2 - x^2)dy}{dt} = av$$

Since $x + y = R$, we have

$$dy = -dx$$

$$\Rightarrow \frac{-\pi(R^2 - x^2)dx}{dt} = av \tag{5.69}$$

Using equations (5.68) and (5.69),

$$-\frac{\pi(R^2 - x^2)dx}{dt} = a\sqrt{2gy},$$

where $y = R - x$

$$\Rightarrow -\frac{\pi(R^2 - x^2)dx}{dt} = a\sqrt{2g(R - x)}$$

Separating the variables,

$$-\pi\frac{(R^2 - x^2)}{\sqrt{R - x}} = a\sqrt{2g}\,dt$$

Integrating both sides,

$$-\pi\int_0^R (\sqrt{R - x})(R + x)\,dx = a\sqrt{2g}\int_0^t dt$$

$$\Rightarrow -\pi\int_0^R (\sqrt{R - x})(R + x)\,dx = a\sqrt{2g}\,t$$

$$\Rightarrow -\pi I = a\sqrt{2g}\,t \tag{5.70}$$

Let $I = \int(R + x)\sqrt{R - x}\,dx$.
  Put $R - x = y$

$$\Rightarrow dx = -dy$$

Then, $I = \int 2R - y)\}\sqrt{y}(-dy)$

$$= -\left[\int_R^0 2R\sqrt{y}\,dy - \int_R^0 y^{\frac{3}{2}}\,dy\right]$$

$$= \left[2R\int_0^R y^{\frac{1}{2}}\,dy - \int_0^R y^{\frac{3}{2}}\,dy\right]$$

$$= 2R\frac{R^{\frac{3}{2}}}{\frac{3}{2}} - \frac{R^{\frac{5}{2}}}{\frac{5}{2}} = R^{\frac{5}{2}}\left(\frac{4}{3} - \frac{2}{5}\right) = \frac{14}{15}R^{\frac{5}{2}}$$

Putting the value of $I = \frac{14}{15}R^{\frac{5}{2}}$ in equation (5.70),

$$\pi\frac{14}{15}R^{\frac{5}{2}} = a\sqrt{2g}\,t$$

$$\Rightarrow t = \frac{14\pi R^2}{15a}\sqrt{\frac{R}{2g}} \quad \text{Ans.}$$

Problem 11 A light cylindrical vessel of mass m and area of cross-section of the base A_2 is filled with a non-viscous liquid of density ρ up to a height H and is placed on a smooth horizontal surface. A small hole of area of A_1 is made at the bottom of the vessel. (a) Find the acceleration of the vessel as the function of the height y of the water column in the vessel. (b) What is the total distance covered by the vessel until the entire liquid will escape from the vessel? Disregard the inclination of the free surface of the liquid due to the accelerating vessel.

Solution

(a) The efflux velocity at any height y of the liquid column in the vessel is

$$v_{rel} = \sqrt{2gy} \qquad (5.71)$$

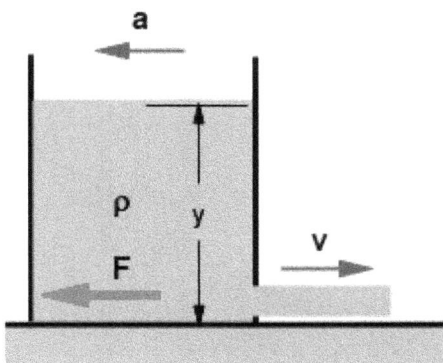

The thrust exerted by the escaping liquid on the vessel is

$$F = v_{rel} \frac{dm}{dt}, \text{ where } \frac{dm}{dt} = A_1 \rho v_{rel}$$

$$\Rightarrow F = v_{rel}(A_1 \rho v_{rel})$$

$$\Rightarrow F = A_1 \rho v_{rel}^2 \qquad (5.72)$$

By using equations (5.71) and (5.72),

$$F = A_1 \rho \left(\sqrt{2gy}\right)^2 = 2A_1 \rho g y \qquad (5.73)$$

The acceleration of the vessel is
$a = \frac{F}{M}$, whose $M = (A_1 \rho y + m)$

$$\Rightarrow a = \frac{F}{m + A_2 \rho y} \qquad (5.74)$$

Using equations (5.73) and (5.74),

$$a = \frac{2A_1 \rho g y}{m + A_2 \rho y} \text{ Ans.}$$

(b) If the mass of the vessel is negligible

$$a \simeq \frac{2A_1 \rho g y}{A_2 \rho y} = 2\frac{A_1}{A_2}g \qquad (5.75)$$

Then, the distance travelled by the vessel is

$$x = \frac{1}{2}at^2 \tag{5.76}$$

By using equations (5.75) and (5.76),

$$x = \frac{A_1}{A_2}gt^2 \tag{5.77}$$

The time of emptying can be obtained (refer to Problem 12) as

$$t = \frac{A_2}{A_1}\sqrt{\frac{2H}{g}} \tag{5.78}$$

Using equations (5.77) and (5.78), we have

$$x = \frac{A_1}{A_2}g\left(\frac{A_2}{A_1}\sqrt{\frac{2H}{g}}\right)^2$$

$$= 2\frac{A_1}{A_2}H \text{ Ans.}$$

Problem 12 A small hole is made at the bottom of a cubical tank of base area A. It is filled with liquid up to a height H. A small hole of area a is made at the bottom of the vessel. Find the (a) height of the liquid column in the vessel as the function of time, (b) time after which η fraction of all the liquid will stay in the vessel, and (c) time after which all the liquid will come out of the vessel.
   Solution

(a) The velocity of efflux at any depth y is

$$v = \sqrt{2gy} \tag{5.79}$$

The volume rate at which the liquid goes out of the vessel is

$$R = -\frac{dV}{dt} = av = a(\sqrt{2gy}), \text{ where } V = Ay$$

$$\Rightarrow -A\frac{dy}{dt} = a\sqrt{2gy} \tag{5.80}$$

Separating the variable and integrating both sides,

$$-A\int_H^y \frac{dy}{\sqrt{y}} = a\sqrt{2g}\int_0^t dt$$

$$\Rightarrow -2A\sqrt{y}\Big|_H^y = a\sqrt{2g}\,t$$

$$\Rightarrow 2A\left(\sqrt{H} - \sqrt{y}\right) = a\sqrt{2g}\,t$$

$$\Rightarrow \sqrt{H} - \sqrt{y} = \frac{a}{A}\sqrt{\frac{g}{2}}\,t$$

Then we have $= \left(\sqrt{H} - \frac{a}{A}\sqrt{\frac{g}{2}}\,t\right)^2$

(b) For emptying $y = \eta H$, we have the time

$$t = \frac{A}{a}\sqrt{\frac{2}{g}}\left\{\sqrt{H} - \sqrt{\eta H}\right\}$$

$$t = \frac{A}{a}\sqrt{\frac{2H}{g}}\left(1 - \sqrt{\eta}\right) \text{ Ans.}$$

(c) At $t = 0$, $y = H$ and at $t = T$, $y = 0$

$$\text{Then, } T = \frac{A}{a}\sqrt{\frac{2H}{g}}\,(1 - 0)$$

$$= \frac{A}{a}\sqrt{\frac{2H}{g}} \text{ Ans.}$$

Problem 13 A horizontal tube of length l and cross-sectional area A is completely filled with water. It is made to rotate with an angular velocity ω in a horizontal plane about a vertical axis passing through one of its ends. A hole of area a is made at the other end. Find the (a) velocity of efflux v of the liquid as the function of the length x of the liquid column in the tube and (b) ratio of the time after all the liquid will escape from the tube and the time period of rotation of the tube.

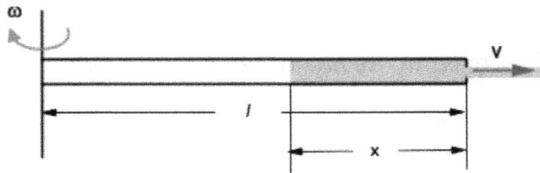

Solution

(a) Referring to Example 16 of chapter 4, the expression of the pressure at the end of the tube is

$$\int_{P_0}^{P} dP = \rho\omega^2 \int_{\ell-x}^{\ell} r\, dr$$

$$\Rightarrow P - P_0 = \rho\omega^2 \left\{ \frac{\ell^2 - (\ell - x)^2}{2} \right\}$$

$$\Rightarrow P = P_0 + \frac{\rho\omega^2}{2}(2\ell - x)x \tag{5.81}$$

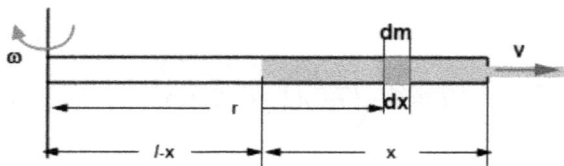

The Bernoulli's equation at 1 (just inside) and 2 (just outside) the tube at its free end is

$$P + \frac{1}{2}\rho v_1^2 = P_0 + \frac{1}{2}\rho v^2$$

Since $v_1 \ll v$ because $A \gg a$, we have

$$P = P_0 + \frac{1}{2}\rho v^2 \tag{5.82}$$

Using equations (5.81) and (5.82),

$$\frac{\rho\omega^2}{2}(2\ell - x)x = \frac{1}{2}\rho v^2 \Rightarrow \omega = \sqrt{\frac{v^2}{(2\ell - x)x}} = \frac{v}{\sqrt{(2\ell - x)x}} \tag{5.83}$$

If $\omega =$ constant, $v = \omega\sqrt{(2\ell - x)x}$.

(b) The velocity of liquid in the tube is

$$v' = v\frac{a}{A} = \frac{a}{A}\omega\sqrt{(2\ell - x)x}$$

The rate of decrease in volume of liquid in the tube is

$$-A\frac{dx}{dt} = Av' = A\frac{a}{A}\omega\sqrt{(2\ell - x)x}$$

$$\Rightarrow -A\frac{dx}{\sqrt{(2\ell - x)x}} = a\omega dt$$

Integrating both sides, we have

$$-A \int_\ell^0 \frac{dx}{(2\ell - x)x} = a\omega \int_0^t dt \qquad (5.84)$$

Put $x = k^2$, then $dx = 2kdk$. Then,

$$I = \int \frac{dx}{\sqrt{(2\ell - x)x}} = \int \frac{2kdk}{\sqrt{(2\ell - k^2)k}}$$

$$= 2 \int \frac{dk}{\sqrt{2\ell - k^2}} = 2\sin^{-1}\frac{k}{\sqrt{2\ell}}, \text{ where } k = \sqrt{x}$$

$$\Rightarrow I = 2\sin^{-1}\frac{\sqrt{x}}{\sqrt{2\ell}}$$

$$\Rightarrow I = 2\sin^{-1}\sqrt{\frac{x}{2\ell}} \qquad (5.85)$$

Using equations (5.84) and (5.85),

$$-A\left\{2\sin^{-1}\sqrt{\frac{x}{2\ell}}\right\}\bigg|_\ell^0 = a\omega t \Rightarrow -2A\left\{0 - \sin^{-1}\sqrt{\frac{1}{2\ell}}\right\} = a\omega t$$

$$\Rightarrow t = \frac{(2A)\left(\frac{\pi}{4}\right)}{a\omega}$$

$$\Rightarrow t = \frac{a\pi}{2a\omega}, \text{ where } \frac{2\pi}{\omega} = T$$

$$\Rightarrow t = \frac{A\pi}{2a\left(\frac{2\pi}{T}\right)} = \frac{AT}{4a}$$

$$\Rightarrow \frac{t}{T} = \frac{A}{4a} \text{ Ans.}$$

Problem 14 In the last problem, let the tube be given an initial angular velocity ω_o so that the tube rotates about the vertical axis freely without any friction. If the mass of the tube is M and the initial mass of the liquid is m, find the velocity of efflux when half of the liquid will escape.

Solution

The initial moment of inertia (M.I.) of the system is

$I_0 = $ M.I. of the tube $+$ M.I. of water

$$\Rightarrow I_0 = \frac{M\ell^2}{3} + I_{\text{water}}, \text{ where } I_{\text{water}} = \frac{m\ell^2}{3}$$

$$\Rightarrow I_0 = (M + m)\frac{\ell^2}{3} \tag{5.86}$$

When half of the liquid escapes, the M.I. of water is

$$I'_\omega = \int dm\, x^2$$

$$= \int_{x_1 = \frac{\ell}{2}}^{x_2 = \ell} \left(\frac{m}{\ell}dx\right) x^2$$

$$= \frac{m}{\ell}\left(\frac{x^3}{3}\right)\Bigg|_{\frac{\ell}{2}}^{\ell}$$

$$= \frac{m}{3\ell}\left[\ell^3 - \left(\frac{\ell}{2}\right)^3\right]$$

$$\Rightarrow I'_\omega = \frac{7m\ell^2}{24}$$

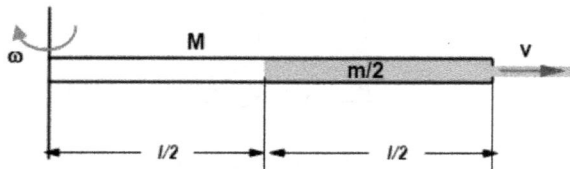

The total M.I. of the system (tube + water) is

$$I = \frac{M\ell^2}{3} + \frac{7m\ell^2}{24} = (8M + 7m)\frac{\ell^2}{24} \tag{5.87}$$

Let ω = angular velocity when half of liquid (water) has escaped.
   Conserving angular momentum,

$$I_0\omega_0 = I\omega$$

$$\Rightarrow \omega = \frac{I_0\omega_0}{I} \tag{5.88}$$

Using equations (5.86)–(5.88),

$$\omega = \frac{(M + m)\frac{\ell^2}{3}\omega_0}{\frac{8M + 7m}{24}\ell^2}$$

$$\Rightarrow \omega = \frac{8(M + m)\omega_0}{8M + 7m} \tag{5.89}$$

Referring to equation (5.83) the previous problem, we have the velocity of efflux is

$$v = \omega\sqrt{(2\ell - x)x} \tag{5.90}$$

Using equations (5.89) and (5.90),

$$v = \frac{8(M + m)\omega_0}{8M + 7m}\sqrt{(2\ell - x)x}, \text{ where } x = \frac{\ell}{2}$$

Then, we have

$$v = \frac{8(M + m)\omega_0}{8M + 7m}\sqrt{\left(2\ell - \frac{\ell}{2}\right)\frac{\ell}{2}}$$

$$\Rightarrow v = \frac{4\sqrt{3}(M + m)\omega_0 l}{8M + 7m} \text{ Ans.}$$

Problem 15 A cylindrical vessel has two immiscible liquids of densities and heights ρ_1 and ρ_2 and h_1 and h_2, respectively. A hole is at a depth y from the interface of the liquids. Find the (a) range R as the function of y, (b) condition for maximum range, and (c) maximum range.

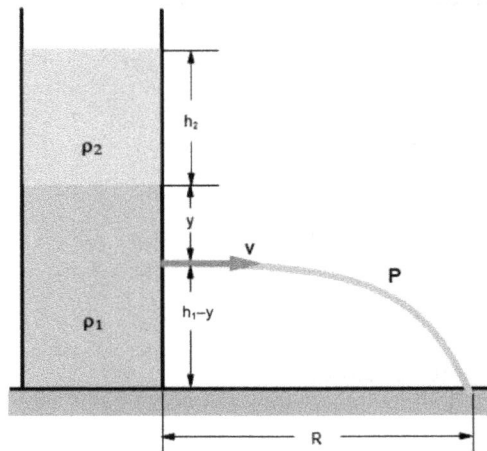

Solution

(a) The velocity of efflux is given by applying Bernoulli's equation between points 1 and 2, where

$$P_1 = P_{adm} = P_0, \quad v_1 = v_{efflux} = v, \quad P_2 = P_0 + \rho_2 g h_2 + \rho_1 g y_1$$

and $v_2 \cong 0$ (because $A \gg a$)

Applying Bernoulli's equation,

$$P_1 + \frac{1}{2}\rho v_1^2 = P_2 + \frac{1}{2}\rho v_2^2 \Rightarrow P_0 + \frac{1}{2}\rho_1(v^2)^2 = P_0 + \rho_2 g h_2 + \rho_1 g y + \frac{1}{2}\rho_2(0)^2$$

$$\Rightarrow v = \sqrt{\frac{2g(\rho_2 h_2 + \rho_1 y)}{\rho_1}} \tag{5.91}$$

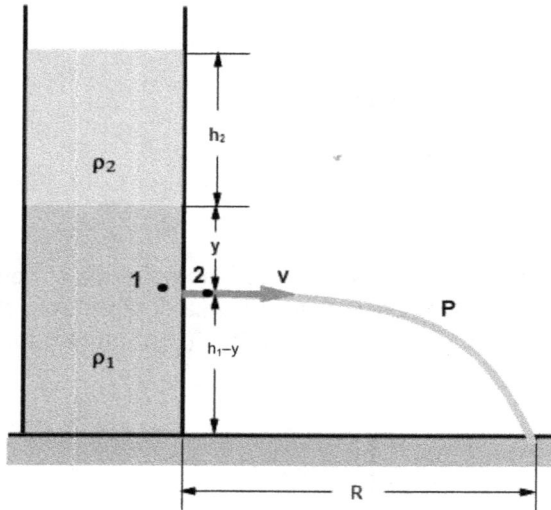

The time of flight of the liquid 1 is

$$t = \sqrt{\frac{2(h_1 - y)}{g}} \tag{5.92}$$

Then the range R is

$$R = vt \tag{5.93}$$

Using the last three equations,

$$R = vt = \left\{ \sqrt{\frac{2g(\rho_2 h_2 + \rho_1 y)}{\rho_1}} \right\} \left\{ \sqrt{\frac{2(h_1 - y)}{g}} \right\}$$

$$= 2\sqrt{\frac{(\rho_2 h_2 + \rho_1 y)(h_1 - y)}{\rho_1}}$$

$$R = 2\sqrt{\left(\frac{\rho_2}{\rho_1} h_2 + y\right)(h_1 - y)} \quad \text{Ans.}$$

(b) For the maximum range $\frac{dR^2}{dy} = 0$

$$\Rightarrow \frac{d}{dy}\left\{\left(\frac{\rho_2}{\rho_1} h_2 + y\right)(h_1 - y)\right\} = 0$$

$$\Rightarrow \left(\frac{\rho_2}{\rho_1} h_2 + y\right)(-1) + (h_1 - y) = 0$$

$$\Rightarrow 2y = h_1 + \frac{\rho_2}{\rho_1} h_2$$

$$y = \frac{1}{2}\left[\frac{\rho_1 h_1 + \rho_2 h_2}{\rho_1}\right] \quad \text{Ans.}$$

(c) Then putting the value of y, we have

$$R_{\max} = 2\sqrt{\left(\frac{\rho_2}{\rho_1} h_2 + \frac{\rho_1 h_1 + \rho_2 h_2}{2\rho_1}\right)\left(h_1 - \frac{\rho_1 h_1 + \rho_2 h_2}{2\rho_1}\right)}$$

$$= 2\sqrt{\frac{(3\rho_2 h_2 + \rho_1 h_1)(\rho_1 h_1 - \rho_2 h_2)}{4\rho_1^2}}$$

$$= \sqrt{\left(3\frac{\rho_2}{\rho_1} h_2 + h_1\right)\left(h_1 - \frac{\rho_2}{\rho_1} h_2\right)} \quad \text{Ans.}$$

Problem 16 A U-tube of area of cross-sections A_1 and A_2 has a liquid of density ρ filled up to heights y_1 and y_2, respectively, and maintained by a stop-cock fitted at the horizontal limb of the U-tube. If we open the stop-cock, find the work done by gravity in equalizing the levels of liquid in both the limbs.

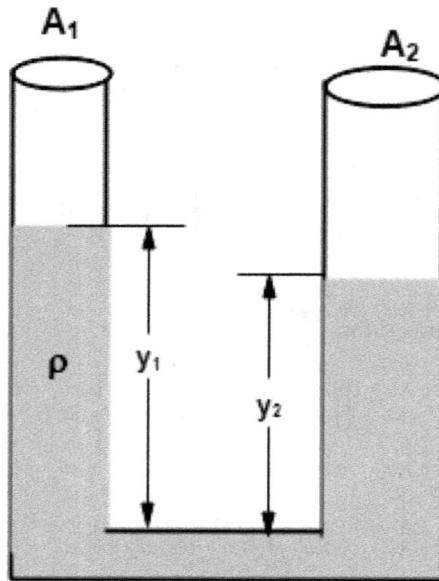

Solution

The initial gravitational potential energy of the liquid in the vertical limb is

$$U_i = m_1 g h_{c1} + m_2 g h_{c2}$$

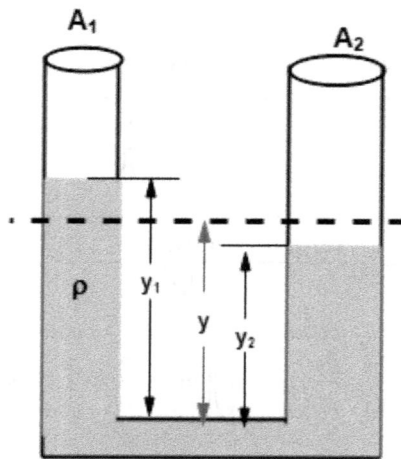

$$= (A_1 \rho y_1)(g)\left(\frac{y_1}{2}\right) + (A_2 \rho y_2)(g)\left(\frac{y_2}{2}\right)$$

$$= \frac{\rho g}{2}\left(A_1 y_1^2 + A_2 y_2^2\right) \tag{5.94}$$

The final gravitational potential energy (after equalization of liquid in both limbs) is

$$U_f = \frac{\rho g}{2}(A_1 y^2 + A_2 y^2) = \left(\frac{A_1 + A_2}{2}\right)\rho g y^2 \tag{5.95}$$

By volume conservation, we have

$$y = \frac{A_1 y_1 + A_2 y_2}{A_1 + A_2} \tag{5.96}$$

The change in gravitational potential energy is

$$\Delta U = U_f - U_i \tag{5.97}$$

Using all four equations,

$$\Delta U = (A_1 + A_2)\rho g \left(\frac{A_1 y_1 + y_2 A_2}{A_1 + A_2}\right)^2 - \frac{\rho g}{2}(A_1 y_1 + A_2 y_2)^2$$

$$\Rightarrow \Delta U = \rho g \left[\frac{(A_1 y_1 + A_2 y_2)^2}{2(A_1 + A_2)} - \frac{(A_1 y_1^2 + A_2 y_2^2)}{2}\right]$$

$$\Rightarrow \Delta U = \rho g \left[\frac{(A_1^2 y_1^2 + A_2^2 y_2^2 + 2A_1 A_2 y_1 y_2) - A_1^2 y_1^2 - A_1 A_2 y_2^2 = A_1 A_2 y_1^2 - A_2 y^2}{2(A_1 + A_2)}\right]$$

$$= -\frac{\rho g}{2(A_1 + A_2)}\left(A_1 A_2 y_1^2 + A_1 A_2 y_2^2 + A_2 y_1 y_2\right)$$

$$= -\frac{\rho g A_1 A_2}{2(A_1 + A_2)}\left(y_1^2 + y^2 - 2y_1 y_2\right)$$

$$= -\frac{A_1 A_2 \rho g}{2(A_1 + A_2)}(y_2 - y_1)^2$$

So, work done by gravity will be

$$\Rightarrow W_{gr} = -\Delta U = \frac{A_1 A_2 \rho g}{2(A_1 + A_2)}(y_2 - y_1)^2 \text{ Ans.}$$

Problem 17 A vertical U-tube of uniform cross-section has a liquid filled up to a height $h = 2a$. The tube is made to rotate by slowly increasing the angular velocity so that the level of liquid in the revolving limb will slowly rise up to its top. In this process, find the work done by the external agent. Assume that the horizontal limb has length $= 2a$. Assume that $M = $ mass of water.

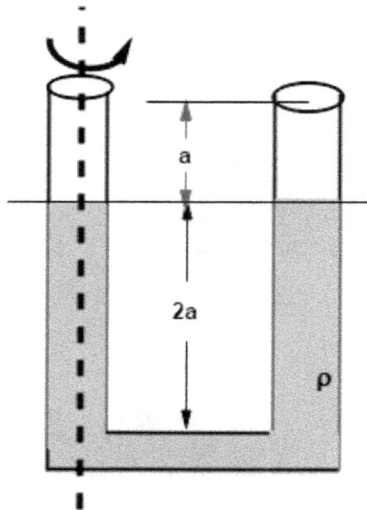

Solution

By slowly increasing the angular velocity, the level of liquid in the right limb will slowly rise and that in the left limb will slowly fall. If the angular velocity is kept constant when the level of liquid in the right limb will attain the top, the liquid level rises by the length a in the right limb and falls by the length a in the left limb. The change in gravitational potential energy of the liquid is

$$\Delta U = \Delta U_{\text{left}} + \Delta U_{\text{right}}$$

$$= \rho A g \frac{(h_2 - h_1)^2}{4},$$

where $h_1 = a$ and $h_2 = 3a$.

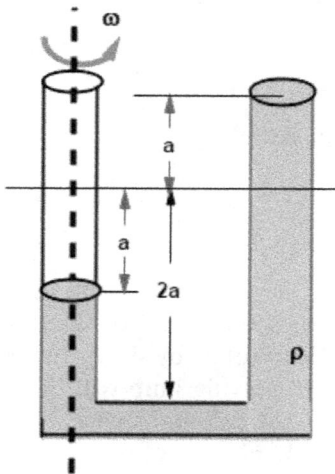

5-67

$$\Rightarrow \Delta U = \frac{1}{4}\rho A g (3a - a)^2 = \rho A g a^2 \tag{5.98}$$

The change in KE is

$$\Delta K = \frac{1}{2}I\omega^2 \tag{5.99}$$

The moment of inertia

$$I = I_{\text{left}} + I_{\text{right}} + I_{\text{horizontal}}$$

$$= 0 + m_r(2a)^2 + m_h\frac{(2a)^2}{3}$$

$$= \{A\rho(3a)\}(2a)^2 + A\rho(2a)\frac{(2a)^2}{3}$$

$$= A\rho a^3\left(12 + \frac{8}{3}\right) = \frac{44}{3}A\rho a^3$$

$$\Rightarrow I = \frac{44}{3}A\rho a^3 \tag{5.100}$$

Let's now calculate ω. We know that the pressure gradient along the radial direction is

$$\frac{dP}{dr} = \rho r \omega^2$$

$$\Rightarrow dP = \rho r \omega^2 dr$$

$$\Rightarrow \int_{P_1}^{P_2} dP = \rho \omega^2 \int_0^{2a} r\, dr$$

$$\Rightarrow P_2 - P_1 = \rho\frac{\omega^2}{2}\{(2a)^2 - (0)^2\}$$

$$\Rightarrow P_2 - P_1 = 2\rho\omega^2 a^2 \tag{5.101}$$

Since according to Pascal's law,
$P_1 = P_0 + \rho g a$ and $P_2 = P_0 + \rho g(3a)$, we have

$$P_2 - P_1 = 2\rho g a \tag{5.102}$$

Using equations (5.101) and (5.102), we have

$$2\rho\omega^2 a^2 = 2\rho g a$$

$$\Rightarrow \omega = \sqrt{\frac{g}{a}}$$

Putting $\omega = \sqrt{\frac{g}{a}}$ in equation (5.99),

$$\Delta K = \frac{1}{2}I\left(\frac{g}{a}\right)$$ (5.103)

Using equations (5.100) and (5.103),

$$\Delta K = \frac{1}{2} \times \frac{44}{3}A\rho a^3\frac{g}{a}$$

$$\Rightarrow \Delta K = \frac{22}{3}A\rho ga^2$$ (5.104)

The work done by the external agent is

$$W_{\text{ext}} = \Delta U + \Delta K$$ (5.105)

Using equations (5.98), (5.104), and (5.105), we have

$$W_{\text{ext}} = A\rho ga^2 + \frac{22}{3}a\rho ga^2 = \frac{25}{3}A\rho ga^2,$$

where $M = A\rho(2a + 2a + 2a) = 6A\rho a$

$$\Rightarrow W_{\text{ext}} = \frac{25}{3}\left(\frac{M}{6}\right)ga = \frac{25}{18}Mga \text{ Ans.}$$

Problem 18 A cylindrical vessel of height h is completely filled with water. It is kept on another cylindrical wooden platform of height h_o. Water escapes through a small hole made at a height y from the bottom of the vessel. Find the (a) range R as the function of y, (b) condition for maximum range, and (c) maximum range.
   Solution

   (a) The velocity of efflux is

$$v = \sqrt{2g(H - y)}$$

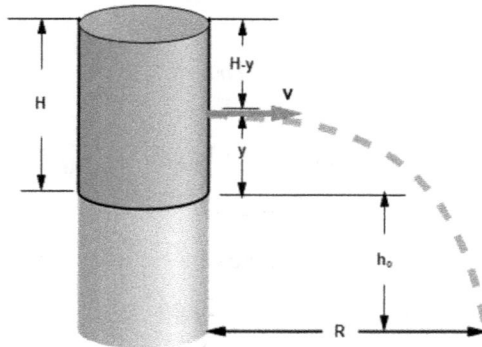

The time taken by the liquid to hit the ground is

$$t = \sqrt{\frac{2(y + h_0)}{g}}$$

Then the range of the liquid is

$$R = vt$$

$$= \sqrt{2g(H - y)} \sqrt{\frac{2(y + h_0)}{g}}$$

$$R = 2\sqrt{(H - y)(y + h_0)} \text{ Ans.}$$

(b) For R_{max}, $\frac{dR^2}{dy} = 0$

$$\Rightarrow (H - y) + (y + h_0)(-1) = 0$$

$$\Rightarrow H - y = y + h_0$$

$$\Rightarrow y = \frac{H - h_0}{2} \text{ Ans.}$$

(c) Then

$$R_{\text{max}} = 2\sqrt{H - \left(\frac{H - h_0}{2}\right)\left(\frac{H - h_0}{2} + h_0\right)}$$

$$= (H + h_0) \text{ Ans.}$$

Problem 19 A water reservoir is fixed with a ceiling so that it maintains a fixed height of water in it. It pours water into the fixed cylindrical vessel of height H at a rate $Q = R$. Find the (a) height of the water column in the vessel as the function of time and (b) height of the water column in the vessel.
   Solution

(a) The velocity of efflux for small hole is

$$v \simeq \sqrt{2gy}$$

The rate of escaping liquid is

$$R_1 = -av$$

$$\Rightarrow R_1 = -a\sqrt{2gy} \tag{5.106}$$

The rate of the liquid entering the vessel is

$$R_2 = +R \tag{5.107}$$

The rate of increase in volume of liquid in the vessel is

$$R_{net} = R = R_2 + R_1$$

$$\frac{d}{dt}(A. y) = R - a\sqrt{2gy}$$

$$\Rightarrow A\frac{dy}{dt} = R - a\sqrt{2gy}$$

$$\Rightarrow \frac{Ady}{R - a\sqrt{2gy}} = dt$$

Integrating both sides,

$$A \int_0^y \frac{dy}{R - a\sqrt{2gy}} = \int_0^t dt \qquad (5.108)$$

Let $R - a\sqrt{2gy} = x$

$$\Rightarrow (R - x)^2 = \left(a\sqrt{2gy}\right)^2$$

$$\Rightarrow (R - x)^2 = a^2(2gy)$$

$$\Rightarrow 2(R - x)(-dx) = 2ga^2 dy$$

$$\Rightarrow dy = -\frac{(R - x)dx}{ga^2}$$

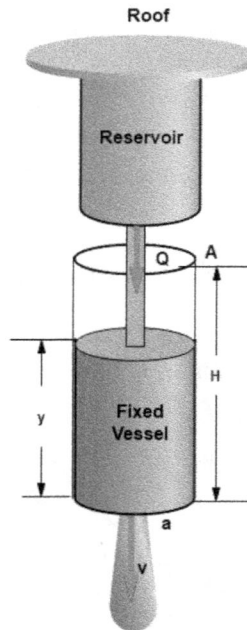

Roof

Reservoir

Q   A

H

y   Fixed
Vessel

a

V

Putting the values of dy and $R - a\sqrt{2gy} = x$, we have

$$I = \int \frac{dy}{R - a\sqrt{2gy}} = \int \frac{\frac{-(R-x)dx}{ga^2}}{x}$$

$$= -\frac{1}{ga^2}\left[R \int \frac{dx}{x} + \int dx\right]$$

$$= -\frac{1}{ga^2}[R \ln x + x]$$

$$= -\frac{1}{ga^2}\left[R \ln\left(R - a\sqrt{2gy}\right) + R - a\sqrt{2gy}\right]$$

Then, $I = \int_0^y \frac{dy}{R - a\sqrt{2gy}}$

$$= -\frac{1}{ga^2}\left[R \ln\left(R - a\sqrt{2gy}\right) + R - a\sqrt{2gy}\right]_0^y$$

$$= -\frac{1}{ga^2}\left[R \ln \frac{R - a\sqrt{2gy}}{R} + R - a\sqrt{2gy} - R\right]$$

$$\Rightarrow I = \frac{t}{A} = -\frac{1}{ga^2}\left[R \ln\left(1 - \frac{a\sqrt{2gy}}{R}\right) - a\sqrt{2gy}\right] \tag{5.109}$$

Using equations (5.108) and (5.109), we have $-\frac{A}{ga^2}[R \ln(1 - \frac{a\sqrt{2gy}}{R}) - a\sqrt{2gy}] = t \Rightarrow R \ln(1 - \frac{a\sqrt{2gy}}{R}) - a\sqrt{2gy} = -\frac{ga^2t}{A}$.

(b) If

$$1 - \frac{a\sqrt{2gy}}{R} \to 0, \, t \to \infty$$

$$\Rightarrow y_{max} = \frac{R^2}{2ga^2} \text{ Ans.}$$

Chapter 6

Properties of matter

6.1 Introduction

We enjoy our lives due to this fascinating property of matter called *elasticity*. Had our body been totally rigid, bending, walking, playing, and any other activities would not have been possible. We cannot produce any sound wave without elasticity of air. Springs act owing to the elasticity. Jumping is possible due to elasticity of matter. At microscopic scale, elasticity is the root cause of reaction force in collision while the bodies experience physical contact.

On the other hand, there is another property of matter called *surface tension*, which forms the basis of soaking of water by trees via capillary action. Cotton dresses are suitable in summer season because the fine pores in cotton act as sweat capillaries. Capillary action is responsible for rising of ink in a pen or oil in a wick in a lamp. Water spreads throughout the soil via capillary rise. Detergents are used to lower the surface tension in water to remove the dirt from clothes.

Viscosity is the third property of matter useful for our practical applications such as testing of composition ratio in food products, liquid doses in the form of oral, injections, etc, in pharmaceutical applications. Viscosity has a major role in coating and spraying, oil blending, inking, painting, gluing, etc. Thus, these three properties of matter have an all-encompassing effect in our daily lives.

6.2 Elastic property

Definition of elasticity: While studying mechanics of rigid body, we assumed that the solids are rigid even though no solid is perfectly rigid. When we press or squeeze or twist a plate of iron hands, we cannot see a noticeable deformation, that is, change in its shape and size. But when we do the same thing by a machine, the deformation of the iron plate will be prominent. So, a large force is required for an iron plate to deform considerably. On the other hand, some solids like rubber, plastic, etc, need small force to deform. For example, iron is far more rigid than rubber. Rubber deforms more under less force and iron deforms much less with moderate force. You may be tempted to say rubber is more elastic than iron, but that is an incorrect statement! Rather, iron is more elastic than rubber. This will be clearer in later sections. Here we need to understand that matter (solid, liquid, and gas) can undergo deformation when an external force acts on it. When we remove the external force, matter regains its original shape and size.

Then, we can state that a solid object deforms when an external force acts on it. If the deformation is too small (within the elastic limit) compared to its size (or the applied force is small), the solid object regains its original shape and size. This property of solids is called elasticity. Liquids and gases also exhibit elasticity.

The above discussion tells us that, elasticity is a property of matter by virtue of which it regains its original shape and size after the removal of the external force acting on it.

Micro-interpretation of elasticity: We have seen in mechanics how a tension or reaction force is developed in a string or rod. Let's restate the same thing here: 'when we deform a rod (or string) by hanging it, accelerating it, or by pulling or pushing it directly at both its ends, we disturb its interatomic or intermolecular spacing (average distance of separation between two adjacent atoms or molecules)'. When the interatomic/molecular spacing increases slightly, each atom pulls the nearest atom; when the interatomic/molecular distance decreases, each atom pushes the other atoms. In other words, an interatomic/molecular force of attraction and repulsion appears when an external force acts on a solid. Hence, microscopic interaction between the nearest atoms or molecules of matter manifests as a macroscopic force called elastic force.

Microscopically, an elastic force is an interatomic (molecular) force. Mechanical forces like tension and reaction are the macroscopic or gross effects of microscopic interatomic (molecular) interactions, which are electromagnetic in origin. Elastic forces can be compressive tensile or shearing.

Reaction forces in matter: The matter responds to an external force. When we push or pull or twist an object, it responds to an external force or torque by either accelerating or by deforming its shape or size. Let's interpret this by considering the variation of interatomic force F with the distance of separation r of two neighbouring atoms.

We can see that at $r = r_0 = 10^{-10}$ m, $F = 0$. At this distance of interatomic separation, no force acts on the atoms. At $r < r_0$, force F becomes positive, which signifies that a repulsion force acts on the atoms. At $r > r_0$, the force becomes negative. This physically signifies an attraction force between the atoms. If $r \to \infty$, the interatomic force of attraction decreases to zero. When $r \to 0$, the interatomic repulsion becomes incredibly large. With the above information let's try to explain the behaviour of solid, liquid, and gas when we apply a force on them.

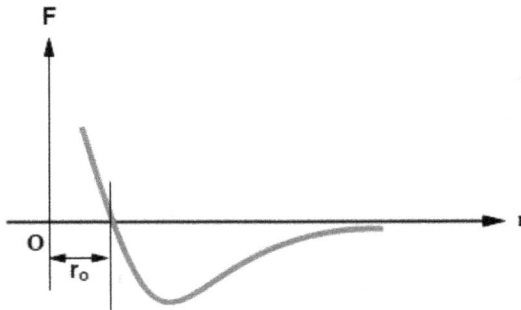

Solid: In solid state, the atoms are strongly attached to neighbouring atoms maintaining an equilibrium (mean) distance $r_0 = 10^{-10}$ m. This means that each atom oscillates about its mean position by small amplitudes. Hence, the molecules and atoms in a solid have strong spring-like interatomic forces. This strong interatomic force in a solid is essentially the cause of a repeating three-dimensional structure, which is known as a lattice structure.

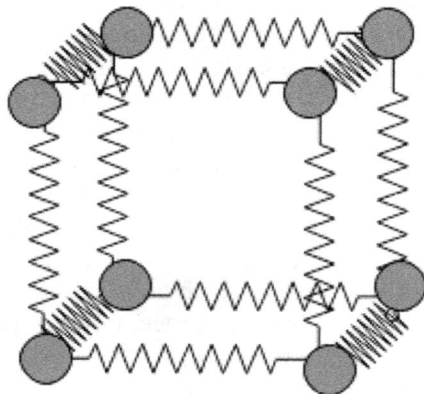

If we compress a solid, the interatomic distance r will be less than r_0. This causes a strong repulsive force between the atoms. This appears as a gross manifestation of a compressive reaction force. Thus, a solid can respond to a compressive or pushing external force.

When you pull a solid, the interatomic separation increases ($r > r_0$). As the force becomes negative in the graph, we can imagine it as a strong attraction force between the atoms. As all atoms start pulling the other atoms, a tensile reaction or tension appears inside the solid.

If you try to twist a solid by applying a torque, a shearing force will appear in the solid. This can also increase the intermolecular separation. As a result, it induces an appropriate shearing force or torque in response to the applied shearing force or torque.

Since a solid can resist (or withstand) the applied tensile and compressive forces with a little deformation compared to its dimensions, we can say that a solid maintains a definite volume (size). Since a solid can withstand a shearing force, it does not change its shape appreciably. Hence, a solid possesses a definite shape.

Recapitulating, a solid can resist external forces (tensile, compressive, and shearing) by exerting an equal and opposite reaction force on the external agent. A solid can provide a reaction force in any direction.

Types of deformation: The explanation of the elastic properties of a solid lies in its atomic structure. The major portion of matter is vacuum as the interatomic spacing is too large compared to the size of an atom; intermolecular distance is also too large compared to a molecule. It is a well-known fact that matter, especially solids, is an aggregation of atoms or molecule organised as a repeating a three-dimensional pattern called a lattice structure as shown in the above figure. The force of interaction between two atoms or molecules can be imagined as the restoring force in tiny (molecular) springs. As the atoms are connected with molecular springs, when we apply the forces F and $-F$, the springs can be compressed or elongated. This means that a solid can be compressed and elongated along its length because of the decrease and increase in interatomic distance, respectively. Furthermore, if we twist a solid, the molecular springs are twisted, inducing internal reactive shearing forces F and $-F$. Thus, a solid can undergo compression, elongation, and twisting or shearing. The compressive or tensile forces can change the size of solids, whereas the shearing forces change the shape of solids.

Recapitulating, application of a force (compressive, tensile, or shearing) deforms a solid by deforming the atomic or molecular springs. The deformed molecular springs induce reaction forces in the form of tensile or shearing forces inside the solids. These internal molecular forces are called elastic forces.

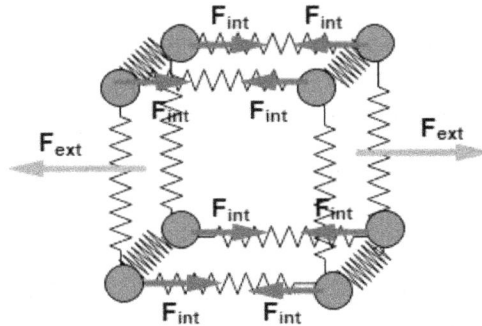

Elastic forces are the internal forces developed (appeared) inside a solid in response to the external forces acting on it. Elastic forces can be tensile, compressive, or shearing. Elastic forces appear due to the molecular springs, which explain the interatomic/molecular force of interaction. Solids have the property of inducing shearing forces, whereas liquids and gases experience very little shearing force in comparison to solids.

Example 1 A horizontal force of magnitude F acts at the end P of a uniform rigid rod, which is welded at point Q. In each case 1 and 2 as shown in the figure, find the reaction force acting at a point C at a distance x from the fixed end Q of the rod.

Solution

Since the shaded segment is in translational equilibrium at rest, $F_{net} = |F_{ext} - F_{int}| = 0$ in both figures. This gives $F_{int} = F_{ext}$.

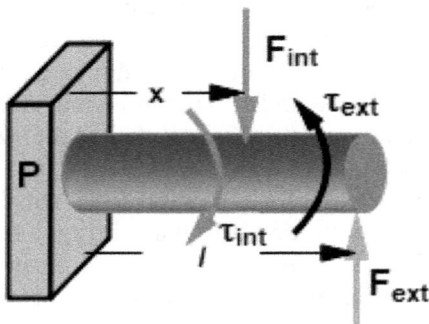

Since the rod is in rotational equilibrium, $\tau_{net} = \tau_{ext} - \tau_{int} = F_{int}x - F_{ext}l = 0$, this gives $F_{int} = \frac{F_{ext}l}{x}$ Ans.

Compressive forces compress the rod; tensile forces elongate the rod and shearing forces twist (bend) the rod. The tensile and compressive forces are equal to the applied force when the rod is in equilibrium horizontally, whereas the shearing force need not be equal to the applied force.

Stress and strain

Stress: When we apply a force on a body, the force is communicated to different parts of the body *via* intermolecular force. As a consequence, an internal (elastic) force is felt at each element of the body. Then, we say that the body is under stress as the elastic force is distributed inside the body. The stress defines the distribution of force F over an area A. If we divide the force by area, the intensity of force, that is, $\frac{F}{A}$, is defined as *stress* or pressure denoted as $p = F/A$.

As we know, force is distributed inside the body. In this case, we imagine an elementary area dA. The force acting on that area is dF, say. Then, dF/dA can be termed as pressure at a point. This is just a rough idea that the ratio of two vector quantities such as force and area is not a vector quantity. So, roughly speaking, pressure is a scalar quantity. However, in a strict sense, it has a certain directional property, as will be explained a little later. At this level, we define pressure or stress as a scalar quantity; stress has no specified direction at a point. You can see that the quantity F/A or dF/dA does not depend on the orientation in a liquid as shown in the figure. But the situation will be different in a solid. This is because a solid can exhibit three types of stresses corresponding to three types of forces described earlier.

Let's consider a horizontal rod of length l and cross-sectional area A welded at one of its ends. If we push the rod by a force F along the rod, the ratio $\frac{F}{A}$ is called compressive stress. If we pull the rod by the force F along its length, $\frac{F}{A}$ is called tensile stress. Since both tensile and compressive stress are generated by the force F

acting along the length of the rod, we can call them linear or longitudinal stress. In linear stress $p = \frac{F}{A}$, F is perpendicular to the plane of the given area A.

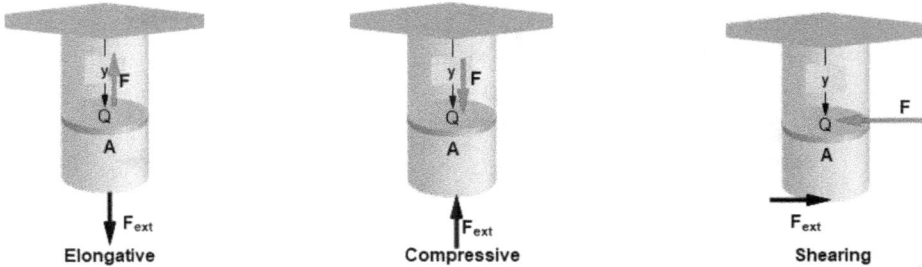

| Elongative | Compressive | Shearing |

When we pull the rod down by applying the force F parallel to the plane of the cross-sectional area, the corresponding stress $\frac{F}{A}$ is called shearing stress.

In the case of liquids and gases, we define the internal stress throughout the volume because the external force is distributed throughout the volume of liquids and gases because of their ability to flow. Hence, we define the stress (pressure) at any point of a fluid as

$$p = \lim_{A \to 0} \frac{F}{A} \frac{dF}{dA}$$

We have explained the pressure in hydrostatics and hydrodynamics. Hence, pressure of the fluids can be defined as a volumetric stress. The unit of stress in N m^{-2}.

Stress is a measure of force, given as $p = \frac{F}{A}$

$$p_{\text{long}} = \frac{F_{\text{long}}}{A}; \quad p_{\text{shearing}} = \frac{F_{\text{tan}}}{A}; \quad p_{\text{volumetric}} = \frac{dF}{dA}$$

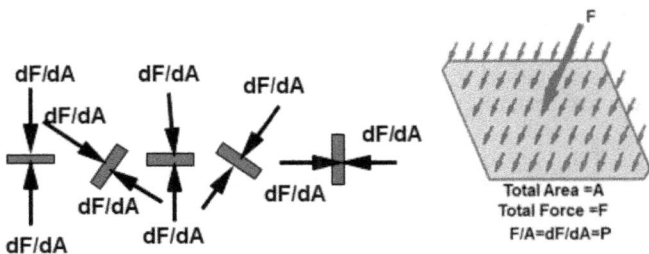

Example 2 Two equal and opposite forces F and $-F$ act on a rod of uniform cross-sectional area A, as shown in the figure. Find the (i) shearing, (ii) longitudinal stress on the section AB, and (iii) longitudinal stress when shearing stress on the area AB is maximum.

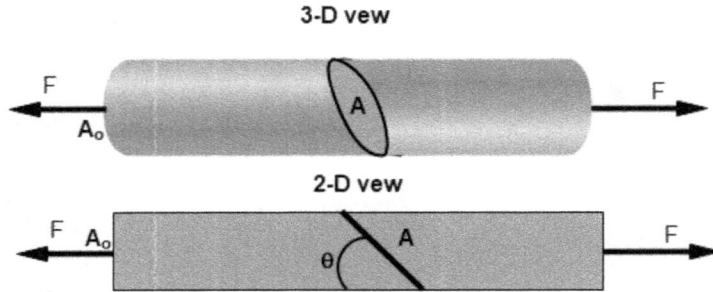

3-D vew

2-D vew

Solution

(i) As the net force acting on the rod is zero, it is in equilibrium. Let the tension in the segment AB be F'. Applying Newton's 2nd law for the segment 1,

$$F_{net} = F' - F = ma,$$

where

$$a = 0$$

This gives $F' = F$.

3-D vew

2-D vew

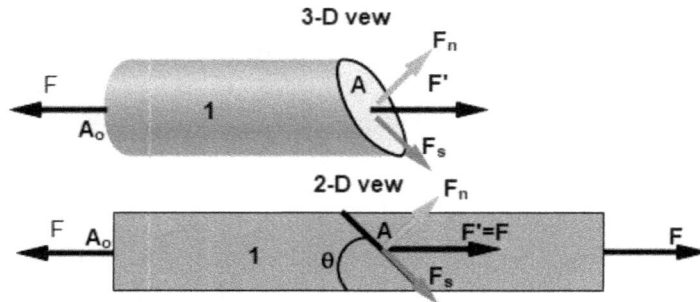

Resolving the force F' parallel and perpendicular to the given area $AB(=A,$ say), we have

$$F_{tan} = F' \cos \theta \text{ and } F_{long} = F' \sin \theta$$

$$\text{Then } p_{shearing} = \frac{F_{tan}}{A}$$

$$= \frac{F' \cos \theta}{A}, \text{ where } A = \frac{A_o}{\sin \theta} \text{ and } F' = F$$

This gives

$$p_{shearing} = \frac{F \sin \theta \cos \theta}{A_o} \quad \text{Ans.}$$

(ii) Similarly, the longitudinal stress is

$$p_{\text{long}} = \frac{F_{\text{long}}}{A}$$

$$= \frac{F' \sin \theta}{A},$$

where $A = \frac{A_o}{\sin \theta}$ and $F' = F$.

This gives $p_{\text{long}} = \frac{F}{A_o} \sin^2 \theta$ Ans.

(iii) $p_{\text{shearing}} = \frac{F \sin \theta \cos \theta}{A_o} = \frac{F \sin 2\theta}{2A_o}$ is maximum when $\sin 2\theta = \text{maximum} = 1$.
This gives $\theta = 45°$.

$$p_{\text{long}} = \frac{F}{A} \sin^2 \theta = \frac{F}{A_o} \sin^2 45° = \frac{F}{2A_o}$$ Ans.

N.B.:

1. When a force F acts at an angle θ with the outward normal $\hat{n}$ to the area A, you should not be tempted to write the stress as

$$p = \frac{F}{A} \text{ or } \frac{F}{A \cos \theta} \text{ or } \frac{F}{A \sin \theta} \text{ or } \frac{F \sin \theta}{A} \text{ or } \frac{F \cos \theta}{A}$$

2. To find the linear (or longitudinal) stress, resolve the force perpendicular to the plane of a given area A, then divide this component $F_\perp$ (or F_{long}) by the area A to obtain

$$p_{long} = \frac{F_\perp}{A} = \frac{F \cos \theta}{A}$$

3. To find shearing stress, resolve F parallel to the plane of the given area and then divide $F_\parallel$ by the area A to obtain

$$p_{\text{shearing}} = \frac{F_\parallel}{A} = \frac{F \sin \theta}{A}$$

Strain: When we pull a flexible rubber rod by a small force F, a small fraction of its length l, breadth b, and height h will change. We say that the rod is strained (or deformed). Let the elongation of the rod be Δl. If we join two such identical rubber rods end to end and then pull them by same force F, we will see that the composite rubber rod will elongate by a length $2\Delta l$. It is thus experimentally verified that the change in length Δl is directly proportional to the length of the material. Hence, it is always logical to take the ratio $\frac{\Delta l}{l}$ as a constant quantity for a given force. We call this unitless and dimensionless quantity 'linear strain'.

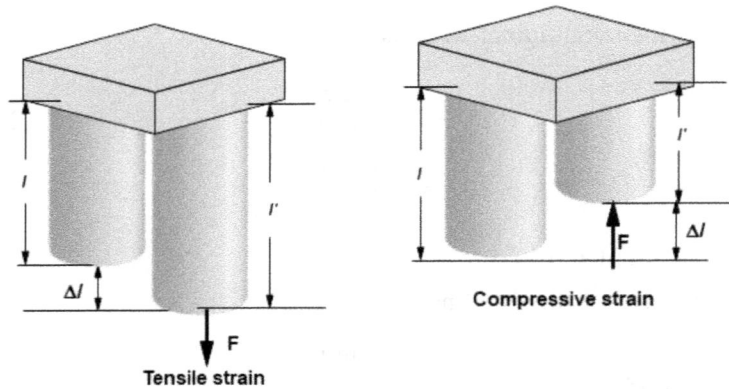

Tensile strain   Compressive strain

Similarly, $\frac{\Delta l}{l}$ and $\frac{\Delta b}{b}$ are called lateral strains, where $\Delta l = l' - l$, $\Delta h = h' - h$, and $\Delta b = b' - b$; the strains can be positive when the length increases (elongative or tensile) and negative when the length decreases (compressive). It is practically evident that when we pull the rubber rod, its length increases and breadth and height decrease. Hence, longitudinal (linear) strain is positive when the stress is tensile and it is positive when the stress is compressive.

Similarly, we can define the quantity $\frac{\Delta A}{A}$ as a real strain and $\frac{\Delta V}{V} =$ volumetric strain, where $\Delta A = A' - A$ and $\Delta V = V' - V$.

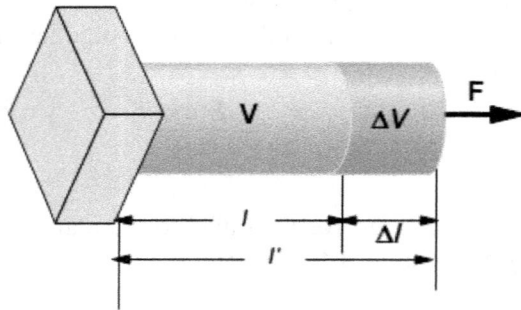

Another type of strain is there, that is, shearing strain. It is defined as the angle of twist ϕ as shown in the figure; $\phi = \frac{\Delta y}{l}$ (in radian).

Strain is the measure of deformation; we have linear strain $\frac{\Delta l}{l}$; lateral strain $\frac{\Delta b}{b}$ or $\frac{\Delta h}{h}$; areal strain $\frac{\Delta A}{A}$; volumetric strain $= \frac{\Delta V}{V}$; and strains can be positive or negative. Shearing strain $=$ angle of twist. Strain has no dimension; except shearing strain, other strains are unitless.

Poisson's ratio: When you pull a cylindrical wire, its length increases. Consequently, its thickness decreases. Similarly, if you compress the wire, its length decreases, but the thickness increases. Hence, when the longitudinal (linear strain) is positive, the transverse (lateral) strain is negative.

The ratio of lateral strain and longitudinal strain is called Poisson's ratio, denoted by σ.

$$\sigma = \frac{\text{lateral strain}}{\text{linear strain}},$$

where lateral strain $= \frac{\Delta r}{r}$ and the linear strain $= \frac{\Delta l}{l}$.

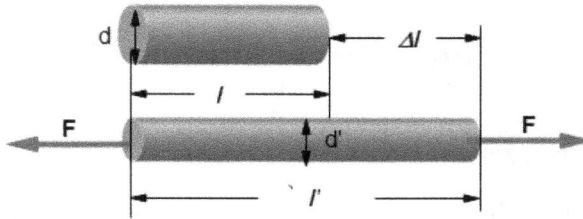

Then, the Poisson's ratio can be given as

$$\sigma = \frac{\frac{\Delta r}{r}}{\frac{\Delta l}{l}}$$

We can see that σ is a unitless and dimensionless quantity. σ must lie between -1 and $+\frac{1}{2}$ theoretically. However, in practice, σ must always be greater than zero.

$$\sigma < \frac{1}{2} \text{ and } \sigma > -1.$$

6.3 Hooke's law

Robert Hooke experimentally proved that for a very small change in length of elastic bodies, the stress is directly proportional to the corresponding strain.

Because the stress is directly proportional to strain, the ratio of stress and strain is a constant quantity within an elastic limit; So, we can write

$$\frac{\text{stress}}{\text{starin}} = E, \text{ a constant, which can be defined as the modulus of elasticity.}$$

Symbolically, $\frac{p}{e} = E$, where $p=$ stress and $e =$ strain.

Types of elastic moduli: We have different types of elastic moduli E as given below.

For solid: $\frac{\text{longitudinal stress}}{\text{longitudinal strain}} = \frac{p}{\Delta l / l} = Y$, where $Y =$ Young's modulus.

For shearing stress: $\frac{\text{shearing (transverse) stress}}{\text{shearing (transverse) strain}} = \frac{p}{\phi} = \eta$, where $\eta=$ modules of rigidity or shear modulus. It is applied mainly for solids.

For volumetric stress: $\frac{\text{volumetric stress}}{\text{volumetric strain}} = \frac{p}{\Delta V / V} = B$, where $B=$ bulk modulus, which is used for fluids.

   1. Hooke's law relates the applied stress to the strain of a body; stress $\propto$ strain for small deformations.

2. The elastic modulus $E = \frac{stress}{strain}$. It can be Young's modulus Y, shear modulus (or modulus of rigidity) η, and bulk modulus B.
3. Shear modulus measures the resistance of a solid to the change in its length (or linear dimensions).
4. Shear modulus measures the resistance of a solid to the change in its shape.
5. Bulk modulus measures the resistance of a solid to the change in size (volume) of the solid.

Physical significance of Hooke's law: Let's logically think why the linear stress should be directly proportional to the linear strain. Hooke's law states that

$p = eE$, where $E = Y$, $p = \frac{F}{A}$ and $e = \frac{\Delta l}{l}$.

This gives $F = \frac{YA}{l}\Delta l$.

We need to prove that $F \propto \frac{\Delta l}{l}$ (for small deformations) and $F \propto A$ by the following logic:

A thin wire can be imagined as a parallel combination of arrays of molecular springs. When we pull a wire, we really pull the springs. If the area of cross-section is more, the numbers of parallel combinations (arrays) will be more. Hence, proportionately more force is required for the same elongation. This means $F \propto A$. You can experiment it by pulling two similar rubber strips joined together. You can see that double amount of force is required to pull them by a given length compared to that for each strip. Similarly, if we pull the strips by a force F, say, when they are joined end to end, you can observe the double elongation compared to that of a single strip. This means that for same force F, $\Delta l \propto l$. If we pull a single strip with more and more force, for small deformation, $F \propto \Delta l$.

Since $k_{eq} = \frac{YA}{l}$, we have $k_{eq}l = YA$.

This means that the product of stiffness of an elastic wire and its length is always constant. If we cut the wire into two equal parts, the stiffness of each part will be double that of the original wire. Following the same logic we can say that if a spring is cut into N equal parts, each part will have stiffness N times the stiffness of the original spring.

Hence, $kl = $ constant, where $k = $ stiffness and $l = $ natural of the body. It is also equally valid for the springs.

Stress-strain curve for a typical metal: When we pull a metallic wire slowly by increasing the stress, the stress and strain go on increasing from point O to A proportionally, as shown in the following. This means that stress-strain graph is a straight line up to A as it obeys Hooke's law. If the stress is increased further, the strain will no longer be linear to the stress. The linearity is lost after the limiting stress at B, which is known as elastic limit. If a material is deformed beyond the elastic limit, it will regain its original shape and size. However, from B to C, more stress is required to elongate the wire more and more (but not proportionately). Attaining the critical load (stress) at E, if we decrease the stress, the body (wire) will still go on elongating. Then, at point E, the corresponding stress is called the ultimate strength. After that, at F, the wire breaks under a reduced stress. Hence, F is called the breaking point and the corresponding stress is called the breaking stress.

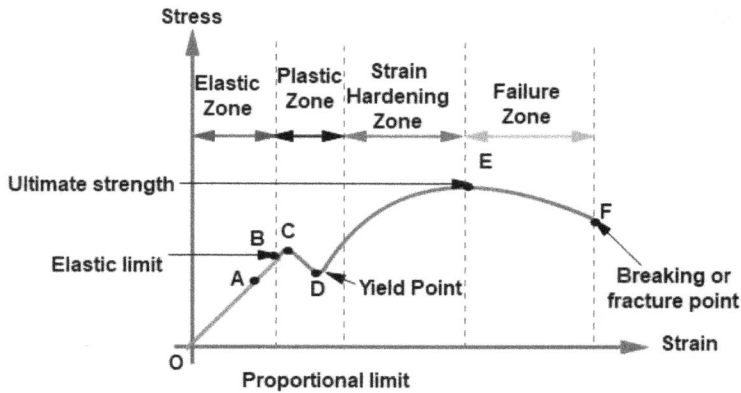

We will see that the area under stress-strain graph gives us the energy per unit volume, that is, volume energy density.

Young's modulus: Thomas Young gave a clear understanding of proportionality of strain with the stress of a solid as suggested by Robert Hooke. He experimentally verified that whenever a force acts along the length of a solid body, its linear deformation is directly proportional to the corresponding linear stress. Using Hooke's law, we have

$$\frac{\text{Linear stress } (p)}{\text{Linear strain } (e)} = Y$$

This gives

$$p = eY$$

Then, within the elastic limit, the slope of stress-strain graph gives us the Young's modulus. This quantifies the stiffness of a solid bar.

N.B.: Although the more conventional symbol for the stress is Greek letter 'sigma', I prefer to use the letter 'p' for stress in this book.

Example 3 Find the stiffness of a thin wire of length l, area of cross-section A, and Young's modulus Y.

Solution

If the force F stretches the wire by a length $x(\ll l)$, the longitudinal stress is

$$p = \frac{F}{A} \tag{6.1}$$

and the linear strain is

$$e = \frac{x}{l} \tag{6.2}$$

Putting the values of p and e from equations (6.1) and (6.2), respectively, in Hooke's law $p = eY$, we have

$$\frac{F}{A} = Y\frac{x}{l}$$

This gives

$$F = \frac{YA}{l}x \qquad (6.3)$$

Comparing equation (6.3) with the equation (Hooke's law for the spring) $\overrightarrow{F} = -k\overrightarrow{x}$, we have $k_{eq} = \frac{YA}{l}$ Ans.

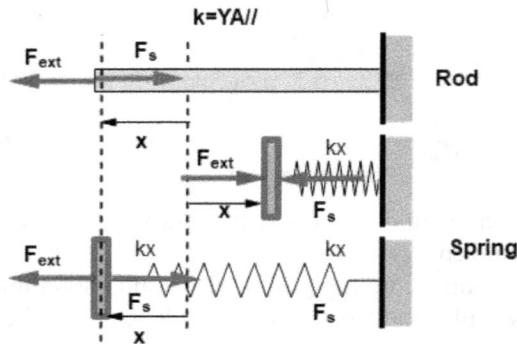

$F = -kx$ is also called Hooke's law. Here, $F =$ spring force F_s (but not F_{ext} or F_{net}). The negative sign signifies that spring force F_s and x (displacement of the tip of the spring from its relaxed or undformed position) are oppositely directed. Although any linear elastic body can be imagined as a spring, a totally different principle works for different materials of phases of matter (solid, liquid and gas). The linear deformation of the rod is due to the compressive or tensile force, whereas the shearing force in the spring is responsible for the transverse (sheering) deformation. Thus, the rod can be made equivalent to a spring obeying the formula $F = -kx$.

Bulk modulus: It is a measure of the ability to resist or withstand the change in volume of a material due to an external pressure to compress it. Let's squeeze a sponge ball by hand; its volume decreases by δV, then the relative change in the volume is known as the volumetric strain given as $e = -\delta V/V/$. The negative sign signifies that the body decreases its volume due to the external pressure or stress P. The ratio of volumetric stress (pressure P) and volumetric strain is defined as the bulk modulus B, given as

$$P/(-\delta V/V) = B$$

This is applied for any fluid (liquid and gas) apart from a solid.

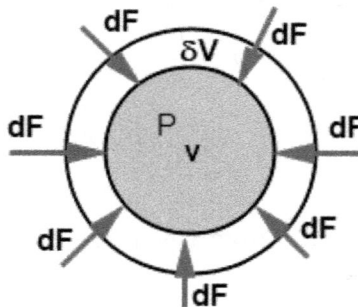

We can see that a distributed compressive force can create a pressure difference $\delta P = P - P_{\text{initial}}$ inside a body decreasing its volume by δV. Then, the volumetric stress is equal to δP but not P (for a fluid) and the volumetric strain is equal to $-\delta V/V$ (but not $\delta V/V$). By dividing the volumetric stress by volumetric strain, we can define the bulk modulus as stated above.

Example 4 A vertical tube contains an ideal gas of adiabatic exponent γ at atmospheric pressure being loaded by a light piston. (a) If we push the piston slowly so that the temperature of the gas remains constant (isothermal process), find the bulk modulus of the gas. (b) If the piston is shifted so as to maintain an adiabatic condition, find the bulk modulus of the gas.

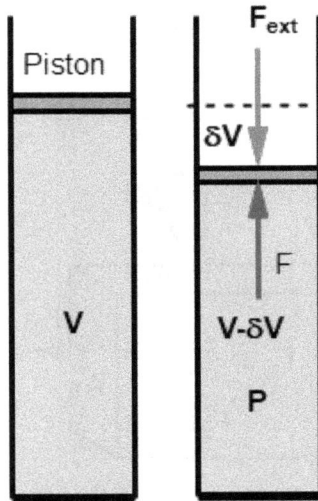

Solution

(a) Let the gas undergo a compression by a volume δV by the applied force F, say. If the displacement of the piston is x, the change in volume of the gas is $\delta V = -Ax$. Since the temperature of the gas remains constant, $PV = nRT = \text{constant}$. Differentiating both sides, we have

$$P\delta V + V\delta P = 0$$

Then, we have, $P/(-\delta V/V) = P = P_{\text{atm}}$.
Since by definition $\delta P/(-\delta V/V) = \text{bulk modulus} = B$, we have $B = P_{\text{atm}}$. Ans.

(b) In adiabatic process, $PV^{\gamma} = \text{constant}$. So, taking the logarithm of both sides, $\log P + \gamma \log V = \text{constant}$. Taking differentials of both sides, we have

$$\gamma \, \delta V/V + \delta P/P = 0. \quad \text{So,} \quad P/(-\delta V/V) = \gamma P = \gamma P_{\text{atm}}$$

Since by definition $\delta P/(-\delta V/V)$ = bulk modulus = B, we have, $B = \gamma\, P_{atm}$.

N.B.: For an ideal gas, isothermal bulk modulus is $B_{isothermal} = P$ (pressure); adiabatic bulk modulus is $B_{adiabatic} = \gamma\, P$.

Example 5 Prove that the relation between Young's and bulk modulus can be given by the expression

$$B = \frac{Y}{3(1 - 2\sigma)}, \quad \text{where } \sigma = \text{Poisson's ratio.}$$

Solution

Let's assume a cubical volume of liquid is subjected to a pressure P. Let the compressive forces F_1, F_2, F_3 act on the three mutually perpendicular faces of the cube. Since the pressure P inside the cube is the same for all forces, we can write

$$P = \frac{F_1}{A_1} = \frac{F_2}{A_2} = \frac{F_3}{A_3}$$

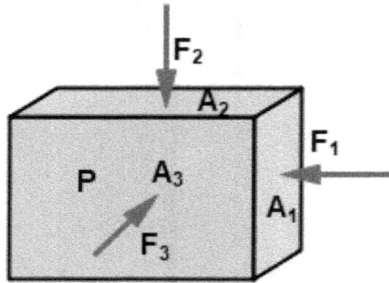

Due to the compressive force F_1, the strain of the length l is

$$e_1 = \left(\frac{\Delta l}{l}\right)_1 = \frac{p}{Y}$$

Due to the compressive force F_2, the strain of the height h is

$$\frac{\Delta h}{h} = -\frac{p}{Y}$$

Then, due to the compressive force F_2, the corresponding strain in length is
$e_2 = \left(\frac{\Delta l}{l}\right)_2 = -\sigma\left(\frac{\Delta h}{h}\right)$, where $\frac{\Delta h}{h} = -\frac{p}{Y}$

This gives $e_2 = \left(\frac{\Delta l}{l}\right)_2 = \frac{\sigma p}{Y}$.

Similarly due to compressive force F_3, the corresponding strain in length l is

$$e_3 = \left(\frac{\Delta l}{l}\right)_3 = \frac{\sigma p}{Y}$$

Since all the compressive forces are acting simultaneously, the net strain of the length l is

$$e = e_1 + e_2 + e_3$$

Substituting the values of e_1, e_2 and e_3, we have

$$e = -\frac{p}{Y} + \frac{\sigma p}{Y} + \frac{\sigma p}{Y}$$

This gives

$$e = \frac{\Delta l}{l} = -\frac{p}{Y}(1 - 2\sigma)$$

Since the pressure is the same and the thing (cube) is symmetrical, the total strain in breadth and height can be given as

$$\frac{\Delta b}{b} = \frac{\Delta h}{h} = -\frac{p}{Y}(1 - 2\sigma)$$

Then, the sum of all three strains is

$$\frac{\Delta l}{l} + \frac{\Delta b}{b} + \frac{\Delta h}{h} = -\frac{3p}{Y}(1 - 2\sigma) \tag{6.4}$$

Since $V = lbh$, taking differentials of both sides, we have

$$\frac{\Delta V}{V} = \frac{\Delta l}{l} + \frac{\Delta b}{b} + \frac{\Delta h}{h} \tag{6.5}$$

Using equations (6.4) and (6.5), we have

$$\frac{\Delta V}{V} = -\frac{3p}{Y}(1 - 2\sigma) \tag{6.6}$$

Substituting $\frac{\Delta V}{V}$ from equation (6.6) in Hooke's law

$$\frac{\Delta V}{V} = -\frac{p}{B},$$

we have

$$B = \frac{Y}{3(1 - 2\sigma)}$$

Since σ and Y are constant for a particular fluid at given condition, its bulk modulus is a constant.

Example 6 A rubber ball of bulk modulus B is taken to a depth h of a liquid of density ρ. Find the fractional change in radius of the ball.

Solution

The volumetric strain of the ball is

$$\frac{\delta V}{V} = -\frac{p}{B},$$

where

$$p = \rho g h$$

Then

$$-\frac{\delta V}{V} = \frac{\rho g h}{B} \tag{6.7}$$

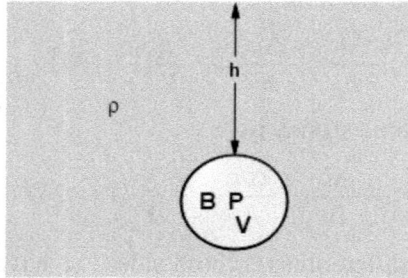

Since the volume of the sphere is $V = \frac{4}{3}\pi r^3$, we have

$$\frac{\delta V}{V} = \frac{3\delta r}{r} \tag{6.8}$$

Using equations (6.7) and (6.8), we have $\frac{\delta r}{r} = \frac{\rho g h}{3B}$ Ans.

Shear modulus: As we described earlier, shear modulus measures the shearing strain of a solid in response to the shearing stress. This means that shear modulus governs the elastic shear stiffness of a body that defines the degree of rigidity of the material. So, this modulus is also known as modulus of rigidity denoted by the symbol η.

By applying Hooke's law of elasticity, the modulus of rigidity is defined as the ratio of shearing stress and shearing strain. It is expressed as

$$\eta = \frac{p_{\text{shear}}}{e_{\text{shear}}} \tag{6.9}$$

If a force F acts on the parallel to the cross-sectional area A of the rectangular vertical bar, it induces a shearing force F_s at the point of application of force F. At equilibrium, $F = F_s$. Then the shearing stress is

$$p_{\text{shear}} = \frac{F}{A} \tag{6.10}$$

As a result, the cross-section A undergoes a small displacement x.

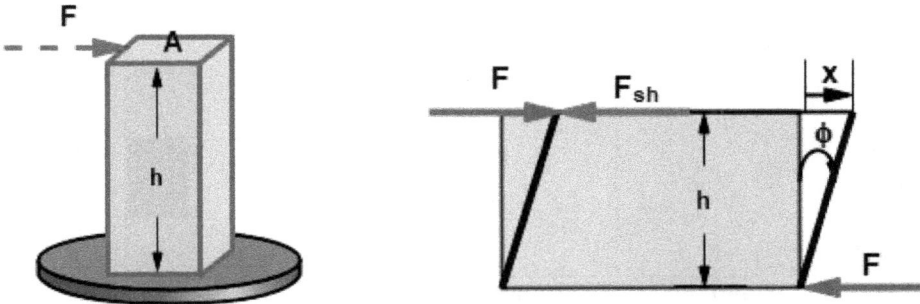

Hence, the shearing strain is

$$e_{shear} = \frac{x}{h}$$

Since $\frac{x}{h} = \phi$ (angle of twist), the shearing strain is given as an angle of twist:

$$e = \phi \qquad (6.11)$$

Using the last two equations, the Hooke's law to define the modulus of rigidity is given as

$$\eta = \frac{p}{\phi}. \text{ where } p = F/A$$

Then, we have

$$\eta = \frac{F}{A\phi} \qquad (6.12)$$

You can prove that the relation between the modulus of rigidity η, which is a constant quantity, and Young's modulus can be given as

$$\eta = \frac{Y}{2(1 + \sigma)}$$

The greater the shear modulus, thre more difficult it will be to twist (or bend) a solid.

Example 7 A horizontal rod fixed at one of its ends has length l, rigidity modulus η, and area of cross-section A. A bob of mass m hangs from the free end of the rod by a light string of effective stiffness k. Find the (a) displacement of the free end of the rod and (b) bob.

Solution

(a) The distance $AB = y$ fallen by the tip of the rod is given as

$$y = \phi l,$$

where

$$\phi = \frac{(T/A)}{\eta} \quad \text{(Hooke's law)}$$

Then

$$y = \frac{Tl}{A\eta}$$

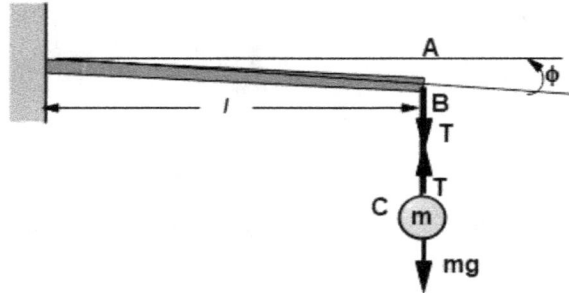

Substituting $F = mg$ (for equilibrium of the hanging mass), we have

$$y = \frac{mgl}{A\eta} \quad \text{Ans.}$$

(b) The distance fallen by the bob from the relaxed position of the string is

$$s = x + y,$$

where $x =$ elongation of the string $= mg/k$. Then, we have

$$s = \frac{mg}{k} + \frac{mgl}{A\eta} = \left(\frac{1}{k} + \frac{l}{A\eta}\right)mg \quad \text{Ans.}$$

6.4 Calculation of deformation

When a tensile or compressive force acts on a linear object, its length changes. To find the change in length, first of all we need to take an elementary segment of length dx. Since for a given force F the area of cross-section of the elementary portion does not vary considerably, we can write the stress as

$$p = \frac{F}{A}$$

Then by applying Hooke's law, the strain of the elementary segment can be given as

$$e = \frac{p}{Y}, \quad \text{where } p = \frac{F}{A}$$

This gives

$$e = \frac{F}{AY} \tag{6.13}$$

Let $dy =$ change in length of the elementary length dx. So, the strain of the element is

$$e = \frac{dy}{dx} \tag{6.14}$$

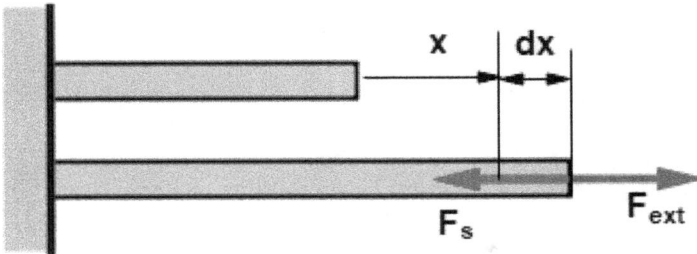

Substituting e from equations (6.14) in (6.13), we have

$$\frac{dy}{dx} = \frac{F}{AY}$$

Then

$$dy = \frac{Fdx}{AY}$$

Integrating both sides, we have

$$\int_0^{\Delta l} dy = \frac{1}{Y} \int_0^l \frac{F}{A} dx,$$

where $\Delta l=$ deformation of the object.

This gives

$$\Delta l = \frac{1}{Y} \int_0^l \frac{F}{A} dx = \frac{1}{Y} \int_0^l p dx,$$

where $p = \frac{F}{A} =$ stress of the object at a distance x from one of its ends.

The above expression gives us the deformation of any linear object when a compressive or tensile force acts on it.

Sometimes the force F or area of cross-section A or both F and A may vary inside the object. In all these cases, generally we can find F and A as the functions of distance x from one end of the linear object (a rod, say). Then we can find the stress $p = f(x)$ and put in the formula

$$\Delta l = \frac{1}{Y} \int_0^l p(x) dx$$

to obtain the total deformation Δl.

Summarizing the above facts, let's note the following points to find the deformation:

1. Take any point P at a distance x from any end of the object.
2. Find the tension F at P as a function of x.
3. Find the cross-sectional area A at P as a function of x.
4. Then find the stress at P as the function of x by using the formula $p = \frac{F}{A}$.
5. Finally by substituting $p = f(x)$ in the formula $\Delta l = \frac{1}{Y} \int_0^l p(x)\, dx$, find the deformation Δl.

Let's look at the following examples to use the above method.

Example 8 A bob of mass m hangs from the ceiling of a smooth trolley car, which is moving with a constant acceleration a. If the Young's modulus, radius, and length of the string are $Y, r,$ and l, respectively, find the (a) stress in the string and (b) extension of the string when it makes a constant angle relative to vertical.

Solution

(a) Relative to the accelerating frame (trolley car), the forces acting on the bob are weight $mg\downarrow$, pseudo-force $F_{ps} = -ma\searrow$, and tension $T\nwarrow$. Since the acceleration of the bob is zero relative to the trolley car, we have

$$\overrightarrow{F_{net}} = m\overrightarrow{g} + (-m\overrightarrow{a}) + \overrightarrow{T} = m\overrightarrow{a}_{rel},$$

where

$$\overrightarrow{a}_{rel} = 0$$

Then, the tension in the string is

$$T = m\,|\overrightarrow{a} - \overrightarrow{g}|$$

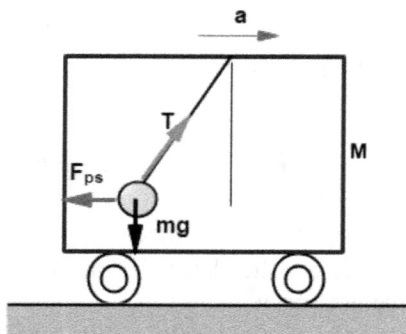

Since the string is light, the same tension is felt at all points of the formula

$$p = \frac{F'}{A}$$

where

$$F' = T = m\,|\overrightarrow{a} - \overrightarrow{g}|$$

Then, we have $p = \dfrac{m\,|\vec{a}-\vec{g}|}{\pi r^2} = \dfrac{m\sqrt{g^2+a^2}}{\pi r^2}$ Ans.

(b) Since p is constant, talking p out of the integral of the formula

$$\Delta l = \frac{1}{Y} \int_0^l p\,dx, \text{ we have}$$

$$\Delta l = \frac{p}{Y} \int_0^l dx = \frac{pl}{Y}$$

Substituting the obtained value of the stress p in the last expression, we have

$$\Delta l = \frac{m\,|\vec{a}-\vec{g}|\,l}{\pi r^2 Y} = \frac{m\left(\sqrt{g^2+a^2}\right)l}{\pi r^2 Y}$$ Ans.

N.B.: In the foregoing example, if an external net horizontal force F acts on the (trolley car + pendulum) system assuming l = length of the string, the elongation of the string can be found by putting $a = F/(M+m)$ in the obtained answer to have

$$\Delta l = \left(\sqrt{g^2 + \frac{F^2}{(M+m)^2}}\right)\frac{ml}{\pi r^2 Y}, \text{ where } M = \text{mass of the trolley car}$$

If tension F remains constant for all points of the wire (or string) and the wire has uniform cross-section, taking F/A out of the integral in the expression $\Delta l = \frac{1}{Y}\int_0^l \frac{F}{A}dx$, we have $\Delta l = \frac{Fl}{AY}$. Hence, you should not misunderstand this expression as a general expression for deformation of any rod.

When the mass of the string is not negligible, there are many cases where the tension in the string varies. For instance, when the string accelerates in horizontal surfaces or rotates or hangs from a rigid support, the tension in the string varies along its length. Let's take the example of a hanging string.

Example 9 A uniform string of modulus Y, density ρ, and length l hangs from a point P. A force F pulls the string vertically down at its lowest point B. Find the:
(a) stress at point A located a distance x from the lowest point B of the string and y from point P, and
(b) elongation of the string.

Solution

(a) Let's take point A at a distance y from point P of suspension of the string. Let the tension in the string at P be T. We can see that the shaded portion of the string is in equilibrium under the action of $m'g\downarrow$ and $T\uparrow$, where $m' =$ mass of the shaded portion of the string.

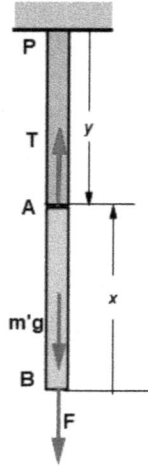

Then applying Newton's second law, $F = ma$, where $a = 0$ and $F = T - m'g$, we have

$$T = m'g$$

Substituting the values of

$$m' = \frac{m}{l}x, \text{ we have}$$

$$T = \frac{m}{l}gx + F, \text{ where } x = (l - y)$$

This gives

$$T = m\left(1 - \frac{y}{l}\right)g + F = mgx/l + F$$

Then, the stress p at point A is

$$p = \frac{T}{A}$$

From the last two expressions, we have

$$p = \frac{m(l - y)g}{Al} + F/A = \frac{mgx}{Al} + F/A$$

Substituting $\frac{m}{Al} = \rho$, we have

$$p = \rho(l - y)g + F/A = \rho gx + F/A \quad \text{Ans.}$$

(b) Now putting $p = \rho g x + F/A$ in expression $\Delta l = \frac{1}{Y} \int_0^l p(x)\, dx$, we have

$$\Delta l = \frac{1}{Y} \int_0^l (\rho g x + F/A)\, dx$$

This gives

$$\Delta l = \frac{\rho g l^2}{2Y} + \frac{Fl}{AY} \quad \text{Ans.}$$

If the area of cross-section of a linear object varies, even though the tension inside the object is uniform, the stress of the object becomes non-uniform. Then, how do we find the stress and elongation? Let's look at the following example.

Example 10 A truncated conical rod of length l is pulled by two equal and opposite forces F and $-F$ as shown in the figure. The radii of cross-section of the rod are r_1 and r_2 and its Young's modulus is Y. Find the extension of the conical rod.

Solution

Let the cross-sectional area of the rod at a distance x from its left-hand side end be A, where

$$A = \pi r^2; \ r = r_1 - \frac{(r_1 - r_2)}{l} \cdot x$$

Then, we have

$$A = \pi \left[r_1 - \frac{(r_1 - r_2)}{l} x \right]^2 \tag{6.15}$$

The tension F' at a distance x from the LHS end of the rod is given as $F' = F$ because the rod is stationary.

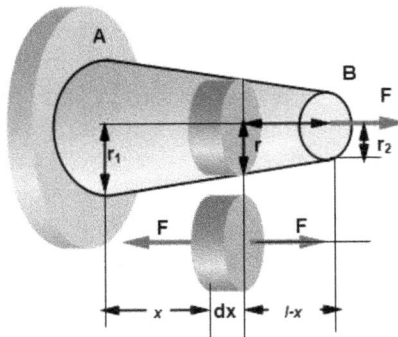

Then, the stress at a distance of x is given as

$$p = F/A = \frac{F}{\pi \left[r_1 - \frac{(r_1 - r_2)}{l} x \right]^2}$$

Now substituting p in the expression

$$\Delta l = \frac{1}{Y} \int_0^l p \, dx, \text{ we have}$$

$$\Delta l = \frac{1}{Y} \frac{F}{\pi} \int_0^l \frac{dx}{\left[r_1 - (r_1 - r_2)\frac{x}{l} \right]^2} \tag{6.16}$$

To evaluate the above integral, put

$$r_1 - \frac{(r_1 - r_2)x}{l} = y$$

This gives

$$dx = -\frac{l \, dy}{(r_1 - r_2)}; \quad \text{when } x = 0, \ y = r_1,$$

when

$$x = l, \ y = r_2$$

Hence, the expression (6.16) can be given as

$$\Delta l = -\frac{Fl}{Y\pi(r_1 - r_2)} \int_{r_1}^{r_2} \frac{dy}{y^2}$$

$$= -\frac{Fl}{Y\pi(r_1 - r_2)} \left[-\frac{1}{y} \right]_{r_1}^{r_2} = \frac{Fl}{Y\pi r_1 r_2} \quad \text{Ans.}$$

6.5 Elastic energy

Whenever we squeeze or twist an object, the interatomic distance either increases to r_1, say, or decreases to r_2, say, from equilibrium distance r_0. Referring to the U–r graph between two atoms given in section 6.2, we can understand that the electrostatic potential energy U between two atoms is minimum at $r = r_0$ (stable equilibrium position). For any other values of r, U will increase. This means that in a deformed molecules the potential energy is greater than the minimum value. The difference in these two values of potential energy, that is, $|U - U_{min}|$ signifies an increase in energy between the atoms or molecules of a deformed body. Macroscopically, this excess energy of interaction between all atoms/molecules of a deformed object is called elastic energy. Hence, elastic energy is a gross (macroscopic) manifestation of microscopic electrostatic energy between the atoms. When we remove the external load (stress), the body becomes undeformed and the elastic energy, in principle will be converted into vibrational energy followed by heat, light, sound, etc.

Recapitulating, when a body is deformed, elastic energy is stored in it. Elastic energy is the increment of electrostatic potential energy of interaction between the atoms of the deformed bodies. After removal of the applied force, the elastic energy will be released in the form of heat, light, sound, etc.

6.6 Calculation of elastic energy

Work done: Let's consider a thin wire of Young's modulus Y, cross-sectional area A, and length l. Its left end is rigidly fixed with the vertical wall. If we slowly pull the other end of the wire, at any elongation x of the wire is, let the force required to pull the wire be F_{ext}. If you pull the wire slowly, it will also pull you with an equal and opposite force due to its elasticity. Therefore, this force is known as elastic force denoted as F_{elastic}

For slow pulling, acceleration $a = 0$, so $F_{\text{ext}} = F_{\text{elastic}}$, where $F_{\text{elastic}} = \frac{YAx}{l}$ as we derived earlier.

This gives

$$F_{\text{ext}} = \frac{YAx}{l} \qquad (6.17)$$

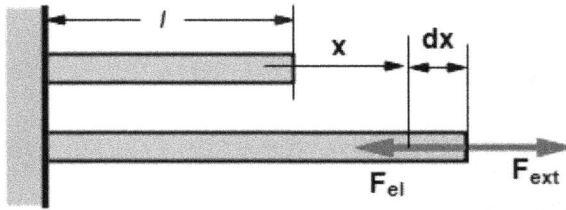

For an elementary displacement dx the work done by the external force is

$$dW_{\text{ext}} = F_{\text{ext}} \, dx$$

Using equation (6.17) and (ii), we have

$$dW_{\text{ext}} = \frac{YA}{l} x \, dx$$

Integrating both sides, the total work by the external agent in pulling the wire by a distance x is

$$W_{\text{ext}} = \int dW_{\text{ext}} = \frac{YA}{l} \int_0^x x \, dx$$

Then, we have

$$W_{\text{ext}} = \frac{YAx^2}{2l}$$

Energy stored:
Since $W_{\text{ext}} = \Delta U$ according to the work–energy theorem, we have

$$\Delta U = \frac{YAx^2}{2l},$$

where $\Delta U = U_f - U_i = U$ (because $U_i = 0$ as the wire was undeformed initially).

Then the elastic potential energy stored in the wire is

$$U = \frac{YAx^2}{2l}$$

Since the volume of the wire is $V = A(l + x) \simeq Al$ because $x \ll l$ the elastic potential energy stored per unit volume of the wire can be given as

$$u_v = \frac{U}{V} = \frac{\frac{1}{2l}YAx^2}{Al}$$

$$\Rightarrow u_v = \frac{Y}{2}\left(\frac{x}{l}\right)^2$$

Energy density: Putting $\frac{x}{l} = e$, we have

$$u_v = \frac{Ye^2}{2}$$

Substituting $e = \frac{p}{Y}$, we have

$$u_v = \frac{p^2}{2Y}$$

The above expression can also be written as

$$u_v = \frac{1}{2}(Ye)e$$

Substituting $Ye = p$, we have

$$u_v = \frac{1}{2}pe = \frac{1}{2} \times \text{stress} \times \text{strain}$$

The above expressions of energy density are applicable for any elementary segment (point) of a deformed elastic material. For the variable stress and strain, using the expression $u_v = \frac{dU}{dV}$, the total energy stored in a linear object can be given by

$$U = \int u_v dV = \int \frac{Ye^2}{2}dV = \int \frac{p^2}{2Y}dV = \int \frac{1}{2}pedV$$

$$\Rightarrow U = \frac{Y}{2}\int e^2 dV = \frac{1}{2Y}\int p^2 dV = \frac{1}{2}\int pedV$$

Now, we have three formulae of elastic energy stored:
1. $U = \frac{Y}{2}\int e^2 dV$
2. $U = \frac{1}{2Y}\int p^2 dV$
3. $U = \frac{1}{2}\int pedV$,

where $e = $ linear strain $= \frac{x}{l}$; $(x = \Delta l)$ and $p = $ linear stress $= \frac{F}{A}$.

Example 11 A smooth uniform string of natural length l, cross-sectional area A, and Young's modulus Y is pulled along its length by a force F on a horizontal surface. Find the elastic potential energy stored in the string.

   Solution

   As discussed earlier, the tension T in the string at a distance x from its free end is given as

$$T = \frac{F}{l}x$$

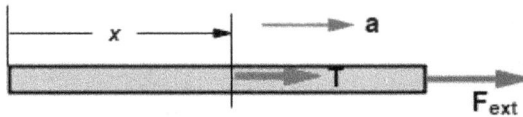

Hence, the stress in the string is

$$p = \frac{T}{A} = \frac{F}{Al}x$$

Substituting (p) in the formula

$$U = \frac{1}{2Y} \int p^2 \, dV,$$

we have

$$U = \frac{1}{2Y} \int_0^l \frac{F^2}{A^2 l^2} x^2 dV, \text{ where } dV = A dx$$

This gives

$$U = \frac{F^2 l}{6AY} \quad \text{Ans.}$$

6.7 Viscosity

You know that an ideal liquid cannot resist any shearing force. However, practically all liquids show a small finite resistance to the shearing forces. For instance, it is more difficult to wave our hands in a liquid than in air due to greater friction in water. If you rotate the water in a bucket and leave it for some time, water will come to rest due to its internal friction. The internal kinetic friction between any two moving layers is called viscous force. Let's interpret it microscopically using the ideas of molecular interaction and momentum transfer.

   Micro-interpretation: Let's imagine two adjacent thin layers of liquid moving with different velocities v_1 and v_2. Let $v_1 = v$ and $v_2 = v - \Delta v$ as shown in the figure. We know that nearby molecules of one layer move to the other layer due to random molecular motion and intermolecular attraction. These two effects retard the relative motion of the molecules, which appears as fluid friction between the liquid layers or viscosity of liquid.

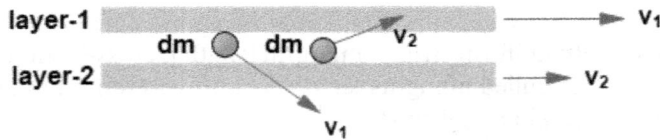

Due to the random motion of the fluid molecules, the exchange of momenta takes place between the layers. The slower molecules from layer 2 will try to slow down the motion of layer 1. Simultaneously the faster molecules from layer 1 accelerate the motion of layer 2. Thus, the momentum exchange between the two layers (1 and 2) reduces the relative velocity (motion) between them. This appears as a resistance to relative motion between the layers. Thus, the frictional (viscous) force between the layers is a macroscopic effect as a shearing (tangential) force that acts on each layer so as to hinder the relative motion between the layers.

Since the molecular attraction depends on temperature, pressure, and nature of fluid, the viscosity is governed by these factors. Recapitulating:

1. Viscosity is attributed to mainly two factors: (a) molecular attraction and (b) momentum transfer.
2. Intermolecular attraction retards the motion of the fluids.
3. Exchange of momenta between the molecules of any two layers reduces the relative motion between them and thus increases the resistance to the relative motion of the fluid layers.

6.8 Newton's law of viscosity

When you look carefully into the flowing water in a canal, you can notice that the things (straws, leaves etc) at the water surface move most rapidly and the particles (dust and sand etc), at the base of the canal do not move noticeably. This tells us that the velocity of water decreases gradually with depth. So, the velocity gradient dv/dy is negative. A velocity profile for a streamline flow or a small height h is shown in the following figure. It is practically evident that the velocity decreases linearly with depth y for a small thickness h of liquid.

Let's take a thin segment of the liquid layer of thickness dy at a depth y from the free surface of the liquid. As a viscous force F_v acts on the thin layer in the backward direction, the shearing stress f (shearing force acting on each unit area) at any point of the liquid is given as

$$f = \frac{F_v}{A}$$

It is experimentally verified that the shearing stress f is directly proportional to the velocity v at that point and inversely proportional to the vertical distance (depth) y from the surface.

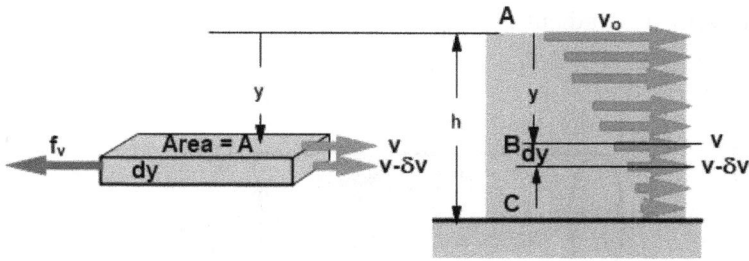

Putting the above two effects together, we have

$$f \propto \frac{v}{y}$$

Equating both sides by a constant of proportionality η, we have

$$f = \eta \frac{v}{y}, \text{ where } \eta = \text{coefficient of viscosity}$$

More generally, substituting $\frac{v}{y} = \frac{dv}{dy}$ and putting a minus sign because viscous force acting on the layer opposes the velocity of the layer relative to that of its adjacent layers, we have

$$f = -\eta \frac{dv}{dy},$$

Hence, the viscous force acting on a layer of area A is given as

$$F = -\eta A \frac{dv}{dy}$$

The above expression is called Newton's law of viscosity, which tells us that shearing stress at any point on a moving layer of a liquid is directly proportional to the velocity gradient (rate of change in velocity with distance perpendicular to velocity) at that point. This relation is strictly valid for streamline flow of viscous liquids, which can be called Newtonian fluids (liquids):
1. $F \propto A$: Viscous force is greater on a layer with more surface area.
2. $F \propto \frac{1}{h}$: Viscous force is greater on a layer with less height.
3. $F \propto v$: Viscous force is greater when a layer moves more rapidly.

Example 12 A cylinder piston of length l and radius R is moved with constant velocity v in a cylindrical tube filled with oil. If the clearance between the piston and tube is $h(\ll R)$, assuming η as the coefficient of viscosity of the oil, find the external force required to drive the piston with constant velocity.
Solution
Since the thickness of the film is small, the velocity profile of the oil can be assumed as linear.

Then the velocity gradient is

$$\frac{dv}{dy} = -\frac{v}{h}$$

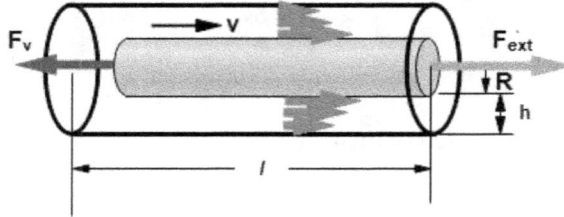

Substituting $\frac{dv}{dy} = \frac{v}{h}$ in Newton's viscosity formula

$$F_v/A = -\eta\frac{dv}{dy}$$

Hence, the shearing force on the curved surface of the piston is

$$F_v = \eta\frac{vA}{h},$$

where

$$A = 2\pi Rl$$

This gives

$$F = 2\pi R\eta\frac{vl}{h} \quad \text{Ans.}$$

6.9 Stoke's law

While a train passes, due to the viscosity of the air, the layers of air in contact with the train move with the train. This means that stationary air gains momentum in the direction of motion of the train. As a result, the velocity of the train tends to decrease. In other words, a viscous force acts on the train so as to retard it. In general, the magnitude of viscous force depends on shape, size, and speed of the body. George Gabriel Stokes studied the viscous force on moving spherical objects in Newtonian fluids. He experimentally showed that the viscous force acting on a spherical body in a Newtonian fluid is directly proportional to the radius and velocity of the body. The viscous force acts backward.

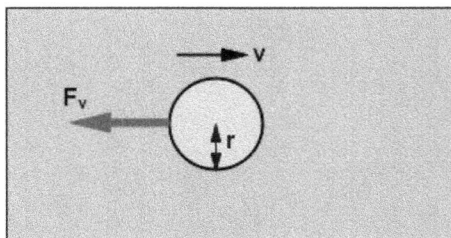

Symbolically, $F_v \propto v$ and $F_v \propto r$.

Combining these two effects, we can write $F \propto vr$. Equating both sides by a constant of proportionality $6\pi\eta$, we have

$$F = 6\pi\eta vr$$

This law is called Stoke's law, which is as follows.

When a spherical body of radius r moves with a speed v in a viscous medium of coefficient of viscosity η, it experiences a viscous force opposing its velocity, which can be vectorially given as $\overrightarrow{F} = -6\pi\eta\overrightarrow{v}r$. The viscous force does not depend on the mass of the body.

Example 13 Terminal velocity: Derive an expression for the velocity of a sphere (a raindrop, say) falling from rest in the viscous medium (air) as the function of time.

Solution

Let the sphere (raindrop) fall from rest ($v = 0$) at $t = 0$. Initially, $v = 0$ so, $F_{\text{viscous}} = 0$. Since the raindrop is denser than air, where $mg > F_b$ is the upward buoyant force acting on the sphere due to the surrounding fluid (air). Then, the sphere will accelerate down at $t = 0$. Let the sphere gain a downward velocity v after a time t. So, an upward viscous force $F_v\uparrow$ will act on the drop in addition to $mg\downarrow$ and $F_b\uparrow$.

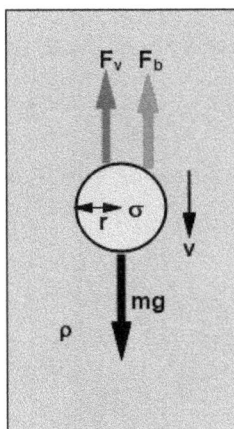

Applying Newton's 2nd law, we have

$$F_{\text{net}} = mg - F_b - F_v = ma,$$

where $F_b = v\rho g = \frac{m}{\sigma}\rho g$, $F_v = 6\pi\eta r v$ and $a = \frac{dv}{dt}$

$$\Rightarrow mg\left(1 - \frac{\rho}{\sigma}\right) - 6\pi\eta r v = m\frac{dv}{dt}$$

$$\Rightarrow \frac{mdv}{mg\left(1 = \frac{\rho}{\sigma}\right) - 6\pi\eta r v} = dt$$

Since the sphere accelerates from $v = 0$ to $v = v$ during the time t, integrating both sides we have

$$\int_0^v \frac{mdv}{mg\left(1 - \frac{\rho}{\sigma}\right) - 6\pi\eta r v} = \int_0^t dt$$

Evaluating the integral, we have

$$\frac{-m}{6\pi\eta r}\ln\left[mg\left(1 - \frac{\rho}{\sigma}\right) - 6\pi\eta r v\right]_0^v = t$$

$$\Rightarrow v = \frac{mg\left(1 - \frac{\rho}{\sigma}\right)}{6\pi\eta r}\left(1 - e^{-\frac{6\pi\eta r t}{m}}\right) \quad \text{Ans.}$$

The sphere moves from rest and after a long time it moves with a constant velocity called terminal velocity.

As the sphere falls from a large height, after a long time its velocity will attain a limiting value.

Since $\lim\limits_{t-\infty} e^{-\frac{6\pi\eta r t}{m}} \to 0$, the terminal velocity is $v_{\text{term}} = v|_{t\to\infty} = \frac{mg(1 - \frac{\rho}{\sigma})}{6\pi\eta r}$, where $m = \frac{4}{3}\pi r^3\sigma$

$$\Rightarrow v_{\text{term}} = v|_{t\to\infty} = \frac{2g(\sigma - \rho)r^2}{9\eta}$$

Hence, terminal velocity is the velocity of the body when the net force acting on the drop is zero. The gravity completely counteracts the combined effect of viscous force and buoyant force so that the body moves down with a constant velocity, called terminal velocity.

The v–t graph is exponential. At $t = 0$, $v = 0$ and at $t \to 0$, $v \to v_{\text{term}}$.

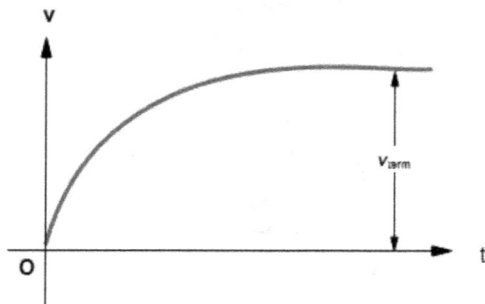

Since $v_t = \frac{2g(\sigma - \rho)r^2}{9\eta}$, v_t is greater if radius and density of the sphere are greater. Furthermore, the terminal velocity will be greater if the coefficient of viscosity is less, at a particular value of g.

6.10 Surface tension

You might have seen a fisherman's raft floating on a lake along with the leaves fallen from the trees near the lake. You may be tempted to think that the underlying principles behind the flotation of both a leaf and a raft or boat would be the same. But, it is interesting to note that the basic principles of floatation of the boat and leaves are different. The boat floats because it is partially immersed in water and the upward buoyant force of water balances the weight of the boat. On the other hand, the leaves do not sink in the water; they just stay on the water surface. Hence, buoyant force does not arise here. Then, what actually pushes a leaf up against gravity? Let's answer the above question by the following logic.

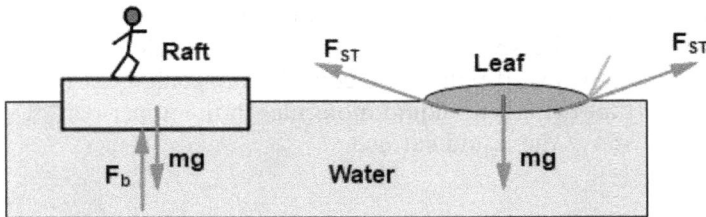

As a leaf touches the water surface, we can say that the water surface pushes the leaf up. We can see that the water surface gets deformed as the leaf stays on it just like a trampoline gets curved (stretched or deformed). So, we can conclude that the water surface behaves like a stretched membrane (rubber band). When a leaf is kept on the water surface, the contacting surface of the water curves down. As a result, the tangential force developed along the water surface becomes inclined, the vertical component of which counteracts the weight of the leaf. This property of water surface is called surface tension.

Surface tension of a liquid is a property of the liquid surface by virtue of which it behaves as a stretched membrane. Due to surface tension, the surface of a liquid can hold lighter objects by countering their weights. For instance, insects can crawl on a liquid surface. However, there is a basic difference between a stretched membrane and a liquid surface. In stretched membrane the tension increases with amount of stretch according to Hooke's law, but the surface tension force on a liquid for a particular length is constant.

The surface tension of liquid arises from the unbalanced molecular cohesive force at or near the surface. So, the free surface of liquid tends to contract and behaves as a stretched membrane.

6.11 Molecular theory of surface tension

Sphere of influence: To understand the cause of surface tension we need to understand the molecular interaction of molecules at different places inside a liquid. Referring to the F–r diagram (refer to section 6.2) we find that the force of attraction between any two molecules decreases to a negligible value at $r = 10^{-7}$ cm.

This means that a molecule A, say, can affect the other molecules with a sphere of radius $r(=10^{-7}$ cm) drawn considering molecule A as the center of the sphere. This is called 'sphere of influence' of molecule A. Using the idea of 'sphere of influence' Laplace explained the cause of surface tension as follows.

Force on different molecules: Let's consider four molecules A, B, C, and D; the sphere of influence of molecule A lies completely inside the liquid. Hence, all surrounding molecules within the sphere of influence attract the molecule equally in all directions. Consequently, the net force acting on molecule A is zero. This means that molecule A is free to move inside the liquid with its thermal speed. Now come to molecule B, which is situated a little below the liquid surface such that its sphere of influence is mostly lying inside the liquid. There are a few more molecules below molecule B than above it.

Hence, the net downward force is little bit more than the net upward force of the molecules lying inside the sphere of influence on molecule B. Then, molecule B experiences a little downward pull. If we consider molecule C at the surface of the liquid, we can see that there is no liquid molecules in the upper half of the sphere of influence, that is, above the liquid surface.

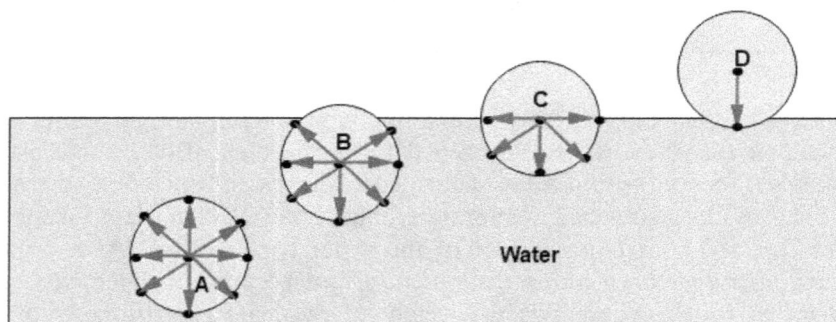

Cohesive pressure: So, no upward cohesive force is there to counteract the downward cohesive force acting on molecule C. Hence, molecule C experiences maximum downward cohesive force (pull). The amount of cohesive force per unit area of the surface of a liquid is called cohesive pressure.

In van der Waals' modified gas equation

$$\left(P + \frac{a}{V^2}\right)(V - b) = nRT,$$

the cohesive pressure is given by the factor $\frac{a}{V^2}$. The cohesive force causes the liquid surface to contract towards the interior of the liquid until the repulsive (upward) collisional force stops further contraction when the area of the surface becomes minimum.

For molecule D, the maximum portion of its sphere of influence is situated above the liquid surface. So, no considerable force acts on it due to the liquid and the molecule D moves haphazardly in vapour state.

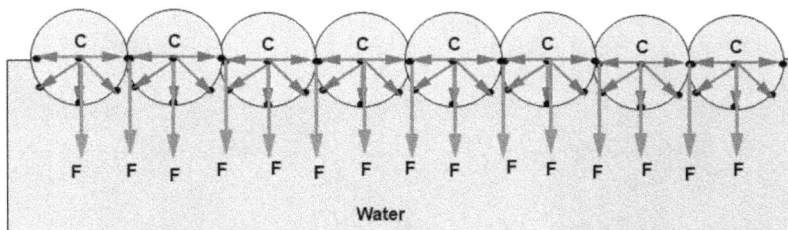

Array of surface molecules create a streched rubberband-like effect

Recapitulating, each surface molecule experiences maximum downward cohesive force (F, say). The cohesive force per unit surface area is called cohesive pressure. In other words, a liquid surface is not free, rather it is under pressure by the net downward cohesive force exerted by the liquid molecules lying within the sphere of influence (a distance of 10^{-7} cm from the surface molecules). As the liquid surface is pulled down it tends to occupy a minimum surface area being balanced by an upward reaction caused by the colliding bulk liquid molecules.

6.12 Measuring surface tension

Molecular theory tells us that the surface molecules (molecules of the skin of liquid surface) of a liquid attract each other more strongly than the molecules in the bulk of liquid. Now, let's draw an imaginary line AB on the free surface of the liquid. It will divide the surface of liquid into two parts. Now, we can imagine two rows 1 and 2 of surface molecules separated by the imaginary line AB. Since each molecule in row 1 attracts the neighbouring molecules in row 2, the surface tension (cohesive) force is uniformly distributed over all molecules of the rows.

If the total surface tension force acting on the rows of a length δl is δF, for uniform distribution of the surface tension force, dividing δF by δl, we have '$\delta F/\delta l$', which is a constant quantity, defined as a measure of surface tension of liquid. We call it surface tension denoted by the letter T. Experimentally it is verified that surface tension of a liquid is constant at given temperature. If temperature increases, cohesive forces decrease. So, surface tension decreases with temperature. It is experimentally verified that surface tension decreases linearly with temperature as $T_\theta = T_0(1 - \alpha\theta)$, where $\alpha = $ temperature coefficient of surface tension.

Since the direction of surface tension is perpendicular to any imaginary line AB, surface tension T has no preferred direction. Therefore, we can treat the surface tension T as a scalar, whereas force is a vector. Its unit is $\mathrm{N\,m^{-1}}$ or $\mathrm{dyne\,cm^{-1}}$ (the same as the unit of stiffness of a spring).

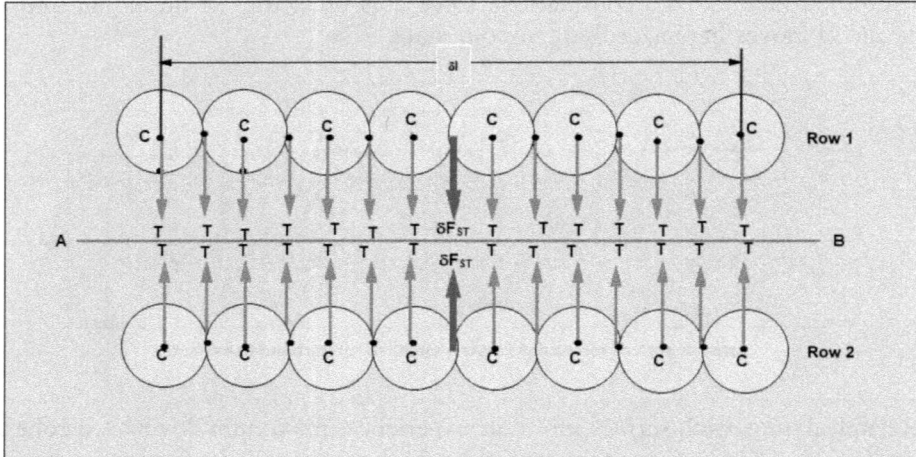

Recapitulating, surface tension of a liquid is given as the force acting on one side of a unit length of an imaginary line drawn on the liquid surface. Symbolically $T = \frac{\delta F}{\delta l} = \frac{dF}{dl}$ or more generally F/l. It is a line (linear) intensity of force of cohesion on the free surface of liquid. It is constant for a given temperature. It varies with temperature as $T_\theta = T_0(1 - \alpha\theta)$, where $\alpha =$ temperature coefficient of surface tension. It is a scalar quantity because it has no preferred (definite) direction. Its unit is $\mathrm{N\,m^{-1}}$ or $\mathrm{dyne\,cm^{-1}}$. Surface tension changes if the liquid is contaminated. For instance, surface tension of water decreases when it is mixed with kerosene.

Example 14 Explain the existence of surface tension in a thin film.
  Solution
  Let's trap a thin soap film by dipping a square wire frame into a soap solution. The frame has a flexible cotton loop fixed between points A and B. You can see that the cotton loop can stay with any shape on the soap. This is because the net force acting on each element of the loop due to the soap film is zero. If you prick the soap film trapped by the loop in region 2, at any point P by a needle, you can see that the loop assumes a circular shape immediately after rapturing the soap film in region 2. This is possible only when the soap film in region 1 pulls each point (element) of the loop radially outward with equal magnitude of force dF. Thus, the loop will remain in equilibrium experiencing zero net force. Furthermore, the soap film will acquire minimum possible area. Hence, the loop has to be stretched equally in radially outward directions so as to enclose a maximum area assuming a circular shape.

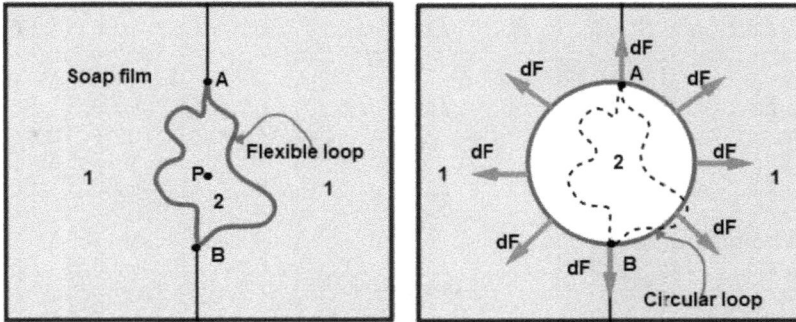

In the foregoing example, if we change the shape and size of the wire frame, in all cases the loop will turn to be same circle after pricking the soap film at P. So, due to surface tension the thin film acquires minimum surface area. Net force acting on the loop is zero but net force dF acting at each elementary segment (point) of the loop by the surrounding thin soap film is not zero. The ratio of elementary force dF and elementary length dl, that is, $\frac{dF}{dl}$, is a constant quantity at a given temperature. We define this quantity as a measure of surface tension.

Example 15 A greased needle of length l is floating on the water surface. Find the maximum possible mass of the needle so that it remains floating. Assume $T =$ surface tension of water.

Solution

If $m =$ mass of the needle is m, the liquid surface gets deformed. The surface tension forces acting on both sides of the needle make an angle θ, say, with vertical. Referring to the free-body diagram, the forces acting on the needle are the surface tension forces on both sides of the needle $F\nwarrow$, $F\nearrow$, and the weight $mg\downarrow$ of the needle. Resolving the forces vertically for its equilibrium, we have

$$\sum F_y = F \cos \theta + F \cos \theta - mg = 0$$

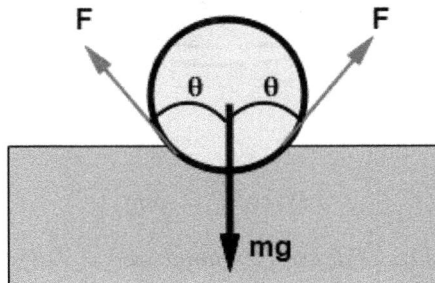

This gives

$$m = \frac{2F \cos \theta}{g},$$

where

$$F = Tl$$

Then, we have

$$m = \frac{2Tl \cos \theta}{g}$$

For m to be maximum,

$$\cos \theta = 1$$

Hence, the maximum possible mass of the needle is

$$m_{max} = \frac{2Tl}{g} \quad \text{Ans.}$$

Example 16 A greased thin ring of mass m and radius R rests on the water surface. Find the (a) maximum value of mass m and (b) minimum force required to lift for a given mass m.

    Solution

(a) If m = mass of the needle is m, the liquid surface gets deformed. The surface tension forces acting on both sides of the needle make an angle θ, say, with vertical. Referring to the free-body diagram, the forces acting on the needle are the surface tension forces on both sides of the needle $dF\nwarrow$, $dF\nearrow$, and the weight $dmg\downarrow$ of the needle. Resolving the forces vertically for its equilibrium, we have

$$dF_y = dF \cos \theta + dF \cos \theta - dmg = 0$$

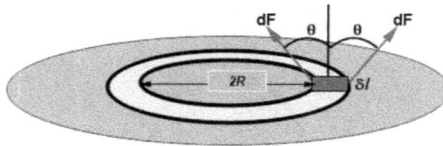

This gives

$$2dF \cos \theta = dmg$$

$$\text{Or, } 2Tdl \cos \theta = dmg (\because dF = Tdl)$$

Integrating both sides, we have

$$2Tl \cos \theta = mg$$

$$\Rightarrow m = \frac{2Tl \cos \theta}{g},$$

where $l = 2\pi R$

$$\Rightarrow m = \frac{2T(2\pi R)\cos \theta}{g} = \frac{4\pi R \cos \theta}{g}$$

For m to be maximum, $\cos \theta = 1$. Hence, $m_{max} = \frac{4\pi R}{g}$ Ans.

(b) For lifting the ring we have to counterbalance the weight of the ring and the surface tension force. At the verge of losing contact, the surface tension force on each elementary length will act vertically downward. The water pulls the ring down with a maximum force

$$F_{ST} = 4\pi RT$$

So, the vertical upward external force required just to lift the ring is

$$F_{ext} = mg + 4\pi RT \quad \text{Ans.}$$

6.13 Surface energy

Origin of surface energy: The surface of a liquid is ideally 10–12 atomic layer. The interatomic spacing is 10^{-10} m and the sphere of influence has radius 10^{-9} m, which is called a molecular range. Up to this depth from the liquid surface, this thickness is known as the skin (or surface layer). As we explained earlier, each surface molecule experiences maximum downward cohesive pull F and this decays to zero within the skin of the liquid surface. Below the skin can be termed as bulk of liquid. So, when we try to bring any atom (or molecule) from the bulk to the surface of liquid, we have to fight against the downward cohesive force. Thus, positive work will be done by the external agent. This work will be stored as the extra potential energy between the surface molecules.

Let's take two molecules 1 and 2 at the surface and molecule 3 at the bottom of the surface layer. If we pull molecule 3 from the bottom to the top of the surface layer, we have to make room between molecules 1 and 2 to accommodate molecule 3. So, the surface molecules 1 and 2 will have to be pushed apart against their mutual force of attraction (cohesion). During this process, the external agent (force) F_{ext} will not be able to do any work because $F_{ext} \perp d\overrightarrow{S_1}$ (and $d\overrightarrow{S_2}$), but ultimately it is responsible for performing work in stretching the surface by increasing the surface molecules. Since a surface molecule experiences more cohesive force than the other deeper molecules in the surface layer, the amount of potential energy of surface molecules is greater than the potential energy of the others. Thus, we can say that while adding surface molecules we really increase the magnitude of potential energy of surface molecules. The increment of potential energy of the additional surface formed is called 'surface energy'.

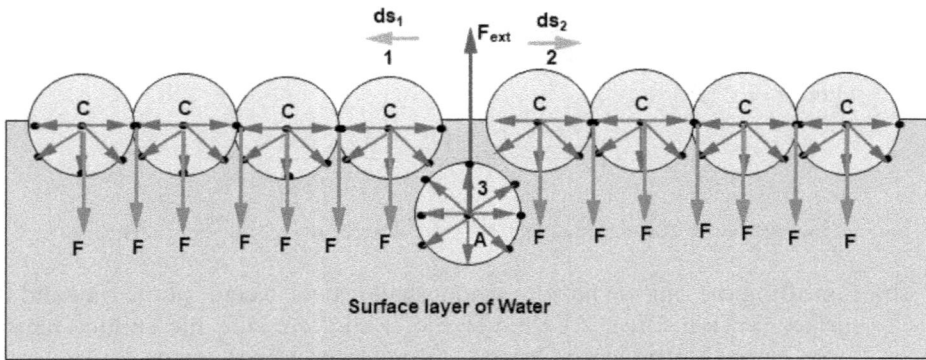

To separate the surface molecules 1 and 2 to accomdate the bulk
molecule 3 , the external agent (F_{ext}) does work

Stability of surface: We know that the total potential energy of the system must be minimum for attaining a stable configuration. Here the system is the 'surface layer'. Since the thickness of the surface layer is fixed (10^{-9} m), to possess minimum potential energy, the number of surface molecules must be minimum. In other words, the surface must be squeezed to have a minimum area. This happens due to the downward cohesive force (sum of all molecular cohesive forces F) acting on the surface molecules. On the other hand, the energetic molecules move from the bulk to the surface in the process of their random motion. Thus, a liquid surface is always agitated by continuously changing molecules. However, the free surface of a liquid tries to occupy the minimum surface area to possess the minimum potential energy.

The above discussion tells us that:

1. The free surface of a liquid possesses the minimum potential energy to remain in stable equilibrium.
2. The potential energy possessed by the surface molecules of liquid per unit surface area is called surface energy.
3. The free surface area of a liquid tends to be minimum to minimise the potential energy (or surface energy).
4. The surface molecules have the maximum amount (magnitude) of potential energy. A negative sign is assigned due to cohesion to possess a minimum potential energy.
5. Deeper molecules in the surface layer have less force and less magnitude of potential energy than the molecules nearer to the free surface of liquid.

Example 17 A soap film is trapped by a rectangular loop. If the surface tension of the soap solution is T, find the minimum external work done in pulling the light rod AB of length l through a distance x.

Solution

When the rod is slowly moved, acceleration $a = 0$; so, $F_{ext} = F$, where $F =$ surface tension force $= Tl + Tl$ (because there are two surfaces of a soap film; upper and lower).

Then $F_{ext} = 2Tl$.

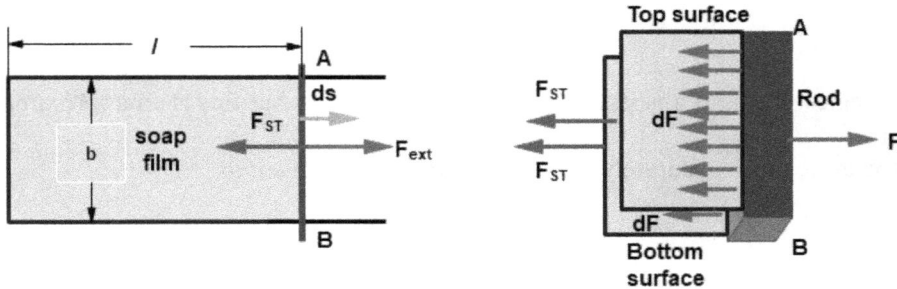

The work done by the external agent in pulling the rod by a distance dx is

$$dW = F_{ext} dx, \quad \text{where } F_{ext} = 2Tl$$

So, we have

$$dW = 2Tldx$$

The total work done by the external agent is

$$W = \int dW = \int_0^x 2Tldx$$

This gives

$$W = 2Tlx$$

N.B.: Surface tension is a surface property. Since we have two surfaces of the soap film, each surface pulls the rod with a force $F' = Tl$. Thus, the total surface tension force acting on the rod is $F = 2F' = 2Tl$.

6.14 Relation between surface tension and surface energy

Referring to the last example we can understand that the work done by the external agent causes an increment of potential energy of the surface because new molecules enter into the surface. As the film is stretched, area increases because its mass is constant. In the process of stretching (pulling the rod) the interior molecules are pulled towards the surface against the cohesive force. Eventually, the molecular agitation (internal kinetic energy) decreases and the surface cools down. So, the surface absorbs heat from the surroundings to maintain its temperature constant. The mechanical work done $W_{ext} = \Delta U$ (energy stored) plus heat absorbed Q is equal to the total energy E of the surface.

$$E = W + Q,$$

where

$$W = (T)2lx = TA \ (A = 2lx = \text{ increase in surface area})$$

Then, we have

$$T = \frac{E - Q}{A}$$

If we neglect the heat absorbed Q, we can write $T = \frac{E}{A}$, but this is an approximated result.

However, more accurately, surface tension can be given as

$$T = \frac{W}{A},$$

where $W =$ external (or mechanical) work done when slowly increasing the surface area by 'A'. If we put $A = 1$, we have

$$T = W$$

This means that the surface tension is numerically equal to the mechanical work done when increasing the surface area of the liquid film (surface) by unity. The energy stored due to the mechanical work in the surface is called the free energy of the surface. In other words, surface tension is equal to the free surface energy of the liquid film.

Summarising the above facts, we can state that surface tension is numerically equal to the external (mechanical) work done in stretching the liquid surface by unity. If we ignore the heat absorbed, surface tension can be defined as the surface energy per unit area. Symbolically

$$T = \frac{dW}{dA}\left(\text{or}\frac{W}{A}\right),$$

where $W =$ mechanical work $=$ potential energy stored in the surface.

Example 18 N number of small droplets each of radius r and surface tension T coalesce to form a bigger drop. Find the change in (a) surface energy and (b) temperature of the water.
Solution

(a) The total surface area of all N droplet is

$$A_1 = 4\pi r^2 N$$

The surface area of the bigger drop is

$$A_2 = 4\pi R^2$$

Then, the change in surface area is

$$\Delta A = A_2 - A_1 = 4\pi(R^2 - r^2 N)$$

The change in surface energy is

$$\Delta E = T\Delta A = 4\pi T(R^2 - r^2 N) \tag{6.18}$$

R can be found by conserving the mass of the system of N droplets. The initial mass = sum of mass of N droplets = mass of the bigger drop. Then

$$N\left(\frac{4}{3}\pi r^3\right)\rho = \frac{4}{3}\pi R^3\rho$$

This gives

$$R = N^{1/3}r \tag{6.19}$$

Using equations (6.18) and (6.19), we have $\Delta E = 4\pi r^2(N^{2/3} - N)T$.

Since $\Delta E < 0$, the surface energy decreases. Hence, heat is evolved, which is equal to change in surface energy. Ans.

(b) The temperature rise is given as

$$\Delta E = 4\pi r^2(N^{2/3} - N)T = mc\Delta\theta$$

Putting $m = N(\frac{4}{3}\pi r^3)\rho$ and simplifying the factors, we have
$\Delta\theta = -\frac{3}{C\rho r}(1 - \frac{1}{N^{1/3}})T$, where $\rho = 1$ gm cc^{-1}, $C = 1$ cal gm^{-1} °C,
$T = 72$ dyne cm^{-1} Ans.

6.15 Pressure difference across a liquid surface

Pressure difference due to curvature of liquid surface: There are three types of surfaces such as plane, concave, and convex as shown the figure. At the plane surface, any surface molecule P, say, is pulled by other surface molecules equally from all sides. So, the net force acting on the surface molecule is zero. Then, the force per unit area at point P is equal to the normal cohesion pressure.

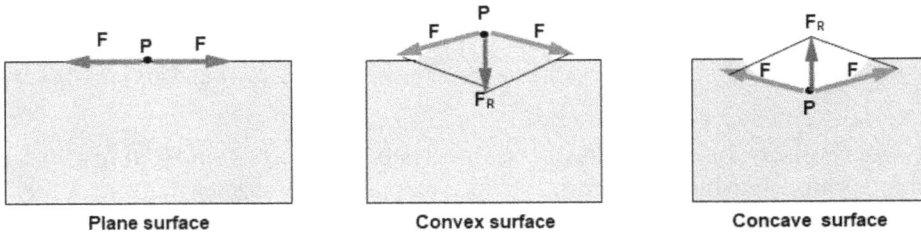

Plane surface   Convex surface   Concave surface

We know that the water meniscus in a capillary tube is concave. If the surface of a liquid is concave, any molecule on the surface is pulled by its surrounding molecules with equal magnitude of forces. As a result, the net force F acting on molecule P due to the concavity is vertically downward. As the surface is pushed up by the additional force F, the resultant surface tension force on the liquid decreases as it opposes the downward cohesion force.

In the case of liquid drops and soap bubbles, the surface is convex. We can see that the net surface tension force F acting on the surface molecule P (due to the other surface molecules) is vertically downward, or radially inward. This increases the downward cohesion force by F. Hence, the cohesion pressure increases.

Let's now find the increase or decrease of cohesion pressure because of the curvature of liquid surface.

Excess pressure inside a liquid drop: Let the pressures inside and outside the liquid drop be P and P_0, respectively. We know that the effective surface area of the liquid drop is equal to πR^2. Now we consider a hemispherical portion of the drop and draw FBD. It is in equilibrium under the forces $P\pi R^2$, $P_0\pi R^2$ and the surface tension force is $2\pi RT$ as shown in the following figure. We have assumed that gravity is not there to give the drop an exact spherical shape.

Equating the upward downward forces, we have

$$P\pi R^2 = P_0\pi R^2 + 2\pi RT$$

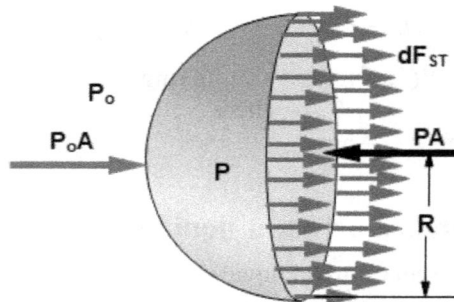

This gives

$$P = P_0 + \frac{2T}{R}$$

The previous expression tells us:

The pressure inside a liquid drop is greater than the outside pressure by an amount $\frac{2T}{R}$, where T = surface tension of the liquid and R = radius of the liquid drop. Hence, the excess pressure inside the drop, that $\Delta P \propto \frac{1}{R}$.

Excess pressure inside a bubble: A bubble has two surfaces 1 and 2 of nearly same radius because the bubble is thin. Hence, we have two surface tension forces of nearly equal magnitude, that is $T(2\pi R)$. Then, the total surface tension force is $4\pi RT$. Following the process described in the previous derivation, we can equate the upward and downward forces to obtain

$$P\pi R^2 = 4\pi RT + P_0\pi R^2$$

This gives

$$P = P_0 + \frac{4T}{R}$$

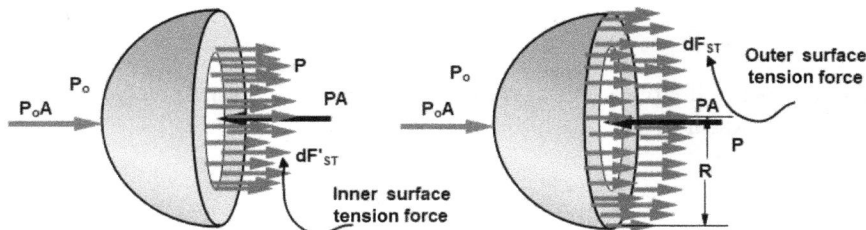

The above expression tells us that:

The excess pressure ΔP inside a bubble placed in air is equal to $\frac{4T}{R}$, where T = surface tension of the bubble and R = radius of the bubble; This means that $\Delta P \propto \frac{1}{R}$. If the bubble is inside water, it will have only one free surface. Hence, the excess pressure $\Delta P = \frac{2T}{R}$, which is equal to the pressure difference inside a liquid drop in air.

Energy method of finding excess pressure

By using the formula, mechanical work done = change in surface energy, we can also find the excess pressure in a liquid drop (or a soap or water bubble). During the drop formation, Let's assume that the radius increases from r to $r + dr$. Hence, the surface area of the bubble increases from $4\pi r^2$ to $4\pi(r + dr)^2$.

Then the change in surface energy is

$\Delta A = 4\pi[(r + dr)^2 - r^2] = 8\pi r^2 dr$, neglecting $(dr)^2$ because dr is very small.

Hence, the increase in surface energy is

$$dE = T\Delta A = 8\pi R^2 T dr \tag{6.20}$$

During the process of formation of the bubble, the work done by the excess pressure $\Delta P(=P'$, say) is

$$dW = P'dV,$$

where $dV = 4\pi r^2 dr$ = increase in volume of the drop.

This gives

$$dW = 4\pi r^2 P' dr \tag{6.21}$$

Then substituting dE from equation (6.20) and dW from equation (6.21) in the expression $dW = dE$, we have

$$8\pi r^2 T dr = 4\pi r^2 P' dr$$

This gives

$$P'(=\Delta P) = \frac{2T}{r}, \text{ where } r = R$$

Similarly, by substituting $dA = 16\pi r^2 dr$ for the bubble (because there are two surfaces; inner and outer) in the expression $dE = TdA$, we have $dE = 16\pi r^2 T dr$. Then equating dE with $dW = 4\pi r^2 P' dr$, we have

$$P' = \frac{4T}{r}, \text{ where } r = R$$

If P_1 and P_2 are the pressures at right side and left side, respectively, of the curved surface S of a liquid drop,

$$P_2 - P_1 = \frac{2T}{R}$$

For the bubble,

$$P_2 - P_1 = \frac{4T}{R}, \text{ where } R = \text{radius of curvature}$$

Hence, $P_2 > P_1$; pressure at the concave side is more.

If the bubble is in water, the outer surface of the bubble is absent. So, we can use the formula $P_2 - P_1 = \frac{2T}{R}$.

Example 19 Two bubbles of radii R_1 and R_2 coalesce to form a complex bubble. Find the radius of curvature R_3, say, of the interface of the bubbles.

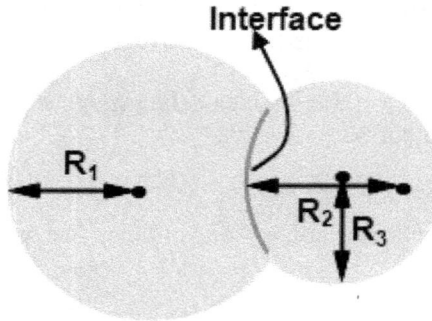

Solution

Let's use the formula $P_2 - P_1 = \frac{2T}{R}$, for the surfaces 1, 2, and 3 to obtain

$$P_1 - P_0 = \frac{2T}{R_1} \tag{6.22}$$

$$P_2 - P_1 = \frac{2T}{R_2} \tag{6.23}$$

and

$$P_2 = P_0 = \frac{2T}{R_3} \qquad (6.24)$$

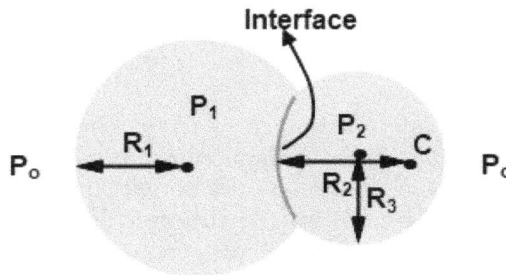

Using equations (6.22)–(6.24), we have

$$-\frac{2T}{R_1} + \frac{2T}{R_3} = \frac{2T}{R_2}$$

This gives

$$R_2 = \frac{R_1 R_3}{R_1 - R_3} \quad \text{Ans.}$$

If R_1 is less than R_3, R_2 will be negative, so the center of curvature of the interface will lie in its left side.

6.16 Angle of contact

Wetting: As you know, each liquid molecules attracts the other surface molecules of the liquid, which is known as cohesion. If the liquid comes in contact with a solid, an additional force comes into play between the molecules of solid and liquid at their interface, which is termed as adhesion. When the adhesive force is stronger than cohesive force, liquid molecules are more attracted towards the glass molecules. So, at the contacting points, liquid molecules rise up little bit on the surface of the solid. Thus, the liquid surface is curved up at the edge forming a concave liquid meniscus. The angle θ made by the meniscus with the downward tangent drawn at the interface of solid and liquid is called 'angle of contact'. As the liquid meniscus is concave, θ is acute. Then, we can call the liquid 'wetting' the solid. For instance, water wets the glass.

Non-wetting: If the cohesion force is stronger than adhesive force, liquid molecules are more attracted by themselves at the edge of the surface than by the solid molecules. Hence, the liquid molecules will shy away from the solid wall. In consequence, the liquid meniscus is curved down at its edge. You can call it concave meniscus, forming an obtuse angle θ of contact. In this case, the liquid is said to be 'non-wetting' the solid. For instance, mercury meniscus is convex in a glass tube and Hence, it does not wet the glass.

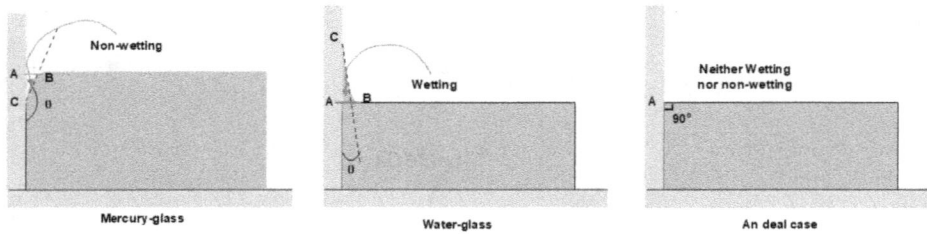

The resulting liquid meniscus is determined by the combined action of cohesive and adhesive forces. If $F_a > F_c$, θ is acute; the liquid wets the solid spreading over the solid surface; If $F_a < F_c$, θ is obtuse; the liquid does not wet the liquid. If $F_a = F_c$, $\theta = 0$; there is perfect wetting (clean glass water). However, if $\theta = 90°$, it will be neither wetting nor non-wetting and the liquid meniscus remains flat at the edge. This is an ideal case of perfect contact, which does not occur in practice. This is because there is no solid and liquid having same potential function that defines the cohesion.

Recapitulating, the angle of contact is determined by the cohesive and adhesive forces.

6.17 Capillary action

Definition: The word capillary is derived from the Latin word 'capillus', which means hair. Then, any hair like thin tube can be called 'capillary tube'. If a glass capillary tube open at both ends is dipped in water, we can see that the water first rises in the tube and finally remains at rest at certain height as shown in the following figure (right side).

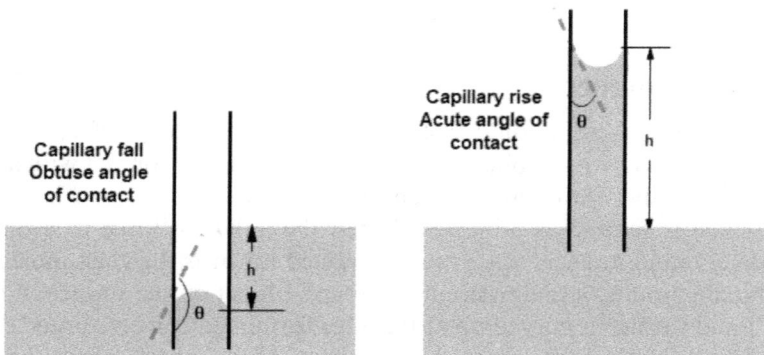

This is called 'capillary rise'. If we insert the glass capillary tube in mercury, the level of mercury gradually comes down and remains at rest at certain depth h, inside the capillary, as shown in figure. The fall of liquid meniscus in a capillary tube is called 'capillary fall'. If we change the angle of orientation of the capillary tube or immerse it more and more into the liquid until the liquid meniscus inside the capillary tube touches the top end of the capillary tube, the height of capillary rise or depth of capillary fall will remain the same.

From the above discussion we define the capillary action as:

The rise or fall of liquid level in a capillary tube relative to the liquid level outside the capillary tube.

Cause of capillary action

Role of adhesion: If the liquid rises just after inserting the capillary tube, we have to conclude that the molecules of the capillary tube pull the molecules of the contacting liquid up against the downward forces such as gravity and cohesion, etc. As the meniscus is curved up (concave), it must stick to the capillary tube (glass). Hence, we imagine a net reaction force F, say, the vertical component of F, that is, F_y holds the peripheral liquid molecules against the downward forces and the horizontal component F_x keeps the liquid molecules in contact with the capillary tube at the interface.

In case of capillary fall, there must be a downward reaction (pressing force) F_y and horizontal reaction F_x keeps the molecules in contact with the tube against other constraint forces.

Role of surface tension: As we see the entire liquid surface moving up like a piston, we can say that surface tension is responsible for it. When the peripheral molecule P is lifted up, surface tension force tries to pull the molecule down along the tangent to the liquid meniscus. Since the surface molecules are tied with each other by surface tension forces, the liquid surface behaves as a stretched membrane. Hence, any attempt in lifting the edge molecules of liquid surface will pull the entire skin (surface) of the liquid. In consequence, the liquid surface is pulled up due to surface tension like a piston, which eventually pulls up the liquid below the skin or surface. If surface tension were not present, only the edge molecules 'like P' would have been lifted up to maximum available height. As the liquid surface rises, the increasing weight of the rising liquid column will stop the liquid surface at certain height. Hence, the surface tension limits the capillary rise. In case of capillary fall, the entire skin of liquid surface presses the liquid against all upward forces like hydrostatic forces.

The capillary action is attributed to both adhesion and cohesion (surface tension) of liquid. Due to adhesion, liquid molecules in contact with the capillary tube will start rising up while sticking to the wall of the capillary tube. Due to surface tension the entire free surface of the liquid is lifted up like a piston instead of only the peripheral liquid molecules of the free liquid surface. The capillary rise occurs when adhesion is stronger than cohesion. In capillary fall, cohesion dominates adhesion. However, in both capillary rise and fall (capillary action), competition between surface tension, cohesion and adhesion takes place.

Measuring capillary rise (Jurin's equation)

Method 1: Let the liquid column be in equilibrium when its height is h. As the liquid is pushed equally from all sides, the resultant of all horizontal forces acting on it is zero. Since the liquid column is in equilibrium, the net vertical force acting on it must be zero. The vertical forces acting on the liquid column are weight $mg\downarrow$, net vertical component of reaction forces, that is, $(T\cos\theta)2\pi R\uparrow$, atmospheric force $P_0 A\downarrow$ at the top and $P_0 A\uparrow$ at the bottom of the liquid column.

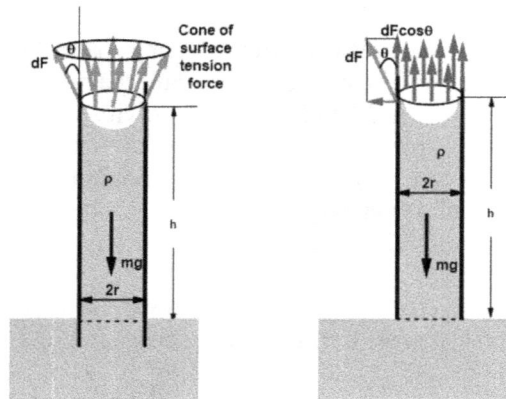

Hence, the net force acting on the liquid column is

$$\sum F_y (T \cos \theta) 2\pi r + P_0 A - P_0 A - mg = 0$$

This gives

$$2\pi T r \cos \theta = mg$$

where $m = \rho(\pi r^2 h)$ (neglecting the curvature of the liquid meniscus).
Then, we have

$$h = \frac{2T \cos \theta}{\rho g r}$$

Method 2: Let's take four points A, B, C, and D as shown in the figure. A and B are the points just above and just below the meniscus, respectively. The pressures at points A and B are given as

$$P_A = P_B + \frac{2T}{R},$$

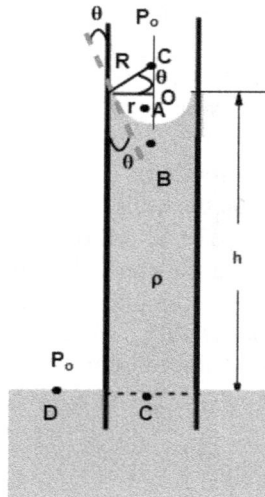

where R = radius of curvature of liquid meniscus. Substituting $P_A = P_0$, we have

$$P_B = P_0 - \frac{2T}{R} \tag{6.25}$$

If C is the point at the bottom of the excess liquid column, we have

$$P_C = P_B + \rho g h$$

where h = height of the bottom of the excess liquid column.

Substituting $P_C = P_D = P_0$, we have

$$P_B = P_0 - \rho g h \tag{6.26}$$

Using equations (6.25) and (6.26), we have

$$P_0 - \rho g h = P_0 = \frac{2T}{R}$$

This gives

$$h = \frac{2T}{\rho g R},$$

where

$$R = \frac{r}{\cos \theta}$$

Then, the height of the capillary rise is

$$h = \frac{2T \cos \theta}{\rho g r} \quad \text{Ans.}$$

N.B.: Let's note the following points from the above expression:
1. h depends on surface tension T, angle of contact θ, radius of the tube, and density ρ of liquid and acceleration due to gravity (or effective g). Hence, the capillary action depends on the nature of liquid and solid (capillary tube) in contact and temperature of the surroundings (because $T = T_0(1 - \alpha\theta)$).
2. If $\theta < 90°$, h is +ve (wetting), which signifies capillary rise and concave meniscus
    $\theta = 90°$, $h = 0$ (neither wetting nor non-wetting); plane meniscus
    $\theta > 90°$, $h = $ −ve (non-wetting), which signifies capillary fall; convex meniscus.
3. Since T, ρ, and θ are constant at given temperature, for any given liquid and capillary tube $hr = c$; this tells us that capillary rise (or fall) will be greater in thinner tubes and vice versa.

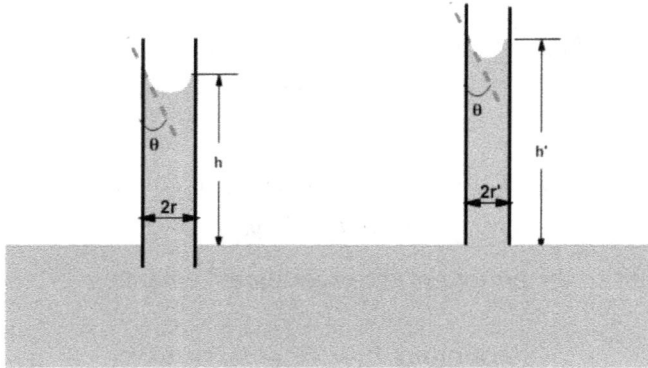

For sufficient height of the capillary tube, greater capillary rise in thinner tube; whereas angle of contact remains constant

4. Since T, ρ, θ, g, and R are constant for a given set of capillary and liquid, we have $h = $ constant. This means that if we change the angle β of orientation from vertical, the capillary rise h (or capillary fall) remains constant even though the length of the excess liquid in the tube increases from h to $h\sec\beta$.

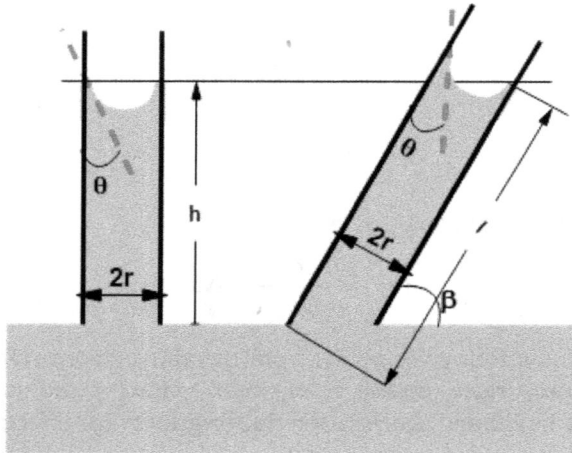

5. If the radius of the liquid meniscus remains constant, the height (but not length) of the liquid column remains the same in capillary tubes of different shapes and sizes.

Rise of liquid in a capillary of insufficient length

The expression of capillary rise is given as

$$h = \frac{2T \cos\theta}{\rho g r}$$

Substituting

$$\frac{r}{\cos \theta} = R \text{ (radius of curvature of the liquid meniscus)}$$

we have

$$hR = \frac{2T}{\rho g}$$

Since T, ρ, and g are constant, we have $hR = \text{constant} = h_0 R_0$, say, where $h =$ free height of the liquid and $R =$ radius of curvature of free meniscus of liquid.

If we slowly push the capillary tube vertically down (into) the liquid, the height of the liquid column remains the same until the length of the tube above the liquid surface is equal to $\frac{2T \cos \theta}{\rho g r}(=h)$. Thereafter, the liquid meniscus gets flattened gradually by increasing its radius of curvature to obey the law $hR = \text{constant}$.

This means that $hR = h'R'$

When the capillary tube just sinks, $h = 0$, we can see that the meniscus becomes flat and radius of curvature $R \to \infty$.

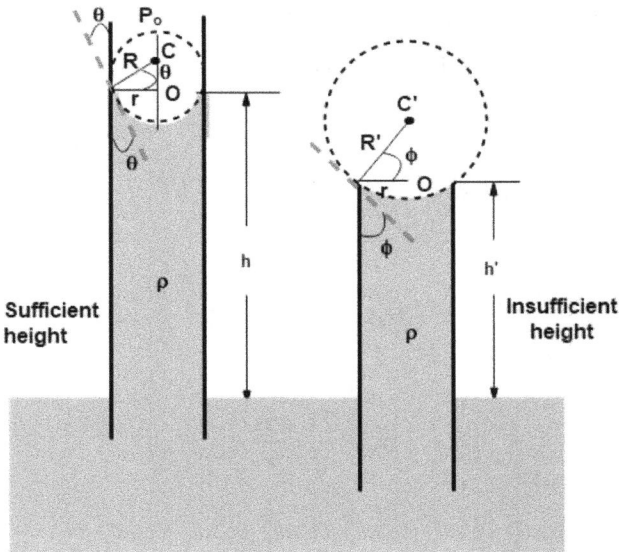

When the length of a capillary above the liquid surface is less than the free height $h = \frac{2T \cos \theta}{\rho g}$, the meniscus assumes a new shape such that the product of its radius of curvature and the new height of the liquid column remains constant, that is, $\frac{2T}{\rho g}$. Hence, the liquid meniscus flattens due to the strong adhesive force between the molecules of liquid and capillary tube. In consequence, the liquid does not ooze (come) out of the tube.

Example 20 In a capillary rise, find the heat developed taking all standard notations as described in the foregoing section.

Solution

As the liquid rises, positive work is done by surface tension in pulling the liquid up by a distance h, which is given as

$W_{ST} = F_y h$, where $F_y = (T \cos \theta) 2\pi r$.

This gives

$$W_{ST}(2\pi r T \cos \theta)h$$

Substituting

$$h = \frac{2T \cos \theta}{\rho g r},$$

we have

$$W_{ST} = \frac{4\pi T^2 \cos^2 \theta}{\rho g}$$

The work done by gravity during the capillary rise is

$$W_{gr} = -\Delta U = mg y_{CM},$$

where

$$y_{CM} = \frac{h}{2} \text{ and } m = (\pi r^2 h)\rho$$

Then

$$W_{gr} = \frac{\pi r^2 \rho g h^2}{2},$$

where

$$h = \frac{2T \cos \theta}{\rho g r}$$

This gives

$$W_{gr} = \frac{2\pi T^2 \cos^2 \theta}{\rho g} \tag{6.27}$$

Applying the first law of thermodynamics, the dissipated energy can be given as

$$Q = W_{ST} = W_{gr} \tag{6.28}$$

Using equations (6.27) and (6.28), we can find the value of Q.

The heat loss is due to the minimization of energy in stable equilibrium of the water column.

Example 21 A vertical U-tube contains a liquid of density ρ and surface tension T. If the radii of the limbs of the U-tube are R and r, find the difference in the heights of liquid column in the limbs. Assume $\theta = $ angle f contact.

Solution

Let the heights of liquid column in the limbs be h_1 and h_2. Using the formula $\Delta P = \frac{2T}{R}$ for meniscus in the limbs, we have the pressures at points A and B given as

$$P_A = P_0 - \frac{2T}{R_1} \tag{6.29}$$

and

$$P_B = P_0 - \frac{2T}{R_2} \tag{6.30}$$

Using the formula $\Delta P = \rho g h$, we have the pressures at A and C

$$P_A(=P_C) - P_B = \rho g(h_2 - h_1) \tag{6.31}$$

Substituting P_A from equation (6.29) and P_B from equation (6.30) in equation (6.31), we have

$$\left(P_0 - \frac{2T}{R_1} \right) - \left(P_0 - \frac{2T}{R_2} \right) = \rho g \Delta h$$

This gives

$$\Delta h = \frac{2T(R_1 - R_2)}{\rho g R_1 R_2}$$

Putting $R_1 = r/\cos\theta$ and $R_2 = r/\cos\theta$, we have

$$\Delta h = \frac{2T(R - r)}{\rho g R r} \cos\theta \quad \text{Ans.}$$

Problem 1 One end of a rod of uniform cross-section A_o is rigidly fixed with a surface. The other end of the rod is pulled along the rod with a force F. Find the maximum possible (a) tangential or shearing stress and (b) longitudinal stress.

Solution

(a) As the body is at rest equal and opposite force F acts on the cross-section A. Resolving the force on F normal to the cross-section A, we have

$$F_n = F \sin \theta$$

So, the tensile stress is

$$p_{\text{ten}} = \frac{F_\perp}{A}$$

$$= \frac{F \sin \theta}{\left(\frac{A_0}{\sin \theta}\right)} = \frac{F \sin^2 \theta}{A_0}$$

p_t is minimum when $\theta = 0$ and maximum when $\theta = 90°$

$$p_t\big|_{\text{min}} = 0 \text{ and } p_t\big|_{\text{max}} = \frac{F}{A_0} \quad \text{Ans.}$$

(b) Resolving the force on F tangential to the cross-section A, we have

$$F_{sh} = F \cos \theta$$

The shearing stress is

$$p_{sh} = \frac{F_{11}}{A} = \frac{F \cos \theta}{\left(\frac{A_0}{\sin \theta}\right)}$$

$$= \frac{F}{A_0} \sin \theta \cos \theta$$

$$= \frac{F}{2A_0} \sin 2\theta$$

$\Rightarrow P_{sh}$ is minimum when $\theta = 0$ and $\theta = 90°$ and p_{sh} is maximum when $\theta = 45°$.
$P_{sh}\big|_{\text{min}} = 0 \text{ and } P_{sh}\big|_{\text{max}} = \frac{F}{2A_0} \quad \text{Ans.}$

Problem 2 A smooth rod of uniform cross-section A is pulled by the force $2F$ and $3F$ as shown in the figure. Find the (a) stress at a point located at a distance $L/3$ from left end, (b) stress at a point located at a distance x, (c) elongation of the rod, and (d) energy stored in the rod.

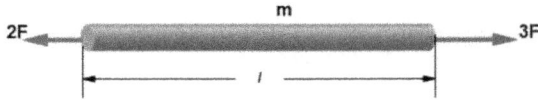

Solution

(a) The stress at the given position is

$$p = \frac{F'}{A} = \frac{m'a + 2F}{A}$$

$$= \frac{\left[\left(\frac{m}{3}\right)\left\{\frac{3F - 2F}{m}\right\} + 2F\right]}{A}$$

$$= \frac{\left(\frac{F}{3} + 2F\right)}{A} = \frac{7F}{3A}$$

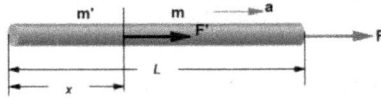

(b) The stress at a distance x is

$$p = \frac{F'}{A} = \frac{m'a + 2F}{A}$$

$$= \frac{1}{A}\left\{\left(\frac{mx}{L}\right)\left(\frac{F}{m}\right) + 2F\right\}$$

$$= \frac{1}{A}\left(\frac{F}{L}x + 2F\right)$$

$$= \frac{F}{A}\left(\frac{x}{L} + 2\right)$$

$$p = \frac{F}{AL}(x + 2L) \quad \text{Ans.}$$

(c) Then, the elongation of the rod is

$$\Delta L = \frac{1}{Y}\int p\, dx$$

$$= \frac{1}{Y} \int_0^L \frac{F}{AL}(x + 2L)dx$$

$$= \frac{F}{YAL}\left(\frac{x^2}{2} + 2Lx\right)\Big|_0^L$$

$$= \frac{F}{YAL}\left(\frac{L^2}{2} + 2L^2\right) = \frac{5FL}{2YA} \quad \text{Ans.}$$

(d) Then, the energy stored is

$$U = \frac{1}{2Y} \int p^2(x)\, dV$$

$$= \frac{1}{2Y} \int_0^L P^2(x)\, A dx$$

$$= \frac{A}{2Y}\left(\frac{F}{AL}\right)^2 \int_0^L (2L + x)^2 dx$$

$$= \frac{F^2}{2AL^2Y}\left\{4L^2 \int_0^L dx + \int_0^L x^2 dx + 4L \int_0^L x\, dx\right\}$$

$$= \frac{F^2}{2AL^2Y}\left\{4L^3 + \frac{L^3}{3} + 2L^3\right\}$$

$$= \frac{F^2L^3}{2AL^2Y}\left(6 + \frac{1}{3}\right)$$

$$= \frac{19F^2L}{6YA} \quad \text{Ans.}$$

Problem 3 A smooth rod of uniform cross-section A is pulled by a force F as shown in the figure. Find the (a) stress at a point located at a distance x, (b) elongation of the rod, and (c) energy stored in the rod.

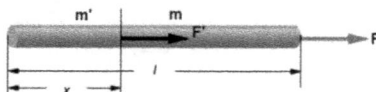

Solution

(a) The stress at a distance x is given as

$$p = \frac{F'}{A} = \frac{m'a}{A}$$

$$\Rightarrow p = \frac{(\rho Ax)a}{A} = \rho ax = \rho \frac{F}{m}x \quad \text{Ans.}$$

(b) Then, the elongation is

$$\Delta l = \frac{1}{Y}\int p(x)\,dx$$

$$= \frac{1}{Y}\int_0^L \left(\frac{\rho Fx}{m}\right) dx$$

$$= \frac{\rho F}{mY}\left(\frac{L^2}{2}\right) = \frac{\rho FL^2}{2mY}$$

(c) The energy stored is

$$U = \frac{1}{2Y}\int P^2\,dx$$

$$= \frac{1}{2Y}\int \left(\frac{\rho F}{m}x\right)^2 dx$$

$$\Rightarrow U = \frac{\rho^2 F^2 L^3}{6m^2 Y} \quad \text{Ans.}$$

Problem 4 A rod of mass m, length l, and density ρ hangs from a celling. Find the (a) stress at a point located at a distance x, (b) elongation of the rod, (c) average stress and strain of the rod, and (d) average energy density of the rod.
Solution

(a) The tension at a distance x is

$$T(x) = m'g = (\rho Ax)g$$

$\Rightarrow$ The stress at a distance x is

$$p(x) = \frac{T(x)}{A} = \frac{\rho g Ax}{A}$$

$$\Rightarrow p(x) = \rho gx \quad \text{Ans.}$$

(b) Then the elongation

$$\Delta L = \frac{\int p(x)\,dx}{lY}$$

$$= \frac{\int_0^L Pgx\, dx}{Y}$$

$$= \frac{\rho g l^2}{2Y} \quad \text{Ans.}$$

(c) The average stress is

$$p(\bar{x}) = \frac{0 + \rho g L}{2} = \frac{\rho g L}{2}(\because p(x) \alpha x)$$

The average strain is

$$\frac{\Delta L}{L} = \frac{\rho g l^2}{2Y}/l = \frac{\rho g l}{2Y} \quad \text{Ans.}$$

(d) The strain energy is

$$U = \frac{1}{2} \int |p(x)|^2 A\, dx$$

$$= \frac{A}{2Y} \int_0^L |p(x)|^2\, dx$$

$$= \frac{A}{2Y} \int_0^L \rho^2 g^2 x^2\, dx$$

$$= \frac{A\rho^2 g^2 l^3}{6Y}$$

⇒ The average elastic energy density is

$$u_{av} = \frac{U}{V} = \frac{\rho^2 g^2 l^2}{6Y} \quad \text{Ans.}$$

Problem 5 Two blocks of mass $3m$ and $5m$ hang from two light metallic strings. The ratio of lengths, area of cross-section, and Young's modulus of the wires are 3:5, 1:3, and 2:1, respectively. Find the ratio of (a) stress, (b) strain, (c) elongation, (d) elastic energy densities, and (e) elastic energy stored of the rods.

Solution

(a) Since the strings are loaded with the blocks of weights $3mg$ and $5mg$, the ratio of the stress is

$$\frac{P_1}{P_2} = \frac{\frac{5mg}{A_1}}{\frac{3mg}{A_2}} = \frac{5}{3}\frac{A_2}{A_1}$$

$$= \frac{5}{3} \times \frac{3}{1} = 5 \quad \text{Ans.}$$

(b) The ratio of strain is $\frac{e_1}{e_2} = \frac{\frac{P_1}{Y_1}}{\frac{P_2}{Y_2}} = (\frac{P_1}{P_2})(\frac{Y_2}{Y_1})$

$$= 5 \times \frac{1}{2} = 2.5 \quad \text{Ans.}$$

(c) The ratio of elongation is $\frac{\Delta L_1}{\Delta L_2} = \frac{e_1 L_1}{e_2 L_2} = (2.5)(\frac{3}{5}) = 1.5 \quad \text{Ans.}$

(d) The ratio of energy densities is $\frac{uv_1}{uv_2} = \frac{Y_1}{Y_2}(\frac{e_1}{e_2})^2 = (2)(2.5)^2 = 12.5 \quad \text{Ans.}$

(e) The ratio of energy stored is $\frac{U_1}{U_2} = \frac{uv_1 \cdot A_1 L_1}{uv_2 A_2 L_2} = (12.5)(\frac{1}{3})(\frac{3}{5}) = 2.5 \quad \text{Ans.}$

Problem 6 A solid uniform cone of slant length L and radius of the base R is resting on a horizontal floor. Find the stress at a vertical distance x from the apex of the cone.
   Solution

(a) Let's assume that at a vertical distance x, the cross-section of the cone is $A = \pi r^2$.
   Then, the stress is $p(x) = \frac{T}{A} = \frac{m'g}{A} = \frac{(\frac{1}{3}\pi r^2 x \rho)g}{\pi r^2} = \frac{\rho g x}{3} \quad \text{Ans.}$

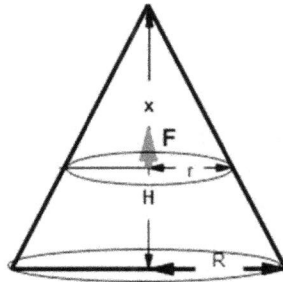

(b) The total elongation is $\Delta L = \frac{1}{Y} \int p(x)\, dx$

$$= \frac{\frac{\rho g}{3}}{Y} \int_0^H x\, dx$$

$= \frac{\rho g H^2}{6Y}$, where $H^2 = L^2 - R^2$ obtained after putting the obtained value of stress.

$$\Rightarrow \Delta L = \frac{\rho g (L^2 - R^2)}{6Y} \quad \text{Ans.}$$

(c) The elastic energy stored is

$$U = \int \frac{p^2(x)}{2Y} A\, dx$$

$$= \frac{1}{2Y} \int \left(\frac{\rho g x}{3}\right)^2 A\, dx$$

$$= \frac{1}{2Y} \int_0^H \left(\frac{\rho g x}{3}\right)^2 \pi r^2\, dx$$

$$= \frac{\pi \rho^2 g^2}{18Y} \int_0^H x^2 r^2\, dx$$

$$= \frac{\pi \rho^2 g^2}{18Y} \int_0^H x^2 \left(\frac{R}{H} x\right)^2 dx$$

$$= \frac{\pi \rho^2 g^2 R^2}{18 Y H^2} \left(\frac{x^5}{5}\right)_0^H = \frac{\pi \rho^2 g^2 R^2 H^3}{90Y}$$

$$= \frac{\pi^2 \rho^2 g^2 R^2 (L^2 - R^2)^{\frac{3}{2}}}{90Y} \quad \text{Ans.}$$

Problem 7 A uniform solid cone is being pulled from both sides with equal magnitude of force F. The radii of cross-sections of the ends and the length of the cone are a, b, and L, respectively. Find the (a) stress at a distance x from the right end, (b) elongation of the cone, and (c) elastic energy stored in the cone. Put $F = Mg$, where $M =$ mass of the cone.

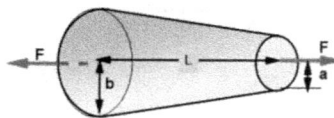

Solution

(a) The tension at a distance x from the right end of the cone is equal to F, because the cone is stationary under the action of equal and opposite force of magnitude F. So, the stress is

$$p(x) = \frac{F}{A} = \frac{F}{\pi r^2},$$

where $r = a + x \tan \theta = a + \frac{x(b-a)}{L}$

$$\Rightarrow p(x) = \frac{F}{\pi \left\{ a + \frac{(b-a)}{L} x \right\}^2} \quad \text{Ans.}$$

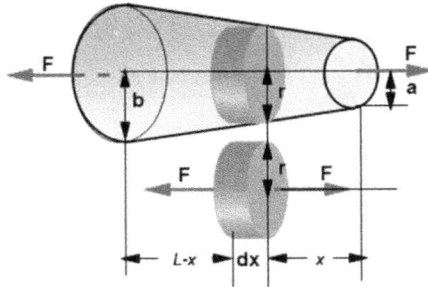

(b) Then the elongation of the truncated rod is

$$\Delta L = \frac{1}{Y} \int p(x) dx$$

$$= \frac{F}{\pi Y} \int_0^L \frac{dx}{\left(a + \frac{b-a}{L} x \right)^2}$$

$$= \frac{Fh}{Y\pi(b-a)} \left(\frac{-1}{a + \frac{b-a}{L} x} \right)_0^L$$

$$= \frac{Fh}{Y\pi(b-a)} \left(\frac{1}{a} - \frac{1}{b} \right)$$

$$= \frac{Fh}{\pi ab Y} = \frac{Mgh}{\pi ab Y} \quad \text{Ans.}$$

(c) The energy stored is

$$U = \int \frac{p^2(x)}{2Y} \frac{A(x)dx}{dv}$$

6-64

$$= \frac{F^2}{2Y} \int \frac{dx}{A(x)} = \frac{F^2}{2Y} \int_0^h \frac{dx}{\left(a + \frac{b-a}{L}x\right)^2}$$

$$= \frac{F^2 h}{2Y\pi(b-a)}\left(\frac{1}{a} - \frac{1}{b}\right)$$

$$\Rightarrow U = \frac{F^2 h}{2\pi ab\, Y} = \frac{M^2 g^2 h}{2\pi ab\, Y} \quad \text{Ans.}$$

Problem 8 Two identical steel rods of radius of cross-section r and one brass rod of radius of cross-section $2r$ are hanging from a ceiling. These three rods are light and their natural lengths are l. A metallic bar of mass M is welded with the rods so that after loading the rod it remains horizontal.

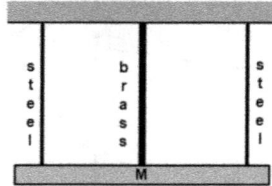

Find the
- (a) elongation of each rod
- (b) tension in each rod
- (c) total energy stored in the rods
- (d) ratio of energy densities in steel and brass rod
- (e) ratio of total energy stored in steel and brass rod.

Solution

(a) Due to the symmetry of load distribution, all the rods have the same elongation, x, say. If the stiffness of the rods are k_1 and k_2, respectively, referring to the free-body diagram,

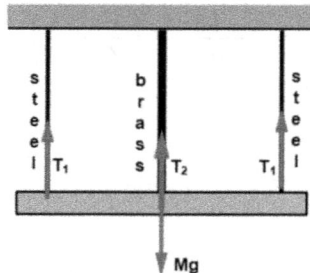

$$\Rightarrow 2T_1 + T_2 = Mg$$

$$\Rightarrow 2k_1 x + k_2 x = Mg$$

$$\Rightarrow x = \frac{Mg}{2k_1 + k_2},$$

where

$$k_1 = \frac{Y_s A_s}{l} \text{ and } k_2 = \frac{Y_b A_b}{l}.$$

$$\Rightarrow x = \frac{Mg}{\left(\frac{2Y_s A_s}{l}\right) + \left(\frac{Y_b A_b}{l}\right)}$$

$$= \frac{Mgl}{(2Y_s A_s + Y_b A_b)} = \frac{Mgl}{2\pi r^2 (Y_s + 2Y_b)} \quad \text{Ans.}$$

(b) Then the tension in the steel rod is

$$T_1 = k_1 x = \frac{Mg k_1}{2K_1 + K_2}$$

$$= \frac{Mg}{2 + \frac{K_2}{K_1}} = \frac{Mg}{2 + \frac{Y_b A_b l_1}{Y_s A_s l_2}}$$

$$= \frac{Mg}{2 + \frac{Y_b A_b l}{Y_s A_s l}} = \frac{Mg}{2 + \frac{Y_b A_b l}{Y_s A_s l}}$$

$$\Rightarrow T_1 = \frac{Mg}{2\left(1 + \frac{2Y_b}{Y_s}\right)} \left(\because \frac{A_b}{A_s} = \frac{r_b^2}{r_s^2} = 4\right) \quad \text{Ans.}$$

The tension in the brass rod is $T_2 = K_2 x = \frac{Mg K_2}{2K_1 + K_2} = \frac{Mg}{2\frac{K_1}{K_2} + 1}$

$$\Rightarrow T_2 = \frac{Mg}{2\frac{Y_1 A_1 l_2}{Y_2 A_2 l_1} + 1} = \frac{Mg}{2\frac{Y_s A_s}{Y_b A_b} + 1}$$

$$= \frac{Mg}{2\frac{Y_s r^2}{Y_b (4r^2)} + 1} = \frac{Mg}{\frac{Y_s}{2Y_b} + 1} \quad \text{Ans.}$$

(c) The total elastic energy stored is

$$U = \frac{F^2}{2k} = \frac{m^2 g^2}{2(2k_1 + k_2)},$$

where $k_1 = \frac{Y_s A_s}{l}$ and $k_2 = \frac{Y_b A_b}{l}$.

$$\Rightarrow U = \frac{M^2 g^2 l}{2(2 Y_s A_s + Y_b A_b)}.$$

$$= \frac{M^2 g^2 l}{2\{2 Y_s \pi r^2 + Y_b(4\pi r^2)\}} = \frac{M^2 g^2 l}{4\pi r^2 (Y_s + 2 Y_b)} \quad \text{Ans.}$$

(d) The ratio of energy densities in the steel and brass rod is

$$\Rightarrow u_s / u_b = \frac{Y_s e_s^2 / 2}{Y_b e_b^2 / 2} = \frac{Y_s}{Y_b} \quad \text{Ans.}$$

(e) The ratio of total energy stored in steel and brass rod

$$\Rightarrow U_s / U_b = u_s A_s l_s / u_b A_b l_b$$

$$= \frac{Y_s}{Y_b} \frac{A_s}{A_b} = \frac{Y_s}{Y_b} \frac{\pi r^2}{4\pi r^2} = \frac{Y_s}{4 Y_b} \quad \text{Ans.}$$

Problem 9 Two wires 1 and 2 of radius of cross-section r and $2r$ are welded with a fixed surface. Their lengths have a slight difference of x_o. First, we pull the string 1 slowly until it touches with the light washer fitted with the wire 2. Then we weld the wire 1 with the washer. Now we pull the washer to the right by a distance x_o. In this process, find the ratio of the average pulling force from x_o to $2x_o$ and 0 to x_o. Assume the ratio of the Young's moduli of the wires is 2:1.

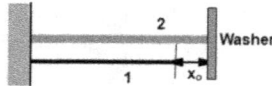

Solution
Let the effective stiffness of the rods be k_1 and k_2. Then the ratio of their stiffness is given as

$$\frac{k_2}{k_1} = \left(\frac{Y_2}{Y_1}\right)\left(\frac{A_2}{A_1}\right) = \frac{Y_2}{Y_1}\left(\frac{r_2}{r_1}\right)^2$$

$$= (1/2)(2)^2$$

$$\Rightarrow \frac{k_2}{k_1} = 2 \tag{6.32}$$

The applied force F_1 at $x = x_0$ is $F_1 = k_1 x_0$. Then the average force between $0 - x_0$ is

$$F_{av_1} = \frac{1}{2}k_1 x_0$$

The first string has elongation $2x_o$ and the 2nd string has elongation x_o. So, the forces at $x = x_o$ and $x = 2x_o$ are $F_1 = k_1 x_0$ and $F_2 = k_1(2x_0) + k_2 x_0$, respectively. Then the average force between $x_0 - 2x_0$ is

$$F_{av_2} = \frac{1}{2}\{(k_1 x_0) + (2k_1 x_0 + k_2 x_0)\}$$

$$\Rightarrow F_{av_2} = \frac{(3k_1 + k_2)x_0}{2}$$

It is given that $\frac{F_{av_2}}{F_{av_1}} = \eta$. Then, by putting the obtained average forces F_{av_1} and F_{av_2}, in the last expression, we have

$$\frac{(3k_1 + k_2)\frac{x_0}{2}}{\frac{k_1 x_0}{2}} = \eta$$

$$\Rightarrow \eta = 3 + \frac{k_2}{k_1} \qquad (6.33)$$

Using the last equations (6.32) and (6.33), we have

$$\eta = 3 + 2 = 5 \quad \text{Ans.}$$

Problem 10 Two uniform rods of lengths $L_1 = 0.5$ m and $L_2 = 0.25$ m area of cross-section $A_1 = 10^{-3}$ m^2 and $A_2 = 0.5 \times 10^{-3}$ m^2 are rigidly joined end to end. This composite rod is made horizontal and welded with the rigid clamps P and Q as shown in the figure. If a horizontal force $F = 100$ KN acts at the interface R of the rods, find the (a) displacement of point R and (b) total elastic energy stored in the rod. Assume the Young's modulus of the rods as $Y_1 = 10^{10}$ N m^{-2} and $Y_2 = 2 \times 10^{10}$ N m^{-2}, respectively.

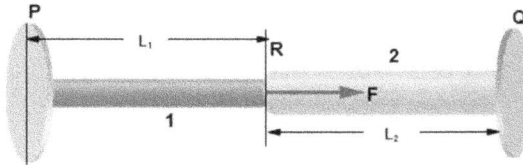

Solution

(a) Rods 1 and 2 can be replaced by two springs of stiffness k_1 and k_2, respectively. As we pull the joint R of these two springs towards the right slowly by a displacement x, each spring will undergo the same deformation (spring 1 is elongated and spring 2 will be compressed). Then the net spring force will oppose the applied force F. At the equilibrium deformation $x = \Delta L$, the net force will be zero. Then, we can write

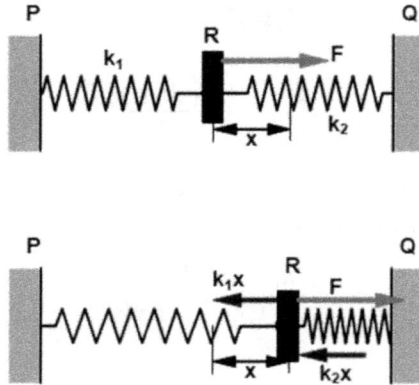

$$(k_1 + k_2)\Delta l = F$$

Then, the displacement Δl of the interface is given as

$$\Delta l = \frac{F}{k_1 + k_2}.$$

where $k_1 = \frac{Y_1 A_1}{L_1}$ and $k_2 = \frac{Y_2 A_2}{L_2}$

$$\Rightarrow \Delta L = \frac{F L_1 L_2}{Y_1 A_1 L_2 + Y_2 A_2 L_1}$$

Putting the given values of the parameters,

$$\Delta L = \frac{(100 \times 10^3 N)\left(\frac{1}{2}\right)\left(\frac{1}{4}\right)}{(10^{10}) \times (10^{-3})\left(\frac{1}{4}\right) + 4 \times 10^{10} \times (0.5 \times 10^{-3})\left(\frac{1}{2}\right)}$$

$$\Rightarrow \Delta L = \frac{\frac{1}{8} \times 10^5}{10^7(0.25 + 1)}$$

$$= \frac{4}{8} \times \frac{10^{-2}}{5} m = 10^{-3} m \quad \text{Ans.}$$

(b) The deformation energy of the rod is

$$U = \frac{(k_1 + k_2)(\Delta L)^2}{2}$$

$$= \frac{1}{2}\left(\frac{Y_1 A_1}{L_1} + \frac{Y_2 A_2}{L_2}\right)\Delta L^2$$

$$= \frac{1}{2}\left\{\frac{10^{10} \times 10^{-3}}{\frac{1}{2}} + \frac{4 \times 10^{10} \times 0.5 \times 10^{-3}}{\frac{1}{4}}\right\}$$

$$= \frac{1}{2}(10^7)(2 + 8) = 0.5 \times 10^8 \text{ J} = 5 \times 10^7 \text{ J} \quad \text{Ans.}$$

Problem 11 Three uniform rods 1, 2, and 3 of lengths l_1, l_2, and l_3, respectively, are rigidly joined end to end. This composite rod is made horizontal and welded with the rigid fixed clamps P and Q as shown in the figure. The area of cross-section of each rod is A and Young's modulus of the rods are Y_1, Y_2, and Y_3, respectively. Find the (a) reactions at the ends of the composite rod and (b) deformation of the rods.

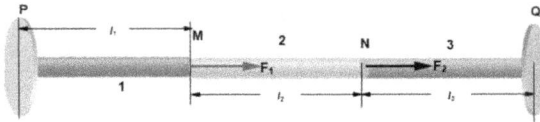

Solution

Method 1: Let R_1 and R_2 be the reactions at P and Q, respectively. Since the composite rod is constrained between two rigid supports, its net deformation is zero; $\Delta L = 0$.

Or, $\Delta l_1 + \Delta l_2 + \Delta l_3 = 0$

Referring to the free-body diagram, rods 1 and 2 are under elongation and rod 3 is compressed. For compression the deformation is negative and for elongation the deformation is positive. Putting the values of individual deformations of the rods, we have

$$\Rightarrow \left(\frac{R_1 l_1}{Y_1 A}\right) + \left(\frac{R_1 l_2}{Y_2 A}\right) + \left(-\frac{R_2 l_3}{Y_3 A}\right) = 0$$

$$\Rightarrow R_1\left(\frac{l_1}{Y_1} + \frac{l_2}{Y_2}\right) - \frac{R_2 l_3}{Y_3} = 0 \tag{6.34}$$

Since the net force on the composite rod is zero,

$$R_1 + R_2 = F_1 + F_2 \tag{6.35}$$

Solving equations (6.34) and (6.35), we have

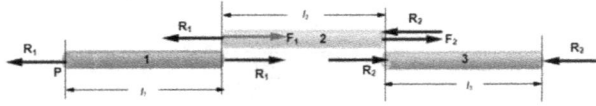

$$R_1\left(\frac{l_1}{Y_1} + \frac{l_2}{Y_2}\right) - \frac{(F_1 + F_2 - R_1)l_3}{Y_3} = 0$$

$$\Rightarrow R_1\left(\frac{l_1}{Y_1} + \frac{l_2}{Y_2} + \frac{l_3}{Y_3}\right) = \frac{(F_1 + F_2)l_3}{Y_3}$$

$$\Rightarrow R_1 = \frac{(F_1 + F_2)l_3}{Y_3\left(\frac{l_1}{Y_1} + \frac{l_2}{Y_2} + \frac{l_3}{Y_3}\right)} \quad \text{Ans.}$$

The other reaction force is

$$R_2 = F_1 + F_2 - R_1$$

$$= F_1 + F_2 - \frac{(F_1 + F_2)l_3}{Y_3\left(\frac{l_1}{Y_1} + \frac{l_2}{Y_2} + \frac{l_3}{Y_3}\right)}$$

$$= (F_1 + F_2)\left\{1 - \frac{l_3}{Y_3\left(\frac{l_1}{Y_1} + \frac{l_2}{Y_2} + \frac{l_3}{Y_3}\right)}\right\}$$

$$= \frac{\left(\frac{l_1}{Y_1} + \frac{l_2}{Y_2}\right)(F_1 + F_2)}{Y_3\left(\frac{l_1}{Y_1} + \frac{l_2}{Y_2} + \frac{l_3}{Y_3}\right)} \quad \text{Ans.}$$

(b) The deformation of rod 1 is

$$\Delta l_1 = \frac{R_1 l_1}{Y_1 A} = \frac{(F_1 + F_2)l_1 l_3}{Y_1 Y_3\left(\frac{l_1}{Y_1} + \frac{l_2}{Y_2} + \frac{l_3}{Y_3}\right)A}$$

The deformation of rod 2 is

$$\Delta l_2 = \frac{R_1 l_2}{Y_2 A} = \frac{(F_1 + F_2)l_2 l_3}{Y_2 Y_3\left(\frac{l_1}{Y_1} + \frac{l_2}{Y_2} + \frac{l_3}{Y_3}\right)A}$$

The deformation of rod 3 is

$$\Delta l_3 = -\frac{R_2 l_3}{Y_3 A} = -\frac{\left(\frac{l_1}{Y_1} + \frac{l_2}{Y_2}\right)(F_1 + F_2)\frac{l_3}{Y_3}}{\left(\frac{l_1}{Y_1} + \frac{l_2}{Y_2} + \frac{l_3}{Y_3}\right)A}$$

$$= -\frac{(F_1 + F_2)(l_1 Y_2 + l_2 Y_1)l_3}{(Y_2 Y_3 l_1 + Y_1 Y_3 l_2 + l_3 Y_1 Y_2)A} \quad \text{Ans.}$$

Problem 12 Three uniform rods 1, 2, and 3 of lengths $l_1 = l, l_2 = 2l$, and $l_3 = l/2$, respectively, are rigidly joined end to end. This composite rod is made horizontal and welded with the rigid fixed clamps at A as shown in the figure. The Young's modulus of the rods are $Y_1 = Y$, $Y_2 = 2Y/3$, and $Y_3 = Y/3$, respectively. The area of cross-section the rods are $A_1 = A$, $A_2 = 3A/2$, and $A_3 = A/2$. Find the (a) deformation of the composite rod, (b) deformation of each rod, (c) displacements of B and C, (d) stress in each rod, and (e) elastic energy stored in the composite rod.

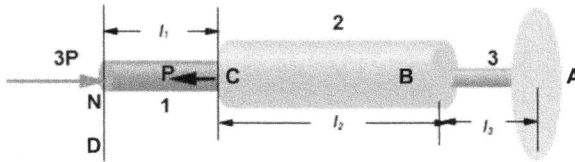

Solution

(a) Since the acceleration of *center of mass* is zero
$a_c = 0$, the rod is at rest. Hence, the net force acting on each rod must be zero.

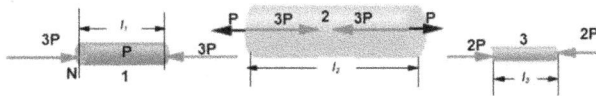

Referring to the free-body diagram, all the rods are under compression with the forces $3P$, $2P$, and $2P$, respectively. Then deformation of the composite rod is equal to the sum of deformation (compression) of each rod, given as

$$\Delta l = \Delta l_1 + \Delta l_2 + \Delta l_3$$

$$= \left(-\frac{3P}{k_1}\right) + \left(-\frac{2P}{K_2}\right) + \left(-\frac{2P}{k_3}\right) = -\frac{3P}{\left(\frac{Y_1 A_1}{l_1}\right)} - \frac{2P}{\frac{Y_2 A_2}{l_2}} - \frac{2P}{\frac{Y_3 A_3}{l_3}}$$

$$= -\frac{Pl_1}{Y_1 A_1}\left\{3 + 2\left(\frac{l_2}{l_1}\right)\left(\frac{A_1}{A_2}\right)\left(\frac{Y_1}{Y_2}\right) + 2\left(\frac{l_3}{l_1}\right)\left(\frac{A_1}{A_3}\right)\left(\frac{Y_1}{Y_3}\right)\right\}$$

$$= -\frac{Pl}{YA}\left\{3 + 2(2)\left(\frac{2}{3}\right)\left(\frac{3}{2}\right) + 2\left(\frac{1}{2}\right)(2)(3)\right\}$$

$$= -\frac{PL}{YA}\{3 + 4 + 6\} = -\frac{13PL}{YA} \quad \text{Ans.}$$

(b) $\Delta l_1 = -\frac{3PL}{YA}$, $\Delta l_2 = -\frac{4PL}{YA}$ and $\Delta l_3 = -\frac{6PL}{YA}$; negative sign signifies compression. Ans.

(c) The displacements of the interfaces are $\frac{-4PL}{YA}$

$$\vec{X_A} = \vec{0}, \ \vec{X_B} = \frac{3Pl}{YA}\hat{i} \ X_c = \Delta l_2 + \Delta l_3 = \frac{4Pl}{YA} + \frac{6Pl}{YA} = \frac{10Pl}{YA}\rightarrow$$

$$\vec{X_D} = \Delta l = \frac{13Pl}{YA}\hat{i} \ \text{Ans.}$$

(d) The stress in the rods can be given as

$$p_1 = \frac{3P}{A_1} = \frac{3P}{A} \quad \text{(compressive) Ans.}$$

$$p_2 = \frac{2P}{A_2} = \frac{2P}{\frac{3A}{2}} = \frac{4P}{3A} \quad \text{(compressive) Ans.}$$

$$p_3 = \frac{2P}{A_3} = \frac{2P}{\frac{A}{2}} = \frac{4P}{A} \quad \text{(compressive) Ans.}$$

(e) The elastic energy stored:

$$U = \frac{1}{2Y_1}\left(\frac{3P}{A}\right)^2 A_1 l_1 + \frac{1}{2Y_2}\left(\frac{4P}{3A}\right)^2 A_2 l_2 + \frac{1}{2Y_3}\left(\frac{4P}{A}\right)^2 A_3 l_3$$

$$= \frac{P^2 A_1}{2A^2 Y_1}\left[9 + \frac{16}{9}\left(\frac{Y_1}{Y_2}\right)\left(\frac{A_2}{A_1}\right)\left(\frac{l_2}{l_1}\right) + \frac{16Y_1}{Y_3}\left(\frac{A_3}{A_1}\right)\left(\frac{l_3}{l_1}\right)\right]$$

$$= \frac{P^2 L}{2YA}\left[9 + \frac{16}{9} \times \frac{3}{2} \times \frac{3}{2} \times \frac{2}{1} + 16 \times \frac{3}{1} \times \frac{1}{2} \times \frac{1}{2}\right]$$

$$= \frac{P^2 L}{2YA}[9 + 8 + 12] = \frac{29P^2 L}{2YA} \quad \text{Ans.}$$

Problem 13 Three uniform rods 1, 2, and 3 of lengths l_1, l_2, and l_3, respectively, are rigidly joined end to end. This composite rod is made horizontal and welded with the rigid fixed clamps P and Q as shown in the figure. The cross-section and Young's modulus of each rod are $Y = 2 \times 10^{11}$ N m^{-2} and $A = 10^{-4}$ m^2, respectively. Three forces $F_1 = 10$ KN, $F_2 = 20$ KN, and $F_3 = 90$ KN act on the composite rod as shown in the figure. Find the (a) deformation of the composite rod the composite rod, (b) deformation of the rods, (c) displacements of the interfaces, (d) stress in each rod, and (e) elastic energy stored in the composite rod. Put $l_1 = 1.5$ m, $l_2 = 1$ m, and $l_3 = 2$ m.

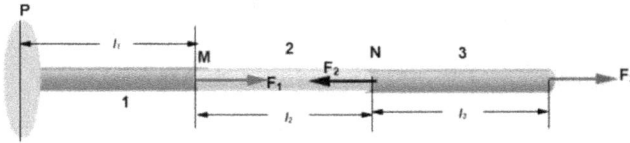

Solution
Method 1: F_1 passes through (acts on) rod 1 only, F_2 passes through rod 1 and 2, and F_3 passes through all three rods. So, the total elongation is given as

$$\Delta L = \left\{ \frac{F_1 l_1}{YA} - \frac{F_2(l_1 + l_2)}{YA} + \frac{F_3(l_1 + l_2 + l_3)}{YA} \right\}$$

Putting the values of the forces,

$$\Delta L = \left\{ \frac{+10 l_1}{YA} - \frac{20(l_1 + l_2)}{YA} + \frac{90(l_1 + l_2 + l_3)}{YA} \right\} \times 10^3 \text{ m}$$

$$= 10^3 \left(\frac{80 l_1 + 70 l_2 + 90 l_3}{YA} \right)$$

$$= \frac{10^4}{YA} (8 l_1 + 7 l_2 + 9 l_3) \quad \text{Ans.}$$

Method 2: Referring to the free-body diagram the net forces acting on the face of the rods are R_1 (assumed), $F_1 - R_1$, and F_3, respectively. For equilibrium of each rod,

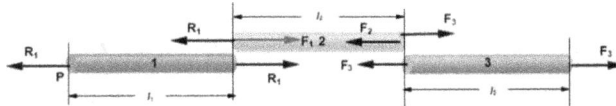

$$F_1 - R_1 = F_2 - F_3$$

Or, $R_1 = F_1 - F_2 + F_3 = (10 - 20 + 90) KN = 80 \ KN$

So, 80 KN (tensile), 70 KN (compressive), and 90 KN (tensile) are the forces acting on the faces of rods 1, 2, and 3, respectively. Then, the net deformation of the composite rod is

$$\Delta l = \left(\frac{80 l_1}{YA} + \frac{70 l_2}{YA} + \frac{90 l_3}{YA} \right) 10^3$$

$$= \frac{10^4}{YA} \{ 8 l_1 + 7 l_2 + 9 l_3 \}$$

$$= \frac{10^4}{2 \times 10^{11} \times 10^{-4}} (8 \times 1.5 + 7 \times 1 + 9 \times 2)$$

$$= 0.5 \times 10^{-3} (12 + 7 + 18)$$

$$= 18.5 \times 10^{-3} \, \text{m}$$

$$= 1.85 \times 10^{-2} \, \text{m}$$

$$= 1.85 \, \text{cm} \quad \text{Ans.}$$

(d) The energy stored in the composite rod is

$$U = \frac{F_1{}^2}{2k_1} + \frac{F_2{}^2}{2k_2} + \frac{F_3{}^2}{2k_3}$$

$$U = \frac{(80 \times 10^3)^2}{\frac{2YA}{l_1}} + \frac{(90 \times 10^3)^2}{\frac{2YA}{l_2}} + \frac{(70 \times 10^3)^2}{\frac{2YA}{l_3}}$$

$$= \frac{10^8}{2YA} (64 l_1 + 81 l_2 + 49 l_3)$$

$$= \frac{10^8}{2 \times (2 \times 10^{11})(10^{-4})} (64 \times 1.5 + 81 \times 1 + 49 \times 2)$$

$$= \frac{2750}{4} = 687.5 J \quad \text{Ans.}$$

Problem 14 A uniform rod of length L, Young's modulus Y, and density ρ rotates in a horizontal plane about a vertical axis passing through one of its ends. (a) If the maximum breaking stress of the rod is p_{max}, find the maximum angular velocity ω and frequency of rotation of the rod. (b) If the angular velocity ω is less than its maximum value, find the elongation of the rod. (c) What is the average energy stored in the rod?

Solution

(a) As the rod rotates with an angular velocity ω, each element revolves with the same angular velocity. So, the tension at any radial distance x is equal to the forces acting on all elements from x to l.

The tension at any distance x is

$$T = \int_x^L dm\, a = \int_x^L dm\,(r\omega^2)(\because a = r\omega^2)$$

$$\Rightarrow T = \int_x^L r\, dm\,\omega^2$$

$$= \rho A \omega^2 \left(\frac{L^2 - x^2}{2}\right)$$

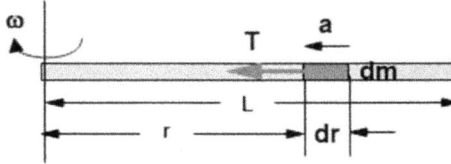

Then the stress at a radial distance x is

$$\Rightarrow \frac{T}{A} = p(x) = \rho\omega^2 \left(\frac{L^2 - x^2}{2}\right)$$

The stress is maximum when $x = 0$ (at the axis of rotation). So maximum stress is

$$p_{max} = \frac{\rho\omega_m^2 L^2}{2}$$

$$\Rightarrow \omega_{max} = \frac{1}{L}\sqrt{\frac{2p_{max}}{\rho}}$$

$$\Rightarrow f_{max} = \frac{1}{2\pi L}\sqrt{\frac{2p_{max}}{\rho}} \quad \text{Ans.}$$

(b) The elongation of the rod is

$$\Delta L = \frac{1}{Y}\int p(x)\, dx$$

$$= \frac{\rho \omega^2}{2Y} \int_0^L (L^2 - x^2) \, dx$$

$$= \frac{\rho \omega^2}{2Y} \left(L^3 - \frac{L^3}{3} \right)$$

$$\Rightarrow \Delta L = \frac{\rho \omega^2 L^3}{3Y} \quad \text{Ans.}$$

(c) The energy stored in the rod is

$$U = \int \frac{p^2}{2Y} A \, dx$$

$$= \frac{A \rho^2 \omega^4}{8Y} \int_0^L (L^4 + x^4 - 2L^2 x^2) \, dx$$

$$= \frac{A \rho^2 \omega^4}{8Y} \left(L^5 + \frac{L^5}{5} - \frac{2L^5}{3} \right)$$

$$= \frac{\rho^2 \omega^4 A L^5}{15Y}$$

The average energy density is

$$\frac{U}{V} = (u_v)_{av} = \frac{\rho^2 \omega^4 L^4}{15Y} \quad \text{Ans.}$$

Problem 15 A uniform flexible circular loop of radius R, mass M, Young's modulus Y, and density ρ rotates in a horizontal plane about a vertical axis passing through the center of loop with an angular velocity ω. Find the (a) tension in the loop, (b) increment of the radius of the loop, (c) energy stored in the loop, and (d) product of maximum frequency of rotation and length of the loop if the braking stress of the loop is σ_b.

Solution

(a) The net radial force acting on the element δm is equal to $2T \sin \frac{\delta \theta}{2}$, which pulls the element towards the center of the circle with an acceleration $a = R \omega^2$. Then, by applying Newton's 2nd law on the element, we have

$$2T \sin \frac{\delta \theta}{2} = \delta m a = \delta m R \omega^2$$

$$\Rightarrow T\delta\theta \simeq \left(\frac{M}{L}R d\theta\right)R\omega^2$$

$$\Rightarrow T = \frac{MR^2\omega^2}{2\pi R} = \frac{MR\omega^2}{2\pi} \quad \text{Ans.}$$

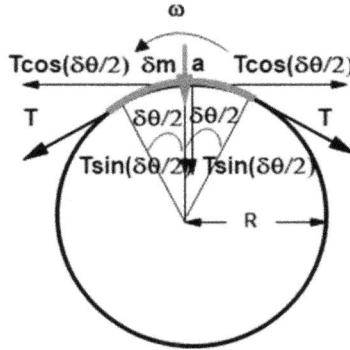

(b) The relative change in length is

$$\frac{\Delta L}{L} = \frac{p}{Y} = \frac{(T/A)}{Y} = \frac{T}{AY}$$

$$\Rightarrow \frac{2\pi dR}{2\pi R} = \frac{MR\omega^2}{2\pi AY}$$

$$\Rightarrow \frac{\Delta R}{R} = \left(\frac{M}{2\pi RA}\right)\left(\frac{R^2\omega^2}{Y}\right)$$

$$\Rightarrow \frac{\Delta R}{R} = \frac{\rho R^2\omega^2}{Y}$$

$$\Rightarrow \Delta R = \frac{\rho R^3\omega^2}{Y} \quad \text{Ans.}$$

(c) The energy stored in the loop is

$$U = \frac{p^2}{2Y}AL = \frac{Y}{2}e^2. \ AL$$

$$\Rightarrow \frac{U}{V} = u_v = \frac{Ye^2}{2}$$

$$= \frac{Y}{2}\left(\frac{\Delta L}{L}\right)^2 = \frac{Y}{2}\left(\frac{\Delta R}{R}\right)^2$$

$$= \frac{Y\rho^2 R^6 \omega^4}{2Y^2} = \frac{\rho^2 R^6 \omega^4}{2Y}$$

(d) Referring to (a), the tension in the wire is $T = \mu R^2 \omega^2$, where $\mu = \frac{M}{L}$

$$\Rightarrow \sigma_b A = \mu R^2 \omega_{max}^2$$

$$\Rightarrow \omega_{max} = \sqrt{\frac{\sigma_b}{\left(\frac{\mu}{A}\right) R^2}}$$

$$\Rightarrow \omega_{max} = \sqrt{\frac{\sigma_b}{\rho R^2}}$$

$$\Rightarrow f_{max} = \frac{1}{2\pi R} \sqrt{\frac{\sigma_b}{\rho}}$$

$$\Rightarrow f_{max} = \frac{1}{L} \sqrt{\frac{\sigma_b}{\rho}}$$

$$\Rightarrow f_{max} L = \sqrt{\frac{\sigma_b}{\rho}} \quad \text{Ans.}$$

Problem 16 A metallic wire of length $2L$ is fastened between two rigid supports A and B such that it is just taut. If a block of mass m is welded at the mod point C of the wire, it comes down by a distance $CD = x$, which is much smaller than the length of the wire. If Young's modulus and density of the wire are Y and ρ, respectively, find the (a) value of x and (b) elastic energy stored in the wire.

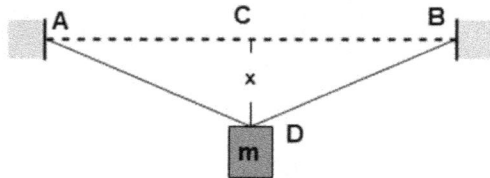

Solution

(a) The strain produced in the wire due to the hanging mass is

$$\frac{\Delta L}{2L} = \frac{2\left(\sqrt{L^2 + x^2} - L\right)}{2L}$$

$$= \left(1 + \frac{x^2}{L^2}\right)^{\frac{1}{2}} - 1$$

$$\simeq 1 + \frac{x^2}{2L^2} - 1$$

$$= \frac{x^2}{2L^2}$$

The tension in the string is given as

$$2T \sin\theta = mg$$

$$\Rightarrow T = \frac{mg}{2} \csc\theta$$

The stress produced in the wire is $p = \frac{T}{A} = \frac{mg}{A \sin\theta}$.
Putting $\sin\theta \simeq \frac{x}{L}$ ($\because x \ll L$), we have

$$p = \frac{mgL}{Ax}$$

Now the stress/strain is

$$\frac{p}{\frac{\Delta l}{L}} (=Y) = \frac{mg\frac{L}{Ax}}{\frac{x^2}{2L^2}}$$

$$\Rightarrow Y = \frac{2mgL^3}{Ax^3}$$

$$\Rightarrow x = \left(\frac{2mgL^2}{AY}\right)^{\frac{1}{3}}$$

(b) The elastic energy stored is

$$U = 2 \times \frac{1}{2}K\left(\frac{1}{2}\Delta L\right)^2 = \frac{K}{4}(\Delta L)^2,$$

where $\Delta L = \frac{x^2}{2L}$ and $K = \frac{YA}{L}$

$$\Rightarrow U = \frac{1}{4}\left(\frac{YA}{L}\right)\left(\frac{x^2}{2L}\right)^2 = \frac{YAx^4}{16L^3}$$

$$= \frac{YA}{16L^3}\left(\frac{2mgL^3}{AY}\right)^{\frac{4}{3}}$$

$$= \frac{m}{8}gL\left(\frac{2mg}{AY}\right)^{\frac{1}{3}} \quad \text{Ans.}$$

Problem 17 Two strings 1 and 2 are hanging from a ceiling. The radii of cross-sections are $2r$ and r. The lowest points of the strings are connected with a light rigid horizontal rod of length l. The ratio of the Young's moduli is 1:2. A weight Mg is loaded at a distance x from the string 1 so that after loading the rod remains horizontal.

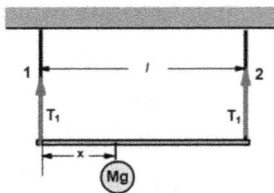

Find the ratio of
- (a) tensions of the strings
- (b) stresses of the strings
- (c) strain of the strings
- (d) elastic energy stored in the strings
- (e) elastic energy densities in the strings.

Solution

(a) As the rod is horizontal and stationary, the torques of the tensions about the point of suspension of the load is zero, which is given as

$$T_1 x = T_2(l - x)$$

$$\Rightarrow \frac{T_1}{T_2} = \frac{l - x}{x}$$

(b) Then the ratio of the stresses is $\frac{p_1}{p_2} = \frac{(T_1/A_1)}{T_2/A_2} = \frac{T_1}{T_2}\frac{A_2}{A_1}$

$$= \left(\frac{l - x}{x}\right)\left(\frac{r_2}{r_1}\right)^2 = \left(\frac{l - x}{x}\right)\left(\frac{1}{2}\right)^2 \quad \text{Ans.}$$

(c) The ratio of the strains is $\frac{e_1}{e_2} = \frac{p_1/Y_1}{p_2/Y_2} = \left(\frac{l-x}{x}\right)\left(\frac{r_2}{r_1}\right)^2\left(\frac{Y_2}{Y_1}\right)$

$$= \left(\frac{l - x}{x}\right)\left(\frac{1}{2}\right)^2 (2) = \frac{l - x}{2x} \quad \text{Ans.}$$

(d) The ratio of the elastic energy is

$$\frac{U_1}{U_2} = \frac{\left(\frac{p_1^2}{2Y_1}\right)(A_1 L_1)}{\left(\frac{p_2^2}{2Y_2}\right)(A_2 L_2)} = \left(\frac{p_1}{p_2}\right)^2 \left(\frac{Y_2}{Y_1}\right)\left(\frac{A_1}{A_2}\right)$$

$$= \left(\frac{l-x}{x}\right)^2 \left(\frac{r_2}{r_1}\right)^4 \left(\frac{Y_2}{Y_1}\right)\left(\frac{r_1}{r_2}\right)^2$$

$$= \left(\frac{l-x}{x}\right)^2 \left(\frac{r_2}{r_1}\right)^2 \left(\frac{Y_2}{Y_1}\right)$$

$$= \left(\frac{l-x}{x}\right)^2 \left(\frac{1}{2}\right)^2 \quad (2)$$

$$= \frac{1}{2}\left(\frac{l-x}{x}\right)^2 \quad \text{Ans.}$$

(e) The ratio of the elastic energy density is

$$\frac{u_{v1}}{u_{v2}} = \frac{\frac{p_1^2}{2Y_1}}{\frac{p_2^2}{2Y_2}} = \left(\frac{p_1}{p_2}\right)^2 \left(\frac{Y_2}{Y_1}\right)$$

$$= \left(\frac{l-x}{x}\right)^2 \left(\frac{r_2}{r_1}\right)^4 \left(\frac{Y_2}{Y_1}\right)$$

$$= \left(\frac{l-x}{x}\right)^2 \left(\frac{1}{2}\right)^4 \quad (2)$$

$$= \frac{1}{8}\left(\frac{l-x}{x}\right)^2 \quad \text{Ans.}$$

Problem 18 A truncated cone of radii r_1 and r_2 and height h is placed on a horizontal floor. Find the y (vertical distance) from the top of the cone vs. the radius of the cone such that the stress remains uniform inside the cone.
Solution
The stress at a distance y is given as

$$p = \frac{mg}{A} = \frac{\int dm\, g}{A}$$

$$p = \frac{\int (\rho \pi r^2 dy) g}{A}$$

$$\Rightarrow \rho \pi g \int r^2 \, dy = pA = p\pi r^2$$

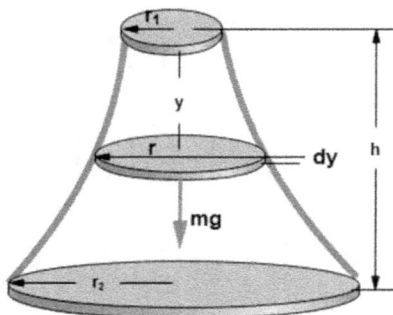

$$\Rightarrow \rho g \, r^2 \, dy = p \, 2r \, dr$$

$$\Rightarrow \rho g dy = 2p \frac{dr}{r}$$

$$\Rightarrow \rho g \int_0^y dy = 2p \int_{r_1}^r \frac{dr}{r}$$

$$\Rightarrow \rho g y = 2p \ln \frac{r}{r_1} \tag{6.36}$$

Putting $y = h$ and $r = r_2$, we have

$$\rho g h = 2p \ln \frac{r_2}{r_1}$$

$$\Rightarrow p = \frac{\rho g h}{2 \ln \frac{r_2}{r_1}} \tag{6.37}$$

By using equations (6.36) and (6.37),

$$\rho g y = 2 \left(\frac{\rho g h}{2 \ln \frac{r_2}{r_1}} \right) \ln \frac{r}{r_1}$$

$$y = \frac{h \ln \frac{r}{r_1}}{\ln \frac{r_2}{r_1}} \quad \text{Ans.}$$

Problem 19 A metallic plate of length L, breadth b, and thickness h is bent into a circular arc of mean radius R. If the Young's modulus of the metal is Y, find the elastic energy stored in the plate.

Solution

The strain of the material (bent strip) at a distance x from the mean radius is given as

$$e = \frac{\Delta l}{l} = \frac{(R+x)\phi - R\phi}{R\phi}$$

$$\Rightarrow e = \frac{x}{R} \tag{6.38}$$

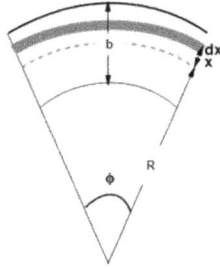

The strain (elastic) energy stored in the elementary (thin) strip (arc) of volume dV is

$$dU = \frac{Y}{2}e^2 dV = \frac{Y}{2}e^2 dV$$

Then, the total elastic energy stored in the bent strip (arc) is

$$U = \int_0^V \frac{Y}{2}e^2 dV = \int_{-b/2}^{+b/2} \frac{Y}{2}e^2\{(R\phi)hdx\} \Rightarrow U = \frac{Y}{2}(R\phi)h \int_{-b/2}^{+b/2} e^2 dx \tag{6.39}$$

Putting $e = \frac{x}{R}$ from equations (6.38) in (6.39),

$$U_{arc} = \frac{YRh\phi}{2} \int_{-\frac{b}{2}}^{\frac{b}{2}} \frac{x^2}{R^2} dx$$

$$= \frac{YRh\phi}{2R^2} \left\{ \frac{\left(\frac{b}{2}\right)^3 - \left(-\frac{b}{2}\right)^3}{3} \right\}$$

$$\Rightarrow U = \frac{YRh\phi}{2R^2}\left(\frac{b^3}{4}\right)\left(\frac{1}{3}\right) = \frac{Yhb^3\phi}{24R},$$

where $\phi = L/R$

$$U = \frac{Yhb^3 L}{24R^2}$$

Problem 20 What is the maximum internal pressure that can be sustained by a glass (a) tube and (b) spherical shell of thickness b and mean radius R? Assume the breaking stress is σ_m.

Solution

(a) Let's take a square element of side δl. As this element is curved subtending an angle $\delta\theta$ at the center of curvature, the net radially inward force acting on the element of area δA is equal to $2T\sin\frac{\delta\theta}{2}$, which is counterbalanced by the pressure force $P\delta A$. Then, by applying Newton's 2nd law on the elementary strip, we have

$$2T\sin\frac{\delta\theta}{2} = P\delta A$$

$$\Rightarrow T\delta\theta \simeq P\delta A \tag{6.40}$$

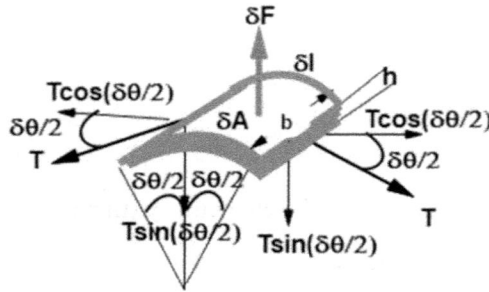

$$T\delta\theta \cong \delta F$$

$$\Rightarrow T\delta\theta = P\delta l.\, h$$

$$\Rightarrow \sigma_m bh\delta\theta = PRd\theta h$$

$$\Rightarrow P = \sigma_m \frac{b}{R} \quad \text{Ans.}$$

(b) Let's take a square patch of side δl. As this element is part of the sphere, each of the four sides of the elementary square patch is curved subtending an angle $\delta\theta$ at the center of curvature O. For two opposite sides, the net radially inward force acting on the element of area δA is equal to $2T\sin\frac{\delta\theta}{2}$. For the other two opposite sides, we have the same radially inward force $2T\sin\frac{\delta\theta}{2}$. So, the net radially inward force is equal to $4T\sin\frac{\delta\theta}{2} \cong 2T\delta\theta$, which is counterbalanced by the pressure force $P\delta A$. Then, by applying Newton's 2nd law on the elementary strip, we have

Applying NSL on the elementary patch,

$$2T\delta\theta = \delta F = P\delta A$$

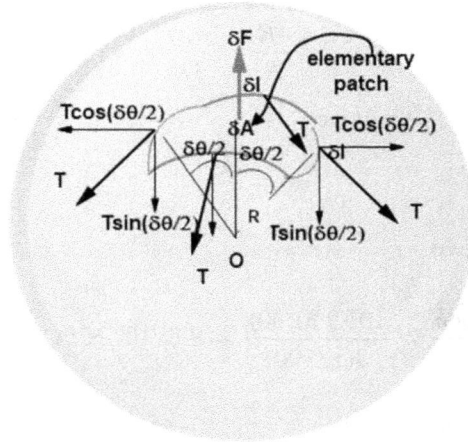

Putting $\delta A = (\delta l)^2$, we have

$$\Rightarrow 2Td\theta = P(\delta l)^2$$

$$\Rightarrow 2(\sigma_m b\delta l)\delta\theta = P(\delta l)^2$$

$$\Rightarrow (2\sigma_m bR\delta\theta)\delta\theta = P(R\delta\theta)^2$$

$$\Rightarrow P = \frac{2\sigma_m b}{R} \quad \text{Ans.}$$

Problem 21 A solid rubber ball is placed $h = 30$ m below the water surface. If the radius of the decreases by $\eta = 1\%$, find the (a) volumetric strain (b) bulk modulus of the ball.

Solution

(a) At the depth of h, the volumetric strain is
$\frac{dV}{V} = \frac{\Delta p}{B}$, where Δp = volumetric stress

$$= P - P_{\text{atm}} = \rho gh$$

$$\Rightarrow \frac{dV}{V} = \frac{\rho gh}{B} \quad \text{Ans.}$$

(b) Since $V = \frac{4}{3}\pi R^3$, taking a log of both sides, $\ln v = \ln\frac{4}{3}\pi + 3\ln R$

$$\Rightarrow \frac{\delta V}{V} = \frac{3\delta R}{R}$$

Putting the obtained value of $\frac{\delta V}{V} = \frac{\rho g h}{B}$, we have

$$\frac{\rho g h}{B} = \frac{3\delta R}{R}$$

$$\Rightarrow \frac{\delta R}{R} = \frac{\rho g h}{3B}$$

Putting $\frac{\delta R}{R} = \eta$, we have

$$\eta = \frac{\rho g h}{3B}$$

Then, we have

$$B = \frac{\rho g h}{3\eta} = \frac{10^3 (9.8)(30)}{3(1/100)} = 9.8 \times 10^6 \ \text{N m}^{-2} \quad \text{Ans.}$$

Problem 22 Derive the expression to relate the (a) Young's modulus in terms of bulk modulus and Poisson's ratio and (b) optimum value of Poisson's ratio.

Solution

(a) The compressive forces are $+PA_1$ and $-PA_1$. They produce a volumetric strain

$$e_1 = \frac{\Delta V}{V} = \frac{P}{B} \tag{6.41}$$

Since $V = l_1 l_2 l_3$

$$\frac{\delta V}{V} = \frac{\delta l_1}{l_1} + \frac{\delta l_2}{l_2} + \frac{\delta l_3}{l_3} \tag{6.42}$$

As we know, the Poisson's ratio is $\frac{\text{Lateral strain}}{\text{Long strain}} = -\sigma$.

Since $\dfrac{\frac{\delta l_1}{l_1}}{\frac{\delta l_3}{l_3}} = -\sigma$ and $\dfrac{\frac{\delta l_2}{l_2}}{\frac{\delta l_3}{l_3}} = -\sigma$, we can write

$$\frac{\delta l_1}{l_1} = -\sigma \frac{\delta l_3}{l_3} \tag{6.43}$$

$$\frac{\delta l_2}{l_2} = -\sigma \frac{\delta l_3}{3} \tag{6.44}$$

Using (6.42), (6.43), and (6.44), we have

$$\frac{\delta V}{V} = \frac{\delta l_3}{l_3}(1 - 2\sigma), \tag{6.45}$$

where $\frac{\delta l_3}{l_3} = \frac{p}{Y}$

$$\Rightarrow e_1 = \frac{\delta v}{v} = \frac{p}{Y}(1 - 2\sigma) \tag{6.46}$$

Since there are three pairs and each face experiences the same pressure, the net volumetric strain is

$$e = 3e_1$$

$$\Rightarrow e = \frac{\Delta V}{V}\bigg|_{\text{total}} = \frac{3p}{Y}(1 - 2\sigma)$$

Putting $e_{\text{vol}} = \frac{p}{B}$, we have

$$\frac{3P}{Y}(1 - 2\sigma) = \frac{P}{B}$$

This gives $Y = 3B(1 - 2\sigma)$ Ans.

(b) Then the compressibility is

$$\beta = \frac{1}{B} = \frac{1}{\left\{\dfrac{Y}{3(1 - 2\sigma)}\right\}}$$

$$\Rightarrow \beta = \frac{3}{Y}(1 - 2\sigma) \text{ Ans.}$$

Since compressibility β cannot be zero if we apply the forces from all sides

$$\beta > 0$$

$$\Rightarrow \frac{3}{Y}(1 - 2\sigma) > 0$$

$$\Rightarrow \frac{3}{Y}(1 - 2\sigma) > 0$$

$$\Rightarrow \sigma < \frac{1}{2}$$

Problem 23 A plate of length L is bent into a hollow cylinder of mean radius R. If the modulus of rigidity is η, find the (a) total energy stored in the cylinders and (b) volume energy density (c) volume energy density at a radial distance x.
   Solution

(a) The shearing strain at a radial distance x is

$$e = \frac{x\phi}{l}$$

Then the energy stored in the thin cylindrical shell of the volume $dv = 2\pi x \, dx \, l$ is

$$dU = \frac{1}{2}\eta e^2 dv$$

Then, the total energy stored in the cylinder is

$$U = \int \frac{1}{2}\eta e^2 dv$$

$$= \frac{\eta}{2}\int e^2 \, dv = \frac{\eta}{2}\int_0^R \left(\frac{x\phi}{l}\right)^2 2\pi x \, dx \, l$$

$$= \frac{\pi\eta\phi^2}{l}\int_0^R x^3 dx$$

$$\Rightarrow U = \frac{\pi\eta\phi^2 R^4}{4l} \quad \text{Ans.}$$

(b) The average volume energy density is

$$\frac{U}{V} = u_{av} = \frac{\pi\eta\phi^2 R^4}{4l\pi R^2 l} = \frac{\eta\phi^2 R^2}{4l^2} \quad \text{Ans.}$$

(c) The volume energy density is $u_v(x) = \frac{1}{2}\eta e^2$

$$= \frac{1}{2}\eta\left(\frac{x\phi}{l}\right)^2$$

$$\Rightarrow u_v(x) = \frac{\eta\phi^2 x^2}{2l^2} \quad \text{Ans.}$$

Problem 24 A (a) disc, (b) annular disc, (c) disc with a square hole, and (d) square plate with a circular hole are placed on the water surface so that they float by surface tension effect. The mass in each case is m. Derive an expression for the minimum upward force required to lift the plates in each case. Assume $T =$ surface tension and $\theta =$ angle of contact. Assume necessary assumptions.

   Solution

   (a) When the bodies are lifted up, the surface tension force acts down. So, in each case the external force must counterbalance the weight and downward surface tension force. The downward surface tension force acting on the periphery of the disc is $2\pi RT \cos\theta$. Then, the net downward force is

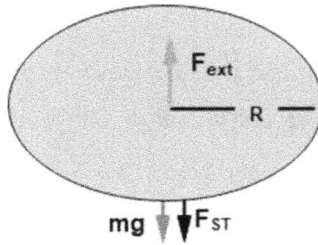

$$F_{ext} = 2\pi RT \cos\theta + mg \quad \text{Ans.}$$

   (b) Annular disc:

   The downward surface tension forces acting on the inner and outer circle are $F_1 = 2\pi RT \cos\theta$ and $F_2 = 2\pi rT \cos\theta$, respectively.

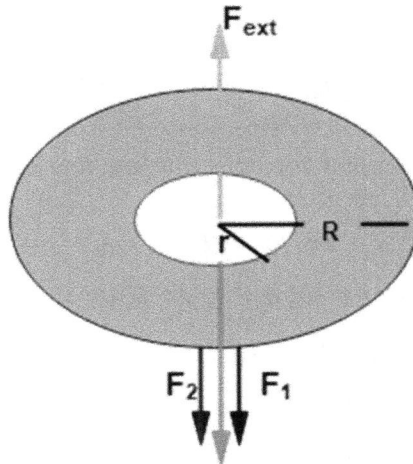

So, $F_{ext} = F_{S.T} + mg$

$$= 2\pi(R + r)T \cos\theta + mg \quad \text{Ans.}$$

(c) Disc with a square hole:

The downward surface tension forces acting on the inner and outer circle are $2\pi RT \cos \theta$ and $4lT \cos \theta$, respectively.

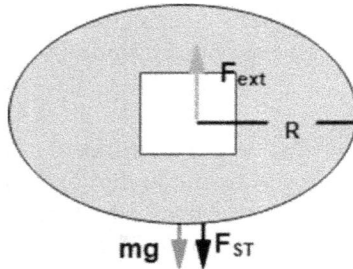

$$mg \quad \downarrow F_{ST}$$

So, $F_{ext} = F_{S.T} + mg$

$$= (2\pi RT + 4lT)\cos \theta + mg$$

$$= T \cos \theta (2\pi R + 4l) + mg$$

$$= 2T \cos \theta (\pi R + 2L) + mg \quad \text{Ans.}$$

(d) Square plate with a circular hole:

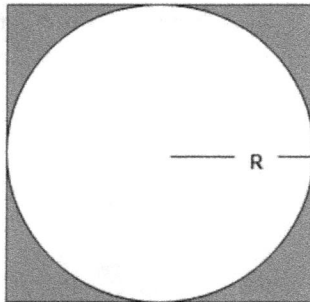

The downward surface tension forces acting on the inner and outer circle are $2\pi RT \cos \theta$ and $4lT \cos \theta$, respectively.

$$\text{So, } F_{ext} = F_{S.T} + mg$$

$$= 4LT \cos \theta + 2\pi RT \cos \theta + mg$$

$$= 2T \cos \theta (\pi R + 2l) + mg$$

If the circle towards the inner sides to be biggest,

$$R_{max} = \frac{L}{2} \text{ Then, } F_{ext} = 2T \cos \theta \left(\frac{\pi L}{2} + 2L \right) + mg,$$

$$= mg + TL \cos \theta (\pi + 4) \quad \text{Ans.}$$

Problem 25 A horizontal cylinder of diameter d, density σ, and length L is floating in a liquid of density ρ such that half of the cylinder is immersed in the liquid. If the θ = angle of contact, find the surface tension of water.

Solution

The cylinder is in equilibrium under the action of net upward surface tension force F_{ST}, buoyant force F_b (up), and downward weight mg of the cylinder.

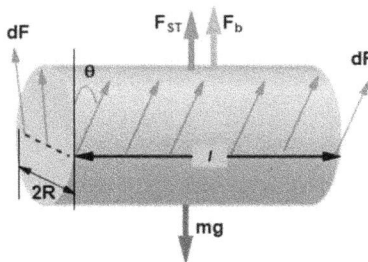

For equilibrium of the cylinder, the net force must be zero:

$$\vec{F_b} + m\vec{g} + \vec{F_{ST}} = 0 \Rightarrow (V/2)\rho g\,\hat{j} - V\sigma g\hat{j} + 2(d+L)T\cos\theta\hat{j} = 0, \qquad \text{where}$$

V = volume of the cylinder and $d = 2R$.

$$\Rightarrow T = \frac{Vg(\sigma - \frac{\rho}{2})}{2(d+L)\cos\theta}, \text{ where } V = \frac{\pi d^2 L}{4}$$

$$\text{So, } T = \left(\frac{\pi d^2 L}{8}\right)\frac{g\left(\sigma - \frac{\rho}{2}\right)}{(d+L)\cos\theta}$$

$$\text{Or, } T = \frac{\pi d^2 Lg\,(2\sigma - \rho)}{16(d+L)\cos\theta} \quad \text{Ans.}$$

Problem 26 A vertical cylinder of diameter d, density σ, and length H is floating in a liquid of density ρ. If the θ = angle of contact and T = surface tension of water, find the depth of immersion of the cylinder.

Solution

The cylinder is in equilibrium under the action of net upward surface tension force F_{ST}, buoyant force F_b (up), and downward weight mg of the cylinder.

For equilibrium of the cylinder,

$$F_{ST} + mg - F_b = 0$$

$$\Rightarrow T\cos\theta(2\pi R) + mg - F_b = 0$$

$$\Rightarrow T = \frac{F_b - mg}{2\pi R\cos\theta}$$

$$=\frac{(\pi R^2 h)\rho g - (\pi R^2 H\sigma)g}{2\pi R \cos\theta}$$

$$\Rightarrow T = \frac{(h\rho - H\sigma)Rg}{2\cos\theta} = \frac{(h\rho - H\sigma)g\ d}{4\cos\theta}$$

Putting $\cos\theta \simeq 1$ for $\theta \simeq 0$, we have

$$h = \frac{1}{\rho}\left(\frac{2T}{RG} + H\sigma\right)$$

$$= \frac{2T}{\rho g R} + H\frac{\sigma}{\rho}$$

$$h = \frac{H\sigma}{\rho} + \frac{4T}{\rho g d} \quad \text{Ans.}$$

Problem 27 A uniform horizontal rod of linear mass density μ and length L is exposed to the film of two liquids of surface tension T_1 and T_2 such that the η fraction of the rod touches the liquid films. Find the (a) net force acting on the rod and (b) acceleration of the rod. Assume $T_2 > T_1$.

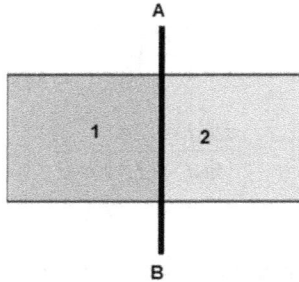

Solution

(a) Both liquids will pull the rod towards them with surface tension forces $2T_1 l$ and $2T_2 l$, respectively. Hence, the frame will experience a net force due to the gap of length l. Then the net force acting on the rod is

$$F = (T_2 - T_1)l \text{ (rightward)} \quad \text{Ans.}$$

(b) The acceleration of the frame is

$$\vec{a} = \frac{\vec{F}_{\text{net}}}{m} = \frac{(T_2 - T_1)l}{\mu L}\hat{i}$$

Since $l = \eta L$, we have

$$\vec{a} = \frac{\eta}{\mu}(T_2 - T_1)\hat{i} \ \text{(rightward)} \quad \text{Ans.}$$

Problem 28 A sector is removed from the equilateral plate ABD of side l so that the right portion is curved with a radius of curvature R. The plate is exposed to the film of two liquids of surface tension T_1 and T_2. Find the net surface tension force acting on the plate. Assume $T_2 > T_1$.

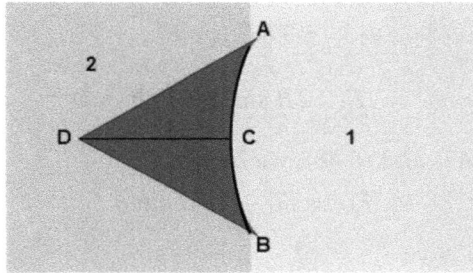

Solution

The free-body diagram shows forces equal to magnitude F act on the sides AD and BD. Let the net force F' act on the curved side ACB of the plate ADB.

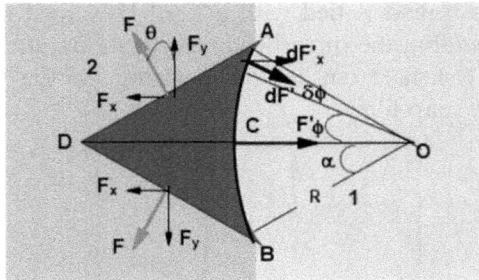

Then the net force acting on the plate is

$$F_{\text{net}} = -F_x - F_x + F' = F' - 2F_x,$$

where $F_x = F \sin \theta = T_2 l \sin \theta$

$$F_{\text{net}} = F' - 2T_2 l \sin \theta \tag{6.47}$$

The net force acting on the arc is

$$F' = \int dF'_x = dF' \cos \phi$$

$$= 2 \int_0^a T_1 R d\phi \cos \phi$$

$$= 2 T_1 R \int_0^\alpha \cos \phi d\phi$$

$$= +2 T_1 R \sin \phi |_0^\alpha$$

$$= 2 T_1 R \sin \alpha$$

$$\Rightarrow F' = l T_1 (\because 2R \sin \alpha = AB = l) \tag{6.48}$$

Using equations (6.47) and (6.48), we have

$$F_{net} = l T_1 - 2 T_2 l \sin \theta$$

Putting $\theta = 30°$, we have

$$F_{net} = l T_1 - 2 T_2 l \sin 30° = (T_1 - T_2) l$$

Since T_2 is greater than T_1, the net force will act to the left. Ans.

Problem 29 A flexible thread is tied with a rigid U-shaped frame. A soap film is trapped in the region so that the thread becomes a circular arc. Find the (a) tension in the thread and (b) surface tension force acting on the thread. Assume that S = surface tension of soap film.

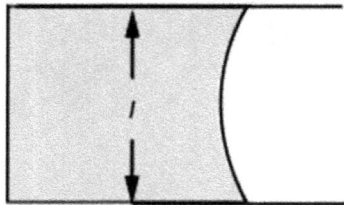

Solution

(a) Consider an element that is pulled up by the surface tension force and down by the components of tension T (but here surface tension is S) in the string.

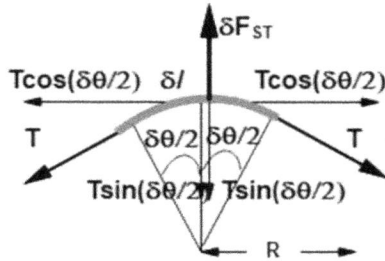

The net force acting on the element of the string is

$$dF_{ST} - 2T \sin \frac{\delta\theta}{2} = 0$$

$$\Rightarrow S(Rd\theta) - T\delta\theta = 0$$

$$\Rightarrow T = SR \quad \text{Ans.}$$

(b) The net force acting on the arc is

$$F_{\text{net}} = F_y = 2 \int dF_{STy}$$

$$= 2 \int dF_{ST} \cos\theta$$

$$= 2 \int_0^\alpha S(Rd\theta)\cos\theta$$

$$= 2SR(\sin\theta)_0^\alpha$$

$$= 2SR \sin\alpha = Sl (\because 2R \sin\alpha = l) \quad \text{Ans.}$$

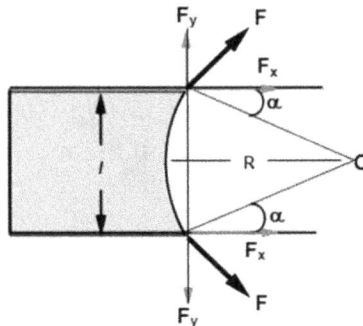

Alternative method: Taking the x-components of tension T and writing the force equation on the arc, we have

$$F_{ST} - 2T \sin\alpha = 0$$

$$\Rightarrow F_{ST} - 2T \sin \alpha = 0$$

$$\Rightarrow F_{ST} - 2(SR)\sin \alpha = 0 (\because T = SR)$$

$$F_{ST} = 2(SR)\sin \alpha = Sl(\because 2R \sin \alpha = l) \quad \text{Ans.}$$

Problem 30 An air bubble of radius R is initially located at a depth h in a lake. If it rises to the surface, find the change in radius of the bubble. Assume $T = $ surface tension of water.

Solution

The pressures inside the bubble at initial position 1 and final position 2 (surface of water) are

$$P_1 = P_0 + \frac{2T}{R_1}$$

$$P_2 = P_0 + \frac{2T}{R_2}$$

Then, $\Delta P = P_2 - P_1$. Using the last three equations, we have

$$\rho g h = 2T \left(\frac{1}{R_1} - \frac{1}{R_2} \right)$$

$$\Rightarrow \rho g h = 2T \left\{ \frac{R_2 - R_1}{R_1 R_2} \right\}$$

Since R_1 and R_2 are nearly equal to R, we can write

$$\Delta R = |R_2 - R_1| = \frac{\rho g h R^2}{2T} \quad \text{Ans.}$$

Problem 31 *Drop count method:* In this method, first of all we can collect and weigh N number of nearly identical liquid drops after detaching from the capillary or radius r by following the *drop weight method*. If the mass of the collected liquid is M, find the surface tension of the liquid. Assume zero angle of contact for perfect wetting.

Solution

If $M = $ total mass of N drops of liquid collected, then the average mass of each drop is

$$m = \frac{M}{N}$$

Using the expression $2\pi r T = mg$, we have

$$T = \frac{mg}{2\pi r} = \frac{Mg}{2\pi N r} \quad \text{Ans.}$$

Problem 32 *Maximum bubble pressure method:* In this method a bubble is formed at the bottom of the capillary when blown from the top with an increasing pressure P. As a result, the bubble grows slowly until it raptures at a critical pressure recorded by height h_m of the monometer as shown in the figure. This is the maximum pressure required to make a bubble at the end of the capillary of radius r. Calculate the surface tension of the liquid.

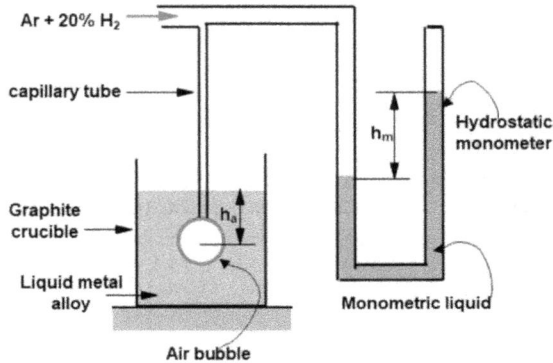

Solution

The pressure inside the bubble can be given as

$$P = P_0 + \rho_a g h_a + \frac{2T}{R} \tag{6.49}$$

$$P = P_0 + \rho_m g h_m \tag{6.50}$$

From (6.49)

$$P - P_0 = \rho_a g h_a = \frac{2T}{R} \tag{6.51}$$

Substituting $P - P_0 = \rho_m g h_m$ from equations (6.50) in (6.51),

$$\frac{2T}{R} = \rho_m g h_m - \rho_a g h_a$$

$$\Rightarrow T = \left(\rho_m h_m - \rho_a h_a\right)\frac{gR}{2} \quad \text{Ans.}$$

Problem 33 *Detachment of a bubble:* When water is heated bubbles are formed at the bottom of the container. Let's consider a bubble newly formed at the bottom of the container, which has a contact circular surface of radius $r \ll R$, where R is the radius of the bubble to be measured. If $T =$ surface tension of water, find the value of R of the bubble at the time of its detaching from the base of the container.

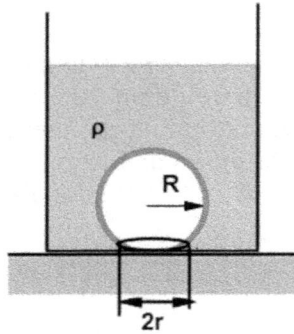

Solution

The bubble is acted on by upward buoyant force F_b and downward net surface tension force. The weight of the air is negligible compared to these two forces, F_{ST}, which must be balanced for the bubble to leave the base.

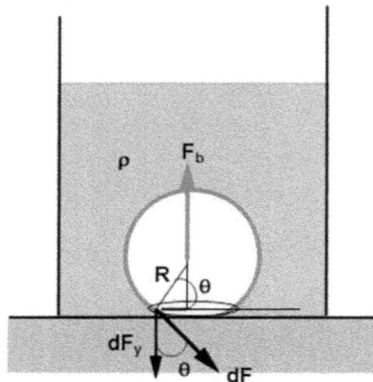

The vertical surface tension force is

$$(F_{S.T})_y = \int dF_y = \int (dF \cos \theta) = \int (T dl \cos \theta)$$

$$= (T \cos \theta) \int dl = (T \cos \theta)(2\pi r) = 2\pi r T \cos \theta,$$

where $r = R \cos \theta$

$$\Rightarrow (F_{S.T})_y = 2\pi r T \left(\frac{r}{R}\right)$$

$$\Rightarrow (F_{S.T})_y = \frac{2\pi r^2 T}{R}$$

For detaching the bottom of the container,

$$F_b \geqslant (F_{S.T})_y$$

$$\frac{4}{3}\pi R^3 \rho g \geqslant \frac{2\pi r^2 T}{R}$$

$$\Rightarrow R \geqslant \left(\frac{3r^2 T}{2\rho g}\right)^{\frac{1}{4}} \quad \text{Ans.}$$

Problem 34 *Effect of surface tension on rising bubble:* An air bubble of radius R rises from the bottom of a lake of height h. Assume that the bubble expands isothermally and its radius at the surface of the lake becomes nR. If the $T = $ surface tension of water, find the depth of the lake.

Solutions At constant temperature, Boyle's law on the bubble between bottom (1) and top (2) positions can be given as

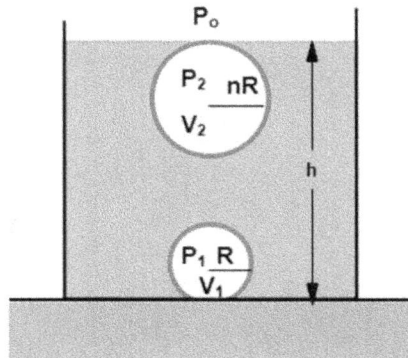

$$p_2 V_2 = P_1 V_1 \tag{6.52}$$

The initial and final pressures inside the bubble are

$$P_1 = P_0 + \rho g h + \frac{2T}{R} \tag{6.53}$$

$$P_2 = P_0 + \frac{2T}{nR} \tag{6.54}$$

Using the last three equations, $\left(P_0 + \frac{2T}{nR}\right)\frac{4}{3}\pi(nR)^3 = \left(P_0 + \rho gh + \frac{2T}{R}\right)\frac{4}{3}\pi R^3$

$$\Rightarrow \left(P_0 + \frac{2T}{nR}\right)n^3 = \left(P_0 + \rho gh + \frac{2T}{R}\right)$$

$$\Rightarrow h = \frac{P_0(n^3 - 1) + 2T\frac{(n^2 - 1)}{R}}{\rho g} \quad \text{Ans.}$$

Problem 35 An air bubble of radius R rises from the bottom of a cylindrical tanker containing two immiscible liquids of densities 3ρ and 2ρ of height $2h$ and h, respectively. Assume that the bubble expands isothermally and its radius at the surface of the lake becomes nR. If T_1 and T_2 are the surface tension of the bottom and top liquids, respectively, find the (a) initial pressure difference across the bubble and (b) change in acceleration of the bubble between its top and bottom positions. Assume $M = $ molar mass of air.

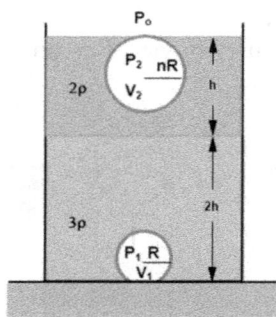

(a) Applying Boyle's law

$$P_f V_f = P_i V_i$$

$$\Rightarrow \left(P_0 + \frac{2T_2}{nR}\right)\frac{4}{3}\pi(nR)^3 = \left\{P_0 + \frac{2T_1}{R} + (2\rho)gh + (3\rho)g(2h)\right\}\frac{4\pi R^3}{3}$$

$$\Rightarrow \left(P_0 + \frac{2T_2}{nR}\right)n^3 = \left(P_0 + \frac{2T_1}{R} + 8\rho gh\right)$$

$$\Rightarrow \Delta P\left(=\frac{2T_1}{R}\right) = \frac{-P_0(n^3 - 1) + 8\rho gh}{\frac{T_2}{T_1}n^2 - 1} \quad \text{Ans.}$$

(b) The difference in accelerations is given as

$$\Delta a = a_f - a_i$$

$$= g\left(\frac{\rho_f}{\sigma_f} - 1\right) - g\left(\frac{\rho_i}{\sigma_i} - 1\right)$$

$$= g\left(\frac{\rho_f}{\sigma_f} - \frac{\rho_i}{\sigma_i}\right) = g\left(\frac{2\rho}{\sigma_f} - \frac{3\rho}{\sigma_i}\right)$$

$$= \frac{\rho g}{\sigma_f}\left(2 - 3\frac{\sigma_f}{\sigma_i}\right) \tag{6.55}$$

Since the density of the air trapped in the bubble is $\sigma = \frac{PM}{RT}$; $M = $ moles mass of the air, we have

$$\frac{\sigma_f}{\sigma_i} = \frac{P_f}{P_i} = \frac{V_i}{V_f} = \left(\frac{R_i}{R_f}\right)^3 = \frac{1}{n^3} \tag{6.56}$$

Finally, using equations (6.55) and (6.56), we have
$\Delta a = \frac{\rho g}{\frac{P_f M}{RT}}\{2 - \frac{3}{n^3}\}$, where $P_f = \frac{2T_2}{nR} + P_0$

$$\Rightarrow \Delta a = \frac{\rho g R T \left(2 - \frac{3}{n^3}\right)}{\left(\frac{2T_2}{nR} + P_0\right)M} \quad \text{Ans.}$$

Problem 36 A soap bubble of radius r is formed over the orifice of a pipe by blowing air in the soap solution. If the radius of the orifice is r, find the minimum velocity of the air at which the bubble will depart from the pipe.

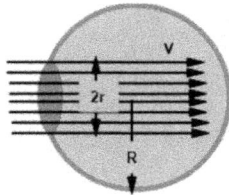

Solution
The impact force of air on the effective area of the orifice is $A = \pi r^2$, given as

$$F_{\text{imp}} = v_{\text{rel}}\frac{dm}{dt} = vA\rho v = \rho A v^2 = \rho \pi r^2 v^2 = \pi \rho r^2 v^2$$

$$\Rightarrow F_{\text{imp}} = \pi \rho r^2 v^2 \tag{6.57}$$

This force must overcome the hydrostatic pressure force

$$F_{\text{hydro}} = (\Delta P)A = \frac{4T}{R}\pi r^2 \tag{6.58}$$

For the bubble to leave the orifice,

$$F_{\text{hydro}} \geqslant F_{\text{imp}}$$

$$\Rightarrow \frac{4T}{R}\pi r^2 \geqslant \pi\rho r^2 v^2$$

$$\Rightarrow v \leqslant 2\sqrt{\frac{T}{\rho R}} \quad \text{Ans.}$$

Problem 37 Two soap bubbles merge at isothermal condition to form a bigger bubble. In this process let the change in surface area and volume of the air be ΔA and ΔV, respectively. Assuming $S =$ surface tension and $P_o =$ atmospheric pressure, prove that

$$3P_0\Delta V + 4S\Delta A = 0$$

(b) If a, b are the radii of the smaller bubbles before merging and c is radius of the bigger bubble after merging, find the temperature of air.

Solution

(a) For no leakage of air, conservation of mass (total numbers of moles) is

$$n = n_1 + n_2$$

For isothermal process $PV = nRT$ because the temperature T is constant in an isothermal process

$$\Rightarrow \frac{P_1 V_1}{RT} + \frac{P_2 V_2}{RT} = \frac{PV}{RT} \Rightarrow P_1 V_1 + P_2 V_2 = PV$$

$$\Rightarrow \left(P_0 + \frac{4S}{r_1}\right)\frac{4\pi}{3}r_1^3 + \left(P_0 + \frac{4S}{r_2}\right)\frac{4\pi}{3}r_2^3 = \left(P_0 + \frac{4S}{r}\right)\frac{4\pi}{3}r^3$$

$$\Rightarrow P_0\left\{\frac{4\pi}{3}r_1^3 + \frac{4\pi}{3}r_2^3 - \frac{4\pi r^3}{3}\right\}$$

$$+ \frac{4S}{3}\{4\pi r_1^2 + 4\pi r_2^2 - 4\pi r^2\} = 0$$

$$\Rightarrow P_0(-\Delta V) + \frac{4S}{3}(-\Delta A) = 0$$

$$\Rightarrow 3P_0\Delta V + 4S\Delta A = 0 \text{ Proved}$$

(b) From the last expression, $T = \dfrac{3P_0\Delta V}{-4\Delta A} = \dfrac{3P_0\frac{4}{3}\pi(r^3 - r_1^3 - r_2^3)}{-4 4\pi(r^2 - r_1^2 - r_2^2)}$

$= \dfrac{P_0(r^3 - r_1^3 - r_2^3)}{4(r_1^2 + r_2^2 - r^2)}$, where $r = c$, $r_1 = a$ and $r_2 = b$

So, $S = \dfrac{P_0(c^3 - a^3 - b^3)}{4(a^2 + b^2 - c^2)}$ Ans.

Problem 38 *Attraction force caused by a wetting liquid droplet:* A liquid drop is trapped between the two glass plates separated by a small distance x. If T is the surface tension of the liquid and gravitational effect is ignored, (a) find the pressure difference across the liquid drop pressed against the plates and (b) prove that the force of attraction (pressing force) between the plates due to surface tension can be given by the following expressions:

$$F = \frac{2TA^2}{V} = \frac{2TV}{x^2} = \frac{2Tm}{\rho x^2}\cos\theta,$$

where m, V, and ρ are the mass, volume, and density of the liquid drop, respectively, and A is the circular area of the liquid in contact with each plate. Assume θ = angle of contact between water and plates.

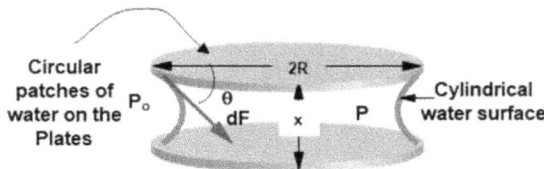

Solution

(a) The pressure difference across the liquid drop pressed against the plates is

$$\Delta P = P - P_0$$

$$= T\left(\frac{1}{R_1} + \frac{1}{R_2}\right) \tag{6.59}$$

Referring to the figure, we have

$$\frac{x}{2\cos\theta} = R_1 \text{ and } R_2 = R \tag{6.60}$$

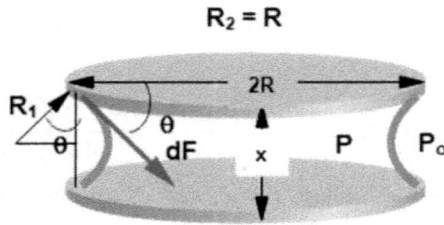

Using the last two equations, we have

$$\Delta P = T\left(\frac{1}{\frac{x}{2\cos\theta}} + \frac{1}{R}\right)$$

Since $x \ll R$, $\frac{1}{R} \ll \frac{2}{x}$

$$\Rightarrow \Delta P \cong \frac{2T}{x}\cos\theta$$

(b) Hence, the attraction (pressing) force between the plates is

$$F = (\Delta P)A_{eH} = \frac{2TA_{eH}\cos\theta}{x} \quad \text{Ans.}$$

$$= \frac{2T}{x} \cdot \pi R^2 \cos\theta$$

$$= \frac{2\pi TR^2 \cos\theta}{x}$$

$$\Rightarrow F = \frac{2T}{x^2}(A_{eH} \cdot x)\cos\theta$$

$$\Rightarrow F = \frac{2T}{x^2}V\cos\theta$$

$$\Rightarrow F = \frac{2Tm\cos\theta}{\rho x^2} \quad \text{Ans.}$$

Alternative (energy) method:
The surface energy of the drop is

$$U = -2(T\cos\theta)A$$

$$= -\left(\frac{T\cos\theta}{x}\right)Ax$$

$$= -\frac{2TV}{x} = -\frac{2Tm}{\rho x}\cos\theta$$

Then, $\overrightarrow{F} = -\frac{ru}{\delta x}\hat{i}$

$$= -\frac{2Tm}{\rho x^2}\cos\theta\hat{i}$$

$$= -\frac{2Tm}{\rho x^2}\cos\theta\hat{i} \quad \text{Ans.}$$

Problem 39 *Height of mercury tablet:* In the previous problem if we remove the plate, the mercury drop of density ρ forms the shape of a tablet of height h on a glass surface. If the top surface of the surface is nearly flat and the angle of contact of glass leaf is $180 - \theta$, find the value of H.

Solution

The mercury drop acquires a shape of a tablet, the upper part is flattened due to the effect of gravity, and the lower part is curved due to the repulsion between molecules of glass and mercury (θ is obtuse). Let $b =$ diameters of the tablet. Writing the force equation for the LHS half segment of the tablet,

$Tb + (T\cos\theta)b = P_{av}.\ Hb$, where the average pressure is

$$P_{av} = \frac{\rho g H + 0}{2}$$

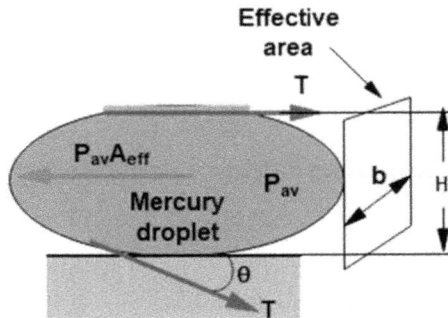

$$\Rightarrow T(1 + \cos\theta)b = \frac{\rho g H^2 b}{2}$$

Then, we have

$$H \simeq \sqrt{\frac{2T(1 + \cos\theta)}{\rho g}}$$

Problem 40 *Drop weight method:* If a liquid of density ρ is made to fall slowly in a capillary tube of radius r, it first forms a drop at the tip of the tube whose size increases gradually until the weight of the drop is equal to the upward surface tension force of the liquid acting on the drop. Let's assume that the drop detaches from the tube when it assumes the shape of a hemi-sphere approximately. Find the maximum height of the column of the liquid.

Solution

Referring to the free-body diagram on the hemispherical waters drop, we have

$$2\pi rT = (P - P_0)\pi r^2$$

$$\Rightarrow \frac{2T}{r} = P - P_0,$$

where $P - P_0 = \rho g(h_m + r) \simeq \rho g h_m$ $(\because h >> r)$.

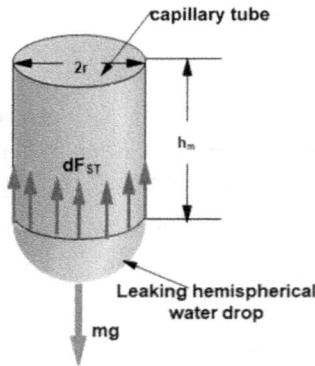

$$\Rightarrow h_m = \frac{2T}{\rho g r} \quad \text{Ans.}$$

Alternative method: The pressure difference across the surface of the water drop is

$$\Delta P = \frac{2T}{r}$$

$$\Rightarrow P - P_0 = \frac{2T}{r} \tag{6.61}$$

Since

$$r \ll H, \quad P \cong P_0 + \rho g h_m \tag{6.62}$$

By using equations (6.61) and (6.62)

$$\frac{2T}{r} = (P - P_0) = \rho g h_{\text{max}}$$

$$\Rightarrow h_{\text{max}} = \frac{2T}{\rho g r} \quad \text{Ans.}$$

Problem 41 In the case of insufficient height of a vertical capillary tube above the liquid level, it may happen that at some stage the height of the liquid column would be equal to the radius of the liquid meniscus. Find that height.

Solution

The radius of the liquid meniscus is $R \cong h$ (given). Then the pressure at A is

$$P_A = P_0 - \frac{2T}{R}, \text{ where } R \simeq h$$

$$\Rightarrow P_A = P_0 - \frac{2T}{h} \tag{6.63}$$

The pressure at B is

$$P_B = P_A = \rho g h \tag{6.64}$$

where $P_B \simeq P_0$.

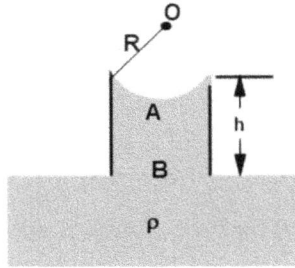

$$\Rightarrow P_0 = \left(P_0 - \frac{2T}{h}\right) + \rho g h$$

$$\Rightarrow h = \sqrt{\frac{2T}{\rho g}} \quad \text{Ans.}$$

Alternative method:

We know that for insufficient height h

$$\Rightarrow hR = \frac{2T}{\rho g}$$

Putting $h = R$, we have

$$h^2 = \frac{2T}{\rho g}$$

$$\Rightarrow h = \sqrt{\frac{2T}{\rho g}} \quad \text{Ans.}$$

Problem 42 A container is partially filled with a liquid of density ρ_2. A capillary tube of radius r is vertically dipped in this liquid. Let's now pour another liquid of lesser density ρ_1 slowly in the container to a height h so that there is only denser liquid column of height h in the capillary tube. Assuming zero angle of contact, find the surface tension of the heavier liquid.

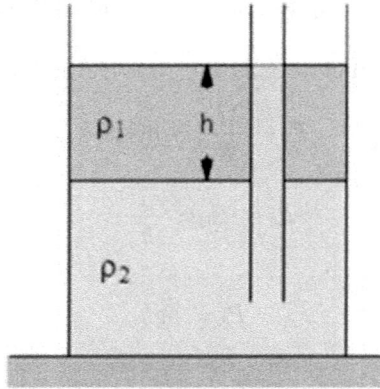

Solution

Pressure equalization at A and B yields

$$P_A = P_0 - \frac{2T}{R} + \rho_2 g(h + h')$$

$$P_B = P_0 + \rho_1 gh + \rho_2 gh'$$

Since $P_A = P_B$

$$P_0 - \frac{2T}{R} + \rho_2 g(h + h') = P_0 + \rho_1 gh + \rho_2 gh'$$

$$\Rightarrow \frac{2T}{R} = (\rho_2 - \rho_1)gh$$

$$\Rightarrow T = \frac{1}{2}(\rho_2 - \rho_1)ghR \quad \text{Ans.}$$

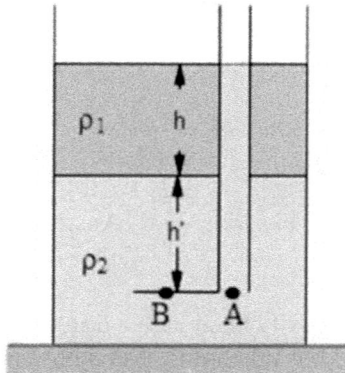

Problem 43 *Accelerating capillary tube*: A vertical communicating U-tube contains a liquid of surface tension T. The liquid meniscus has radii r_1, r_2 in the left and right limbs, respectively, while the tube moves with a horizontal acceleration a as shown in the figure. (a) Find the height difference $h = (h_1-h_2)$ of the liquid in two communicating capillary tubes. (b) What must the be acceleration so that both the limbs have the same height of liquid column? (c) What is the height difference if the tube is either at rest or moves with a constant velocity? (d) If we neglect the effect of surface tension, find the value of h. (e) If the radii of the limbs are the same, find the value of h.

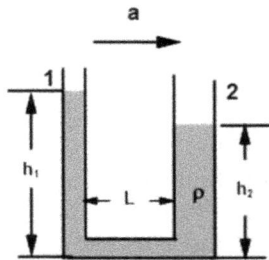

Solution

(a) Equating the difference of pressures at A and B with $\rho a L$,

$$P_A - P_B = \rho a L$$

$$\Rightarrow \left(P_0 - \frac{2T}{r_1} + \rho g h_1 \right) - \left(P_0 - \frac{2T}{r_2} + \rho g h_2 \right) = \rho a L$$

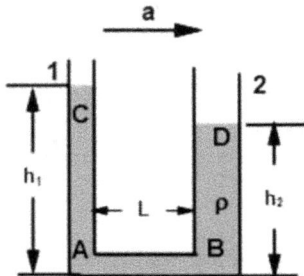

$$\Rightarrow (h_1 - h_2) - \frac{2T}{\rho g}\left(\frac{1}{r_1} - \frac{1}{r_2} \right) = a L$$

$$\Rightarrow h = \frac{2T}{\rho g}\left(\frac{1}{r_1} - \frac{1}{r_2}\right) + \frac{a}{g}L$$

$$\Rightarrow h = \frac{2T(r_2 - r_1)}{\rho r_1 r_2 g} + \frac{aL}{g} \quad \text{Ans.}$$

(b) If $h = 0$, we have

$$\vec{a} = -\frac{2T(r_2 - r_1)}{\rho r_1 r_2 L}\hat{i}$$

$\Rightarrow$ The U-tube must have this acceleration towards the left. Ans.

(c) If $a = 0$, we have

$$h = \frac{2T(r_2 - r_1)}{\rho r_1 r_2 g} \quad \text{Ans.}$$

(d) If $T = 0$, $h = \frac{aL}{g}$. Ans.

(e) If $r_1 = r_2$, $h = \frac{aL}{g}$ Ans.

Problem 44 A U-tube having the limbs of diameters d_1 and d_2 contain a liquid of density ρ and surface tension is T. The tube rotates about a vertical axis so that the limbs are situated at distances b and c as shown in the figure. Find the angular velocity ω of rotation of the tube if the difference of heights of the liquid columns in the limbs is (a) h and (b) $h = 0$.

Solution

(a) The pressure equation along both limbs gives

$$P_A = P_0 + \rho g h_1 - \frac{2T}{R_1} - \frac{1}{2}\rho r_1^2 \omega^2 = P_0 + \rho g h_2 - \frac{2T}{R_2} - \frac{1}{2}\rho r_2^2 \omega^2$$

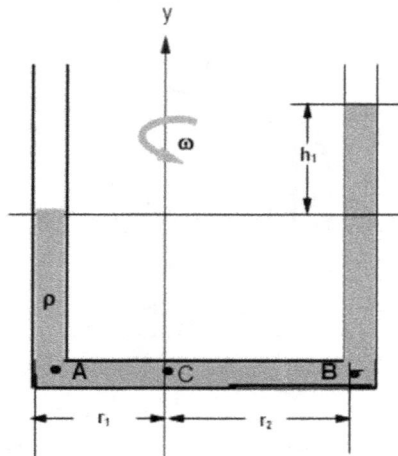

$$\Rightarrow \rho\omega(r_2^2 - r_1^2) = \rho g(h_2 - h_1) + 2T\left(\frac{1}{R_1} - \frac{1}{R_2}\right)$$

$$\Rightarrow \omega = \sqrt{2\frac{\left\{\rho g h + 4T\left(\frac{1}{d_1} - \frac{1}{d_2}\right)\right\}}{\rho(c^2 - b^2)}}$$

(b) If $h = 0$, the angular speed becomes

$$\Rightarrow \omega = \sqrt{\frac{8T(d_2 - d_1)}{\rho d_1 d_2(c^2 - b^2)}}$$

Problem 45 The L-shaped vessel moves with a horizontal acceleration a in the vertical plane as shown in the figure. If the pressure at the closed end A of the L-shaped tube is η times the atmospheric pressure P_o, find the surface tension of the liquid. Assume θ = wetting angle of contact between liquid and glass, ρ = density of the liquid.

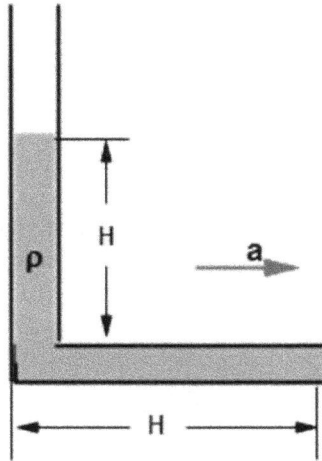

Solution
The pressure equation gives

$$P_C - P_A = \rho a L$$

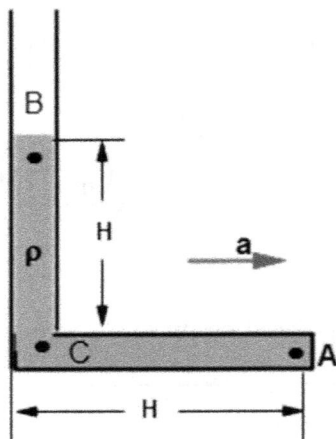

Or, $P_0 - \dfrac{2T \cos \theta}{r} + \rho g H - \eta P_0 = \rho a H$

$\Rightarrow P_0 (1 - \eta) + \rho H (g - a) = \dfrac{2T \cos \theta}{r}$

$\Rightarrow T = \dfrac{r}{2 \cos \theta} \{ \rho H (g - a) - P_0 (\eta - 1) \}$ Ans.

Problem 46 *Rise of liquid in conical capillary tube*

A truncated glass capillary tube with semivertical apex angle α is immersed vertically in water so that water rises to a height h. Assume θ = wetting angle of contact between liquid and glass, ρ = density of the liquid, average radius of cross-section is r, and surface tension of water-glass is T. (a) Find the height h in terms of the above given parameters. (b) What will be the answer if the capillary tube is inverted?

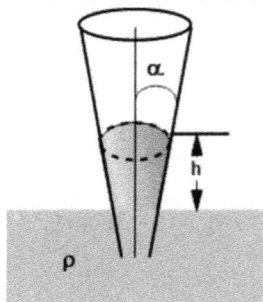

Solution

(a) Writing the force equation on the liquid column, we have

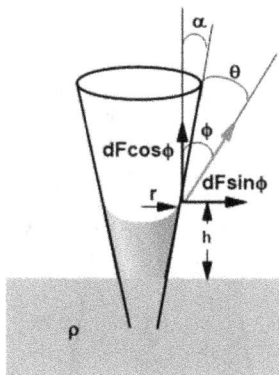

$$\int dF \cos \phi = \pi r^2 h \rho g$$

$$\Rightarrow \int T dl \cos \phi = \pi r^2 h \rho g$$

$$\Rightarrow T \cos \phi \int dl = \pi r^2 h \rho g$$

$$\Rightarrow (T \cos \phi) 2\pi r = \pi r^2 h \rho g$$

$$\Rightarrow \rho r = \frac{2T \cos \phi}{\rho g r}, \text{ where } \phi = \theta + \alpha$$

$$\Rightarrow h = \frac{2T \cos(\theta + \alpha)}{\rho g r} \quad \text{Ans.}$$

(b) Writing the force equation on the liquid column, we have

$$(T \cos \phi) 2\pi r = \pi r^2 h \rho g$$

$$\Rightarrow h = \frac{2T \cos \phi}{\rho g r} \text{ where } \phi = \theta - \alpha,$$

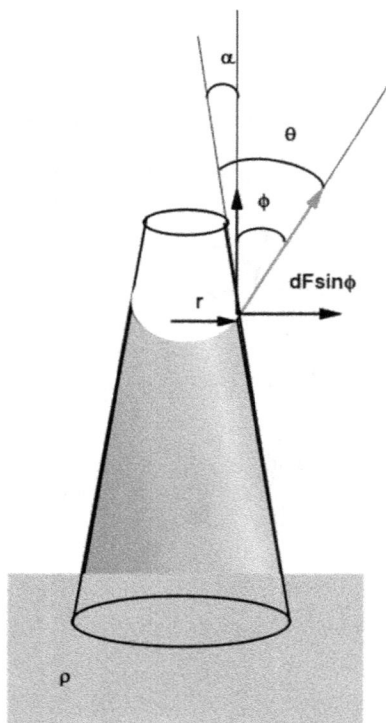

$$\Rightarrow h = \frac{2T \cos(\theta - \alpha)}{\rho g r} \quad \text{Ans.}$$

Alternative method: Applying Pascal's law, we have

$$P_c + \rho g h = P_B$$

$$\Rightarrow P_A - \frac{2T}{R_A} + \rho g h = P_B$$

$$\Rightarrow P_0 - \frac{2T}{R_A} + \lg h = P_0$$

$$\Rightarrow h = \frac{2T}{\rho g R_A}, \quad \text{where } R_A = \frac{r}{\cos \phi}$$

$$\Rightarrow h = \frac{2T \cos \phi}{\rho g r}, \quad \text{where } \phi = \theta \pm \alpha$$

Problem 47 A long capillary tube is dipped vertically in a liquid so that the length of capillary above the liquid surface is equal to h, which is insufficient for free rise of liquid. If we go on pulling the capillary up, find the maximum change in gravitational potential energy of the liquid column in the capillary. Assume $T = $ surface tension of water, $r = $ radius of the capillary tube, and ρ is the density of water.

Solution

The liquid level does not increase beyond force length (height)

$$h_f = \frac{2T \cos \theta}{\rho g r} \simeq \frac{2T}{\rho g} \tag{6.65}$$

The change in gravitational potential energy of the liquid columns is

$$\Delta U_{qr} = U_f - U_i$$

$$= \frac{1}{2} m_f g h_f - \frac{1}{2} m_i g h_i$$

$$= \frac{1}{2} m_i g h_i \left(\frac{m_f h_f}{m_i h_i} - 1 \right)$$

Since $m \, \alpha$ length, $\frac{m_f}{m_i} = \frac{h_f}{h_i}$

$$. \Rightarrow \Delta U_{qr} = \frac{1}{2} m_i g h_i \left\{ \left(\frac{h_f}{h_i} \right)^2 - 1 \right\} \tag{6.66}$$

Putting $m_i = \pi r^2 h \rho$, $h_i = h$, $h_f = \frac{2T}{\rho g r}$ from equation (6.65) in equation (6.66), we obtain

$$\Delta U_{gr} = \frac{\pi r^2 \rho g h^2}{2} \left\{ \left(\frac{2T}{\rho g r h} \right)^2 - 1 \right\} \quad \text{Ans.}$$

Problem 48 A vertical capillary tube of insufficient length (for free capillary rise) just touches the surface of water in a pond. If we incline the tube at an angle θ with horizontal, find the fraction change of gravitational potential energy of the water column inside the capillary tube. Assume that the lowest point of the tube always touches the liquid surface.

Solution

Since the water cannot come out of the tube, the same mass of water remains in the tube during the tilting. Hence, the potential energy changes by

$$\Delta U = U_f - U_i = mg \Delta h$$

$$=mg\frac{(h)(\cos\theta)}{2} - mg\frac{h}{2}$$

$$=mg\frac{h}{2}(\cos\theta - 1)$$

$$\Rightarrow \Delta U = U_i(\cos\theta - 1)$$

$$\Rightarrow \frac{\Delta U}{U_i} = \cos\theta - 1 = -(1 - \cos\theta) \quad \text{Ans.}$$

Problem 49 A glass capillary tube sealed at the top having length L and radius r ($r \ll L$) is dipped vertically into water. To what height h does the capillary need to be submersed so that the water level inside and outside of the capillary will coincide? Assume that the surface tension of water is T, atmospheric pressure is P_o, and angle of contact between glass-water interface $= 0°$.

Solution

Let h = depth of immersion of the tube. Applying Boyle's law between initial and final portion,

$$P_0 V_0 = PV$$

$$\Rightarrow P_0 AL = PA(L - h)$$

$$\Rightarrow P_0 L = P(L - h) \tag{6.67}$$

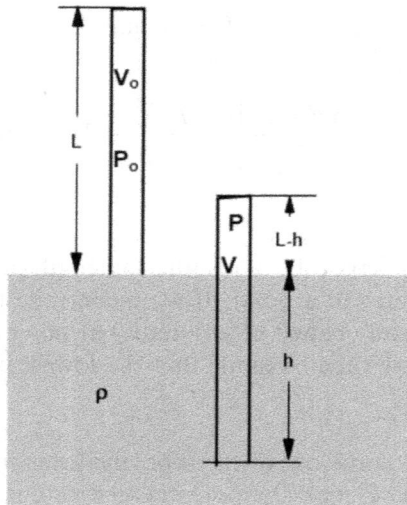

Pressure calculation:

$$P = P_0 + \frac{2T}{R},$$

where $R = \frac{r}{\cos\theta} \simeq r(\because \cos\theta \simeq 1)$

$$\Rightarrow P = P_0 + \frac{2T}{r} \tag{6.68}$$

Using equations (6.67) and (6.68),

$$P_0 L = \left(P_0 + \frac{2T}{r}\right)(L - h)$$

$$\Rightarrow P_0 L = P_0 L + \frac{2T}{r}L - P_0 h - \frac{2T}{r}h \Rightarrow h\left[P_0 + \frac{2T}{r}\right] = \frac{2T}{r}L$$

$$\Rightarrow h\left[\frac{P_0 r}{2T} + 1\right] = L$$

$$\Rightarrow h = \frac{L}{\left(1 + \frac{P_0 r}{2T}\right)} \quad \text{Ans.}$$

Problem 50 A thin plate of mass m and area A is dragged on a thin film of oil of thickness h coefficient of viscosity η by a constant force F. Find the (a) expression of velocity vs. time and (b) force required to drag it with a constant velocity v_o. (c) Derive velocity vs. time if the external force F is withdrawn. (d) What is the maximum distance covered by the plate after the withdrawal of the force? Assume a linear velocity profile for small thickness of oil.
   Solution

(a) The net force acting on the plate is

$$F_{\text{net}} = F - f_v$$

$$\Rightarrow ma = F - \eta\frac{v}{h}A$$

$$\Rightarrow a = \frac{dv}{dt} = \frac{F}{m} - \frac{\eta A}{mh}v$$

$$\Rightarrow \frac{dv}{dt} = k_1 - k_2 v, \text{ where } k_1 = \frac{F}{m} \text{ and } k_2 = \frac{\eta A}{mh}$$

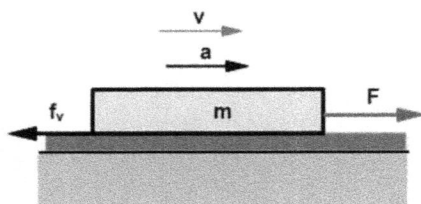

Solving this equation, we have

$$v = \frac{k_1}{k_2}(1 - e^{-k_2 t})$$

$$= \frac{Fh}{\eta A}\left(1 - e^{-\frac{\eta A}{mh}t}\right) \quad \text{Ans.}$$

(b) $F = f_v = \frac{\eta v_0 A}{h}$ Ans.

(c) If F is withdrawn, $F_{\text{net}} = -\eta \frac{v}{h} A$

Or, $m\frac{dv}{dt} = -\eta \frac{v}{h} A$

Or, $\frac{dv}{v} = -\frac{\eta A}{mh} dt$

Or, $\int_{v_0}^{v} \frac{dv}{v} = -\eta \frac{At}{mh}$

Or, $v = v_0 e^{-\frac{\eta A t}{mh}}$ Ans.

(d) $m\frac{u dv}{dx} = -\frac{\eta A}{h} v$

Or, $v \, dv = -\frac{\eta A}{mh} v \, dx$

Or, $\int_{v_0}^{v} dv = -\frac{\eta A}{mh} \int_{0}^{x} dx$

$$\Rightarrow \theta = v_0 - \frac{\eta A}{mh} x$$

Or, $x_{\text{max}} = \frac{m v_0 h}{\eta A}$ Ans.

Problem 51 *Rotating disc in a viscous liquid:* A disc of mass m and radius R is welded with a smooth and rigid vertical axle so that it can rotate about the vertical axis. Now, let the disc be placed inside a liquid layer of thickness $h \ll R$ and coefficient of viscosity η and given an initial spin angular velocity ω_o. (a) Find the retarding torque acting on the disc. (b) How does the angular speed of the disc vary as the function of (i) time and (ii) angular distance θ? (c) After how many rotations will the disc stop?

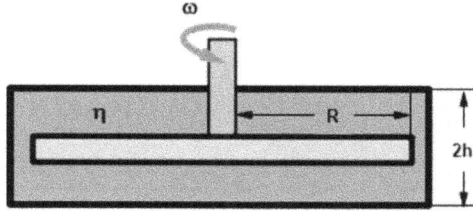

Solution

(a) Taking a small segment of area dA on a thin ring of radius r and thickness dr, let's find the elementary viscous force δf_v:

$$\delta f_v = \eta \left(\frac{r\omega}{\frac{h}{2}} \delta A \right)$$

This force produces a retarding torque

$$\delta \tau = r \delta f$$

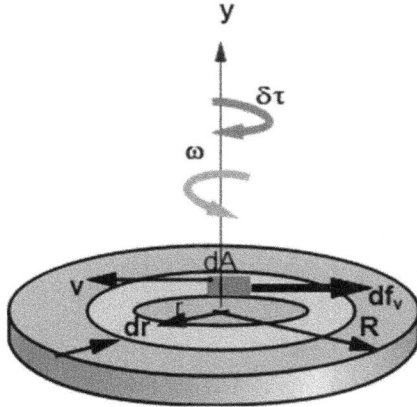

Then, the total torque exerted by the elementary viscous forces acting on every segment of the ring is $\tau_{\text{ring}} = \int d\tau = \int r\, dF$, where $df = \frac{2\eta r\omega}{h} \delta A$.

Then, $\tau_{\text{ring}} = \int r\{2\eta \frac{r\omega}{h} dA\}$, where $\int \delta A = A_{\text{ring}}$

$$\Rightarrow \tau_{\text{ring}} = 2\eta \frac{r^2 \omega}{h} A_{\text{ring}}$$

Integrating τ_{ring} we can get the restoring torque acting on the top surface of the disc, which is given as

$$\tau'_{\text{disc}} = \int \tau_{\text{ring}} = \frac{2\eta\omega}{h} \int r^2 A_{\text{ring}}, \quad \text{where } A_{\text{ring}} = 2\pi r dr$$

Hence, $\tau'_{\text{disc}} = \frac{2\eta\omega}{h} \int_0^{R_2} r^2 2\pi r dr$

$$= 4\pi\eta \frac{\omega}{h} \int_0^R r^3 dr = \frac{4\pi\eta\omega R^4}{4h}$$

$$\Rightarrow \tau'_{\text{disc}} = \frac{\pi\eta\omega R^4}{h}$$

Since there are two surfaces (top and bottom) of the disc exposed to liquid, total torque acting on the disc is

$$\tau_{\text{disc}} = 2\tau'_{\text{disc}} = \frac{2\pi\eta\omega R^4}{h} \quad \text{Ans.}$$

(b) (i) This retarding torque will decrease its angular velocity at the rate

$$-\frac{d\omega}{dt} = -\frac{\omega d\omega}{R d\theta}$$

Then, $\tau_{\text{disc}} = -I\frac{d\omega}{dt} = -I\frac{\omega d\omega}{R d\theta}$

$$\frac{2\pi\eta\omega R^4}{h} = -\frac{mR^2}{2}\frac{d\omega}{dt} = -\frac{mR^2}{2}\frac{\omega d\omega}{d\theta}$$

Or, $\int_{\omega_0}^{\omega} \frac{d\omega}{\omega} = -\frac{4\pi\eta R^2}{mh} \int_0^t dt$

Or, $\omega = \omega_0 e^{-\frac{4\pi\eta R^2}{mh}t} \quad \text{Ans.}$

(ii) $\int_{\omega_0}^{\omega} d\omega = -\frac{4\pi\eta R^2}{mh} \int_0^\theta d\theta$

$$\Rightarrow \omega = \omega_0 - \frac{4\pi\eta R^2}{mh}\theta \quad \text{Ans.}$$

(c) The total angle covered θ is obtained by putting $\omega = 0$

$$\theta_{\text{max}} = \frac{mh\omega_0}{4\pi\eta R^2}$$

Hence, the total number of rotations is

$$N = \frac{\theta_{\text{max}}}{2\pi} = \frac{mh\omega_0}{8\pi^2\eta R^2} \quad \text{Ans.}$$

Problem 52 *Flow between two rotating coaxial cylinders (Rotation viscometer)*: Two coaxial cylinders can rotate freely about their common vertical axis. A viscous liquid of coefficient of viscosity η is present in between the cylinders. The gap between the cylinders is too small compared to their radii. The inner and outer cylinders are given constant angular velocities ω_1 and ω_2, respectively, as shown in the figure. In steady state, calculate the (a) torque exerted on the cylinders by the liquid, (b) angular velocity of the rotating liquid as the function of radial distance, and (c) radial distance, where $\omega = 0$ if the cylinders rotate opposite to each other.

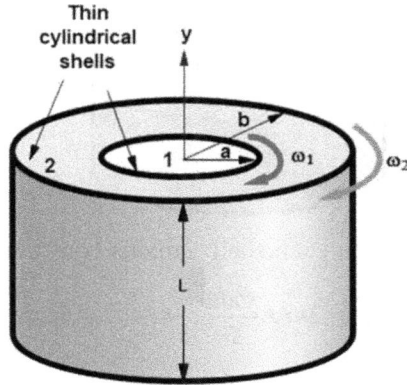

Solution

(a) The viscous force df acting on the element of area dA is

$$\delta f_v = \eta \frac{dv}{dr} dA$$

This force δf_v produces a torque $\delta \tau$ about the axis of rotation, given as

$$\delta \tau = \int r \delta F_v$$

$$\Rightarrow \delta \tau = r\eta \frac{dv}{dr} dA$$

The net torque acting on the thin cylindrical shell is

$$\tau_{\text{shell}} = \int d\tau = r\eta \frac{dv}{dr} \int dA$$

$$= \eta r \frac{dv}{dr} A_{\text{shell}}$$

$$\tau_{\text{shell}} = \eta r \frac{dv}{dr} 2\pi r L$$

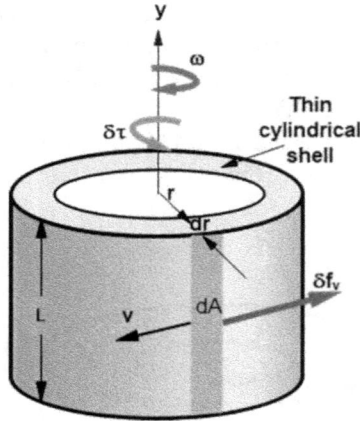

Then, the net torque acting on each shell remains constant

$$\tau_{\text{shell}} = 2\pi\eta L \frac{r^2 dv}{dr}, \text{ where } v = r\omega$$

$$\Rightarrow \tau_{\text{shell}} = 2\pi\eta L \, r^3 \frac{d\omega}{dr}$$

$$\Rightarrow 2\pi\eta L \int_{\omega_1}^{\omega} d\omega = \tau_{\text{shell}} \int_{r=a}^{r} \frac{dr}{r^3}$$

$$\Rightarrow 2\pi\eta L(\omega - \omega_1) = \tau_{\text{shell}} \left(-\frac{1}{r^2} \right)_{r=a}^{r=r} \qquad (6.69)$$

$$\Rightarrow 2\pi\eta L(\omega - \omega_1) = \tau_{\text{shell}} \left(\frac{1}{a^2} - \frac{1}{r^2} \right)$$

Putting $r = b$ and $\omega = \omega_2$, we have

$$\tau_{\text{shell}} = \frac{2\pi\eta L(\omega_2 - \omega_1)}{\frac{1}{a^2} - \frac{1}{b^2}} \quad \text{Ans.}$$

(b) By putting the value of τ_{shell} in equation (6.69),

$$2\pi\eta L(\omega - \omega_1) = \frac{2\pi\eta L(\omega_2 - \omega_1)}{\left(\frac{1}{a^2} - \frac{1}{b^2} \right)} \left(\frac{1}{a^2} - \frac{1}{r^2} \right)$$

$$\Rightarrow \omega = \omega_1 + \frac{(\omega_2 - \omega_1)a^2 b^2}{b^2 - a^2} \left(\frac{1}{a^2} - \frac{1}{r^2} \right)$$

Putting $\omega_2 = 0$, we have

$$\omega = \omega_1 \left(\frac{\frac{1}{b^2} - \frac{1}{r^2}}{\frac{1}{a^2} - \frac{1}{b^2}} \right)$$

$$= \frac{\omega_1(r^2 - b^2)a^2}{(b^2 - a^2)r^2} = \frac{\omega_1\left(1 - \frac{b^2}{r^2}\right)}{\left(\frac{b^2}{a^2} - 1\right)} \quad \text{Ans.}$$

(c) Rearranging the terms, $\omega = \frac{a^2 b^2}{b^2 - a^2}[\omega_1(\frac{1}{a^2} - \frac{1}{b^2}) + (\omega_2 - \omega_1)(\frac{1}{a^2} - \frac{1}{r^2})]$

$$= \frac{a^2 b^2}{b^2 - a^2}[\omega_1(\frac{1}{a^2} - \frac{1}{b^2} - \frac{1}{a^2} + \frac{1}{r^2}) + \omega_2(\frac{1}{a^2} - \frac{1}{r^2})]$$

$$\omega = \frac{a^2 b^2}{b^2 - a^2}\left[\omega_1\left(\frac{1}{r^2} - \frac{1}{b^2}\right) + \omega_2\left(\frac{1}{a^2} - \frac{1}{r^2}\right)\right]$$

If $\omega_2 = -\omega_1$, putting $\omega = 0$ in the expression $\omega = f(r)$, we have

$$\omega_1\left(\frac{1}{r^2} - \frac{1}{b^2}\right) = \omega_2\left(\frac{1}{a^2} - \frac{1}{r^2}\right)$$

$$\Rightarrow r = \sqrt{\frac{(\omega_1 + \omega_2)a^2 b^2}{(a^2\omega + b^2\omega_2)}} \quad \text{Ans.}$$

Problem 53 Let the viscous liquid in the last problem be stationary in the gap between the cylinders. If the inner cylinder moves with a velocity v_0 along its axis, assuming the laminar flow, find the velocity of the liquid as the function of radial distance r.

Solution

The viscous force acting at a radial distance on the thin cylindrical shell is

$$f_v = -\frac{dv}{dr}A$$

$$\Rightarrow f_v = -\eta\frac{dv}{dr}2\pi r L$$

$$\Rightarrow f_v \int_a^r \frac{dr}{r} = -2\pi\eta L \int_{v_0}^v dv$$

$$\Rightarrow f_v \ln\frac{r}{a} = 2\pi\eta L(v - v_0) \tag{6.70}$$

Putting $r = b$ and $v = 0$, we get

$$f_v = \frac{2\pi\eta L v_0}{\ln\left(\frac{b}{a}\right)}$$

(6.71)

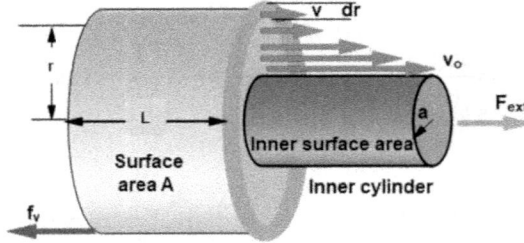

Putting f_v from equations (6.71) in (6.70),

$$-2\pi\eta L(v - v_0) = \frac{2\pi\eta L v_0}{\ln\frac{b}{a}} \ln\frac{r}{a}$$

$$\Rightarrow v_0\left(1 - \frac{\ln\frac{r}{a}}{\ln\frac{b}{a}}\right) = v$$

$$\Rightarrow v_0 \frac{\ln\frac{b}{a} - \ln\frac{r}{a}}{\ln\frac{b}{a}} = v$$

$$\Rightarrow v = v_0 \frac{\ln\frac{b}{r}}{\ln\frac{b}{a}} = v_0 \frac{\ln\frac{r}{b}}{\ln\frac{a}{b}} \quad \text{Ans.}$$

Alternative method: Referring to equation (6.60), we have

$$fv \int \frac{dr}{r} = -2\pi\eta L \int_0^v dv$$

$$\Rightarrow f_v \ln\frac{r}{b} = -2\pi\eta L v,$$

where $f_v = \frac{2\pi\eta L v_0}{\ln\frac{b}{a}}$

$$\Rightarrow v = v_0 \frac{\ln\frac{r}{b}}{\ln\frac{a}{b}} \quad \text{Ans.}$$

Problem 54 In a steady laminar flow in a tube radius R and length L, let's maintain a pressure difference ΔP between the ends of the tube. (a) Show that the velocity profile of the flow is given by the expression for the laminar flow of a viscous liquid as

$$v = v_o\left(1 - \frac{r^2}{R^2}\right),$$

where $v_o = -(\frac{\partial p}{\partial x})\frac{R^2}{4\eta} = \frac{R^2\Delta p}{4\eta L}$.

In terms of the given parameters (v_o, R, ρ, R, and L), find the (b) mass of liquid flowing per second in the tube in terms of pressure difference along the tube, (c) average momentum of the liquid per unit volume, (d) average kinetic energy of the liquid in the tube per unit volume, (e) viscous force of the liquid on the tube, (f) pressure difference along the tube of length L, (g) shear stress acting on the internal wall of the tube.

Solution

    (a) Applying Newton's law of viscosity on the thin cylindrical shell of viscous flow, we have

$$f_v = -\eta\frac{\delta v}{\delta r}A$$

$$f_v = -\eta\frac{dv}{dr}2\pi rL \tag{6.72}$$

Since $\frac{f_v}{A_{eH}} = \Delta P = P_1 - P_2$, we have

$$f_v = (P_1 - P_2)A_{eH} = (P_1 - P_2)\pi r^2 \tag{6.73}$$

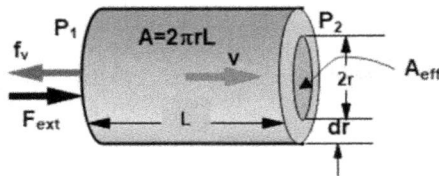

Eliminating f_v in equations (6.72) by (6.73), we have

$$\pi r^2(p_1 - p_2) = -\eta\frac{dv}{dr}2\pi rL$$

$$\Rightarrow -\frac{p_1 - p_2}{2\eta L}rdr = dv$$

$$\Rightarrow \int_{v=0}^{v} dv = \frac{(p_2 - p_1)}{L}\frac{1}{2\eta}\int_{r=R}^{r} rdr$$

$$\Rightarrow v = \frac{1}{2\eta} \frac{\delta P}{\delta x} \left(\frac{r^2 - R^2}{2} \right)$$

$$\Rightarrow v = -\frac{1}{4\eta} \frac{\delta P}{\delta x} (R^2 - r^2),$$

where $-\frac{\delta p}{\delta x} = -(\frac{p_1 - p_2}{L}) = \frac{\Delta p}{L}$

Then, $v = \frac{\Delta p}{4\eta L} (R^2 - r^2)$

$$\Rightarrow v = \frac{(\Delta p) R^2}{4\eta L} \left(1 - \frac{r^2}{R^2} \right) \quad \text{Ans.}$$

(b) The mass rate of flow is

$$Q = \rho \int v dA$$

$$= \rho \int_0^R \frac{(\Delta p)(R^2 - r^2)}{4\eta L} 2\pi r dr$$

$$= \rho \frac{\Delta p}{4\eta L} 2\pi \left\{ \int_0^R R^2 r dr - \int_0^R r^3 dr \right\}$$

$$= \frac{\rho \pi (\Delta p)}{2\eta L} \left(R^2 \frac{r^2}{2} - \frac{r^4}{4} \right)_0^R$$

$$= \frac{\rho \pi (\Delta p)}{2\eta L} \left(\frac{R^4}{2} - \frac{R^4}{4} \right)$$

$$\Rightarrow Q = \frac{\pi \rho R^4 (\Delta p)}{8\eta L} \quad \text{Ans.}$$

(c) We have obtained the expression

$$v = \frac{(\Delta p) R^2}{4\eta L} \left(1 - \frac{r^2}{R^2} \right) = v_0 \left(1 - \frac{r^2}{R^2} \right),$$

where $v_0 = \frac{\Delta p R^2}{4\eta L}$.

Then, the momentum of the liquid is

$$P = \int dP = \int dm \, v$$

$$= \int (2\pi r L dr \rho) \, v_0 \left(1 - \frac{r^2}{R^2} \right)$$

$$= 2\pi \rho v_0 L \int_0^R \left(1 - \frac{r^2}{R^2} \right) r \, dr$$

$$= 2\pi \rho v_0 L \left(\frac{R^2}{2} - \frac{R^2}{4} \right)$$

$$= \frac{\pi}{2} \rho v_0 R^2 L$$

$$\Rightarrow \frac{P}{Al} = P_v = \frac{\rho v_0}{2} \quad \text{Ans.}$$

(d) The *KE* of the liquid in the tube is $K = \int dK = \frac{1}{2} \int v^2 dm$

$$= \frac{1}{2} \int v_0^2 \left(1 - \frac{r^2}{R^2} \right) (2\pi r dr L \rho)$$

$$= \pi L \rho v_0^2 \int_0^R \left(1 - \frac{r^2}{R^2} \right) r \, dr$$

$$= \pi L \rho v_0^2 \int_0^R \left(1 + \frac{r^4}{R^4} - \frac{2r^2}{R^2} \right) r \, dr$$

$$= \pi L \rho v_0^2 \left(\frac{R^2}{2} + \frac{R^2}{6} - \frac{2 \times R^4}{4R^2} \right)$$

$$= \pi L \rho v_0^2 R^2$$

(e) The viscous force $f_v = \frac{\pi \eta v_0 r^2 L}{R^2}$ Ans.

      Putting $r = R$, $(f_v)_{\text{tube}} = 4\pi \eta v_0 L$ Ans.

(f) The pressure difference is

$$\Delta p = \frac{f}{\pi r^2} = \frac{-\eta \frac{\delta v}{do - r} \cdot 2\pi r L}{\pi r^2}$$

$$= \frac{-\eta \frac{\delta}{\delta r} \left\{ v_0 \left(1 - \frac{r^2}{R^2} \right) \right\} 2\pi r L}{\pi r^2}$$

$$= \eta v_0 \frac{2r}{R^2} \frac{2\pi r L}{\pi r^2}$$

$$\Rightarrow \Delta p = \frac{4 \eta v_0 L}{R^2} \quad \text{Ans.}$$

(g) The shear stress is

$$\sigma_{sh} = \left. \frac{f}{A_{surface}} \right|_{r=R}$$

$$= \frac{\frac{4 \eta v_0 \pi r^2 L}{R^2}}{2\pi r L}$$

$$= 2 \eta v_0 \left. \frac{r}{R^2} \right|_{r=R} = \frac{2 \eta v_0}{R} \quad \text{Ans.}$$

Problem 55 *Steady annular flow in between concentric cylinders:* Let's consider a steady laminar flow through the annular space between two coaxial thin cylinders as shown in the figure. The flow of liquid is along the axis of the tubes and is maintained by a constant pressure gradient $-\frac{\partial p}{\partial z}$. The radii of the inner and outer cylinders are $R_i = r_1$ and $R_o = r_2$, respectively. Assuming η as the coefficient of viscosity of the liquid, (a) prove that the velocity of the liquid varies with the radial distance r as

$$v_z = \frac{1}{4\eta} \left(-\frac{\partial p}{\partial z} \right) \left\{ \frac{\ln \frac{r}{r_1}}{\ln \frac{r_2}{r_1}} \left(r_2^2 - r_1^2 \right) - \left(r^2 - r_1^2 \right) \right\}.$$

(b) Find the position where the flow velocity is maximum.

Solution

The viscous force at any radial distance r is

$$f = -\eta \frac{dv}{dr} A$$

$$= -\eta \frac{dv}{dr} 2\pi r L$$

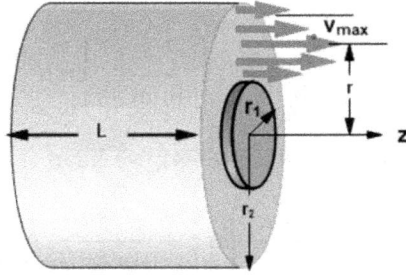

Then, the pressure difference is

$$\Delta p = \frac{f}{A_{eH}} = \frac{-\eta \frac{dv}{dr} 2\pi r L}{\pi (r^2 - a^2)}$$

$$\Rightarrow -\frac{\Delta p}{2\eta L}\left(\frac{r^2 - a^2}{r}\right) dr = +dv$$

$$\Rightarrow \int dv = \frac{\Delta p}{2\eta L} \int \left(r - \frac{a^2}{r}\right) dr$$

$$\Rightarrow v = -\frac{\Delta p r^2}{4\eta L} + C_1 \ln r + C_2 \qquad (6.74)$$

When $r = r_1$, $v = 0$

$$\Rightarrow 0 = -\frac{\Delta p}{4\eta L}(r_1)^2 + C_1 \ln r_1 + C_2 \qquad (6.75)$$

When $r = r_2$, $v = 0$

$$\Rightarrow 0 = -\frac{\Delta p}{4\eta L}(r_2)^2 + C_1 \ln r_2 + C_2 \qquad (6.76)$$

Subtracting equation (6.76) from equation (6.75), we have

$$\Rightarrow -\frac{\Delta p}{4\eta L}(r_1^2 - r_2^2) + C_1 \ln \frac{r_1}{r_2}$$

$$\Rightarrow C_1 = -\frac{\Delta p(r_2^2 - r_1^2)}{4\eta L \ln \frac{r_1}{r_2}} = \left(\frac{\delta p}{\delta z}\right)\left(\frac{r_2^2 - r_1^2}{4\eta \ln \frac{r_2}{r_1}}\right)$$

Putting C_1 in equation (6.75), we have

$$0 = -\frac{\delta p}{\delta z}\frac{1}{4\eta}r_1^2 + \left(\frac{\delta p}{\delta z}\right)\left(\frac{r_2^2 - r_1^2}{4\eta \ln \frac{r_2}{r_1}}\right)\ln r_1 + C_2$$

$$\Rightarrow C_2 = \frac{r_1^2}{4\eta}\frac{\delta p}{\delta z} - \left(\frac{r_2^2 - r_1^2}{4\eta \ln \frac{r2}{r1}}\right)\ln r_1 \frac{\delta p}{\delta z}$$

Then putting C_1 and C_2 in equation (6.74), we have

$$v = -\frac{\delta p}{\delta z}\frac{r^2}{4\eta} + \frac{\delta p}{\delta z}\left(\frac{r_2^2 - r_1^2}{4\eta \ln \frac{r2}{r1}}\right)\ln r + \frac{r_1^2}{4\eta}\frac{\delta p}{\delta z} - \frac{(r_2^2 - r_1^2)}{4\eta \ln \frac{r2}{r1}}\ln r_1 \frac{\delta p}{\delta z}$$

$$= \left(+\frac{\delta p}{\delta z}\right)\left\{\frac{(r_2^2 - r_1^2)\ln\left(\frac{r}{r_1}\right)}{4\eta \ln\left(\frac{r2}{r1}\right)} - \frac{(r^2 - r_1^2)}{4\eta}\right\}$$

$$\Rightarrow v = \frac{1}{4\eta}\left(\frac{\delta p}{\delta z}\right)\left\{(r_2^2 - r_1^2)\frac{\ln \frac{r}{r1}}{\ln \frac{r2}{r1}} - (r^2 - r_1^2)\right\} \text{ Proved}$$

(b) $\frac{dv_z}{dr} = \frac{1}{4\eta}\left(-\frac{\delta p}{\delta z}\right)\left\{\frac{\eta}{r}\frac{1}{r_1}\left(\frac{r_2^2 - r_1^2}{\ln \frac{2}{\eta}}\right) - 2r\right\} = 0$

$$\Rightarrow r = \sqrt{\frac{r_2^2 - r_1^2}{2 \ln \frac{r2}{r1}}} \quad \text{Ans.}$$

Problem 56 *Lifetime of a soap bubble:* A soap bubble is blown at the end of a capillary tube of radius a and length L as shown in the figure. When the other end is left open, the bubble begins to deflate. Find the (a) radius of the bubble as a function of time and (b) lifetime of the bubble. Assume that the initial radius of the bubble was R and surface tension of soap solution is T.

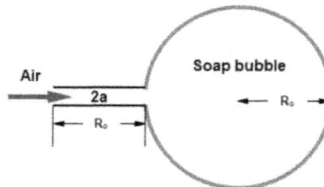

Solution

(a) Let the radius shrinks by dr. The corresponding decrease in volume is $dv = 4\pi r^2 dr$ during time dt. Then the rate of change in volume is equal to the rate $-Q$ of flow of a through the tube.

$Q = \frac{dV}{dt}$, where $\frac{dV}{dt} = -4\pi r^2 \frac{dr}{dt}$

Putting $Q = \frac{\pi a^4}{8\eta L}\Delta P$ (Poiseuille's equation), we have

$-4\pi r^2 \frac{dr}{dt} = \frac{\pi a^4}{8\eta L}\Delta p$, where $\Delta P = \frac{4T}{r}$

$$\Rightarrow -4\pi r^2 \frac{dr}{dt} = \frac{\pi a^4}{8\eta L}\frac{4\pi}{r}$$

$$\Rightarrow -4\int_R^r r^3 \, dr = \frac{a^4 T}{2\eta L}\int_0^t dt$$

$$\Rightarrow R^4 - r^4 = \frac{a^4 Tt}{2\eta L}$$

$$\text{Or, } r = \left(R^4 - \frac{a^4 Tt}{2\eta L}\right)^{\frac{1}{4}} \quad \text{Ans.}$$

(b) Putting $r = 0$, we have

$$T = \frac{2\eta L R^4}{a^4 T} \quad \text{Ans.}$$

Problem 57 *Tubes in series:* The tubes of lengths L_1, L_2, and L_3 and radii a, b, and c, respectively, are connected in series. If the pressure difference across the first capillary tube is p, find the pressure difference across the (a) 2nd and (b) 3rd capillary tubes. (Assume $a{:}b{:}c = 3{:}2{:}1$ and $L_1{:}L_2{:}L_3 = 9{:}8{:}1$.)
Solution
(a) and (b) since the tubes are connected in series, the rate of flow Q remains the same in all values. Since $Q = \frac{\pi \Delta p r^4}{8\eta l}$,

$$Q_1 = Q_2 = Q_3$$

$$Q = \frac{\Delta p_1 r_1^4}{L_1} = \frac{\Delta p_2 r_2^4}{L_2} = \frac{\Delta p_3 r_3^4}{L_3} \tag{6.77}$$

$$\Delta p_2 = \frac{L_2}{L_1}\Delta p_1 \left(\frac{r_1}{r_2}\right)^4 \text{ and } \Delta p_3 = \Delta p_1 \frac{L_3}{L_1}\left(\frac{r_1}{r_3}\right)^4$$

$$\Rightarrow \Delta p_2 = \left(\frac{L_2}{L_1}\right) p \left(\frac{a}{b}\right)^4 = \frac{9}{2}p, \quad \Delta p_3 = p\left(\frac{L_3}{L_1}\right)\left(\frac{b}{c}\right)^2 = 9p$$

(c) The total pressure difference is

$$\Delta p = \Delta p_1 + \Delta p_2 + \Delta p_3$$

$$= \Delta p_1 + \Delta p_1 \left(\frac{r_1}{r_2}\right)^4 \frac{L_2}{L_1} + \Delta p_1 \frac{L_3}{L_1}\left(\frac{r_1}{r_3}\right)$$

$$= p\left[1 + \left(\frac{a}{b}\right)^4 \frac{L_2}{L_1} + \left(\frac{a}{c}\right)^4 \frac{L_3}{L_1}\right]$$

$$= p\left[1 + \left(\frac{3}{2}\right)^4\left(\frac{8}{9}\right) + \left(\frac{3}{1}\right)^4\left(\frac{1}{9}\right)\right]$$

$$= \left(1 + \frac{81}{16} \times \frac{8}{9} + \frac{81}{9}\right)p$$

$$= \left(\frac{29}{2}\right)p \quad \text{Ans.}$$

Problem 58 A vertical tank A contains a viscous liquid of density $\rho = 1.0 \text{ g cc}^{-1}$ that flows along a horizontal tube fitted with two vertical monometers. If the liquid head in the monometer tubes are $h_1 = 10$ cm and $h_2 = 20$ cm and the tank maintains a constant height $h_3 = 35$ cm, find the velocity of the liquid flow in the tube.

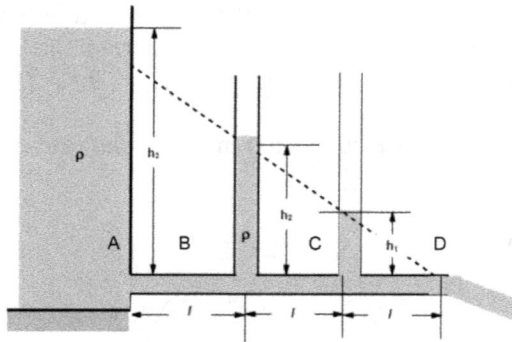

Solution

Applying Bernoulli's equation between C and D,

$$R_0 + \rho g h_1 + \frac{1}{2}\rho v^2 = P_0 + \rho\frac{v^2}{2} + Q_{CD}$$

$$Q_{CD} = \rho g h_1 \tag{6.78}$$

Applying BE between B and D,

$$P_0 + \rho g h_2 + \frac{1}{2}\rho v^2 = P_0 + \frac{1}{2}\rho v^2 + Q_{BD}$$

$$Q_{BD} = \rho g h_2 \tag{6.79}$$

From equations (6.78) and (6.79) it is obvious that

$$Q \propto h(\because Q = \rho g h)$$

$$\Rightarrow Q_{AD} = \rho g y \tag{6.80}$$

The applying BE between A and D,

$$P_0 + \rho g h_3 = P_0 + \frac{1}{2}\rho v^2 + Q_{AD}$$

$$\Rightarrow \rho g h_3 = \frac{1}{2}\rho v^2 + Q_{AD} \tag{6.81}$$

Putting Q_{AD} from equations (6.80) in (6.81),

$$\rho g h_3 = \frac{1}{2}\rho v^2 + egy$$

$$\Rightarrow v = \sqrt{2g(h_3 - y)}, \text{ where}$$

$$y = AD \tan\theta = (l_1 + l_2 + l_3)\left(\frac{h_2 - h_1}{l_2}\right)$$

$$= \frac{(l + l + l)(h_2 - h_1)}{l} = 3(h_2 - h_1)$$

Then, $v = \sqrt{2g\{h_3 - 3(h_2 - h_1)\}}$ Ans.

Problem 59 A steel ball of radius r and density σ is released from rest at $t = 0$ in a liquid column of sufficient height to attain terminal velocity. If the coefficient of viscosity of liquid is η and the density of liquid is ρ, find the time after which the ball attains a speed which is equal to a fraction n of the terminal speed.

Solution

The net force acting on the ball is

$$F_{\text{net}} = mg - F_b - F_v$$

$$= mg - mg\frac{p}{\sigma} - 6\pi\eta ru$$

$$\Rightarrow m\frac{dv}{dt} = mg\left(\frac{\sigma - p}{\sigma}\right) - 6\pi\eta ru$$

$\Rightarrow \frac{dv}{dt} = g(\frac{\sigma - P}{\sigma}) - \frac{6\pi\eta ru}{m}$, where $m = \frac{4}{3}\pi r^3\sigma$

$\Rightarrow \frac{dv}{dt} = k_1 - k_2v$, where $k_1 = g(\frac{\sigma - P}{\sigma})$ and $k_2 = \frac{9\eta}{2r^2\sigma}$

$$\Rightarrow \int_0^v \frac{dv}{k_1 - k_2v} = \int dt$$

$$\Rightarrow -\frac{1}{k_2}\ln\frac{(k_1 - k_2v)}{k_1} = t$$

$\Rightarrow -k_2t = \ln(1 - \frac{k_2}{k_1}v)$, where $\frac{k_1}{k_2} = v_{\text{terminal}} = v_{\text{max}}$

$$\Rightarrow t = -\frac{1}{k_2}\ln\left(1 - \frac{v}{v_{\text{max}}}\right)$$

$$t = -\frac{2r^2a}{9\eta}\ln(1 - n) \quad \text{Ans.}$$

Problem 60 Consider two solid spheres P and Q each of density σ ($= 8$ g cm^{-3}) and radii $r_1 = 1$ cm and $r_2 = 0.5$ cm, respectively. Sphere P is dropped into a liquid of density $\rho_1 = 1$ g cm^{-3} and viscosity $\eta_1(= 3$ poise). Sphere Q is dropped into a liquid of density $\rho_2 = 2$ g cm^{-3} and viscosity η_2 ($= 2$ poise). Find the ratio of their (a) velocities, (b) momenta, and (c) KE in steady state.

Solution

The terminal velocity is given as

$$v_t = \frac{2r^2}{9\eta}(\sigma - \rho)$$

(a) $\frac{v_1}{v_2} = (\frac{\eta_2}{\eta_1})(\frac{r_1}{r_2})^2(\frac{\sigma_1 - \rho_1}{\sigma_2 - \rho_2})$

$$= \frac{8}{3} \times \frac{7}{6} = \frac{28}{9} \quad \text{Ans.}$$

(b) $\frac{P_1}{P_2} = (\frac{\eta_2}{\eta_1})(\frac{r_1}{r_2})^5\frac{(\sigma_1 - \rho_1)}{(\sigma_2 - \rho_2)}(\frac{\sigma_1}{\sigma_2})$

$$= \left(\frac{2}{3}\right)(2)^5\left(\frac{7}{6}\right) = \frac{224}{9}$$

$$= \frac{2 \times 32 \times 7 \times 7}{3 \times 6 \times 6} = \frac{784}{27} \quad \text{Ans.}$$

(c) $\frac{K_1}{K_2} = (\frac{\eta_2}{\eta_1})^2(\frac{r_1}{r_2})^5(\frac{\sigma_1-\rho_1}{\sigma_2-\rho_1})^2(\frac{\sigma_1}{\sigma_2})$

$$= \frac{1568}{27} \quad \text{Ans.}$$

Problem 61 Two spheres P and Q of equal radii r and densities ρ and 2.5 ρ, respectively, are placed in two immiscible viscous liquids L_1 and L_2 being connected by a light inextensible string as shown in the figure. The densities and viscosities of liquids are $\rho/2$, ρ, and η, $\eta/3$, respectively. If the string is taut and spheres move with common velocity v as shown in the figure, find the (a) terminal velocity and (b) tension in the string acceleration of the balls while moving in their respective liquids.

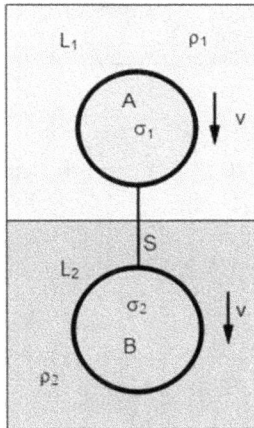

Solution

(a) The force equations for the balls are given as

$$T + gm_1\left(1 - \frac{\rho_1}{\sigma_1}\right) - F_{v_1} = 0 \tag{6.82}$$

$$m_2g\left(1 - \frac{\rho_2}{\sigma_2}\right) - T - F_{v_2} = 0 \tag{6.83}$$

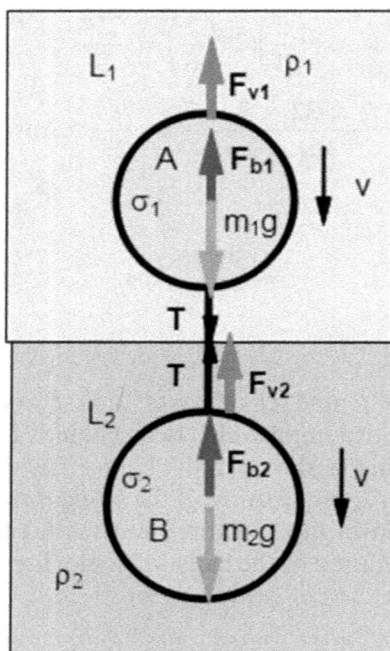

Solving these two equations,

$$F_{v_1} + F_{v_2} = m_1 g\left(1 - \frac{\rho_1}{\sigma_1}\right) + m_2 g\left(1 - \frac{\rho_2}{\sigma_2}\right)$$

$$\Rightarrow 6\pi r v(\eta_1 + \eta_2) = \frac{4}{3}\pi r^3\{(\sigma_1 - \rho_1) + (\sigma_2 - \rho_2)\}g$$

$$\Rightarrow v = \frac{2r^2 g}{9(\eta_1 + \eta_2)}\left[(\sigma_1 + \sigma_2) - (\rho_1 + \rho_2)\right]$$

$$= \frac{2r^2 g}{9(\eta_1 + \eta_2)}\left[(\rho + 2.5\rho) - \left(\frac{\rho}{2} + \rho\right)\right]$$

$$= \frac{4\rho r^2 g}{9\left(\eta + \frac{\eta}{3}\right)} = \frac{\rho r^2 g}{3\eta} \quad \text{Ans.}$$

(b) The tension in the string can be calculated by using equations (6.82) or (6.83). Putting $v = \frac{\rho r^2 g}{3\eta m}$ equation (6.82) after proper evaluation we have,

$$T = 6\pi\eta_1 v - \frac{4}{3}\pi r^3(\sigma_1 - \rho_1)g$$

$$= 6\pi\eta\left(\frac{\rho r^2 g}{3\eta}\right) - \frac{4}{3}\pi r^3\left(\rho - \frac{\rho}{2}\right)g$$

$$= 2\pi\rho r^3 g\left(1 - \frac{1}{3}\right)$$

$$= \left(\frac{4\pi}{3}r^3\rho\right)g \quad \text{Ans.}$$

www.ingramcontent.com/pod-product-compliance
Lightning Source LLC
Chambersburg PA
CBHW082114210326
41599CB00031B/5767